Student's Solutions Manual
Part 1

Thomas/Finney

Calculus and
Analytic Geometry
9TH Edition

Student's Solutions Manual
Part 1

Thomas/Finney

Calculus and
Analytic Geometry
9TH Edition

Maurice D. Weir
U.S. Naval Postgraduate School

Addison-Wesley Publishing Company

Reading, Massachusetts • Menlo Park, California • New York
Don Mills, Ontario • Wokingham, England • Amsterdam • Bonn
Sydney • Singapore • Tokyo • Madrid • San Juan • Milan • Paris

Reproduced by Addison-Wesley from camera-ready copy supplied by the author.

Copyright © 1996 by Addison-Wesley Publishing Company, Inc.

ISBN 0-201-53179-8
     6 7 8 9 10- CRS-99

PREFACE TO THE STUDENT

This Student's Solutions Manual contains the solutions to all of the odd-numbered exercises in the 9th Edition of CALCULUS AND ANALYTIC GEOMETRY by Ross L. Finney and George B. Thomas, Jr., excluding the Computer Algebra System (CAS) exercises. We have worked each solution to ensure that it

- conforms exactly to the methods, procedures and steps presented in the text

- is mathematically correct

- includes all of the steps necessary so you can follow the logical argument and algebra

- includes a graph or figure whenever called for by the exercise, or if needed to help with the explanation

- is formatted in an appropriate style to aid you in its understanding

How to use a solution's manual

- solve the assigned problem yourself

- if you get stuck along the way, refer to the solution in the manual as an aid but continue to solve the problem on your own

- if you cannot continue, reread the textbook section, or work through that section in the Student Study Guide, or consult your instructor

- after solving the problem (if odd-numbered), carefully compare your solution procedure to the one in the manual

- if your answer is correct but your solution procedure seems to differ from the one in the manual, and you are unsure your method is correct, consult your instructor

- if your answer is incorrect and you cannot find your error, consult your instructor

TABLE OF CONTENTS

7 Techniques of Integration 307

8 Infinite Series 359

9 Conic Sections, Parametrized Curves, and Polar Coordinates 407

Student's Solutions Manual
Part 1

Thomas/Finney

Calculus and
Analytic Geometry
9TH Edition

PRELIMINARIES

P.1 REAL NUMBERS AND THE REAL LINE

1. Executing long division, $\frac{1}{9} = 0.\overline{1}$, $\frac{2}{9} = 0.\overline{2}$, $\frac{3}{9} = 0.\overline{3}$, $\frac{8}{9} = 0.\overline{8}$

3. (a) False   (b) True   (c) True   (d) True
    (e) True   (f) True   (g) True   (h) True

5. $-2x > 4 \Rightarrow x < -2$

7. $5x - 3 \le 7 - 3x \Rightarrow 8x \le 10 \Rightarrow x \le \frac{5}{4}$

9. $2x - \frac{1}{2} \ge 7x + \frac{7}{6} \Rightarrow -\frac{1}{2} - \frac{7}{6} \ge 5x$

    $\Rightarrow \frac{1}{5}\left(-\frac{10}{6}\right) \ge x$ or $-\frac{1}{3} \ge x$

11. $\frac{4}{5}(x-2) < \frac{1}{3}(x-6) \Rightarrow 12(x-2) < 5(x-6)$

    $\Rightarrow 12x - 24 < 5x - 30 \Rightarrow 7x < -6$ or $x < -\frac{6}{7}$

13. $y = 3$ or $y = -3$

15. $2t + 5 = 4$ or $2t + 5 = -4 \Rightarrow 2t = -1$ or $2t = -9 \Rightarrow t = -\frac{1}{2}$ or $t = -\frac{9}{2}$

17. $8 - 3s = \frac{9}{2}$ or $8 - 3s = -\frac{9}{2} \Rightarrow -3s = -\frac{7}{2}$ or $-3s = -\frac{25}{2} \Rightarrow s = \frac{7}{6}$ or $s = \frac{25}{6}$

19. $-2 < x < 2$; solution interval $(-2, 2)$

21. $-3 \le t - 1 \le 3 \Rightarrow -2 \le t \le 4$; solution interval $[-2, 4]$

23. $-4 < 3y - 7 < 4 \Rightarrow 3 < 3y < 11 \Rightarrow 1 < y < \frac{11}{3}$;

    solution interval $\left(1, \frac{11}{3}\right)$

25. $-1 \le \frac{z}{5} - 1 \le 1 \Rightarrow 0 \le \frac{z}{5} \le 2 \Rightarrow 0 \le z \le 10$;

    solution interval $[0, 10]$

27. $-\frac{1}{2} < 3 - \frac{1}{x} < \frac{1}{2} \Rightarrow -\frac{7}{2} < -\frac{1}{x} < -\frac{5}{2} \Rightarrow \frac{7}{2} > \frac{1}{x} < \frac{5}{2} \Rightarrow \frac{2}{7} < x < \frac{2}{5}$;

    solution interval $\left(\frac{2}{7}, \frac{2}{5}\right)$

29. $2s \ge 4$ or $-2s \ge 4 \Rightarrow s \ge 2$ or $s \le -2$;

    solution intervals $(-\infty, -2] \cup [2, \infty)$

31. $1 - x > 1$ or $-(1 - x) > 1 \Rightarrow -x > 0$ or $x > 2 \Rightarrow x < 0$ or $x > 2$;

    solution intervals $(-\infty, 0) \cup (2, \infty)$

33. $\frac{r+1}{2} \geq 1$ or $-\left(\frac{r+1}{2}\right) \geq 1 \Rightarrow r + 1 \geq 2$ or $r + 1 \leq -2 \Rightarrow r \geq 1$ or $r \leq -3$;

    solution intervals $(-\infty, -3] \cup [1, \infty)$

35. $x^2 < 2 \Rightarrow |x| < \sqrt{2} \Rightarrow -\sqrt{2} < x < \sqrt{2}$;

    solution interval $\left(-\sqrt{2}, \sqrt{2}\right)$

37. $4 < x^2 < 9 \Rightarrow 2 < |x| < 3 \Rightarrow 2 < x < 3$ or $2 < -x < 3 \Rightarrow 2 < x < 3$ or $-3 < x < -2$;

    solution intervals $(-3, -2) \cup (2, 3)$

39. $(x - 1)^2 < 4 \Rightarrow |x - 1| < 2 \Rightarrow -2 < x - 1 < 2 \Rightarrow -1 < x < 3$;

    solution interval $(-1, 3)$

41. $x^2 - x < 0 \Rightarrow x(x - 1) < 0 \Rightarrow$ solution interval $(0, 1)$

43. True if $a \geq 0$; False if $a < 0$.

45. (1) $|a + b| = (a + b)$ or $|a + b| = -(a + b)$;

    both squared equal $(a + b)^2$

    (2) $ab \leq |ab| = |a||b|$

    (3) $|a| = a$ or $|a| = -a$, so $|a|^2 = a^2$; likewise, $|b|^2 = b^2$

    (4) $x^2 \leq y^2$ implies $\sqrt{x^2} \leq \sqrt{y^2}$ or $x \leq y$ for all nonnegative real numbers x and y. Let $x = |a + b|$ and $y = |a| + |b|$ so that $|a + b|^2 \leq (|a| + |b|)^2 \Rightarrow |a + b| \leq |a| + |b|$.

47. $-3 \leq x \leq 3$ and $x > -\frac{1}{2} \Rightarrow -\frac{1}{2} < x \leq 3$.

49. (a) From the graph, $\frac{x}{2} > 1 + \frac{4}{x} \Rightarrow x \in (-2, 0) \cup (4, \infty)$

    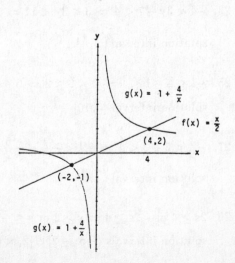

    (b) $\frac{x}{2} > 1 + \frac{4}{x} \Rightarrow \frac{x}{2} - 1 - \frac{4}{x} > 0$

    $x > 0$: $\frac{x}{2} - 1 - \frac{4}{x} > 0 \Rightarrow \frac{x^2 - 2x - 8}{2x} > 0 \Rightarrow \frac{(x - 4)(x + 2)}{2x} > 0$

    $\Rightarrow x > 4$ since x is positive;

    $x < 0$: $\frac{x}{2} - 1 - \frac{4}{x} > 0 \Rightarrow \frac{x^2 - 2x - 8}{2x} < 0 \Rightarrow \frac{(x - 4)(x + 2)}{2x} < 0$

    $\Rightarrow x < -2$ since x is negative;

    sign of $(x - 4)(x + 2)$

    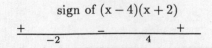

    Solution interval: $(-2, 0) \cup (4, \infty)$

P.2 COORDINATES, LINES, AND INCREMENTS

1. $\Delta x = -1 - (-3) = 2,\ \Delta y = -2 - 2 = -4;\ d = \sqrt{(\Delta x)^2 + (\Delta y)^2} = \sqrt{4 + 16} = 2\sqrt{5}$

3. $\Delta x = -8.1 - (-3.2) = -4.9,\ \Delta y = -2 - (-2) = 0;\ d = \sqrt{(-4.9)^2 + 0^2} = 4.9$

5. Circle with center $(0,0)$ and radius 1.

7. Disk (i.e., circle together with its interior points) with center $(0,0)$ and radius $\sqrt{3}$.

9. $m = \dfrac{\Delta y}{\Delta x} = \dfrac{-1 - 2}{-2 - (-1)} = 3$

    perpendicular slope $= -\dfrac{1}{3}$

11. $m = \dfrac{\Delta y}{\Delta x} = \dfrac{3 - 3}{-1 - 2} = 0$

    perpendicular slope does not exist

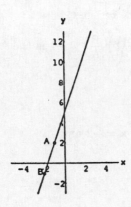

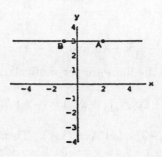

13. (a) $x = -1$ (b) $y = \dfrac{4}{3}$ 15. (a) $x = 0$ (b) $y = -\sqrt{2}$

17. $P(-1,1),\ m = -1 \Rightarrow y - 1 = -1(x - (-1)) \Rightarrow y = -x$

19. $P(3,4),\ Q(-2,5) \Rightarrow m = \dfrac{\Delta y}{\Delta x} = \dfrac{5 - 4}{-2 - 3} = -\dfrac{1}{5} \Rightarrow y - 4 = -\dfrac{1}{5}(x - 3) \Rightarrow y = -\dfrac{1}{5}x + \dfrac{23}{5}$

21. $m = -\dfrac{5}{4},\ b = 6 \Rightarrow y = -\dfrac{5}{4}x + 6$ 23. $m = 0,\ P(-12,-9) \Rightarrow y = -9$

25. $a = -1,\ b = 4 \Rightarrow (0,4)$ and $(-1,0)$ are on the line $\Rightarrow m = \dfrac{\Delta y}{\Delta x} = \dfrac{0 - 4}{-1 - 0} = 4 \Rightarrow y = 4x + 4$

27. $P(5,-1),\ L:\ 2x + 5y = 15 \Rightarrow m_L = -\dfrac{2}{5} \Rightarrow$ parallel line is $y - (-1) = -\dfrac{2}{5}(x - 5) \Rightarrow y = -\dfrac{2}{5}x + 1$

29. $P(4,10),\ L:\ 6x - 3y = 5 \Rightarrow m_L = 2 \Rightarrow m_\perp = -\dfrac{1}{2} \Rightarrow$ perpendicular line is $y - 10 = -\dfrac{1}{2}(x - 4) \Rightarrow y = -\dfrac{1}{2}x + 12$

31.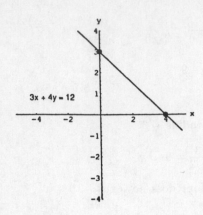

33.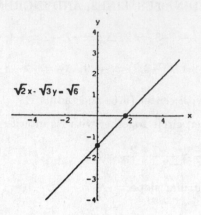

35. $Ax + By = C_1 \Leftrightarrow y = -\dfrac{A}{B}x + \dfrac{C_1}{B}$ and $Bx - Ay = C_2 \Leftrightarrow y = \dfrac{B}{A}x - \dfrac{C_2}{A}$. Since $\left(-\dfrac{A}{B}\right)\left(\dfrac{B}{A}\right) = -1$ is the product of the slopes, the lines are perpendicular.

37. New position $= \left(x_{old} + \Delta x, y_{old} + \Delta y\right) = (-2 + 5, 3 + (-6)) = (3, -3)$.

39. $\Delta x = 5$, $\Delta y = 6$, $B(3, -3)$. Let $A = (x, y)$. Then $\Delta x = x_2 - x_1 \Rightarrow 5 = 3 - x \Rightarrow x = -2$ and $\Delta y = y_2 - y_1 \Rightarrow 6 = -3 - y \Rightarrow y = -9$. Therefore, $A = (-2, -9)$.

41. (a) $A \approx (69°, 0 \text{ in})$, $B \approx (68°, .4 \text{ in}) \Rightarrow m = \dfrac{68° - 69°}{.4 - 0} \approx -2.5°/\text{in}$

    (b) $A \approx (68°, .4 \text{ in})$, $B \approx (10°, 4 \text{ in}) \Rightarrow m = \dfrac{10° - 68°}{4 - .4} \approx -16.1°/\text{in}$

    (c) $A \approx (10°, 4 \text{ in})$, $B \approx (5°, 4.6 \text{ in}) \Rightarrow m = \dfrac{5° - 10°}{4.6 - 4} \approx -8.3°/\text{in}$

43. $p = kd + 1$ and $p = 10.94$ at $d = 100 \Rightarrow k = \dfrac{10.94 - 1}{100} = 0.0994$. Then $p = 0.0994d + 1$ is the diver's pressure equation so that $d = 50 \Rightarrow p = (0.0994)(50) + 1 = 5.97$ atmospheres.

45. $C = \dfrac{5}{9}(F - 32)$ and $C = F \Rightarrow F = \dfrac{5}{9}F - \dfrac{160}{9} \Rightarrow \dfrac{4}{9}F = -\dfrac{160}{9}$ or $F = -40°$ gives the same numerical reading.

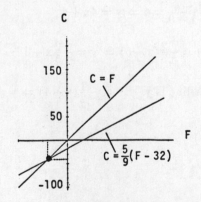

47. length $AB = \sqrt{(5 - 1)^2 + (5 - 2)^2} = \sqrt{16 + 9} = 5$

    length $AC = \sqrt{(4 - 1)^2 + (-2 - 2)^2} = \sqrt{9 + 16} = 5$

    length $BC = \sqrt{(4 - 5)^2 + (-2 - 5)^2} = \sqrt{1 + 49} = \sqrt{50} = 5\sqrt{2} \neq 5$

49. Length $AB = \sqrt{(\Delta x)^2 + (\Delta y)^2} = \sqrt{1^2 + 4^2} = \sqrt{17}$ and length $BC = \sqrt{(\Delta x)^2 + (\Delta y)^2} = \sqrt{4^2 + 1^2} = \sqrt{17}$.

Also, slope $AB = \frac{4}{-1}$ and slope $BC = \frac{1}{4}$, so $AB \perp BC$. Thus, the points are vertices of a square. The coordinate increments from the fourth vertex $D(x, y)$ to A must equal the increments from C to $B \Rightarrow 2 - x = \Delta x = 4$ and $-1 - y = \Delta y = 1 \Rightarrow x = -2$ and $y = -2$. Thus $D(-2, -2)$ is the fourth vertex.

51. Let $A(-1, 1)$, $B(2, 3)$, and $C(2, 0)$ denote the points. Since BC is vertical and has length $|BC| = 3$, let $D_1(-1, 4)$ be located vertically upward from A and $D_2(-1, -2)$ be located vertically downward from A so that $|BC| = |AD_1| = |AD_2| = 3$. Denote the point $D_3(x, y)$. Since the slope of AB equals the slope of CD_3 we have $\frac{y - 3}{x - 2} = -\frac{1}{3} \Rightarrow 3y - 9 = -x + 2$ or $x + 3y = 11$. Likewise, the slope of AC equals the slope of BD_3 so that $\frac{y - 0}{x - 2} = \frac{2}{3} \Rightarrow 3y = 2x - 4$ or $2x - 3y = 4$.

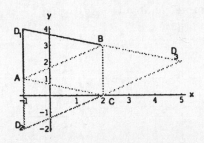

Solving the system of equations $\left. \begin{array}{r} x + 3y = 11 \\ 2x - 3y = \ \ 4 \end{array} \right\}$ we find $x = 5$ and $y = 2$ yielding the vertex $D_3(5, 2)$.

53. $2x + ky = 3$ has slope $-\frac{2}{k}$ and $4x + y = 1$ has slope -4. The lines are perpendicular when $-\frac{2}{k}(-4) = -1$ or $k = -8$ and parallel when $-\frac{2}{k} = -4$ or $k = \frac{1}{2}$.

55. Let $M(a, b)$ be the midpoint. Since the two triangles shown in the figure are congruent, the value a must lie midway between x_1 and x_2, so $a = \frac{x_1 + x_2}{2}$.

Similarly, $b = \frac{y_1 + y_2}{2}$.

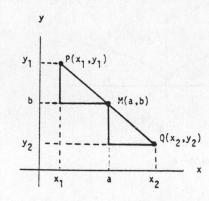

P.3 FUNCTIONS

1. domain $= (-\infty, \infty)$; range $= [1, \infty)$

3. domain $= (0, \infty)$; y in range $\Rightarrow y = \frac{1}{\sqrt{t}}$, $t > 0 \Rightarrow y^2 = \frac{1}{t}$ and $y > 0 \Rightarrow$ y can be any positive real number $\Rightarrow$ range $= (0, \infty)$.

5. $4 - z^2 = (2 - z)(2 + z) \geq 0 \Leftrightarrow z \in [-2, 2] =$ domain. Largest value is $g(0) = \sqrt{4} = 2$ and smallest value is $g(-2) = g(2) = \sqrt{0} = 0 \Rightarrow$ range $= [0, 2]$.

7. (a) Not the graph of a function of x since it fails the vertical line test.
   (b) Is the graph of a function of x since any vertical line intersects the graph at most once.

9. base $= x$; $(\text{height})^2 + \left(\dfrac{x}{2}\right)^2 = x^2 \Rightarrow$ height $= \dfrac{\sqrt{3}}{2}x$; area is $a(x) = \dfrac{1}{2}(\text{base})(\text{height}) = \dfrac{1}{2}(x)\left(\dfrac{\sqrt{3}}{2}\right) = \dfrac{\sqrt{3}}{4}x^2$; perimeter is $p(x) = x + x + x = 3x$.

11. Let $D =$ diagonal of a face of the cube and $\ell =$ the length of an edge. Then $\ell^2 + D^2 = d^2$ and (by Exercise 10) $D^2 = 2\ell^2 \Rightarrow 3\ell^2 = d^2 \Rightarrow \ell = \dfrac{d}{\sqrt{3}}$. The surface area is $6\ell^2 = \dfrac{6d^2}{3} = 2d^2$ and the volume is $\ell^3 = \left(\dfrac{d^2}{3}\right)^{3/2} = \dfrac{d^3}{3\sqrt{3}}$.

13. Symmetric about
    the origin

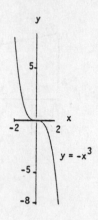

15. Symmetric about
    the origin

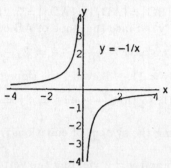

17. Symmetric about
    the y-axis

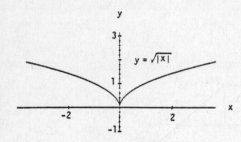

19. Symmetric about
    the origin

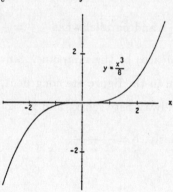

21. No symmetry

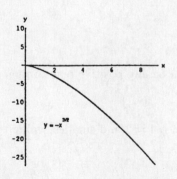

23. Symmetric about
    the y-axis

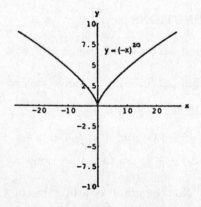

25. Neither graph passes the vertical line test
    (a)   (b)

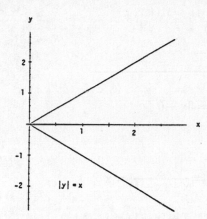

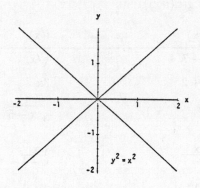

27. even   29. even   31. odd   33. even

35. neither   37. neither

39. D_f: $-\infty < x < \infty$, D_g: $x \geq 1 \Rightarrow D_{f+g} = D_{fg}$: $x \geq 1$. R_f: $-\infty < y < \infty$, R_g: $y \geq 0$, R_{f+g}: $y \geq 1$, R_{fg}: $y \geq 0$

41. D_f: $-\infty < x < \infty$, D_g: $-\infty < x < \infty \Rightarrow D_{f/g}$: $-\infty < x < \infty$ since $g(x) \neq 0$ for any x; $D_{g/f}$: $-\infty < x < \infty$

   since $f(x) \neq 0$ for any x. R_f: $y = 2$, R_g: $y \geq 1$, $R_{f/g}$: $0 < y \leq 2$, $R_{g/f}$: $y \geq \frac{1}{2}$

43. (a) $f(g(0)) = f(-3) = 2$

    (b) $g(f(0)) = g(5) = 22$

    (c) $f(g(x)) = f(x^2 - 3) = x^2 - 3 + 5 = x^2 + 2$

    (d) $g(f(x)) = g(x + 5) = (x + 5)^2 - 3 = x^2 + 10x + 22$

    (e) $f(f(-5)) = f(0) = 5$

    (f) $g(g(2)) = g(1) = -2$

    (g) $f(f(x)) = f(x + 5) = (x + 5) + 5 = x + 10$

    (h) $g(g(x)) = g(x^2 - 3) = (x^2 - 3)^2 - 3 = x^4 - 6x^2 + 6$

45. (a) $u(v(f(x))) = u\left(v\left(\frac{1}{x}\right)\right) = u\left(\frac{1}{x^2}\right) = 4\left(\frac{1}{x}\right)^2 - 5 = \frac{4}{x^2} - 5$

    (b) $u(f(v(x))) = u\left(f\left(x^2\right)\right) = u\left(\frac{1}{x^2}\right) = 4\left(\frac{1}{x^2}\right) - 5 = \frac{4}{x^2} - 5$

    (c) $v(u(f(x))) = v\left(u\left(\frac{1}{x}\right)\right) = v\left(4\left(\frac{1}{x}\right) - 5\right) = \left(\frac{4}{x} - 5\right)^2$

    (d) $v(f(u(x))) = v(f(4x - 5)) = v\left(\frac{1}{4x - 5}\right) = \left(\frac{1}{4x - 5}\right)^2$

    (e) $f(u(v(x))) = f\left(u\left(x^2\right)\right) = f\left(4\left(x^2\right) - 5\right) = \frac{1}{4x^2 - 5}$

    (f) $f(v(u(x))) = f(v(4x - 5)) = f\left((4x - 5)^2\right) = \frac{1}{(4x - 5)^2}$

47. (a) $y = f(g(x))$ (b) $y = j(g(x))$
(c) $y = g(g(x))$ (d) $y = j(j(x))$
(e) $y = g(h(f(x)))$ (f) $y = h(j(f(x)))$

49.

	$g(x)$	$f(x)$	$(f \circ g)(x)$
(a)	$x - 7$	$\sqrt{x}$	$\sqrt{x - 7}$
(b)	$x + 2$	$3x$	$3(x + 2) = 3x + 6$
(c)	x^2	$\sqrt{x - 5}$	$\sqrt{x^2 - 5}$
(d)	$\dfrac{x}{x-1}$	$\dfrac{x}{x-1}$	$\dfrac{\frac{x}{x-1}}{\frac{x}{x-1}-1} = \dfrac{x}{x-(x-1)} = x$
(e)	$\dfrac{1}{x-1}$	$1 + \dfrac{1}{x}$	$1 + \dfrac{1}{\frac{1}{x-1}} = 1 + (x-1) = x$
(f)	$\dfrac{1}{x}$	$\dfrac{1}{x}$	$\dfrac{1}{\frac{1}{x}} = x$

51.

x	0	1	2
y	0	1	0

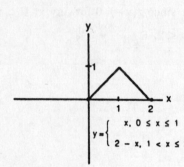

$y = \begin{cases} x, & 0 \le x \le 1 \\ 2 - x, & 1 < x \le 2 \end{cases}$

53. $y = \begin{cases} 3 - x, & x \le 1 \\ 2x, & 1 < x \end{cases}$

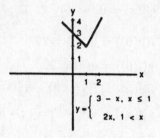

$y = \begin{cases} 3 - x, & x \le 1 \\ 2x, & 1 < x \end{cases}$

55. (a) $f(x) = \begin{cases} x, & 0 \le x \le 1 \\ 2 - x, & 1 < x \le 2 \end{cases}$ (b) $f(x) = \begin{cases} 2, & x \in [0,1] \cup [2,3] \\ 0, & x \in [1,2) \cup [3,4] \end{cases}$

57. (a) $\lfloor x \rfloor = 0$ for $x \in [0,1)$ (b) $\lceil x \rceil = 0$ for $x \in (-1, 0]$

59. Yes. The reflection of $y = \lceil x \rceil$ across the y-axis (to get the graph of $y = \lceil -x \rceil$) and of $y = \lfloor x \rfloor$ across the x-axis (to get the graph of $y = -\lfloor x \rfloor$) are the same.

61. (a) $(fg)(-x) = f(-x)g(-x) = f(x)(-g(x)) = -(fg)(x)$, odd

(b) $\left(\dfrac{f}{g}\right)(-x) = \dfrac{f(-x)}{g(-x)} = \dfrac{f(x)}{-g(x)} = -\left(\dfrac{f}{g}\right)(x)$, odd

(c) $\left(\dfrac{g}{f}\right)(-x) = \dfrac{g(-x)}{f(-x)} = \dfrac{-g(x)}{f(x)} = -\left(\dfrac{g}{f}\right)(x)$, odd

(d) $f^2(-x) = f(-x)f(-x) = f(x)f(x) = f^2(x)$, even

(e) $g^2(-x) = (g(-x))^2 = (-g(x))^2 = g^2(x)$, even

(f) $(f \circ g)(-x) = f(g(-x)) = f(-g(x)) = f(g(x)) = (f \circ g)(x)$, even

(g) $(g \circ f)(-x) = g(f(-x)) = g(f(x)) = (g \circ f)(x)$, even

(h) $(f \circ f)(-x) = f(f(-x)) = f(f(x)) = (f \circ f)(x)$, even

(i) $(g \circ g)(-x) = g(g(-x)) = g(-g(x)) = -g(g(x)) = -(g \circ g)(x)$, odd

63. (a)   (b)

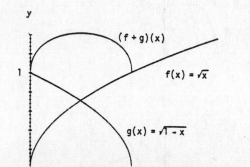

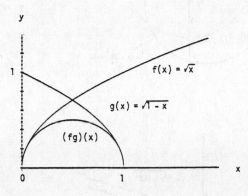

(c)   (d)

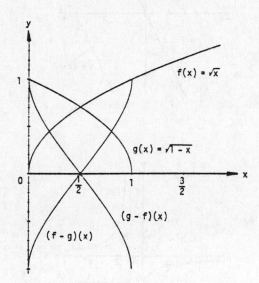

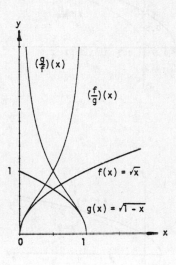

P.4 SHIFTING GRAPHS

1. (a) $y = -(x + 7)^2$ (b) $y = -(x - 4)^2$

3. (a) Position 4   (b) Position 1   (c) Position 2   (d) Position 3

5.

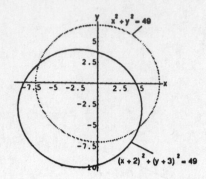

7.

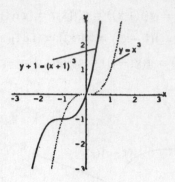

9.

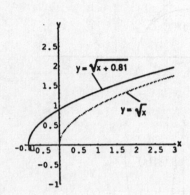

11.

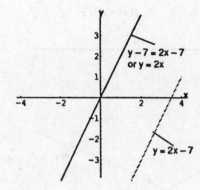

13.

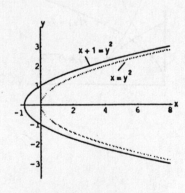

15.

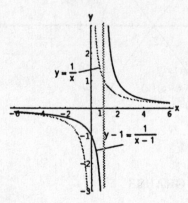

17.

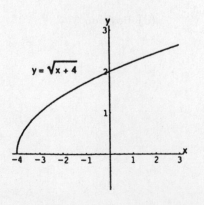

19.

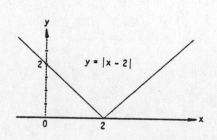

21.

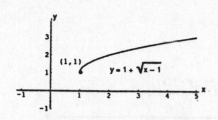

23.

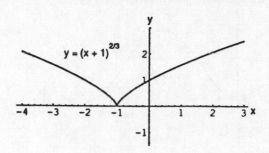

25.

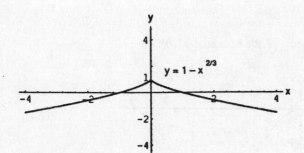

27.

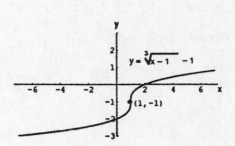

29.

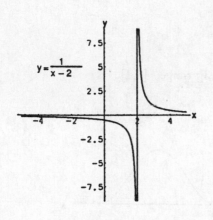

31.

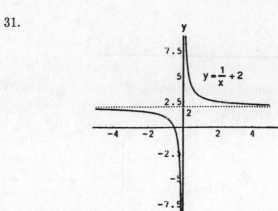

33.

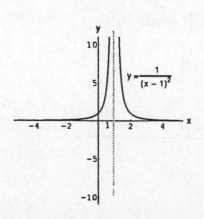

35.

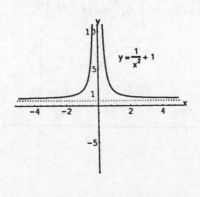

37. (a) domain: $[0,2]$; range: $[2,3]$

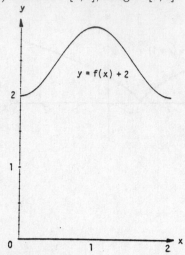

$y = f(x) + 2$

(b) domain: $[0,2]$; range: $[-1,0]$

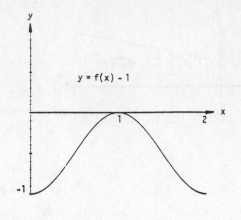

$y = f(x) - 1$

(c) domain: $[0,2]$; range: $[0,2]$

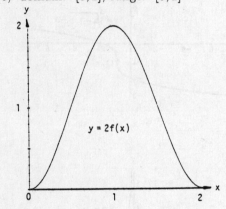

$y = 2f(x)$

(d) domain: $[0,2]$; range: $[-1,0]$

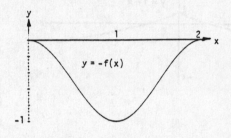

$y = -f(x)$

(e) domain: $[-2,0]$; range: $[0,1]$

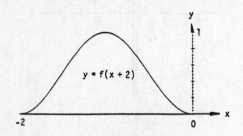

$y = f(x + 2)$

(f) domain: $[1,3]$; range: $[0,1]$

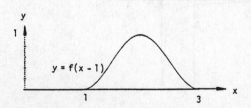

$y = f(x - 1)$

(g) domain: $[-2,0]$; range: $[0,1]$

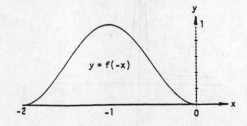

$y = f(-x)$

(h) domain: $[-1,1]$; range: $[0,1]$

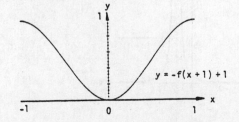

$y = -f(x + 1) + 1$

39. $C(0,2)$, $a = 2 \Rightarrow x^2 + (y-2)^2 = 4$

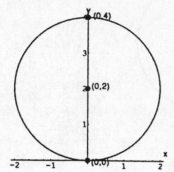

41. $C(-1,5)$, $a = \sqrt{10} \Rightarrow (x+1)^2 + (y-5)^2 = 10$

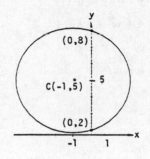

43. $C\left(-\sqrt{3},-2\right)$, $a = 2 \Rightarrow \left(x+\sqrt{3}\right)^2 + (y+2)^2 = 4$,

$x = 0 \Rightarrow \left(0+\sqrt{3}\right)^2 + (y+2)^2 = 4 \Rightarrow (y+2)^2 = 1$

$\Rightarrow y+2 = \pm 1 \Rightarrow y = -1$ or $y = -3$. Also, $y = 0$

$\Rightarrow \left(x+\sqrt{3}\right)^2 + (0+2)^2 = 4 \Rightarrow \left(x+\sqrt{3}\right)^2 = 0$

$\Rightarrow x = -\sqrt{3}$

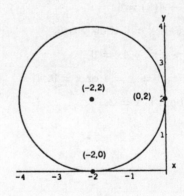

45. $x^2 + y^2 + 4x - 4y + 4 = 0$

$\Rightarrow x^2 + 4x + y^2 - 4y = -4$

$\Rightarrow x^2 + 4x + 4 + y^2 - 4y + 4 = 4$

$\Rightarrow (x+2)^2 + (y-2)^2 = 4 \Rightarrow C = (-2,2)$, $a = 2$.

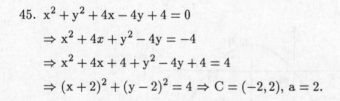

47. $x^2 + y^2 - 3y - 4 = 0 \Rightarrow x^2 + y^2 - 3y = 4$

$\Rightarrow x^2 + y^2 - 3y + \frac{9}{4} = \frac{25}{4}$

$\Rightarrow x^2 + \left(y - \frac{3}{2}\right)^2 = \frac{25}{4} \Rightarrow C = \left(0, \frac{3}{2}\right)$,

$a = \frac{5}{2}$.

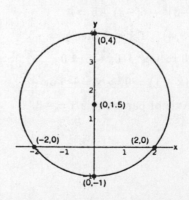

49. $x^2 + y^2 - 4x + 4y = 0$

$\Rightarrow x^2 - 4x + y^2 + 4y = 0$

$\Rightarrow x^2 - 4x + 4 + y^2 + 4y + 4 = 8$

$\Rightarrow (x-2)^2 + (y+2)^2 = 8$

$\Rightarrow C(2, -2), a = \sqrt{8}.$

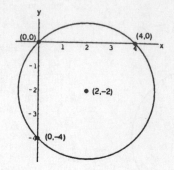

51. $x = -\dfrac{b}{2a} = -\dfrac{-2}{2(1)} = 1$

$\Rightarrow y = (1)^2 - 2(1) - 3 = -4$

$\Rightarrow V = (1, -4).$ If $x = 0$ then $y = -3.$

Also, $y = 0 \Rightarrow x^2 - 2x - 3 = 0$

$\Rightarrow (x-3)(x+1) = 0 \Rightarrow x = 3$ or

$x = -1.$ Axis of parabola is $x = 1.$

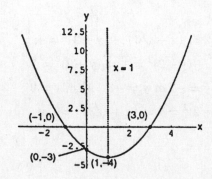

53. $x = -\dfrac{b}{2a} = -\dfrac{4}{2(-1)} = 2$

$\Rightarrow y = -(2)^2 + 4(2) = 4$

$\Rightarrow V = (2, 4).$ If $x = 0$ then $y = 0.$

Also, $y = 0 \Rightarrow -x^2 + 4x = 0$

$\Rightarrow -x(x-4) = 0 \Rightarrow x = 4$ or $x = 0.$

Axis of parabola is $x = 2.$

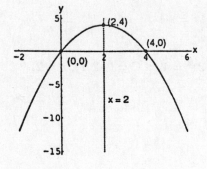

55. $x = -\dfrac{b}{2a} = -\dfrac{-6}{2(-1)} = -3$

$\Rightarrow y = -(-3)^2 - 6(-3) - 5 = 4$

$\Rightarrow V = (-3, 4).$ If $x = 0$ then $y = -5.$

Also, $y = 0 \Rightarrow -x^2 - 6x - 5 = 0$

$\Rightarrow (x+5)(x+1) = 0 \Rightarrow x = -5$ or

$x = -1.$ Axis of parabola is $x = -3.$

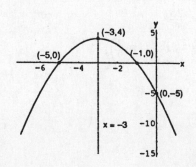

57. $x = -\dfrac{b}{2a} = -\dfrac{1}{2(1/2)} = -1$

$\Rightarrow y = \dfrac{1}{2}(-1)^2 + (-1) + 4 = \dfrac{7}{2}$

$\Rightarrow V = \left(-1, \dfrac{7}{2}\right)$. If $x = 0$ then $y = 4$.

Also, $y = 0 \Rightarrow \dfrac{1}{2}x^2 + x + 4 = 0$

$\Rightarrow x = \dfrac{-1 \pm \sqrt{-7}}{1} \Rightarrow$ no x intercepts.

Axis of parabola is $x = -1$.

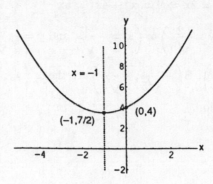

59. $x = -\dfrac{1}{2(-1)} = \dfrac{1}{2} \Rightarrow y = \dfrac{1}{2} - \left(\dfrac{1}{2}\right)^2 = \dfrac{1}{4}$

$\Rightarrow V = \left(\dfrac{1}{2}, \dfrac{1}{4}\right)$. If $x = 0$ then $y = 0$.

If $y = 0$ then $x - x^2 = 0 \Rightarrow x = 0$ or $x = 1$.

Domain of f: $x - x^2 \geq 0 \Rightarrow 0 \leq x \leq 1$,

the portion of the parabola above the x-axis

(or on it). Range of f: $0 \leq f(x) \leq \dfrac{1}{2}$.

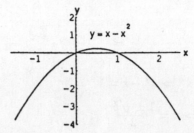

61. The points that lie outside the circle with center $(0,0)$ and radius $\sqrt{7}$.

63. The points that lie on or inside the circle with center $(1,0)$ and radius 2.

65. The points lying outside the circle with center $(0,0)$ and radius 1, but inside the circle with center $(0,0)$, and radius 2 (i.e., a washer).

67. $x^2 + y^2 + 6y < 0 \Rightarrow x^2 + (y+3)^2 < 9$.
The interior points of the circle centered at $(0,-3)$ with radius 3, but above the line $y = -3$.

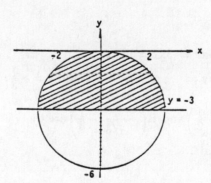

69. $(x+2)^2 + (y-1)^2 < 6$ 71. $x^2 + y^2 \leq 2$, $x \geq 1$

73. $y - y_0 = m(x - x_0)$, the point-slope form of the equation of a line.

75. $x^2 + y^2 = 1$ and $y = 2x \Rightarrow 1 = x^2 + 4x^2 = 5x^2$

$\Rightarrow \left(x = \dfrac{1}{\sqrt{5}} \text{ and } y = \dfrac{2}{\sqrt{5}} \right) \text{ or } \left(x = -\dfrac{1}{\sqrt{5}} \text{ and } y = -\dfrac{2}{\sqrt{5}} \right).$

Thus, $A\left(\dfrac{1}{\sqrt{5}}, \dfrac{2}{\sqrt{5}} \right)$, $B\left(-\dfrac{1}{\sqrt{5}}, -\dfrac{2}{\sqrt{5}} \right)$ are the

points of intersection.

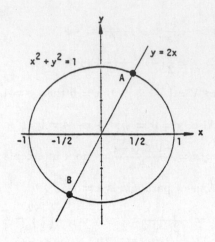

77. $y - x = 1$ and $y = x^2 \Rightarrow x^2 - x = 1$

$\Rightarrow x^2 - x - 1 = 0 \Rightarrow x = \dfrac{1 \pm \sqrt{5}}{2}.$

If $x = \dfrac{1 + \sqrt{5}}{2}$, then $y = x + 1 = \dfrac{3 + \sqrt{5}}{2}.$

If $x = \dfrac{1 - \sqrt{5}}{2}$, then $y = x + 1 = \dfrac{3 - \sqrt{5}}{2}.$

Thus, $A\left(\dfrac{1 + \sqrt{5}}{2}, \dfrac{3 + \sqrt{5}}{2} \right)$ and $B\left(\dfrac{1 - \sqrt{5}}{2}, \dfrac{3 - \sqrt{5}}{2} \right)$

are the intersection points.

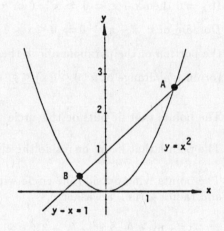

79. $y = 2x^2 - 1 = -x^2 \Rightarrow 3x^2 = 1$

$\Rightarrow x = \dfrac{1}{\sqrt{3}} \text{ and } y = -\dfrac{1}{3} \text{ or } x = -\dfrac{1}{\sqrt{3}} \text{ and } y = -\dfrac{1}{3}.$

Thus, $A\left(\dfrac{1}{\sqrt{3}}, -\dfrac{1}{3} \right)$ and $B\left(-\dfrac{1}{\sqrt{3}}, -\dfrac{1}{3} \right)$ are the

intersection points.

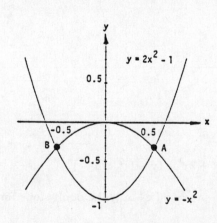

81. $x^2 + y^2 = 1 = (x-1)^2 + y^2$

$\Rightarrow x^2 = (x-1)^2 = x^2 - 2x + 1$

$\Rightarrow 0 = -2x + 1 \Rightarrow x = \frac{1}{2}.$ Hence

$y^2 = 1 - x^2 = \frac{3}{4}$ or $y = \pm\frac{\sqrt{3}}{2}.$ Thus,

$A\left(\frac{1}{2}, \frac{\sqrt{3}}{2}\right)$ and $B\left(\frac{1}{2}, -\frac{\sqrt{3}}{2}\right)$ are the

intersection points.

P.5 TRIGONOMETRIC FUNCTIONS

1. (a) $s = r\theta = (10)\left(\frac{4\pi}{5}\right) = 8\pi$ m

   (b) $s = r\theta = (10)(110°)\left(\frac{\pi}{180°}\right) = \frac{110\pi}{18} = \frac{55\pi}{9}$ m

3. $\theta = 80° \Rightarrow \theta = 80°\left(\frac{\pi}{180°}\right) = \frac{4\pi}{9} \Rightarrow s = (6)\left(\frac{4\pi}{9}\right) = 8.4$ in. (since the diameter = 12 in. $\Rightarrow$ radius = 6 in.)

5.

θ	$-\pi$	$-\frac{2\pi}{3}$	0	$\frac{\pi}{2}$	$\frac{3\pi}{4}$
$\sin\theta$	0	$-\frac{\sqrt{3}}{2}$	0	1	$\frac{1}{\sqrt{2}}$
$\cos\theta$	-1	$-\frac{1}{2}$	1	0	$-\frac{1}{\sqrt{2}}$
$\tan\theta$	0	$\sqrt{3}$	0	und.	-1
$\cot\theta$	und.	$\frac{1}{\sqrt{3}}$	und.	0	-1
$\sec\theta$	-1	-2	1	und.	$-\sqrt{2}$
$\csc\theta$	und.	$-\frac{2}{\sqrt{3}}$	und.	1	$\sqrt{2}$

7. $\cos x = -\frac{4}{5}, \tan x = -\frac{3}{4}$

9. $\sin x = -\frac{\sqrt{8}}{3}, \tan x = -\sqrt{8}$

11. $\sin x = -\frac{1}{\sqrt{5}}, \cos x = -\frac{2}{\sqrt{5}}$

13.

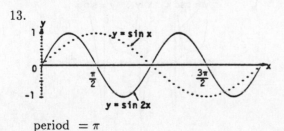

period $= \pi$

15.

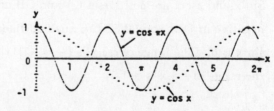

period $= 2$

17.

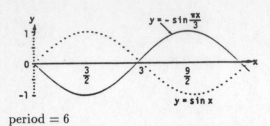

period = 6

19.

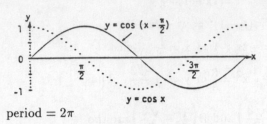

period = 2π

21.

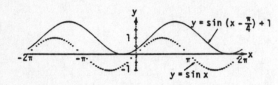

period = 2π

23. period = $\frac{\pi}{2}$, symmetric about the origin

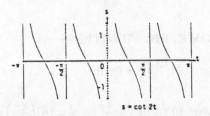

25. period = 4, symmetric about the y-axis

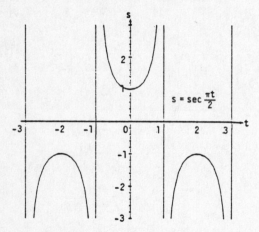

27. (a) Cos x and sec x are positive in Q1 and QIV and negative in QII and QIII. Sec x is undefined when cos x is 0. The range of sec x is $(-\infty, -1] \cup [1, \infty)$; the range of cos x is $[-1, 1]$.

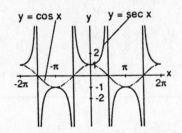

(b) Sin x and csc x are positive in Q1 and QII and negative in QIII and QIV. Csc x is undefined when sin x is 0. The range of csc x is $(-\infty, -1] \cup [1, \infty)$; the range of sin x is $[-1, 1]$.

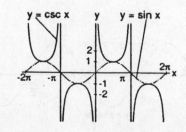

29. D: $-\infty < x < \infty$; R: $y = -1, 0, 1$

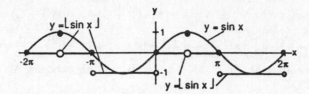

31. $\cos\left(x - \dfrac{\pi}{2}\right) = \cos x \cos\left(-\dfrac{\pi}{2}\right) - \sin x \sin\left(-\dfrac{\pi}{2}\right) = (\cos x)(0) - (\sin x)(-1) = \sin x$

33. $\sin\left(x + \dfrac{\pi}{2}\right) = \sin x \cos\left(\dfrac{\pi}{2}\right) + \cos x \sin\left(\dfrac{\pi}{2}\right) = (\sin x)(0) + (\cos x)(1) = \cos x$

35. $\cos(A - B) = \cos(A + (-B)) = \cos A \cos(-B) - \sin A \sin(-B) = \cos A \cos B - \sin A (-\sin B)$
$\qquad = \cos A \cos B + \sin A \sin B$

37. If $B = A$, $A - B = 0 \Rightarrow \cos(A - B) = \cos 0 = 1$. Also $\cos(A - B) = \cos(A - A) = \cos A \cos A + \sin A \sin A$
$\qquad = \cos^2 A + \sin^2 A$. Therefore, $\cos^2 A + \sin^2 A = 1$.

39. $\cos(\pi + x) = \cos \pi \cos x - \sin \pi \sin x = (-1)(\cos x) - (0)(\sin x) = -\cos x$

41. $\sin\left(\dfrac{3\pi}{2} - x\right) = \sin\left(\dfrac{3\pi}{2}\right)\cos(-x) + \cos\left(\dfrac{3\pi}{2}\right)\sin(-x) = (-1)(\cos x) + (0)(\sin(-x)) = -\cos x$

43. $\sin \dfrac{7\pi}{12} = \sin\left(\dfrac{\pi}{4} + \dfrac{\pi}{3}\right) = \sin \dfrac{\pi}{4}\cos \dfrac{\pi}{3} + \cos \dfrac{\pi}{4}\sin \dfrac{\pi}{3} = \left(\dfrac{\sqrt{2}}{2}\right)\left(\dfrac{1}{2}\right) + \left(\dfrac{\sqrt{2}}{2}\right)\left(\dfrac{\sqrt{3}}{2}\right) = \dfrac{\sqrt{6} + \sqrt{2}}{4}$

45. $\cos \dfrac{\pi}{12} = \cos\left(\dfrac{\pi}{3} - \dfrac{\pi}{4}\right) = \cos \dfrac{\pi}{3}\cos\left(-\dfrac{\pi}{4}\right) - \sin \dfrac{\pi}{3}\sin\left(-\dfrac{\pi}{4}\right) = \left(\dfrac{1}{2}\right)\left(\dfrac{\sqrt{2}}{2}\right) - \left(\dfrac{\sqrt{3}}{2}\right)\left(-\dfrac{\sqrt{2}}{2}\right) = \dfrac{1 + \sqrt{3}}{2\sqrt{2}}$

47. $\cos^2 \dfrac{\pi}{8} = \dfrac{1 + \cos\left(\dfrac{2\pi}{8}\right)}{2} = \dfrac{1 + \dfrac{\sqrt{2}}{2}}{2} = \dfrac{2 + \sqrt{2}}{4}$
$\qquad\qquad$ 49. $\sin^2 \dfrac{\pi}{12} = \dfrac{1 - \cos\left(\dfrac{2\pi}{12}\right)}{2} = \dfrac{1 - \dfrac{\sqrt{3}}{2}}{2} = \dfrac{2 - \sqrt{3}}{4}$

51. $\tan(A + B) = \dfrac{\sin(A + B)}{\cos(A + B)} = \dfrac{\sin A \cos B + \cos A \cos B}{\cos A \cos B - \sin A \sin B} = \dfrac{\dfrac{\sin A \cos B}{\cos A \cos B} + \dfrac{\cos A \sin B}{\cos A \cos B}}{\dfrac{\cos A \cos B}{\cos A \cos B} - \dfrac{\sin A \sin B}{\cos A \cos B}} = \dfrac{\tan A + \tan B}{1 - \tan A \tan B}$

53. According to the figure in the text, we have the following: By the law of cosines, $c^2 = a^2 + b^2 - 2ab \cos \theta$
$\qquad = 1^2 + 1^2 - 2\cos(A - B) = 2 - 2\cos(A - B)$. By distance formula, $c^2 = (\cos A - \cos B)^2 + (\sin A - \sin B)^2$
$\qquad = \cos^2 A - 2\cos A \cos B + \cos^2 B + \sin^2 A - 2\sin A \sin B + \sin^2 B = 2 - 2(\cos A \cos B + \sin A \sin B)$. Thus
$\qquad c^2 = 2 - 2\cos(A - B) = 2 - 2(\cos A \cos B + \sin A \sin B) \Rightarrow \cos(A - B) = \cos A \cos B + \sin A \sin B$.

55. $c^2 = a^2 + b^2 - 2ab \cos C = 2^2 + 3^2 - 2(2)(3)\cos(60°) = 4 + 9 - 12\cos(60°) = 13 - 12\left(\dfrac{1}{2}\right) = 7$.
$\qquad$ Thus, $c = \sqrt{7} \approx 2.65$.

57. From the figures in the text, we see that $\sin B = \dfrac{h}{c}$. If C is an acute angle, then $\sin C = \dfrac{h}{b}$. On the other hand,
$\qquad$ if C is obtuse (as in the figure on the right), then $\sin C = \sin(\pi - C) = \dfrac{h}{b}$. Thus, in either case,
$\qquad h = b \sin C = c \sin B \Rightarrow ah = ab \sin C = ac \sin B$.

By the law of cosines, $\cos C = \dfrac{a^2 + b^2 - c^2}{2ab}$ and $\cos B = \dfrac{a^2 + c^2 - b^2}{2ac}$. Moreover, since the sum of the

interior angles of a triangle is π, we have $\sin A = \sin(\pi - (B+C)) = \sin(B+C) = \sin B \cos C + \cos B \sin C$

$= \left(\dfrac{h}{c}\right)\left[\dfrac{a^2+b^2-c^2}{2ab}\right] + \left[\dfrac{a^2+c^2-b^2}{2ac}\right]\left(\dfrac{h}{b}\right) = \left(\dfrac{h}{2abc}\right)(2a^2 + b^2 - c^2 + c^2 - b^2) = \dfrac{ah}{bc} \Rightarrow ah = bc \sin A.$

Combining our results we have $ah = ab \sin C$, $ah = ac \sin B$, and $ah = bc \sin A$. Dividing by abc gives

$\dfrac{h}{bc} = \underbrace{\dfrac{\sin A}{a} = \dfrac{\sin C}{c} = \dfrac{\sin B}{b}}_{\text{law of sines}}.$

59. From the figure at the right and the law of cosines,

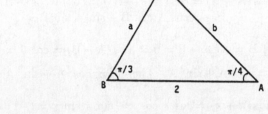

$b^2 = a^2 + 2^2 - 2(2a) \cos B = a^2 + 4 - 4a\left(\dfrac{1}{2}\right) = a^2 - 2a + 4.$

Applying the law of sines to the figure, $\dfrac{\sin A}{a} = \dfrac{\sin B}{b}$

$\Rightarrow \dfrac{\sqrt{2}/2}{a} = \dfrac{\sqrt{3}/2}{b} \Rightarrow b = \sqrt{\dfrac{3}{2}}a.$ Thus, combining results,

$a^2 - 2a + 4 = b^2 = \dfrac{3}{2}a^2 \Rightarrow 0 = \dfrac{1}{2}a^2 + 2a - 4$

$\Rightarrow 0 = a^2 + 4a - 8.$ From the quadratic formula and the

fact that $a > 0$, we have $a = \dfrac{-4 + \sqrt{4^2 - 4(1)(-8)}}{2} = \dfrac{4\sqrt{3} - 4}{2} \simeq 1.464.$

61. $A = 2$, $B = 2\pi$, $C = -\pi$, $D = -1$

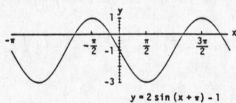

$y = 2\sin(x + \pi) - 1$

63. $A = -\dfrac{2}{\pi}$, $B = 4$, $C = 0$, $D = \dfrac{1}{\pi}$

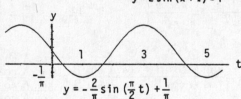

$y = -\dfrac{2}{\pi}\sin\left(\dfrac{\pi}{2}t\right) + \dfrac{1}{\pi}$

65. (a) amplitude $= |A| = 37$ (b) period $= |B| = 365$
    (c) right horizontal shift $= C = 101$ (d) upward vertical shift $= D = 25$

PRELIMINARIES PRACTICE EXERCISES

1. Since the particle moved to the y-axis, $-2 + \Delta x = 0 \Rightarrow \Delta x = 2$. Since $\Delta y = 3\Delta x = 6$, the new coordinates are $(x + \Delta x, y + \Delta y) = (-2 + 2, 5 + 6) = (0, 11)$.

3. The triangle ABC is neither an isosceles triangle nor is it a right triangle. The lengths of AB, BC and AC are $\sqrt{53}$, $\sqrt{72}$ and $\sqrt{65}$, respectively. The slopes of AB, BC and AC are $\dfrac{7}{2}$, -1 and $\dfrac{1}{8}$, respectively.

5. The area is $A = \pi r^2$ and the circumference is $C = 2\pi r$. Thus, $r = \dfrac{C}{2\pi} \Rightarrow A = \pi\left(\dfrac{C}{2\pi}\right)^2 = \dfrac{C^2}{4\pi}.$

7. The coordinates of a point on the parabola are (x, x^2). The angle of inclination θ joining this point to the origin satisfies the equation $\tan \theta = \frac{x^2}{x} = x$. Thus the point has coordinates $(x, x^2) = (\tan \theta, \tan^2 \theta)$.

9.

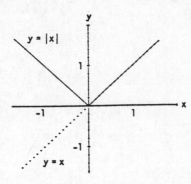

The graph of $f_2(x) = f_1(|x|)$ is the same as the graph of $f_1(x)$ to the right of the y-axis. The graph of $f_2(x)$ to the left of the y-axis is the reflection of $y = f_1(x)$, $x \geq 0$ across the y-axis.

11.

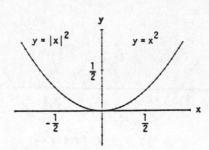

It does not change the graph.

13.

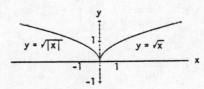

The graph of $f_2(x) = f_1(|x|)$ is the same as the graph of $f_1(x)$ to the right of the y-axis. The graph of $f_2(x)$ to the left of the y-axis is the reflection of $y = f_1(x)$, $x \geq 0$ across the y-axis.

15.

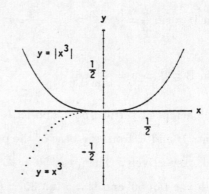

Whenever $g_1(x)$ is positive, the graph of $y = g_2(x) = |g_1(x)|$ is the same as the graph of $y = g_1(x)$. When $g_1(x)$ is negative, the graph of $y = g_2(x)$ is the reflection of the graph of $y = g_1(x)$ across the x-axis.

17.

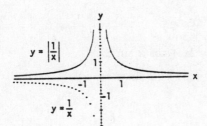

Whenever $g_1(x)$ is positive, the graph of $y = g_2(x) = |g_1(x)|$ is the same as the graph of $y = g_1(x)$. When $g_1(x)$ is negative, the graph of $y = g_2(x)$ is the reflection of the graph of $y = g_1(x)$ across the x-axis.

19.

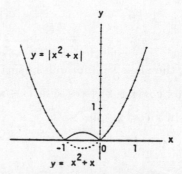

Whenever $g_1(x)$ is positive, the graph of $y = g_2(x) = |g_1(x)|$ is the same as the graph of $y = g_1(x)$. When $g_1(x)$ is negative, the graph of $y = g_2(x)$ is the reflection of the graph of $y = g_1(x)$ across the x-axis.

21.

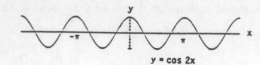

period $= \pi$

23.

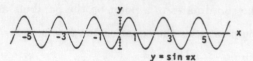

period $= 2$

25.

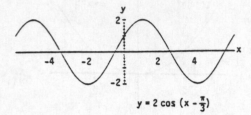

period $= 2\pi$

27. (a) $\sin B = \sin \frac{\pi}{3} = \frac{b}{c} = \frac{b}{2} \Rightarrow b = 2 \sin \frac{\pi}{3} = 2\left(\frac{\sqrt{3}}{2}\right) = \sqrt{3}$. By the theorem of Pythagoras,

$a^2 + b^2 = c^2 \Rightarrow a = \sqrt{c^2 - b^2} = \sqrt{4-3} = 1$.

(b) $\sin B = \sin \frac{\pi}{3} = \frac{b}{c} = \frac{2}{c} \Rightarrow c = \frac{2}{\sin \frac{\pi}{3}} = \frac{2}{\left(\frac{\sqrt{3}}{2}\right)} = \frac{4}{\sqrt{3}}$. Thus, $a = \sqrt{c^2 - b^2} = \sqrt{\left(\frac{4}{\sqrt{3}}\right)^2 - (2)^2} = \sqrt{\frac{4}{3}} = \frac{2}{\sqrt{3}}$.

29. (a) $\tan B = \frac{b}{a} \Rightarrow a = \frac{b}{\tan B}$ (b) $\sin A = \frac{a}{c} \Rightarrow c = \frac{a}{\sin A}$

31. Let $h =$ height of vertical pole, and let b and c denote the distances
of points B and C from the base of the pole, measured along the flat
ground, respectively. Then, $\tan 50° = \frac{h}{c}$, $\tan 35° = \frac{h}{b}$, and $b - c = 10$.
Thus, $h = c \tan 50°$ and $h = b \tan 35° = (c + 10) \tan 35°$
$\Rightarrow c \tan 50° = (c + 10) \tan 35° \Rightarrow c (\tan 50° - \tan 35°) = 10 \tan 35°$
$\Rightarrow c = \dfrac{10 \tan 35°}{\tan 50° - \tan 35°} \Rightarrow h = c \tan 50° = \dfrac{10 \tan 35° \tan 50°}{\tan 50° - \tan 35°}$
≈ 16.98 m.

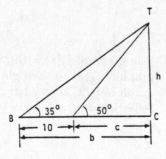

33. Using the angle sum formulas: $\sin (3x) = \sin (x + 2x) = \sin x \cos 2x + \cos x \sin 2x$

$= \sin x \cos (x + x) + \cos x \sin (x + x) = (\sin x)\left(\cos^2 x - \sin^2 x\right) + (\cos x)(2 \sin x \cos x)$

$= 3 \sin x \cos^2 x - \sin^3 x$.

35. (a)

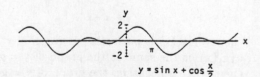

(b) The period appears to be 4π.

(c) $f(x + 4\pi) = \sin(x + 4\pi) + \cos\left(\dfrac{x + 4\pi}{2}\right) = \sin(x + 2\pi) + \cos\left(\dfrac{x}{2} + 2\pi\right) = \sin x + \cos\dfrac{x}{2}$

since the period of sine and cosine is 2π. Thus, $f(x)$ has period 4π.

PRELIMINARIES ADDITIONAL EXERCISES–THEORY, EXAMPLES, APPLICATIONS

1. Each of the triangles pictured has the same base

$b = v\Delta t = v(1 \text{ sec})$. Moreover, the height of each

triangle is the same value h. Thus $\dfrac{1}{2}$(base)(height) $= \dfrac{1}{2}$bh

$= A_1 = A_2 = A_3 = \dots$. In conclusion, the object sweeps

out equal areas in each one second interval.

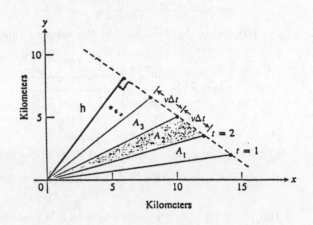

3. There are (infinitely) many such function pairs. For example, $f(x) = 3x$ and $g(x) = 4x$ satisfy

$f(g(x)) = f(4x) = 3(4x) = 12x = 4(3x) = g(3x) = g(f(x))$.

5. If f is odd and defined at x, then $f(-x) = -f(x)$. Thus $g(-x) = f(-x) - 2 = -f(x) - 2$ whereas

$-g(x) = -(f(x) - 2) = -f(x) + 2$. Then g cannot be odd because $g(-x) = -g(x) \Rightarrow -f(x) - 2 = -f(x) + 2$

$\Rightarrow 4 = 0$, which is a contradiction. Also, $g(x)$ is not even unless $f(x) = 0$ for all x. On the other hand, if f is

even, then $g(x) = f(x) - 2$ is also even: $g(-x) = f(-x) - 2 = f(x) - 2 = g(x)$.

7. For (x, y) in the 1st quadrant, $|x| + |y| = 1 + x$

$\Leftrightarrow x + y = 1 + x \Leftrightarrow y = 1$. For (x, y) in the 2nd

quadrant, $|x| + |y| = x + 1 \Leftrightarrow -x + y = x + 1$

$\Leftrightarrow y = 2x + 1$. In the 3rd quadrant, $|x| + |y| = x + 1$

$\Leftrightarrow -x - y = x + 1 \Leftrightarrow y = -2x - 1$. In the 4th

quadrant, $|x| + |y| = x + 1 \Leftrightarrow x + (-y) = x + 1$

$\Leftrightarrow y = -1$. The graph is given at the right.

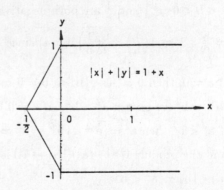

9. By the law of sines, $\dfrac{\sin\dfrac{\pi}{3}}{\sqrt{3}} = \dfrac{\sin A}{a} = \dfrac{\sin B}{b} = \dfrac{\sin\dfrac{\pi}{4}}{b} \Rightarrow b = \dfrac{\sqrt{3}\sin(\pi/4)}{\sin(\pi/3)} = \dfrac{\sqrt{3}\left(\dfrac{\sqrt{2}}{2}\right)}{\dfrac{\sqrt{3}}{2}} = \sqrt{2}$.

11. By the law of cosines, $a^2 = b^2 + c^2 - 2bc\cos A \Rightarrow \cos A = \dfrac{b^2 + c^2 - a^2}{2bc} = \dfrac{2^2 + 3^2 - 2^2}{2(2)(3)} = \dfrac{3}{4}$.

13. By the law of cosines, $b^2 = a^2 + c^2 - 2ac \cos B \Rightarrow \cos B = \dfrac{a^2 + c^2 - b^2}{2ac} = \dfrac{2^2 + 4^2 - 3^2}{(2)(2)(4)} = \dfrac{4 + 16 - 9}{16}$

$= \dfrac{11}{16}$. Since $0 < B < \pi$, $\sin B = \sqrt{1 - \cos^2 B} = \sqrt{1 - \dfrac{121}{256}} = \dfrac{\sqrt{135}}{16} = \dfrac{3\sqrt{15}}{16}$.

15. (a) $\sin^2 x + \cos^2 x = 1 \Rightarrow \sin^2 x = 1 - \cos^2 x = (1 - \cos x)(1 + \cos x) \Rightarrow (1 - \cos x) = \dfrac{\sin^2 x}{1 + \cos x}$

$\Rightarrow \dfrac{1 - \cos x}{\sin x} = \dfrac{\sin x}{1 + \cos x}$

(b) Using the definition of the tangent function and the double angle formulas, we have

$$\tan^2\left(\frac{x}{2}\right) = \frac{\sin^2\left(\frac{x}{2}\right)}{\cos^2\left(\frac{x}{2}\right)} = \frac{\dfrac{1 - \cos\left(2\left(\frac{x}{2}\right)\right)}{2}}{\dfrac{1 + \cos\left(2\left(\frac{x}{2}\right)\right)}{2}} = \frac{1 - \cos x}{1 + \cos x}.$$

17. As in the proof of the law of sines of Section P.5, Exercise 57, $ah = bc \sin A = ab \sin C = ac \sin B$

$\Rightarrow$ the area of ABC $= \frac{1}{2}$(base)(height) $= \frac{1}{2}ah = \frac{1}{2}bc \sin A = \frac{1}{2}ab \sin C = \frac{1}{2}ac \sin B$.

19. 1. $b + c - (a + c) = b - a$, which is positive since $a < b$. Thus, $a + c < b + c$.

2. $b - c - (a - c) = b - a$, which is positive since $a < b$. Thus, $a - c < b - c$.

3. $c > 0$ and $a < b \Rightarrow c - 0 = c$ and $b - a$ are positive $\Rightarrow (b - a)c = bc - ac$ is positive $\Rightarrow ac < bc$.

4. $a < b$ and $c < 0 \Rightarrow b - a$ and $-c$ are positive $\Rightarrow (b - a)(-c) = ac - bc$ is positive $\Rightarrow bc < ac$.

5.   Since $a > 0$, a and $\frac{1}{a}$ are positive $\Rightarrow \frac{1}{a} > 0$.

6.   Since $0 < a < b$, both $\frac{1}{a}$ and $\frac{1}{b}$ are positive. By (3), $a < b$ and $\frac{1}{a} > 0 \Rightarrow a\left(\frac{1}{a}\right) < b\left(\frac{1}{a}\right)$ or $1 < \frac{b}{a}$

$\Rightarrow 1\left(\frac{1}{b}\right) < \frac{b}{a}\left(\frac{1}{b}\right)$ by (3) since $\frac{1}{b} > 0 \Rightarrow \frac{1}{b} < \frac{1}{a}$.

7. $a < b < 0 \Rightarrow \frac{1}{a}$ and $\frac{1}{b}$ are both negative, i.e., $\frac{1}{a} < 0$ and $\frac{1}{b} < 0$. By (4), $a < b$ and $\frac{1}{a} < 0 \Rightarrow b\left(\frac{1}{a}\right) < a\left(\frac{1}{a}\right)$

$\Rightarrow \frac{b}{a} < 1 \Rightarrow 1\left(\frac{1}{b}\right) < \frac{b}{a}\left(\frac{1}{b}\right)$ by (4) since $\frac{1}{b} < 0 \Rightarrow \frac{1}{b} < \frac{1}{a}$.

21. (a) If $a = 0$, then $0 = |a| < |b| \Leftrightarrow b \neq 0 \Leftrightarrow 0 = |a|^2 < |b|^2$. Since $|a|^2 = |a||a| = |a^2| = a^2$ and

$|b|^2 = b^2$ we obtain $a^2 < b^2$. If $a \neq 0$ then $|a| > 0$ and $|a| < |b| \Rightarrow a^2 < b^2$. On the other hand,

if $a^2 < b^2$ then $a^2 = |a|^2 < |b|^2 = b^2 \Rightarrow 0 < |b|^2 - |a|^2 = (|b| - |a|)(|b| + |a|)$. Since $(|b| + |a|) > 0$

and the product $(|b| - |a|)(|b| + |a|)$ is positive, we must have $(|b| - |a|) > 0 \Rightarrow |b| > |a|$. Thus

$|a| < |b| \Leftrightarrow a^2 < b^2$.

(b) $ab \leq |ab| \Rightarrow -ab \geq -2|ab|$ by Exercise 19(4) above $\Rightarrow a^2 - 2ab + b^2 \geq |a|^2 - 2|a||b| + |b|^2$, since

$|a|^2 = a^2$ and $|b|^2 = b^2$. Factoring both sides, $(a - b)^2 \geq (|a| - |b|)^2 \Rightarrow |a - b| \geq ||a| - |b||$, by part (a).

23. If f is even and odd, then $f(-x) = -f(x)$ and $f(-x) = f(x) \Rightarrow f(x) = -f(x)$ for all x in the domain of f.

Thus $2f(x) = 0 \Rightarrow f(x) = 0$.

25. $y = ax^2 + bx + c = a\left(x^2 + \frac{b}{a}x + \frac{b^2}{4a^2}\right) - \frac{b^2}{4a} + c = a\left(x + \frac{b}{2a}\right)^2 - \frac{b^2}{4a} + c$

   (a) If $a > 0$ the graph is a parabola that opens upward. Increasing a causes a vertical stretching and a shift of the vertex toward the y-axis and upward. If $a < 0$ the graph is a parabola that opens downward. Decreasing a causes a vertical stretching and a shift of the vertex toward the y-axis and downward.

   (b) If $a > 0$ the graph is a parabola that opens upward. If also $b > 0$, then increasing b causes a shift of the graph downward to the left; if $b < 0$, then decreasing b causes a shift of the graph downward and to the right.

       If $a < 0$ the graph is a parabola that opens downward. If $b > 0$, increasing b shifts the graph upward to the right. If $b < 0$, decreasing b shifts the graph upward to the left.

   (c) Changing c (for fixed a and b) by Δc shifts the graph upward Δc units if $\Delta c > 0$, and downward $-\Delta c$ units if $\Delta c < 0$.

27. If $m > 0$, the x-intercept of $y = mx + 2$ must be negative. If $m < 0$, then the x-intercept exceeds $\frac{1}{2}$

   $\Rightarrow 0 = mx + 2$ and $x > \frac{1}{2} \Rightarrow x = -\frac{2}{m} > \frac{1}{2} \Rightarrow 0 > m > -4$.

NOTES:

CHAPTER 1 LIMITS AND CONTINUITY

1.1 RATE OF CHANGE AND LIMITS

1. (a) Does not exist. As x approaches 1 from the right, g(x) approaches 0. As x approaches 1 from the left, g(x) approaches 1. There is no single number L that all the values g(x) get arbitrarily close to as $x \to 1$.
   (b) 1
   (c) 0

3. (a) True   (b) True   (c) False
   (d) False   (e) False   (f) True

5. $\lim\limits_{x \to 0} \frac{x}{|x|}$ does not exist because $\frac{x}{|x|} = \frac{x}{x} = 1$ if $x > 0$ and $\frac{x}{|x|} = \frac{x}{-x} = -1$ if $x < 0$. As x approaches 0 from the left, $\frac{x}{|x|}$ approaches -1. As x approaches 0 from the right, $\frac{x}{|x|}$ approaches 1. There is no single number L that all the function values get arbitrarily close to as $x \to 0$.

7. Nothing can be said about f(x) because the existence of a limit as $x \to x_0$ does not depend on how the function is defined at x_0. In order for a limit to exist, f(x) must be arbitrarily close to a single real number L when x is close enough to x_0. That is, the existence of a limit depends on the values of f(x) for x <u>near</u> x_0, not on the definition of f(x) at x_0 itself.

9. No, the definition does not require that f be defined at $x = 1$ in order for a limiting value to exist there. If f(1) is defined, it can be any real number, so we can conclude nothing about f(1) from $\lim\limits_{x \to 1} f(x) = 5$.

11. (a) $f(x) = (x^2 - 9)/(x + 3)$

x	−3.1	−3.01	−3.001	−3.0001	−3.00001	−3.000001
f(x)	−6.1	−6.01	−6.001	−6.0001	−6.00001	−6.000001

x	−2.9	−2.99	−2.999	−2.9999	−2.99999	−2.999999
f(x)	−5.9	−5.99	−5.999	−5.9999	−5.99999	−5.999999

The estimate is $\lim\limits_{x \to -3} f(x) = -6$.

(b)

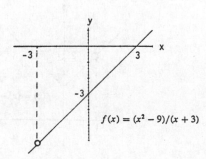

$f(x) = (x^2 - 9)/(x + 3)$

(c) $f(x) = \frac{x^2 - 9}{x + 3} = \frac{(x + 3)(x - 3)}{x + 3} = x - 3$ if $x \neq -3$, and $\lim\limits_{x \to -3} (x - 3) = -3 - 3 = -6$.

13. (a) $G(x) = (x + 6)/(x^2 + 4x - 12)$

x	−5.9	−5.99	−5.999	−5.9999	−5.99999	−5.999999
G(x)	−.126582	−.1251564	−.1250156	−.1250015	−.1250001	−.1250000

x	−6.1	−6.01	−6.001	−6.0001	−6.00001	−6.000001
G(x)	−.123456	−.124843	−.124984	−.124998	−.124999	−.124999

(b)

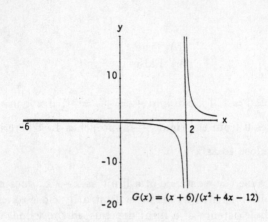

(c) $G(x) = \dfrac{x + 6}{(x^2 + 4x - 12)} = \dfrac{x + 6}{(x + 6)(x - 2)} = \dfrac{1}{x - 2}$ if $x \neq -6$, and $\displaystyle\lim_{x \to -6} \dfrac{1}{x - 2} = \dfrac{1}{-6 - 2} = -\dfrac{1}{8} = -0.125.$

15. (a) $f(x) = (x^2 - 1)/(|x| - 1)$

x	−1.1	−1.01	−1.001	−1.0001	−1.00001	−1.000001
f(x)	2.1	2.01	2.001	2.0001	2.00001	2.000001

x	−.9	−.99	−.999	−.9999	−.99999	−.999999
f(x)	1.9	1.99	1.999	1.9999	1.99999	1.999999

(b)

$f(x) = (x^2 - 1)/(|x| - 1)$

(c) $f(x) = \dfrac{x^2 - 1}{|x| - 1} = \begin{cases} \dfrac{(x + 1)(x - 1)}{x - 1} = x + 1, & x \geq 0 \text{ and } x \neq 1 \\[3mm] \dfrac{(x + 1)(x - 1)}{-(x + 1)} = 1 - x, & x < 0 \text{ and } x \neq -1 \end{cases}$, and $\displaystyle\lim_{x \to -1} (1 - x) = 1 - (-1) = 2.$

17. (a) $g(\theta) = (\sin \theta)/\theta$

θ	.1	.01	.001	.0001	.00001	.000001
$g(\theta)$	.998334	.999983	.999999	.999999	.999999	.999999

θ	−.1	−.01	−.001	−.0001	−.00001	−.000001
$g(\theta)$	.998334	.999983	.999999	.999999	.999999	.999999

$$\lim_{\theta \to 0} g(\theta) = 1$$

(b)

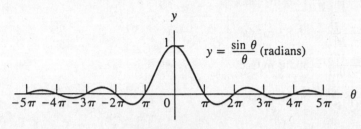

$$y = \frac{\sin \theta}{\theta} \text{ (radians)}$$

NOT TO SCALE

19. (a) $f(x) = x^{1/(1-x)}$

x	.9	.99	.999	.9999	.99999	.999999
f(x)	.348678	.366032	.367695	.367861	.367877	.367879

x	1.1	1.01	1.001	1.0001	1.00001	1.000001
f(x)	.385543	.369711	.368063	.367897	.367881	.367878

$$\lim_{x \to 1} f(x) \approx 0.36788$$

(b)

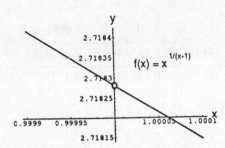

$$f(x) = x^{1/(x-1)}$$

Graph is NOT TO SCALE. Also the intersection of the axes is not the origin: the axes intersect at the point $(1, 2.71820)$.

21. $\lim\limits_{x \to 2} 2x = 2(2) = 4$

23. $\lim\limits_{x \to \frac{1}{3}} (3x - 1) = 3\left(\frac{1}{3}\right) - 1 = 0$

25. $\lim\limits_{x \to -1} 3x(2x - 1) = 3(-1)(2(-1) - 1) = 9$

27. $\lim\limits_{x \to \frac{\pi}{2}} x \sin x = \frac{\pi}{2} \sin \frac{\pi}{2} = \frac{\pi}{2}$

29. (a) $\dfrac{\Delta f}{\Delta x} = \dfrac{f(3) - f(2)}{3 - 2} = \dfrac{28 - 9}{1} = 19$

(b) $\dfrac{\Delta f}{\Delta x} = \dfrac{f(1) - f(-1)}{1 - (-1)} = \dfrac{2 - 0}{2} = 1$

31. (a) $\dfrac{\Delta h}{\Delta t} = \dfrac{h\left(\frac{3\pi}{4}\right) - h\left(\frac{\pi}{4}\right)}{\frac{3\pi}{4} - \frac{\pi}{4}} = \dfrac{-1-1}{\frac{\pi}{2}} = -\dfrac{4}{\pi}$ (b) $\dfrac{\Delta h}{\Delta t} = \dfrac{h\left(\frac{\pi}{2}\right) - h\left(\frac{\pi}{6}\right)}{\frac{\pi}{2} - \frac{\pi}{6}} = \dfrac{0 - \sqrt{3}}{\frac{\pi}{3}} = \dfrac{-3\sqrt{3}}{\pi}$

33. $\dfrac{\Delta R}{\Delta \theta} = \dfrac{R(2) - R(0)}{2 - 0} = \dfrac{\sqrt{8+1} - \sqrt{1}}{2} = \dfrac{3-1}{2} = 1$

35. (a)

Q	Slope of PQ $= \dfrac{\Delta p}{\Delta t}$
$Q_1(10, 225)$	$\dfrac{650 - 225}{20 - 10} = 42.5$ m/sec
$Q_2(14, 375)$	$\dfrac{650 - 375}{20 - 14} = 45.83$ m/sec
$Q_3(16.5, 475)$	$\dfrac{650 - 475}{20 - 16.5} = 50.00$ m/sec
$Q_4(18, 550)$	$\dfrac{650 - 550}{20 - 18} = 50.00$ m/sec

(b) At $t = 20$, the Cobra was traveling approximately 50 m/sec or 180 km/h.

37. (b) $\dfrac{\Delta p}{\Delta t} = \dfrac{174 - 62}{1994 - 1992} = \dfrac{112}{2} = 56$ thousand dollars per year

(c) Approximately \$42,000/yr

39. (a) $\dfrac{\Delta g}{\Delta x} = \dfrac{g(2) - g(1)}{2 - 1} = \dfrac{\sqrt{2} - 1}{2 - 1} \approx 0.414213$ $\dfrac{\Delta g}{\Delta x} = \dfrac{g(1.5) - g(1)}{1.5 - 1} = \dfrac{\sqrt{1.5} - 1}{0.5} \approx 0.449489$

$\dfrac{\Delta g}{\Delta x} = \dfrac{g(1+h) - g(1)}{(1+h) - 1} = \dfrac{\sqrt{1+h} - 1}{h}$

(b) $g(x) = \sqrt{x}$

$1 + h$	1.1	1.01	1.001	1.0001	1.00001	1.000001
$\sqrt{1+h}$	1.04880	1.004987	1.0004998	1.0000499	1.000005	1.0000005
$(\sqrt{1+h} - 1)/h$	0.4880	0.4987	0.4998	0.499	0.5	0.5

(c) The rate of change of $g(x)$ at $x = 1$ is 0.5.

(d) The calculator gives $\displaystyle\lim_{h \to 0} \dfrac{\sqrt{1+h} - 1}{h} = \dfrac{1}{2}$.

1.2 RULES FOR FINDING LIMITS

1. $\displaystyle\lim_{x \to -7} (2x + 5) = 2(-7) + 5 = -14 + 5 = -9$

3. $\displaystyle\lim_{x \to 2} (-x^2 + 5x - 2) = -(2)^2 + 5(2) - 2 = -4 + 10 - 2 = 4$

5. $\displaystyle\lim_{t \to 6} 8(t - 5)(t - 7) = 8(6 - 5)(6 - 7) = -8$ 7. $\displaystyle\lim_{x \to 2} \dfrac{x + 3}{x + 6} = \dfrac{2 + 3}{2 + 6} = \dfrac{5}{8}$

9. $\displaystyle\lim_{y \to -5} \dfrac{y^2}{5 - y} = \dfrac{(-5)^2}{5 - (-5)} = \dfrac{25}{10} = \dfrac{5}{2}$

11. $\lim\limits_{x \to -1} 3(2x-1)^2 = 3(2(-1)-1)^2 = 3(-3)^2 = 27$

13. $\lim\limits_{y \to -3} (5-y)^{4/3} = [5-(-3)]^{4/3} = (8)^{4/3} = \left((8)^{1/3}\right)^4 = 2^4 = 16$

15. $\lim\limits_{h \to 0} \dfrac{3}{\sqrt{3h+1}+1} = \dfrac{3}{\sqrt{3(0)+1}+1} = \dfrac{3}{\sqrt{1}+1} = \dfrac{3}{2}$

17. $\lim\limits_{x \to 5} \dfrac{x-5}{x^2-25} = \lim\limits_{x \to 5} \dfrac{x-5}{(x+5)(x-5)} = \lim\limits_{x \to 5} \dfrac{1}{x+5} = \dfrac{1}{5+5} = \dfrac{1}{10}$

19. $\lim\limits_{x \to -5} \dfrac{x^2+3x-10}{x+5} = \lim\limits_{x \to -5} \dfrac{(x+5)(x-2)}{x+5} = \lim\limits_{x \to -5} (x-2) = -5-2 = -7$

21. $\lim\limits_{t \to 1} \dfrac{t^2+t-2}{t^2-1} = \lim\limits_{t \to 1} \dfrac{(t+2)(t-1)}{(t-1)(t+1)} = \lim\limits_{t \to 1} \dfrac{t+2}{t+1} = \dfrac{1+2}{1+1} = \dfrac{3}{2}$

23. $\lim\limits_{x \to -2} \dfrac{-2x-4}{x^3+2x^2} = \lim\limits_{x \to -2} \dfrac{-2(x+2)}{x^2(x+2)} = \lim\limits_{x \to -2} \dfrac{-2}{x^2} = \dfrac{-2}{4} = -\dfrac{1}{2}$

25. $\lim\limits_{u \to 1} \dfrac{u^4-1}{u^3-1} = \lim\limits_{u \to 1} \dfrac{(u^2+1)(u+1)(u-1)}{(u^2+u+1)(u-1)} = \lim\limits_{u \to 1} \dfrac{(u^2+1)(u+1)}{u^2+u+1} = \dfrac{(1+1)(1+1)}{1+1+1} = \dfrac{4}{3}$

27. $\lim\limits_{x \to 9} \dfrac{\sqrt{x}-3}{x-9} = \lim\limits_{x \to 9} \dfrac{\sqrt{x}-3}{(\sqrt{x}-3)(\sqrt{x}+3)} = \lim\limits_{x \to 9} \dfrac{1}{\sqrt{x}+3} = \dfrac{1}{\sqrt{9}+3} = \dfrac{1}{6}$

29. $\lim\limits_{x \to 1} \dfrac{x-1}{\sqrt{x+3}-2} = \lim\limits_{x \to 1} \dfrac{(x-1)(\sqrt{x+3}+2)}{(\sqrt{x+3}-2)(\sqrt{x+3}+2)} = \lim\limits_{x \to 1} \dfrac{(x-1)(\sqrt{x+3}+2)}{(x+3)-4} = \lim\limits_{x \to 1} \left(\sqrt{x+3}+2\right)$

$$= \sqrt{4}+2 = 4$$

31. (a) quotient rule
    (b) difference and power rules
    (c) sum and constant multiple rules

33. (a) $\lim\limits_{x \to c} f(x)\, g(x) = \left[\lim\limits_{x \to c} f(x)\right]\left[\lim\limits_{x \to c} g(x)\right] = (5)(-2) = -10$

    (b) $\lim\limits_{x \to c} 2f(x)\, g(x) = 2\left[\lim\limits_{x \to c} f(x)\right]\left[\lim\limits_{x \to c} g(x)\right] = 2(5)(-2) = -20$

    (c) $\lim\limits_{x \to c} [f(x)+3g(x)] = \lim\limits_{x \to c} f(x) + 3\lim\limits_{x \to c} g(x) = 5+3(-2) = -1$

    (d) $\lim\limits_{x \to c} \dfrac{f(x)}{f(x)-g(x)} = \dfrac{\lim\limits_{x \to c} f(x)}{\lim\limits_{x \to c} f(x) - \lim\limits_{x \to c} g(x)} = \dfrac{5}{5-(-2)} = \dfrac{5}{7}$

35. (a) $\lim\limits_{x \to b} [f(x)+g(x)] = \lim\limits_{x \to b} f(x) + \lim\limits_{x \to b} g(x) = 7+(-3) = 4$

    (b) $\lim\limits_{x \to b} f(x) \cdot g(x) = \left[\lim\limits_{x \to b} f(x)\right]\left[\lim\limits_{x \to b} g(x)\right] = (7)(-3) = -21$

    (c) $\lim\limits_{x \to b} 4g(x) = \left[\lim\limits_{x \to b} 4\right]\left[\lim\limits_{x \to b} g(x)\right] = (4)(-3) = -12$

    (d) $\lim\limits_{x \to b} f(x)/g(x) = \lim\limits_{x \to b} f(x) / \lim\limits_{x \to b} g(x) = \dfrac{7}{-3} = -\dfrac{7}{3}$

37. $\lim\limits_{h \to 0} \dfrac{(1+h)^2 - 1^2}{h} = \lim\limits_{h \to 0} \dfrac{1 + 2h + h^2 - 1}{h} = \lim\limits_{h \to 0} \dfrac{h(2+h)}{h} = \lim\limits_{h \to 0} (2+h) = 2$

39. $\lim\limits_{h \to 0} \dfrac{[3(2+h) - 4] - [3(2) - 4]}{h} = \lim\limits_{h \to 0} \dfrac{3h}{h} = 3$

41. $\lim\limits_{h \to 0} \dfrac{\sqrt{7+h} - \sqrt{7}}{h} = \lim\limits_{h \to 0} \dfrac{\left(\sqrt{7+h} - \sqrt{7}\right)\left(\sqrt{7+h} + \sqrt{7}\right)}{h\left(\sqrt{7+h} + \sqrt{7}\right)} = \lim\limits_{h \to 0} \dfrac{(7+h) - 7}{h\left(\sqrt{7+h} + \sqrt{7}\right)}$

$= \lim\limits_{h \to 0} \dfrac{h}{h\left(\sqrt{7+h} + \sqrt{7}\right)} = \lim\limits_{h \to 0} \dfrac{1}{\sqrt{7+h} + \sqrt{7}} = \dfrac{1}{2\sqrt{7}}$

43. $\lim\limits_{x \to 0} \sqrt{5 - 2x^2} = \sqrt{5 - 2(0)^2} = \sqrt{5}$ and $\lim\limits_{x \to 0} \sqrt{5 - x^2} = \sqrt{5 - (0)^2} = \sqrt{5}$; by the sandwich theorem,

$\lim\limits_{x \to 0} f(x) = \sqrt{5}$

45. (a) $\lim\limits_{x \to 0} \left(1 - \dfrac{x^2}{6}\right) = 1 - \dfrac{0}{6} = 1$ and $\lim\limits_{x \to 0} 1 = 1$; by the sandwich theorem, $\lim\limits_{x \to 0} \dfrac{x \sin x}{2 - 2 \cos x} = 1$

(b) For $x \neq 0$, $y = (x \sin x)/(2 - 2 \cos x)$ lies between the other two graphs in the figure, and the graphs converge as $x \to 0$.

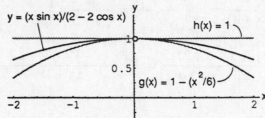

47. $\lim\limits_{x \to c} f(x)$ exists at those points c where $\lim\limits_{x \to c} x^4 = \lim\limits_{x \to c} x^2$. Thus, $c^4 = c^2 \Rightarrow c^2(1 - c^2) = 0$

$\Rightarrow c = 0, 1,$ or -1. Moreover, $\lim\limits_{x \to 0} f(x) = \lim\limits_{x \to 0} x^2 = 0$ and $\lim\limits_{x \to -1} f(x) = \lim\limits_{x \to 1} f(x) = 1$.

49. $1 = \lim\limits_{x \to 4} \dfrac{f(x) - 5}{x - 2} = \dfrac{\lim\limits_{x \to 4} f(x) - \lim\limits_{x \to 4} 5}{\lim\limits_{x \to 4} x - \lim\limits_{x \to 4} 2} = \dfrac{\lim\limits_{x \to 4} f(x) - 5}{4 - 2} \Rightarrow \lim\limits_{x \to 4} f(x) - 5 = 2(1) \Rightarrow \lim\limits_{x \to 4} f(x) = 2 + 5 = 7.$

51. (a) $0 = 3 \cdot 0 = \left[\lim\limits_{x \to 2} \dfrac{f(x) - 5}{x - 2}\right]\left[\lim\limits_{x \to 2} (x - 2)\right] = \lim\limits_{x \to 2} \left[\left(\dfrac{f(x) - 5}{x - 2}\right)(x - 2)\right] = \lim\limits_{x \to 2} [f(x) - 5] = \lim\limits_{x \to 2} f(x) - 5$

$\Rightarrow \lim\limits_{x \to 2} f(x) = 5.$

(b) $0 = 4 \cdot 0 = \left[\lim\limits_{x \to 2} \dfrac{f(x) - 5}{x - 2}\right]\left[\lim\limits_{x \to 2} (x - 2)\right] \Rightarrow \lim\limits_{x \to 2} f(x) = 5$ as in part (a).

53. (a) $\lim\limits_{x\to0} x \sin\frac{1}{x} = 0$

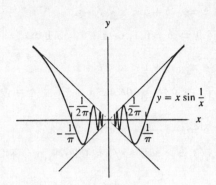

(b) $-1 \le \sin\frac{1}{x} \le 1$ for $x \ne 0$:

$x > 0 \Rightarrow -x \le x \sin\frac{1}{x} \le x \Rightarrow \lim\limits_{x\to0} x \sin\frac{1}{x} = 0$ by the sandwich theorem;

$x < 0 \Rightarrow -x \ge x \sin\frac{1}{x} \ge x \Rightarrow \lim\limits_{x\to0} x \sin\frac{1}{x} = 0$ by the sandwich theorem.

1.3 TARGET VALUES AND FORMAL LIMITS

1.

Step 1: $|x - 5| < \delta \Rightarrow -\delta < x - 5 < \delta \Rightarrow -\delta + 5 < x < \delta + 5$

Step 2: $\delta + 5 = 7 \Rightarrow \delta = 2$, or $-\delta + 5 = 1 \Rightarrow \delta = 4$.

The value of δ which assures $|x - 5| < \delta \Rightarrow 1 < x < 7$ is the smaller value, $\delta = 2$.

3.

Step 1: $\left|x - (-3)\right| < \delta \Rightarrow -\delta < x + 3 < \delta \Rightarrow -\delta - 3 < x < \delta - 3$

Step 2: $-\delta - 3 = -\frac{7}{2} \Rightarrow \delta = \frac{1}{2}$, or $\delta - 3 = -\frac{1}{2} \Rightarrow \delta = \frac{5}{2}$.

The value of δ which assures $\left|x - (-3)\right| < \delta \Rightarrow -\frac{7}{2} < x < -\frac{1}{2}$ is the smaller value, $\delta = \frac{1}{2}$.

5.

Step 1: $\left|x - \frac{1}{2}\right| < \delta \Rightarrow -\delta < x - \frac{1}{2} < \delta \Rightarrow -\delta + \frac{1}{2} < x < \delta + \frac{1}{2}$

Step 2: $-\delta + \frac{1}{2} = \frac{4}{9} \Rightarrow \delta = \frac{1}{18}$, or $\delta + \frac{1}{2} = \frac{4}{7} \Rightarrow \delta = \frac{1}{14}$.

The value of δ which assures $\left|x - \frac{1}{2}\right| < \delta \Rightarrow \frac{4}{9} < x < \frac{4}{7}$ is the smaller value, $\delta = \frac{1}{18}$.

7. Step 1: $|x - 5| < \delta \Rightarrow -\delta < x - 5 < \delta \Rightarrow -\delta + 5 < x < \delta + 5$

   Step 2: From the graph, $-\delta + 5 = 4.9 \Rightarrow \delta = 0.1$, or $\delta + 5 = 5.1 \Rightarrow \delta = 0.1$; thus $\delta = 0.1$ in either case.

9. Step 1: $|x - 1| < \delta \Rightarrow -\delta < x - 1 < \delta \Rightarrow -\delta + 1 < x < \delta + 1$

   Step 2: From the graph, $-\delta + 1 = \frac{9}{16} \Rightarrow \delta = \frac{7}{16}$, or $\delta + 1 = \frac{25}{16} \Rightarrow \delta = \frac{9}{16}$; thus $\delta = \frac{7}{16}$.

11. Step 1: $|x - 2| < \delta \Rightarrow -\delta < x - 2 < \delta \Rightarrow -\delta + 2 < x < \delta + 2$

    Step 2: From the graph, $-\delta + 2 = \sqrt{3} \Rightarrow \delta = 2 - \sqrt{3} \approx 0.2679$, or $\delta + 2 = \sqrt{5} \Rightarrow \delta = \sqrt{5} - 2 \approx 0.2361$;

    thus $\delta = \sqrt{5} - 2$.

13. Step 1: $\left|x - (-1)\right| < \delta \Rightarrow -\delta < x + 1 < \delta \Rightarrow -\delta - 1 < x < \delta - 1$

    Step 2: From the graph, $-\delta - 1 = -\frac{16}{9} \Rightarrow \delta = \frac{7}{9} \approx 0.77$, or $\delta - 1 = -\frac{16}{25} \Rightarrow \frac{9}{25} = 0.36$; thus $\delta = \frac{9}{25} = 0.36$.

15. Step 1: $\left|(x + 1) - 5\right| < 0.01 \Rightarrow |x - 4| < 0.01 \Rightarrow -0.01 < x - 4 < 0.01 \Rightarrow 3.99 < x < 4.01$

    Step 2: $|x - 4| < \delta \Rightarrow -\delta < x - 4 < \delta \Rightarrow -\delta + 4 < x < \delta + 4 \Rightarrow \delta = 0.01$.

17. Step 1: $\left|\sqrt{x + 1} - 1\right| < 0.1 \Rightarrow -0.1 < \sqrt{x + 1} - 1 < 0.1 \Rightarrow 0.9 < \sqrt{x + 1} < 1.1 \Rightarrow 0.81 < x + 1 < 1.21$

    $\Rightarrow -0.19 < x < 0.21$

    Step 2: $|x - 0| < \delta \Rightarrow -\delta < x < \delta \Rightarrow \delta = 0.019$.

19. Step 1: $\left|\sqrt{19 - x} - 3\right| < 1 \Rightarrow -1 < \sqrt{19 - x} - 3 < 1 \Rightarrow 2 < \sqrt{19 - x} < 4 \Rightarrow 4 < 19 - x < 16$

    $\Rightarrow -4 > x - 19 > -16 \Rightarrow 15 > x > 3$ or $3 < x < 15$

    Step 2: $|x - 10| < \delta \Rightarrow -\delta < x - 10 < \delta \Rightarrow -\delta + 10 < x < \delta + 10$.

    Then $-\delta + 10 = 3 \Rightarrow \delta = 7$, or $\delta + 10 = 15 \Rightarrow \delta = 5$; thus $\delta = 5$.

21. Step 1: $\left|\frac{1}{x} - \frac{1}{4}\right| < 0.05 \Rightarrow -0.05 < \frac{1}{x} - \frac{1}{4} < 0.05 \Rightarrow 0.2 < \frac{1}{x} < 0.3 \Rightarrow \frac{10}{2} > x > \frac{10}{3}$ or $\frac{10}{3} < x < 5$.

    Step 2: $|x - 4| < \delta \Rightarrow -\delta < x - 4 < \delta \Rightarrow -\delta + 4 < x < \delta + 4$.

    Then $-\delta + 4 = \frac{10}{3}$ or $\delta = \frac{2}{3}$, or $\delta + 4 = 5$ or $\delta = 1$; thus $\delta = \frac{2}{3}$.

23. Step 1: $\left|x^2 - 4\right| < 0.5 \Rightarrow -0.5 < x^2 - 4 < 0.5 \Rightarrow 3.5 < x^2 < 4.5 \Rightarrow \sqrt{3.5} < |x| < \sqrt{4.5} \Rightarrow -\sqrt{4.5} < x < -\sqrt{3.5}$,

    for x near -2.

    Step 2: $\left|x - (-2)\right| < \delta \Rightarrow -\delta < x + 2 < \delta \Rightarrow -\delta - 2 < x < \delta - 2$.

    Then $-\delta - 2 = -\sqrt{4.5} \Rightarrow \delta = \sqrt{4.5} - 2 \approx 0.1213$, or $\delta - 2 = -\sqrt{3.5} \Rightarrow \delta = 2 - \sqrt{3.5} \approx 0.1292$;

    thus $\delta = \sqrt{4.5} - 2 \approx 0.12$.

25. Step 1: $\left|(x^2 - 5) - 11\right| < 1 \Rightarrow |x^2 - 16| < 1 \Rightarrow -1 < x^2 - 16 < 1 \Rightarrow 15 < x^2 < 17 \Rightarrow \sqrt{15} < x < \sqrt{17}$.

    Step 2: $|x - 4| < \delta \Rightarrow -\delta < x - 4 < \delta \Rightarrow -\delta + 4 < x < \delta + 4$.

    Then $-\delta + 4 = \sqrt{15} \Rightarrow \delta = 4 - \sqrt{15} \approx 0.1270$, or $\delta + 4 = \sqrt{17} \Rightarrow \delta = \sqrt{17} - 4 \approx 0.1231$;

    thus $\delta = \sqrt{17} - 4 \approx 0.12$.

27. Step 1: $|mx - 2m| < 0.03 \Rightarrow -0.03 < mx - 2m < 0.03 \Rightarrow -0.03 + 2m < mx < 0.03 + 2m \Rightarrow$
$2 - \frac{0.03}{m} < x < 2 + \frac{0.03}{m}.$

Step 2: $|x - 2| < \delta \Rightarrow -\delta < x - 2 < \delta \Rightarrow -\delta + 2 < x < \delta + 2.$
Then $-\delta + 2 = 2 - \frac{0.03}{m} \Rightarrow \delta = \frac{0.03}{m}$, or $\delta + 2 = 2 + \frac{0.03}{m} \Rightarrow \delta = \frac{0.03}{m}$. In either case, $\delta = \frac{0.03}{m}$.

29. Step 1: $\left|(mx + b) - \left(\frac{m}{2} + b\right)\right| < c \Rightarrow -c < mx - \frac{m}{2} < c \Rightarrow -c + \frac{m}{2} < mx < c + \frac{m}{2} \Rightarrow \frac{1}{2} - \frac{c}{m} < x < \frac{1}{2} + \frac{c}{m}.$

Step 2: $\left|x - \frac{1}{2}\right| < \delta \Rightarrow -\delta < x - \frac{1}{2} < \delta \Rightarrow -\delta + \frac{1}{2} < x < \delta + \frac{1}{2}.$

Then $-\delta + \frac{1}{2} = \frac{1}{2} - \frac{c}{m} \Rightarrow \delta = \frac{c}{m}$, or $\delta + \frac{1}{2} = \frac{1}{2} + \frac{c}{m} \Rightarrow \delta = \frac{c}{m}$. In either case, $\delta = \frac{c}{m}$.

31. $\lim_{x \to 3} (3 - 2x) = 3 - 2(3) = -3$

Step 1: $|3 - 2x - (-3)| < 0.02 \Rightarrow -0.02 < 6 - 2x < 0.02 \Rightarrow -6.02 < -2x < -5.98 \Rightarrow 3.01 > x > 2.99$ or
$2.99 < x < 3.01.$

Step 2: $0 < |x - 3| < \delta \Rightarrow -\delta < x - 3 < \delta \Rightarrow -\delta + 3 < x < \delta + 3.$
Then $-\delta + 3 = 2.99 \Rightarrow \delta = 0.01$, or $\delta + 3 = 3.01 \Rightarrow \delta = 0.01$; thus $\delta = 0.01$.

33. $\lim_{x \to 2} \frac{x^2 - 4}{x - 2} = \lim_{x \to 2} \frac{(x + 2)(x - 2)}{(x - 2)} = \lim_{x \to 2} (x + 2) = 2 + 2 = 4, x \neq 2$

Step 1: $\left|\left(\frac{x^2 - 4}{x - 2}\right) - 4\right| < 0.05 \Rightarrow -0.05 < \frac{(x + 2)(x - 2)}{(x - 2)} - 4 < 0.05 \Rightarrow 3.95 < x + 2 < 4.05, x \neq 2$
$\Rightarrow 1.95 < x < 2.05, x \neq 2.$

Step 2: $|x - 2| < \delta \Rightarrow -\delta < x - 2 < \delta \Rightarrow -\delta + 2 < x < \delta + 2.$
Then $-\delta + 2 = 1.95 \Rightarrow \delta = 0.05$, or $\delta + 2 = 2.05 \Rightarrow \delta = 0.05$; thus $\delta = 0.05$.

35. $\lim_{x \to -3} \sqrt{1 - 5x} = \sqrt{1 - 5(-3)} = \sqrt{16} = 4$

Step 1: $\left|\sqrt{1 - 5x} - 4\right| < 0.5 \Rightarrow -0.5 < \sqrt{1 - 5x} - 4 < 0.5 \Rightarrow 3.5 < \sqrt{1 - 5x} < 4.5 \Rightarrow 12.25 < 1 - 5x < 20.25$
$\Rightarrow 11.25 < -5x < 19.25 \Rightarrow -3.85 < x < -2.25.$

Step 2: $|x - (-3)| < \delta \Rightarrow -\delta < x + 3 < \delta \Rightarrow -\delta - 3 < x < \delta - 3.$
Then $-\delta - 3 = -3.85 \Rightarrow \delta = 0.85$, or $\delta - 3 = -2.25 \Rightarrow 0.75$; thus $\delta = 0.75$.

37. Step 1: $|(9 - x) - 5| < \epsilon \Rightarrow -\epsilon < 4 - x < \epsilon \Rightarrow -\epsilon - 4 < -x < \epsilon - 4 \Rightarrow \epsilon + 4 > x > 4 - \epsilon \Rightarrow 4 - \epsilon < x < 4 + \epsilon.$
Step 2: $|x - 4| < \delta \Rightarrow -\delta < x - 4 < \delta \Rightarrow -\delta + 4 < x < \delta + 4.$
Then $-\delta + 4 = -\epsilon + 4 \Rightarrow \delta = \epsilon$, or $\delta + 4 = \epsilon + 4 \Rightarrow \delta = \epsilon$. Thus choose $\delta = \epsilon$.

39. Step 1: $\left|\sqrt{x - 5} - 2\right| < \epsilon \Rightarrow -\epsilon < \sqrt{x - 5} - 2 < \epsilon \Rightarrow 2 - \epsilon < \sqrt{x - 5} < 2 + \epsilon \Rightarrow (2 - \epsilon)^2 < x - 5 < (2 + \epsilon)^2$
$\Rightarrow (2 - \epsilon)^2 + 5 < x < (2 + \epsilon)^2 + 5.$
Step 2: $|x - 9| < \delta \Rightarrow -\delta < x - 9 < \delta \Rightarrow -\delta + 9 < x < \delta + 9.$

Then $-\delta + 9 = \epsilon^2 - 2\epsilon + 9 \Rightarrow \delta = 2\epsilon - \epsilon^2$, or $\delta + 9 = \epsilon^2 + 2\epsilon + 9 \Rightarrow \delta = 2\epsilon + \epsilon^2$. Thus choose the smaller distance, $\delta = 2\epsilon - \epsilon^2$.

41. Step 1:   For $x \neq 1$, $\left| x^2 - 1 \right| < \epsilon \Rightarrow -\epsilon < x^2 - 1 < \epsilon \Rightarrow 1 - \epsilon < x^2 < 1 + \epsilon \Rightarrow \sqrt{1-\epsilon} < |x| < \sqrt{1+\epsilon}$
$\Rightarrow \sqrt{1-\epsilon} < x < \sqrt{1+\epsilon}$ near $x = 1$.

Step 2: $|x - 1| < \delta \Rightarrow -\delta < x - 1 < \delta \Rightarrow -\delta + 1 < x < \delta + 1$.
Then $-\delta + 1 = \sqrt{1-\epsilon} \Rightarrow \delta = 1 - \sqrt{1-\epsilon}$, or $\delta + 1 = \sqrt{1+\epsilon} \Rightarrow \delta = \sqrt{1+\epsilon} - 1$. Choose $\delta = \min\left\{ 1 - \sqrt{1-\epsilon}, \sqrt{1+\epsilon} - 1 \right\}$, that is, the smaller of the two distances.

43. Step 1: $\left| \frac{1}{x} - 1 \right| < \epsilon \Rightarrow -\epsilon < \frac{1}{x} - 1 < \epsilon \Rightarrow 1 - \epsilon < \frac{1}{x} < 1 + \epsilon \Rightarrow \frac{1}{1+\epsilon} < x < \frac{1}{1-\epsilon}$.

Step 2: $|x - 1| < \delta \Rightarrow -\delta < x - 1 < \delta \Rightarrow 1 - \delta < x < 1 + \delta$.
Then $1 - \delta = \frac{1}{1+\epsilon} \Rightarrow \delta = 1 - \frac{1}{1+\epsilon} = \frac{\epsilon}{1+\epsilon}$, or $1 + \delta = \frac{1}{1-\epsilon} \Rightarrow \delta = \frac{1}{1-\epsilon} - 1 = \frac{\epsilon}{1-\epsilon}$.
Choose $\delta = \frac{\epsilon}{1+\epsilon}$, the smaller of the two distances.

45. Step 1: $\left| \left(\frac{x^2 - 9}{x + 3} \right) - (-6) \right| < \epsilon \Rightarrow -\epsilon < (x - 3) + 6 < \epsilon, \; x \neq -3 \Rightarrow -\epsilon < x + 3 < \epsilon \Rightarrow -\epsilon - 3 < x < \epsilon - 3$.

Step 2: $\left| x - (-3) \right| < \delta \Rightarrow -\delta < x + 3 < \delta \Rightarrow -\delta - 3 < x < \delta - 3$.
Then $-\delta - 3 = -\epsilon - 3 \Rightarrow \delta = \epsilon$, or $\delta - 3 = \epsilon - 3 \Rightarrow \delta = \epsilon$. Choose $\delta = \epsilon$.

47. Step 1: $x < 1$: $(4 - 2x) - 2 < \epsilon \Rightarrow 2 - 2x < \epsilon \Rightarrow x \geq \frac{2 - \epsilon}{2} = 1 - \frac{\epsilon}{2}$;
$x \geq 1$: $(6x - 4) - 2 < \epsilon \Rightarrow 6x - 6 < \epsilon \Rightarrow x < \frac{6 + \epsilon}{6} = 1 + \frac{\epsilon}{6}$.

Step 2: $|x - 1| < \delta \Rightarrow -\delta < x - 1 < \delta \Rightarrow 1 - \delta < x < 1 + \delta$.
Then $1 - \delta = 1 - \frac{\epsilon}{2} \Rightarrow \delta = \frac{\epsilon}{2}$, or $1 + \delta = 1 + \frac{\epsilon}{6} \Rightarrow \delta = \frac{\epsilon}{6}$. Choose $\delta = \frac{\epsilon}{6}$.

49. By the figure, $-x \leq x \sin \frac{1}{x} \leq x$ for all $x > 0$ and $-x \geq x \sin \frac{1}{x} \geq x$ for $x < 0$. Since $\lim_{x \to 0} (-x) = \lim_{x \to 0} x = 0$, then by the sandwich theorem, in either case, $\lim_{x \to 0} x \sin \frac{1}{x} = 0$.

51. As x approaches the value 2, the values of $f(x)$ approach 5. Thus for every number $\epsilon > 0$, there exists a $\delta > 0$ such that $0 < |x - 2| < \delta \Rightarrow \left| f(x) - 5 \right| < \epsilon$.

53. Let $f(x) = x^2$. The function values do get closer to -1 as x approaches 0, but $\lim_{x \to 0} f(x) = 0$, not -1. The function $f(x) = x^2$ never gets <u>arbitrarily</u> <u>close</u> to -1 for x near 0.

55. $|A - 9| \leq 0.01 \Rightarrow -0.01 \leq \pi \left(\frac{x}{2} \right)^2 - 9 \leq 0.01 \Rightarrow 8.99 \leq \frac{\pi x^2}{4} \leq 9.01 \Rightarrow \frac{4}{\pi}(8.99) \leq x^2 \leq \frac{4}{\pi}(9.01)$
$\Rightarrow 2\sqrt{\frac{8.99}{\pi}} \leq x \leq 2\sqrt{\frac{9.01}{\pi}}$ or $3.384 \leq x \leq 3.387$. To be safe, the left endpoint was rounded up and the right endpoint was rounded down.

57. (a) $-\delta < x - 1 < 0 \Rightarrow 1 - \delta < x < 1 \Rightarrow f(x) = x.$ Then $|f(x) - 2| = |x - 2| = 2 - x > 2 - 1 = 1.$ That is,

   $|f(x) - 2| \geq 1 \geq \frac{1}{2}$ no matter how small δ is taken when $1 - \delta < x < 1 \Rightarrow \lim_{x \to 1} f(x) \neq 2.$

   (b) $0 < x - 1 < \delta \Rightarrow 1 < x < 1 + \delta \Rightarrow f(x) = x + 1.$ Then $|f(x) - 1| = |(x + 1) - 1| = |x| = x > 1.$ That is,

   $|f(x) - 1| \geq 1$ no matter how small δ is taken when $1 < x < 1 + \delta \Rightarrow \lim_{x \to 1} f(x) \neq 1.$

   (c) $-\delta < x - 1 < 0 \Rightarrow 1 - \delta < x < 1 \Rightarrow f(x) = x.$ Then $|f(x) - 1.5| = |x - 1.5| = 1.5 - x > 1.5 - 1 = 0.5.$

   Also, $0 < x - 1 < \delta \Rightarrow 1 < x < 1 + \delta \Rightarrow f(x) = x + 1.$ Then $|f(x) - 1.5| = |(x + 1) - 1.5| = |x - 0.5|$

   $= x - 0.5 > 1 - 0.5 = 0.5.$ Thus, no matter how small δ is taken, there exists a value of x such that

   $-\delta < x - 1 < \delta$ but $|f(x) - 1.5| \geq \frac{1}{2} \Rightarrow \lim_{x \to 1} f(x) \neq 1.5.$

59. (a) For $3 - \delta < x < 3 + \delta \Rightarrow f(x) > 4.8 \Rightarrow |f(x) - 4| \geq 0.8.$ Thus for $\epsilon < 0.8, |f(x) - 4| \geq \epsilon$ whenever

   $3 - \delta < x < 3$ no matter how small we choose $\delta > 0 \Rightarrow \lim_{x \to 3} f(x) \neq 4.$

   (b) For $3 < x < 3 + \delta \Rightarrow f(x) < 3 \Rightarrow |f(x) - 4.8| \geq 1.8.$ Thus for $\epsilon < 1.8, |f(x) - 4.8| \geq \epsilon$ whenever $3 < x < 3 + \delta$

   no matter how small we choose $\delta > 0 \Rightarrow \lim_{x \to 3} f(x) \neq 4.8.$

   (c) For $3 - \delta < x < 3 \Rightarrow f(x) > 4.8 \Rightarrow |f(x) - 3| \geq 1.8.$ Again, for $\epsilon < 1.8, |f(x) - 3| \geq \epsilon$ whenever $3 - \delta < x < 3$

   no matter how small we choose $\delta > 0 \Rightarrow \lim_{x \to 3} f(x) \neq 3.$

1.4 EXTENSIONS OF THE LIMIT CONCEPT

1. (a) True (b) True (c) False (d) True
   (e) True (f) True (g) False (h) False
   (i) False (j) False (k) True (l) False

3. (a) $\lim_{x \to 2^+} f(x) = \frac{2}{2} + 1 = 2, \ \lim_{x \to 2^-} f(x) = 3 - 2 = 1$

   (b) No, $\lim_{x \to 2} f(x)$ does not exist because $\lim_{x \to 2^+} f(x) \neq \lim_{x \to 2^-} f(x)$

   (c) $\lim_{x \to 4^-} f(x) = \frac{4}{2} + 1 = 3, \ \lim_{x \to 4^+} f(x) = \frac{4}{2} + 1 = 3$

   (d) Yes, $\lim_{x \to 4} f(x) = 3$ because $3 = \lim_{x \to 4^-} f(x) = \lim_{x \to 4^+} f(x)$

5. (a) No, $\lim_{x \to 0^+} f(x)$ does not exist since $\sin\left(\frac{1}{x}\right)$ does not approach any single value as x approaches 0

   (b) $\lim_{x \to 0^-} f(x) = \lim_{x \to 0^+} 0 = 0$

   (c) $\lim_{x \to 0} f(x)$ does not exist because $\lim_{x \to 0^+} f(x)$ does not exist

7. (a)

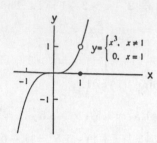

(b) $\lim\limits_{x \to 1^-} f(x) = 1 = \lim\limits_{x \to 1^+} f(x)$

(c) Yes, $\lim\limits_{x \to 1} f(x) = 1$ since the right-hand and left-hand limits exist and equal 1

9. (a) domain: $0 \le x \le 2$
   range: $0 < y \le 1$ and $y = 2$

   (b) $\lim\limits_{x \to c} f(x)$ exists for c belonging to

   $(0, 1) \cup (1, 2)$

   (c) $x = 2$

   (d) $x = 0$

11. $\lim\limits_{x \to -0.5^-} \sqrt{\dfrac{x+2}{x-1}} = \sqrt{\dfrac{-0.5+2}{-0.5+1}} = \sqrt{\dfrac{3/2}{1/2}} = \sqrt{3}$

13. $\lim\limits_{x \to -2^+} \left(\dfrac{x}{x+1}\right)\left(\dfrac{2x+5}{x^2+x}\right) = \left(\dfrac{-2}{-2+1}\right)\left(\dfrac{2(-2)+5}{(-2)^2+(-2)}\right) = (2)\left(\dfrac{1}{2}\right) = 1$

15. $\lim\limits_{h \to 0^+} \dfrac{\sqrt{h^2+4h+5} - \sqrt{5}}{h} = \lim\limits_{h \to 0^+} \left(\dfrac{\sqrt{h^2+4h+5} - \sqrt{5}}{h}\right)\left(\dfrac{\sqrt{h^2+4h+5} + \sqrt{5}}{\sqrt{h^2+4h+5} + \sqrt{5}}\right)$

$= \lim\limits_{h \to 0^+} \dfrac{(h^2+4h+5) - 5}{h\left(\sqrt{h^2+4h+5} + \sqrt{5}\right)} = \lim\limits_{h \to 0^+} \dfrac{h(h+4)}{h\left(\sqrt{h^2+4h+5} + \sqrt{5}\right)} = \dfrac{0+4}{\sqrt{5}+\sqrt{5}} = \dfrac{2}{\sqrt{5}}$

17. (a) $\lim\limits_{x \to -2^+} (x+3)\dfrac{|x+2|}{x+2} = \lim\limits_{x \to -2^+} (x+3)\dfrac{(x+2)}{(x+2)}$ $(|x+2| = x+2 \text{ for } x > -2)$

$= \lim\limits_{x \to -2^+} (x+3) = (-2) + 3 = 1$

(b) $\lim\limits_{x \to -2^-} (x+3)\dfrac{|x+2|}{x+2} = \lim\limits_{x \to -2^-} (x+3)\left[\dfrac{-(x+2)}{(x+2)}\right]$ $(|x+2| = -(x+2) \text{ for } x < -2)$

$= \lim\limits_{x \to -2^-} (x+3)(-1) = -(-2+3) = -1$

19. (a) $\lim\limits_{\theta \to 3^+} \dfrac{\lfloor \theta \rfloor}{\theta} = \dfrac{3}{3} = 1$

(b) $\lim\limits_{\theta \to 3^-} \dfrac{\lfloor \theta \rfloor}{\theta} = \dfrac{2}{3}$

21. $\lim\limits_{x \to 0^+} \dfrac{1}{3x} = \infty$ $\left(\dfrac{\text{positive}}{\text{positive}}\right)$

23. $\lim\limits_{x \to 2^-} \dfrac{3}{x-2} = -\infty$ $\left(\dfrac{\text{positive}}{\text{negative}}\right)$

25. $\displaystyle\lim_{x\to-8^+} \frac{2x}{x+8} = -\infty$ $\left(\dfrac{\text{negative}}{\text{positive}}\right)$ 27. $\displaystyle\lim_{x\to7} \frac{4}{(x-7)^2} = \infty$ $\left(\dfrac{\text{positive}}{\text{positive}}\right)$

29. (a) $\displaystyle\lim_{x\to0^+} \frac{2}{3x^{1/3}} = \infty$ (b) $\displaystyle\lim_{x\to0^-} \frac{2}{3x^{1/3}} = -\infty$

31. $\displaystyle\lim_{x\to0} \frac{4}{x^{2/5}} = \lim_{x\to0} \frac{4}{\left(x^{1/5}\right)^2} = \infty$ 33. $\displaystyle\lim_{x\to\left(\frac{\pi}{2}\right)^-} \tan x = \infty$

35. $\displaystyle\lim_{\theta\to0^-} (1 + \csc\theta) = -\infty$

37. (a) $\displaystyle\lim_{x\to2^+} \frac{1}{x^2-4} = \lim_{x\to2^+} \frac{1}{(x+2)(x-2)} = \infty$ $\left(\dfrac{1}{\text{positive}\cdot\text{positive}}\right)$

(b) $\displaystyle\lim_{x\to2^-} \frac{1}{x^2-4} = \lim_{x\to2^-} \frac{1}{(x+2)(x-2)} = -\infty$ $\left(\dfrac{1}{\text{positive}\cdot\text{negative}}\right)$

(c) $\displaystyle\lim_{x\to-2^+} \frac{1}{x^2-4} = \lim_{x\to-2^+} \frac{1}{(x+2)(x-2)} = -\infty$ $\left(\dfrac{1}{\text{positive}\cdot\text{negative}}\right)$

(d) $\displaystyle\lim_{x\to-2^-} \frac{1}{x^2-4} = \lim_{x\to-2^-} \frac{1}{(x+2)(x-2)} = \infty$ $\left(\dfrac{1}{\text{negative}\cdot\text{negative}}\right)$

39. (a) $\displaystyle\lim_{x\to0^+} \frac{x^2}{2} - \frac{1}{x} = 0 + \lim_{x\to0^+} \frac{1}{-x} = -\infty$ $\left(\dfrac{1}{\text{negative}}\right)$

(b) $\displaystyle\lim_{x\to0^-} \frac{x^2}{2} - \frac{1}{x} = 0 + \lim_{x\to0^-} \frac{1}{-x} = \infty$ $\left(\dfrac{1}{\text{positive}}\right)$

(c) $\displaystyle\lim_{x\to\sqrt[3]{2}} \frac{x^2}{2} - \frac{1}{x} = \frac{2^{2/3}}{2} - \frac{1}{2^{1/3}} = 2^{-1/3} - 2^{-1/3} = 0$

(d) $\displaystyle\lim_{x\to-1} \frac{x^2}{2} - \frac{1}{x} = \frac{1}{2} - \left(\frac{1}{-1}\right) = \frac{3}{2}$

41. (a) $\displaystyle\lim_{x\to0^+} \frac{x^2-3x+2}{x^3-2x^2} = \lim_{x\to0^+} \frac{(x-2)(x-1)}{x^2(x-2)} = -\infty$ $\left(\dfrac{\text{negative}\cdot\text{negative}}{\text{positive}\cdot\text{negative}}\right)$

(b) $\displaystyle\lim_{x\to2^+} \frac{x^2-3x+2}{x^3-2x^2} = \lim_{x\to2^+} \frac{(x-2)(x-1)}{x^2(x-2)} = \lim_{x\to2^+} \frac{x-1}{x^2} = \frac{1}{4},\ x\neq2$

(c) $\displaystyle\lim_{x\to2^-} \frac{x^2-3x+2}{x^3-2x^2} = \lim_{x\to2^-} \frac{(x-2)(x-1)}{x^2(x-2)} = \lim_{x\to2^-} \frac{x-1}{x^2} = \frac{1}{4},\ x\neq2$

(d) $\displaystyle\lim_{x\to2} \frac{x^2-3x+2}{x^3-2x^2} = \lim_{x\to2} \frac{(x-2)(x-1)}{x^2(x-2)} = \lim_{x\to2^+} \frac{x-1}{x^2} = \frac{1}{4},\ x\neq2$

(e) $\displaystyle\lim_{x\to0} \frac{x^2-3x+2}{x^3-2x^2} = \lim_{x\to0} \frac{(x-2)(x-1)}{x^2(x-2)} = -\infty$ $\left(\dfrac{\text{negative}\cdot\text{negative}}{\text{positive}\cdot\text{negative}}\right)$

43. (a) $\lim\limits_{t\to 0^+} \left[2 - \dfrac{3}{t^{1/3}}\right] = -\infty$ (b) $\lim\limits_{t\to 0^-} \left[2 - \dfrac{3}{t^{1/3}}\right] = \infty$

45. (a) $\lim\limits_{x\to 0^+} \left[\dfrac{1}{x^{2/3}} + \dfrac{2}{(x-1)^{2/3}}\right] = \infty$ (b) $\lim\limits_{x\to 0^-} \left[\dfrac{1}{x^{2/3}} + \dfrac{2}{(x-1)^{2/3}}\right] = \infty$

(c) $\lim\limits_{x\to 1^+} \left[\dfrac{1}{x^{2/3}} + \dfrac{2}{(x-1)^{2/3}}\right] = \infty$ (d) $\lim\limits_{x\to 1^-} \left[\dfrac{1}{x^{2/3}} + \dfrac{2}{(x-1)^{2/3}}\right] = \infty$

47. Yes. If $\lim\limits_{x\to a^+} f(x) = L = \lim\limits_{x\to a^-} f(x)$, then $\lim\limits_{x\to a} f(x) = L$. If $\lim\limits_{x\to a^+} f(x) \neq \lim\limits_{x\to a^-} f(x)$, then $\lim\limits_{x\to a} f(x)$ does not exist.

49. If f is an odd function of x, then $f(-x) = -f(x)$. Given $\lim\limits_{x\to 0^+} f(x) = 3$, then $\lim\limits_{x\to 0^-} f(x) = -3$.

51. $I = (5, 5 + \delta) \Rightarrow 5 < x < 5 + \delta$. Also, $\sqrt{x-5} < \epsilon \Rightarrow x - 5 < \epsilon^2 \Rightarrow x < 5 + \epsilon^2$. Choose $\delta = \epsilon^2$
$\Rightarrow \lim\limits_{x\to 5^+} \sqrt{x-5} = 0$.

53. As $x \to 0^-$ the number x is always negative. Thus, $\left|\dfrac{x}{|x|} - (-1)\right| < \epsilon \Rightarrow \left|\dfrac{x}{-x} + 1\right| < \epsilon \Rightarrow 0 < \epsilon$ which is always

true independent of the value of x. Hence we can choose any $\delta > 0$ with $-\delta < x < 0 \Rightarrow \lim\limits_{x\to 0^-} \dfrac{x}{|x|} = -1$.

55. (a) $\lim\limits_{x\to 400^+} |x| = 400; \, \big||x| - 400\big| < \epsilon \Rightarrow -\epsilon < |x| - 400 < \epsilon \Rightarrow -\epsilon + 400 < |x| < \epsilon + 400$. For x near 400, $|x| = x$
$\Rightarrow -\epsilon + 400 < x < \epsilon + 400$. Choose $\delta = \epsilon$ and $400 < x < 400 + \delta \Rightarrow \lim\limits_{x\to 400^+} |x| = 400$.

(b) $\lim\limits_{x\to 400^-} |x| = 400$; as in part (a), choose $\delta = \epsilon$ and $400 - \delta < x < 400 \Rightarrow \lim\limits_{x\to 400^-} |x| = 400$.

(c) $\lim\limits_{x\to 400} |x| = 400$, since f has limit L at x_0 if and only if f has right-hand limit L and left-hand limit L at x_0.

57. For every real number $B > 0$, we must find a $\delta > 0$ such that for all x, $0 < |x - 0| < \delta \Rightarrow \dfrac{1}{x^2} > B$. Now

$\dfrac{1}{x^2} > B > 0 \Leftrightarrow \dfrac{1}{B} > x^2 \Leftrightarrow \dfrac{1}{\sqrt{B}} > |x|$. Thus, choose $\delta = \dfrac{1}{\sqrt{B}}$. Then $0 < |x| < \delta \Rightarrow |x| < \dfrac{1}{\sqrt{B}} \Rightarrow \dfrac{1}{x^2} > B$ so that

$\lim\limits_{x\to 0} \dfrac{1}{x^2} = \infty$.

59. For every real number $-B < 0$, we must find a $\delta > 0$ such that for all x, $0 < |x - 3| < \delta \Rightarrow \dfrac{-2}{(x-3)^2} < -B$.

Now, $\dfrac{-2}{(x-3)^2} < -B < 0 \Leftrightarrow \dfrac{2}{(x-3)^2} > B > 0 \Leftrightarrow \dfrac{(x-3)^2}{2} < \dfrac{1}{B} \Leftrightarrow (x-3)^2 < \dfrac{2}{B} \Leftrightarrow 0 < |x-3| < \sqrt{\dfrac{2}{B}}$. Choose

$\delta = \sqrt{\dfrac{2}{B}}$, then $0 < |x-3| < \delta \Rightarrow \dfrac{-2}{(x-3)^2} < -B < 0$ so that $\lim\limits_{x\to 3} \dfrac{-2}{(x-3)^2} = -\infty$.

61. (a) We say that f(x) approaches infinity as x approaches x_0 from the left, and write $\lim\limits_{x\to x_0^-} f(x) = \infty$, if

for every positive number B, there exists a corresponding number $\delta > 0$ such that for all x,
$x_0 - \delta < x < x_0 \Rightarrow f(x) > B$.

(b) We say that f(x) approaches minus infinity as x approaches x_0 from the right, and write $\lim\limits_{x \to x_0^+} f(x) = -\infty$, if for every positive number B (or negative number −B) there exists a corresponding number $\delta > 0$ such that for all x, $x_0 < x < x_0 + \delta \Rightarrow f(x) < -B$.

(c) We say that f(x) approaches minus infinity as x approaches x_0 from the left, and write $\lim\limits_{x \to x_0^-} f(x) = -\infty$, if for every positive number B (or negative number −B) there exists a corresponding number $\delta > 0$ such that for all x, $x_0 - \delta < x < x_0 \Rightarrow f(x) < -B$.

63. For $B > 0$, $\frac{1}{x} < -B < 0 \Leftrightarrow -\frac{1}{x} > B > 0 \Leftrightarrow -x < \frac{1}{B} \Leftrightarrow -\frac{1}{B} < x$. Choose $\delta = \frac{1}{B}$. Then $-\delta < x < 0$

$\Rightarrow -\frac{1}{B} < x \Rightarrow \frac{1}{x} < -B$ so that $\lim\limits_{x \to 0^-} \frac{1}{x} = -\infty$.

65. For $B > 0$, $\frac{1}{x-2} > B \Leftrightarrow 0 < x - 2 < \frac{1}{B}$. Choose $\delta = \frac{1}{B}$. Then $2 < x < 2 + \delta \Rightarrow 0 < x - 2 < \delta \Rightarrow 0 < x - 2 < \frac{1}{B}$

$\Rightarrow \frac{1}{x-2} > B > 0$ so that $\lim\limits_{x \to 2^+} \frac{1}{x-2} = \infty$.

67. For $B > 0$ and $0 < x < 1$, $\frac{1}{1-x^2} > B \Leftrightarrow 1 - x^2 < \frac{1}{B} \Leftrightarrow (1-x)(1+x) < \frac{1}{B}$. Now $\frac{1+x}{2} < 1$ since $x < 1$. Choose

$\delta < \frac{1}{2B}$. Then $1 - \delta < x < 1 \Rightarrow -\delta < x - 1 < 0 \Rightarrow 1 - x < \delta < \frac{1}{2B} \Rightarrow (1-x)(1+x) < \frac{1}{B}\left(\frac{1+x}{2}\right) < \frac{1}{B}$

$\Rightarrow \frac{1}{1-x^2} > B$ for $0 < x < 1$ and x near $1 \Rightarrow \lim\limits_{x \to 1^-} \frac{1}{1-x^2} = \infty$.

1.5 CONTINUITY

1. No, discontinuous at $x = 2$, not defined at $x = 2$

3. Continuous on $[-1, 3]$

5. (a) Yes

   (b) Yes, $\lim\limits_{x \to -1^+} f(x) = 0$

   (c) Yes

   (d) Yes

7. (a) No

   (b) No

9. $f(2) = 0$, since $\lim\limits_{x \to 2^-} f(x) = -2(2) + 4 = 0 = \lim\limits_{x \to 2^+} f(x)$

11. Nonremovable discontinuity at $x = 1$ because $\lim\limits_{x \to 1} f(x)$ fails to exist ($\lim\limits_{x \to 1^-} f(x) = 1$ and $\lim\limits_{x \to 1^+} f(x) = 0$). Removable discontinuity at $x = 0$ by assigning the number $\lim\limits_{x \to 0} f(x) = 0$ to be the value of $f(0)$ rather than $f(0) = 1$.

13. Discontinuous only when $x - 2 = 0 \Rightarrow x = 2$

15. Discontinuous only when $x^2 - 4x + 3 = 0 \Rightarrow (x-3)(x-1) = 0 \Rightarrow x = 3$ or $x = 1$

17. Continuous everywhere. ($|x-1| + \sin x$ defined for all x; limits exist and are equal to function values.)

19. Discontinuous only at $x = 0$

21. Discontinuous when 2x is an integer multiple of π, i.e., $2x = n\pi$, n an integer $\Rightarrow x = \frac{n\pi}{2}$, n an integer, but continuous at all other x.

23. Discontinuous at odd integer multiples of $\frac{\pi}{2}$, i.e., $x = (2n-1)\frac{\pi}{2}$, n an integer, but continuous at all other x.

25. Discontinuous when $2x + 3 < 0$ or $x < -\frac{3}{2} \Rightarrow$ continuous on the interval $\left[-\frac{3}{2}, \infty\right)$.

27. Continuous everywhere: $(2x-1)^{1/3}$ is defined for all x; limits exist and are equal to function values.

29. $\lim\limits_{x \to \pi} \sin(x - \sin x) = \sin(\pi - \sin \pi) = \sin(\pi - 0) = \sin \pi = 0$

31. $\lim\limits_{y \to 1} \sec\left(y \sec^2 y - \tan^2 y - 1\right) = \lim\limits_{y \to 1} \sec\left(y \sec^2 y - \sec^2 y\right) = \lim\limits_{y \to 1} \sec\left((y-1)\sec^2 y\right) = \sec\left((1-1)\sec^2 1\right)$
    $= \sec 0 = 1$

33. $\lim\limits_{t \to 0} \cos\left[\dfrac{\pi}{\sqrt{19 - 3 \sec 2t}}\right] = \cos\left[\dfrac{\pi}{\sqrt{19 - 3 \sec 0}}\right] = \cos\dfrac{\pi}{\sqrt{16}} = \cos\dfrac{\pi}{4} = \dfrac{\sqrt{2}}{2}$

35. $g(x) = \dfrac{x^2 - 9}{x - 3} = \dfrac{(x+3)(x-3)}{(x-3)} = x + 3, \ x \neq 3 \Rightarrow g(3) = \lim\limits_{x \to 3}(x+3) = 6$

37. $f(s) = \dfrac{s^3 - 1}{s^2 - 1} = \dfrac{(s^2 + s + 1)(s-1)}{(s+1)(s-1)} = \dfrac{s^2 + s + 1}{s+1}, \ s \neq 1 \Rightarrow f(1) = \lim\limits_{s \to 1}\left(\dfrac{s^2 + s + 1}{s+1}\right) = \dfrac{3}{2}$

39. As defined, $\lim\limits_{x \to 3^-} f(x) = (3)^2 - 1 = 8$ and $\lim\limits_{x \to 3^+} (2a)(3) = 6a$. For f(x) to be continuous we must have
    $6a = 8 \Rightarrow a = \dfrac{4}{3}$.

41. The function can be extended: $f(0) \approx 2.3$.

43. The function cannot be extended to be continuous at $x = 0$. If $f(0) = 1$, it will be continuous from the right. Or if $f(0) = -1$, it will be continuous from the left.

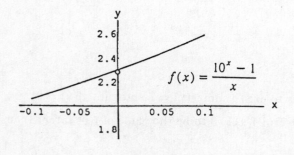

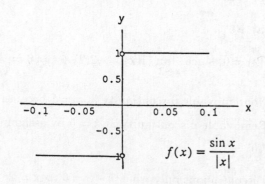

45. f(x) is continuous on $[0,1]$ and $f(0) < 0$, $f(1) > 0$
$\Rightarrow$ by the Intermediate Value Theorem f(x) takes
on every value between $f(0)$ and $f(1)$ $\Rightarrow$ the
equation $f(x) = 0$ has at least one solution between
$x = 0$ and $x = 1$.

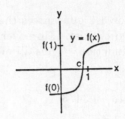

47. Let $f(x) = x^3 - 15x + 1$ which is continuous on $[-4, 4]$. Then $f(-4) = -3$, $f(-1) = 15$, $f(1) = -13$, and $f(4) = 5$.

By the Intermediate Value Theorem $f(x) = 0$ for some x in each of the intervals $-4 < x < -1$, $-1 < x < 1$, and

$1 < x < 4$. That is, $x^3 - 15x + 1 = 0$ has three solutions in $[-4, 4]$.

49. Answers may vary. Note that f is continuous for every value of x.

    (a) $f(0) = 10$, $f(1) = 1^3 - 8(1) + 10 = 3$. Since $3 < \pi < 10$, by the Intermediate Value Theorem, there exists a c
    so that $0 < c < 1$ and $f(c) = \pi$.

    (b) $f(0) = 10$, $f(-4) = (-4)^3 - 8(-4) + 10 = -22$. Since $-22 < -\sqrt{3} < 10$, by the Intermediate Value
    Theorem, there exists a c so that $-4 < c < 0$ and $f(c) = -\sqrt{3}$.

    (c) $f(0) = 10$, $f(1000) = (1000)^3 - 8(1000) + 10 = 999{,}992{,}010$. Since $10 < 5{,}000{,}000 < 999{,}992{,}010$, by the
    Intermediate Value Theorem, there exists a c so that $0 < c < 1000$ and $f(c) = 5{,}000{,}000$.

51. Answers may vary. For example, $f(x) = \dfrac{\sin(x-2)}{x-2}$ is discontinuous at $x = 2$ because it is not defined there.
However, the discontinuity can be removed because f has a limit (namely 1) as $x \to 2$.

53. (a) Suppose x_0 is rational $\Rightarrow f(x_0) = 1$. Choose $\epsilon = \frac{1}{2}$. For any $\delta > 0$ there is an irrational number x (actually
    infinitely many) in the interval $(x_0 - \delta, x_0 + \delta) \Rightarrow f(x) = 0$. Then $0 < |x - x_0| < \delta$ but $\big|f(x) - f(x_0)\big|$
    $= 1 > \frac{1}{2} = \epsilon$, so $\lim\limits_{x \to x_0} f(x)$ fails to exist $\Rightarrow$ f is discontinuous at x_0 rational.
        On the other hand, x_0 irrational $\Rightarrow f(x_0) = 0$ and there is a rational number x in $(x_0 - \delta, x_0 + \delta) \Rightarrow f(x)$
    $= 1$. Again $\lim\limits_{x \to x_0} f(x)$ fails to exist $\Rightarrow$ f is discontinuous at x_0 irrational. That is, f is discontinuous at
    every point.

    (b) f is neither right-continuous nor left-continuous at any point x_0 because in every interval $(x_0 - \delta, x_0)$ or
    $(x_0, x_0 + \delta)$ there exist both rational and irrational real numbers. Thus neither limits $\lim\limits_{x \to x_0^-} f(x)$ and
    $\lim\limits_{x \to x_0^+} f(x)$ exist by the same arguments used in part (a).

55. No. For instance, if $f(x) = 0$, $g(x) = \lceil x \rceil$, then $h(x) = 0\big(\lceil x \rceil\big) = 0$ is continuous at $x = 0$ and g(x) is not.

57. Yes, because of the Intermediate Value Theorem. If $f(a)$ and $f(b)$ did have different signs then f would have to
equal zero at some point between a and b since f is continuous on $[a, b]$.

59. If $f(0) = 0$ or $f(1) = 1$, we are done (i.e., $c = 0$ or $c = 1$ in those cases). Then let $f(0) = a > 0$ and $f(1) = b < 1$
because $0 \le f(x) \le 1$. Define $g(x) = f(x) - x \Rightarrow$ g is continuous on $[0, 1]$. Moreover, $g(0) = f(0) - 0 = a > 0$ and
$g(1) = f(1) - 1 = b - 1 < 0 \Rightarrow$ by the Intermediate Value Theorem there is a number c in $(0, 1)$ such that
$g(c) = 0 \Rightarrow f(c) - c = 0$ or $f(c) = c$.

61. Theorem 1 in Section 1.2 guarantees that all limits exist for the algebraic combinations in Theorem 6 whenever $\lim\limits_{x\to c} f(x)$ and $\lim\limits_{x\to c} g(x)$ exist. Moreover, $f(c)$ and $g(c) \neq 0$, both exist and $\lim\limits_{x\to c} f(x) = f(c)$, $\lim\limits_{x\to c} g(x) = g(c)$ by continuity of each function. Thus all of the continuity conditions hold for each algebraic combination in Theorem 6.

63. $x \approx 1.8794, -1.5321, -0.3473$ 65. $x \approx 1.7549$

67. $x \approx 3.5156$ 69. $x \approx 0.7391$

1.6 TANGENT LINES

1. P_1: $m_1 = 1$, P_2: $m_2 = 5$ 3. P_1: $m_1 = \dfrac{5}{2}$, P_2: $m_2 = -\dfrac{1}{2}$

5. $m = \lim\limits_{h\to 0} \dfrac{\left[4 - (-1+h)^2\right] - \left(4 - (-1)^2\right)}{h}$

$\quad = \lim\limits_{h\to 0} \dfrac{-\left(1 - 2h + h^2\right) + 1}{h} = \lim\limits_{h\to 0} \dfrac{h(2 - h)}{h} = 2;$

at $(-1, 3)$: $y = 3 + 2(x - (-1)) \Rightarrow y = 2x + 5$,

tangent line

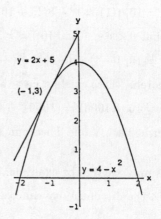

7. $m = \lim\limits_{h\to 0} \dfrac{2\sqrt{1+h} - 2\sqrt{1}}{h} = \lim\limits_{h\to 0} \dfrac{2\sqrt{1+h} - 2}{h} \cdot \dfrac{2\sqrt{1+h} + 2}{2\sqrt{1+h} + 2}$

$\quad = \lim\limits_{h\to 0} \dfrac{4(1+h) - 4}{2h\left(\sqrt{1+h} + 1\right)} = \lim\limits_{h\to 0} \dfrac{2}{\sqrt{1+h} + 1} = 1;$

at $(1, 2)$: $y = 2 + 1(x - 1) \Rightarrow y = x + 1$, tangent line

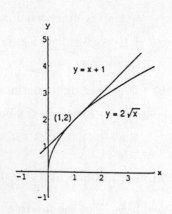

9. $m = \lim\limits_{h \to 0} \dfrac{(-2+h)^3 - (-2)^3}{h} = \lim\limits_{h \to 0} \dfrac{-8+12h-6h^2+h^3+8}{h}$

     $= \lim\limits_{h \to 0} \left(12 - 6h + h^2\right) = 12;$

at $(-2,-8):$ $y = -8 + 12(x - (-2)) \Rightarrow y = 12x + 16,$

tangent line

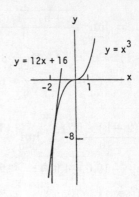

11. $m = \lim\limits_{h \to 0} \dfrac{\left[(2+h)^2 + 1\right] - 5}{h} = \lim\limits_{h \to 0} \dfrac{\left(5 + 4h + h^2\right) - 5}{h} = \lim\limits_{h \to 0} \dfrac{h(4+h)}{h} = 4;$

at $(2,5):$ $y - 5 = 4(x - 2),$ tangent line

13. $m = \lim\limits_{h \to 0} \dfrac{\dfrac{3+h}{(3+h)-2} - 3}{h} = \lim\limits_{h \to 0} \dfrac{(3+h) - 3(h+1)}{h(h+1)} = \lim\limits_{h \to 0} \dfrac{-2h}{h(h+1)} = -2;$

at $(3,3):$ $y - 3 = -2(x - 3),$ tangent line

15. $m = \lim\limits_{h \to 0} \dfrac{(2+h)^3 - 8}{h} = \lim\limits_{h \to 0} \dfrac{\left(8 + 12h + 6h^2 + h^3\right) - 8}{h} = \lim\limits_{h \to 0} \dfrac{h\left(12 + 6h + h^2\right)}{h} = 12;$

at $(2,8):$ $y - 8 = 12(t - 2),$ tangent line

17. $m = \lim\limits_{h \to 0} \dfrac{\sqrt{4+h} - 2}{h} = \lim\limits_{h \to 0} \dfrac{\sqrt{4+h} - 2}{h} \cdot \dfrac{\sqrt{4+h} + 2}{\sqrt{4+h} + 2} = \lim\limits_{h \to 0} \dfrac{(4+h) - 4}{h\left(\sqrt{4+h} + 2\right)} = \lim\limits_{h \to 0} \dfrac{h}{h\left(\sqrt{4+h} + 2\right)} = \dfrac{1}{\sqrt{4} + 2}$

     $= \dfrac{1}{4};$ at $(4,2):$ $y - 2 = \dfrac{1}{4}(x - 4),$ tangent line

19. At $x = -1,$ $y = 5 \Rightarrow m = \lim\limits_{h \to 0} \dfrac{5(-1+h)^2 - 5}{h} = \lim\limits_{h \to 0} \dfrac{5\left(1 - 2h + h^2\right) - 5}{h} = \lim\limits_{h \to 0} \dfrac{5h(-2+h)}{h} = -10,$ slope

21. At $x = 3,$ $y = \dfrac{1}{2} \Rightarrow m = \lim\limits_{h \to 0} \dfrac{\dfrac{1}{(3+h)-1} - \dfrac{1}{2}}{h} = \lim\limits_{h \to 0} \dfrac{2 - (2+h)}{2h(2+h)} = \lim\limits_{h \to 0} \dfrac{-h}{2h(2+h)} = -\dfrac{1}{4},$ slope

23. At a horizontal tangent the slope $m = 0 \Rightarrow 0 = m = \lim\limits_{h \to 0} \dfrac{\left[(x+h)^2 + 4(x+h) - 1\right] - \left(x^2 + 4x - 1\right)}{h}$

     $= \lim\limits_{h \to 0} \dfrac{\left(x^2 + 2xh + h^2 + 4x + 4h - 1\right) - \left(x^2 + 4x - 1\right)}{h} = \lim\limits_{h \to 0} \dfrac{\left(2xh + h^2 + 4h\right)}{h} = \lim\limits_{h \to 0} (2x + h + 4) = 2x + 4;$

$2x + 4 = 0 \Rightarrow x = -2.$ Then $f(-2) = 4 - 8 - 1 = -5 \Rightarrow (-2,-5)$ is the point on the graph where there is a horizontal tangent.

25. $-1 = m = \lim\limits_{h \to 0} \dfrac{\frac{1}{(x+h)-1} - \frac{1}{x-1}}{h} = \lim\limits_{h \to 0} \dfrac{(x-1)-(x+h-1)}{h(x-1)(x+h-1)} = \lim\limits_{h \to 0} \dfrac{-h}{h(x-1)(x+h-1)} = -\dfrac{1}{(x-1)^2}$

$\Rightarrow (x-1)^2 = 1 \Rightarrow x^2 - 2x = 0 \Rightarrow x(x-2) = 0 \Rightarrow x = 0$ or $x = 2$. If $x = 0$, then $y = -1$ and $m = -1$

$\Rightarrow y = -1 - (x-0) = -(x+1)$. If $x = 2$, then $y = 1$ and $m = -1 \Rightarrow y = 1 - (x-2) = -(x-3)$.

27. $\lim\limits_{h \to 0} \dfrac{f(2+h) - f(2)}{h} = \lim\limits_{h \to 0} \dfrac{\left(100 - 4.9(2+h)^2\right) - \left(100 - 4.9(2)^2\right)}{h} = \lim\limits_{h \to 0} \dfrac{-4.9\left(4 + 4h + h^2\right) + 4.9(4)}{h}$

$= \lim\limits_{h \to 0} (-19.6 - 4.9h) = -19.6$. The minus sign indicates the object is falling <u>downward</u> at a speed of 19.6 m/sec.

29. $\lim\limits_{h \to 0} \dfrac{f(3+h) - f(3)}{h} = \lim\limits_{h \to 0} \dfrac{\pi(3+h)^2 - \pi(3)^2}{h} = \lim\limits_{h \to 0} \dfrac{\pi\left[9 + 6h + h^2 - 9\right]}{h} = \lim\limits_{h \to 0} \pi(6+h) = 6\pi$

31. Slope at origin $= \lim\limits_{h \to 0} \dfrac{f(0+h) - f(0)}{h} = \lim\limits_{h \to 0} \dfrac{h^2 \sin\left(\frac{1}{h}\right)}{h} = \lim\limits_{h \to 0} h \sin\left(\frac{1}{h}\right) = 0 \Rightarrow$ yes, f(x) does have a tangent at the origin with slope 0.

33. $\lim\limits_{h \to 0^-} \dfrac{f(0+h) - f(0)}{h} = \lim\limits_{h \to 0^-} \dfrac{-1-0}{h} = \infty$, and $\lim\limits_{h \to 0^+} \dfrac{f(0+h) - f(0)}{h} = \lim\limits_{h \to 0^+} \dfrac{1-0}{h} = \infty$. Therefore,

$\lim\limits_{h \to 0} \dfrac{f(0+h) - f(0)}{h} = \infty \Rightarrow$ yes, the graph of f has a vertical tangent at the origin.

35. (a) The graph appears to have a cusp at $x = 0$.

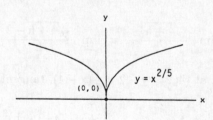

(b) $\lim\limits_{h \to 0^-} \dfrac{f(0+h) - f(0)}{h} = \lim\limits_{h \to 0^-} \dfrac{h^{2/5} - 0}{h} = \lim\limits_{h \to 0^-} \dfrac{1}{h^{3/5}} = -\infty$ and $\lim\limits_{h \to 0^+} \dfrac{1}{h^{3/5}} = \infty \Rightarrow$ limit does not exist

$\Rightarrow$ the graph of $y = x^{2/5}$ does not have a vertical tangent at $x = 0$.

37. (a) The graph appears to have a vertical tangent at $x = 0$.

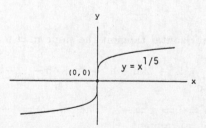

(b) $\lim\limits_{h \to 0} \dfrac{f(0+h) - f(0)}{h} = \lim\limits_{h \to 0} \dfrac{h^{1/5} - 0}{h} = \lim\limits_{h \to 0} \dfrac{1}{h^{4/5}} = \infty \Rightarrow y = x^{1/5}$ has a vertical tangent at $x = 0$.

39. (a) The graph appears to have a cusp at $x = 0$.

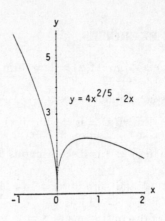

$y = 4x^{2/5} - 2x$

(b) $\displaystyle\lim_{h\to 0^-} \frac{f(0+h)-f(0)}{h} = \lim_{h\to 0^-} \frac{4h^{2/5}-2h}{h} = \lim_{h\to 0^-} \frac{4h^{2/5}-2h}{h} = \lim_{h\to 0^-} \frac{4}{h^{3/5}} - 2 = -\infty$ and $\displaystyle\lim_{h\to 0^+} \frac{4}{h^{3/5}} - 2$

$= \infty \Rightarrow$ limit does not exist $\Rightarrow$ the graph of $y = 4x^{2/5} - 2x$ does not have a vertical tangent at $x = 0$.

41. (a) The graph appears to have a vertical tangent at $x = 1$ and a cusp at $x = 0$.

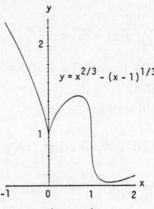

$y = x^{2/3} - (x-1)^{1/3}$

(b) $x = 1$: $\displaystyle\lim_{h\to 0} \frac{(1+h)^{2/3} - (1+h-1)^{1/3} - 1}{h} = \lim_{h\to 0} \frac{(1+h)^{2/3} - h^{1/3} - 1}{h} = -\infty$

$\Rightarrow y = x^{2/3} - (x-1)^{1/3}$ has a vertical tangent at $x = 1$;

$x = 0$: $\displaystyle\lim_{h\to 0} \frac{f(0+h)-f(0)}{h} = \lim_{h\to 0} \frac{h^{2/3} - (h-1)^{1/3} - (-1)^{1/3}}{h} = \lim_{h\to 0} \left[\frac{1}{h^{1/3}} - \frac{(h-1)^{1/3}}{h} + \frac{1}{h} \right]$

does not exist $\Rightarrow y = x^{2/3} - (x-1)^{1/3}$ does not have a vertical tangent at $x = 0$.

43. (a) The graph appears to have a vertical tangent at $x = 0$.

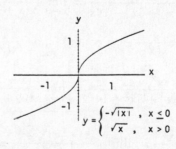

$y = \begin{cases} -\sqrt{|x|}, & x \le 0 \\ \sqrt{x}, & x > 0 \end{cases}$

(b) $\displaystyle\lim_{h\to 0^+} \frac{f(0+h)-f(0)}{h} = \lim_{x\to 0^+} \frac{\sqrt{h}-0}{h} = \lim_{h\to 0} \frac{1}{\sqrt{h}} = \infty;$

$\displaystyle\lim_{h\to 0^-} \frac{f(0+h)-f(0)}{h} = \lim_{h\to 0^-} \frac{-\sqrt{|h|}-0}{h} = \lim_{h\to 0^-} \frac{-\sqrt{|h|}}{-|h|} = \lim_{h\to 0^-} \frac{1}{\sqrt{|h|}} = \infty$

CHAPTER 1 PRACTICE EXERCISES

1. At $x = -1$: $\lim\limits_{x \to -1^-} f(x) = \lim\limits_{x \to -1^+} f(x) = 1 \Rightarrow \lim\limits_{x \to -1} f(x)$

   $= 1 = f(-1) \Rightarrow f$ is continuous at $x = -1$.

   At $x = 0$: $\lim\limits_{x \to 0^-} f(x) = \lim\limits_{x \to 0^+} f(x) = 0 \Rightarrow \lim\limits_{x \to 0} f(x) = 0$.

   But $f(0) = 1 \neq \lim\limits_{x \to 0} f(x) \Rightarrow f$ is discontinuous at $x = 0$.

   At $x = 1$: $\lim\limits_{x \to 1^-} f(x) = -1$ and $\lim\limits_{x \to 1^+} f(x) = 1 \Rightarrow \lim\limits_{x \to 1} f(x)$

   does not exist $\Rightarrow f$ is discontinuous at $x = 1$.

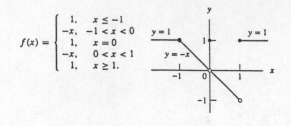

$$f(x) = \begin{cases} 1, & x \le -1 \\ -x, & -1 < x < 0 \\ 1, & x = 0 \\ -x, & 0 < x < 1 \\ 1, & x \ge 1. \end{cases}$$

3. (a) $\lim\limits_{x \to c} 3f(x) = 3 \lim\limits_{x \to c} f(x) = 3(-7) = -21$

   (b) $\lim\limits_{x \to c} (f(x))^2 = \lim\limits_{x \to c} (f(x) \cdot f(x)) = \lim\limits_{x \to c} f(x) \cdot \lim\limits_{x \to c} f(x) = 49$

   (c) $\lim\limits_{x \to 1} (f(x) \cdot g(x)) = \lim\limits_{x \to c} f(x) \cdot \lim\limits_{x \to c} g(x) = (-7)(0) = 0$

   (d) $\lim\limits_{x \to c} \dfrac{f(x)}{g(x) - 7} = \dfrac{\lim\limits_{x \to c} f(x)}{\lim\limits_{x \to c} g(x) - 7} = \dfrac{-7}{0 - 7} = 1$

   (e) $\lim\limits_{x \to c} \cos(g(x)) = \cos\left(\lim\limits_{x \to c} g(x) \right) = \cos(0) = 1$

   (f) $\lim\limits_{x \to c} \left| f(x) \right| = \left| \lim\limits_{x \to c} f(x) \right| = |-7| = 7$

5. Since $\lim\limits_{x \to 0} x = 0$ we must have that $\lim\limits_{x \to 0} (4 - g(x)) = 0$. Otherwise, if $\lim\limits_{x \to 0} (4 - g(x))$ is a finite positive

   number, we would have $\lim\limits_{x \to 0^-} \left[\dfrac{4 - g(x)}{x} \right] = -\infty$ and $\lim\limits_{x \to 0^+} \left[\dfrac{4 - g(x)}{x} \right] = \infty$ so the limit could not equal 1 as

   $x \to 0$. Similar reasoning holds if $\lim\limits_{x \to 0} (4 - g(x))$ is a finite negative number. We conclude that $\lim\limits_{x \to 0} g(x) = 4$.

7. $\lim\limits_{x \to 0^+} [4\, g(x)]^{1/3} = 2 \Rightarrow \left[\lim\limits_{x \to 0^+} 4\, g(x) \right]^{1/3} = 2 \Rightarrow \lim\limits_{x \to 0^+} 4\, g(x) = 8$, since $2^3 = 8$. Then $\lim\limits_{x \to 0^+} g(x) = 2$.

9. $\lim\limits_{x \to 1} \dfrac{3x^2 + 1}{g(x)} = \infty \Rightarrow \lim\limits_{x \to 1} g(x) = 0$ since $\lim\limits_{x \to 1} (3x^2 + 1) = 4$

11. (a) $\lim\limits_{x \to 0} \dfrac{x^2 - 4x + 4}{x^3 + 5x^2 - 14x} = \lim\limits_{x \to 0} \dfrac{(x - 2)(x - 2)}{x(x + 7)(x - 2)} = \lim\limits_{x \to 0} \dfrac{x - 2}{x(x + 7)}, x \neq 2$; the limit does not exist because

   $\lim\limits_{x \to 0^-} \dfrac{x - 2}{x(x + 7)} = \infty$ and $\lim\limits_{x \to 0^+} \dfrac{x - 2}{x(x + 7)} = -\infty$

   (b) $\lim\limits_{x \to 2} \dfrac{x^2 - 4x + 4}{x^3 + 5x^2 - 14x} = \lim\limits_{x \to 2} \dfrac{(x - 2)(x - 2)}{x(x + 7)(x - 2)} = \lim\limits_{x \to 2} \dfrac{x - 2}{x(x + 7)}, x \neq 2 = \dfrac{0}{2(9)} = 0$

13. $\lim\limits_{x \to 1} \dfrac{1 - \sqrt{x}}{1 - x} = \lim\limits_{x \to 1} \dfrac{1 - \sqrt{x}}{(1 - \sqrt{x})(1 + \sqrt{x})} = \lim\limits_{x \to 1} \dfrac{1}{1 + \sqrt{x}} = \dfrac{1}{2}$

15. $\lim\limits_{h\to 0} \dfrac{(x+h)^2 - x^2}{h} = \lim\limits_{h\to 0} \dfrac{(x^2 + 2hx + h^2) - x^2}{h} = \lim\limits_{h\to 0} (2x + h) = 2x$

17. $\lim\limits_{x\to 0} \dfrac{\frac{1}{2+x} - \frac{1}{2}}{x} = \lim\limits_{x\to 0} \dfrac{2 - (2+x)}{2x(2+x)} = \lim\limits_{x\to 0} \dfrac{-1}{4 + 2x} = -\dfrac{1}{4}$

19. (a) $\lim\limits_{x\to c} f(x) = \lim\limits_{x\to c} x^{1/3} = c^{1/3} = f(c)$ for every real number c $\Rightarrow$ f is continuous on $(-\infty, \infty)$

   (b) $\lim\limits_{x\to c} g(x) = \lim\limits_{x\to c} x^{3/4} = c^{3/4} = g(c)$ for every nonnegative real number c $\Rightarrow$ g is continuous on $[0, \infty)$

   (c) $\lim\limits_{x\to c} h(x) = \lim\limits_{x\to c} x^{-2/3} = \dfrac{1}{c^{2/3}} = h(c)$ for every nonzero real number c $\Rightarrow$ h is continuous on $(-\infty, 0)$ and

   $(0, \infty)$

   (d) $\lim\limits_{x\to c} k(x) = \lim\limits_{x\to c} x^{-1/6} = \dfrac{1}{c^{1/6}} = k(c)$ for every positive real number c $\Rightarrow$ h is continuous on $(0, \infty)$

21. Yes, f does have a continuous extension to a = 1:

   define $f(1) = \lim\limits_{x\to 1} \dfrac{x-1}{x - \sqrt[4]{x}} = \dfrac{4}{3}$.

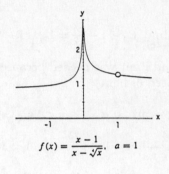

$f(x) = \dfrac{x-1}{x - \sqrt[4]{x}}, \quad a = 1$

23. From the graph we see that $\lim\limits_{t\to 0^-} h(t) \neq \lim\limits_{t\to 0^+} h(t)$

   so h <u>cannot</u> be extended to a continuous function at a = 0.

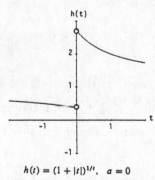

$h(t) = (1 + |t|)^{1/t}, \quad a = 0$

25. (a) $f(-1) = -1$ and $f(2) = 5 \Rightarrow$ f has a root between -1 and 2 by the Intermediate Value Theorem.

   (b) root is 1.32471795724

CHAPTER 1 ADDITIONAL EXERCISES–THEORY, EXAMPLES, APPLICATIONS

1. (a) No, since $\lim\limits_{x\to c} f(x) = 5$ means the function f(x) approaches the value 5 as x gets arbitrarily close to c, but

   not that f(x) takes on the value 5 at x = c.

   (b) Although $f(c) = 5$, the function f(x) could approach some other value as x $\to$ c, so $\lim\limits_{x\to c} f(x)$ does not have
   to equal 5.

3. (a)

x	0.1	0.01	0.001	0.0001	0.00001
x^x	0.7943	0.9550	0.9931	0.9991	0.9999

Apparently, $\lim\limits_{x \to 0^+} x^x = 1$

(b)

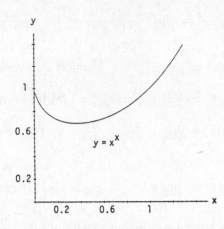

$y = x^x$

5. $\lim\limits_{v \to c^-} L = \lim\limits_{v \to c^-} L_0 \sqrt{1 - \dfrac{v^2}{c^2}} = L_0 \sqrt{1 - \dfrac{\lim\limits_{v \to c^-} v^2}{c^2}} = L_0 \sqrt{1 - \dfrac{c^2}{c^2}} = 0$

The left-hand limit was needed because the function L is undefined if $v > c$ (the rocket cannot move faster than the speed of light).

7. (a) Since $x \to 0^+$, $0 < x^3 < x < 1 \Rightarrow (x^3 - x) \to 0^- \Rightarrow \lim\limits_{x \to 0^+} f(x^3 - x) = \lim\limits_{y \to 0^-} f(y) = B$ where $y = x^3 - x$.

(b) Since $x \to 0^-$, $-1 < x < x^3 < 0 \Rightarrow (x^3 - x) \to 0^+ \Rightarrow \lim\limits_{x \to 0^-} f(x^3 - x) = \lim\limits_{y \to 0^+} f(y) = A$ where $y = x^3 - x$.

(c) Since $x \to 0^+$, $0 < x^4 < x^2 < 1 \Rightarrow (x^2 - x^4) \to 0^+ \Rightarrow \lim\limits_{x \to 0^+} f(x^2 - x^4) = \lim\limits_{y \to 0^+} f(y) = A$ where $y = x^2 - x^4$.

(d) Since $x \to 0^-$, $-1 < x < 0 \Rightarrow 0 < x^4 < x^2 < 1 \Rightarrow (x^2 - x^4) \to 0^+ \Rightarrow \lim\limits_{x \to 0^+} f(x^2 - x^4) = A$ as in part (c).

9. $f(x) = x + 2 \cos x \Rightarrow f(0) = 0 + 2 \cos 0 = 2 > 0$ and $f(-\pi) = -\pi + 2 \cos(-\pi) = -\pi - 2 < 0$. Since $f(x)$ is continuous on $[-\pi, 0]$, by the Intermediate Value Theorem, $f(x)$ must take on every value between $[-\pi - 2, 2]$. Thus there is some number c in $[-\pi, 0]$ such that $f(c) = 0$; i.e., c is a solution to $x + 2 \cos x = 0$.

11. $\left| \dfrac{\sqrt{x}}{2} - 1 \right| < 0.2 \Rightarrow -0.2 < \dfrac{\sqrt{x}}{2} - 1 < 0.2 \Rightarrow 0.8 < \dfrac{\sqrt{x}}{2} < 1.2 \Rightarrow 1.6 < \sqrt{x} < 2.4 \Rightarrow 2.56 < x < 5.76$.

$\left| \dfrac{\sqrt{x}}{2} - 1 \right| < 0.1 \Rightarrow -0.1 < \dfrac{\sqrt{x}}{2} - 1 < 0.1 \Rightarrow 0.9 < \dfrac{\sqrt{x}}{2} < 1.1 \Rightarrow 1.8 < \sqrt{x} < 2.2 \Rightarrow 3.24 < x < 4.84$.

13. Yes. Let R be the radius of the equator (earth) and suppose at a fixed instant of time we label noon as the zero point, 0, on the equator $\Rightarrow 0 + \pi R$ represents the midnight point (at the same exact time). Suppose x_1 is a point on the equator "just after" noon $\Rightarrow x_1 + \pi R$ is simultaneously "just after" midnight. It seems reasonable that the temperature T at a point just after noon is hotter than it would be at the diametrically opposite point just after midnight: That is, $T(x_1) - T(x_1 + \pi R) > 0$. At exactly the same moment in time pick x_2 to be a point just before midnight $\Rightarrow x_2 + \pi R$ is just before noon. Then $T(x_2) - T(x_2 + \pi R) < 0$. Assuming the temperature function T is continuous along the equator (which is reasonable), the Intermediate Value Theorem says there is a point c between 0 (noon) and πR (simultaneously midnight) such that $T(c) - T(c + \pi R) = 0$; i.e., there is always a pair of antipodal points on the earth's equator where the temperatures are the same.

15. Show $\lim\limits_{x \to 1} f(x) = \lim\limits_{x \to 1} \left(x^2 - 7\right) = -6 = f(1).$

Step 1: $\left|\left(x^2 - 7\right) + 6\right| < \epsilon \Rightarrow -\epsilon < x^2 - 1 < \epsilon \Rightarrow 1 - \epsilon < x^2 < 1 + \epsilon \Rightarrow \sqrt{1 - \epsilon} < x < \sqrt{1 + \epsilon}.$

Step 2: $|x - 1| < \delta \Rightarrow -\delta < x - 1 < \delta \Rightarrow -\delta + 1 < x < \delta + 1.$

Then $-\delta + 1 = \sqrt{1 - \epsilon}$ or $\delta + 1 = \sqrt{1 + \epsilon}$. Choose $\delta = \min\left\{1 - \sqrt{1 - \epsilon}, \sqrt{1 + \epsilon} - 1\right\}$, then

$0 < |x - 1| < \delta \Rightarrow \left|\left(x^2 - 7\right) - 6\right| < \epsilon$ and $\lim\limits_{x \to 1} f(x) = -6.$ By the continuity test, $f(x)$ is continuous at $x = 1.$

17. Show $\lim\limits_{x \to 2} h(x) = \lim\limits_{x \to 2} \sqrt{2x - 3} = 1 = h(2).$

Step 1: $\left|\sqrt{2x - 3} - 1\right| < \epsilon \Rightarrow -\epsilon < \sqrt{2x - 3} - 1 < \epsilon \Rightarrow 1 - \epsilon < \sqrt{2x - 3} < 1 + \epsilon \Rightarrow \dfrac{(1 - \epsilon)^2 + 3}{2} < x < \dfrac{(1 + \epsilon)^2 + 3}{2}.$

Step 2: $|x - 2| < \delta \Rightarrow -\delta < x - 2 < \delta$ or $-\delta + 2 < x < \delta + 2.$

Then $-\delta + 2 = \dfrac{(1 - \epsilon)^2 + 3}{2} \Rightarrow \delta = 2 - \dfrac{(1 - \epsilon)^2 + 3}{2} = \dfrac{1 - (1 - \epsilon)^2}{2} = \epsilon - \dfrac{\epsilon^2}{2}$, or $\delta + 2 = \dfrac{(1 + \epsilon)^2 + 3}{2}$

$\Rightarrow \delta = \dfrac{(1 + \epsilon)^2 + 3}{2} - 2 = \dfrac{(1 + \epsilon)^2 - 1}{2} = \epsilon + \dfrac{\epsilon^2}{2}.$ Choose $\delta = \epsilon - \dfrac{\epsilon^2}{2}$, the smaller of the two values. Then,

$0 < |x - 2| < \delta \Rightarrow \left|\sqrt{2x - 3} - 1\right| < \epsilon$, so $\lim\limits_{x \to 2} \sqrt{2x - 3} = 1.$ By the continuity test, $h(x)$ is continuous at $x = 2.$

19. Show $\lim\limits_{x \to -1} f(x) = \lim\limits_{x \to -1} \dfrac{x^2 - 1}{x + 1} = \lim\limits_{x \to -1} \dfrac{(x + 1)(x - 1)}{(x + 1)} = -2, x \neq -1.$

Define the continuous extension of $f(x)$ as $F(x) = \begin{cases} \dfrac{x^2 - 1}{x + 1}, & x \neq -1 \\ -2 & , \ x = -1 \end{cases}$. We now prove the limit of $f(x)$ as $x \to -1$

exists and has the correct value.

Step 1: $\left|\dfrac{x^2 - 1}{x + 1} - (-2)\right| < \epsilon \Rightarrow -\epsilon < \dfrac{(x + 1)(x - 1)}{(x + 1)} + 2 < \epsilon \Rightarrow -\epsilon < (x - 1) + 2 < \epsilon, x \neq -1 \Rightarrow -\epsilon - 1 < x < \epsilon - 1.$

Step 2: $|x - (-1)| < \delta \Rightarrow -\delta < x + 1 < \delta \Rightarrow -\delta - 1 < x < \delta - 1.$

Then $-\delta - 1 = -\epsilon - 1 \Rightarrow \delta = \epsilon$, or $\delta - 1 = \epsilon - 1 \Rightarrow \delta = \epsilon.$ Choose $\delta = \epsilon.$ Then $0 < |x - (-1)| < \delta$

$\Rightarrow \left|\dfrac{x^2 - 1}{x + 1}\right| < \epsilon \Rightarrow \lim\limits_{x \to -1} F(x) = -2.$ Since the conditions of the continuity test are met by $F(x)$, then $f(x)$ has a

continuous extension to $F(x)$ at $x = -1.$

21. (a) If $a \geq b$, then $a - b \geq 0 \Rightarrow |a - b| = a - b \Rightarrow \max(a, b) = \dfrac{a + b}{2} + \dfrac{|a - b|}{2} = \dfrac{a + b}{2} + \dfrac{a - b}{2} = \dfrac{2a}{2} = a.$

If $a \leq b$, then $a - b \leq 0 \Rightarrow |a - b| = -(a - b) = b - a \Rightarrow \max(a, b) = \dfrac{a + b}{2} + \dfrac{|a - b|}{2} = \dfrac{a + b}{2} + \dfrac{b - a}{2}$

$= \dfrac{2b}{2} = b.$

(b) Let $\min(a, b) = \dfrac{a + b}{2} - \dfrac{|a - b|}{2}.$

23. (a) The function f is bounded on D if $f(x) \geq M$ and $f(x) \leq N$ for all x in D. This means $M \leq f(x) \leq N$ for all x in D. Choose B to be $\max\{|M|, |N|\}$. Then $|f(x)| \leq B$. On the other hand, if $|f(x)| \leq B$, then

$-B \leq f(x) \leq B \Rightarrow f(x) \geq -B$ and $f(x) \leq B \Rightarrow f(x)$ is bounded on D with $N = B$ an upper bound and $M = -B$ a lower bound.

(b) Assume $f(x) \leq N$ for all x and that $L > N$. Let $\epsilon = \dfrac{L - N}{2}$. Since $\lim\limits_{x \to x_0} f(x) = L$ there is a $\delta > 0$ such that

$0 < |x - x_0| < \delta \Rightarrow |f(x) - L| < \epsilon \Leftrightarrow L - \epsilon < f(x) < L + \epsilon \Leftrightarrow L - \dfrac{L - N}{2} < f(x) < L + \dfrac{L - N}{2}$

$\Leftrightarrow \dfrac{L + N}{2} < f(x) < \dfrac{3L - N}{2}$. But $L > N \Rightarrow \dfrac{L + N}{2} > N \Rightarrow N < f(x)$ contrary to the boundedness assumption

$f(x) \leq N$. This contradiction proves $L \leq N$.

(c) Assume $M \leq f(x)$ for all x and that $L < M$. Let $\epsilon = \dfrac{M - L}{2}$. As in part (b), $0 < |x - x_0| < \delta$

$\Rightarrow L - \dfrac{M - L}{2} < f(x) < L + \dfrac{M - L}{2} \Leftrightarrow \dfrac{3L - M}{2} < f(x) < \dfrac{M + L}{2} < M$, a contradiction.

NOTES:

CHAPTER 2 DERIVATIVES

2.1 THE DERIVATIVE OF A FUNCTION

1. Step 1: $f(x) = 4 - x^2$ and $f(x+h) = 4 - (x+h)^2$

   Step 2: $\dfrac{f(x+h) - f(x)}{h} = \dfrac{[4 - (x+h)^2] - (4 - x^2)}{h} = \dfrac{(4 - x^2 - 2xh - h^2) - 4 + x^2}{h} = \dfrac{-2xh - h^2}{h} = \dfrac{h(-2x - h)}{h}$
   $= -2x - h$

   Step 3: $f'(x) = \lim\limits_{h \to 0} (-2x - h) = -2x$; $f'(-3) = 6$, $f'(0) = 0$, $f'(1) = -2$

3. Step 1: $g(t) = \dfrac{1}{t^2}$ and $g(t+h) = \dfrac{1}{(t+h)^2}$

   Step 2: $\dfrac{g(t+h) - g(t)}{h} = \dfrac{\dfrac{1}{(t+h)^2} - \dfrac{1}{t^2}}{h} = \dfrac{\left(\dfrac{t^2 - (t+h)^2}{(t+h)^2 \cdot t^2}\right)}{h} = \dfrac{t^2 - (t^2 + 2th + h^2)}{(t+h)^2 \cdot t^2 \cdot h} = \dfrac{-2th - h^2}{(t+h)^2 t^2 h}$
   $= \dfrac{h(-2t - h)}{(t+h)^2 t^2 h} = \dfrac{-2t - h}{(t+h)^2 t^2}$

   Step 3: $g'(t) = \lim\limits_{h \to 0} \dfrac{-2t - h}{(t+h)^2 t^2} = \dfrac{-2t}{t^2 \cdot t^2} = \dfrac{-2}{t^3}$; $g'(-1) = 2$, $g'(2) = -\dfrac{1}{4}$, $g'(\sqrt{3}) = -\dfrac{2}{3\sqrt{3}}$

5. Step 1: $p(\theta) = \sqrt{3\theta}$ and $p(\theta + h) = \sqrt{3(\theta + h)}$

   Step 2: $\dfrac{p(\theta + h) - p(\theta)}{h} = \dfrac{\sqrt{3(\theta + h)} - \sqrt{3\theta}}{h} = \dfrac{\left(\sqrt{3\theta + 3h} - \sqrt{3\theta}\right)}{h} \cdot \dfrac{\left(\sqrt{3\theta + 3h} + \sqrt{3\theta}\right)}{\left(\sqrt{3\theta + 3h} + \sqrt{3\theta}\right)} = \dfrac{(3\theta + 3h) - 3\theta}{h\left(\sqrt{3\theta + 3h} + \sqrt{3\theta}\right)}$
   $= \dfrac{3h}{h\left(\sqrt{3\theta + 3h} + \sqrt{3\theta}\right)} = \dfrac{3}{\sqrt{3\theta + 3h} + \sqrt{3\theta}}$

   Step 3: $p'(\theta) = \lim\limits_{h \to 0} \dfrac{3}{\sqrt{3\theta + 3h} + \sqrt{3\theta}} = \dfrac{3}{\sqrt{3\theta} + \sqrt{3\theta}} = \dfrac{3}{2\sqrt{3\theta}}$; $p'(1) = \dfrac{3}{2\sqrt{3}}$, $p'(3) = \dfrac{1}{2}$, $p'\left(\dfrac{2}{3}\right) = \dfrac{3}{2\sqrt{2}}$

7. $y = f(x) = 2x^3$ and $f(x+h) = 2(x+h)^3 \Rightarrow \dfrac{dy}{dx} = \lim\limits_{h \to 0} \dfrac{2(x+h)^3 - 2x^3}{h} = \lim\limits_{h \to 0} \dfrac{2\left(x^3 + 3x^2 h + 3xh^2 + h^3\right) - 2x^3}{h}$

   $= \lim\limits_{h \to 0} \dfrac{6x^2 h + 6xh^2 + 2h^3}{h} = \lim\limits_{h \to 0} \dfrac{h\left(6x^2 + 6xh + 2h^2\right)}{h} = \lim\limits_{h \to 0} \left(6x^2 + 6xh + 2h^2\right) = 6x^2$

9. $s = r(t) = \dfrac{t}{2t + 1}$ and $r(t+h) = \dfrac{t+h}{2(t+h) + 1} \Rightarrow \dfrac{ds}{dt} = \lim\limits_{h \to 0} \dfrac{\left(\dfrac{t+h}{2(t+h) + 1}\right) - \left(\dfrac{t}{2t + 1}\right)}{h}$

   $= \lim\limits_{h \to 0} \dfrac{\left(\dfrac{(t+h)(2t+1) - t(2t + 2h + 1)}{(2t + 2h + 1)(2t + 1)}\right)}{h} = \lim\limits_{h \to 0} \dfrac{(t+h)(2t+1) - t(2t + 2h + 1)}{(2t + 2h + 1)(2t + 1)h}$

$$= \lim_{h \to 0} \frac{2t^2 + t + 2ht + h - 2t^2 - 2ht - t}{(2t + 2h + 1)(2t + 1)h} = \lim_{h \to 0} \frac{h}{(2t + 2h + 1)(2t + 1)h} = \lim_{h \to 0} \frac{1}{(2t + 2h + 1)(2t + 1)}$$

$$= \frac{1}{(2t + 1)(2t + 1)} = \frac{1}{(2t + 1)^2}$$

11. $p = f(q) = \dfrac{1}{\sqrt{q + 1}}$ and $f(q + h) = \dfrac{1}{\sqrt{(q + h) + 1}} \Rightarrow \dfrac{dp}{dq} = \lim\limits_{h \to 0} \dfrac{\left(\dfrac{1}{\sqrt{(q + h) + 1}}\right) - \left(\dfrac{1}{\sqrt{q + 1}}\right)}{h}$

$$= \lim_{h \to 0} \frac{\left(\dfrac{\sqrt{q + 1} - \sqrt{q + h + 1}}{\sqrt{q + h + 1}\sqrt{q + 1}}\right)}{h} = \lim_{h \to 0} \frac{\sqrt{q + 1} - \sqrt{q + h + 1}}{h\sqrt{q + h + 1}\sqrt{q + 1}}$$

$$= \lim_{h \to 0} \frac{\left(\sqrt{q + 1} - \sqrt{q + h + 1}\right)}{h\sqrt{q + h + 1}\sqrt{q + 1}} \cdot \frac{\left(\sqrt{q + 1} + \sqrt{q + h + 1}\right)}{\left(\sqrt{q + 1} + \sqrt{q + h + 1}\right)} = \lim_{h \to 0} \frac{(q + 1) - (q + h + 1)}{h\sqrt{q + h + 1}\sqrt{q + 1}\left(\sqrt{q + 1} + \sqrt{q + h + 1}\right)}$$

$$= \lim_{h \to 0} \frac{-h}{h\sqrt{q + h + 1}\sqrt{q + 1}\left(\sqrt{q + 1} + \sqrt{q + h + 1}\right)} = \lim_{h \to 0} \frac{-1}{\sqrt{q + h + 1}\sqrt{q + 1}\left(\sqrt{q + 1} + \sqrt{q + h + 1}\right)}$$

$$= \frac{-1}{\sqrt{q + 1}\sqrt{q + 1}\left(\sqrt{q + 1} + \sqrt{q + 1}\right)} = \frac{-1}{2(q + 1)\sqrt{q + 1}}$$

13. $f(x) = x + \dfrac{9}{x}$ and $f(x + h) = (x + h) + \dfrac{9}{(x + h)} \Rightarrow \dfrac{f(x + h) - f(x)}{h} = \dfrac{\left[(x + h) + \dfrac{9}{(x + h)}\right] - \left[x + \dfrac{9}{x}\right]}{h}$

$$= \frac{x(x + h)^2 + 9x - x^2(x + h) - 9(x + h)}{x(x + h)h} = \frac{x^3 + 2x^2h + xh^2 + 9x - x^3 - x^2h - 9x - 9h}{x(x + h)h} = \frac{x^2h + xh^2 - 9h}{x(x + h)h}$$

$$= \frac{h(x^2 + xh - 9)}{x(x + h)h} = \frac{x^2 + xh - 9}{x(x + h)}; \ f'(x) = \lim_{h \to 0} \frac{x^2 + xh - 9}{x(x + h)} = \frac{x^2 - 9}{x^2} = 1 - \frac{9}{x^2}; \ m = f'(-3) = 0$$

15. $\dfrac{ds}{dt} = \lim\limits_{h \to 0} \dfrac{\left[(t + h)^3 - (t + h)^2\right] - \left(t^3 - t^2\right)}{h} = \lim\limits_{h \to 0} \dfrac{\left(t^3 + 3t^2h + 3th^2 + h^3\right) - \left(t^2 + 2th + h^2\right) - t^3 + t^2}{h}$

$$= \lim_{h \to 0} \frac{3t^2h + 3th^2 + h^3 - 2th - h^2}{h} = \lim_{h \to 0} \frac{h\left(3t^2 + 3th + h^2 - 2t - h\right)}{h} = \lim_{h \to 0} \left(3t^2 + 3th + h^2 - 2t - h\right)$$

$$= 3t^2 - 2t; \ m = \left.\frac{ds}{dt}\right|_{t = -1} = 5$$

17. $f(x) = \dfrac{8}{\sqrt{x - 2}}$ and $f(x + h) = \dfrac{8}{\sqrt{(x + h) - 2}} \Rightarrow \dfrac{f(x + h) - f(x)}{h} = \dfrac{\dfrac{8}{\sqrt{(x + h) - 2}} - \dfrac{8}{\sqrt{x - 2}}}{h}$

$$= \frac{8\left(\sqrt{x - 2} - \sqrt{x + h - 2}\right)}{h\sqrt{x + h - 2}\sqrt{x - 2}} \cdot \frac{\left(\sqrt{x - 2} + \sqrt{x + h - 2}\right)}{\left(\sqrt{x - 2} + \sqrt{x + h - 2}\right)} = \frac{8[(x - 2) - (x + h - 2)]}{h\sqrt{x + h - 2}\sqrt{x - 2}\left(\sqrt{x - 2} + \sqrt{x + h - 2}\right)}$$

$$= \frac{-8h}{h\sqrt{x + h - 2}\sqrt{x - 2}\left(\sqrt{x - 2} + \sqrt{x + h - 2}\right)} \Rightarrow f'(x) = \lim_{h \to 0} \frac{-8}{\sqrt{x + h - 2}\sqrt{x - 2}\left(\sqrt{x - 2} + \sqrt{x + h - 2}\right)}$$

$$= \frac{-8}{\sqrt{x-2}\,\sqrt{x-2}\,(\sqrt{x-2}+\sqrt{x-2})} = \frac{-4}{(x-2)\sqrt{x-2}};\ m = f'(6) = \frac{-4}{4\sqrt{4}} = -\frac{1}{2} \Rightarrow \text{the equation of the tangent}$$

line at $(6,4)$ is $y - 4 = -\frac{1}{2}(x-6)$.

19. $s = f(t) = 1 - 3t^2$ and $f(t+h) = 1 - 3(t+h)^2 = 1 - 3t^2 - 6th - 3h^2 \Rightarrow \dfrac{ds}{dt} = \lim\limits_{h\to 0} \dfrac{f(t+h) - f(t)}{h}$

$$= \lim_{h\to 0} \frac{(1 - 3t^2 - 6th - 3h^2) - (1 - 3t^2)}{h} = \lim_{h\to 0} (-6t - 3h) = -6t \Rightarrow \frac{ds}{dt}\bigg|_{t=-1} = 6$$

21. $r = f(\theta) = \dfrac{2}{\sqrt{4-\theta}}$ and $f(\theta+h) = \dfrac{2}{\sqrt{4-(\theta+h)}} \Rightarrow \dfrac{dr}{d\theta} = \lim\limits_{h\to 0} \dfrac{f(\theta+h) - f(\theta)}{h} = \lim\limits_{h\to 0} \dfrac{\dfrac{2}{\sqrt{4-\theta-h}} - \dfrac{2}{\sqrt{4-\theta}}}{h}$

$$= \lim_{h\to 0} \frac{2\sqrt{4-\theta} - 2\sqrt{4-\theta-h}}{h\sqrt{4-\theta}\,\sqrt{4-\theta-h}} = \lim_{h\to 0} \frac{2\sqrt{4-\theta} - 2\sqrt{4-\theta-h}}{h\sqrt{4-\theta}\,\sqrt{4-\theta-h}} \cdot \frac{(2\sqrt{4-\theta} + 2\sqrt{4-\theta-h})}{(2\sqrt{4-\theta} + 2\sqrt{4-\theta-h})}$$

$$= \lim_{h\to 0} \frac{4(4-\theta) - 4(4-\theta-h)}{2h\sqrt{4-\theta}\,\sqrt{4-\theta-h}(\sqrt{4-\theta}+\sqrt{4-\theta-h})} = \lim_{h\to 0} \frac{2}{\sqrt{4-\theta}\,\sqrt{4-\theta-h}(\sqrt{4-\theta}+\sqrt{4-\theta-h})}$$

$$= \frac{2}{(4-\theta)(2\sqrt{4-\theta})} = \frac{1}{(4-\theta)\sqrt{4-\theta}} \Rightarrow \frac{dr}{d\theta}\bigg|_{\theta=0} = \frac{1}{8}$$

23. $f(x) = \dfrac{1}{x+2}$ and $f(-1) = 1 \Rightarrow f'(-1) = \lim\limits_{x\to -1} \dfrac{f(x) - f(-1)}{x - (-1)} = \lim\limits_{x\to -1} \dfrac{\left(\dfrac{1}{x+2} - 1\right)}{x+1} = \lim\limits_{x\to -1} \dfrac{\left(\dfrac{1-(x+2)}{x+2}\right)}{x+1}$

$$= \lim_{x\to -1} \frac{-(x+1)}{(x+2)(x+1)} = \lim_{x\to -1} \frac{-1}{(x+2)} = -1$$

25. $g(t) = \dfrac{t}{t-1}$ and $g(3) = \dfrac{3}{2} \Rightarrow g'(3) = \lim\limits_{t\to 3} \dfrac{g(t) - g(3)}{t-3} = \lim\limits_{t\to 3} \dfrac{\left(\dfrac{t}{t-1} - \dfrac{3}{2}\right)}{t-3} = \lim\limits_{t\to 3} \dfrac{2t - 3(t-1)}{2(t-1)(t-3)}$

$$= \lim_{t\to 3} \frac{-t+3}{2(t-1)(t-3)} = \lim_{t\to 3} \frac{-1}{2(t-1)} = -\frac{1}{4}$$

27. Note that as x increases, the slope of the tangent line to the curve is first negative, then zero (when $x = 0$), then positive $\Rightarrow$ the slope is always increasing which matches (b).

29. $f_3(x)$ is an oscillating function like the cosine. Everywhere that the graph of f_3 has a horizontal tangent we expect f_3' to be zero, and (d) matches this condition.

31. (a) f' is not defined at $x = 0, 1, 4$. At these points, the left-hand and right-hand derivatives do not agree.

For example, $\lim\limits_{x\to 0^-} \dfrac{f(x) - f(0)}{x - 0} = $ slope of line joining $(-4,0)$ and $(0,2) = \dfrac{1}{2}$ but $\lim\limits_{x\to 0^+} \dfrac{f(x) - f(0)}{x - 0} = $ slope of

line joining $(0,2)$ and $(1,-2) = -4$. Since these values are not equal, $f'(0) = \lim\limits_{x\to 0} \dfrac{f(x) - f(0)}{x - 0}$ does not exist.

(b)

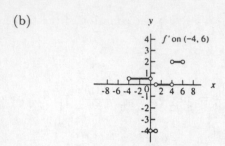

33.

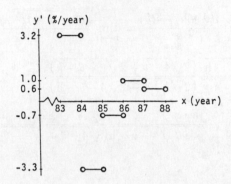

35. Left-hand derivative: For $h < 0$, $f(0 + h) = f(h) = h^2$ (using $y = x^2$ curve) $\Rightarrow \lim\limits_{h \to 0^-} \dfrac{f(0 + h) - f(0)}{h}$

$= \lim\limits_{h \to 0^-} \dfrac{h^2 - 0}{h} = \lim\limits_{h \to 0^-} h = 0$;

Right-hand derivative: For $h > 0$, $f(0 + h) = f(h) = h$ (using $y = x$ curve) $\Rightarrow \lim\limits_{h \to 0^+} \dfrac{f(0 + h) - f(0)}{h}$

$= \lim\limits_{h \to 0^+} \dfrac{h - 0}{h} = \lim\limits_{h \to 0^+} 1 = 1$;

Then $\lim\limits_{h \to 0^-} \dfrac{f(0 + h) - f(0)}{h} \neq \lim\limits_{h \to 0^+} \dfrac{f(0 + h) - f(0)}{h} \Rightarrow$ the derivative $f'(0)$ does not exist.

37. Left-hand derivative: When $h < 0$, $1 + h < 1 \Rightarrow f(1 + h) = \sqrt{1 + h} \Rightarrow \lim\limits_{h \to 0^-} \dfrac{f(1 + h) - f(1)}{h}$

$= \lim\limits_{h \to 0^-} \dfrac{\sqrt{1 + h} - 1}{h} = \lim\limits_{h \to 0^-} \dfrac{\left(\sqrt{1 + h} - 1\right) \cdot \left(\sqrt{1 + h} + 1\right)}{\left(\sqrt{1 + h} + 1\right)} = \lim\limits_{h \to 0^-} \dfrac{(1 + h) - 1}{h\left(\sqrt{1 + h} + 1\right)} = \lim\limits_{h \to 0^-} \dfrac{1}{\sqrt{1 + h} + 1} = \dfrac{1}{2}$;

Right-hand derivative: When $h > 0$, $1 + h > 1 \Rightarrow f(1 + h) = 2(1 + h) - 1 = 2h + 1 \Rightarrow \lim\limits_{h \to 0^+} \dfrac{f(1 + h) - f(1)}{h}$

$= \lim\limits_{h \to 0^+} \dfrac{(2h + 1) - 1}{h} = \lim\limits_{h \to 0^+} 2 = 2$;

Then $\lim\limits_{h \to 0^-} \dfrac{f(1 + h) - f(1)}{h} \neq \lim\limits_{h \to 0^+} \dfrac{f(1 + h) - f(1)}{h} \Rightarrow$ the derivative $f'(1)$ does not exist.

39. (a) The function is differentiable on its domain $-3 \leq x \leq 2$ (it is smooth)
    (b) none
    (c) none

41. (a) The function is differentiable on $-3 \leq x < 0$ and $0 < x \leq 3$
   (b) none
   (c) The function is neither continuous nor differentiable at $x = 0$ since $\lim\limits_{x \to 0^-} f(x) \neq \lim\limits_{x \to 0^+} f(x)$

43. (a) f is differentiable on $-1 \leq x < 0$ and $0 < x \leq 2$
   (b) f is continuous but not differentiable at $x = 0$: $\lim\limits_{x \to 0} f(x) = 0$ exists but there is a cusp at $x = 0$, so

   $$f'(0) = \lim_{h \to 0} \frac{f(0 + h) - f(0)}{h} \text{ does not exist}$$

   (c) none

45. (a) $f'(x) = \lim\limits_{h \to 0} \dfrac{f(x + h) - f(x)}{h} = \lim\limits_{h \to 0} \dfrac{-(x + h)^2 - (-x^2)}{h} = \lim\limits_{h \to 0} \dfrac{-x^2 - 2xh - h^2 + x^2}{h} = \lim\limits_{h \to 0} (-2x - h) = -2x$

   (b)

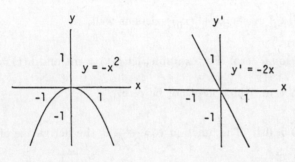

   (c) $y' = -2x$ is positive for $x < 0$, y' is zero when $x = 0$, y' is negative when $x > 0$
   (d) $y = -x^2$ is increasing for $-\infty < x < 0$ and decreasing for $0 < x < \infty$; the function is increasing on intervals where $y' > 0$ and decreasing on intervals where $y' < 0$

47. (a) Using the alternate formula for calculating derivatives: $f'(c) = \lim\limits_{x \to c} \dfrac{f(x) - f(c)}{x - c} = \lim\limits_{x \to c} \dfrac{\left(\dfrac{x^3}{3} - \dfrac{c^3}{3}\right)}{x - c}$

   $$= \lim_{x \to c} \frac{x^3 - c^3}{3(x - c)} = \lim_{x \to c} \frac{(x - c)(x^2 + xc + c^2)}{3(x - c)} = \lim_{x \to c} \frac{x^2 + xc + c^2}{3} = c^2 \Rightarrow f'(x) = x^2$$

   (b)

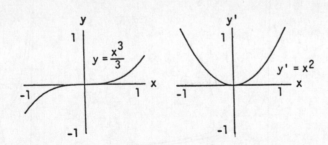

   (c) y' is positive for all $x \neq 0$, and $y' = 0$ when $x = 0$; y' is never negative
   (d) $y = \dfrac{x^3}{3}$ is increasing for all $x \neq 0$ (the graph is horizontal at $x = 0$) because y is increasing where $y' > 0$; y is never decreasing

49. $y' = \lim\limits_{x \to c} \dfrac{f(x) - f(c)}{x - c} = \lim\limits_{x \to c} \dfrac{x^3 - c^3}{x - c} = \lim\limits_{x \to c} \dfrac{(x - c)(x^2 + xc + c^2)}{x - c} = \lim\limits_{x \to c} (x^2 + xc + c^2) = 3c^2.$

   The slope of the curve $y = x^3$ at $x = c$ is $y' = 3c^2$. Notice that $3c^2 \geq 0$ for all $c \Rightarrow y = x^3$ never has a negative slope.

51. $y' = \lim\limits_{h \to 0} \dfrac{\left(2(x+h)^2 - 13(x+h) + 5\right) - \left(2x^2 - 13x + 5\right)}{h} = \lim\limits_{h \to 0} \dfrac{2x^2 + 4xh + 2h^2 - 13x - 13h + 5 - 2x^2 + 13x - 5}{h}$

$= \lim\limits_{h \to 0} \dfrac{4xh + 2h^2 - 13h}{h} = \lim\limits_{h \to 0} (4x + 2h - 13) = 4x - 13$, slope at x. The slope is -1 when $4x - 13 = -1$

$\Rightarrow 4x = 12 \Rightarrow x = 3 \Rightarrow y = 2 \cdot 3^2 - 13 \cdot 3 + 5 = -16$. Thus the tangent line is $y + 16 = (-1)(x - 3)$ and the point of tangency is $(3, -16)$.

53. No. Derivatives of functions have the intermediate value property. The function $f(x) = \lfloor x \rfloor$ satisfies $f(0) = 0$ and $f(1) = 1$ but does not take on the value $\frac{1}{2}$ anywhere in $[0, 1] \Rightarrow$ f does not have the intermediate value property. Thus f cannot be the derivative of any function on $[0, 1] \Rightarrow$ f cannot be the derivative of any function on $(-\infty, \infty)$.

55. Yes; the derivative of $-f$ is $-f'$ so that $f'(x_0)$ exists $\Rightarrow -f'(x_0)$ exists as well.

57. Yes, $\lim\limits_{t \to 0} \dfrac{g(t)}{h(t)}$ can exist but it need not equal zero. For example, let $g(t) = mt$ and $h(t) = t$. Then $g(0) = h(0)$ $= 0$, but $\lim\limits_{t \to 0} \dfrac{g(t)}{h(t)} = \lim\limits_{t \to 0} \dfrac{mt}{t} = \lim\limits_{t \to 0} m = m$, which need not be zero.

59. The graphs are shown below for $h = 1, 0.5, 0.1$. The function $y = \dfrac{1}{2\sqrt{x}}$ is the derivative of the function $y = \sqrt{x}$ so that $\dfrac{1}{2\sqrt{x}} = \lim\limits_{h \to 0} \dfrac{\sqrt{x+h} - \sqrt{x}}{h}$. The graphs reveal that $y = \dfrac{\sqrt{x+h} - \sqrt{x}}{h}$ gets closer to $y = \dfrac{1}{2\sqrt{x}}$ as h gets smaller and smaller.

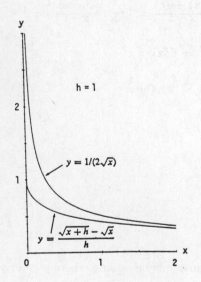
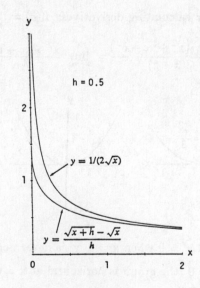
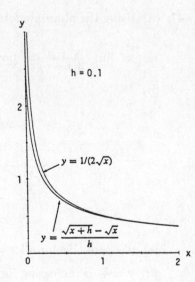

61. Weierstrass's nowhere differentiable
    continuous function.

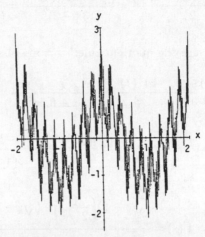

$$g(x) = \cos(\pi x) + \left(\frac{2}{3}\right)^1 \cos(9\pi x) + \left(\frac{2}{3}\right)^2 \cos(9^2\pi x) + \left(\frac{2}{3}\right)^3 \cos(9^3\pi x)$$

$$+ \cdots + \left(\frac{2}{3}\right)^7 \cos(9^7\pi x)$$

2.2 DIFFERENTIATION RULES

1. $y = -x^2 + 3 \Rightarrow \dfrac{dy}{dx} = \dfrac{d}{dx}(-x^2) + \dfrac{d}{dx}(3) = -2x + 0 = -2x \Rightarrow \dfrac{d^2y}{dx^2} = -2$

3. $s = 5t^3 - 3t^5 \Rightarrow \dfrac{ds}{dt} = \dfrac{d}{dt}(5t^3) - \dfrac{d}{dt}(3t^5) = 15t^2 - 15t^4 \Rightarrow \dfrac{d^2s}{dt^2} = \dfrac{d}{dt}(15t^2) - \dfrac{d}{dt}(15t^4) = 30t - 60t^3$

5. $y = \dfrac{4}{3}x^3 - x \Rightarrow \dfrac{dy}{dx} = 4x^2 - 1 \Rightarrow \dfrac{d^2y}{dx^2} = 8x$

7. $w = 3z^{-2} - z^{-1} \Rightarrow \dfrac{dw}{dz} = -6z^{-3} + z^{-2} = -6z^{-3} + \dfrac{1}{z^2} \Rightarrow \dfrac{d^2w}{dz^2} = 18z^{-4} - 2z^{-3} = 18z^{-4} - \dfrac{2}{z^3}$

9. $y = 6x^2 - 10x - 5x^{-2} \Rightarrow \dfrac{dy}{dx} = 12x - 10 + 10x^{-3} \Rightarrow \dfrac{d^2y}{dx^2} = 12 - 0 - 30x^{-4} = 12 - 30x^{-4}$

11. $r = \dfrac{1}{3}s^{-2} - \dfrac{5}{2}s^{-1} \Rightarrow \dfrac{dr}{ds} = -\dfrac{2}{3}s^{-3} + \dfrac{5}{2}s^{-2} = \dfrac{-2}{3s^3} + \dfrac{5}{2s^2} \Rightarrow \dfrac{d^2r}{ds^2} = 2s^{-4} - 5s^{-3} = \dfrac{2}{s^4} - \dfrac{5}{s^3}$

13. (a) $y = (3 - x^2)(x^3 - x + 1) \Rightarrow y' = (3 - x^2) \cdot \dfrac{d}{dx}(x^3 - x + 1) + (x^3 - x + 1) \cdot \dfrac{d}{dx}(3 - x^2)$

    $= (3 - x^2)(3x^2 - 1) + (x^3 - x + 1)(-2x) = -5x^4 + 12x^2 - 2x - 3$

    (b) $y = -x^5 + 4x^3 - x^2 - 3x + 3 \Rightarrow y' = -5x^4 + 12x^2 - 2x - 3$

15. (a) $y = (x^2 + 1)\left(x + 5 + \dfrac{1}{x}\right) \Rightarrow y' = (x^2 + 1) \cdot \dfrac{d}{dx}\left(x + 5 + \dfrac{1}{x}\right) + \left(x + 5 + \dfrac{1}{x}\right) \cdot \dfrac{d}{dx}(x^2 + 1)$

    $= (x^2 + 1)(1 - x^{-2}) + (x + 5 + x^{-1})(2x) = (x^2 - 1 + 1 - x^{-2}) + (2x^2 + 10x + 2) = 3x^2 + 10x + 2 - \dfrac{1}{x^2}$

    (b) $y = x^3 + 5x^2 + 2x + 5 + \dfrac{1}{x} \Rightarrow y' = 3x^2 + 10x + 2 - \dfrac{1}{x^2}$

17. $y = \dfrac{2x+5}{3x-2}$; use the quotient rule: $u = 2x + 5$ and $v = 3x - 2 \Rightarrow u' = 2$ and $v' = 3 \Rightarrow y' = \dfrac{vu' - uv'}{v^2}$

$$= \frac{(3x-2)(2) - (2x+5)(3)}{(3x-2)^2} = \frac{6x - 4 - 6x - 15}{(3x-2)^2} = \frac{-19}{(3x-2)^2}$$

19. $g(x) = \dfrac{x^2 - 4}{x + 0.5}$; use the quotient rule: $u = x^2 - 4$ and $v = x + 0.5 \Rightarrow u' = 2x$ and $v' = 1 \Rightarrow g'(x) = \dfrac{vu' + uv'}{v^2}$

$$= \frac{(x+0.5)(2x) - (x^2 - 4)(1)}{(x+0.5)^2} = \frac{2x^2 + x - x^2 + 4}{(x+0.5)^2} = \frac{x^2 + x + 4}{(x+0.5)^2}$$

21. $v = (1-t)(1+t^2)^{-1} = \dfrac{1-t}{1+t^2} \Rightarrow \dfrac{dv}{dt} = \dfrac{(1+t^2)(-1) - (1-t)(2t)}{(1+t^2)^2} = \dfrac{-1 - t^2 - 2t + 2t^2}{(1+t^2)^2} = \dfrac{t^2 - 2t - 1}{(1+t^2)^2}$

23. $f(s) = \dfrac{\sqrt{s} - 1}{\sqrt{s} + 1} \Rightarrow f'(s) = \dfrac{(\sqrt{s}+1)\left(\frac{1}{2\sqrt{s}}\right) - (\sqrt{s}-1)\left(\frac{1}{2\sqrt{s}}\right)}{(\sqrt{s}+1)^2} = \dfrac{(\sqrt{s}+1) - (\sqrt{s}-1)}{2\sqrt{s}(\sqrt{s}+1)^2} = \dfrac{1}{\sqrt{s}(\sqrt{s}+1)^2}$

NOTE: $\dfrac{d}{ds}(\sqrt{s}) = \dfrac{1}{2\sqrt{s}}$ from Example 2 in Section 2.1

25. $v = \dfrac{1 + x - 4\sqrt{x}}{x} \Rightarrow v' = \dfrac{x\left(1 - \frac{2}{\sqrt{x}}\right) - (1 + x - 4\sqrt{x})}{x^2} = \dfrac{2\sqrt{x} - 1}{x^2} = -\dfrac{1}{x^2} + 2x^{-3/2}$

27. $y = \dfrac{1}{(x^2 - 1)(x^2 + x + 1)}$; use the quotient rule: $u = 1$ and $v = (x^2 - 1)(x^2 + x + 1) \Rightarrow u' = 0$ and

$v' = (x^2 - 1)(2x + 1) + (x^2 + x + 1)(2x) = 2x^3 + x^2 - 2x - 1 + 2x^3 + 2x^2 + 2x = 4x^3 + 3x^2 - 1$

$\Rightarrow \dfrac{dy}{dx} = \dfrac{vu' - uv'}{v^2} = \dfrac{0 - 1(4x^3 + 3x^2 - 1)}{(x^2 - 1)^2(x^2 + x + 1)^2} = \dfrac{-4x^3 - 3x^2 + 1}{(x^2 - 1)^2(x^2 + x + 1)^2}$

29. $y = \frac{1}{2}x^4 - \frac{3}{2}x^2 - x \Rightarrow y' = 2x^3 - 3x - 1 \Rightarrow y'' = 6x^2 - 3 \Rightarrow y''' = 12x \Rightarrow y^{(4)} = 12 \Rightarrow y^{(n)} = 0$ for all $n \geq 5$

31. $y = \dfrac{x^3 + 7}{x} = x^2 + 7x^{-1} \Rightarrow \dfrac{dy}{dx} = 2x - 7x^{-2} \Rightarrow \dfrac{d^2y}{dx^2} = 2 + 14x^{-3}$

33. $r = \dfrac{(\theta - 1)(\theta^2 + \theta + 1)}{\theta^3} = \dfrac{\theta^3 - 1}{\theta^3} = 1 - \dfrac{1}{\theta^3} = 1 - \theta^{-3} \Rightarrow \dfrac{dr}{d\theta} = 0 + 3\theta^{-4} = 3\theta^{-4} \Rightarrow \dfrac{d^2r}{d\theta^2} = -12\theta^{-5}$

35. $w = \left(\dfrac{1+3z}{3z}\right)(3-z) = \left(\frac{1}{3}z^{-1} + 1\right)(3 - z) = z^{-1} - \frac{1}{3} + 3 - z = z^{-1} + \frac{8}{3} - z \Rightarrow \dfrac{dw}{dz} = -z^{-2} + 0 - 1 = -z^{-2} - 1$

$\Rightarrow \dfrac{d^2w}{dz^2} = 2z^{-3} - 0 = 2z^{-3}$

37. $p = \left(\dfrac{q^2 + 3}{12q}\right)\left(\dfrac{q^4 - 1}{q^3}\right) = \dfrac{q^6 - q^2 + 3q^4 - 3}{12q^4} = \frac{1}{12}q^2 - \frac{1}{12}q^{-2} + \frac{1}{4} - \frac{1}{4}q^{-4} \Rightarrow \dfrac{dp}{dq} = \frac{1}{6}q + \frac{1}{6}q^{-3} + q^{-5}$

$\Rightarrow \dfrac{d^2p}{dq^2} = \frac{1}{6} - \frac{1}{2}q^{-4} - 5q^{-6}$

39. $u(0) = 5$, $u'(0) = -3$, $v(0) = -1$, $v'(0) = 2$

(a) $\frac{d}{dx}(uv) = uv' + vu' \Rightarrow \frac{d}{dx}(uv)\Big|_{x=0} = u(0)v'(0) + v(0)u'(0) = 5 \cdot 2 + (-1)(-3) = 13$

(b) $\frac{d}{dx}\left(\frac{u}{v}\right) = \frac{vu' - uv'}{v^2} \Rightarrow \frac{d}{dx}\left(\frac{u}{v}\right)\Big|_{x=0} = \frac{v(0)u'(0) - u(0)v'(0)}{(v(0))^2} = \frac{(-1)(-3) - (5)(2)}{(-1)^2} = -7$

(c) $\frac{d}{dx}\left(\frac{v}{u}\right) = \frac{uv' - vu'}{u^2} \Rightarrow \frac{d}{dx}\left(\frac{v}{u}\right)\Big|_{x=0} = \frac{u(0)v'(0) - v(0)u'(0)}{(u(0))^2} = \frac{(5)(2) - (-1)(-3)}{(5)^2} = \frac{7}{25}$

(d) $\frac{d}{dx}(7v - 2u) = 7v' - 2u' \Rightarrow \frac{d}{dx}(7v - 2u)\Big|_{x=0} = 7v'(0) - 2u'(0) = 7 \cdot 2 - 2(-3) = 20$

41. $y = x^3 - 4x + 1$. Note that $(2, 1)$ is on the curve: $1 = 2^3 - 4(2) + 1$

(a) Slope of the tangent at (x, y) is $y' = 3x^2 - 4 \Rightarrow$ slope of the tangent at $(2, 1)$ is $y'(2) = 3(2)^2 - 4 = 8$. Thus the slope of the line perpendicular to the tangent at $(2, 1)$ is $-\frac{1}{8} \Rightarrow$ the equation of the line perpendicular to to the tangent line at $(2, 1)$ is $y - 1 = -\frac{1}{8}(x - 2)$ or $y = -\frac{x}{8} + \frac{5}{4}$.

(b) The slope of the curve at x is $m = 3x^2 - 4$ and the smallest value for m is -4 when $x = 0$ and $y = 1$.

(c) We want the slope of the curve to be $8 \Rightarrow y' = 8 \Rightarrow 3x^2 - 4 = 8 \Rightarrow 3x^2 = 12 \Rightarrow x^2 = 4 \Rightarrow x = \pm 2$. When $x = 2$, $y = 1$ and the tangent line has equation $y - 1 = 8(x - 2)$ or $y = 8x - 15$; when $x = -2$, $y = (-2)^3 - 4(-2) + 1 = 1$, and the tangent line has equation $y - 1 = 8(x + 2)$ or $y = 8x + 17$.

43. $y = \frac{4x}{x^2 + 1} \Rightarrow \frac{dy}{dx} = \frac{(x^2 + 1)(4) - (4x)(2x)}{(x^2 + 1)^2} = \frac{4x^2 + 4 - 8x^2}{(x^2 + 1)^2} = \frac{4(-x^2 + 1)}{(x^2 + 1)^2}$. When $x = 0$, $y = 0$ and $y' = \frac{4(0 + 1)}{1}$

$= 4$, so the tangent to the curve at $(0, 0)$ is the line $y = 4x$. When $x = 1$, $y = 2 \Rightarrow y' = 0$, so the tangent to the curve at $(1, 2)$ is the line $y = 2$.

45. $y = ax^2 + bx + c$ passes through $(0, 0) \Rightarrow 0 = a(0) + b(0) + c \Rightarrow c = 0$; $y = ax^2 + bx$ passes through $(1, 2)$ $\Rightarrow 2 = a + b$; $y' = 2ax + b$ and since the curve is tangent to $y = x$ at the origin, its slope is 1 at $x = 0$ $\Rightarrow y' = 1$ when $x = 0 \Rightarrow 1 = 2a(0) + b \Rightarrow b = 1$. Then $a + b = 2 \Rightarrow a = 1$. In summary $a = b = 1$ and $c = 0$ so the curve is $y = x^2 + x$.

47. (a) $y = x^3 - x \Rightarrow y' = 3x^2 - 1$. When $x = -1$, $y = 0$ and $y' = 2 \Rightarrow$ the tangent line to the curve at $(-1, 0)$ is $y = 2(x + 1)$ or $y = 2x + 2$.

(c) $\left.\begin{array}{l} y = x^3 - x \\ y = 2x + 2 \end{array}\right\} \Rightarrow x^3 - x = 2x + 2 \Rightarrow x^3 - 3x - 2 = (x - 2)(x + 1)^2 = 0 \Rightarrow x = 2$ or $x = -1$. Therefore the other intersection point is $(2, 6)$

49. $p = \frac{nRT}{V - nb} - \frac{an^2}{V^2}$. We are holding T constant, and a, b, n, R are also constant so their derivatives are zero

$\Rightarrow \frac{dP}{dV} = \frac{(V - nb) \cdot 0 - (nRT)(1)}{(V - nb)^2} - \frac{V^2(0) - (an^2)(2V)}{(V^2)^2} = \frac{-nRT}{(V - nb)^2} - \frac{2an^2}{V^3}$

51. Let c be a constant $\Rightarrow \frac{dc}{dx} = 0 \Rightarrow \frac{d}{dx}(u \cdot c) = u \cdot \frac{dc}{dx} + c \cdot \frac{du}{dx} = u \cdot 0 + c\frac{du}{dx} = c\frac{du}{dx}$. Thus when one of the functions is a constant, the Product Rule is just the Constant Multiple Rule $\Rightarrow$ the Constant Multiple Rule is a special case of the Product Rule.

53. Let $f(x) = x^n$. Then $f'(c) = \lim\limits_{x \to c} \dfrac{f(x) - f(c)}{x - c} = \lim\limits_{x \to c} \dfrac{x^n - c^n}{x - c} = \lim\limits_{x \to c} \dfrac{(x - c)\left(x^{n-1} + x^{n-2}c + \ldots + xc^{n-2} + c^{n-1}\right)}{x - c}$

$= \lim\limits_{x \to c} \left(x^{n-1} + x^{n-2}c + \ldots + x \cdot c^{n-2} + c^{n-1}\right) = c^{n-1} + c^{n-2}c + \ldots + c \cdot c^{n-2} + c^{n-1}$

$= \underbrace{c^{n-1} + c^{n-1} + \ldots + c^{n-1} + c^{n-1}}_{n \text{ terms}} = n \cdot c^{n-1}$. Thus $f'(c) = n \cdot c^{n-1} \Rightarrow f'(x) = n \cdot x^{n-1} \Rightarrow \dfrac{d}{dx}\left(x^n\right) = nx^{n-1}$.

55. In this problem we don't know the Power Rule works with fractional powers so we can't use it. Remember $\dfrac{d}{dx}\left(\sqrt{x}\right) = \dfrac{1}{2\sqrt{x}}$ (from Example 2 in Section 2.1)

(a) $\dfrac{d}{dx}\left(x^{3/2}\right) = \dfrac{d}{dx}\left(x \cdot x^{1/2}\right) = x \cdot \dfrac{d}{dx}\left(\sqrt{x}\right) + \sqrt{x}\,\dfrac{d}{dx}(x) = x \cdot \dfrac{1}{2\sqrt{x}} + \sqrt{x} \cdot 1 = \dfrac{\sqrt{x}}{2} + \sqrt{x} = \dfrac{3\sqrt{x}}{2} = \dfrac{3}{2}x^{1/2}$

(b) $\dfrac{d}{dx}\left(x^{5/2}\right) = \dfrac{d}{dx}\left(x^2 \cdot x^{1/2}\right) = x^2\,\dfrac{d}{dx}\left(\sqrt{x}\right) + \sqrt{x}\,\dfrac{d}{dx}\left(x^2\right) = x^2 \cdot \left(\dfrac{1}{2\sqrt{x}}\right) + \sqrt{x} \cdot 2x = \dfrac{1}{2}x^{3/2} + 2x^{3/2} = \dfrac{5}{2}x^{3/2}$

(c) $\dfrac{d}{dx}\left(x^{7/2}\right) = \dfrac{d}{dx}\left(x^3 \cdot x^{1/2}\right) = x^3\,\dfrac{d}{dx}\left(\sqrt{x}\right) + \sqrt{x}\,\dfrac{d}{dx}\left(x^3\right) = x^3 \cdot \left(\dfrac{1}{2\sqrt{x}}\right) + \sqrt{x} \cdot 3x^2 = \dfrac{1}{2}x^{5/2} + 3x^{5/2} = \dfrac{7}{2}x^{5/2}$

(d) We have $\dfrac{d}{dx}\left(x^{3/2}\right) = \dfrac{3}{2}x^{1/2}$, $\dfrac{d}{dx}\left(x^{5/2}\right) = \dfrac{5}{2}x^{3/2}$, $\dfrac{d}{dx}\left(x^{7/2}\right) = \dfrac{7}{2}x^{5/2}$ so it appears that $\dfrac{d}{dx}\left(x^{n/2}\right) = \dfrac{n}{2}x^{(n/2)-1}$

whenever n is an odd positive integer ≥ 3.

2.3 RATES OF CHANGE

1. $s = 0.8t^2$, $0 \leq t \leq 10$

(a) displacement $= \Delta s = s(10) - s(0) = 80 - 0 = 80$ m, average velocity $= \dfrac{\Delta s}{\Delta t} = \dfrac{80}{10} = 8$ m/sec

(b) $v = \dfrac{ds}{dt} = 1.6t \Rightarrow$ speed at $t = 0$ is $\left|v(0)\right| = 0$ m/sec and speed at $t = 10$ is $\left|v(10)\right| = 16$ m/sec;

$a = \dfrac{d^2s}{dt^2} = 1.6 \Rightarrow a(0) = 1.6$ m/sec^2 and $a(10) = 1.6$ m/sec^2

(c) The body changes direction when $v = 0 \Rightarrow t = 0 \Rightarrow$ the body never changes direction in the interval

3. $s = -t^3 + 3t^2 - 3t$, $0 \leq t \leq 3$

(a) displacement $= \Delta s = s(3) - s(0) = -9$ m, $v_{av} = \dfrac{\Delta s}{\Delta t} = \dfrac{-9}{3} = -3$ m/sec

(b) $v = \dfrac{ds}{dt} = -3t^2 + 6t - 3 \Rightarrow \left|v(0)\right| = \left|-3\right| = 3$ m/sec and $\left|v(3)\right| = \left|-12\right| = 12$ m/sec; $a = \dfrac{d^2s}{dt^2} = -6t + 6$

$\Rightarrow a(0) = 6$ m/sec^2 and $a(3) = -12$ m/sec^2

(c) $v = 0 \Rightarrow -3t^2 + 6t - 3 = 0 \Rightarrow t^2 - 2t + 1 = 0 \Rightarrow (t - 1)^2 = 0 \Rightarrow t = 1$. For all other values of t in the

interval the velocity v is negative (the graph of $v = -3t^2 + 6t - 3$ is a parabola with vertex at $t = 1$ which

opens downward $\Rightarrow$ the body never changes direction.

5. $s = \dfrac{25}{t^2} - \dfrac{5}{t}$, $1 \leq t \leq 5$

(a) $\Delta s = s(5) - s(1) = -20$ m, $v_{av} = \dfrac{-20}{4} = -5$ m/sec

(b) $v = \frac{-50}{t^3} + \frac{5}{t^2} \Rightarrow |v(1)| = 45$ m/sec and $|v(5)| = \frac{1}{5}$ m/sec; $a = \frac{150}{t^4} - \frac{10}{t^3} \Rightarrow a(1) = 140$ m/sec^2 and

$a(5) = \frac{4}{25}$ m/sec^2

(c) $v = 0 \Rightarrow \frac{-50 + 5t}{t^3} = 0 \Rightarrow -50 + 5t = 0 \Rightarrow t = 10 \Rightarrow$ the body does not change direction in the interval

7. $s = t^3 - 6t^2 + 9t$ and let the positive direction be to the right on the s-axis.
   (a) $v = 3t^2 - 12t + 9$ so that $v = 0 \Rightarrow t^2 - 4t + 3 = (t - 3)(t - 1) = 0 \Rightarrow t = 1$ or 3; $a = 6t - 12 \Rightarrow a(1)$
   $= -6$ m/sec^2 and $a(3) = 6$ m/sec^2. Thus the body is motionless but being accelerated left when $t = 1$, and
   motionless but being accelerated right when $t = 3$.
   (b) $a = 0 \Rightarrow 6t - 12 = 0 \Rightarrow t = 2$ with speed $|v(2)| = |12 - 24 + 9| = 3$ m/sec
   (c) The body moves to the right or forward on $0 \le t < 1$, and to the left or backward on $1 < t < 2$. The
   positions are $s(0) = 0$, $s(1) = 4$ and $s(2) = 2 \Rightarrow$ total distance $= |s(1) - s(0)| + |s(2) - s(1)| = |4| + |-2|$
   $= 6$ m.

9. $s_m = 1.86t^2 \Rightarrow v_m = 3.72t$ and solving $3.72t = 27.8 \Rightarrow t \approx 7.5$ sec on Mars; $s_j = 11.44t^2 \Rightarrow v_j = 22.88t$ and

   solving $22.88t = 27.8 \Rightarrow t \approx 1.2$ sec on Jupiter.

11. (a) $v(t) = s'(t) = 24 - 9.8t$ m/sec, and $a(t) = v'(t) = s''(t) = -9.8$ m/sec^2

    (b) Solve $v(t) = 0 \Rightarrow 24 - 9.8t = 0 \Rightarrow t \approx 2.4$ sec

    (c) $s(2.4) \approx 29.4$

    (d) Solve $s(t) = 14.7 \Rightarrow -4.9t^2 + 24t - 14.7 = 0 \Rightarrow t \approx \frac{-24 \pm \sqrt{287.9}}{-9.8} \Rightarrow t \approx 0.7$ sec going up and

    ≈ 4.2 sec going down

    (e) Twice the time it took to reach its highest point or ≈ 4.9 sec

13. Solving $s_m = 832t - 2.6t^2 = 0 \Rightarrow t(832 - 2.6t) = 0 \Rightarrow t = 0$ or $320 \Rightarrow 320$ sec on the moon; solving

    $s_e = 832t - 16t^2 = 0 \Rightarrow t(832 - 16t) = 0 \Rightarrow t = 0$ or $52 \Rightarrow 52$ sec on the earth. Also, $v_m = 832 - 5.2t = 0$

    $\Rightarrow t = 160$ and $s_m(160) \approx 66,560$ ft, the height it reaches above the moon's surface; $v_e = 832 - 32t = 0$

    $\Rightarrow t = 26$ and $s_e(26) \approx 10,816$ ft, the height it reaches above the earth's surface.

15. (a) $\lim\limits_{\theta \to \frac{\pi}{2}} v = \lim\limits_{\theta \to \frac{\pi}{2}} 9.8(\sin \theta)t = 9.8t$ so we expect $v = 9.8t$ m/sec in free fall

    (b) $a = \frac{dv}{dt} = 9.8$ m/sec^2

17. (a) at 2 and 7 seconds   (b) between 3 and 6 seconds: $3 \le t \le 6$

    (c)   (d)

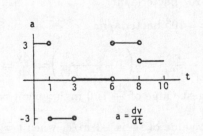

19. (a) 190 ft/sec   (b) 2 sec
    (c) at 8 sec, 0 ft/sec   (d) 10.8 sec, 90 ft/sec
    (e) From $t = 8$ until $t = 10.8$ sec, a total of 2.8 sec
    (f) Greatest acceleration happens 2 sec after launch

(g) From t = 2 to t = 10.8 sec; during this period, $a = \dfrac{v(10.8) - v(2)}{10.8 - 2} \approx -32 \text{ ft/sec}^2$

21. (a)

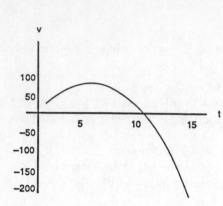

(b)

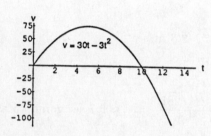

23. C = position, A = velocity, and B = acceleration. Neither A nor C can be the derivative of B because B's derivative is constant. Graph C cannot be the derivative of A either, because A has some negative slopes while C has only positive values. So, C, being the derivative of neither A nor B must be the graph of position. Curve C has both positive and negative slopes, so its derivative, the velocity, must be A and not B. That leaves B for acceleration.

25. (a) $c(100) = 11{,}000 \Rightarrow c_{av} = \dfrac{11{,}000}{100} = \$110$; $c(x) = 2000 + 100x - .1x^2 \Rightarrow c'(x) = 100 - .2x$

(b) Marginal cost $= c'(x) \Rightarrow$ the marginal cost of producing 100 machines is $c'(100) = \$80$

(c) The cost of producing the 101st machine is $c(101) - c(100) = 100 - \dfrac{201}{10} = \$79.90$

27. $b(t) = 10^6 + 10^4 t - 10^3 t^2 \Rightarrow b'(t) = 10^4 - (2)(10^3 t) = 10^3(10 - 2t)$

(a) $b'(0) = 10^4$ bacteria/hr   (b) $b'(5) = 0$ bacteria/hr

(c) $b'(10) = -10^4$ bacteria/hr

29. (a) $y = 6\left(1 - \dfrac{t}{12}\right)^2 = 6\left(1 - \dfrac{t}{6} + \dfrac{t^2}{144}\right) \Rightarrow \dfrac{dy}{dt} = \dfrac{t}{12} - 1$

(b) The largest value of $\dfrac{dy}{dt}$ is 0 m/h when t = 12 and the fluid level is falling the slowest at that time. The smallest value of $\dfrac{dy}{dt}$ is −1 m/h, when t = 0, and the fluid level is falling the fastest at that time.

(c) In this situation, $\frac{dy}{dt} \leq 0 \Rightarrow$ the graph of y is

always decreasing. As $\frac{dy}{dt}$ increases in value,

the slope of the graph of y increases from -1

to 0 over the interval $0 \leq t \leq 12$.

31. $200 \text{ km/hr} = 55\frac{5}{9} = \frac{500}{9}$ m/sec, and $D = \frac{10}{9}t^2 \Rightarrow V = \frac{20}{9}t$. Thus $V = \frac{500}{9} \Rightarrow \frac{20}{9}t = \frac{500}{9} \Rightarrow t = 25$ sec. When

$t = 25$, $D = \frac{10}{9}(25)^2 = \frac{6250}{9}$ m

33.

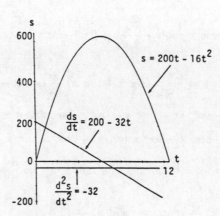

(a) $v = 0$ when $t = 6.25$ sec
(b) $v > 0$ when $0 \leq t < 6.25 \Rightarrow$ body moves up; $v < 0$ when $6.25 < t \leq 12.5 \Rightarrow$ body moves down
(c) body changes direction at $t = 6.25$ sec
(d) body speeds up on $(6.25, 12.5]$ and slows down on $[0, 6.25)$
(e) The body is moving fastest at the endpoints $t = 0$ and $t = 12.5$ when it is traveling 200 ft/sec. It's moving slowest at $t = 6.25$ when the speed is 0.
(f) When $t = 6.25$ the body is $s = 625$ m from the origin and farthest away.

35.

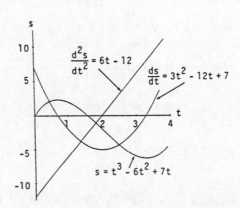

(a) $v = 0$ when $t = \dfrac{6 \pm \sqrt{15}}{3}$ sec

(b) $v < 0$ when $\dfrac{6 - \sqrt{15}}{3} < t < \dfrac{6 + \sqrt{15}}{3} \Rightarrow$ body moves left; $v > 0$ when $0 \leq t < \dfrac{6 - \sqrt{15}}{3}$ or $\dfrac{6 + \sqrt{15}}{3} < t \leq 4$

   $\Rightarrow$ body moves right

(c) body changes direction at $t = \dfrac{6 \pm \sqrt{15}}{3}$ sec

(d) body speeds up on $\left(\dfrac{6 - \sqrt{15}}{3}, 2 \right) \cup \left(\dfrac{6 + \sqrt{15}}{3}, 4 \right]$ and slows down on $\left[0, \dfrac{6 - \sqrt{15}}{3} \right) \cup \left(2, \dfrac{6 + \sqrt{15}}{3} \right)$.

(e) The body is moving fastest at $t = 0$ and $t = 4$ when it is moving 7 units/sec and slowest at $t = \dfrac{6 \pm \sqrt{15}}{3}$ sec

(f) When $t = \dfrac{6 + \sqrt{15}}{3}$ the body is at position $s \approx -6.303$ units and farthest from the origin.

2.4 DERIVATIVES OF TRIGONOMETRIC FUNCTIONS

1. $y = -10x + 3 \cos x \Rightarrow \dfrac{dy}{dx} = -10 + 3 \dfrac{d}{dx}(\cos x) = -10 - 3 \sin x$

3. $y = \csc x - 4\sqrt{x} + 7 \Rightarrow \dfrac{dy}{dx} = -\csc x \cot x - \dfrac{4}{2\sqrt{x}} + 0 = -\csc x \cot x - \dfrac{2}{\sqrt{x}}$

5. $y = (\sec x + \tan x)(\sec x - \tan x) \Rightarrow \dfrac{dy}{dx} = (\sec x + \tan x) \dfrac{d}{dx}(\sec x - \tan x) + (\sec x - \tan x) \dfrac{d}{dx}(\sec x + \tan x)$

$= (\sec x + \tan x)(\sec x \tan x - \sec^2 x) + (\sec x - \tan x)(\sec x \tan x + \sec^2 x)$

$= (\sec^2 x \tan x + \sec x \tan^2 x - \sec^3 x - \sec^2 x \tan x) + (\sec^2 x \tan x - \sec x \tan^2 x + \sec^3 x - \tan x \sec^2 x) = 0.$

$\left(\text{Note also that } y = \sec^2 x - \tan^2 x = (\tan^2 x + 1) - \tan^2 x = 1 \Rightarrow \dfrac{dy}{dx} = 0. \right)$

7. $y = \dfrac{\cot x}{1 + \cot x} \Rightarrow \dfrac{dy}{dx} = \dfrac{(1 + \cot x) \dfrac{d}{dx}(\cot x) - (\cot x) \dfrac{d}{dx}(1 + \cot x)}{(1 + \cot x)^2} = \dfrac{(1 + \cot x)(-\csc^2 x) - (\cot x)(-\csc^2 x)}{(1 + \cot x)^2}$

$= \dfrac{-\csc^2 x - \csc^2 x \cot x + \csc^2 x \cot x}{(1 + \cot x)^2} = \dfrac{-\csc^2 x}{(1 + \cot x)^2}$

9. $y = \dfrac{4}{\cos x} + \dfrac{1}{\tan x} = 4 \sec x + \cot x \Rightarrow \dfrac{dy}{dx} = 4 \sec x \tan x - \csc^2 x$

11. $y = x^2 \sin x + 2x \cos x - 2 \sin x \Rightarrow \dfrac{dy}{dx} = \left(x^2 \cos x + (\sin x)(2x) \right) + \left((2x)(-\sin x) + (\cos x)(2) \right) - 2 \cos x$

$= x^2 \cos x + 2x \sin x - 2x \sin x + 2 \cos x - 2 \cos x = x^2 \cos x$

13. $s = \tan t - t \Rightarrow \dfrac{ds}{dt} = \dfrac{d}{dt}(\tan t) - 1 = \sec^2 t - 1$

15. $s = \dfrac{1 + \csc t}{1 - \csc t} \Rightarrow \dfrac{ds}{dt} = \dfrac{(1 - \csc t)(-\csc t \cot t) - (1 + \csc t)(\csc t \cot t)}{(1 - \csc t)^2}$

$= \dfrac{-\csc t \cot t + \csc^2 t \cot t - \csc t \cot t - \csc^2 t \cot t}{(1 - \csc t)^2} = \dfrac{-2 \csc t \cot t}{(1 - \csc t)^2}$

17. $r = 4 - \theta^2 \sin \theta \Rightarrow \dfrac{dr}{d\theta} = -\left(\theta^2 \dfrac{d}{d\theta}(\sin \theta) + (\sin \theta)(2\theta) \right) = -\left(\theta^2 \cos \theta + 2\theta \sin \theta \right) = -\theta(\theta \cos \theta + 2 \sin \theta)$

19. $r = \sec \theta \csc \theta \Rightarrow \dfrac{dr}{d\theta} = (\sec \theta)(-\csc \theta \cot \theta) + (\csc \theta)(\sec \theta \tan \theta)$

$= \left(\dfrac{-1}{\cos \theta}\right)\left(\dfrac{1}{\sin \theta}\right)\left(\dfrac{\cos \theta}{\sin \theta}\right) + \left(\dfrac{1}{\sin \theta}\right)\left(\dfrac{1}{\cos \theta}\right)\left(\dfrac{\sin \theta}{\cos \theta}\right) = \dfrac{-1}{\sin^2 \theta} + \dfrac{1}{\cos^2 \theta} = \sec^2 \theta - \csc^2 \theta$

21. $p = 5 + \dfrac{1}{\cot q} = 5 + \tan q \Rightarrow \dfrac{dp}{dq} = \sec^2 q$

23. $p = \dfrac{\sin q + \cos q}{\cos q} \Rightarrow \dfrac{dp}{dq} = \dfrac{(\cos q)(\cos q - \sin q) - (\sin q + \cos q)(-\sin q)}{\cos^2 q}$

$= \dfrac{\cos^2 q - \cos q \sin q + \sin^2 q + \cos q \sin q}{\cos^2 q} = \dfrac{1}{\cos^2 q} = \sec^2 q$

25. (a) $y = \csc x \Rightarrow y' = -\csc x \cot x \Rightarrow y'' = -\Big((\csc x)(-\csc^2 x) + (\cot x)(-\csc x \cot x)\Big) = \csc^3 x + \csc x \cot^2 x$

$= (\csc x)(\csc^2 x + \cot^2 x) = (\csc x)(\csc^2 x + \csc^2 x - 1) = 2 \csc^3 x - \csc x$

(b) $y = \sec x \Rightarrow y' = \sec x \tan x \Rightarrow y'' = (\sec x)(\sec^2 x) + (\tan x)(\sec x \tan x) = \sec^3 x + \sec x \tan^2 x$

$= (\sec x)(\sec^2 x + \tan^2 x) = (\sec x)(\sec^2 x + \sec^2 x - 1) = 2 \sec^3 x - \sec x$

27. $\lim\limits_{x \to 2} \sin\left(\dfrac{1}{x} - \dfrac{1}{2}\right) = \sin\left(\dfrac{1}{2} - \dfrac{1}{2}\right) = \sin 0 = 0$

29. $\lim\limits_{x \to 0} \sec\left[\cos x + \pi \tan\left(\dfrac{\pi}{4 \sec x}\right) - 1\right] = \sec\left[\cos 0 + \pi \tan\left(\dfrac{\pi}{4 \sec 0}\right) - 1\right] = \sec\left[1 + \pi \tan\left(\dfrac{\pi}{4}\right) - 1\right] = \sec \pi = -1$

31. $\lim\limits_{t \to 0} \tan\left(1 - \dfrac{\sin t}{t}\right) = \tan\left(1 - \lim\limits_{t \to 0} \dfrac{\sin t}{t}\right) = \tan(1 - 1) = 0$

33. $\lim\limits_{\theta \to 0} \dfrac{\sin \sqrt{2\theta}}{\sqrt{2\theta}} = \lim\limits_{x \to 0} \dfrac{\sin x}{x} = 1 \qquad$ (where $x = \sqrt{2\theta}$)

35. $\lim\limits_{y \to 0} \dfrac{\sin 3y}{4y} = \dfrac{1}{4} \lim\limits_{y \to 0} \dfrac{3 \sin 3y}{3y} = \dfrac{3}{4} \lim\limits_{y \to 0} \dfrac{\sin 3y}{3y} = \dfrac{3}{4} \lim\limits_{\theta \to 0} \dfrac{\sin \theta}{\theta} = \dfrac{3}{4} \qquad$ (where $\theta = 3y$)

37. $\lim\limits_{x \to 0} \dfrac{\tan 2x}{x} = \lim\limits_{x \to 0} \dfrac{\left(\dfrac{\sin 2x}{\cos 2x}\right)}{x} = \lim\limits_{x \to 0} \dfrac{\sin 2x}{x \cos 2x} = \left(\lim\limits_{x \to 0} \dfrac{1}{\cos 2x}\right)\left(\lim\limits_{x \to 0} \dfrac{2 \sin 2x}{2x}\right) = 1 \cdot 2 = 2$

39. $\lim\limits_{x \to 0} \dfrac{x \csc 2x}{\cos 5x} = \lim\limits_{x \to 0} \left(\dfrac{x}{\sin 2x} \cdot \dfrac{1}{\cos 5x}\right) = \left(\dfrac{1}{2} \lim\limits_{x \to 0} \dfrac{2x}{\sin 2x}\right)\left(\lim\limits_{x \to 0} \dfrac{1}{\cos 5x}\right) = \left(\dfrac{1}{2} \cdot 1\right)(1) = \dfrac{1}{2}$

41. $\lim\limits_{x \to 0} \dfrac{x + x \cos x}{\sin x \cos x} = \lim\limits_{x \to 0} \left(\dfrac{x}{\sin x \cos x} + \dfrac{x \cos x}{\sin x \cos x}\right) = \lim\limits_{x \to 0} \left(\dfrac{x}{\sin x} \cdot \dfrac{1}{\cos x}\right) + \lim\limits_{x \to 0} \dfrac{x}{\sin x}$

$= \lim\limits_{x \to 0} \left(\dfrac{1}{\frac{\sin x}{x}}\right) \cdot \lim\limits_{x \to 0} \left(\dfrac{1}{\cos x}\right) + \lim\limits_{x \to 0} \left(\dfrac{1}{\frac{\sin x}{x}}\right) = (1)(1) + 1 = 2$

43. $\lim\limits_{t \to 0} \dfrac{\sin(1 - \cos t)}{1 - \cos t} = \lim\limits_{\theta \to 0} \dfrac{\sin \theta}{\theta} = 1$ since $\theta = 1 - \cos t \to 0$ as $t \to 0$

45. $\lim\limits_{\theta \to 0} \dfrac{\sin \theta}{\sin 2\theta} = \lim\limits_{\theta \to 0} \left(\dfrac{\sin \theta}{\sin 2\theta} \cdot \dfrac{2\theta}{2\theta}\right) = \dfrac{1}{2} \lim\limits_{\theta \to 0} \left(\dfrac{\sin \theta}{\theta} \cdot \dfrac{2\theta}{\sin 2\theta}\right) = \dfrac{1}{2} \cdot 1 \cdot 1 = \dfrac{1}{2}$

47. $\lim\limits_{x \to 0} \dfrac{\tan 3x}{\sin 8x} = \lim\limits_{x \to 0} \left(\dfrac{\sin 3x}{\cos 3x} \cdot \dfrac{1}{\sin 8x} \right) = \lim\limits_{x \to 0} \left(\dfrac{\sin 3x}{\cos 3x} \cdot \dfrac{1}{\sin 8x} \cdot \dfrac{8x}{3x} \cdot \dfrac{3}{8} \right)$

$= \dfrac{3}{8} \lim\limits_{x \to 0} \left(\dfrac{1}{\cos 3x} \right) \left(\dfrac{\sin 3x}{3x} \right) \left(\dfrac{8x}{\sin 8x} \right) = \dfrac{3}{8} \cdot 1 \cdot 1 \cdot 1 = \dfrac{3}{8}$

49. $y = \sin x \Rightarrow y' = \cos x \Rightarrow$ slope of tangent at

    $x = -\pi$ is $y'(-\pi) = \cos(-\pi) = -1$; slope of

    tangent at $x = 0$ is $y'(0) = \cos(0) = 1$; and

    slope of tangent at $x = \dfrac{3\pi}{2}$ is $y'\left(\dfrac{3\pi}{2}\right) = \cos\dfrac{3\pi}{2}$

    $= 0$. The tangent at $(-\pi, 0)$ is $y - 0 = -1(x + \pi)$,

    or $y = -x - \pi$; the tangent at $(0,0)$ is

    $y - 0 = 1(x - 0)$, or $y = x$; and the tangent at

    $\left(\dfrac{3\pi}{2}, -1\right)$ is $y = -1$.

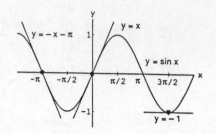

51. $y = \sec x \Rightarrow y' = \sec x \tan x \Rightarrow$ slope of tangent at $x = -\dfrac{\pi}{3}$

    is $\sec\left(-\dfrac{\pi}{3}\right) \tan\left(-\dfrac{\pi}{3}\right) = -2\sqrt{3}$; slope of tangent at $x = \dfrac{\pi}{4}$

    is $\sec\left(\dfrac{\pi}{4}\right) \tan\left(\dfrac{\pi}{4}\right) = \sqrt{2}$. The tangent at the point

    $\left(-\dfrac{\pi}{3}, \sec\left(-\dfrac{\pi}{3}\right)\right) = \left(-\dfrac{\pi}{3}, 2\right)$ is $y - 2 = -2\sqrt{3}\left(x + \dfrac{\pi}{3}\right)$; the

    tangent at the point $\left(\dfrac{\pi}{4}, \sec\left(\dfrac{\pi}{4}\right)\right) = \left(\dfrac{\pi}{4}, \sqrt{2}\right)$ is $y - \sqrt{2}$

    $= \sqrt{2}\left(x - \dfrac{\pi}{4}\right)$.

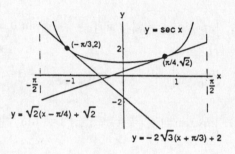

53. Yes, $y = x + \sin x \Rightarrow y' = 1 + \cos x$; horizontal tangent occurs where $1 + \cos x = 0 \Rightarrow \cos x = -1$
    $\Rightarrow x = \pi$

55. No, $y = x - \cot x \Rightarrow y' = 1 + \csc^2 x$; horizontal tangent occurs where $1 + \csc^2 x = 0 \Rightarrow \csc^2 x = -1$. But there
    are no x-values for which $\csc^2 x = -1$.

57. We want all points on the curve where the tangent

    line has slope 2. Thus, $y = \tan x \Rightarrow y' = \sec^2 x$ so

    that $y' = 2 \Rightarrow \sec^2 x = 2 \Rightarrow \sec x = \pm\sqrt{2}$

    $\Rightarrow x = \pm\dfrac{\pi}{4}$. Then the tangent line at $\left(\dfrac{\pi}{4}, 1\right)$ has

    equation $y - 1 = 2\left(x - \dfrac{\pi}{4}\right)$; the tangent line at

    $\left(-\dfrac{\pi}{4}, -1\right)$ has equation $y + 1 = 2\left(x + \dfrac{\pi}{4}\right)$.

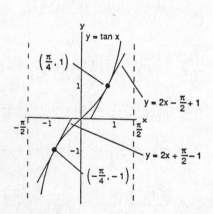

59. $y = 4 + \cot x - 2 \csc x \Rightarrow y' = -\csc^2 x + 2 \csc x \cot x = -\left(\dfrac{1}{\sin x}\right)\left(\dfrac{1 - 2\cos x}{\sin x}\right)$

    (a) When $x = \dfrac{\pi}{2}$, then $y' = -1$; the tangent line is $y = -x + \dfrac{\pi}{2} + 2$.

(b) To find the location of the horizontal tangent set $y' = 0 \Rightarrow 1 - 2\cos x = 0 \Rightarrow x = \frac{\pi}{3}$ radians. When $x = \frac{\pi}{3}$, then $y = 4 - \sqrt{3}$ is the horizontal tangent.

61. $s = 2 - 2\sin t \Rightarrow v = \frac{ds}{dt} = -2\cos t \Rightarrow a = \frac{dv}{dt} = 2\sin t \Rightarrow j = \frac{da}{dt} = 2\cos t.$ Therefore, velocity $= v\left(\frac{\pi}{4}\right)$
$= -\sqrt{2}$ m/sec; speed $= \left|v\left(\frac{\pi}{4}\right)\right| = \sqrt{2}$ m/sec; acceleration $= a\left(\frac{\pi}{4}\right) = \sqrt{2}$ m/sec^2; jerk $= j\left(\frac{\pi}{4}\right) = \sqrt{2}$ m/sec^3.

63. $\lim\limits_{x \to 0} f(x) = \lim\limits_{x \to 0} \frac{\sin^2 3x}{x^2} = \lim\limits_{x \to 0} 9\left(\frac{\sin 3x}{3x}\right)\left(\frac{\sin 3x}{3x}\right) = 9$ so that f is continuous at $x = 0 \Rightarrow \lim\limits_{x \to 0} f(x) = f(0)$
$\Rightarrow 9 = c.$

65. $\frac{d^{999}}{dx^{999}}(\cos x) = \sin x$ because $\frac{d^4}{dx^4}(\cos x) = \cos x \Rightarrow$ the derivative of $\cos x$ any number of times that is a multiple of 4 is $\cos x$. Thus, dividing 999 by 4 gives $999 = 249 \cdot 4 + 3 \Rightarrow \frac{d^{999}}{dx^{999}}(\cos x)$
$= \frac{d^3}{dx^3}\left[\frac{d^{249 \cdot 4}}{dx^{249 \cdot 4}}(\cos x)\right] = \frac{d^3}{dx^3}(\cos x) = \sin x.$

67. (a) $y = \sec x = \frac{1}{\cos x} \Rightarrow \frac{dy}{dx} = \frac{(\cos x)(0) - (1)(-\sin x)}{(\cos x)^2} = \frac{\sin x}{\cos^2 x} = \left(\frac{1}{\cos x}\right)\left(\frac{\sin x}{\cos x}\right) = \sec x \tan x$

$\Rightarrow \frac{d}{dx}(\sec x) = \sec x \tan x$

(b) $y = \csc x = \frac{1}{\sin x} \Rightarrow \frac{dy}{dx} = \frac{(\sin x)(0) - (1)(\cos x)}{(\sin x)^2} = \frac{-\cos x}{\sin^2 x} = \left(\frac{-1}{\sin x}\right)\left(\frac{\cos x}{\sin x}\right) = -\csc x \cot x$

$\Rightarrow \frac{d}{dx}(\csc x) = -\csc x \cot x$

69.

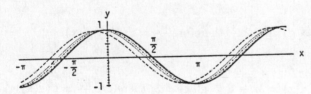 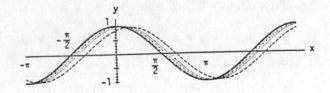

As h takes on the values of 1, 0.5, 0.3 and 0.1 the corresponding dashed curves of $y = \dfrac{\sin(x + h) - \sin x}{h}$ get

closer and closer to the black curve $y = \cos x$ because $\frac{d}{dx}(\sin x) = \lim\limits_{h \to 0} \dfrac{\sin(x + h) - \sin x}{h} = \cos x.$ The same

is true as h takes on the values of $-1, -0.5, -0.3$ and $-0.1.$

71. This is a grapher exercise. Compare your graphs with Exercises 69 and 70.

73. $y = \tan x \Rightarrow y' = \sec^2 x$, so the smallest value
$y' = \sec^2 x$ takes on is $y' = 1$ when $x = 0$;
y' has no maximum value since $\sec^2 x$ has no
largest value on $\left(-\frac{\pi}{2}, \frac{\pi}{2}\right)$; y' is never negative
since $\sec^2 x \geq 1$.

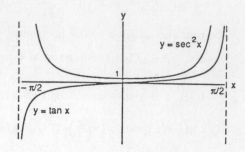

75. $y = \frac{\sin x}{x}$ appears to cross the y-axis at $y = 1$, since

$\lim\limits_{x \to 0} \frac{\sin x}{x} = 1$; $y = \frac{\sin 2x}{x}$ appears to cross the y-axis

at $y = 2$, since $\lim\limits_{x \to 0} \frac{\sin 2x}{x} = 2$; $y = \frac{\sin 4x}{x}$ appears to

cross the y-axis at $y = 4$, since $\lim\limits_{x \to 0} \frac{\sin 4x}{x} = 4$.

However, none of these graphs actually cross the y-axis

since $x = 0$ is not in the domain of the functions. Also,

$\lim\limits_{x \to 0} \frac{\sin 5x}{x} = 5$, $\lim\limits_{x \to 0} \frac{\sin(-3x)}{x} = -3$, and $\lim\limits_{x \to 0} \frac{\sin kx}{x}$

$= k \Rightarrow$ the graphs of $y = \frac{\sin 5x}{x}$, $y = \frac{\sin(-3x)}{x}$, and

$y = \frac{\sin kx}{x}$ approach 5, -3, and k, respectively, as

$x \to 0$. However, the graphs do not actually cross the y-axis.

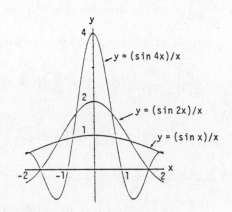

2.5 THE CHAIN RULE

1. $f(u) = 6u - 9 \Rightarrow f'(u) = 6 \Rightarrow f'(g(x)) = 6$; $g(x) = \frac{1}{2}x^4 \Rightarrow g'(x) = 2x^3$; therefore $\frac{dy}{dx} = f'(g(x))g'(x)$
$= 6 \cdot 2x^3 = 12x^3$

3. $f(u) = \sin u \Rightarrow f'(u) = \cos u \Rightarrow f'(g(x)) = \cos(3x + 1)$; $g(x) = 3x + 1 \Rightarrow g'(x) = 3$; therefore $\frac{dy}{dx} = f'(g(x))g'(x)$
$= (\cos(3x + 1))(3) = 3 \cos(3x + 1)$

5. $f(u) = \cos u \Rightarrow f'(u) = -\sin u \Rightarrow f'(g(x)) = -\sin(\sin x)$; $g(x) = \sin x \Rightarrow g'(x) = \cos x$; therefore
$\frac{dy}{dx} = f'(g(x))g'(x) = -(\sin(\sin x)) \cos x$

7. $f(u) = \tan u \Rightarrow f'(u) = \sec^2 u \Rightarrow f'(g(x)) = \sec^2(10x - 5)$; $g(x) = 10x - 5 \Rightarrow g'(x) = 10$; therefore
$\frac{dy}{dx} = f'(g(x))g'(x) = (\sec^2(10x - 5))(10) = 10 \sec^2(10x - 5)$

9. With $u = (2x + 1)$, $y = u^5$: $\frac{dy}{dx} = \frac{dy}{du}\frac{du}{dx} = 5u^4 \cdot 2 = 10(2x + 1)^4$

11. With $u = \left(1 - \frac{x}{7}\right)$, $y = u^{-7}$: $\frac{dy}{dx} = \frac{dy}{du}\frac{du}{dx} = -7u^{-8} \cdot \left(-\frac{1}{7}\right) = \left(1 - \frac{x}{7}\right)^{-8}$

13. With $u = \left(\frac{x^2}{8} + x - \frac{1}{x}\right)$, $y = u^4$: $\frac{dy}{dx} = \frac{dy}{du}\frac{du}{dx} = 4u^3 \cdot \left(\frac{x}{4} + 1 + \frac{1}{x^2}\right) = 4\left(\frac{x^2}{8} + x - \frac{1}{x}\right)^3\left(\frac{x}{4} + 1 + \frac{1}{x^2}\right)$

15. With $u = \tan x$, $y = \sec u$: $\frac{dy}{dx} = \frac{dy}{du}\frac{du}{dx} = (\sec u \tan u)(\sec^2 x) = (\sec(\tan x)\tan(\tan x))\sec^2 x$

17. With $u = \sin x$, $y = u^3$: $\frac{dy}{dx} = \frac{dy}{du}\frac{du}{dx} = 3u^2 \cos x = 3(\sin^2 x)(\cos x)$

19. $p = \sqrt{3-t} = (3-t)^{1/2} \Rightarrow \frac{dp}{dt} = \frac{1}{2}(3-t)^{-1/2}\cdot\frac{d}{dt}(3-t) = -\frac{1}{2}(3-t)^{-1/2} = \frac{-1}{2\sqrt{3-t}}$

21. $s = \frac{4}{3\pi}\sin 3t + \frac{4}{5\pi}\cos 5t \Rightarrow \frac{ds}{dt} = \frac{4}{3\pi}\cos 3t\cdot\frac{d}{dt}(3t) + \frac{4}{5\pi}(-\sin 5t)\cdot\frac{d}{dt}(5t) = \frac{4}{\pi}\cos 3t - \frac{4}{\pi}\sin 5t$

$= \frac{4}{\pi}(\cos 3t - \sin 5t)$

23. $r = (\csc\theta + \cot\theta)^{-1} \Rightarrow \frac{dr}{d\theta} = -(\csc\theta + \cot\theta)^{-2}\frac{d}{d\theta}(\csc\theta + \cot\theta) = \frac{\csc\theta\cot\theta + \csc^2\theta}{(\csc\theta + \cot\theta)^2} = \frac{\csc\theta(\cot\theta + \csc\theta)}{(\csc\theta + \cot\theta)^2}$

$= \frac{\csc\theta}{\csc\theta + \cot\theta}$

25. $y = x^2\sin^4 x + x\cos^{-2}x \Rightarrow \frac{dy}{dx} = x^2\frac{d}{dx}(\sin^4 x) + \sin^4 x\cdot\frac{d}{dx}(x^2) + x\frac{d}{dx}(\cos^{-2}x) + \cos^{-2}x\cdot\frac{d}{dx}(x)$

$= x^2\left(4\sin^3 x\frac{d}{dx}(\sin x)\right) + 2x\sin^4 x + x\left(-2\cos^{-3}x\cdot\frac{d}{dx}(\cos x)\right) + \cos^{-2}x$

$= x^2(4\sin^3 x\cos x) + 2x\sin^4 x + x\left((-2\cos^{-3}x)(-\sin x)\right) + \cos^{-2}x$

$= 4x^2\sin^3 x\cos x + 2x\sin^4 x + 2x\sin x\cos^{-3}x + \cos^{-2}x$

27. $y = \frac{1}{21}(3x-2)^7 + \left(4 - \frac{1}{2x^2}\right)^{-1} \Rightarrow \frac{dy}{dx} = \frac{7}{21}(3x-2)^6\cdot\frac{d}{dx}(3x-2) + (-1)\left(4 - \frac{1}{2x^2}\right)^{-2}\cdot\frac{d}{dx}\left(4 - \frac{1}{2x^2}\right)$

$= \frac{7}{21}(3x-2)^6\cdot 3 + (-1)\left(4 - \frac{1}{2x^2}\right)^{-2}\left(\frac{1}{x^3}\right) = (3x-2)^6 - \frac{1}{x^3\left(4 - \frac{1}{2x^2}\right)^2}$

29. $y = (4x+3)^4(x+1)^{-3} \Rightarrow \frac{dy}{dx} = (4x+3)^4(-3)(x+1)^{-4}\cdot\frac{d}{dx}(x+1) + (x+1)^{-3}(4)(4x+3)^3\cdot\frac{d}{dx}(4x+3)$

$= (4x+3)^4(-3)(x+1)^{-4}(1) + (x+1)^{-3}(4)(4x+3)^3(4) = -3(4x+3)^4(x+1)^{-4} + 16(4x+3)^3(x+1)^{-3}$

$= \frac{(4x+3)^3}{(x+1)^4}[-3(4x+3) + 16(x+1)] = \frac{(4x+3)^3(4x+7)}{(x+1)^4}$

31. $h(x) = x\tan(2\sqrt{x}) + 7 \Rightarrow h'(x) = x\frac{d}{dx}(\tan(2x^{1/2})) + \tan(2x^{1/2})\cdot\frac{d}{dx}(x) + 0$

$= x\sec^2(2x^{1/2})\cdot\frac{d}{dx}(2x^{1/2}) + \tan(2x^{1/2}) = x\sec^2(2\sqrt{x})\cdot\frac{1}{\sqrt{x}} + \tan(2\sqrt{x}) = \sqrt{x}\sec^2(2\sqrt{x}) + \tan(2\sqrt{x})$

33. $f(\theta) = \left(\frac{\sin\theta}{1+\cos\theta}\right)^2 \Rightarrow f'(\theta) = 2\left(\frac{\sin\theta}{1+\cos\theta}\right)\cdot\frac{d}{d\theta}\left(\frac{\sin\theta}{1+\cos\theta}\right) = \frac{2\sin\theta}{1+\cos\theta}\cdot\frac{(1+\cos\theta)(\cos\theta) - (\sin\theta)(-\sin\theta)}{(1+\cos\theta)^2}$

$= \frac{(2\sin\theta)(\cos\theta + \cos^2\theta + \sin^2\theta)}{(1+\cos\theta)^3} = \frac{(2\sin\theta)(\cos\theta + 1)}{(1+\cos\theta)^3} = \frac{2\sin\theta}{(1+\cos\theta)^2}$

35. $r = \sin(\theta^2)\cos(2\theta) \Rightarrow \frac{dr}{d\theta} = \sin(\theta^2)(-\sin 2\theta)\frac{d}{d\theta}(2\theta) + \cos(2\theta)(\cos(\theta^2))\cdot\frac{d}{d\theta}(\theta^2)$

$= \sin(\theta^2)(-\sin 2\theta)(2) + (\cos 2\theta)(\cos(\theta^2))(2\theta) = -2\sin(\theta^2)\sin(2\theta) + 2\theta\cos(2\theta)\cos(\theta^2)$

37. $q = \sin\left(\frac{t}{\sqrt{t+1}}\right) \Rightarrow \frac{dq}{dt} = \cos\left(\frac{t}{\sqrt{t+1}}\right)\cdot\frac{d}{dt}\left(\frac{t}{\sqrt{t+1}}\right) = \cos\left(\frac{t}{\sqrt{t+1}}\right)\cdot\frac{\sqrt{t+1}\,(1) - t\cdot\frac{d}{dt}(\sqrt{t+1})}{(\sqrt{t+1})^2}$

$= \cos\left(\frac{t}{\sqrt{t+1}}\right)\cdot\frac{\sqrt{t+1} - \frac{t}{2\sqrt{t+1}}}{t+1} = \cos\left(\frac{t}{\sqrt{t+1}}\right)\left(\frac{2(t+1)-t}{2(t+1)^{3/2}}\right) = \left(\frac{t+2}{2(t+1)^{3/2}}\right)\cos\left(\frac{t}{\sqrt{t+1}}\right)$

39. $y = \sin^2(\pi t - 2) \Rightarrow \frac{dy}{dt} = 2\sin(\pi t - 2)\cdot\frac{d}{dt}\sin(\pi t - 2) = 2\sin(\pi t - 2)\cdot\cos(\pi t - 2)\cdot\frac{d}{dt}(\pi t - 2)$

$= 2\pi\sin(\pi t - 2)\cos(\pi t - 2)$

41. $y = (1 + \cos 2t)^{-4} \Rightarrow \frac{dy}{dt} = -4(1 + \cos 2t)^{-5}\cdot\frac{d}{dt}(1 + \cos 2t) = -4(1 + \cos 2t)^{-5}(-\sin 2t)\cdot\frac{d}{dt}(2t) = \frac{8\sin 2t}{(1+\cos 2t)^5}$

43. $y = \sin(\cos(2t - 5)) \Rightarrow \frac{dy}{dt} = \cos(\cos(2t - 5))\cdot\frac{d}{dt}\cos(2t - 5) = \cos(\cos(2t - 5))\cdot(-\sin(2t - 5))\cdot\frac{d}{dt}(2t - 5)$

$= -2\cos(\cos(2t - 5))(\sin(2t - 5))$

45. $y = \left[1 + \tan^4\left(\frac{t}{12}\right)\right]^3 \Rightarrow \frac{dy}{dt} = 3\left[1 + \tan^4\left(\frac{t}{12}\right)\right]^2\cdot\frac{d}{dt}\left[1 + \tan^4\left(\frac{t}{12}\right)\right] = 3\left[1 + \tan^4\left(\frac{t}{12}\right)\right]^2\left[4\tan^3\left(\frac{t}{12}\right)\cdot\frac{d}{dt}\tan\left(\frac{t}{12}\right)\right]$

$= 12\left[1 + \tan^4\left(\frac{t}{12}\right)\right]^2\left[\tan^3\left(\frac{t}{12}\right)\sec^2\left(\frac{t}{12}\right)\cdot\frac{1}{12}\right] = \left[1 + \tan^4\left(\frac{t}{12}\right)\right]^2\left[\tan^3\left(\frac{t}{12}\right)\sec^2\left(\frac{t}{12}\right)\right]$

47. $y = (1 + \cos(t^2))^{1/2} \Rightarrow \frac{dy}{dt} = \frac{1}{2}(1 + \cos(t^2))^{-1/2}\cdot\frac{d}{dt}(1 + \cos(t^2)) = \frac{1}{2}(1 + \cos(t^2))^{-1/2}\left(-\sin(t^2)\cdot\frac{d}{dt}(t^2)\right)$

$= -\frac{1}{2}(1 + \cos(t^2))^{-1/2}(\sin(t^2))\cdot 2t = -\frac{t\sin(t^2)}{\sqrt{1 + \cos(t^2)}}$

49. $y = \left(1 + \frac{1}{x}\right)^3 \Rightarrow y' = 3\left(1 + \frac{1}{x}\right)^2\left(-\frac{1}{x^2}\right) = -\frac{3}{x^2}\left(1 + \frac{1}{x}\right)^2 \Rightarrow y'' = \left(-\frac{3}{x^2}\right)\cdot\frac{d}{dx}\left(1 + \frac{1}{x}\right)^2 - \left(1 + \frac{1}{x}\right)^2\cdot\frac{d}{dx}\left(\frac{3}{x^2}\right)$

$= \left(-\frac{3}{x^2}\right)\left(2\left(1 + \frac{1}{x}\right)\left(-\frac{1}{x^2}\right)\right) + \left(\frac{6}{x^3}\right)\left(1 + \frac{1}{x}\right)^2 = \frac{6}{x^4}\left(1 + \frac{1}{x}\right) + \frac{6}{x^3}\left(1 + \frac{1}{x}\right)^2 = \frac{6}{x^3}\left(1 + \frac{1}{x}\right)\left(\frac{1}{x} + 1 + \frac{1}{x}\right)$

$= \frac{6}{x^3}\left(1 + \frac{1}{x}\right)\left(1 + \frac{2}{x}\right)$

51. $y = \frac{1}{9}\cot(3x - 1) \Rightarrow y' = -\frac{1}{9}\csc^2(3x - 1)(3) = -\frac{1}{3}\csc^2(3x - 1) \Rightarrow y'' = \left(-\frac{2}{3}\right)\left(\csc(3x - 1)\cdot\frac{d}{dx}\csc(3x - 1)\right)$

$= -\frac{2}{3}\csc(3x - 1)(-\csc(3x - 1)\cot(3x - 1)\cdot\frac{d}{dx}(3x - 1)) = 2\csc^2(3x - 1)\cot(3x - 1)$

53. $g(x) = \sqrt{x} \Rightarrow g'(x) = \frac{1}{2\sqrt{x}} \Rightarrow g(1) = 1$ and $g'(1) = \frac{1}{2}$; $f(u) = u^5 + 1 \Rightarrow f'(u) = 5u^4 \Rightarrow f'(g(1)) = f'(1) = 5$;

therefore, $(f \circ g)'(1) = f'(g(1))\cdot g'(1) = 5\cdot\frac{1}{2} = \frac{5}{2}$

55. $g(x) = 5\sqrt{x} \Rightarrow g'(x) = \dfrac{5}{2\sqrt{x}} \Rightarrow g(1) = 5$ and $g'(1) = \dfrac{5}{2}$; $f(u) = \cot\left(\dfrac{\pi u}{10}\right) \Rightarrow f'(u) = -\csc^2\left(\dfrac{\pi u}{10}\right)\left(\dfrac{\pi}{10}\right)$

$= \dfrac{-\pi}{10}\csc^2\left(\dfrac{\pi u}{10}\right) \Rightarrow f'(g(1)) = f'(5) = -\dfrac{\pi}{10}\csc^2\left(\dfrac{\pi}{2}\right) = -\dfrac{\pi}{10}$; therefore, $(f \circ g)'(1) = f'(g(1))g'(1) = -\dfrac{\pi}{10}\cdot\dfrac{5}{2}$

$= -\dfrac{\pi}{4}$

57. $g(x) = 10x^2 + x + 1 \Rightarrow g'(x) = 20x + 1 \Rightarrow g(0) = 1$ and $g'(0) = 1$; $f(u) = \dfrac{2u}{u^2+1} \Rightarrow f'(u) = \dfrac{(u^2+1)(2) - (2u)(2u)}{(u^2+1)^2}$

$= \dfrac{-2u^2+2}{(u^2+1)^2} \Rightarrow f'(g(0)) = f'(1) = 0$; therefore, $(f \circ g)'(0) = f'(g(0))g'(0) = 0 \cdot 1 = 0$

59. (a) $y = 2f(x) \Rightarrow \dfrac{dy}{dx} = 2f'(x) \Rightarrow \left.\dfrac{dy}{dx}\right|_{x=2} = 2f'(2) = 2\left(\dfrac{1}{3}\right) = \dfrac{2}{3}$

(b) $y = f(x) + g(x) \Rightarrow \dfrac{dy}{dx} = f'(x) + g'(x) \Rightarrow \left.\dfrac{dy}{dx}\right|_{x=3} = f'(3) + g'(3) = 2\pi + 5$

(c) $y = f(x) \cdot g(x) \Rightarrow \dfrac{dy}{dx} = f(x)g'(x) + g(x)f'(x) \Rightarrow \left.\dfrac{dy}{dx}\right|_{x=3} = f(3)g'(3) + g(3)f'(3) = 3 \cdot 5 + (-4)(2\pi) = 15 - 8\pi$

(d) $y = \dfrac{f(x)}{g(x)} \Rightarrow \dfrac{dy}{dx} = \dfrac{g(x)f'(x) - f(x)g'(x)}{[g(x)]^2} \Rightarrow \left.\dfrac{dy}{dx}\right|_{x=2} = \dfrac{g(2)f'(2) - f(2)g'(2)}{[g(2)]^2} = \dfrac{(2)\left(\frac{1}{3}\right) - (8)(-3)}{2^2} = \dfrac{37}{6}$

(e) $y = f(g(x)) \Rightarrow \dfrac{dy}{dx} = f'(g(x))g'(x) \Rightarrow \left.\dfrac{dy}{dx}\right|_{x=2} = f'(g(2))g'(2) = f'(2)(-3) = \dfrac{1}{3}(-3) = -1$

(f) $y = (f(x))^{1/2} \Rightarrow \dfrac{dy}{dx} = \dfrac{1}{2}(f(x))^{-1/2} \cdot f'(x) = \dfrac{f'(x)}{2\sqrt{f(x)}} \Rightarrow \left.\dfrac{dy}{dx}\right|_{x=2} = \dfrac{f'(2)}{2\sqrt{f(2)}} = \dfrac{\left(\frac{1}{3}\right)}{2\sqrt{8}} = \dfrac{1}{6\sqrt{8}} = \dfrac{1}{12\sqrt{2}} = \dfrac{\sqrt{2}}{24}$

(g) $y = (g(x))^{-2} \Rightarrow \dfrac{dy}{dx} = -2(g(x))^{-3} \cdot g'(x) \Rightarrow \left.\dfrac{dy}{dx}\right|_{x=3} = -2(g(3))^{-3}g'(3) = -2(-4)^{-3} \cdot 5 = \dfrac{5}{32}$

(h) $y = \left((f(x))^2 + (g(x))^2\right)^{1/2} \Rightarrow \dfrac{dy}{dx} = \dfrac{1}{2}\left((f(x))^2 + (g(x))^2\right)^{-1/2}\left(2f(x) \cdot f'(x) + 2g(x) \cdot g'(x)\right)$

$\Rightarrow \left.\dfrac{dy}{dx}\right|_{x=2} = \dfrac{1}{2}\left((f(2))^2 + (g(2))^2\right)^{-1/2}\left(2f(2)f'(2) + 2g(2)g'(2)\right) = \dfrac{1}{2}\left(8^2 + 2^2\right)^{-1/2}\left(2 \cdot 8 \cdot \dfrac{1}{3} + 2 \cdot 2 \cdot (-3)\right)$

$= -\dfrac{5}{3\sqrt{17}}$

61. $\dfrac{ds}{dt} = \dfrac{ds}{d\theta} \cdot \dfrac{d\theta}{dt}$; $s = \cos\theta \Rightarrow \dfrac{ds}{d\theta} = -\sin\theta \Rightarrow \left.\dfrac{ds}{d\theta}\right|_{\theta=\frac{3\pi}{2}} = -\sin\left(\dfrac{3\pi}{2}\right) - 1$ so that $\dfrac{ds}{dt} = \dfrac{ds}{d\theta} \cdot \dfrac{d\theta}{dt} = 1 \cdot 5 = 5$

63. With $y = x$, we should get $\dfrac{dy}{dx} = 1$ for both (a) and (b):

(a) $y = \dfrac{u}{5} + 7 \Rightarrow \dfrac{dy}{du} = \dfrac{1}{5}$; $u = 5x - 35 \Rightarrow \dfrac{du}{dx} = 5$; therefore, $\dfrac{dy}{dx} = \dfrac{dy}{du} \cdot \dfrac{du}{dx} = \dfrac{1}{5} \cdot 5 = 1$, as expected

(b) $y = 1 + \dfrac{1}{u} \Rightarrow \dfrac{dy}{du} = -\dfrac{1}{u^2}$; $u = (x-1)^{-1} \Rightarrow \dfrac{du}{dx} = -(x-1)^{-2}(1) = \dfrac{-1}{(x-1)^2}$; therefore $\dfrac{dy}{dx} = \dfrac{dy}{du} \cdot \dfrac{du}{dx}$

$= \dfrac{-1}{u^2} \cdot \dfrac{-1}{(x-1)^2} = \dfrac{-1}{\left((x-1)^{-1}\right)^2} \cdot \dfrac{-1}{(x-1)^2} = (x-1)^2 \cdot \dfrac{1}{(x-1)^2} = 1$, again as expected

65. $y = 2 \tan\left(\frac{\pi x}{4}\right) \Rightarrow \frac{dy}{dx} = \left(2 \sec^2 \frac{\pi x}{4}\right)\left(\frac{\pi}{4}\right) = \frac{\pi}{2} \sec^2 \frac{\pi x}{4}$

(a) $\left.\frac{dy}{dx}\right|_{x=1} = \frac{\pi}{2} \sec^2\left(\frac{\pi}{4}\right) = \pi \Rightarrow$ slope of tangent is 2; thus, $y(1) = 2 \tan\left(\frac{\pi}{4}\right) = 2$ and $y'(1) = \pi \Rightarrow$ tangent line is

given by $y - 2 = \pi(x - 1) \Rightarrow y = \pi x + 2 - \pi$

(b) $y' = \frac{\pi}{2} \sec^2\left(\frac{\pi x}{4}\right)$ and the smallest value the secant function can have in $-2 < x < 2$ is $1 \Rightarrow$ the minimum

value of y' is $\frac{\pi}{2}$ and that occurs when $\frac{\pi}{2} = \frac{\pi}{2} \sec^2\left(\frac{\pi x}{4}\right) \Rightarrow 1 = \sec^2\left(\frac{\pi x}{4}\right) \Rightarrow \pm 1 = \sec\left(\frac{\pi x}{4}\right) \Rightarrow x = 0$.

67. $s = A \cos(2\pi bt) \Rightarrow v = \frac{ds}{dt} = -A \sin(2\pi bt)(2\pi b) = -2\pi bA \sin(2\pi bt)$. If we replace b with 2b to double the

frequency, the velocity formula gives $v = -4\pi bA \sin(4\pi bt) \Rightarrow$ doubling the frequency causes the velocity to

double. Also $v = -2\pi bA \sin(2\pi bt) \Rightarrow a = \frac{dv}{dt} = -4\pi^2 b^2 A \cos(2\pi bt)$. If we replace b with 2b in the

acceleration formula, we get $a = -16\pi^2 b^2 A \cos(4\pi bt) \Rightarrow$ doubling the frequency causes the acceleration to

quadruple. Finally, $a = -4\pi^2 b^2 A \cos(2\pi bt) \Rightarrow j = \frac{da}{dt} = 8\pi^3 b^3 A \sin(2\pi bt)$. If we replace b with 2b in the jerk

formula, we get $j = 64\pi^3 b^3 A \sin(2\pi bt) \Rightarrow$ doubling the frequency multiplies the jerk by a factor of 8.

69. $s = (1 + 4t)^{1/2} \Rightarrow v = \frac{ds}{dt} = \frac{1}{2}(1 + 4t)^{-1/2}(4) = 2(1 + 4t)^{-1/2} \Rightarrow v(6) = 2(1 + 4 \cdot 6)^{-1/2} = \frac{2}{5}$ m/sec;

$v = 2(1 + 4t)^{-1/2} \Rightarrow a = \frac{dv}{dt} = -\frac{1}{2} \cdot 2(1 + 4t)^{-3/2}(4) = -4(1 + 4t)^{-3/2} \Rightarrow a(6) = -4(1 + 4 \cdot 6)^{-3/2} = -\frac{4}{125}$ m/sec^2

71. v proportional to $\frac{1}{\sqrt{s}} \Rightarrow v = \frac{k}{\sqrt{s}}$ for some constant $k \Rightarrow \frac{dv}{ds} = -\frac{k}{2s^{3/2}}$. Thus, $a = \frac{dv}{dt} = \frac{dv}{ds} \cdot \frac{ds}{dt} = \frac{dv}{ds} \cdot v$

$= -\frac{k}{2s^{3/2}} \cdot \frac{k}{\sqrt{s}} = -\frac{k^2}{2}\left(\frac{1}{s^2}\right) \Rightarrow$ acceleration is a constant times $\frac{1}{s^2}$ so a is proportional to $\frac{1}{s^2}$.

73. $T = 2\pi\sqrt{\frac{L}{g}} \Rightarrow \frac{dT}{dL} = 2\pi \cdot \frac{1}{2\sqrt{\frac{L}{g}}} \cdot \frac{1}{g} = \frac{\pi}{g\sqrt{\frac{L}{g}}} = \frac{\pi}{\sqrt{gL}}$. Therefore, $\frac{dT}{du} = \frac{dT}{dL} \cdot \frac{dL}{du} = \frac{\pi}{\sqrt{gL}} \cdot kL = \frac{\pi k\sqrt{L}}{\sqrt{g}} = \frac{1}{2} \cdot 2\pi k\sqrt{\frac{L}{g}}$

$= \frac{kT}{2}$, as required.

75. The graph of $y = (f \circ g)(x)$ has a horizontal tangent at $x = 1$ provided that $(f \circ g)'(1) = 0 \Rightarrow f'(g(1))g'(1) = 0$ $\Rightarrow$ either $f'(g(1)) = 0$ or $g'(1) = 0$ (or both) $\Rightarrow$ either the graph of f has a horizontal tangent at $u = g(1)$, or the graph of g has a horizontal tangent at $x = 1$ (or both).

77. From the power rule, with $y = x^{1/4}$, we get $\frac{dy}{dx} = \frac{1}{4}x^{-3/4}$. From the chain rule, $y = \sqrt{\sqrt{x}}$

$\Rightarrow \frac{dy}{dx} = \frac{1}{2\sqrt{\sqrt{x}}} \cdot \frac{d}{dx}(\sqrt{x}) = \frac{1}{2\sqrt{\sqrt{x}}} \cdot \frac{1}{2\sqrt{x}} = \frac{1}{4}x^{-3/4}$, in agreement.

79. As $h \to 0$, the graph of $y = \dfrac{\sin 2(x + h) - \sin 2x}{h}$

approaches the graph of $y = 2 \cos 2x$ because

$$\lim_{h \to 0} \frac{\sin 2(x + h) - \sin 2x}{h} = \frac{d}{dx}(\sin 2x) = 2 \cos 2x.$$

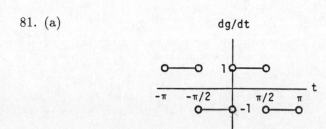

81. (a)

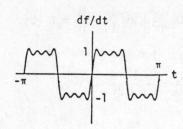

(b) $\dfrac{df}{dt} = 1.27324 \sin 2t + 0.42444 \sin 6t + 0.2546 \sin 10t + 0.18186 \sin 14t$

(c) The curve of $y = \dfrac{df}{dt}$ approximates $y = \dfrac{dg}{dt}$

the best when t is not $-\pi$, $-\dfrac{\pi}{2}$, 0, $\dfrac{\pi}{2}$, nor π.

2.6 IMPLICIT DIFFERENTIATION AND RATIONAL EXPONENTS

1. $y = x^{9/4} \Rightarrow \dfrac{dy}{dx} = \dfrac{9}{4}x^{5/4}$

3. $y = \sqrt[3]{2x} = (2x)^{1/3} \Rightarrow \dfrac{dy}{dx} = \dfrac{1}{3}(2x)^{-2/3} \cdot 2 = \dfrac{2^{1/3}}{3x^{2/3}}$

5. $y = 7\sqrt{x + 6} = 7(x + 6)^{1/2} \Rightarrow \dfrac{dy}{dx} = \dfrac{7}{2}(x + 6)^{-1/2} = \dfrac{7}{2\sqrt{x + 6}}$

7. $y = (2x + 5)^{-1/2} \Rightarrow \dfrac{dy}{dx} = -\dfrac{1}{2}(2x + 5)^{-3/2} \cdot 2 = -(2x + 5)^{-3/2}$

9. $y = x(x^2 + 1)^{1/2} \Rightarrow y' = (1)(x^2 + 1)^{1/2} + \left(\dfrac{x}{2}\right)(x^2 + 1)^{-1/2}(2x) = \dfrac{2x^2 + 1}{\sqrt{x^2 + 1}}$

11. $s = \sqrt[7]{t^2} = t^{2/7} \Rightarrow \dfrac{ds}{dt} = \dfrac{2}{7}t^{-5/7}$

13. $y = \sin\left((2t+5)^{-2/3}\right) \Rightarrow \dfrac{dy}{dt} = \cos\left((2t+5)^{-2/3}\right) \cdot \left(-\dfrac{2}{3}\right)(2t+5)^{-5/3} \cdot 2 = -\dfrac{4}{3}(2t+5)^{-5/3}\cos\left((2t+5)^{-2/3}\right)$

15. $f(x) = \sqrt{1-\sqrt{x}} = \left(1-x^{1/2}\right)^{1/2} \Rightarrow f'(x) = \dfrac{1}{2}\left(1-x^{1/2}\right)^{-1/2}\left(-\dfrac{1}{2}x^{-1/2}\right) = \dfrac{-1}{4\left(\sqrt{1-\sqrt{x}}\right)\sqrt{x}} = \dfrac{-1}{4\sqrt{x\left(1-\sqrt{x}\right)}}$

17. $h(\theta) = \sqrt[3]{1+\cos(2\theta)} = (1+\cos 2\theta)^{1/3} \Rightarrow h'(\theta) = \dfrac{1}{3}(1+\cos 2\theta)^{-2/3} \cdot (-\sin 2\theta)\cdot 2 = -\dfrac{2}{3}(\sin 2\theta)(1+\cos 2\theta)^{-2/3}$

19. $x^2y + xy^2 = 6$:

    Step 1: $\left(x^2\dfrac{dy}{dx} + y \cdot 2x\right) + \left(x \cdot 2y\dfrac{dy}{dx} + y^2 \cdot 1\right) = 0$

    Step 2: $x^2\dfrac{dy}{dx} + 2xy\dfrac{dy}{dx} = -2xy - y^2$

    Step 3: $\dfrac{dy}{dx}\left(x^2 + 2xy\right) = -2xy - y^2$

    Step 4: $\dfrac{dy}{dx} = \dfrac{-2xy - y^2}{x^2 + 2xy}$

21. $2xy + y^2 = x + y$:

    Step 1: $\left(2x\dfrac{dy}{dx} + 2y\right) + 2y\dfrac{dy}{dx} = 1 + \dfrac{dy}{dx}$

    Step 2: $2x\dfrac{dy}{dx} + 2y\dfrac{dy}{dx} - \dfrac{dy}{dx} = 1 - 2y$

    Step 3: $\dfrac{dy}{dx}(2x + 2y - 1) = 1 - 2y$

    Step 4: $\dfrac{dy}{dx} = \dfrac{1 - 2y}{2x + 2y - 1}$

23. $x^2(x-y)^2 = x^2 - y^2$:

    Step 1: $x^2\left[2(x-y)\left(1-\dfrac{dy}{dx}\right)\right] + (x-y)^2(2x) = 2x - 2y\dfrac{dy}{dx}$

    Step 2: $-2x^2(x-y)\dfrac{dy}{dx} + 2y\dfrac{dy}{dx} = 2x - 2x^2(x-y) - 2x(x-y)^2$

    Step 3: $\dfrac{dy}{dx}\left[-2x^2(x-y) + 2y\right] = 2x\left[1 - x(x-y) - (x-y)^2\right]$

    Step 4: $\dfrac{dy}{dx} = \dfrac{2x\left[1 - x(x-y) - (x-y)^2\right]}{-2x^2(x-y) + 2y} = \dfrac{x\left[1 - x(x-y) - (x-y)^2\right]}{y - x^2(x-y)} = \dfrac{x\left(1 - x^2 + xy - x^2 + 2xy - y^2\right)}{x^2y - x^3 + y}$

             $= \dfrac{x - 2x^3 + 3x^2y - xy^2}{x^2y - x^3 + y}$

25. $y^2 = \dfrac{x-1}{x+1} \Rightarrow 2y\dfrac{dy}{dx} = \dfrac{(x+1) - (x-1)}{(x+1)^2} = \dfrac{2}{(x+1)^2} \Rightarrow \dfrac{dy}{dx} = \dfrac{1}{y(x+1)^2}$

27. $x = \tan y \Rightarrow 1 = (\sec^2 y)\dfrac{dy}{dx} \Rightarrow \dfrac{dy}{dx} = \dfrac{1}{\sec^2 y} = \cos^2 y$

29. $x + \tan(xy) = 0 \Rightarrow 1 + \left[\sec^2(xy)\right]\left(y + x\,\frac{dy}{dx}\right) = 0 \Rightarrow x\sec^2(xy)\,\frac{dy}{dx} = -1 - y\sec^2(xy) \Rightarrow \frac{dy}{dx} = \frac{-1 - y\sec^2(xy)}{x\sec^2(xy)}$

$= \frac{-1}{x\sec^2(xy)} - \frac{y}{x} = \frac{-\cos^2(xy)}{x} - \frac{y}{x} = \frac{-\cos^2(xy) - y}{x}$

31. $y\sin\left(\frac{1}{y}\right) = 1 - xy \Rightarrow y\left[\cos\left(\frac{1}{y}\right)\cdot(-1)\,\frac{1}{y^2}\cdot\frac{dy}{dx}\right] + \sin\left(\frac{1}{y}\right)\cdot\frac{dy}{dx} = -x\,\frac{dy}{dx} - y \Rightarrow \frac{dy}{dx}\left[-\frac{1}{y}\cos\left(\frac{1}{y}\right) + \sin\left(\frac{1}{y}\right) + x\right] = -y$

$\Rightarrow \frac{dy}{dx} = \frac{-y}{-\frac{1}{y}\cos\left(\frac{1}{y}\right) + \sin\left(\frac{1}{y}\right) + x} = \frac{-y^2}{y\sin\left(\frac{1}{y}\right) - \cos\left(\frac{1}{y}\right) + xy}$

33. $\theta^{1/2} + r^{1/2} = 1 \Rightarrow \frac{1}{2}\theta^{-1/2} + \frac{1}{2}r^{-1/2}\cdot\frac{dr}{d\theta} = 0 \Rightarrow \frac{dr}{d\theta}\left[\frac{1}{2\sqrt{r}}\right] = \frac{-1}{2\sqrt{\theta}} \Rightarrow \frac{dr}{d\theta} = -\frac{2\sqrt{r}}{2\sqrt{\theta}} = -\frac{\sqrt{r}}{\sqrt{\theta}}$

35. $\sin(r\theta) = \frac{1}{2} \Rightarrow \left[\cos(r\theta)\right]\left(r + \theta\,\frac{dr}{d\theta}\right) = 0 \Rightarrow \frac{dr}{d\theta}\left[\theta\cos(r\theta)\right] = -r\cos(r\theta) \Rightarrow \frac{dr}{d\theta} = \frac{-r\cos(r\theta)}{\theta\cos(r\theta)} = -\frac{r}{\theta},$

$\cos(r\theta) \neq 0$

37. $x^2 + y^2 = 1 \Rightarrow 2x + 2yy' = 0 \Rightarrow 2yy' = -2x \Rightarrow \frac{dy}{dx} = y' = -\frac{x}{y};$ now to find $\frac{d^2y}{dx^2}, \frac{d}{dx}(y') = \frac{d}{dx}\left(-\frac{x}{y}\right)$

$\Rightarrow y'' = \frac{y(-1) + xy'}{y^2} = \frac{-y + x\left(-\frac{x}{y}\right)}{y^2}$ since $y' = -\frac{x}{y} \Rightarrow \frac{d^2y}{dx^2} = y'' = \frac{-y^2 - x^2}{y^3}$

39. $y^2 = x^2 + 2x \Rightarrow 2yy' = 2x + 2 \Rightarrow y' = \frac{2x+2}{2y} = \frac{x+1}{y};$ then $y'' = \frac{y - (x+1)y'}{y^2} = \frac{y - (x+1)\left(\frac{x+1}{y}\right)}{y^2}$

$\Rightarrow \frac{d^2y}{dx^2} = y'' = \frac{y^2 - (x+1)^2}{y^3}$

41. $2\sqrt{y} = x - y \Rightarrow y^{-1/2}y' = 1 - y' \Rightarrow y'\left(y^{-1/2} + 1\right) = 1 \Rightarrow \frac{dy}{dx} = y' = \frac{1}{y^{-1/2} + 1} = \frac{\sqrt{y}}{\sqrt{y} + 1};$ we can

differentiate the equation $y'\left(y^{-1/2} + 1\right) = 1$ again to find y'': $y'\left(-\frac{1}{2}y^{-3/2}y'\right) + \left(y^{-1/2} + 1\right)y'' = 0$

$\Rightarrow \left(y^{-1/2} + 1\right)y'' = \frac{1}{2}[y']^2 y^{-3/2} \Rightarrow \frac{d^2y}{dx^2} = y'' = \frac{\frac{1}{2}\left(\frac{1}{y^{-1/2} + 1}\right)^2 y^{-3/2}}{\left(y^{-1/2} + 1\right)} = \frac{1}{2y^{3/2}\left(y^{-1/2} + 1\right)^3} = \frac{1}{2\left(1 + \sqrt{y}\right)^3}$

43. $x^3 + y^3 = 16 \Rightarrow 3x^2 + 3y^2y' = 0 \Rightarrow 3y^2y' = -3x^2 \Rightarrow y' = -\frac{x^2}{y^2};$ we differentiate $y^2y' = -x^2$ to find y'':

$y^2y'' + y'[2y\cdot y'] = -2x \Rightarrow y^2y'' = -2x - 2y[y']^2 \Rightarrow y'' = \frac{-2x - 2y\left(-\frac{x^2}{y^2}\right)^2}{y^2} = \frac{-2x - \frac{2x^4}{y^3}}{y^2}$

$= \frac{-2xy^3 - 2x^4}{y^5} \Rightarrow \frac{d^2y}{dx^2}\bigg|_{(2,2)} = \frac{-32 - 32}{32} = -2$

45. $y^2 + x^2 = y^4 - 2x$ at $(-2, 1)$ and $(-2, -1) \Rightarrow 2y\dfrac{dy}{dx} + 2x = 4y^3\dfrac{dy}{dx} - 2 \Rightarrow 2y\dfrac{dy}{dx} - 4y^3\dfrac{dy}{dx} = -2 - 2x$

$\Rightarrow \dfrac{dy}{dx}(2y - 4y^3) = -2 - 2x \Rightarrow \dfrac{dy}{dx} = \dfrac{x+1}{2y^3 - y} \Rightarrow \dfrac{dy}{dx}\bigg|_{(-2,1)} = -1$ and $\dfrac{dy}{dx}\bigg|_{(-2,-1)} = 1$

47. $x^2 + xy - y^2 = 1 \Rightarrow 2x + y + xy' - 2yy' = 0 \Rightarrow (x - 2y)y' = -2x - y \Rightarrow y' = \dfrac{2x+y}{2y-x}$;

   (a) the slope of the tangent line $m = y'\big|_{(2,3)} = \dfrac{7}{4} \Rightarrow$ the tangent line is $y - 3 = \dfrac{7}{4}(x-2) \Rightarrow y = \dfrac{7}{4}x - \dfrac{1}{2}$

   (b) the normal line is $y - 3 = -\dfrac{4}{7}(x-2) \Rightarrow y = -\dfrac{4}{7}x + \dfrac{29}{7}$

49. $x^2y^2 = 9 \Rightarrow 2xy^2 + 2x^2yy' = 0 \Rightarrow x^2yy' = -xy^2 \Rightarrow y' = -\dfrac{y}{x}$;

   (a) the slope of the tangent line $m = y'\big|_{(-1,3)} = -\dfrac{y}{x}\big|_{(-1,3)} = 3 \Rightarrow$ the tangent line is $y - 3 = 3(x+1)$

   $\Rightarrow y = 3x + 6$

   (b) the normal line is $y - 3 = -\dfrac{1}{3}(x+1) \Rightarrow y = -\dfrac{1}{3}x + \dfrac{8}{3}$

51. $6x^2 + 3xy + 2y^2 + 17y - 6 = 0 \Rightarrow 12x + 3y + 3xy' + 4yy' + 17y' = 0 \Rightarrow y'(3x + 4y + 17) = -12x - 3y$

   $\Rightarrow y' = \dfrac{-12x - 3y}{3x + 4y + 17}$;

   (a) the slope of the tangent line $m = y'\big|_{(-1,0)} = \dfrac{-12x - 3y}{3x + 4y + 17}\bigg|_{(-1,0)} = \dfrac{6}{7} \Rightarrow$ the tangent line is $y - 0 = \dfrac{6}{7}(x+1)$

   $\Rightarrow y = \dfrac{6}{7}x + \dfrac{6}{7}$

   (b) the normal line is $y - 0 = -\dfrac{7}{6}(x+1) \Rightarrow y = -\dfrac{7}{6}x - \dfrac{7}{6}$

53. $2xy + \pi \sin y = 2\pi \Rightarrow 2xy' + 2y + \pi(\cos y)y' = 0 \Rightarrow y'(2x + \pi \cos y) = -2y \Rightarrow y' = \dfrac{-2y}{2x + \pi \cos y}$;

   (a) the slope of the tangent line $m = y'\big|_{(1,\frac{\pi}{2})} = \dfrac{-2y}{2x + \pi \cos y}\bigg|_{(1,\frac{\pi}{2})} = -\dfrac{\pi}{2} \Rightarrow$ the tangent line is

   $y - \dfrac{\pi}{2} = -\dfrac{\pi}{2}(x-1) \Rightarrow y = -\dfrac{\pi}{2}x + \pi$

   (b) the normal line is $y - \dfrac{\pi}{2} = \dfrac{2}{\pi}(x-1) \Rightarrow y = \dfrac{2}{\pi}x - \dfrac{2}{\pi} + \dfrac{\pi}{2}$

55. $y = 2\sin(\pi x - y) \Rightarrow y' = 2[\cos(\pi x - y)] \cdot (\pi - y') \Rightarrow y'[1 + 2\cos(\pi x - y)] = 2\pi \cos(\pi x - y)$

   $\Rightarrow y' = \dfrac{2\pi \cos(\pi x - y)}{1 + 2\cos(\pi x - y)}$;

   (a) the slope of the tangent line $m = y'\big|_{(1,0)} = \dfrac{2\pi \cos(\pi x - y)}{1 + 2\cos(\pi x - y)}\bigg|_{(1,0)} = 2\pi \Rightarrow$ the tangent line is

   $y - 0 = 2\pi(x-1) \Rightarrow y = 2\pi x - 2\pi$

   (b) the normal line is $y - 0 = -\dfrac{1}{2\pi}(x-1) \Rightarrow y = -\dfrac{x}{2\pi} + \dfrac{1}{2\pi}$

57. Solving $x^2 + xy + y^2 = 7$ and $y = 0 \Rightarrow x^2 = 7 \Rightarrow x = \pm\sqrt{7} \Rightarrow (-\sqrt{7}, 0)$ and $(\sqrt{7}, 0)$ are the points where the

   curve crosses the x-axis. Now $x^2 + xy + y^2 = 7 \Rightarrow 2x + y + xy' + 2yy' = 0 \Rightarrow (x + 2y)y' = -2x - y$

$\Rightarrow y' = -\dfrac{2x+y}{x+2y} \Rightarrow m = -\dfrac{2x+y}{x+2y} \Rightarrow$ the slope at $\left(-\sqrt{7},0\right)$ is $m = -\dfrac{-2\sqrt{7}}{-\sqrt{7}} = -2$ and the slope at $\left(\sqrt{7},0\right)$ is

$m = -\dfrac{2\sqrt{7}}{\sqrt{7}} = -2$. Since the slope is -2 in each case, the corresponding tangents must be parallel.

59. $y^4 = y^2 - x^2 \Rightarrow 4y^3 y' = 2yy' - 2x \Rightarrow 2(2y^3 - y)y' = -2x \Rightarrow y' = \dfrac{x}{y - 2y^3}$; the slope of the tangent line at

$\left(\dfrac{\sqrt{3}}{4}, \dfrac{\sqrt{3}}{2}\right)$ is $\dfrac{x}{y - 2y^3}\bigg|_{\left(\frac{\sqrt{3}}{4}, \frac{\sqrt{3}}{2}\right)} = \dfrac{\frac{\sqrt{3}}{4}}{\frac{\sqrt{3}}{2} - \frac{6\sqrt{3}}{8}} = \dfrac{\frac{1}{4}}{\frac{1}{2} - \frac{3}{4}} = \dfrac{1}{2-3} = -1$; the slope of the tangent line at $\left(\dfrac{\sqrt{3}}{4}, \dfrac{1}{2}\right)$

is $\dfrac{x}{y - 2y^3}\bigg|_{\left(\frac{\sqrt{3}}{4}, \frac{1}{2}\right)} = \dfrac{\frac{\sqrt{3}}{4}}{\frac{1}{2} - \frac{2}{8}} = \dfrac{2\sqrt{3}}{4-2} = \sqrt{3}$

61. $y^4 - 4y^2 = x^4 - 9x^2 \Rightarrow 4y^3 y' - 8yy' = 4x^3 - 18x \Rightarrow y'(4y^3 - 8y) = 4x^3 - 18x \Rightarrow y' = \dfrac{4x^3 - 18x}{4y^3 - 8y} = \dfrac{2x^3 - 9x}{2y^3 - 4y}$

$= \dfrac{x(2x^2 - 9)}{y(2y^2 - 4)} = m$; $(-3,2)$: $m = \dfrac{(-3)(18-9)}{2(8-4)} = -\dfrac{27}{8}$; $(-3,-2)$: $m = \dfrac{27}{8}$; $(3,2)$: $m = \dfrac{27}{8}$; $(3,-2)$: $m = -\dfrac{27}{8}$

63. (a) if $f(x) = \dfrac{3}{2}x^{2/3} - 3$, then $f'(x) = x^{-1/3}$ and $f''(x) = -\dfrac{1}{3}x^{-4/3}$ so the claim $f''(x) = x^{-1/3}$ is false

   (b) if $f(x) = \dfrac{9}{10}x^{5/3} - 7$, then $f'(x) = \dfrac{3}{2}x^{2/3}$ and $f''(x) = x^{-1/3}$ is true

   (c) $f''(x) = x^{-1/3} \Rightarrow f'''(x) = -\dfrac{1}{3}x^{-4/3}$ is true

   (d) if $f'(x) = \dfrac{3}{2}x^{2/3} + 6$, then $f''(x) = x^{-1/3}$ is true

65. $x^2 + 2xy - 3y^2 = 0 \Rightarrow 2x + 2xy' + 2y - 6yy' = 0 \Rightarrow y'(2x - 6y) = -2x - 2y \Rightarrow y' = \dfrac{x+y}{3y-x} \Rightarrow$ the slope of the

tangent line $m = y'\big|_{(1,1)} = \dfrac{x+y}{3y-x}\bigg|_{(1,1)} = 1 \Rightarrow$ the equation of the normal line at $(1,1)$ is $y - 1 = -1(x-1)$

$\Rightarrow y = -x + 2$. To find where the normal line intersects the curve we substitute into its equation:

$x^2 + 2x(2-x) - 3(2-x)^2 = 0 \Rightarrow x^2 + 4x - 2x^2 - 3(4 - 4x + x^2) = 0 \Rightarrow -4x^2 + 16x - 12 = 0 \Rightarrow x^2 - 4x + 3 = 0$

$\Rightarrow (x-3)(x-1) = 0 \Rightarrow x = 3$ and $y = -x + 2 = -1$. Therefore, the normal to the curve at $(1,1)$ intersects the

curve at the point $(3,-1)$. Note that it also intersects the curve at $(1,1)$.

67. $y^2 = x \Rightarrow \dfrac{dy}{dx} = \dfrac{1}{2y}$. If a normal is drawn from $(a,0)$ to (x_1, y_1) on the curve its slope satisfies $\dfrac{y_1 - 0}{x_1 - a} = -2y_1$

$\Rightarrow y_1 = -2y_1(x_1 - a)$ or $a = x_1 + \dfrac{1}{2}$. Since $x_1 \geq 0$ on the curve, we must have that $a \geq \dfrac{1}{2}$. By symmetry, the

two points on the parabola are $\left(x_1, \sqrt{x_1}\right)$ and $\left(x_1, -\sqrt{x_1}\right)$. For the normal to be perpendicular,

$\left(\dfrac{\sqrt{x_1}}{x_1 - a}\right)\left(\dfrac{\sqrt{x_1}}{a - x_1}\right) = -1 \Rightarrow \dfrac{x_1}{(a - x_1)^2} = 1 \Rightarrow x_1 = (a - x_1)^2 \Rightarrow x_1 = \left(x_1 + \dfrac{1}{2} - x_1\right)^2 \Rightarrow x_1 = \dfrac{1}{4}$ and $y_1 = \pm\dfrac{1}{2}$.

Therefore, $\left(\dfrac{1}{4}, \pm\dfrac{1}{2}\right)$ and $a = \dfrac{3}{4}$.

69. $xy^3 + x^2y = 6 \Rightarrow x\left(3y^2\dfrac{dy}{dx}\right) + y^3 + x^2\dfrac{dy}{dx} + 2xy = 0 \Rightarrow \dfrac{dy}{dx}(3xy^2 + x^2) = -y^3 - 2xy \Rightarrow \dfrac{dy}{dx} = \dfrac{-y^3 - 2xy}{3xy^2 + x^2}$

$= -\dfrac{y^3 + 2xy}{3xy^2 + x^2}$; also, $xy^3 + x^2y = 6 \Rightarrow x(3y^2) + y^3\dfrac{dx}{dy} + x^2 + y\left(2x\dfrac{dx}{dy}\right) = 0 \Rightarrow \dfrac{dx}{dy}(y^3 + 2xy) = -3xy^2 - x^2$

$\Rightarrow \dfrac{dx}{dy} = -\dfrac{3xy^2 + x^2}{y^3 + 2xy}$; thus $\dfrac{dx}{dy}$ appears to equal $\dfrac{1}{\frac{dy}{dx}}$. The two different treatments view the graphs as functions

symmetric across the line $y = x$, so their slopes are reciprocals of one another at the corresponding points (a, b) and (b, a).

71. $x^4 + 4y^2 = 1$:

(a) $y^2 = \dfrac{1 - x^4}{4} \Rightarrow y = \pm\dfrac{1}{2}\sqrt{1 - x^4} \Rightarrow \dfrac{dy}{dx} = \pm\dfrac{1}{4}(1 - x^4)^{-1/2}(-4x^3) = \dfrac{\pm x^3}{(1 - x^4)^{1/2}}$; differentiating implicitly, we

find, $4x^3 + 8y\dfrac{dy}{dx} = 0 \Rightarrow \dfrac{dy}{dx} = \dfrac{-4x^3}{8y} = \dfrac{-4x^3}{8\left(\pm\frac{1}{2}\sqrt{1 - x^4}\right)} = \dfrac{\pm x^3}{(1 - x^4)^{1/2}}.$

(b)

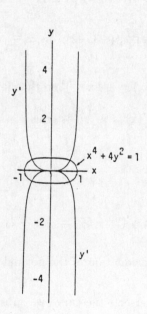

2.7 RELATED RATES OF CHANGE

1. $A = \pi r^2 \Rightarrow \dfrac{dA}{dt} = 2\pi r\dfrac{dr}{dt}$

3. (a) $V = \pi r^2 h \Rightarrow \dfrac{dV}{dt} = \pi r^2\dfrac{dh}{dt}$ (b) $V = \pi r^2 h \Rightarrow \dfrac{dV}{dt} = 2\pi rh\dfrac{dr}{dt}$

   (c) $V = \pi r^2 h \Rightarrow \dfrac{dV}{dt} = \pi r^2\dfrac{dh}{dt} + 2\pi rh\dfrac{dr}{dt}$

5. (a) $\dfrac{dV}{dt} = 1$ volt/sec   (b) $\dfrac{dI}{dt} = -\dfrac{1}{3}$ amp/sec

   (c) $\dfrac{dV}{dt} = R\left(\dfrac{dI}{dt}\right) + I\left(\dfrac{dR}{dt}\right) \Rightarrow \dfrac{dR}{dt} = \dfrac{1}{I}\left(\dfrac{dV}{dt} - R\dfrac{dI}{dt}\right) \Rightarrow \dfrac{dR}{dt} = \dfrac{1}{I}\left(\dfrac{dV}{dt} - \dfrac{V}{I}\dfrac{dI}{dt}\right)$

   (d) $\dfrac{dR}{dt} = \dfrac{1}{2}\left[1 - \dfrac{12}{2}\left(-\dfrac{1}{3}\right)\right] = \left(\dfrac{1}{2}\right)(3) = \dfrac{3}{2}$ ohms/sec, R is increasing

7. (a) $s = \sqrt{x^2 + y^2} = (x^2 + y^2)^{1/2} \Rightarrow \frac{ds}{dt} = \frac{x}{\sqrt{x^2+y^2}} \frac{dx}{dt}$

  (b) $s = \sqrt{x^2 + y^2} = (x^2 + y^2)^{1/2} \Rightarrow \frac{ds}{dt} = \frac{x}{\sqrt{x^2+y^2}} \frac{dx}{dt} + \frac{y}{\sqrt{x^2+y^2}} \frac{dy}{dt}$

  (c) $s = \sqrt{x^2 + y^2} \Rightarrow s^2 = x^2 + y^2 \Rightarrow 2s \frac{ds}{dt} = 2x \frac{dx}{dt} + 2y \frac{dy}{dt} \Rightarrow 2s \cdot 0 = 2x \frac{dx}{dt} + 2y \frac{dy}{dt} \Rightarrow \frac{dx}{dt} = -\frac{y}{x} \frac{dy}{dt}$

9. (a) $A = \frac{1}{2} ab \sin\theta \Rightarrow \frac{dA}{dt} = \frac{1}{2} ab \cos\theta \frac{d\theta}{dt}$ (b) $A = \frac{1}{2} ab \sin\theta \Rightarrow \frac{dA}{dt} = \frac{1}{2} ab \cos\theta \frac{d\theta}{dt} + \frac{1}{2} b \sin\theta \frac{da}{dt}$

  (c) $A = \frac{1}{2} ab \sin\theta \Rightarrow \frac{dA}{dt} = \frac{1}{2} ab \cos\theta \frac{d\theta}{dt} + \frac{1}{2} b \sin\theta \frac{da}{dt} + \frac{1}{2} a \sin\theta \frac{db}{dt}$

11. Given $\frac{d\ell}{dt} = -2$ cm/sec, $\frac{dw}{dt} = 2$ cm/sec, $\ell = 12$ cm and $w = 5$ cm.

  (a) $A = \ell w \Rightarrow \frac{dA}{dt} = \ell \frac{dw}{dt} + w \frac{d\ell}{dt} \Rightarrow \frac{dA}{dt} = 12(2) + 5(-2) = 14$ cm^2/sec, increasing

  (b) $P = 2\ell + 2w \Rightarrow \frac{dP}{dt} = 2 \frac{d\ell}{dt} + 2 \frac{dw}{dt} = 2(-2) + 2(2) = 0$ cm/sec, constant

  (c) $D = \sqrt{w^2 + \ell^2} = (w^2 + \ell^2)^{1/2} \Rightarrow \frac{dD}{dt} = \frac{1}{2}(w^2 + \ell^2)^{-1/2}\left(2w \frac{dw}{dt} + 2\ell \frac{d\ell}{dt}\right) \Rightarrow \frac{dD}{dt} = \frac{w \frac{dw}{dt} + \ell \frac{d\ell}{dt}}{\sqrt{w^2 + \ell^2}}$

  $= \frac{(5)(2) + (12)(-2)}{\sqrt{25 + 144}} = -\frac{14}{13}$ cm/sec, decreasing

13. Given: $\frac{dx}{dt} = 5$ ft/sec, the ladder is 13 ft long, and $x = 12$, $y = 5$ at the instant of time

  (a) Since $x^2 + y^2 = 169 \Rightarrow \frac{dy}{dt} = -\frac{x}{y} \frac{dx}{dt} = -\left(\frac{12}{5}\right)(5) = -12$ ft/sec, the ladder is sliding down the wall

  (b) The area of the triangle formed by the ladder and walls is $A = \frac{1}{2} xy \Rightarrow \frac{dA}{dt} = \left(\frac{1}{2}\right)\left(x \frac{dy}{dt} + y \frac{dx}{dt}\right)$. The area

  is changing at $\frac{1}{2}[12(-12) + 5(5)] = -\frac{119}{2} = -59.5$ ft^2/sec.

  (c) $\cos\theta = \frac{x}{13} \Rightarrow -\sin\theta \frac{d\theta}{dt} = \frac{1}{13} \cdot \frac{dx}{dt} \Rightarrow \frac{d\theta}{dt} = -\frac{1}{13 \sin\theta} \cdot \frac{dx}{dt} = -\left(\frac{1}{5}\right)(5) = -1$ rad/sec

15. Let s represent the distance between the girl and the kite and x represents the horizontal distance between the

  girl and kite $\Rightarrow s^2 = (300)^2 + x^2 \Rightarrow \frac{ds}{dt} = \frac{x}{s} \frac{dx}{dt} = \frac{400(25)}{500} = 20$ ft/sec.

17. $V = \frac{1}{3} \pi r^2 h$, $h = \frac{3}{8}(2r) = \frac{3r}{4} \Rightarrow r = \frac{4h}{3} \Rightarrow V = \frac{1}{3} \pi \left(\frac{4h}{3}\right)^2 h = \frac{16\pi h^3}{27} \Rightarrow \frac{dV}{dt} = \frac{16\pi h^2}{9} \frac{dh}{dt}$

  (a) $\left.\frac{dh}{dt}\right|_{h=4} = \left(\frac{9}{16\pi 4^2}\right)(10) = \frac{90}{256\pi} \approx 0.1119$ m/sec $= 11.19$ cm/sec

  (b) $r = \frac{4h}{3} \Rightarrow \frac{dr}{dt} = \frac{4}{3} \frac{dh}{dt} = \frac{4}{3}\left(\frac{90}{256\pi}\right) = \frac{15}{32\pi} \approx 0.1492$ m/sec $= 14.92$ cm/sec

19. (a) $V = \frac{\pi}{3} y^2 (3R - y) \Rightarrow \frac{dV}{dt} = \frac{\pi}{3}[2y(3R - y) + y^2(-1)] \frac{dy}{dt} \Rightarrow \frac{dy}{dt} = \left[\frac{\pi}{3}(6Ry - 3y^2)\right]^{-1} \frac{dV}{dt} \Rightarrow$ at $R = 13$ and

  $y = 8$ we have $\frac{dy}{dt} = \frac{1}{144\pi}(-6) = \frac{-1}{24\pi}$ m/min

  (b) The hemisphere is on the circle $r^2 + (13 - y)^2 = 169 \Rightarrow r = \sqrt{26y - y^2}$ m

(c) $r = \left(26y - y^2\right)^{1/2} \Rightarrow \dfrac{dr}{dt} = \dfrac{1}{2}\left(26y - y^2\right)^{-1/2}(26 - 2y)\dfrac{dy}{dt} \Rightarrow \dfrac{dr}{dt} = \dfrac{13 - y}{\sqrt{26y - y^2}}\dfrac{dy}{dt} \Rightarrow \dfrac{dr}{dt}\bigg|_{y=8} = \dfrac{13 - 8}{\sqrt{26 \cdot 8 - 64}}\left(\dfrac{-1}{24\pi}\right)$

$= \dfrac{-5}{288\pi}$ m/min

21. If $V = \dfrac{4}{3}\pi r^3$, $r = 5$, and $\dfrac{dV}{dt} = 100\pi$ ft^3/min, then $\dfrac{dV}{dt} = 4\pi r^2 \dfrac{dr}{dt} \Rightarrow \dfrac{dr}{dt} = 1$ ft/min. · Then $S = 4\pi r^2 \Rightarrow \dfrac{dS}{dt}$

$= 8\pi r \dfrac{dr}{dt} = 8\pi(5)(1) = 40\pi$ ft^2/min, the rate at which the area is increasing.

23. Let s represent the distance between the bicycle and balloon, h the height of the balloon and x the horizontal distance between the balloon and the bicycle. The relationship between the variables is $s^2 = h^2 + x^2$

$\Rightarrow \dfrac{ds}{dt} = \dfrac{1}{s}\left(h\dfrac{dh}{dt} + x\dfrac{dx}{dt}\right) \Rightarrow \dfrac{ds}{dt} = \dfrac{1}{85}[68(1) + 51(17)] = 11$ ft/sec.

25. $y = QD^{-1} \Rightarrow \dfrac{dy}{dt} = D^{-1}\dfrac{dQ}{dt} - QD^{-2}\dfrac{dD}{dt} = \dfrac{1}{41}(0) - \dfrac{233}{(41)^2}(-2) = \dfrac{466}{1681}$ L/min $\Rightarrow$ increasing about 0.2772 L/min

27. Let $P(x,y)$ represent a point on the curve $y = x^2$ and θ the angle of inclination of a line containing P and the origin. Consequently, $\tan\theta = \dfrac{y}{x} \Rightarrow \tan\theta = \dfrac{x^2}{x} = x \Rightarrow \sec^2\theta\dfrac{d\theta}{dt} = \dfrac{dx}{dt} \Rightarrow \dfrac{d\theta}{dt} = \cos^2\theta\dfrac{dx}{dt}$. Since $\dfrac{dx}{dt} = 10$ m/sec

and $\cos^2\theta\big|_{x=3} = \dfrac{x^2}{y^2 + x^2} = \dfrac{3^2}{9^2 + 3^2} = \dfrac{1}{10}$, we have $\dfrac{d\theta}{dt}\bigg|_{x=3} = 1$ rad/sec.

29. The distance from the origin is $s = \sqrt{x^2 + y^2}$ and we wish to find $\dfrac{ds}{dt}\bigg|_{(5,12)}$

$= \dfrac{1}{2}\left(x^2 + y^2\right)^{-1/2}\left(2x\dfrac{dx}{dt} + 2y\dfrac{dy}{dt}\right)\bigg|_{(5,12)} = \dfrac{(5)(-1) + (12)(-5)}{\sqrt{25 + 144}} = -5$ m/sec

31. Let $s = 16t^2$ represent the distance the ball has fallen, h the distance between the ball and the ground, and I the distance between the shadow and the point directly beneath the ball. Accordingly, $s + h = 50$ and since the triangle LOQ and triangle PRQ are similar we have

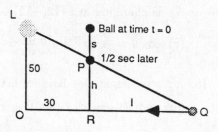

$I = \dfrac{30h}{50 - h} \Rightarrow h = 50 - 16t^2$ and $I = \dfrac{30\left(50 - 16t^2\right)}{50 - \left(50 - 16t^2\right)}$

$= \dfrac{1500}{16t^2} - 30 \Rightarrow \dfrac{dI}{dt} = -\dfrac{1500}{8t^3} \Rightarrow \dfrac{dI}{dt}\bigg|_{t=\frac{1}{2}} = -1500$ ft/sec.

33. The volume of the ice is $V = \dfrac{4}{3}\pi r^3 - \dfrac{4}{3}\pi 4^3 \Rightarrow \dfrac{dV}{dt} = 4\pi r^2 \dfrac{dr}{dt} \Rightarrow \dfrac{dr}{dt}\bigg|_{r=6} = \dfrac{5}{72\pi}$ in/min when $\dfrac{dV}{dt} = 10$ in^3/min.

The surface area is $S = 4\pi r^2 \Rightarrow \dfrac{dS}{dt} = 8\pi r \dfrac{dr}{dt} \Rightarrow \dfrac{dS}{dt}\bigg|_{r=6} = 48\pi\left(\dfrac{5}{72\pi}\right) = \dfrac{10}{3}$ in^2/min.

35. When x represents the length of the shadow, then $\tan\theta = \dfrac{80}{x} \Rightarrow \sec^2\theta\dfrac{d\theta}{dt} = -\dfrac{80}{x^2}\dfrac{dx}{dt} \Rightarrow \dfrac{dx}{dt} = \dfrac{-x^2\sec^2\theta}{80}\dfrac{d\theta}{dt}$.

We are given that $\dfrac{d\theta}{dt} = 0.27° = \dfrac{3\pi}{2000}$ rad/min. At $x = 60$, $\cos\theta = \dfrac{3}{5} \Rightarrow$

$\left|\dfrac{dx}{dt}\right| = \left|\dfrac{-x^2\sec^2\theta}{80}\dfrac{d\theta}{dt}\right|_{\left(\frac{d\theta}{dt} = \frac{3\pi}{2000} \text{ and } \sec\theta = \frac{5}{3}\right)} = \dfrac{3\pi}{16}$ ft/min ≈ 0.589 ft/min ≈ 7.1 in/min.

37. Let x represent distance of the player from second base and s the distance to third base. Then $\frac{dx}{dt} = -16$ ft/sec

(a) $s^2 = x^2 + 8100 \Rightarrow 2s\frac{ds}{dt} = 2x\frac{dx}{dt} \Rightarrow \frac{ds}{dt} = \frac{x}{s}\frac{dx}{dt}$. When the player is 30 ft from first base, $x = 60$

$\Rightarrow s = 30\sqrt{13}$ and $\frac{ds}{dt} = \frac{60}{30\sqrt{13}}(-16) = \frac{-32}{\sqrt{13}} \approx -8.875$ ft/sec

(b) $\cos\theta_1 = \frac{90}{s} \Rightarrow -\sin\theta_1\frac{d\theta_1}{dt} = -\frac{90}{s^2}\cdot\frac{ds}{dt} \Rightarrow \frac{d\theta_1}{dt} = \frac{90}{s^2\sin\theta_1}\cdot\frac{ds}{dt} = \frac{90}{sx}\cdot\frac{ds}{dt}$. Therefore, $x = 60$ and $s = 30\sqrt{13}$

$\Rightarrow \frac{d\theta_1}{dt} = \frac{90}{(30\sqrt{13})(60)}\cdot\left(\frac{-32}{\sqrt{13}}\right) = \frac{-8}{65}$ rad/sec; $\sin\theta_2 = \frac{90}{s} \Rightarrow \cos\theta_2\frac{d\theta_2}{dt} = -\frac{90}{s^2}\cdot\frac{ds}{dt} \Rightarrow \frac{d\theta_2}{dt} = \frac{-90}{s^2\cos\theta_2}\cdot\frac{ds}{dt}$

$= \frac{-90}{sx}\cdot\frac{ds}{dt}$. Therefore, $x = 60$ and $s = 30\sqrt{13} \Rightarrow \frac{d\theta_2}{dt} = \frac{8}{65}$ rad/sec.

(c) $\frac{d\theta_1}{dt} = \frac{90}{s^2\sin\theta_1}\cdot\frac{ds}{dt} = \frac{90}{\left(s^2\cdot\frac{x}{s}\right)}\cdot\left(\frac{x}{s}\right)\cdot\left(\frac{dx}{dt}\right) = \left(\frac{90}{s^2}\right)\left(\frac{dx}{dt}\right) = \left(\frac{90}{x^2+8100}\right)\frac{dx}{dt} \Rightarrow \lim_{x\to 0}\frac{d\theta_1}{dt}$

$= \lim_{x\to 0}\left(\frac{90}{x^2+8100}\right)(-15) = -\frac{1}{6}$ rad/sec; $\frac{d\theta_2}{dt} = \frac{-90}{s^2\cos\theta_2}\cdot\frac{ds}{dt} = \left(\frac{-90}{s^2\cdot\frac{x}{s}}\right)\left(\frac{x}{s}\right)\left(\frac{dx}{dt}\right) = \left(\frac{-90}{s^2}\right)\left(\frac{dx}{dt}\right)$

$= \left(\frac{-90}{x^2+8100}\right)\frac{dx}{dt} \Rightarrow \lim_{x\to 0}\frac{d\theta_2}{dt} = \frac{1}{6}$ rad/sec

39. Let a represent the distance between point O and ship A, b the distance between point O and ship B, and D the distance between the ships. By the Law of Cosines, $D^2 = a^2 + b^2 - 2ab\cos 120°$

$\Rightarrow \frac{dD}{dt} = \frac{1}{2D}\left[2a\frac{da}{dt} + 2b\frac{db}{dt} + a\frac{db}{dt} + b\frac{da}{dt}\right]$. When $a = 5$, $\frac{da}{dt} = 14$, $b = 3$, and $\frac{db}{dt} = 21$, then $\frac{dD}{dt} = \frac{413}{2D}$

where $D = 7$. The ships are moving $\frac{dD}{dt} \approx 29.5$ knots apart.

CHAPTER 2 PRACTICE EXERCISES

1. $y = x^5 - 0.125x^2 + 0.25x \Rightarrow \frac{dy}{dx} = 5x^4 - 0.25x + 0.25$

3. $y = x^3 - 3\left(x^2 + \pi^2\right) \Rightarrow \frac{dy}{dx} = 3x^2 - 3(2x + 0) = 3x^2 - 6x = 3x(x - 2)$

5. $y = (x+1)^2\left(x^2 + 2x\right) \Rightarrow \frac{dy}{dx} = (x+1)^2(2x+2) + \left(x^2+2x\right)(2(x+1)) = 2(x+1)\left[(x+1)^2 + x(x+2)\right]$

$= 2(x+1)\left(2x^2 + 4x + 1\right)$

7. $y = \left(\theta^2 + \sec\theta + 1\right)^3 \Rightarrow \frac{dy}{d\theta} = 3\left(\theta^2 + \sec\theta + 1\right)^2(2\theta + \sec\theta\tan\theta)$

9. $s = \frac{\sqrt{t}}{1+\sqrt{t}} \Rightarrow \frac{ds}{dt} = \frac{(1+\sqrt{t})\cdot\frac{1}{2\sqrt{t}} - \sqrt{t}\left(\frac{1}{2\sqrt{t}}\right)}{(1+\sqrt{t})^2} = \frac{(1+\sqrt{t}) - \sqrt{t}}{2\sqrt{t}(1+\sqrt{t})^2} = \frac{1}{2\sqrt{t}(1+\sqrt{t})^2}$

11. $y = 2\tan^2 x - \sec^2 x \Rightarrow \frac{dy}{dx} = (4\tan x)(\sec^2 x) - (2\sec x)(\sec x\tan x) = 2\sec^2 x\tan x$

13. $s = \cos^4(1-2t) \Rightarrow \frac{ds}{dt} = 4\cos^3(1-2t)(-\sin(1-2t))(-2) = 8\cos^3(1-2t)\sin(1-2t)$

15. $s = (\sec t + \tan t)^5 \Rightarrow \dfrac{ds}{dt} = 5(\sec t + \tan t)^4(\sec t \tan t + \sec^2 t) = 5(\sec t)(\sec t + \tan t)^5$

17. $r = \sqrt{2\theta \sin \theta} = (2\theta \sin \theta)^{1/2} \Rightarrow \dfrac{dr}{d\theta} = \frac{1}{2}(2\theta \sin \theta)^{-1/2}(2\theta \cos \theta + 2 \sin \theta) = \dfrac{\theta \cos \theta + \sin \theta}{\sqrt{2\theta \sin \theta}}$

19. $r = \sin \sqrt{2\theta} = \sin (2\theta)^{1/2} \Rightarrow \dfrac{dr}{d\theta} = \cos (2\theta)^{1/2}\left(\frac{1}{2}(2\theta)^{-1/2}(2)\right) = \dfrac{\cos \sqrt{2\theta}}{\sqrt{2\theta}}$

21. $y = \frac{1}{2}x^2 \csc \frac{2}{x} \Rightarrow \dfrac{dy}{dx} = \frac{1}{2}x^2\left(-\csc \frac{2}{x} \cot \frac{2}{x}\right)\left(\dfrac{-2}{x^2}\right) + \left(\csc \frac{2}{x}\right)\left(\frac{1}{2} \cdot 2x\right) = \csc \frac{2}{x} \cot \frac{2}{x} + x \csc \frac{2}{x}$

23. $y = x^{-1/2} \sec (2x)^2 \Rightarrow \dfrac{dy}{dx} = x^{-1/2} \sec (2x)^2 \tan (2x)^2(2(2x) \cdot 2) + \sec (2x)^2\left(-\frac{1}{2}x^{-3/2}\right)$

$\qquad = 8x^{1/2} \sec (2x)^2 \tan (2x)^2 - \frac{1}{2}x^{-3/2} \sec (2x)^2 = \frac{1}{2}x^{1/2} \sec (2x)^2[16 \tan (2x)^2 - x^{-2}]$

25. $y = 5 \cot x^2 \Rightarrow \dfrac{dy}{dx} = 5(-\csc^2 x^2)(2x) = -10x \csc^2 (x^2)$

27. $y = x^2 \sin^2(2x^2) \Rightarrow \dfrac{dy}{dx} = x^2(2 \sin(2x^2))(\cos(2x^2))(4x) + \sin^2(2x^2)(2x) = 8x^3 \sin(2x^2) \cos(2x^2) + 2x \sin^2(2x^2)$

29. $s = \left(\dfrac{4t}{t+1}\right)^{-2} \Rightarrow \dfrac{ds}{dt} = -2\left(\dfrac{4t}{t+1}\right)^{-3}\left(\dfrac{(t+1)(4) - (4t)(1)}{(t+1)^2}\right) = -2\left(\dfrac{4t}{t+1}\right)^{-3}\dfrac{4}{(t+1)^2} = -\dfrac{(t+1)}{8t^3}$

31. $y = \left(\dfrac{\sqrt{x}}{x+1}\right)^2 \Rightarrow \dfrac{dy}{dx} = 2\left(\dfrac{\sqrt{x}}{x+1}\right) \cdot \dfrac{(x+1)\left(\dfrac{1}{2\sqrt{x}}\right) - (\sqrt{x})(1)}{(x+1)^2} = \dfrac{(x+1) - 2x}{(x+1)^3} = \dfrac{1-x}{(x+1)^3}$

33. $y = \sqrt{\dfrac{x^2 + x}{x^2}} = \left(1 + \dfrac{1}{x}\right)^{1/2} \Rightarrow \dfrac{dy}{dx} = \frac{1}{2}\left(1 + \dfrac{1}{x}\right)^{-1/2}\left(-\dfrac{1}{x^2}\right) = -\dfrac{1}{2x^2 \sqrt{1 + \dfrac{1}{x}}}$

35. $r = \left(\dfrac{\sin \theta}{\cos \theta - 1}\right)^2 \Rightarrow \dfrac{dr}{d\theta} = 2\left(\dfrac{\sin \theta}{\cos \theta - 1}\right)\left[\dfrac{(\cos \theta - 1)(\cos \theta) - (\sin \theta)(-\sin \theta)}{(\cos \theta - 1)^2}\right]$

$\qquad = 2\left(\dfrac{\sin \theta}{\cos \theta - 1}\right)\left(\dfrac{\cos^2 \theta - \cos \theta + \sin^2 \theta}{(\cos \theta - 1)^2}\right) = \dfrac{(2 \sin \theta)(1 - \cos \theta)}{(\cos \theta - 1)^3} = \dfrac{-2 \sin \theta}{(\cos \theta - 1)^2}$

37. $y = (2x + 1) \sqrt{2x + 1} = (2x + 1)^{3/2} \Rightarrow \dfrac{dy}{dx} = \frac{3}{2}(2x + 1)^{1/2}(2) = 3\sqrt{2x + 1}$

39. $y = 3(5x^2 + \sin 2x)^{-3/2} \Rightarrow \dfrac{dy}{dx} = 3\left(-\frac{3}{2}\right)(5x^2 + \sin 2x)^{-5/2}[10x + (\cos 2x)(2)] = \dfrac{-9(5x + \cos 2x)}{(5x^2 + \sin 2x)^{5/2}}$

41. $xy + 2x + 3y = 1 \Rightarrow (xy' + y) + 2 + 3y' = 0 \Rightarrow xy' + 3y' = -2 - y \Rightarrow y'(x + 3) = -2 - y \Rightarrow y' = -\dfrac{y + 2}{x + 3}$

43. $x^3 + 4xy - 3y^{4/3} = 2x \Rightarrow 3x^2 + \left(4x\dfrac{dy}{dx} + 4y\right) - 4y^{1/3}\dfrac{dy}{dx} = 2 \Rightarrow 4x\dfrac{dy}{dx} - 4y^{1/3}\dfrac{dy}{dx} = 2 - 3x^2 - 4y$

$\Rightarrow \dfrac{dy}{dx}\left(4x - 4y^{1/3}\right) = 2 - 3x^2 - 4y \Rightarrow \dfrac{dy}{dx} = \dfrac{2 - 3x^2 - 4y}{4x - 4y^{1/3}}$

45. $(xy)^{1/2} = 1 \Rightarrow \dfrac{1}{2}(xy)^{-1/2}\left(x\dfrac{dy}{dx} + y\right) = 0 \Rightarrow x^{1/2}y^{-1/2}\dfrac{dy}{dx} = -x^{-1/2}y^{1/2} \Rightarrow \dfrac{dy}{dx} = -x^{-1}y \Rightarrow \dfrac{dy}{dx} = -\dfrac{y}{x}$

47. $y^2 = \dfrac{x}{x+1} \Rightarrow 2y\dfrac{dy}{dx} = \dfrac{(x+1)(1) - (x)(1)}{(x+1)^2} \Rightarrow \dfrac{dy}{dx} = \dfrac{1}{2y(x+1)^2}$

49. $p^3 + 4pq - 3q^2 = 2 \Rightarrow 3p^2\dfrac{dp}{dq} + 4\left(p + q\dfrac{dp}{dq}\right) - 6q = 0 \Rightarrow 3p^2\dfrac{dp}{dq} + 4q\dfrac{dp}{dq} = 6q - 4p \Rightarrow \dfrac{dp}{dq}\left(3p^2 + 4q\right) = 6q - 4p$

$\Rightarrow \dfrac{dp}{dq} = \dfrac{6q - 4p}{3p^2 + 4q}$

51. $r\cos 2s + \sin^2 s = \pi \Rightarrow r(-\sin 2s)(2) + (\cos 2s)\left(\dfrac{dr}{ds}\right) + 2\sin s\cos s = 0 \Rightarrow \dfrac{dr}{ds}(\cos 2s) = 2r\sin 2s - 2\sin s\cos s$

$\Rightarrow \dfrac{dr}{ds} = \dfrac{2r\sin 2s - \sin 2s}{\cos 2s} = \dfrac{(2r-1)(\sin 2s)}{\cos 2s} = (2r-1)(\tan 2s)$

53. (a) $x^3 + y^3 = 1 \Rightarrow 3x^2 + 3y^2\dfrac{dy}{dx} = 0 \Rightarrow \dfrac{dy}{dx} = -\dfrac{x^2}{y^2} \Rightarrow \dfrac{d^2y}{dx^2} = \dfrac{y^2(-2x) - (-x^2)\left(2y\dfrac{dy}{dx}\right)}{y^4}$

$\Rightarrow \dfrac{d^2y}{dx^2} = \dfrac{-2xy^2 + (2yx^2)\left(-\dfrac{x^2}{y^2}\right)}{y^4} = \dfrac{-2xy^2 - \dfrac{2x^4}{y}}{y^4} = \dfrac{-2xy^3 - 2x^4}{y^5}$

(b) $y^2 = 1 - \dfrac{2}{x} \Rightarrow 2y\dfrac{dy}{dx} = \dfrac{2}{x^2} \Rightarrow \dfrac{dy}{dx} = \dfrac{1}{yx^2} \Rightarrow \dfrac{dy}{dx} = (yx^2)^{-1} \Rightarrow \dfrac{d^2y}{dx^2} = -(yx^2)^{-2}\left[y(2x) + x^2\dfrac{dy}{dx}\right]$

$\Rightarrow \dfrac{d^2y}{dx^2} = \dfrac{-2xy - x^2\left(\dfrac{1}{yx^2}\right)}{y^2x^4} = \dfrac{-2xy^2 - 1}{y^3x^4}$

55. (a) Let $h(x) = 5f(x) - g(x) \Rightarrow h'(x) = 5f'(x) - g'(x) \Rightarrow h'(1) = 5f'(1) - g'(1) = 5\left(-\dfrac{1}{3}\right) - \left(-\dfrac{8}{3}\right) = 1$

(b) Let $h(x) = f(x)g^3(x) \Rightarrow h'(x) = f(x)\left(3g^2(x)\right)g'(x) + g^3(x)f'(x) \Rightarrow h'(0) = 3f(0)g^2(0)g'(0) + g^3(0)f'(0)$

$= 3(1)(1)^2\left(\dfrac{1}{3}\right) + (1)^3(5) = 6$

(c) Let $h(x) = \dfrac{f(x)}{g(x) + 1} \Rightarrow h'(x) = \dfrac{(g(x) + 1)f'(x) - f(x)g'(x)}{(g(x) + 1)^2} \Rightarrow h'(1) = \dfrac{(g(1) + 1)f'(1) - f(1)g'(1)}{(g(1) + 1)^2}$

$= \dfrac{(-4 + 1)\left(-\dfrac{1}{3}\right) - 3\left(-\dfrac{8}{3}\right)}{(-4 + 1)^2} = 1$

(d) Let $h(x) = f(g(x)) \Rightarrow h'(x) = f'(g(x))g'(x) \Rightarrow h'(0) = f'(g(0))g'(0) = f'(1)\left(\dfrac{1}{3}\right) = \left(-\dfrac{1}{3}\right)\left(\dfrac{1}{3}\right) = -\dfrac{1}{9}$

(e) Let $h(x) = g(f(x)) \Rightarrow h'(x) = g'(f(x))f'(x) \Rightarrow h'(0) = g'(f(0))f'(0) = g'(1)\cdot 5 = \left(-\dfrac{8}{3}\right)(5) = -\dfrac{40}{3}$

(f) Let $h(x) = (x + f(x))^{3/2} \Rightarrow h'(x) = \frac{3}{2}(x + f(x))^{1/2}(1 + f'(x)) \Rightarrow h'(1) = \frac{3}{2}(1 + f(1))^{1/2}(1 + f'(1))$

$= \frac{3}{2}(1 + 3)^{1/2}\left(1 - \frac{1}{3}\right) = 2$

(g) Let $h(x) = f(x + g(x)) \Rightarrow h'(x) = f'(x + g(x))(1 + g'(x)) \Rightarrow h'(0) = f'(g(0))(1 + g'(0))$

$= f'(1)\left(1 + \frac{1}{3}\right) = \left(-\frac{1}{3}\right)\left(\frac{4}{3}\right) = -\frac{4}{9}$

57. $x = t^2 + \pi \Rightarrow \frac{dx}{dt} = 2t$; $y = 3 \sin 2x \Rightarrow \frac{dy}{dx} = 3(\cos 2x)(2) = 6 \cos 2x = 6 \cos(2t^2 + 2\pi) = 6 \cos(2t^2)$; thus,

$\frac{dy}{dt} = \frac{dy}{dx} \cdot \frac{dx}{dt} = 6 \cos(2t^2) \cdot 2t \Rightarrow \left.\frac{dy}{dt}\right|_{t=0} = 6 \cos(0) \cdot 0 = 0$

59. $r = 8 \sin\left(s + \frac{\pi}{6}\right) \Rightarrow \frac{dr}{ds} = 8 \cos\left(s + \frac{\pi}{6}\right)$; $w = \sin(\sqrt{r} - 2) \Rightarrow \frac{dw}{dr} = \cos(\sqrt{r} - 2)\left(\frac{1}{2\sqrt{r}}\right)$

$= \dfrac{\cos\sqrt{8 \sin\left(s + \frac{\pi}{6}\right)} - 2}{2\sqrt{8 \sin\left(s + \frac{\pi}{6}\right)}}$; thus, $\dfrac{dw}{ds} = \dfrac{dw}{dr} \cdot \dfrac{dr}{ds} = \dfrac{\cos\left(\sqrt{8 \sin\left(s + \frac{\pi}{6}\right)} - 2\right)}{2\sqrt{8 \sin\left(s + \frac{\pi}{6}\right)}} \cdot \left[8 \cos\left(s + \frac{\pi}{6}\right)\right]$

$\Rightarrow \left.\dfrac{dw}{ds}\right|_{s=0} = \dfrac{\cos\left(\sqrt{8 \sin\left(\frac{\pi}{6}\right)} - 2\right) \cdot 8 \cos\left(\frac{\pi}{6}\right)}{2\sqrt{8 \sin\left(\frac{\pi}{6}\right)}} = \dfrac{(\cos 0)(8)\left(\frac{\sqrt{3}}{2}\right)}{2\sqrt{4}} = \sqrt{3}$

61. $y^3 + y = 2 \cos x \Rightarrow 3y^2 \frac{dy}{dx} + \frac{dy}{dx} = -2 \sin x \Rightarrow \frac{dy}{dx}(3y^2 + 1) = -2 \sin x \Rightarrow \frac{dy}{dx} = \frac{-2 \sin x}{3y^2 + 1} \Rightarrow \left.\frac{dy}{dx}\right|_{(0,1)}$

$= \dfrac{-2 \sin(0)}{3 + 1} = 0$; $\dfrac{d^2y}{dx^2} = \dfrac{(3y^2 + 1)(-2 \cos x) - (-2 \sin x)\left(6y \frac{dy}{dx}\right)}{(3y^2 + 1)^2}$

$\Rightarrow \left.\dfrac{d^2y}{dx^2}\right|_{(0,1)} = \dfrac{(3 + 1)(-2 \cos 0) - (-2 \sin 0)(6 \cdot 0)}{(3 + 1)^2} = -\dfrac{1}{2}$

63. $f(t) = \frac{1}{2t + 1}$ and $f(t + h) = \frac{1}{2(t + h) + 1} \Rightarrow \dfrac{f(t + h) - f(t)}{h} = \dfrac{\frac{1}{2(t + h) + 1} - \frac{1}{2t + 1}}{h} = \dfrac{2t + 1 - (2t + 2h + 1)}{(2t + 2h + 1)(2t + 1)h}$

$= \dfrac{-2h}{(2t + 2h + 1)(2t + 1)h} = \dfrac{-2}{(2t + 2h + 1)(2t + 1)} \Rightarrow f'(t) = \lim_{h \to 0} \dfrac{f(t + h) - f(t)}{h} = \lim_{h \to 0} \dfrac{-2}{(2t + 2h + 1)(2t + 1)}$

$= \dfrac{-2}{(2t + 1)^2}$

65. (a)

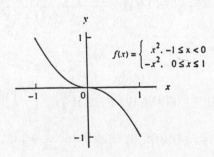

$f(x) = \begin{cases} x^2, & -1 \le x < 0 \\ -x^2, & 0 \le x \le 1 \end{cases}$

(b) $\lim_{x\to 0^-} f(x) = \lim_{x\to 0^-} x^2 = 0$ and $\lim_{x\to 0^+} f(x) = \lim_{x\to 0^+} -x^2 = 0 \Rightarrow \lim_{x\to 0} f(x) = 0$. Since $\lim_{x\to 0} f(x) = 0 = f(0)$ it

follows that f is continuous at $x = 0$.

(c) $\lim_{x\to 0^-} f'(x) = \lim_{x\to 0^-} (2x) = 0$ and $\lim_{x\to 0^+} f'(x) = \lim_{x\to 0^+} (-2x) = 0 \Rightarrow \lim_{x\to 0} f'(x) = 0$. Since this limit exists, it

follows that f is differentiable at $x = 0$.

67. (a)

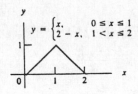

(b) $\lim_{x\to 1^-} f(x) = \lim_{x\to 1^-} x = 1$ and $\lim_{x\to 1^+} f(x) = \lim_{x\to 1^+} (2 - x) = 1 \Rightarrow \lim_{x\to 1} f(x) = 1$. Since $\lim_{x\to 1} f(x) = 1 = f(1)$, it

follows that f is continuous at $x = 1$.

(c) $\lim_{x\to 1^-} f'(x) = \lim_{x\to 1^-} 1 = 1$ and $\lim_{x\to 1^+} f'(x) = \lim_{x\to 1^+} -1 = -1 \Rightarrow \lim_{x\to 1^-} f'(x) \neq \lim_{x\to 1^+} f'(x)$, so $\lim_{x\to 1} f'(x)$ does

not exist $\Rightarrow$ f is not differentiable at $x = 1$.

69. $y = \frac{x}{2} + \frac{1}{2x-4} = \frac{1}{2}x + (2x-4)^{-1} \Rightarrow \frac{dy}{dx} = \frac{1}{2} - 2(2x-4)^{-2}$; the slope of the tangent is $-\frac{3}{2} \Rightarrow -\frac{3}{2}$

$= \frac{1}{2} - 2(2x-4)^{-2} \Rightarrow -2 = -2(2x-4)^{-2} \Rightarrow 1 = \frac{1}{(2x-4)^2} \Rightarrow (2x-4)^2 = 1 \Rightarrow 4x^2 - 16x + 16 = 1$

$\Rightarrow 4x^2 - 16x + 15 = 0 \Rightarrow (2x-5)(2x-3) = 0 \Rightarrow x = \frac{5}{2}$ or $x = \frac{3}{2} \Rightarrow \left(\frac{5}{2}, \frac{9}{4}\right)$ and $\left(\frac{3}{2}, -\frac{1}{4}\right)$ are points on the

curve where the slope is $-\frac{3}{2}$.

71. $y = 2x^3 - 3x^2 - 12x + 20 \Rightarrow \frac{dy}{dx} = 6x^2 - 6x - 12$; the tangent is parallel to the x-axis when $\frac{dy}{dx} = 0$

$\Rightarrow 6x^2 - 6x - 12 = 0 \Rightarrow x^2 - x - 2 = 0 \Rightarrow (x-2)(x+1) = 0 \Rightarrow x = 2$ or $x = -1 \Rightarrow (2, 0)$ and $(-1, 27)$ are

points on the curve where the tangent is parallel to the x-axis.

73. $y = 2x^3 - 3x^2 - 12x + 20 \Rightarrow \frac{dy}{dx} = 6x^2 - 6x - 12$

(a) The tangent is perpendicular to the line $y = 1 - \frac{x}{24}$ when $\frac{dy}{dx} = -\left(\frac{1}{-\left(\frac{1}{24}\right)}\right) = 24$; $6x^2 - 6x - 12 = 24$

$\Rightarrow x^2 - x - 2 = 4 \Rightarrow x^2 - x - 6 = 0 \Rightarrow (x-3)(x+2) = 0 \Rightarrow x = -2$ or $x = 3 \Rightarrow (-2, 16)$ and $(3, 11)$ are

points where the tangent is perpendicular to $y = 1 - \frac{x}{24}$.

(b) The tangent is parallel to the line $y = \sqrt{2} - 12x$ when $\frac{dy}{dx} = -12 \Rightarrow 6x^2 - 6x - 12 = -12 \Rightarrow x^2 - x = 0$

$\Rightarrow x(x - 1) = 0 \Rightarrow x = 0$ or $x = 1 \Rightarrow (0, 20)$ and $(1, 7)$ are points where the tangent is parallel to

$y = \sqrt{2} - 12x$.

75. $y = \tan x$, $-\frac{\pi}{2} < x < \frac{\pi}{2} \Rightarrow \frac{dy}{dx} = \sec^2 x$; now the slope

of $y = -\frac{x}{2}$ is $-\frac{1}{2} \Rightarrow$ the normal line is parallel to

$y = -\frac{x}{2}$ when $\frac{dy}{dx} = 2$. Thus, $\sec^2 x = 2 \Rightarrow \frac{1}{\cos^2 x} = 2$

$\Rightarrow \cos^2 x = \frac{1}{2} \Rightarrow \cos x = \frac{\pm 1}{\sqrt{2}} \Rightarrow x = -\frac{\pi}{4}$ and $x = \frac{\pi}{4}$

for $-\frac{\pi}{2} < x < \frac{\pi}{2} \Rightarrow \left(-\frac{\pi}{4}, -1\right)$ and $\left(\frac{\pi}{4}, 1\right)$ are points

where the normal is parallel to $y = -\frac{x}{2}$.

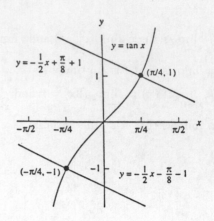

77. $y = x^2 + C \Rightarrow \frac{dy}{dx} = 2x$ and $y = x \Rightarrow \frac{dy}{dx} = 1$; the parabola is tangent to $y = x$ when $2x = 1 \Rightarrow x = \frac{1}{2} \Rightarrow y = \frac{1}{2}$;

thus, $\frac{1}{2} = \left(\frac{1}{2}\right)^2 + C \Rightarrow C = \frac{1}{4}$.

79. The line through $(0, 3)$ and $(5, -2)$ has slope $m = \frac{3 - (-2)}{0 - 5} = -1 \Rightarrow$ the line through $(0, 3)$ and $(5, -2)$ is

$y = -x + 3$; $y = \frac{c}{x+1} \Rightarrow \frac{dy}{dx} = \frac{-c}{(x+1)^2}$, so the curve is tangent to $y = -x + 3 \Rightarrow \frac{dy}{dx} = -1 = \frac{-c}{(x+1)^2}$

$\Rightarrow (x+1)^2 = c$, $x \neq -1$. Moreover, $y = \frac{c}{x+1}$ intersects $y = -x + 3 \Rightarrow \frac{c}{x+1} = -x + 3$, $x \neq -1$

$\Rightarrow c = (x+1)(-x+3)$, $x \neq -1$. Thus $c = c \Rightarrow (x+1)^2 = (x+1)(-x+3) \Rightarrow (x+1)[x+1-(-x+3)]$

$= 0$, $x \neq -1 \Rightarrow (x+1)(2x-2) = 0 \Rightarrow x = 1$ (since $x \neq -1$) $\Rightarrow c = 4$.

81. $x^2 + 2y^2 = 9 \Rightarrow 2x + 4y \frac{dy}{dx} = 0 \Rightarrow \frac{dy}{dx} = -\frac{x}{2y} \Rightarrow \left.\frac{dy}{dx}\right|_{(1,2)} = -\frac{1}{4} \Rightarrow$ the tangent line is $y = 2 - \frac{1}{4}(x-1)$

$= -\frac{1}{4}x + \frac{9}{4}$ and the normal line is $y = 2 + 4(x-1) = 4x - 2$.

83. $xy + 2x - 5y = 2 \Rightarrow \left(x\frac{dy}{dx} + y\right) + 2 - 5\frac{dy}{dx} = 0 \Rightarrow \frac{dy}{dx}(x-5) = -y - 2 \Rightarrow \frac{dy}{dx} = \frac{-y-2}{x-5} \Rightarrow \left.\frac{dy}{dx}\right|_{(3,2)} = 2$

$\Rightarrow$ the tangent line is $y = 2 + 2(x-3) = 2x - 4$ and the normal line is $y = 2 + \frac{-1}{2}(x-3) = -\frac{1}{2}x + \frac{7}{2}$.

85. $x + \sqrt{xy} = 6 \Rightarrow 1 + \frac{1}{2\sqrt{xy}}\left(x\frac{dy}{dx} + y\right) = 0 \Rightarrow x\frac{dy}{dx} + y = -2\sqrt{xy} \Rightarrow \frac{dy}{dx} = \frac{-2\sqrt{xy} - y}{x} \Rightarrow \left.\frac{dy}{dx}\right|_{(4,1)} = \frac{-5}{4}$

$\Rightarrow$ the tangent line is $y = 1 - \frac{5}{4}(x-4) = -\frac{5}{4}x + 6$ and the normal line is $y = 1 + \frac{4}{5}(x-4) = \frac{4}{5}x - \frac{11}{5}$.

87. $x^3y^3 + y^2 = x + y \Rightarrow \left[x^3\left(3y^2\frac{dy}{dx}\right) + y^3(3x^2)\right] + 2y\frac{dy}{dx} = 1 + \frac{dy}{dx} \Rightarrow 3x^3y^2\frac{dy}{dx} + 2y\frac{dy}{dx} - \frac{dy}{dx} = 1 - 3x^2y^3$

$\Rightarrow \frac{dy}{dx}(3x^3y^2 + 2y - 1) = 1 - 3x^2y^3 \Rightarrow \frac{dy}{dx} = \frac{1 - 3x^2y^3}{3x^3y^2 + 2y - 1} \Rightarrow \left.\frac{dy}{dx}\right|_{(1,1)} = -\frac{2}{4}$, but $\left.\frac{dy}{dx}\right|_{(1,-1)}$ is undefined.

Therefore, the curve has slope $-\frac{1}{2}$ at $(1, 1)$ but the slope is undefined at $(1, -1)$.

89. B = graph of f, A = graph of f′. Curve B cannot be the derivative of A because A has only negative slopes while some of B's values are positive.

91.

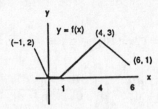

93. (a) 0, 0 $\qquad$ (b) largest 1700, smallest about 1400

95. $\lim\limits_{s\to 0} \dfrac{\sin\left(\frac{s}{2}\right)}{\left(\frac{s}{3}\right)} = \lim\limits_{s\to 0} \dfrac{\sin\left(\frac{s}{2}\right)}{\left(\frac{s}{2}\right)}\cdot\dfrac{1}{\left(\frac{2}{3}\right)} = 1\cdot\dfrac{3}{2} = \dfrac{3}{2}$

97. $\lim\limits_{x\to 0} \dfrac{\sin x}{2x^2 - x} = \lim\limits_{x\to 0}\left[\left(\dfrac{\sin x}{x}\right)\cdot\dfrac{1}{(2x-1)}\right] = (1)\left(\dfrac{1}{-1}\right) = -1$

99. $\lim\limits_{r\to 0} \dfrac{\sin r}{\tan 2r} = \lim\limits_{r\to 0}\left(\dfrac{\sin r}{r}\cdot\dfrac{2r}{\tan 2r}\cdot\dfrac{1}{2}\right) = \left(\dfrac{1}{2}\right)(1)\ \lim\limits_{r\to 0}\dfrac{\cos 2r}{\left(\frac{\sin 2r}{2r}\right)} = \left(\dfrac{1}{2}\right)(1)\left(\dfrac{1}{1}\right) = \dfrac{1}{2}$

101. $\lim\limits_{\theta\to\left(\frac{\pi}{2}\right)^-} \dfrac{4\tan^2\theta + \tan\theta + 1}{\tan^2\theta + 5} = \lim\limits_{\theta\to\left(\frac{\pi}{2}\right)^-} \dfrac{\left(4 + \frac{1}{\tan\theta} + \frac{1}{\tan^2\theta}\right)}{\left(1 + \frac{5}{\tan^2\theta}\right)} = \dfrac{(4+0+0)}{(1+0)} = 4$

103. $\lim\limits_{x\to 0} \dfrac{x\sin x}{2 - 2\cos x} = \lim\limits_{x\to 0} \dfrac{x\sin x}{2(1-\cos x)} = \lim\limits_{x\to 0} \dfrac{x\sin x}{2\left(2\sin^2\left(\frac{x}{2}\right)\right)} = \lim\limits_{x\to 0}\left[\dfrac{\frac{x}{2}\cdot\frac{x}{2}}{\sin^2\left(\frac{x}{2}\right)}\cdot\dfrac{\sin x}{x}\right]$

$= \lim\limits_{x\to 0}\left[\dfrac{\left(\frac{x}{2}\right)}{\sin\left(\frac{x}{2}\right)}\cdot\dfrac{\left(\frac{x}{2}\right)}{\sin\left(\frac{x}{2}\right)}\cdot\dfrac{\sin x}{x}\right] = (1)(1)(1) = 1$

105. $\lim\limits_{x\to 0} \dfrac{\tan x}{x} = \lim\limits_{x\to 0}\left(\dfrac{1}{\cos x}\cdot\dfrac{\sin x}{x}\right) = 1$; let $\theta = \tan x \Rightarrow \theta \to 0$ as $x \to 0 \Rightarrow \lim\limits_{x\to 0} g(x) = \lim\limits_{x\to 0} \dfrac{\tan(\tan x)}{\tan x}$

$= \lim\limits_{\theta\to 0} \dfrac{\tan\theta}{\theta} = 1$. Therefore, to make g continuous at the origin, define $g(0) = 1$.

107. $\lim\limits_{x\to 0} f(x) = \lim\limits_{x\to 0} \dfrac{\sin x}{2x} = \dfrac{1}{2}\lim\limits_{x\to 0}\dfrac{\sin x}{x} = \dfrac{1}{2}$ so that defining $f(0) = \dfrac{1}{2} \Rightarrow$ f is continuous at the origin: $f(0) = \dfrac{1}{2}$

$\Rightarrow k = \dfrac{1}{2}$.

109. (a) $S = 2\pi r^2 + 2\pi rh$ and h constant $\Rightarrow \dfrac{dS}{dt} = 4\pi r\dfrac{dr}{dt} + 2\pi h\dfrac{dr}{dt} = (4\pi r + 2\pi h)\dfrac{dr}{dt}$

   (b) $S = 2\pi r^2 + 2\pi rh$ and r constant $\Rightarrow \dfrac{dS}{dt} = 2\pi r\dfrac{dh}{dt}$

(c) $S = 2\pi r^2 + 2\pi rh \Rightarrow \frac{dS}{dt} = 4\pi r \frac{dr}{dt} + 2\pi\left(r \frac{dh}{dt} + h \frac{dr}{dt}\right) = (4\pi r + 2\pi h)\frac{dr}{dt} + 2\pi r \frac{dh}{dt}$

(d) S constant $\Rightarrow \frac{dS}{dt} = 0 \Rightarrow 0 = (4\pi r + 2\pi h)\frac{dr}{dt} + 2\pi r \frac{dh}{dt} \Rightarrow (2r + h)\frac{dr}{dt} = -r \frac{dh}{dt} \Rightarrow \frac{dr}{dt} = \frac{-r}{2r+h}\frac{dh}{dt}$

111. $A = \pi r^2 \Rightarrow \frac{dA}{dt} = 2\pi r \frac{dr}{dt}$; so $r = 10$ and $\frac{dr}{dt} = -\frac{2}{\pi}$ m/sec $\Rightarrow \frac{dA}{dt} = (2\pi)(10)\left(-\frac{2}{\pi}\right) = -40$ m^2/sec

113. $\frac{dR_1}{dt} = -1$ ohm/sec, $\frac{dR_2}{dt} = 0.5$ ohm/sec; and $\frac{1}{R} = \frac{1}{R_1} + \frac{1}{R_2} \Rightarrow \frac{-1}{R^2}\frac{dR}{dt} = \frac{-1}{R_1^2}\frac{dR_1}{dt} - \frac{1}{R_2^2}\frac{dR_2}{dt}$. Also,

$R_1 = 75$ ohms and $R_2 = 50$ ohms $\Rightarrow \frac{1}{R} = \frac{1}{75} + \frac{1}{50} \Rightarrow R = 30$ ohms. Therefore, from the derivative equation,

$\frac{-1}{(30)^2}\frac{dR}{dt} = \frac{-1}{(75)^2}(-1) - \frac{1}{(50)^2}(0.5) = \left(\frac{1}{5625} - \frac{1}{5000}\right) \Rightarrow \frac{dR}{dt} = (-900)\left(\frac{5000 - 5625}{5625 \cdot 5000}\right) = \frac{9(625)}{50(5625)} = \frac{1}{50}$

$= 0.02$ ohm/sec.

115. Given $\frac{dx}{dt} = -1$ m/sec and $\frac{dy}{dt} = -5$ m/sec, let D be the distance from the origin $\Rightarrow D^2 = x^2 + y^2 \Rightarrow 2D \frac{dD}{dt}$

$= 2x \frac{dx}{dt} + 2y \frac{dy}{dt} \Rightarrow D \frac{dD}{dt} = x \frac{dx}{dt} + y \frac{dy}{dt}$. When $(x, y) = (5, 12)$, $D = \sqrt{5^2 + 12^2} = 13$ and

$13 \frac{dD}{dt} = (5)(-1) + (12)(-5) \Rightarrow \frac{dD}{dt} = -\frac{65}{13} = -5$. Therefore, the particle is <u>approaching</u> the origin at 5 m/sec
(because the distance D is decreasing).

117. (a) From the diagram we have $\frac{10}{h} = \frac{4}{r} \Rightarrow r = \frac{2}{5} h$.

(b) $V = \frac{1}{3}\pi r^2 h = \frac{1}{3}\pi\left(\frac{2}{5}h\right)^2 h = \frac{4\pi h^3}{75} \Rightarrow \frac{dV}{dt} = \frac{4\pi h^2}{25}\frac{dh}{dt}$, so $\frac{dV}{dt} = -5$ and $h = 6 \Rightarrow \frac{dh}{dt} = -\frac{125}{144\pi}$ ft/min.

119. (a) From the sketch in the text, $\frac{d\theta}{dt} = -0.6$ rad/sec and $x = \tan\theta$. Also $x = \tan\theta \Rightarrow \frac{dx}{dt} = \sec^2\theta \frac{d\theta}{dt}$; at

point A, $x = 0 \Rightarrow \theta = 0 \Rightarrow \frac{dx}{dt} = (\sec^2 0)(-0.6) = -0.6$. Therefore the speed of the light is $0.6 = \frac{3}{5}$ km/sec

when it reaches point A.

(b) $\frac{(3/5) \text{ rad}}{\text{sec}} \cdot \frac{1 \text{ rev}}{2\pi \text{ rad}} \cdot \frac{60 \text{ sec}}{\text{min}} = \frac{18}{\pi}$ revs/min

CHAPTER 2   ADDITIONAL EXERCISES–THEORY, EXAMPLES, APPLICATIONS

1. (a) $\sin 2\theta = 2\sin\theta\cos\theta \Rightarrow \frac{d}{d\theta}(\sin 2\theta) = \frac{d}{d\theta}(2\sin\theta\cos\theta) \Rightarrow 2\cos 2\theta = 2[(\sin\theta)(-\sin\theta) + (\cos\theta)(\cos\theta)]$
$\Rightarrow \cos 2\theta = \cos^2\theta - \sin^2\theta$

(b) $\cos 2\theta = \cos^2\theta - \sin^2\theta \Rightarrow \frac{d}{d\theta}(\cos 2\theta) = \frac{d}{d\theta}(\cos^2\theta - \sin^2\theta) \Rightarrow -2\sin 2\theta = (2\cos\theta)(-\sin\theta) - (2\sin\theta)(\cos\theta)$
$\Rightarrow \sin 2\theta = \cos\theta\sin\theta + \sin\theta\cos\theta \Rightarrow \sin 2\theta = 2\sin\theta\cos\theta$

3. (a) $f(x) = \cos x \Rightarrow f'(x) = -\sin x \Rightarrow f''(x) = -\cos x$, and $g(x) = a + bx + cx^2 \Rightarrow g'(x) = b + 2cx \Rightarrow g''(x) = 2c$;
also, $f(0) = g(0) \Rightarrow \cos(0) = a \Rightarrow a = 1$; $f'(0) = g'(0) \Rightarrow -\sin(0) = b \Rightarrow b = 0$; $f''(0) = g''(0)$
$\Rightarrow -\cos(0) = 2c \Rightarrow c = -\frac{1}{2}$. Therefore, $g(x) = 1 - \frac{1}{2}x^2$.

(b) $f(x) = \sin(x+a) \Rightarrow f'(x) = \cos(x+a)$, and $g(x) = b \sin x + c \cos x \Rightarrow g'(x) = b \cos x - c \sin x$; also,

$f(0) = g(0) \Rightarrow \sin(a) = b \sin(0) + c \cos(0) \Rightarrow c = \sin a$; $f'(0) = g'(0) \Rightarrow \cos(a) = b \cos(0) - c \sin(0)$

$\Rightarrow b = \cos a$. Therefore, $g(x) = \sin x \cos a + \cos x \sin a$.

(c) When $f(x) = \cos x$, $f'''(x) = \sin x$ and $f^{(4)}(x) = \cos x$; when $g(x) = 1 - \frac{1}{2}x^2$, $g'''(x) = 0$ and $g^{(4)}(x) = 0$.

Thus $f'''(0) = 0 = g'''(0)$ so the third derivatives agree at $x = 0$. However, the fourth derivatives do not

agree since $f^{(4)}(0) = 1$ but $g^{(4)}(0) = 0$. In case (b), when $f(x) = \sin(x+a)$ and $g(x)$

$= \sin x \cos a + \cos x \sin a$, notice that $f(x) = g(x)$ for all x, not just $x = 0$. Since this is an identity, we

have $f^{(n)}(x) = g^{(n)}(x)$ for any x and any positive integer n.

5. If the circle $(x-h)^2 + (y-k)^2 = a^2$ and $y = x^2 + 1$ are tangent at $(1,2)$, then the slope of this tangent is

$m = 2x|_{(1,2)} = 2$ and the tangent line is $y = 2x$. The line containing (h,k) and $(1,2)$ is perpendicular to

$y = 2x \Rightarrow \frac{k-2}{h-1} = -\frac{1}{2} \Rightarrow h = 5 - 2k \Rightarrow$ the location of the center is $(5-2k, k)$. Also, $(x-h)^2 + (y-k)^2 = a^2$

$\Rightarrow x - h + (y-k)y' = 0 \Rightarrow 1 + (y')^2 + (y-k)y'' = 0 \Rightarrow y'' = \frac{1 + (y')^2}{k-y}$. At the point $(1,2)$ we know

$y' = 2$ from the tangent line and that $y'' = 2$ from the parabola. Since the second derivatives are equal at $(1,2)$

we obtain $2 = \frac{1 + (2)^2}{k-2} \Rightarrow k = \frac{9}{2}$. Then $h = 5 - 2k = -4 \Rightarrow$ the circle is $(x+4)^2 + \left(y - \frac{9}{2}\right)^2 = a^2$. Since $(1,2)$

lies on the circle we have that $a = \frac{5\sqrt{5}}{2}$.

7. (a) $y = uv \Rightarrow \frac{dy}{dt} = \frac{du}{dt}v + u\frac{dv}{dt} = (0.04u)v + u(0.05v) = 0.09uv = 0.09y$

(b) If $\frac{du}{dt} = -0.02u$ and $\frac{dv}{dt} = 0.03v$, then $\frac{dy}{dt} = (-0.02u)v + (0.03v)u = 0.01uv = 0.01y$, increasing at 1% per

year.

9.

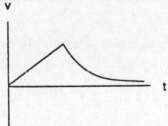

11. (a) $s(t) = 64t - 16t^2 \Rightarrow v(t) = \frac{ds}{dt} = 64 - 32t = 32(2-t)$. The maximum height is reached when $v(t) = 0$

$\Rightarrow t = 2$ sec. The velocity when it leaves the hand is $v(0) = 64$ ft/sec.

(b) $s(t) = 64t - 2.6t^2 \Rightarrow v(t) = \frac{ds}{dt} = 64 - 5.2t$. The maximum height is reached when $v(t) = 0 \Rightarrow t \approx 12.31$ sec.

The maximum height is about $s(12.31) = 393.85$ ft.

13. $m\left(v^2 - v_0^2\right) = k\left(x_0^2 - x^2\right) \Rightarrow m\left(2v\frac{dv}{dt}\right) = k\left(-2x\frac{dx}{dt}\right) \Rightarrow m\frac{dv}{dt} = k\left(-\frac{2x}{2v}\right)\frac{dx}{dt} \Rightarrow m\frac{dv}{dt} = -kx\left(\frac{1}{v}\right)\frac{dx}{dt}$. Then

substituting $\frac{dx}{dt} = v \Rightarrow m\frac{dv}{dt} = -kx$, as claimed.

15. (a) To be continuous at $x = \pi$ requires that $\lim\limits_{x \to \pi^-} \sin x = \lim\limits_{x \to \pi^+} (mx + b) \Rightarrow 0 = m\pi + b \Rightarrow m = -\frac{b}{\pi}$;

(b) If $y' = \begin{cases} \cos x, & x < \pi \\ m, & x \geq \pi \end{cases}$ is differentiable at $x = \pi$, then $\lim\limits_{x \to \pi^-} \cos x = m \Rightarrow m = -1$ and $b = \pi$.

17. For all a, b and for all $x \neq 2$, f is differentiable at x. Next, f differentiable at $x = 2 \Rightarrow$ f continuous at $x = 2$

$\Rightarrow \lim\limits_{x \to 2^-} f(x) = f(2) \Rightarrow 2a = 4a - 2b + 3 \Rightarrow 2a - 2b + 3 = 0$. Also, f differentiable at $x \neq 2$

$\Rightarrow f'(x) = \begin{cases} a, & x < 2 \\ 2ax - b, & x > 2 \end{cases}$. In order that $f'(2)$ exist we must have $a = 2a(2) - b \Rightarrow a = 4a - b \Rightarrow 3a = b$.

Then $2a - 2b + 3 = 0$ and $3a = b \Rightarrow a = \frac{3}{4}$ and $b = \frac{9}{4}$.

19. f odd $\Rightarrow f(-x) = -f(x) \Rightarrow \frac{d}{dx}(f(-x)) = \frac{d}{dx}(-f(x)) \Rightarrow f'(-x)(-1) = -f'(x) \Rightarrow f'(-x) = f'(x) \Rightarrow$ f' is even.

21. Let $h(x) = (fg)(x) = f(x) g(x) \Rightarrow h'(x) = \lim\limits_{x \to x_0} \frac{h(x) - h(x_0)}{x - x_0} = \lim\limits_{x \to x_0} \frac{f(x) g(x) - f(x_0) g(x_0)}{x - x_0}$

$= \lim\limits_{x \to x_0} \frac{f(x) g(x) - f(x) g(x_0) + f(x) g(x_0) - f(x_0) g(x_0)}{x - x_0} = \lim\limits_{x \to x_0} \left[f(x) \left[\frac{g(x) - g(x_0)}{x - x_0} \right] \right] + \lim\limits_{x \to x_0} \left[g(x_0) \left[\frac{f(x) - f(x_0)}{x - x_0} \right] \right]$

$= f(x_0) \lim\limits_{x \to x_0} \left[\frac{g(x) - g(x_0)}{x - x_0} \right] + g(x_0) f'(x_0) = 0 \cdot \lim\limits_{x \to x_0} \left[\frac{g(x) - g(x_0)}{x - x_0} \right] + g(x_0) f'(x_0) = g(x_0) f'(x_0)$, if g is

continuous at x_0. Therefore $(fg)(x)$ is differentiable at x_0 if $f(x_0) = 0$, and $(fg)'(x_0) = g(x_0) f'(x_0)$.

23. If $f(x) = x$ and $g(x) = x \sin\left(\frac{1}{x}\right)$, then $x^2 \sin\left(\frac{1}{x}\right)$ is differentiable at $x = 0$ because $f'(0) = 1$, $f(0) = 0$ and

$\lim\limits_{x \to 0} x \sin\left(\frac{1}{x}\right) = \lim\limits_{x \to 0} \frac{\sin\left(\frac{1}{x}\right)}{\frac{1}{x}} = \lim\limits_{t \to \infty} \frac{\sin t}{t} = 0$ (so g is continuous at $x = 0$). In fact, from Exercise 21,

$h'(0) = g(0) f'(0) = 0$. However, for $x \neq 0$, $h'(x) = \left[x^2 \cos\left(\frac{1}{x}\right) \right]\left(-\frac{1}{x^2}\right) + 2x \sin\left(\frac{1}{x}\right)$. But

$\lim\limits_{x \to 0} h'(x) = \lim\limits_{x \to 0} \left[-\cos\left(\frac{1}{x}\right) + 2x \sin\left(\frac{1}{x}\right) \right]$ does not exist because $\cos\left(\frac{1}{x}\right)$ has no limit as $x \to 0$. Therefore,
the derivative is not continuous at $x = 0$ because it has no limit there.

25. Step 1: The formula holds for $n = 2$ (a single product) since $y = u_1 u_2 \Rightarrow \frac{dy}{dx} = \frac{du_1}{dx} u_2 + u_1 \frac{du_2}{dx}$.

Step 2: Assume the formula holds for $n = k$:

$$y = u_1 u_2 \cdots u_k \Rightarrow \frac{dy}{dx} = \frac{du_1}{dx} u_2 u_3 \cdots u_k + u_1 \frac{du_2}{dx} u_3 \cdots u_k + \ldots + u_1 u_2 \cdots u_{k-1} \frac{du_k}{dx}.$$

If $y = u_1 u_2 \cdots u_k u_{k+1} = (u_1 u_2 \cdots u_k) u_{k+1}$, then $\frac{dy}{dx} = \frac{d(u_1 u_2 \cdots u_k)}{dx} u_{k+1} + u_1 u_2 \cdots u_k \frac{du_{k+1}}{dx}$

$= \left(\frac{du_1}{dx} u_2 u_3 \cdots u_k + u_1 \frac{du_2}{dx} u_3 \cdots u_k + \cdots + u_1 u_2 \cdots u_{k-1} \frac{du_k}{dx} \right) u_{k+1} + u_1 u_2 \cdots u_k \frac{du_{k+1}}{dx}$

$= \frac{du_1}{dx} u_2 u_3 \cdots u_{k+1} + u_1 \frac{du_2}{dx} u_3 \cdots u_{k+1} + \cdots + u_1 u_2 \cdots u_{k-1} \frac{du_k}{dx} u_{k+1} + u_1 u_2 \cdots u_k \frac{du_{k+1}}{dx}$.

Thus the original formula holds for $n = (k+1)$ whenever it holds for $n = k$.

CHAPTER 3 APPLICATIONS OF DERIVATIVES

3.1 EXTREME VALUES OF FUNCTIONS

1. An absolute minimum at $x = C_2$, an absolute maximum at $x = b$. Theorem 1 guarantees the existence of such extreme values because h is continuous on $[a, b]$.

3. No absolute minimum. An absolute maximum at $x = c$. Since the function's domain is an open interval, the function does not satisfy the hypotheses of Theorem 1 and need not have absolute extreme values.

5. An absolute minimum at $x = a$ and an absolute maximum at $x = c$. Note that $y = g(x)$ is not continuous but still has extrema. When the hypothesis of Theorem 1 is satisfied then extrema are guaranteed, but when the hypothesis is not satisfied, absolute extrema may or may not occur.

7. $f(x) = \frac{2}{3}x - 5 \Rightarrow f'(x) = \frac{2}{3} \Rightarrow$ no critical points;

$f(-2) = -\frac{19}{3}$, $f(3) = -3 \Rightarrow$ the absolute maximum is -3 at $x = 3$ and the absolute minimum is $-\frac{19}{3}$ at $x = -2$

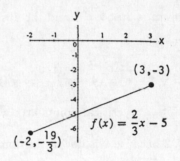

9. $f(x) = x^2 - 1 \Rightarrow f'(x) = 2x \Rightarrow$ a critical point at $x = 0$; $f(-1) = 0$, $f(0) = -1$, $f(2) = 3 \Rightarrow$ the absolute maximum is 3 at $x = 2$ and the absolute minimum is -1 at $x = 0$

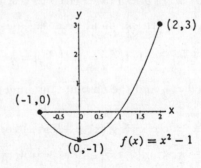

11. $F(x) = -\dfrac{1}{x^2} = -x^{-2} \Rightarrow F'(x) = 2x^{-3} = \dfrac{2}{x^3}$, however

   $x = 0$ is not a critical point since 0 is not in the domain;

   $F(0.5) = -4$, $F(2) = -0.25 \Rightarrow$ the absolute maximum is

   -0.25 at $x = 2$ and the absolute minimum is -4 at

   $x = 0.5$

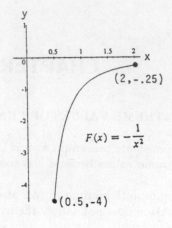

13. $h(x) = \sqrt[3]{x} = x^{1/3} \Rightarrow h'(x) = \dfrac{1}{3}x^{-2/3} \Rightarrow$ a critical point

   at $x = 0$; $h(-1) = -1$, $h(0) = 0$, $h(8) = 2 \Rightarrow$ the absolute

   maximum is 2 at $x = 8$ and the absolute minimum is -1

   at $x = -1$

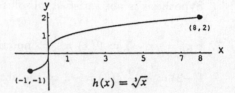

15. $g(x) = \sqrt{4 - x^2} = \left(4 - x^2\right)^{1/2} \Rightarrow g'(x) = \dfrac{1}{2}\left(4 - x^2\right)^{-1/2}(-2x)$

   $= \dfrac{-x}{\sqrt{4 - x^2}} \Rightarrow$ critical points at $x = -2$ and $x = 0$, but not

   at $x = 2$ because 2 is not in the domain; $g(-2) = 0$, $g(0) = 2$,

   $g(1) = \sqrt{3} \Rightarrow$ the absolute maximum is 2 at $x = 0$ and the

   absolute minimum is 0 at $x = -2$

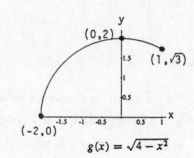

17. $f(\theta) = \sin\theta \Rightarrow f'(\theta) = \cos\theta \Rightarrow \theta = \dfrac{\pi}{2}$ is a critical point, but $\theta = \dfrac{-\pi}{2}$

   is not a critical point because $\dfrac{-\pi}{2}$ is not interior to the domain;

   $f\left(\dfrac{-\pi}{2}\right) = -1$, $f\left(\dfrac{\pi}{2}\right) = 1$, $f\left(\dfrac{5\pi}{6}\right) = \dfrac{1}{2} \Rightarrow$ the absolute maximum is 1

   at $\theta = \dfrac{\pi}{2}$ and the absolute minimum is -1 at $\theta = \dfrac{-\pi}{2}$

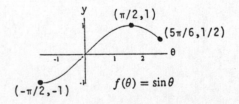

19. $g(x) = \csc x \Rightarrow g'(x) = -(\csc x)(\cot x) \Rightarrow$ a critical point at

   $x = \dfrac{\pi}{2}$; $g\left(\dfrac{\pi}{3}\right) = \dfrac{2}{\sqrt{3}}$, $g\left(\dfrac{\pi}{2}\right) = 1$, $g\left(\dfrac{2\pi}{3}\right) = \dfrac{2}{\sqrt{3}} \Rightarrow$ the absolute

   maximum is $\dfrac{2}{\sqrt{3}}$ at $x = \dfrac{\pi}{3}$ and $x = \dfrac{2\pi}{3}$, and the absolute

   minimum is 1 at $x = \dfrac{\pi}{2}$

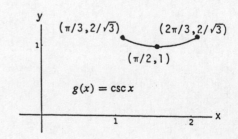

21. $f(t) = 2 - |t| = 2 - \sqrt{t^2} = 2 - (t^2)^{1/2} \Rightarrow f'(t) = -\frac{1}{2}(t^2)^{-1/2}(2t)$

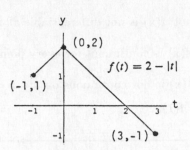

$= -\frac{t}{\sqrt{t^2}} = -\frac{t}{|t|} \Rightarrow$ a critical point at $t = 0$; $f(-1) = 1$,

$f(0) = 2$, $f(3) = -1 \Rightarrow$ the absolute maximum is 2 at $t = 0$

and the absolute minimum is -1 at $t = 3$

23. $f(x) = x^{4/3} \Rightarrow f'(x) = \frac{4}{3}x^{1/3} \Rightarrow$ a critical point at $x = 0$; $f(-1) = 1$, $f(0) = 0$, $f(8) = 16 \Rightarrow$ the absolute

maximum is 16 at $x = 8$ and the absolute minimum is 0 at $x = 0$

25. $g(\theta) = \theta^{3/5} \Rightarrow g'(\theta) = \frac{3}{5}\theta^{-2/5} \Rightarrow$ a critical point at $\theta = 0$; $g(-32) = -8$, $g(0) = 0$, $g(1) = 1 \Rightarrow$ the absolute

maximum is 1 at $\theta = 1$ and the absolute minimum is -8 at $\theta = -32$

27. (a) $x = 0$ is a critical point; $f(-2) = 0$, $f(0) = -4$, $f(2) = 0 \Rightarrow$ local maximum is 0 at $x = \pm 2$ and local

    minimum is -4 at $x = 0$; the absolute maximum is 0 and the absolute minimum is -4

    (b) $x = 0$ is a critical point; $g(-2) = 0$, $g(0) = -4$, $\lim_{x \to 2^-} g(x) = 0 \Rightarrow$ local maximum is 0 at $x = -2$, local

    minimum is -4 at $x = 0$; the absolute maximum is 0 and the absolute minimum is -4

    (c) $x = 0$ is a critical point; $\lim_{x \to -2^+} h(x) = 0$, $h(0) = -4$, $\lim_{x \to 2} h(x) = 0 \Rightarrow$ no local maximum, but there is a

    local minimum of -4 at $x = 0$; the absolute minimum is -4, but there is no absolute maximum

    (d) $x = 0$ is a critical point; $k(-2) = 0$, $k(0) = -4$, $\lim_{x \to \infty} k(x) = \infty \Rightarrow$ local maximum is 0 at $x = -2$ and local

    minimum is -4 at $x = 0$; the absolute minimum is -4, but there is no absolute maximum

    (e) $x = 0$ is not a critical point since 0 is not interior to the domain; $\lim_{x \to 0^+} \ell(x) = -4$ and $\lim_{x \to \infty} \ell(x) = \infty$

    $\Rightarrow$ there are no local or absolute maximum and minimum values

29. Yes, since $f(x) = |x| = \sqrt{x^2} = (x^2)^{1/2} \Rightarrow f'(x) = \frac{1}{2}(x^2)^{-1/2}(2x) = \frac{x}{(x^2)^{1/2}} = \frac{x}{|x|}$ is not defined at $x = 0$. Thus it

is not required that f' be zero at a local extreme point since f' may be undefined there.

31. If $f(c)$ is a local maximum value of f, then $f(x) \le f(c)$ for all x in some open interval (a, b) containing c. Since
f is even, $f(-x) = f(x) \le f(c) = f(-c)$ for all $-x$ in the open interval $(-b, -a)$ containing $-c$. That is, f assumes
a local maximum at the point $-c$. This is also clear from the graph of f because the graph of an even function
is symmetric about the y-axis.

33. If there are no boundary points or critical points the function will have no extreme values in its domain. Such
functions do indeed exist, for example $f(x) = x$ for $-\infty < x < \infty$. (Any other linear function $f(x) = mx + b$
with $m \ne 0$ will do as well.)

3.2 THE MEAN VALUE THEOREM

1. When $f(x) = x^2 + 2x - 1$ for $0 \le x \le 1$, then $\frac{f(1) - f(0)}{1 - 0} = f'(c) \Rightarrow 3 = 2c + 2 \Rightarrow c = \frac{1}{2}$.

3. When $f(x) = x + \frac{1}{x}$ for $\frac{1}{2} \le x \le 2$, then $\frac{f(2) - f(1/2)}{2 - 1/2} = f'(c) \Rightarrow 0 = 1 - \frac{1}{c^2} \Rightarrow c = 1$.

5. Does not; f(x) is not differentiable at x = 0 in (−1, 8).

7. Does; f(x) is continuous for every point of [0, 1] and differentiable for every point in (0, 1).

9. Since f(x) is not continuous on $0 \leq x < 1$, Rolle's Theorem does not apply: $\lim\limits_{x \to 1^-} f(x) = \lim\limits_{x \to 1^-} x = 1$ $\neq 0 = f(1)$.

11. (a) i

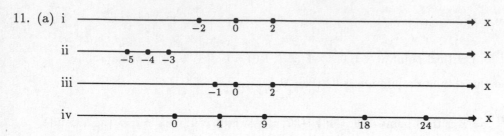

(b) Let r_1 and r_2 be zeros of the polynomial $P(x) = x^n + a_{n-1}x^{n-1} + \ldots + a_1 x + a_0$, then $P(r_1) = P(r_2) = 0$. Since polynomials are everywhere continuous and differentiable, by Rolle's Theorem $P'(r) = 0$ for some r between r_1 and r_2, where $P'(x) = nx^{n-1} + (n-1)a_{n-1}x^{n-2} + \ldots + a_1$.

13. Since f'' exists throughout [a, b] the derivative function f' is continuous there. If f' has more than one zero in [a, b], say $f'(r_1) = f'(r_2) = 0$ for $r_1 \neq r_2$, then by Rolle's Theorem there is a c between r_1 and r_2 such that $f''(c) = 0$, contrary to $f'' > 0$ throughout [a, b]. Therefore f' has at most one zero in [a, b]. The same argument holds if $f'' < 0$ throughout [a, b].

15. Let d(t) represent the distance the automobile traveled in time t. The average speed over $0 \leq t \leq 2$ is $\dfrac{d(2) - d(0)}{2 - 0}$. The Mean Value Theorem says that for some $0 < t_0 < 2$, $d'(t_0) = \dfrac{d(2) - d(0)}{2 - 0}$. The value $d(t_0)$ is the speed of the automobile at time t_0 (which is read on the speedometer).

17. Suppose that $f'(x) \neq 0$ for $0 \leq x \leq 1$. From the Mean Value Theorem there exists a c between 0 and 1 such that $\dfrac{f(1) - f(0)}{1 - 0} = f'(c)$, but since $f'(x) \neq 0$ for $0 \leq x \leq 1$ we have that $f(1) \neq f(0)$.

19. Yes; f' is negative at some point c between a and b. By the Mean Value Theorem, $\dfrac{f(b) - f(a)}{b - a} = f'(c)$ for some point c between a and b. Since $b - a > 0$ and $f(b) < f(a)$, we have $f(b) - f(a) < 0 \Rightarrow f'(c) < 0$.

21. (a) Suppose $x < 1$, then by the Mean Value Theorem $\dfrac{f(x) - f(1)}{x - 1} < 0 \Rightarrow f(x) > f(1)$. Suppose $x > 1$, then by the Mean Value Theorem $\dfrac{f(x) - f(1)}{x - 1} > 0 \Rightarrow f(x) > f(1)$. Therefore $f(x) \geq 1$ for all x since $f(1) = 1$.

(b) Yes. From part (a), $\lim\limits_{x \to 1^-} \dfrac{f(x) - f(1)}{x - 1} \leq 0$ and $\lim\limits_{x \to 1^+} \dfrac{f(x) - f(1)}{x - 1} \geq 0$. Since $f'(1)$ exists, these two one-sided limits are equal and have the value $f'(1) \Rightarrow f'(1) \leq 0$ and $f'(1) \geq 0 \Rightarrow f'(1) = 0$.

23. $f'(x) = [\cos x \sin(x + 2) + \sin x \cos(x + 2)] - 2 \sin(x + 1) \cos(x + 1) = \sin(x + x + 2) - \sin 2(x + 1)$ $= \sin(2x + 2) - \sin(2x + 2) = 0$. Therefore, the function has the constant value $f(0) = -\sin^2 1 \approx -0.7081$ which explains why the graph is a horizontal line.

25. (a) $g(x) = \frac{1}{x} \Rightarrow g'(x) = -\frac{1}{x^2} \Rightarrow g'(x) < 0$ for all $x \neq 0 \Rightarrow g(x)$ is always decreasing on every interval in its domain.

   (b) No interval in the domain of $g(x)$ can contain both positive and negative numbers since 0 does not belong to the domain of g. Therefore g is not defined on an interval containing both $x = -1$ and $x = 1$.

27. $f'(x) = \left(1 + x^4 \cos x\right)^{-1} \Rightarrow f''(x) = -\left(1 + x^4 \cos x\right)^{-2}\left(4x^3 \cos x - x^4 \sin x\right)$

   $= -x^3\left(1 + x^4 \cos x\right)^{-2}(4 \cos x - x \sin x) < 0$ for $0 \leq x \leq 0.1 \Rightarrow f'(x)$ is decreasing when $0 \leq x \leq 0.1$

   $\Rightarrow$ min $f' \approx 0.9999$ and max $f' = 1$. Now we have $0.9999 \leq \dfrac{f(0.1) - 1}{0.1} \leq 1 \Rightarrow 0.09999 \leq f(0.1) - 1 \leq 0.1$

   $\Rightarrow 1.09999 \leq f(0.1) \leq 1.1$.

29. The conclusion of the Mean Value Theorem yields $\dfrac{\frac{1}{b} - \frac{1}{a}}{b - a} = -\dfrac{1}{c^2} \Rightarrow c^2\left(\dfrac{a - b}{ab}\right) = a - b \Rightarrow c = \sqrt{ab}$.

31. By Corollary 1, $f'(x) = 0$ for all $x \Rightarrow f(x) = C$, where C is a constant. Since $f(-1) = 3$ we have $C = 3$

   $\Rightarrow f(x) = 3$ for all x.

33. $g(x) = x^2 \Rightarrow g'(x) = 2x = f'(x)$ for all x. By Corollary 2, $f(x) = g(x) + C$.

   (a) $f(0) = 0 \Rightarrow 0 = g(0) + C = 0 + C \Rightarrow C = 0 \Rightarrow f(x) = x^2 \Rightarrow f(2) = 4$

   (b) $f(1) = 0 \Rightarrow 0 = g(1) + C = 1 + C \Rightarrow C = -1 \Rightarrow f(x) = x^2 - 1 \Rightarrow f(2) = 3$

   (c) $f(-2) = 3 \Rightarrow 3 = g(-2) + C \Rightarrow 3 = 4 + C \Rightarrow C = -1 \Rightarrow f(x) = x^2 - 1 \Rightarrow f(2) = 3$

35. (a) $y = \frac{x^2}{2} + C$ (b) $y = \frac{x^3}{3} + C$ (c) $y = \frac{x^4}{4} + C$

37. (a) $y' = -x^{-2} \Rightarrow y = \frac{1}{x} + C$ (b) $y = x + \frac{1}{x} + C$ (c) $y = 5x + \frac{1}{x} + C$

39. (a) $y = -\frac{1}{2} \cos 2t + C$ (b) $y = 2 \sin \frac{t}{2} + C$

   (c) $y = -\frac{1}{2} \cos 2t + 2 \sin \frac{t}{2} + C$

41. $f(x) = x^2 - x + C; \ 0 = f(0) = 0^2 - 0 + C \Rightarrow C = 0 \Rightarrow f(x) = x^2 - x$

43. $r(\theta) = 8\theta + \cot \theta + C; \ 0 = r\left(\frac{\pi}{4}\right) = 8\left(\frac{\pi}{4}\right) + \cot\left(\frac{\pi}{4}\right) + C \Rightarrow 0 = 2\pi + 1 + C \Rightarrow C = -2\pi - 1$

   $\Rightarrow r(\theta) = 8\theta + \cot \theta - 2\pi - 1$

45. With $f(-2) = 11 > 0$ and $f(-1) = -1 < 0$ we conclude from the Intermediate Value Theorem that

   $f(x) = x^4 + 3x + 1$ has at least one zero between -2 and -1. Then $-2 < x < -1 \Rightarrow -8 < x^3 < -1$

   $\Rightarrow -32 < 4x^3 < -4 \Rightarrow -29 < 4x^3 + 3 < -1 \Rightarrow f'(x) < 0$ for $-2 < x < -1 \Rightarrow f(x)$ is decreasing on $[-2, -1]$

   $\Rightarrow f(x) = 0$ has exactly one solution in the interval $(-2, -1)$.

47. $g(t) = \sqrt{t} + \sqrt{t + 1} - 4 \Rightarrow g'(t) = \dfrac{1}{2\sqrt{t}} + \dfrac{1}{2\sqrt{t + 1}} > 0 \Rightarrow g(t)$ is increasing for t in $(0, \infty)$; $g(3) = \sqrt{3} - 2 < 0$

   and $g(15) = \sqrt{15} > 0 \Rightarrow g(t)$ has exactly one zero in $(0, \infty)$.

49. $r(\theta) = \theta + \sin^2\left(\frac{\theta}{3}\right) - 8 \Rightarrow r'(\theta) = 1 + \frac{2}{3}\sin\left(\frac{\theta}{3}\right)\cos\left(\frac{\theta}{3}\right) = 1 + \frac{1}{3}\sin\left(\frac{2\theta}{3}\right) > 0$ on $(-\infty, \infty) \Rightarrow r(\theta)$ is

    increasing on $(-\infty, \infty)$; $r(0) = -8$ and $r(8) = \sin^2\left(\frac{8}{3}\right) > 0 \Rightarrow r(\theta)$ has exactly one zero in $(-\infty, \infty)$.

51. $r(\theta) = \sec\theta - \frac{1}{\theta^3} + 5 \Rightarrow r'(\theta) = (\sec\theta)(\tan\theta) + \frac{3}{\theta^4} > 0$ on $\left(0, \frac{\pi}{2}\right) \Rightarrow r(\theta)$ is increasing on $\left(0, \frac{\pi}{2}\right)$;

    $r(0.1) \approx -994$ and $r(1.57) \approx 1260.5 \Rightarrow r(\theta)$ has exactly one zero in $\left(0, \frac{\pi}{2}\right)$.

3.3 THE FIRST DERIVATIVE TEST FOR LOCAL EXTREME VALUES

1. (a) $f'(x) = x(x-1) \Rightarrow$ critical points at 0 and 1
   (b) $f' = +++ \mid --- \mid +++ \Rightarrow$ increasing on $(-\infty, 0)$ and $(1, \infty)$, decreasing on $(0, 1)$
       ${}_{0}{}_{1}$
   (c) Local maximum at $x = 0$ and a local minimum at $x = 1$

3. (a) $f'(x) = (x-1)^2(x+2) \Rightarrow$ critical points at -2 and 1
   (b) $f' = --- \mid +++ \mid +++ \Rightarrow$ increasing on $(-2, 1)$ and $(1, \infty)$, decreasing on $(-\infty, -2)$
       ${}_{-2}{}_{1}$
   (c) No local maximum and a local minimum at $x = -2$

5. (a) $f'(x) = (x-1)(x+2)(x-3) \Rightarrow$ critical points at -2, 1 and 3
   (b) $f' = --- \mid +++ \mid --- \mid +++ \Rightarrow$ increasing on $(-2, 1)$ and $(3, \infty)$, decreasing on $(-\infty, -2)$ and $(1, 3)$
       ${}_{-2}{}_{1}{}_{3}$
   (c) Local maximum at $x = 1$, local minima at $x = -2$ and $x = 3$

7. (a) $f'(x) = x^{-1/3}(x+2) \Rightarrow$ critical points at -2 and 0
   (b) $f' = +++ \mid ---)(+++ \Rightarrow$ increasing on $(-\infty, -2)$ and $(0, \infty)$, decreasing on $(-2, 0)$
       ${}_{-2}{}_{0}$
   (c) Local maximum at $x = -2$, local minimum at $x = 0$

9. (a) $g(t) = -t^2 - 3t + 3 \Rightarrow g'(t) = -2t - 3 \Rightarrow$ a critical point at $t = -\frac{3}{2}$; $g' = +++ \mid ---$, increasing on
    ${}_{-3/2}$
    $\left(-\infty, -\frac{3}{2}\right)$, decreasing on $\left(-\frac{3}{2}, \infty\right)$
   (b) local maximum value of $g\left(-\frac{3}{2}\right) = \frac{21}{4}$ at $t = -\frac{3}{2}$
   (c) absolute maximum is $\frac{21}{4}$ at $t = -\frac{3}{2}$

11. (a) $h(x) = -x^3 + 2x^2 \Rightarrow h'(x) = -3x^2 + 4x = x(4 - 3x) \Rightarrow$ critical points at $x = 0, \frac{4}{3}$
    $\Rightarrow h' = --- \mid +++ \mid ---$, increasing on $\left(0, \frac{4}{3}\right)$, decreasing on $(-\infty, 0)$ and $\left(\frac{4}{3}, \infty\right)$
    ${}_{0}{}_{4/3}$
    (b) local maximum value of $h\left(\frac{4}{3}\right) = \frac{32}{27}$ at $x = \frac{4}{3}$; local minimum value of $h(0) = 0$ at $x = 0$
    (c) no absolute extrema

13. (a) $f(\theta) = 3\theta^2 - 4\theta^3 \Rightarrow f'(\theta) = 6\theta - 12\theta^2 = 6\theta(1 - 2\theta) \Rightarrow$ critical points at $\theta = 0, \frac{1}{2} \Rightarrow f' = --- \mid +++ \mid ---$,
    $\phantom{f(\theta) = 3\theta^2 - 4\theta^3 \Rightarrow f'(\theta) = 6\theta - 12\theta^2 = 6\theta(1 - 2\theta) \Rightarrow critical points at \theta = 0, \frac{1}{2} \Rightarrow f' = ---}{}_{0}{}_{1/2}$
    increasing on $\left(0, \frac{1}{2}\right)$, decreasing on $(-\infty, 0)$ and $\left(\frac{1}{2}, \infty\right)$
    (b) a local maximum is $f\left(\frac{1}{2}\right) = \frac{1}{4}$ at $\theta = \frac{1}{2}$, a local minimum is $f(0) = 0$ at $\theta = 0$

(c) no absolute extrema

15. (a) $f(r) = 3r^3 + 16r \Rightarrow f'(r) = 9r^2 + 16 \Rightarrow$ no critical points $\Rightarrow f' = +++++$, increasing on $(-\infty, \infty)$, never decreasing

(b) no local extrema

(c) no absolute extrema

17. (a) $f(x) = x^4 - 8x^2 + 16 \Rightarrow f'(x) = 4x^3 - 16x = 4x(x+2)(x-2) \Rightarrow$ critical points at $x = 0$ and $x = \pm 2$

$\Rightarrow f' = \underset{-2}{---|} +++ \underset{0}{|---} \underset{2}{|} +++$, increasing on $(-2, 0)$ and $(2, \infty)$, decreasing on $(-\infty, -2)$ and $(0, 2)$

(b) a local maximum is $f(0) = 16$ at $x = 0$, local minima are $f(\pm 2) = 0$ at $x = \pm 2$

(c) no absolute maximum; absolute minimum is 0 at $x = \pm 2$

19. (a) $H(t) = \frac{3}{2}t^4 - t^6 \Rightarrow H'(t) = 6t^3 - 6t^5 = 6t^3(1+t)(1-t) \Rightarrow$ critical points at $t = 0, \pm 1$

$\Rightarrow H' = +++ \underset{-1}{|} --- \underset{0}{|} +++ \underset{1}{|} ---$, increasing on $(-\infty, -1)$ and $(0, 1)$, decreasing on $(-1, 0)$ and $(1, \infty)$

(b) the local maxima are $H(-1) = \frac{1}{2}$ at $t = -1$ and $H(1) = \frac{1}{2}$ at $t = 1$, the local minimum is $H(0) = 0$ at $t = 0$

(c) absolute maximum is $\frac{1}{2}$ at $t = \pm 1$; no absolute minimum

21. (a) $g(x) = x\sqrt{8 - x^2} = x(8 - x^2)^{1/2} \Rightarrow g'(x) = (8 - x^2)^{1/2} + x\left(\frac{1}{2}\right)(8 - x^2)^{-1/2}(-2x) = \dfrac{2(2 - x)(2 + x)}{\sqrt{(2\sqrt{2} - x)(2\sqrt{2} + x)}}$

$\Rightarrow$ critical points at $x = \pm 2, \pm 2\sqrt{2} \Rightarrow g' = (\underset{-2\sqrt{2}}{---|} \underset{-2}{+++|} \underset{2}{---|} \underset{2\sqrt{2}}{})$, increasing on $(-2, 2)$, decreasing on $(-2\sqrt{2}, -2)$ and $(2, 2\sqrt{2})$

(b) local maxima are $g(2) = 4$ at $x = 2$ and $g(-2\sqrt{2}) = 0$ at $x = -2\sqrt{2}$, local minima are $g(-2) = -4$ at $x = -2$ and $g(2\sqrt{2}) = 0$ at $x = 2\sqrt{2}$

(c) absolute maximum is 4 at $x = 2$; absolute minimum is -4 at $x = -2$

23. (a) $f(x) = \dfrac{x^2 - 3}{x - 2} \Rightarrow f'(x) = \dfrac{2x(x - 2) - (x^2 - 3)(1)}{(x - 2)^2} = \dfrac{(x - 3)(x - 1)}{(x - 2)^2} \Rightarrow$ critical points at $x = 1, 3$

$\Rightarrow f' = +++ \underset{1}{|} ---)(\underset{2}{} --- \underset{3}{|} +++$, increasing on $(-\infty, 1)$ and $(3, \infty)$, decreasing on $(1, 2)$ and $(2, 3)$, discontinuous at $x = 2$

(b) a local maximum is $f(1) = 2$ at $x = 1$, a local minimum is $f(3) = 6$ at $x = 3$

(c) no absolute extrema

25. (a) $f(x) = x^{1/3}(x + 8) = x^{4/3} + 8x^{1/3} \Rightarrow f'(x) = \frac{4}{3}x^{1/3} + \frac{8}{3}x^{-2/3} = \dfrac{4(x + 2)}{3x^{2/3}} \Rightarrow$ critical points at $x = 0, -2$

$\Rightarrow f' = --- \underset{-2}{|} +++)(\underset{0}{} +++$, increasing on $(-2, 0) \cup (0, \infty)$, decreasing on $(-\infty, -2)$

(b) no local maximum, a local minimum is $f(-2) = -6\sqrt[3]{2} \approx -7.56$ at $x = -2$

(c) no absolute maximum; absolute minimum is $-6\sqrt[3]{2}$ at $x = -2$

27. (a) $h(x) = x^{1/3}(x^2 - 4) = x^{7/3} - 4x^{1/3} \Rightarrow h'(x) = \frac{7}{3}x^{4/3} - \frac{4}{3}x^{-2/3} = \frac{(\sqrt{7}x + 2)(\sqrt{7}x - 2)}{3\sqrt[3]{x^2}} \Rightarrow$ critical points at

$x = 0, \frac{\pm 2}{\sqrt{7}} \Rightarrow h' = +++ \mid \underset{-2/\sqrt{7}}{---})(\underset{0}{---} \mid \underset{2/\sqrt{7}}{+++}$, increasing on $\left(-\infty, \frac{-2}{\sqrt{7}}\right)$ and $\left(\frac{2}{\sqrt{7}}, \infty\right)$, decreasing on

$\left(\frac{-2}{\sqrt{7}}, 0\right)$ and $\left(0, \frac{2}{\sqrt{7}}\right)$

(b) local maximum is $h\left(\frac{-2}{\sqrt{7}}\right) = \frac{24\sqrt[3]{2}}{7^{7/6}} \approx 3.12$ at $x = \frac{-2}{\sqrt{7}}$, the local minimum is $h\left(\frac{2}{\sqrt{7}}\right) = -\frac{24\sqrt[3]{2}}{7^{7/6}} \approx -3.12$

(c) no absolute extrema

29. (a) $f(x) = 2x - x^2 \Rightarrow f'(x) = 2 - 2x = 2(1 - x) \Rightarrow$ a critical point at $x = 1 \Rightarrow f' = +++ \mid \underset{1}{} \underset{2}{---}]$ and $f(1) = 1$,

$f(2) = 0 \Rightarrow$ a local maximum is 1 at $x = 1$, a local minimum is 0 at $x = 2$

(b) absolute maximum is 1 at $x = 1$; no absolute minimum

31. (a) $g(x) = x^2 - 4x + 4 \Rightarrow g'(x) = 2x - 4 = 2(x - 2) \Rightarrow$ a critical point at $x = 2 \Rightarrow g' = [\underset{1}{---} \mid \underset{2}{+++}$ and

$g(1) = 1$, $g(2) = 0 \Rightarrow$ a local maximum is 1 at $x = 1$, a local minimum is $g(2) = 0$ at $x = 2$

(b) no absolute maximum; absolute minimum is 0 at $x = 2$

33. (a) $f(t) = 12t - t^3 \Rightarrow f'(t) = 12 - 3t^2 = 3(2 + t)(2 - t) \Rightarrow$ critical points at $t = \pm 2 \Rightarrow f' = [\underset{-3}{---} \mid \underset{-2}{+++} \mid \underset{2}{---}$

and $f(-3) = -9$, $f(-2) = -16$, $f(2) = 16 \Rightarrow$ local maxima are -9 at $t = -3$ and 16 at $t = -2$, a local

minimum is -16 at $t = -2$

(b) absolute maximum is 16 at $t = 2$; no absolute minimum

35. (a) $h(x) = \frac{x^3}{3} - 2x^2 + 4x \Rightarrow h'(x) = x^2 - 4x + 4 = (x - 2)^2 \Rightarrow$ a critical point at $x = 2 \Rightarrow h' = [\underset{0}{+++} \mid \underset{2}{+++}$ and

$h(0) = 0 \Rightarrow$ no local maximum, a local minimum is 0 at $x = 0$

(b) no absolute maximum; absolute minimum is 0 at $x = 0$

37. (a) $f(x) = \frac{x}{2} - 2\sin\left(\frac{x}{2}\right) \Rightarrow f'(x) = \frac{1}{2} - \cos\left(\frac{x}{2}\right)$, $f'(x) = 0 \Rightarrow \cos\left(\frac{x}{2}\right) = \frac{1}{2} \Rightarrow$ a critical point at $x = \frac{2\pi}{3}$

$\Rightarrow f' = [\underset{0}{---} \mid \underset{2\pi/3}{+++} \underset{2\pi}{]}$ and $f(0) = 0$, $f\left(\frac{2\pi}{3}\right) = \frac{\pi}{3} - \sqrt{3}$, $f(2\pi) = \pi \Rightarrow$ local maxima are 0 at $x = 0$ and π

at $x = 2\pi$, a local minimum is $\frac{\pi}{3} - \sqrt{3}$ at $x = \frac{2\pi}{3}$

(b) The graph of f rises when $f' > 0$, falls when $f' < 0$,
and has a local minimum value at the point where f'
changes from negative to positive.

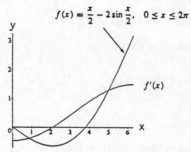

$f(x) = \frac{x}{2} - 2\sin\frac{x}{2}$, $0 \le x \le 2\pi$

39. (a) $f(x) = \csc^2 x - 2\cot x \Rightarrow f'(x) = 2(\csc x)(-\csc x)(\cot x) - 2(-\csc^2 x) = -2(\csc^2 x)(\cot x - 1) \Rightarrow$ a critical

point at $x = \frac{\pi}{4} \Rightarrow f' = (\underset{0}{---} \mid \underset{\pi/4}{+++} \underset{\pi}{)}$ and $f\left(\frac{\pi}{4}\right) = 0 \Rightarrow$ no local maximum, a local minimum is 0 at $x = \frac{\pi}{4}$

(b) The graph of f rises when f' > 0, falls when f' < 0, and has a local minimum value at the point where f' = 0 and the values of f' change from negative to positive. The graph of f steepens as $f'(x) \to \pm \infty$.

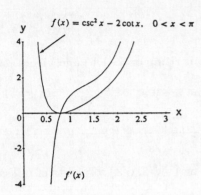

41. $h(\theta) = 3 \cos\left(\frac{\theta}{2}\right) \Rightarrow h'(\theta) = -\frac{3}{2} \sin\left(\frac{\theta}{2}\right) \Rightarrow h' = \underset{0}{[}\text{---}\underset{2\pi}{]}, (0,3)$ and $(2\pi, -3) \Rightarrow$ a local maximum is 3 at $\theta = 0$,

a local minimum is -3 at $\theta = 2\pi$

43. (a)   (b)   (c)   (d)

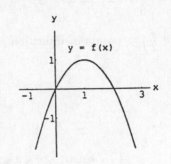
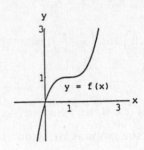
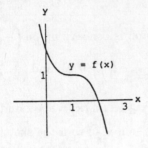

45. (a)   (b)

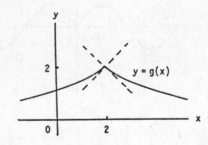

47. $f(x) = x^3 - 3x + 2 \Rightarrow f'(x) = 3x^2 - 3 = 3(x-1)(x+1) \Rightarrow f' = +++ \underset{-1}{|} --- \underset{1}{|} +++ \Rightarrow$ rising since $f'(x) > 0$

3.4 GRAPHING WITH y' AND y''

1. $y = \frac{x^3}{3} - \frac{x^2}{2} - 2x + \frac{1}{3} \Rightarrow y' = x^2 - x - 2 = (x-2)(x+1) \Rightarrow y'' = 2x - 1 = 2\left(x - \frac{1}{2}\right)$. The graph is rising on

$(-\infty, -1)$ and $(2, \infty)$, falling on $(-1, 2)$, concave up on $\left(\frac{1}{2}, \infty\right)$ and concave down on $\left(-\infty, \frac{1}{2}\right)$. Consequently,

a local maximum is $\frac{3}{2}$ at $x = -1$, a local minimum is -3 at $x = 2$, and $\left(\frac{1}{2}, -\frac{3}{4}\right)$ is a point of inflection.

3. $y = \frac{3}{4}(x^2 - 1)^{2/3} \Rightarrow y' = \left(\frac{3}{4}\right)\left(\frac{2}{3}\right)(x^2 - 1)^{-1/3}(2x) = x(x^2 - 1)^{-1/3}$, $y' = ---)(+++|---)(+++ \Rightarrow$ the

$$$-101$

graph is rising on $(-1, 0)$ and $(1, \infty)$, falling on $(-\infty, -1)$ and $(0, 1) \Rightarrow$ a local maximum is $\frac{3}{4}$ at $x = 0$, local

minima are 0 at $x = \pm 1$; $y'' = (x^2 - 1)^{-1/3} + (x)\left(-\frac{1}{3}\right)(x^2 - 1)^{-4/3}(2x) = \dfrac{x^2 - 3}{3\sqrt[3]{(x^2 - 1)^4}}$,

$y'' = +++|---)(---)(---|+++ \Rightarrow$ the graph is concave up on $\left(-\infty, -\sqrt{3}\right)$ and $\left(\sqrt{3}, \infty\right)$, concave

$-\sqrt{3}-11\sqrt{3}$

down on $\left(-\sqrt{3}, \sqrt{3}\right) \Rightarrow$ points of inflection at $\left(\pm\sqrt{3}, \dfrac{3}{2\sqrt[3]{2}}\right)$

5. $y = x + \sin 2x \Rightarrow y' = 1 + 2\cos 2x$, $y' = [\,---|+++|---\,] \Rightarrow$ the graph is rising on $\left(-\frac{\pi}{3}, \frac{\pi}{3}\right)$, falling on

$-2\pi/3-\pi/3\pi/32\pi/3$

$\left(-\frac{2\pi}{3}, -\frac{\pi}{3}\right)$ and $\left(\frac{\pi}{3}, \frac{2\pi}{3}\right) \Rightarrow$ local maxima are $-\frac{2\pi}{3} + \frac{\sqrt{3}}{2}$ at $x = -\frac{2\pi}{3}$ and $\frac{\pi}{3} + \frac{\sqrt{3}}{2}$ at $x = \frac{\pi}{3}$, local minima are

$-\frac{\pi}{3} - \frac{\sqrt{3}}{2}$ at $x = -\frac{\pi}{3}$ and $\frac{2\pi}{3} - \frac{\sqrt{3}}{2}$ at $x = \frac{2\pi}{3}$; $y'' = -4\sin 2x$, $y'' = [\,---|+++|---|+++\,] \Rightarrow$ the

$-2\pi/3-\pi/20\pi/22\pi/3$

graph is concave up on $\left(-\frac{\pi}{2}, 0\right)$ and $\left(\frac{\pi}{2}, \frac{2\pi}{3}\right)$, concave down on $\left(-\frac{2\pi}{3}, -\frac{\pi}{2}\right)$ and $\left(0, \frac{\pi}{2}\right) \Rightarrow$ points of inflection at

$\left(-\frac{\pi}{2}, -\frac{\pi}{2}\right)$, $(0, 0)$, and $\left(\frac{\pi}{2}, \frac{\pi}{2}\right)$

7. If $x \geq 0$, $\sin |x| = \sin x$ and if $x < 0$, $\sin |x| = \sin(-x)$

$= -\sin x$. From the sketch the graph is rising on

$\left(-\frac{3\pi}{2}, -\frac{\pi}{2}\right)$, $\left(0, \frac{\pi}{2}\right)$ and $\left(\frac{3\pi}{2}, 2\pi\right)$, falling on $\left(-2\pi, -\frac{3\pi}{2}\right)$,

$\left(-\frac{\pi}{2}, 0\right)$ and $\left(\frac{\pi}{2}, \frac{3\pi}{2}\right)$; local maxima are 1 at $x = \pm\frac{3\pi}{2}$

and 0 at $x = \pm 2\pi$; local minima are -1 at $x = \pm\frac{3\pi}{2}$ and

0 at $x = 0$; concave up on $(-2\pi, -\pi)$ and $(\pi, 2\pi)$, and concave

down on $(-\pi, 0)$ and $(0, \pi) \Rightarrow$ points of inflection are $(-\pi, 0)$

and $(\pi, 0)$

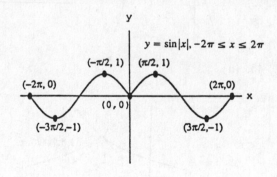

9. When $y = x^2 - 4x + 3$, then $y' = 2x - 4 = 2(x - 2)$ and

$y'' = 2$. The curve rises on $(2, \infty)$ and falls on $(-\infty, 2)$.

At $x = 2$ there is a minimum. Since $y'' > 0$, the curve is

concave up for all x.

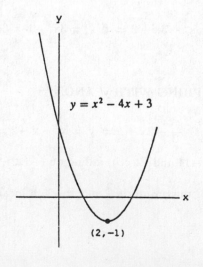

11. When $y = x^3 - 3x + 3$, then $y' = 3x^2 - 3 = 3(x-1)(x+1)$ and $y'' = 6x$. The curve rises on $(-\infty, -1) \cup (1, \infty)$ and falls on $(-1, 1)$. At $x = -1$ there is a local maximum and at $x = 1$ a local minimum. The curve is concave down on $(-\infty, 0)$ and concave up on $(0, \infty)$. There is a point of inflection at $x = 0$.

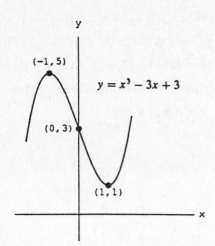

13. When $y = -2x^3 + 6x^2 - 3$, then $y' = -6x^2 + 12x$ $= -6x(x-2)$ and $y'' = -12x + 12 = -12(x-1)$. The curve rises on $(0, 2)$ and falls on $(-\infty, 0)$ and $(2, \infty)$. At $x = 0$ there is a local minimum and at $x = 2$ a local maximum. The curve is concave up on $(-\infty, 1)$ and concave down on $(1, \infty)$. At $x = 1$ there is a point of inflection.

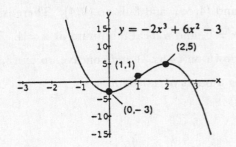

15. When $y = (x-2)^3 + 1$, then $y' = 3(x-2)^2$ and $y'' = 6(x-2)$. The curve never falls and there are no local extrema. The curve is concave down on $(-\infty, 2)$ and concave up on $(2, \infty)$. At $x = 2$ there is a point of inflection.

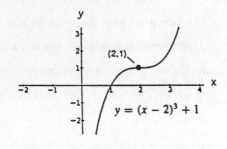

17. When $y = x^4 - 2x^2$, then $y' = 4x^3 - 4x = 4x(x+1)(x-1)$ and $y'' = 12x^2 - 4 = 12\left(x + \dfrac{1}{\sqrt{3}}\right)\left(x - \dfrac{1}{\sqrt{3}}\right)$. The curve rises on $(-1, 0)$ and $(1, \infty)$ and falls on $(-\infty, -1)$ and $(0, 1)$. At $x = \pm 1$ there are local minima and at $x = 0$ a local maximum. The curve is concave up on $\left(-\infty, -\dfrac{1}{\sqrt{3}}\right)$ and $\left(\dfrac{1}{\sqrt{3}}, \infty\right)$ and concave down on $\left(-\dfrac{1}{\sqrt{3}}, \dfrac{1}{\sqrt{3}}\right)$. At $x = \dfrac{\pm 1}{\sqrt{3}}$ there are points of inflection.

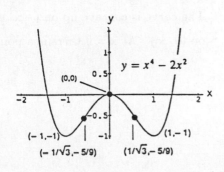

19. When $y = 4x^3 - x^4$, then $y' = 12x^2 - 4x^3 = 4x^2(3-x)$ and $y'' = 24x - 12x^2 = 12x(2-x)$. The curve rises on $(-\infty, 3)$ and falls on $(3, \infty)$. At $x = 3$ there is a local maximum, but there is no local minimum. There are points of inflection at $x = 0$ and $x = 2$.

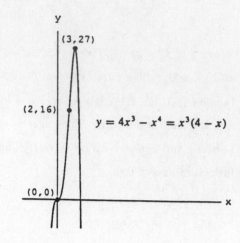

21. When $y = x^5 - 5x^4$, then $y' = 5x^4 - 20x^3 = 5x^3(x-4)$ and $y'' = 20x^3 - 60x^2 = 20x^2(x-3)$. The curve rises on $(-\infty, 0)$ and $(4, \infty)$, and falls on $(0, 4)$. There is a local maximum at $x = 0$, and a local minimum at $x = 4$. The curve is concave down on $(-\infty, 3)$ and concave up on $(3, \infty)$. At $x = 3$ there is a point of inflection.

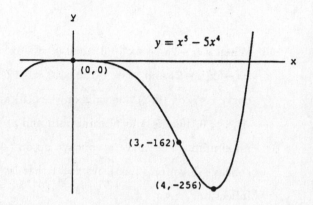

23. When $y = x + \sin x$, then $y' = 1 + \cos x$ and $y'' = -\sin x$. The curve rises on $(0, 2\pi)$. At $x = 0$ there is a local and absolute minimum and at $x = 2\pi$ there is a local and absolute maximum. The curve is concave down on $(0, \pi)$ and concave up on $(\pi, 2\pi)$. At $x = \pi$ there is a point of inflection.

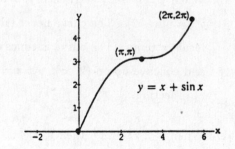

25. When $y = x^{1/5}$, then $y' = \frac{1}{5}x^{-4/5}$ and $y'' = -\frac{4}{25}x^{-9/5}$. The curve rises on $(-\infty, \infty)$ and there are no extrema. The curve is concave up on $(-\infty, 0)$ and concave down on $(0, \infty)$. At $x = 0$ there is a point of inflection.

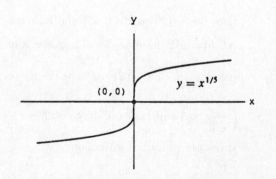

27. When $y = x^{2/5}$, then $y' = \frac{2}{5}x^{-3/5}$ and $y'' = -\frac{6}{25}x^{-8/5}$.
The curve is rising on $(0,\infty)$ and falling on $(-\infty,0)$. At $x = 0$ there is a local and absolute minimum. There is no local or absolute maximum. The curve is concave down on $(-\infty,0)$ and $(0,\infty)$. There are no points of inflection, but a cusp exists at $x = 0$.

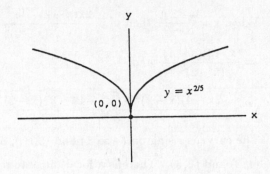

29. When $y = 2x - 3x^{2/3}$, then $y' = 2 - 2x^{-1/3}$ and $y'' = \frac{2}{3}x^{-4/3}$. The curve is rising on $(-\infty,0)$ and $(1,\infty)$, and falling on $(0,1)$. There is a local maximum at $x = 0$ and a local minimum at $x = 1$. The curve is concave up on $(-\infty,0)$ and $(0,\infty)$. There are no points of inflection, but a cusp exists at $x = 0$.

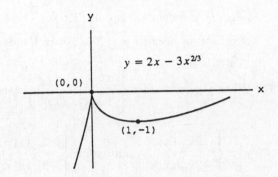

31. When $y = x^{2/3}\left(\frac{5}{2} - x\right) = \frac{5}{2}x^{2/3} - x^{5/3}$, then
$y' = \frac{5}{3}x^{-1/3} - \frac{5}{3}x^{2/3} = \frac{5}{3}x^{-1/3}(1 - x)$ and
$y'' = -\frac{5}{9}x^{-4/3} - \frac{10}{9}x^{-1/3} = -\frac{5}{9}x^{-4/3}(1 + 2x)$.
The curve is rising on $(0,1)$ and falling on $(-\infty,0)$ and $(1,\infty)$. There is a local minimum at $x = 0$ and a local maximum at $x = 1$. The curve is concave up on $\left(-\infty,-\frac{1}{2}\right)$ and concave down on $\left(-\frac{1}{2},0\right)$ and $(0,\infty)$. There is a point of inflection at $x = -\frac{1}{2}$ and a cusp at $x = 0$.

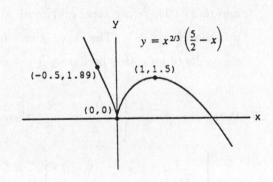

33. When $y = x\sqrt{8 - x^2} = x\left(8 - x^2\right)^{1/2}$, then
$y' = \left(8 - x^2\right)^{1/2} + (x)\left(\frac{1}{2}\right)\left(8 - x^2\right)^{-1/2}(-2x)$
$= \left(8 - x^2\right)^{-1/2}\left(8 - 2x^2\right) = \frac{2(2 - x)(2 + x)}{\sqrt{\left(2\sqrt{2} + x\right)\left(2\sqrt{2} - x\right)}}$ and
$y'' = \left(-\frac{1}{2}\right)\left(8 - x^2\right)^{-3/2}(-2x)\left(8 - 2x^2\right) + \left(8 - x^2\right)^{-1/2}(-4x)$
$= \frac{2x\left(x^2 - 12\right)}{\sqrt{\left(8 - x^2\right)^3}}$. The curve is rising on $(-2,2)$, and falling

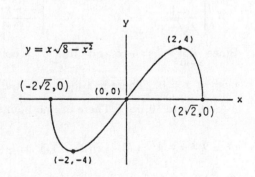

on $\left(-2\sqrt{2},-2\right)$ and $\left(2,2\sqrt{2}\right)$. There are local minima at $x = -2$ and $x = 2\sqrt{2}$, and local maxima at $x = -2\sqrt{2}$ and $x = 2$. The curve is concave up on $\left(-2\sqrt{2},0\right)$ and concave down on $\left(0,2\sqrt{2}\right)$. There is a point of inflection at $x = 0$.

35. When $y = \dfrac{x^2-3}{x-2}$, then $y' = \dfrac{2x(x-2)-\left(x^2-3\right)(1)}{(x-2)^2}$

$= \dfrac{(x-3)(x-1)}{(x-2)^2}$ and

$y'' = \dfrac{(2x-4)(x-2)^2-\left(x^2-4x+3\right)(x-2)}{(x-2)^4} = \dfrac{2}{(x-2)^3}$.

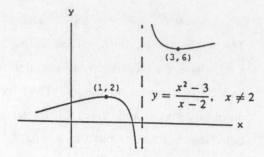

The curve is rising on $(-\infty,1)$ and $(3,\infty)$, and falling on

$(1,2)$ and $(2,3)$. There is a local maximum at $x=1$ and a

local minimum at $x=3$. The curve is concave down on

$(-\infty,2)$ and concave up on $(2,\infty)$. There are no points

of inflection because $x=2$ is not in the domain.

37. When $y = \left|x^2-1\right| = \begin{cases} x^2-1, & |x|\geq 1 \\ 1-x^2, & |x|<1 \end{cases}$, then

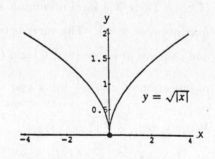

$y' = \begin{cases} 2x, & |x|>1 \\ -2x, & |x|<1 \end{cases}$ and $y'' = \begin{cases} 2, & |x|>1 \\ -2, & |x|<1 \end{cases}$. The

curve rises on $(-1,0)$ and $(1,\infty)$ and falls on $(-\infty,-1)$

and $(0,1)$. There is a local maximum at $x=0$ and local

minima at $x=\pm 1$. The curve is concave up on $(-\infty,-1)$ and $(1,\infty)$, and concave down on $(-1,1)$. There are

no points of inflection because y is not differentiable at $x=\pm 1$ (so there is no tangent line at those points).

39. When $y = \sqrt{|x|} = \begin{cases} \sqrt{x}, & x\geq 0 \\ \sqrt{-x}, & x<0 \end{cases}$, then

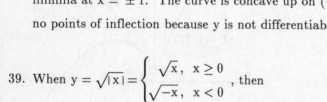

$y' = \begin{cases} \dfrac{1}{2\sqrt{x}}, & x>0 \\ \dfrac{-1}{2\sqrt{-x}}, & x<0 \end{cases}$ and $y'' = \begin{cases} \dfrac{-x^{-3/2}}{4}, & x>0 \\ \dfrac{-(-x)^{-3/2}}{4}, & x<0 \end{cases}$.

Since $\lim\limits_{x\to 0^-} y' = -\infty$ and $\lim\limits_{x\to 0^+} y' = \infty$ there is a

cusp at $x=0$. There is a local minimum at $x=0$, but no local maximum. The curve is concave down on

$(-\infty,0)$ and $(0,\infty)$. There are no points of inflection.

41. $y' = 2+x-x^2 = (1+x)(2-x)$, $y' = ---\underset{-1}{|}+++\underset{2}{|}---$

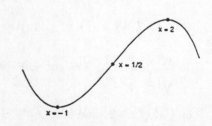

$\Rightarrow$ rising on $(-1,2)$, falling on $(-\infty,-1)$ and $(2,\infty)\Rightarrow$ there is

a local maximum at $x=2$ and a local minimum at $x=-1$;

$y'' = 1-2x$, $y'' = +++\underset{1/2}{|}--- \Rightarrow$ concave up on $\left(-\infty,\tfrac{1}{2}\right)$,

concave down on $\left(\tfrac{1}{2},\infty\right)\Rightarrow$ a point of inflection at $x=\tfrac{1}{2}$

43. $y' = x(x-3)^2$, $y' = ---|+++|+++ \Rightarrow$ rising on $(0, \infty)$,
     $\quad\quad\quad\quad\quad\quad\quad\quad\quad\;\; 0\quad\;\; 3$

    falling on $(-\infty, 0) \Rightarrow$ no local maximum, but there is a local

    minimum at $x = 0$; $y'' = (x-3)^2 + x(2)(x-3)$

    $= 3(x-3)(x-1)$, $y'' = +++|---|+++ \Rightarrow$ concave up
    $\quad\quad\quad\quad\quad\quad\quad\quad\quad\quad\;\; 1\quad\;\; 3$

    on $(-\infty, 1)$ and $(3, \infty)$, concave down on $(1, 3) \Rightarrow$ points of

    inflection at $x = 1$ and $x = 3$

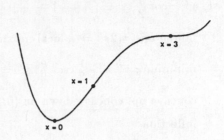

45. $y' = x(x^2 - 12) = x(x - 2\sqrt{3})(x + 2\sqrt{3})$,

    $y' = ---|+++|---|+++ \Rightarrow$ rising on $(-2\sqrt{3}, 0)$
    $\quad\quad\;\; -2\sqrt{3}\quad\; 0\quad\; 2\sqrt{3}$

    and $(2\sqrt{3}, \infty)$, falling on $(-\infty, -2\sqrt{3})$ and $(0, 2\sqrt{3}) \Rightarrow$ a

    local maximum at $x = 0$, local minima at $x = \pm 2\sqrt{3}$;

    $y'' = (1)(x^2 - 12) + (x)(2x) = 3(x - 2)(x + 2)$,

    $y'' = +++|---|+++ \Rightarrow$ concave up on $(-\infty, -2)$
    $\quad\quad\quad\;\; -2\quad\;\; 2$

    and $(2, \infty)$, concave down on $(-2, 2) \Rightarrow$ points of inflection at $x = \pm 2$

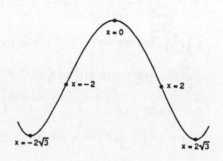

47. $y' = (8x - 5x^2)(4 - x)^2 = x(8 - 5x)(4 - x)^2$,

    $y' = ---|+++|---|--- \Rightarrow$ rising on $\left(0, \frac{8}{5}\right)$,
    $\quad\quad\;\; 0\quad\; 8/5\quad\; 4$

    falling on $(-\infty, 0)$ and $\left(\frac{8}{5}, \infty\right) \Rightarrow$ a local maximum at

    $x = \frac{8}{5}$, a local minimum at $x = 0$;

    $y'' = (8 - 10x)(4 - x)^2 + (8x - 5x^2)(2)(4 - x)(-1)$

    $= 4(4 - x)(5x^2 - 16x + 8)$,

    $y'' = +++|---|+++|--- \Rightarrow$ concave up on
    $\quad\quad\quad\;\; \frac{8-2\sqrt{6}}{5}\quad\; \frac{8+2\sqrt{6}}{5}\quad\; 4$

    $\left(-\infty, \frac{8 - 2\sqrt{6}}{5}\right)$ and $\left(\frac{8 + 2\sqrt{6}}{5}, 4\right)$, concave down on $\left(\frac{8 - 2\sqrt{6}}{5}, \frac{8 + 2\sqrt{6}}{5}\right)$ and $(4, \infty) \Rightarrow$ points of inflection at

    $x = \frac{8 \pm 2\sqrt{6}}{5}$ and $x = 4$

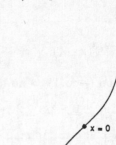

49. $y' = \sec^2 x$, $y' = (+++) \Rightarrow$ rising on $\left(-\frac{\pi}{2}, \frac{\pi}{2}\right)$, never
    $\quad\quad\quad\quad\quad\quad\; -\pi/2\quad\; \pi/2$

    falling $\Rightarrow$ no local extrema;

    $y'' = 2(\sec x)(\sec x)(\tan x) = 2(\sec^2 x)(\tan x)$,

    $y'' = (---|+++) \Rightarrow$ a point of inflection at $x = 0$
    $\quad\;\; -\pi/2\quad\; 0\quad\; \pi/2$

51. $y' = \cot \frac{\theta}{2}$, $y' = (\underset{0}{+++} | \underset{\pi}{---} \underset{2\pi}{)} \Rightarrow$ rising on $(0, \pi)$,

falling on $(\pi, 2\pi) \Rightarrow$ a local maximum at $\theta = \pi$, no local

minimum; $y'' = -\frac{1}{2} \csc^2 \frac{\theta}{2}$, $y'' = (\underset{0}{---}) \underset{2\pi}{} \Rightarrow$ never

concave up, concave down on $(0, 2\pi) \Rightarrow$ no points of

inflection

53. $y' = \tan^2 \theta - 1 = (\tan \theta - 1)(\tan \theta + 1)$,

$y' = (\underset{-\pi/2}{+++} | \underset{-\pi/4}{---} | \underset{\pi/4}{+++} \underset{\pi/2}{)} \Rightarrow$ rising on $\left(-\frac{\pi}{2}, -\frac{\pi}{4}\right)$ and

$\left(\frac{\pi}{4}, \frac{\pi}{2}\right)$, falling on $\left(-\frac{\pi}{4}, \frac{\pi}{4}\right) \Rightarrow$ a local maximum at $\theta = -\frac{\pi}{4}$,

a local minimum at $\theta = \frac{\pi}{4}$; $y'' = 2 \tan \theta \sec^2 \theta$,

$y'' = (\underset{-\pi/2}{---} | \underset{0}{+++} \underset{\pi/2}{)} \Rightarrow$ concave up on $\left(0, \frac{\pi}{2}\right)$, concave

down on $\left(-\frac{\pi}{2}, 0\right) \Rightarrow$ a point of inflection at $\theta = 0$

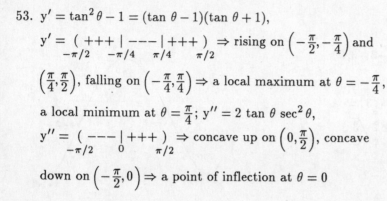

55. $y' = \cos t$, $y' = [\underset{0}{+++} | \underset{\pi/2}{---} | \underset{3\pi/2}{+++}] \underset{2\pi}{} \Rightarrow$ rising on $\left(0, \frac{\pi}{2}\right)$

and $\left(\frac{3\pi}{2}, 2\pi\right)$, falling on $\left(\frac{\pi}{2}, \frac{3\pi}{2}\right) \Rightarrow$ local maxima at $t = \frac{\pi}{2}$ and

$t = 2\pi$, local minima at $t = 0$ and $t = \frac{3\pi}{2}$; $y'' = -\sin t$,

$y'' = [\underset{0}{---} | \underset{\pi}{+++}] \underset{2\pi}{} \Rightarrow$ concave up on $(\pi, 2\pi)$, concave down

on $(0, \pi) \Rightarrow$ a point of inflection at $t = \pi$

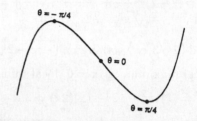

57. $y' = (x + 1)^{-2/3}$, $y' = \underset{-1}{+++})(+++ \Rightarrow$ rising on $(-\infty, \infty)$,

never falling $\Rightarrow$ no local extrema; $y'' = -\frac{2}{3}(x + 1)^{-5/3}$,

$y'' = \underset{-1}{+++})(--- \Rightarrow$ concave up on $(-\infty, -1)$, concave down

on $(-1, \infty) \Rightarrow$ a point of inflection and vertical tangent at

$x = -1$

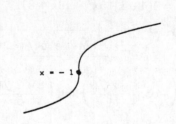

59. $y' = x^{-2/3}(x - 1)$, $y' = \underset{0}{---})(\underset{1}{---} | +++ \Rightarrow$ rising on $(1, \infty)$,

falling on $(-\infty, 1) \Rightarrow$ no local maximum, but a local minimum at

$x = 1$; $y'' = \frac{1}{3}x^{-2/3} + \frac{2}{3}x^{-5/3} = \frac{1}{3}x^{-5/3}(x + 2)$,

$y'' = \underset{-2}{+++} | \underset{0}{---})(+++ \Rightarrow$ concave up on $(-\infty, -2)$ and

$(0, \infty)$, concave down on $(-2, 0) \Rightarrow$ points of inflection at $x = -2$

and $x = 0$, and a vertical tangent at $x = 0$

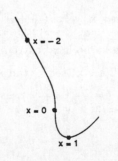

61. $y' = \begin{cases} -2x, & x \le 0 \\ 2x, & x > 0 \end{cases}$, $y' = {+}{+}{+} \underset{0}{|} {+}{+}{+} \Rightarrow$ rising on $(-\infty, \infty)$

$\Rightarrow$ no local extrema; $y'' = \begin{cases} -2, & x < 0 \\ 2, & x > 0 \end{cases}$, $y'' = {-}{-}{-} \underset{0}{)(} {+}{+}{+}$

$\Rightarrow$ concave up on $(0, \infty)$, concave down on $(-\infty, 0) \Rightarrow$ a point of

inflection at $x = 0$

63. The graph of $y = f''(x) \Rightarrow$ the graph of $y = f(x)$ is concave up

on $(0, \infty)$, concave down on $(-\infty, 0) \Rightarrow$ a point of inflection at

$x = 0$; the graph of $y = f'(x) \Rightarrow y' = {+}{+}{+} | {-}{-}{-} | {+}{+}{+} \Rightarrow$

the graph $y = f(x)$ has both a local maximum and a local minimum

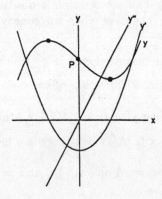

65. The graph of $y = f''(x) \Rightarrow y'' = {-}{-}{-} | {+}{+}{+} | {-}{-}{-} \Rightarrow$ the

graph of $y = f(x)$ has two points of inflection, the graph of

$y = f'(x) \Rightarrow y' = {-}{-}{-} | {+}{+}{+} \Rightarrow$ the graph of $y = f(x)$

has a local minimum

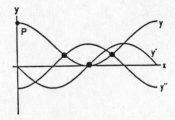

67.

Point	y'	y''
P	$-$	$+$
Q	$+$	0
R	$+$	$-$
S	0	$-$
T	$-$	$-$

69.

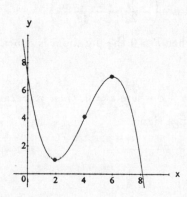

71. Graphs printed in color can shift during a press run, so your values may differ somewhat from those given here.
    (a) The body is moving away from the origin when |displacement| is increasing as t increases, $0 < t < 2$ and $6 < t < 9.5$; the body is moving toward the origin when |displacement| is decreasing as t increases, $2 < t < 6$ and $9.5 < t < 15$
    (b) The velocity will be zero when the slope of the tangent line for $y = s(t)$ is horizontal. The velocity is zero when t is approximately 2, 6, or 9.5 sec.
    (c) The acceleration will be zero at those values of t where the curve $y = s(t)$ has points of inflection. The acceleration is zero when t is approximately 4, 7.5, or 12.5 sec.
    (d) The acceleration is positive when the concavity is up, $4 < t < 7.5$ and $12.5 < t < 15$; the acceleration is negative when the concavity is down, $0 < t < 4$ and $7.5 < t < 12.5$

73. The marginal cost is $\frac{dc}{dx}$ which changes from decreasing to increasing when its derivative $\frac{d^2c}{dx^2}$ is zero. This is a point of inflection of the cost curve and occurs when the production level x is approximately 60 thousand units.

75. When $y' = (x-1)^2(x-2)$, then $y'' = 2(x-1)(x-2) + (x-1)^2$. The curve falls on $(-\infty, 2)$ and rises on $(2, \infty)$. At $x = 2$ there is a local minimum. The curve is concave upward on $(-\infty, 1)$ and $\left(\frac{5}{3}, \infty\right)$, and concave downward on $\left(1, \frac{5}{3}\right)$. At $x = 1$ or $x = \frac{5}{3}$ there are inflection points.

77. The graph must be concave down for $x > 0$ because
    $$f''(x) = -\frac{1}{x^2} < 0.$$

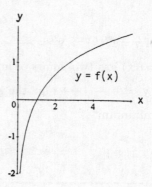

79. The curve will have a point of inflection at $x = 1$ if 1 is a solution of $y'' = 0$; $y = x^3 + bx^2 + cx + d$
    $\Rightarrow y' = 3x^2 + 2bx + c \Rightarrow y'' = 6x + 2b$ and $6(1) + 2b = 0 \Rightarrow b = -3$.

81. (a) $f(x) = ax^2 + bx + c = a\left(x^2 + \frac{b}{a}x\right) + c = a\left(x^2 + \frac{b}{a}x + \frac{b^2}{4a^2}\right) - \frac{b^2}{4a} + c = a\left(x + \frac{b}{2a}\right)^2 - \frac{b^2 - 4ac}{4a}$ a parabola
    whose vertex is at $x = -\frac{b}{2a} \Rightarrow$ the coordinates of the vertex are $\left(-\frac{b}{2a}, -\frac{b^2 - 4ac}{4a}\right)$

    (b) The second derivative, $f''(x) = 2a$, describes concavity $\Rightarrow$ when $a > 0$ the parabola is concave up and when $a < 0$ the parabola is concave down.

83. A quadratic curve never has an inflection point. If $y = ax^2 + bx + c$ where $a \neq 0$, then $y' = 2ax + b$ and $y'' = 2a$. Since 2a is a constant, it is not possible for y'' to change signs.

85. If $y = x^5 - 5x^4 - 240$, then $y' = 5x^3(x-4)$ and
$y'' = 20x^2(x-3)$. The zeros of y′ and y″ are
extrema and points of inflection, respectively.

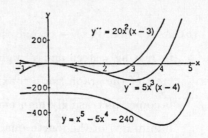

87. If $y = \frac{4}{5}x^5 + 16x^2 - 25$, then $y' = 4x(x^3 + 8)$ and
$y'' = 16(x^3 + 2)$. The zeros of y′ and y″ are
extrema and points of inflection, respectively.

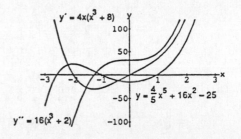

89. The graph of f falls where $f' < 0$, rises where $f' > 0$,
and has horizontal tangents where $f' = 0$. It has local
minima at points where f′ changes from negative to
positive and local maxima where f′ changes from
positive to negative. The graph of f is concave down
where $f'' < 0$ and concave up where $f'' > 0$. It has
points of inflection at values of x where f″ changes
sign and a tangent line exists.

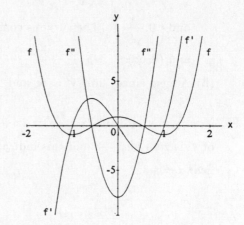

91. (a) It appears to control the number and magnitude of the
local extrema. If $k < 0$, there is a local maximum to the
left of the origin and a local minimum to the right. The
larger the magnitude of k $(k < 0)$, the greater the
magnitude of the extrema. If $k > 0$, the graph has only
positive slopes and lies entirely in the first and third
quadrants with no local extrema. The graph becomes
increasingly steep and straight as $k \to \infty$.

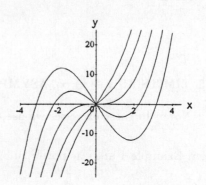

(b) $f'(x) = 3x^2 + k \Rightarrow$ the discriminant $0^2 - 4(3)(k) = -12k$ is positive for $k < 0$, zero for $k = 0$, and
negative for $k > 0$; f′ has two zeros $x = \pm\sqrt{-\frac{k}{3}}$ when $k < 0$, one zero $x = 0$ when $k = 0$ and no real zeros
when $k > 0$; the sign of k controls the number of local extrema.

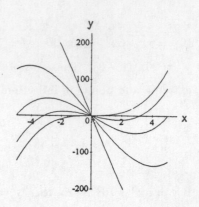

(c) As $k \to \infty$, $f'(x) \to \infty$ and the graph becomes increasingly steep and straight. As $k \to -\infty$, the crest of the graph (local maximum) in the second quadrant becomes increasingly high and the trough (local minimum) in the fourth quadrant becomes increasingly deep.

93. (a) If $y = x^{2/3}(x^2 - 2)$, then $y' = \frac{4}{3}x^{-1/3}(2x^2 - 1)$ and

$y'' = \frac{4}{9}x^{-4/3}(10x^2 + 1)$. The curve rises on

$\left(-\frac{1}{\sqrt{2}}, 0\right)$ and $\left(\frac{1}{\sqrt{2}}, \infty\right)$ and falls on $\left(-\infty, -\frac{1}{\sqrt{2}}\right)$

and $\left(0, \frac{1}{\sqrt{2}}\right)$. The curve is concave up on $(-\infty, 0)$

and $(0, \infty)$.

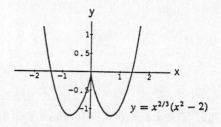

(b) A cusp since $\lim\limits_{x \to 0^-} y' = \infty$ and $\lim\limits_{x \to 0^+} y' = -\infty$.

95. Yes: $y = x^2 + 3\sin 2x \Rightarrow y' = 2x + 6\cos 2x$. The graph of y' is zero near -3 and this indicates a horizontal tangent near $x = -3$.

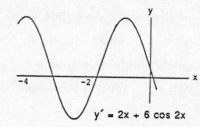

3.5 LIMITS AS $x \to \pm\infty$, ASYMPTOTES AND DOMINANT TERMS

Note: In these exercises we use the result $\lim\limits_{x \to \pm\infty} \frac{1}{x^{m/n}} = 0$ whenever $\frac{m}{n} > 0$. This result follows immediately

from Example 1 and the power rule in Theorem 6: $\lim\limits_{x \to \pm\infty} \left(\frac{1}{x^{m/n}}\right) = \lim\limits_{x \to \pm\infty} \left(\frac{1}{x}\right)^{m/n} = \left(\lim\limits_{x \to \pm\infty} \frac{1}{x}\right)^{m/n} = 0^{m/n}$

$= 0$.

1. (a) -3 (b) -3

3. (a) $\frac{1}{2}$ (b) $\frac{1}{2}$

5. (a) $-\frac{5}{3}$ (b) $-\frac{5}{3}$

7. $-\frac{1}{x} \leq \frac{\sin 2x}{x} \leq \frac{1}{x} \Rightarrow \lim\limits_{x \to \infty} \frac{\sin 2x}{x} = 0$ by the Sandwich Theorem

9. $\displaystyle\lim_{t\to\infty} \frac{2-t+\sin t}{t+\cos t} = \lim_{t\to\infty} \frac{\frac{2}{t}-1+\left(\frac{\sin t}{t}\right)}{1+\left(\frac{\cos t}{t}\right)} = \frac{0-1+0}{1+0} = -1$

11. (a) $\displaystyle\lim_{x\to\infty} \frac{2x+3}{5x+7} = \lim_{x\to\infty} \frac{2+\frac{3}{x}}{5+\frac{7}{x}} = \frac{2}{5}$ (b) $\frac{2}{5}$ (same process as part (a))

13. (a) $\displaystyle\lim_{x\to\infty} \frac{x+1}{x^2+3} = \lim_{x\to\infty} \frac{\frac{1}{x}+\frac{1}{x^2}}{1+\frac{3}{x^2}} = 0$ (b) 0 (same process as part (a))

15. (a) $\displaystyle\lim_{x\to\infty} \frac{1-12x^3}{4x^2+12} = \lim_{x\to\infty} \frac{\frac{1}{x^2}-12x}{4+\frac{12}{x^2}} = -\infty$ (b) $\displaystyle\lim_{x\to-\infty} \frac{1-12x^3}{4x^2+12} = \lim_{x\to-\infty} \frac{\frac{1}{x^2}-12x}{4+\frac{12}{x^2}} = \infty$

17. (a) $\displaystyle\lim_{x\to\infty} \frac{7x^3}{x^3-3x^2+6x} = \lim_{x\to\infty} \frac{7}{1-\frac{3}{x}+\frac{6}{x^2}} = 7$ (b) 7 (same process as part (a))

19. (a) $\displaystyle\lim_{x\to\infty} \frac{2x^5+3}{-x^2+x} = \lim_{x\to\infty} \frac{2x^3+\frac{3}{x^2}}{-1+\frac{1}{x}} = -\infty$ (b) $\displaystyle\lim_{x\to-\infty} \frac{2x^5+3}{-x^2+x} = \lim_{x\to-\infty} \frac{2x^3+\frac{3}{x^2}}{-1+\frac{1}{x}} = \infty$

21. (a) $\displaystyle\lim_{x\to\infty} \frac{x^4}{x^3+1} = \lim_{x\to\infty} \frac{x}{1+\frac{1}{x^3}} = \infty$ (b) $\displaystyle\lim_{x\to-\infty} \frac{x^4}{x^3+1} = \lim_{x\to-\infty} \frac{x}{1+\frac{1}{x^3}} = -\infty$

23. (a) $\displaystyle\lim_{x\to\infty} \frac{-2x^3-2x+3}{3x^3+3x^2-5x} = \lim_{x\to\infty} \frac{-2-\frac{2}{x^2}+\frac{3}{x^3}}{3+\frac{3}{x}-\frac{5}{x^2}} = -\frac{2}{3}$

   (b) $-\frac{2}{3}$ (same process as part (a))

25. $\displaystyle\lim_{x\to\infty} \frac{2\sqrt{x}+x^{-1}}{3x-7} = \lim_{x\to\infty} \frac{\left(\frac{2}{x^{1/2}}\right)+\left(\frac{1}{x^2}\right)}{3-\frac{7}{x}} = 0$

27. $\displaystyle\lim_{x\to-\infty} \frac{\sqrt[3]{x}-\sqrt[5]{x}}{\sqrt[3]{x}+\sqrt[5]{x}} = \lim_{x\to-\infty} \frac{1-x^{(1/5)-(1/3)}}{1+x^{(1/5)-(1/3)}} = \lim_{x\to-\infty} \frac{1-\left(\frac{1}{x^{2/15}}\right)}{1+\left(\frac{1}{x^{2/15}}\right)} = 1$

29. $\displaystyle\lim_{x\to\infty} \frac{2x^{5/3}-x^{1/3}+7}{x^{8/5}+3x+\sqrt{x}} = \lim_{x\to\infty} \frac{2x^{1/15}-\frac{1}{x^{19/15}}+\frac{7}{x^{8/5}}}{1+\frac{3}{x^{3/5}}+\frac{1}{x^{11/10}}} = \infty$

31. Here is one possibility.

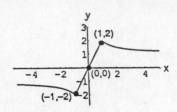

33. Here is one possibility.

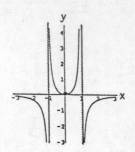

35. Here is one possibility.

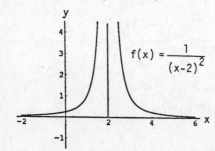

37. Here is one possibility.

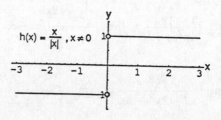

39. $y = \dfrac{1}{x-1}$

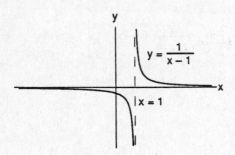

41. $y = \dfrac{1}{2x+4}$

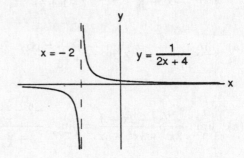

43. $y = \dfrac{x+3}{x+2} = 1 + \dfrac{1}{x+2}$

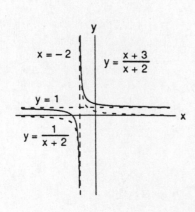

45. $y = \dfrac{2x^2+x-1}{x^2-1}$

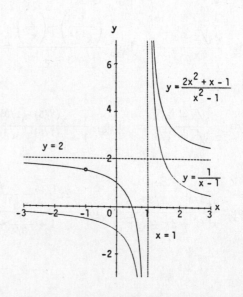

47. $y = \dfrac{x^2 - 1}{x} = x - \dfrac{1}{x}$

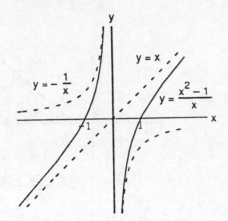

49. $y = \dfrac{x^4 + 1}{x^2} = x^2 + \dfrac{1}{x^2}$

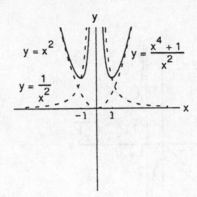

51. $y = \dfrac{1}{x^2 - 1}$

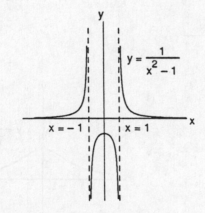

53. $y = -\dfrac{x^2 - 2}{x^2 - 1} = -1 + \dfrac{1}{x^2 - 1}$

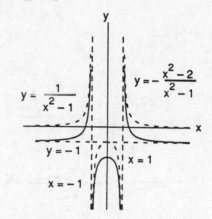

55. $y = \dfrac{x^2}{x - 1} = x + 1 + \dfrac{1}{x - 1}$

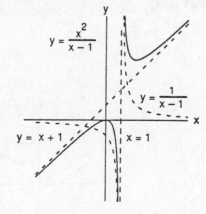

57. $y = \dfrac{x^2 - 4}{x - 1} = x + 1 - \dfrac{3}{x - 1}$

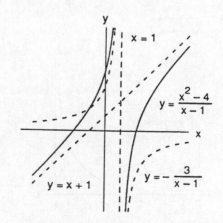

59. $y = \dfrac{x^2 - x + 1}{x - 1} = x + \dfrac{1}{x - 1}$

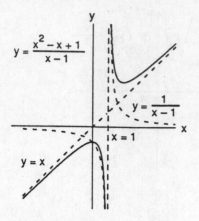

61. $y = \dfrac{x^3 - 3x^2 + 3x - 1}{x^2 + x + 2} = x - 4 + \dfrac{5x + 7}{x^2 + x + 2}$

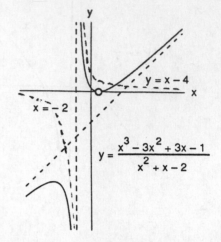

63. $y = \dfrac{x}{x^2 - 1}$

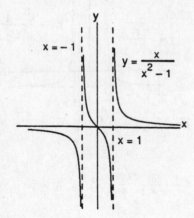

65. $y = \dfrac{8}{x^2 + 4}$

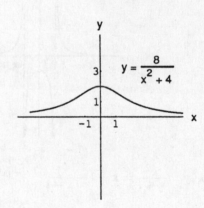

67. $y = \dfrac{x}{\sqrt{4 - x^2}}$

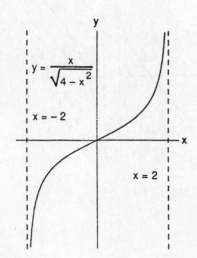

69. $y = x^{2/3} + \dfrac{1}{x^{1/3}}$

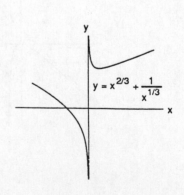

71. $y = \sin\left(\dfrac{\pi}{x^2 + 1}\right)$

73. $y = \sec x + \dfrac{1}{x}$

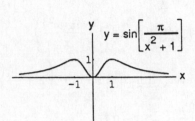

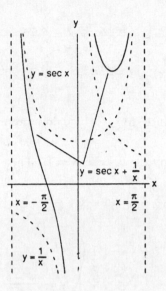

75. $y = \tan x + \dfrac{1}{x^2}$

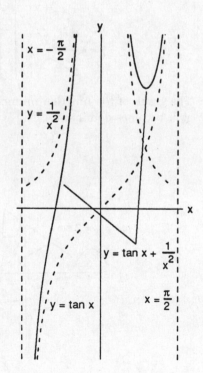

77. f(x) is continuous for all x. $\displaystyle\lim_{x\to\infty} \dfrac{x^3 + x^2}{x^2 + 1} = \infty \Rightarrow$ there exists some b such that $f(b) = \dfrac{b^3 + b^2}{b^2 + 1} > 5{,}000{,}000.$

Also, $f(10) = \dfrac{-900}{101} < -2 < \cos 3 < 5{,}000{,}000 < f(b).$ The result follows from the Intermediate Value Theorem.

79. Since the graphs of odd functions are symmetric about the origin, the function is also increasing when x < 0.

81. Yes. If $\lim\limits_{x\to\infty} \dfrac{f(x)}{g(x)} = 2$ then the ratio of the polynomials' leading coefficients is 2, so $\lim\limits_{x\to-\infty} \dfrac{f(x)}{g(x)} = 2$ as well.

83. At most 2 horizontal asymptotes: one for $x \to \infty$ and possibly another for $x \to -\infty$.

85. (a) $y = 2 + \dfrac{\sin x}{x} \Rightarrow y' = \dfrac{x \cos x - \sin x}{x^2} \Rightarrow \lim\limits_{x\to\infty} y' = \lim\limits_{x\to\infty} \left(\dfrac{\cos x}{x} - \dfrac{\sin x}{x^2} \right) = 0 - 0 = 0$

    (b) $f(x) = 2 + \dfrac{\sin(x^2)}{x} \Rightarrow f'(x) = \dfrac{2x^2 \cos(x^2) - \sin(x^2)}{x^2} \Rightarrow \lim\limits_{x\to\infty} f'(x) = \lim\limits_{x\to\infty} \left(\cos(x^2) - \dfrac{\sin(x^2)}{x^2} \right).$

    Since $\lim\limits_{x\to\infty} \cos(x^2)$ does not exist, $\lim\limits_{x\to\infty} f'(x)$ does not exist. However, $\lim\limits_{x\to\infty} f(x) = \lim\limits_{x\to\infty} \left(2 + \dfrac{\sin(x^2)}{x} \right) = 2.$

87. For any $\epsilon > 0$, take $N = 1$. Then for all $x > N$ we have that $|f(x) - k| = |k - k| = 0 < \epsilon$.

89. $y = -\dfrac{x^2 - 4}{x + 1} = 1 - x + \dfrac{3}{x + 1}$ 91. $y = \dfrac{x^3 - x^2 - 1}{x^2 - 1} = x - 1 + \dfrac{x - 2}{x^2 - 1}$

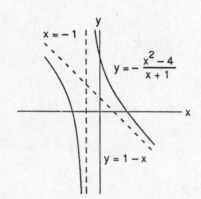   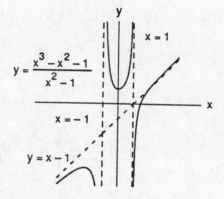

93. The graph of the function mimics each term as it becomes dominant.

95. The graph of the function mimics each term as it becomes dominant.

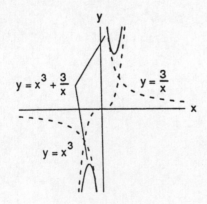   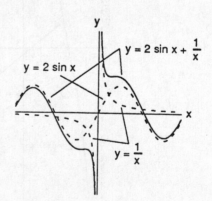

97. The graph of the function mimics each term
    as it becomes dominant.

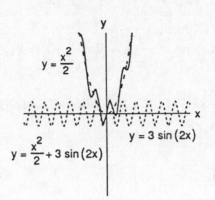

99. (a) $y \to \infty$ (see the accompanying graph)

    (b) $y \to \infty$ (see the accompanying graph)

    (c) cusps at $x = \pm 1$ (see the accompanying graph)

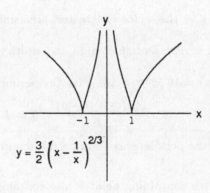

101. (a)

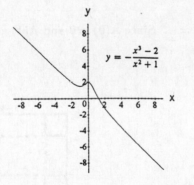

(b)

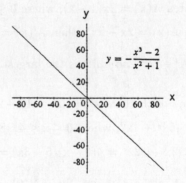

(c) The distance in part (c) is so great that small
    movements are not visible. The activity at the
    origin takes place within a single pixel on the
    screen.

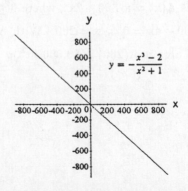

103. $\lim\limits_{x \to \pm\infty} x \sin \frac{1}{x} = \lim\limits_{\theta \to 0} \frac{1}{\theta} \sin \theta = 1,\ \left(\theta = \frac{1}{x}\right)$

105. $\lim\limits_{x \to \pm\infty} \frac{3x + 4}{2x - 5} = \lim\limits_{x \to \pm\infty} \frac{3 + \frac{4}{x}}{2 - \frac{5}{x}} = \lim\limits_{t \to 0} \frac{3 + 4t}{2 - 5t} = \frac{3}{2},\ \left(t = \frac{1}{x}\right)$

107. $\lim\limits_{x \to \pm\infty} \left(3 + \frac{2}{x}\right)\left(\cos \frac{1}{x}\right) = \lim\limits_{\theta \to 0} (3 + 2\theta)(\cos \theta) = (3)(1) = 3,\ \left(\theta = \frac{1}{x}\right)$

3.6 OPTIMIZATION

1. $A = \frac{1}{2}rs$ and $P = s + 2r = 100 \Rightarrow A(r) = \frac{1}{2}r(100 - 2r) = 50r - r^2$; $A'(r) = 50 - 2r = 0 \Rightarrow r = 25$ is a critical point. At $r = 25$ there is a local maximum because $A''(r) < 0$ for all possible r. The values $r = 25$ m and $s = 50$ m give the sector its greatest area since at the endpoints $r = 0$ and $r = 50$ and the area is 0.

3.   Let ℓ and w represent the length and width of the rectangle, respectively. With an area of 16 in.^{2}, we have that $(\ell)(w) = 16 \Rightarrow w = 16\ell^{-1} \Rightarrow$ the perimeter is $P = 2\ell + 2w = 2\ell + 32\ell^{-1}$ and $P'(\ell) = 2 - \frac{32}{\ell^2} = \frac{2(\ell^2 - 16)}{\ell^2}$. Solving $P'(\ell) = 0 \Rightarrow \frac{2(\ell + 4)(\ell - 4)}{\ell^2} = 0 \Rightarrow \ell = -4, 4$. Since $\ell > 0$ for the length of a rectangle, ℓ must be 4 and $w = 4 \Rightarrow$ the perimeter is 16 in., a minimum since $P''(\ell) = \frac{16}{\ell^3} > 0$.

5.   (a) The line containing point P also contains the points $(0, 1)$ and $(1, 0) \Rightarrow$ the line containing P is $y = 1 - x$ $\Rightarrow$ a general point on that line is $(x, 1 - x)$.

   (b) The area $A(x) = 2x(1 - x)$, where $0 \le x \le 1$.

   (c) When $A(x) = 2x - 2x^2$, then $A'(x) = 0 \Rightarrow 2 - 4x = 0 \Rightarrow x = \frac{1}{2}$. Since $A(0) = 0$ and $A(1) = 0$, we conclude that $A\left(\frac{1}{2}\right) = \frac{1}{2}$ sq units is the largest area.

7.   The volume of the box is $V(x) = x(15 - 2x)(8 - 2x)$ $= 120x - 46x^2 + 4x^3$, where $0 \le x \le 4$. Solving $V'(x) = 0$ $\Rightarrow 120 - 92x + 12x^2 = 4(6 - x)(5 - 3x) = 0 \Rightarrow x = \frac{5}{3}$ or 6, but 6 is not in the domain. Since $V(0) = V(4) = 0$, $V\left(\frac{5}{3}\right)$ must be the maximum volume of the box with dimensions $\frac{14}{3} \times \frac{35}{3} \times \frac{5}{3}$ inches.

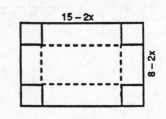

9.   The area is $A(x) = x(800 - 2x)$, where $0 \le x \le 400$. Solving $A'(x) = 800 - 4x = 0 \Rightarrow x = 200$. With $A(0) = A(400) = 0$, the maximum area is $A(200) = 80,000$ m^2.

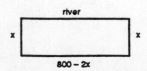

11. We must minimize the weight $= \frac{1}{24}S$, where S is the surface

   area, and $\frac{1}{24}$ ft the thickness of the steel in the tank. The surface

   area is $S(x) = x^2 + 4x\left(\frac{500}{x^2}\right)$, where $0 < x \Rightarrow S'(x) = 2x - \frac{2000}{x^2}$.

   The critical points are 0 and 10, but 0 is not in the domain. Since

   $S''(10) > 0$, at $x = 10$ there is a minimum $\Rightarrow$ the dimensions

   should be 10 ft on the base edge and 5 ft for the height.

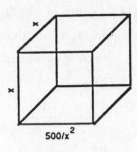

13. The area of the printing is $(y-4)(x-8) = 50$. Consequently,

   $y = \left(\frac{50}{x-8}\right) + 4$. The area of the paper is $A(x) = x\left(\frac{50}{x-8} + 4\right)$,

   where $8 < x$. Then $A'(x) = \left(\frac{50}{x-8} + 4\right) - x\left(\frac{50}{(x-8)^2}\right)$

   $= \frac{4(x-8)^2 - 400}{(x-8)^2} = 0 \Rightarrow$ the critical points are -2 and 18, but

   -2 is not in the domain. Thus $A''(18) > 0 \Rightarrow$ at $x = 18$ we have

   a minimum. Therefore the dimensions 18 by 9 inches minimize

   the amount of paper.

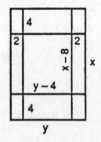

15. The area of the triangle is $A(\theta) = \frac{ab \sin \theta}{2}$, where $0 < \theta < \pi$.

   Solving $A'(\theta) = 0 \Rightarrow \frac{ab \cos \theta}{2} = 0 \Rightarrow \theta = \frac{\pi}{2}$. Since $A''(\theta)$

   $= -\frac{ab \sin \theta}{2} \Rightarrow A''\left(\frac{\pi}{2}\right) < 0$, there is a maximum at $\theta = \frac{\pi}{2}$.

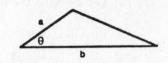

17. A volume $V = \pi r^2 h = 1000 \Rightarrow h = \frac{1000}{\pi r^2}$. The amount of material

   is the surface area given by the sides and bottom of the can

   $\Rightarrow S = 2\pi rh + \pi r^2 = \frac{2000}{r} + \pi r^2, 0 < r$. Then $\frac{dS}{dr} = -\frac{2000}{r^2} + 2\pi r$

   $= 0 \Rightarrow \frac{\pi r^3 - 1000}{r^2} = 0$. The critical points are 0 and $\frac{10}{\sqrt[3]{\pi}}$, but 0

   is not in the domain. Since $\frac{d^2 S}{dr^2} = \frac{4000}{r^3} > 0$, we have a minimum

   surface area when $r = \frac{10}{\sqrt[3]{\pi}}$ cm and $h = \frac{1000}{\pi r^2} = \frac{10}{\sqrt[3]{\pi}}$ cm.

19. (a) From the diagram we have $4x + \ell = 108$ and $V = x^2\ell$. The

   volume of the box is $V(x) = x^2(108 - 4x)$, where $0 \leq x < 27$.

   Then $V'(x) = 216x - 12x^2 = 12x(18 - x) = 0 \Rightarrow$ the critical

   points are 0 and 18, but $x = 0$ results in no box. Since

   $V''(x) = 216 - 24x < 0$ at $x = 18$ we have a maximum. The

   dimensions of the box are $18 \times 18 \times 36$ in.

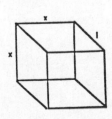

(b) In terms of length, $V(\ell) = x^2\ell = \left(\frac{108-\ell}{4}\right)^2\ell$. The graph indicates that the maximum volume occurs near $\ell = 36$, which is consistent with the result of part (a).

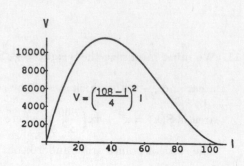

21. (a) From the figure in the text we have $P = 2x + 2y \Rightarrow y = \frac{P}{2} - x$. If $P = 36$, then $y = 18 - x$. When the cylinder is formed, $x = 2\pi r \Rightarrow r = \frac{x}{2\pi}$ and $h = y \Rightarrow h = 18 - x$. The volume of the cylinder is $V = \pi r^2 h$ $\Rightarrow V(x) = \frac{18x^2 - x^3}{4\pi}$. Solving $V'(x) = \frac{3x(12-x)}{4\pi} = 0 \Rightarrow x = 0$ or 12; but when $x = 0$, there is no cylinder. Then $V''(x) = \frac{3}{\pi}\left(3 - \frac{x}{2}\right) \Rightarrow V''(12) < 0 \Rightarrow$ there is a maximum at $x = 12$. The values of $x = 12$ cm and $y = 6$ cm give the largest volume.

(b) In this case $V(x) = \pi x^2(18 - x)$. Solving $V'(x) = 3\pi x(12 - x) = 0 \Rightarrow x = 0$ or 12; but $x = 0$ would result in no cylinder. Then $V''(x) = 6\pi(6 - x) \Rightarrow V''(12) < 0 \Rightarrow$ there is a maximum at $x = 12$. The values of $x = 12$ cm and $y = 6$ cm give the largest volume.

23. (a) Let $4 - x$ be the perimeter of the square and x the circumference of the circle. Then $\frac{x}{2\pi}$ is the radius of the circle and $1 - \frac{x}{4}$ is the side length of the square. The enclosed area is $A = \left(1 - \frac{x}{4}\right)^2 + \pi\left(\frac{x}{2\pi}\right)^2$ $= \left(1 - \frac{x}{4}\right)^2 + \frac{x^2}{4\pi}$ where $0 \leq x \leq 4$, and $\frac{dA}{dx} = 2\left(1 - \frac{x}{4}\right)\left(-\frac{1}{4}\right) + \frac{2x}{4\pi} = 0 \Rightarrow \frac{x}{8} - \frac{1}{2} + \frac{x}{2\pi} = 0 \Rightarrow x = \frac{4\pi}{\pi + 4}$. Since $A(0) = 1$, $A\left(\frac{4\pi}{\pi + 4}\right) = \frac{4}{\pi + 4}$ and $A(4) = \frac{4}{\pi}$, then at $x = 4$ there is a maximum sum of areas.

(b) The radius of the circle is $r = \frac{x}{2\pi} \Rightarrow x = 2\pi r \Rightarrow$ $A = \left(1 - \frac{\pi r}{2}\right)^2 + \pi r^2$, $0 \leq r \leq \frac{2}{\pi}$. The graph indicates the same maximum at $r = \frac{2}{\pi} \Rightarrow x = 4$ consistent with part (a).

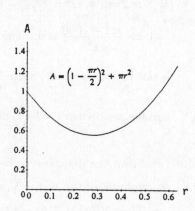

(c) The side length is $s = 1 - \frac{x}{4} \Rightarrow x = 4(1 - s)$

$\Rightarrow A = s^2 + \frac{4}{\pi}(1 - s)^2$, $0 \leq s \leq 1$. The graph indicates

the same maximum at $s = 0 \Rightarrow x = 4$ consistent with

part (a).

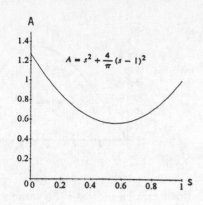

25. From the diagram the perimeter is $P = 2r + 2h + \pi r$,

where r is the radius of the semicircle and h is the

height of the rectangle. The amount of light transmitted

is $A = 2rh + \frac{1}{4}\pi r^2 = r(P - 2r - \pi r) + \frac{1}{4}\pi r^2 = rP - 2r^2 - \frac{3}{4}\pi r^2$.

Then $\frac{dA}{dr} = P - 4r - \frac{3}{2}\pi r = 0 \Rightarrow r = \frac{2P}{8 + 3\pi} \Rightarrow$

$2h = P - \frac{4P}{8 + 3\pi} - \frac{2\pi P}{8 + 3\pi} = \frac{(4 + \pi)P}{8 + 3\pi}$. Therefore,

$\frac{2r}{h} = \frac{8}{4 + \pi}$ gives the proportions that admit the <u>most</u> light since

$\frac{d^2A}{dr^2} = -4 - \frac{3}{2}\pi < 0$.

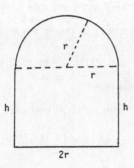

27. The volume of the trough is maximized when the area of the cross section is maximized. From the diagram

the area of the cross section is $A(\theta) = \cos \theta + \sin \theta \cos \theta$, $0 < \theta < \frac{\pi}{2}$. Then $A'(\theta) = -\sin \theta + \cos^2 \theta - \sin^2 \theta$

$= -(2 \sin^2\theta + \sin \theta - 1) = -(2 \sin \theta - 1)(\sin \theta + 1)$ so $A'(\theta) = 0 \Rightarrow \sin \theta = \frac{1}{2}$ or $\sin \theta = -1 \Rightarrow \theta = \frac{\pi}{6}$ because

$\sin \theta \neq -1$ when $0 < \theta < \frac{\pi}{2}$. Also, $A'(\theta) > 0$ for $0 < \theta < \frac{\pi}{6}$ and $A'(\theta) < 0$ for $\frac{\pi}{6} < \theta < \frac{\pi}{2}$. Therefore, at $\theta = \frac{\pi}{6}$

there is a maximum.

29. $s = -\frac{1}{2}gt^2 + v_0t + s_0 \Rightarrow \frac{ds}{dt} = -gt + v_0 = 0 \Rightarrow t = \frac{v_0}{g}$. Then $s\left(\frac{v_0}{g}\right) = -\frac{1}{2}g\left(\frac{v_0}{g}\right)^2 + v_0\left(\frac{v_0}{g}\right) + s_0$

$= \frac{v_0^2}{2g} + s_0$ is the maximum height since $\frac{d^2s}{dt^2} = -g < 0$.

31. (a) From the diagram we have $d^2 = 4r^2 - w^2$. The strength of the beam is $S = kwd^2 = kw(4r^2 - w^2)$. When

$r = 6$, then $S = 144kw - kw^3$. Also, $S'(w) = 144k - 3kw^2 = 3k(48 - w^2)$ so $S'(w) = 0 \Rightarrow w = \pm 4\sqrt{3}$;

$S''(4\sqrt{3}) < 0$ and $-4\sqrt{3}$ is not acceptable. Therefore $S(4\sqrt{3})$ is the maximum strength. The dimensions

of the strongest beam are $4\sqrt{3}$ by $4\sqrt{6}$ inches.

(b)

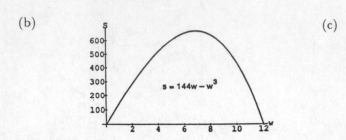

(c)

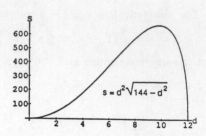

Both graphs indicate the same maximum value and are consistent with each other. Changing k does not change the dimensions that give the strongest beam (i.e., do not change the values of w and d that produce the strongest beam).

33. $\frac{di}{dt} = -2 \sin t + 2 \cos t$, solving $\frac{di}{dt} = 0 \Rightarrow \tan t = 1 \Rightarrow t = \frac{\pi}{4} + n\pi$ where n is a nonnegative integer (in this Exercise t is never negative) $\Rightarrow$ the peak current is $2\sqrt{2}$ amps

35. (a) $2 \sin t = \sin 2t \Rightarrow 2 \sin t - 2 \sin t \cos t = 0 \Rightarrow (2 \sin t)(1 - \cos t) = 0 \Rightarrow t = k\pi$ where k is a positive integer

   (b) The vertical distance between the masses is $s(t) = |s_1 - s_2| = \left((s_1 - s_2)^2\right)^{1/2} = \left((\sin 2t - 2 \sin t)^2\right)^{1/2}$

   $\Rightarrow s'(t) = \left(\frac{1}{2}\right)\left((\sin 2t - 2 \sin t)^2\right)^{-1/2}(2)(\sin 2t - 2 \sin t)(2 \cos 2t - 2 \cos t)$

   $= \frac{2(\cos 2t - \cos t)(\sin 2t - 2 \sin t)}{|\sin 2t - 2 \sin t|} = \frac{4(2 \cos t + 1)(\cos t - 1)(\sin t)(\cos t - 1)}{|\sin 2t - 2 \sin t|} \Rightarrow$ critical times at

   $0, \frac{2\pi}{3}, \pi, \frac{4\pi}{3}, 2\pi$; then $s(0) = 0$, $s\left(\frac{2\pi}{3}\right) = \left|\sin\left(\frac{4\pi}{3}\right) - 2 \sin\left(\frac{2\pi}{3}\right)\right| = \frac{3\sqrt{3}}{2}$, $s(\pi) = 0$, $s\left(\frac{4\pi}{3}\right)$

   $= \left|\sin\left(\frac{8\pi}{3}\right) - 2 \sin\left(\frac{4\pi}{3}\right)\right| = \frac{3\sqrt{3}}{2}$, $s(2\pi) = 0 \Rightarrow$ the greatest distance is $\frac{3\sqrt{3}}{2}$ at $t = \frac{2\pi}{3}$ and $\frac{4\pi}{3}$

37. If $x = (t - 1)(t - 4)^4$, then $v(t) = \frac{dx}{dt} = (t - 4)^3(5t - 8)$ and $v'(t) = 4(t - 4)^2(5t - 11)$.

   (a) The particle is at rest when $v(t) = 0 \Rightarrow t = \frac{8}{5}$ or 4.

   (b) For $\frac{8}{5} < t < 4$, $v(t)$ is negative which indicates that the particle moves to the left.

   (c) Solving $v'(t) = 0 \Rightarrow t = 4$ or $\frac{11}{5}$. At $t = \frac{11}{5}$ there is a maximum since $\left|v\left(\frac{8}{5}\right)\right| = 0$, $\left|v\left(\frac{11}{5}\right)\right| = \frac{2187}{125}$, and $\left|v(4)\right| = 0$. The fastest the particle moves to the left is $\frac{2187}{125}$ units per time.

   (d)

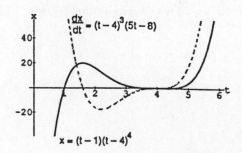

39. The distance $\overline{OT} + \overline{TB}$ is minimized when $\overline{OB}$ is a straight line. Hence $\angle\alpha = \angle\beta \Rightarrow \theta_1 = \theta_2$.

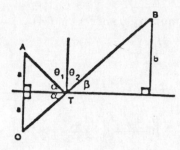

41. $f(x) = x^2 - x + 1 \Rightarrow f'(x) = 2x - 1 = 0 \Rightarrow x = \frac{1}{2}$. Since $f''(x) = 2 > 0$, the function has a minimum value of $f\left(\frac{1}{2}\right) = \frac{1}{4} - \frac{1}{2} + 1 = \frac{3}{4}$, so the function is never negative.

43. (a) The distance is given by $d(x) = \sqrt{(x - c)^2 + x}$ where $x \geq 0$. Also, $d'(x) = \dfrac{2x - 2c + 1}{2\sqrt{(x - c)^2 + x}} = 0 \Rightarrow$ there is a

   minimum at $x = c - \frac{1}{2}$ and $\left(c - \frac{1}{2}, \sqrt{c - \frac{1}{2}}\right)$ is nearest $(c, 0)$.

   (b) If $c < \frac{1}{2}$, then $d'(x)$ cannot equal 0 so the minimum must occur at the endpoint $x = 0 \Rightarrow (0, 0)$ is nearest $(c, 0)$

45. If $f(x) = x^3 + ax^2 + bx$, then $f'(x) = 3x^2 + 2ax + b$ and $f''(x) = 6x + 2a$.
   (a) A local maximum at $x = -1$ and local minimum at $x = 3 \Rightarrow f'(-1) = 0$ and $f'(3) = 0 \Rightarrow 3 - 2a + b = 0$ and $27 + 6a + b = 0 \Rightarrow a = -3$ and $b = -9$.
   (b) A local minimum at $x = 4$ and a point of inflection at $x = 1 \Rightarrow f'(4) = 0$ and $f''(1) = 0 \Rightarrow 48 + 8a + b = 0$ and $6 + 2a = 0 \Rightarrow a = -3$ and $b = -24$.

47. (a) If $y = \cot x - \sqrt{2}\csc x$ where $0 < x < \pi$, then $y' = (\csc x)(\sqrt{2}\cot x - \csc x)$. Solving $y' = 0$

   $\Rightarrow \cos x = \dfrac{1}{\sqrt{2}} \Rightarrow x = \frac{\pi}{4}$. For $0 < x < \frac{\pi}{4}$ we have $y' > 0$, and $y' < 0$ when $\frac{\pi}{4} < x < \pi$. Therefore, at $x = \frac{\pi}{4}$ there is a maximum value of $y = -1$.

   (b)

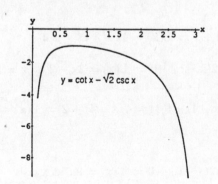

49. The distance between a point on the curve $y = \sqrt{x}$ and $\left(\frac{1}{2}, 16\right)$ is given by $D(x) = \sqrt{\left(\frac{1}{2} - x\right)^2 + \left(16 - \sqrt{x}\right)^2}$.

   Now $D(x)$ is minimized when $f(x) = \left(\frac{1}{2} - x\right)^2 + \left(16 - \sqrt{x}\right)^2$ is minimized. Then $f'(x)$

   $= 2\left(\frac{1}{2} - x\right)(-1) + 2\left(16 - \sqrt{x}\right)\left(-\frac{1}{2\sqrt{x}}\right) = 2x - \dfrac{16}{\sqrt{x}} = 0 \Rightarrow x^{3/2} - 8 = 0 \Rightarrow x = 4$, and $f''(x) = 2 + 8x^{-3/2} > 0$.

Therefore the minimum value of f(x) is $f(4) = \left(-\frac{7}{2}\right)^2 + (16-2)^2 = \frac{49}{4} + 196 = \frac{17 \cdot 49}{4} \Rightarrow$ the minimum distance is $D(4) = \sqrt{\frac{17 \cdot 49}{4}} = \frac{7\sqrt{17}}{2}$.

51. If $x > 0$, then $(x-1)^2 \geq 0 \Rightarrow x^2 + 1 \geq 2x \Rightarrow \frac{x^2+1}{x} \geq 2$. In particular if a, b, c and d are positive integers, then $\left(\frac{a^2+1}{a}\right)\left(\frac{b^2+1}{b}\right)\left(\frac{c^2+1}{c}\right)\left(\frac{d^2+1}{d}\right) \geq 16$.

53. From Exercise 50, Section 2.2 we have $\frac{dR}{dM} = CM - M^2$. Solving $\frac{d^2R}{dM^2} = C - 2M = 0 \Rightarrow M = \frac{C}{2}$. Also, $\frac{d^3R}{dM^3} = -2 \Rightarrow$ at $M = \frac{C}{2}$ there is a maximum.

55. The profit is $p = nx - nc = n(x-c) = \left[a(x-c)^{-1} + b(100-x)\right](x-c) = a + b(100-x)(x-c)$
    $= a + (bc + 100b)x - 100bc - bx^2$. Then $p'(x) = bc + 100b - 2bx$ and $p''(x) = -2b$. Solving $p'(x) = 0 \Rightarrow$ $x = \frac{c}{2} + 50$. At $x = \frac{c}{2} + 50$ there is a maximum profit since $p''(x) = -2b < 0$ for all x.

57. $A(q) = kmq^{-1} + cm + \frac{h}{2}q$, where $q > 0 \Rightarrow A'(q) = -kmq^{-2} + \frac{h}{2} = \frac{hq^2 - 2km}{2q^2}$ and $A''(q) = 2kmq^{-3}$. The critical points are $-\sqrt{\frac{2km}{h}}$, 0, and $\sqrt{\frac{2km}{h}}$, but only $\sqrt{\frac{2km}{h}}$ is in the domain. Then $A''\left(\sqrt{\frac{2km}{h}}\right) > 0 \Rightarrow$ at $q = \sqrt{\frac{2km}{h}}$ there is a minimum average weekly cost.

59. The profit $p(x) = r(x) - c(x) = 6x - \left(x^3 - 6x^2 + 15x\right) = -x^3 + 6x^2 - 9x$, where $x \geq 0$. Then $p'(x) = -3x^2 + 12x - 9 = -3(x-3)(x-1)$ and $p''(x) = -6x + 12$. The critical points are 1 and 3. Thus $p''(1) = 6 > 0 \Rightarrow$ at $x = 1$ there is a local minimum, and $p''(3) = -6 < 0 \Rightarrow$ at $x = 3$ there is a local maximum. But $p(3) = 0 \Rightarrow$ the best you can do is break even.

3.7 LINEARIZATION AND DIFFERENTIALS

1. $f(x) = x^4 \Rightarrow f'(x) = 4x^3 \Rightarrow L(x) = f'(1)(x-1) + f(1) = 4(x-1) + 1 \Rightarrow L(x) = 4x - 3$ at $x = 1$

3. $f(x) = x^3 - x \Rightarrow f'(x) = 3x^2 - 1 \Rightarrow L(x) = f'(1)(x-1) + f(1) = 2(x-1) + 0 \Rightarrow L(x) = 2x - 2$ at $x = 1$

5. $f(x) = \sqrt{x} = x^{1/2} \Rightarrow f'(x) = \left(\frac{1}{2}\right)x^{-1/2} = \frac{1}{2\sqrt{x}} \Rightarrow L(x) = f'(4)(x-4) + f(4) = \frac{1}{4}(x-4) + 2 \Rightarrow L(x) = \frac{1}{4}x + 1$ at $x = 4$

7. $f(x) = x^2 + 2x \Rightarrow f'(x) = 2x + 2 \Rightarrow L(x) = f'(0)(x-0) + f(0) = 2(x-0) + 0 \Rightarrow L(x) = 2x$ at $x = 0$

9. $f(x) = 2x^2 + 4x - 3 \Rightarrow f'(x) = 4x + 4 \Rightarrow L(x) = f'(-1)(x+1) + f(-1) = 0(x+1) + (-5) \Rightarrow L(x) = -5$ at $x = -1$

11. $f(x) = \sqrt[3]{x} = x^{1/3} \Rightarrow f'(x) = \left(\frac{1}{3}\right)x^{-2/3} \Rightarrow L(x) = f'(8)(x-8) + f(8) = \frac{1}{12}(x-8) + 2 \Rightarrow L(x) = \frac{1}{12}x + \frac{4}{3}$ at $x = 8$

13. $f(x) = \sin x \Rightarrow f'(x) = \cos x$

   (a) $L(x) = f'(0)(x - 0) + f(0) = 1(x - 0) + 0$

      $\Rightarrow L(x) = x$ at $x = 0$

   (b) $L(x) = f'(\pi)(x - \pi) + f(\pi) = (-1)(x - \pi) + 0$

      $\Rightarrow L(x) = \pi - x$ at $x = \pi$

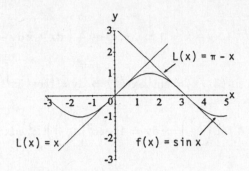

15. $f(x) = \sec x \Rightarrow f'(x) = \sec x \tan x$

   (a) $L(x) = f'(0)(x - 0) + f(0) = 0(x - 0) + 1$

      $\Rightarrow L(x) = 1$ at $x = 0$

   (b) $L(x) = f'\left(-\frac{\pi}{3}\right)\left(x + \frac{\pi}{3}\right) + f\left(-\frac{\pi}{3}\right)$

      $= -2\sqrt{3}\left(x + \frac{\pi}{3}\right) + 2 \Rightarrow L(x) = 2 - 2\sqrt{3}\left(x + \frac{\pi}{3}\right)$

      at $x = -\frac{\pi}{3}$

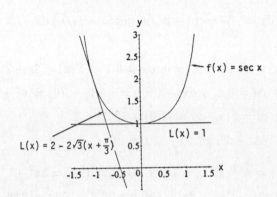

17.

	$f(x)$	$(1 + x)^k \approx 1 + kx$
(a)	$(1 + x)^2$	$1 + 2x$
(b)	$\dfrac{1}{(1 + x)^5} = (1 + x)^{-5}$	$1 + (-5)x = 1 - 5x$
(c)	$\dfrac{2}{1 - x} = 2(1 + (-x))^{-1}$	$2[1 + (-1)(-x)] = 2 + 2x$
(d)	$(1 - x)^6 = (1 + (-x))^6$	$1 + (6)(-x) = 1 - 6x$
(e)	$3(1 + x)^{1/3}$	$3\left[1 + \left(\frac{1}{3}\right)(x)\right] = 3 + x$
(f)	$\dfrac{1}{\sqrt{1 + x}} = (1 + x)^{-1/2}$	$1 + \left(-\frac{1}{2}\right)(x) = 1 - \frac{x}{2}$

19. $f(x) = \sqrt{x + 1} + \sin x = (x + 1)^{1/2} + \sin x \Rightarrow f'(x) = \left(\frac{1}{2}\right)(x + 1)^{-1/2} + \cos x \Rightarrow L_f(x) = f'(0)(x - 0) + f(0)$

   $= \frac{3}{2}(x - 0) + 1 \Rightarrow L_f(x) = \frac{3}{2}x + 1$, the linearization of $f(x)$; $g(x) = \sqrt{x + 1} = (x + 1)^{1/2} \Rightarrow g'(x)$

   $= \left(\frac{1}{2}\right)(x + 1)^{-1/2} \Rightarrow L_g(x) = g'(0)(x - 0) + g(0) = \frac{1}{2}(x - 0) + 1 \Rightarrow L_g(x) = \frac{1}{2}x + 1$, the linearization of $g(x)$;

   $h(x) = \sin x \Rightarrow h'(x) = \cos x \Rightarrow L_h(x) = h'(0)(x - 0) + h(0) = (1)(x - 0) + 0 \Rightarrow L_h(x) = x$, the linearization of

   $h(x)$. $L_f(x) = L_g(x) + L_h(x)$ implies that the linearization of a sum is equal to the sum of the linearizations.

21. $y = x^3 - 3\sqrt{x} = x^3 - 3x^{1/2} \Rightarrow dy = \left(3x^2 - \frac{3}{2}x^{-1/2}\right)dx \Rightarrow dy = \left(3x^2 - \frac{3}{2\sqrt{x}}\right)dx$

23. $y = \dfrac{2x}{1 + x^2} \Rightarrow dy = \left(\dfrac{(2)\left(1 + x^2\right) - (2x)(2x)}{\left(1 + x^2\right)^2}\right)dx = \dfrac{2 - 2x^2}{\left(1 + x^2\right)^2}dx$

25. $2y^{3/2} + xy - x = 0 \Rightarrow 3y^{1/2}\,dy + y\,dx + x\,dy - dx = 0 \Rightarrow \left(3y^{1/2} + x\right)dy = (1-y)\,dx \Rightarrow dy = \dfrac{1-y}{3\sqrt{y}+x}\,dx$

27. $y = \sin\left(5\sqrt{x}\right) = \sin\left(5x^{1/2}\right) \Rightarrow dy = \left(\cos\left(5x^{1/2}\right)\right)\left(\frac{5}{2}x^{-1/2}\right)dx \Rightarrow dy = \dfrac{5\,\cos\left(5\sqrt{x}\right)}{2\sqrt{x}}\,dx$

29. $y = 4\tan\left(\frac{x^3}{3}\right) \Rightarrow dy = 4\left(\sec^2\left(\frac{x^3}{3}\right)\right)\left(x^2\right)dx \Rightarrow dy = 4x^2\,\sec^2\left(\frac{x^3}{3}\right)dx$

31. $y = 3\csc\left(1 - 2\sqrt{x}\right) = 3\csc\left(1 - 2x^{1/2}\right) \Rightarrow dy = 3\left(-\csc\left(1 - 2x^{1/2}\right)\right)\cot\left(1 - 2x^{1/2}\right)\left(-x^{-1/2}\right)dx$

$\Rightarrow dy = \dfrac{3}{\sqrt{x}}\csc\left(1 - 2\sqrt{x}\right)\cot\left(1 - 2\sqrt{x}\right)dx$

33. $f(x) = x^2 + 2x,\ x_0 = 0,\ dx = 0.1 \Rightarrow f'(x) = 2x + 2$

   (a) $\Delta f = f(x_0 + dx) - f(x_0) = f(0.1) - f(0) = .01 + .2 = .21$

   (b) $df = f'(x_0)\,dx = [2(0) + 2](.1) = 2$

   (c) $|\Delta f - df| = |.21 - .20| = .01$

35. $f(x) = x^3 - x,\ x_0 = 1,\ dx = 0.1 \Rightarrow f'(x) = 3x^2 - 1$

   (a) $\Delta f = f(x_0 + dx) - f(x_0) = f(1.1) - f(1) = .231$

   (b) $df = f'(x_0)\,dx = [3(1)^2 - 1](.1) = .2$

   (c) $|\Delta f - df| = |.231 - .2| = .031$

37. $f(x) = x^{-1},\ x_0 = 0.5,\ dx = 0.1 \Rightarrow f'(x) = -x^{-2}$

   (a) $\Delta f = f(x_0 + dx) - f(x_0) = f(.6) - f(.5) = -\frac{1}{3}$

   (b) $df = f'(x_0)\,dx = (-4)\left(\frac{1}{10}\right) = -\frac{2}{5}$

   (c) $|\Delta f - df| = \left|-\frac{1}{3} + \frac{2}{5}\right| = \frac{1}{15}$

39. $V = \frac{4}{3}\pi r^3 \Rightarrow dV = 4\pi r_0^2\,dr$ 41. $S = 6x^2 \Rightarrow dS = 12x_0\,dx$

43. $V = \pi r^2 h$, height constant $\Rightarrow dV = 2\pi r_0 h\,dr$

45. Given $r = 2$ m, $dr = .02$ m

   (a) $A = \pi r^2 \Rightarrow dA = 2\pi r\,dr = 2\pi(2)(.02) = .08\pi$ m^2

   (b) $\left(\frac{.08\pi}{4\pi}\right)(100\%) = 2\%$

47. The error in measurement $dx = (1\%)(10) = 0.1$ cm; $V = x^3 \Rightarrow dV = 3x^2\,dx = 3(10)^2(0.1) = 30$ cm$^3 \Rightarrow$ the percentage error in the volume calculation is $\left(\frac{30}{1000}\right)(100\%) = 3\%$

49. Given $D = 100$ cm, $dD = 1$ cm, $V = \frac{4}{3}\pi\left(\frac{D}{2}\right)^3 = \frac{\pi D^3}{6} \Rightarrow dV = \frac{\pi}{2}D^2\,dD = \frac{\pi}{2}(100)^2(1) = \frac{10^4\pi}{2}$. Then $\frac{dV}{V}(100\%)$

$= \left[\dfrac{\frac{10^4\pi}{2}}{\frac{10^6\pi}{6}}\right](10^2\%) = \left[\dfrac{\frac{10^6\pi}{2}}{\frac{10^6\pi}{6}}\right]\% = 3\%$

51. $V = \pi h^3 \Rightarrow dV = 3\pi h^2$ dh; recall that $\Delta V \approx dV$. Then $|\Delta V| \le (1\%)(V) = \frac{(1)(\pi h^3)}{100} \Rightarrow |dV| \le \frac{(1)(\pi h^3)}{100}$

$\Rightarrow |3\pi h^2 \, dh| \le \frac{(1)(\pi h^3)}{100} \Rightarrow |dh| \le \frac{1}{300} \, h = \left(\frac{1}{3}\%\right) h$. Therefore the greatest tolerated error in the measurement

of h is $\frac{1}{3}\%$.

53. $V = \pi r^2 h$, h is constant $\Rightarrow dV = 2\pi rh$ dr; recall that $\Delta V \approx dV$. We want $|\Delta V| \le \frac{1}{1000} V \Rightarrow |dV| \le \frac{\pi r^2 h}{1000}$

$\Rightarrow |2\pi rh \, dr| \le \frac{\pi r^2 h}{1000} \Rightarrow |dr| \le \frac{r}{2000} = (.05\%)r \Rightarrow$ a .05% variation in the radius can be tolerated.

55. A 5% error in measuring t $\Rightarrow$ dt = (5%)t = $\frac{t}{20}$. Then s = $16t^2 \Rightarrow$ ds = 32t dt = $32t\left(\frac{t}{20}\right) = \frac{32t^2}{20} = \frac{16t^2}{10} = \left(\frac{1}{10}\right)$s

$= (10\%)$s $\Rightarrow$ a 10% error in the calculation of s.

57. Volume = $(x + \Delta x)^3 = x^3 + 3x^2(\Delta x) + 3x(\Delta x)^2 + (\Delta x)^3$

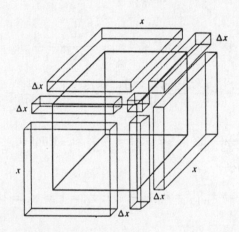

59. $\lim\limits_{x \to 0} \dfrac{\sqrt{1+x}}{1 + \frac{x}{2}} = \dfrac{\sqrt{1+0}}{1 + \frac{0}{2}} = 1$

61. If f has a horizontal tangent at x = a, then f'(a) = 0 and the linearization of f at x = a is

L(x) = f(a) + f'(a)(x − a) = f(a) + 0 · (x − a) = f(a). The linearization is a constant.

63. $f(x) = \frac{4x}{x^2+1} \Rightarrow f'(x) = \frac{4(1-x^2)}{(x^2+1)^2}$;

At x = 0: L(x) = f'(0)(x − 0) + f(0) = 4x;

At x = $\sqrt{3}$: L(x) = f'($\sqrt{3}$)(x − $\sqrt{3}$) + f($\sqrt{3}$)

$= \left(-\frac{1}{2}\right)(x - \sqrt{3}) + \sqrt{3} \Rightarrow L(x) = \frac{1}{2}(3\sqrt{3} - x)$

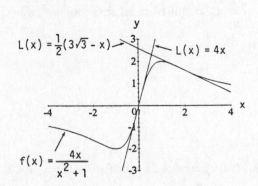

65. $\sqrt{1+x} \approx 1+\frac{x}{2}$ gives the following: $\sqrt{1+1} \approx 1+\frac{1}{2} \Rightarrow \sqrt{\sqrt{1+1}} \approx \sqrt{1+\frac{1}{2}} \approx 1+\frac{1}{4} \Rightarrow \sqrt{\sqrt{\sqrt{1+1}}} \approx \sqrt{1+\frac{1}{4}}$

$\approx 1+\frac{1}{8}$, and so forth. That is, $\underbrace{\sqrt{\ldots\sqrt{\sqrt{1+1}}}}_{\text{n square roots}} \approx 1+\frac{1}{2n} \to 1$ as n $\to \infty$.

For successive tenth roots we obtain the approximation $1+\frac{1}{10n} \to 1$ as n $\to \infty$.

3.8 NEWTON'S METHOD

1. $y = x^2 + x - 1 \Rightarrow y' = 2x + 1 \Rightarrow x_{n+1} = x_n - \dfrac{x_n^2 + x_n - 1}{2x_n + 1}$; $x_0 = 1 \Rightarrow x_1 = 1 - \dfrac{1+1-1}{2+1} = \dfrac{2}{3}$

$\Rightarrow x_2 = \dfrac{2}{3} - \dfrac{\frac{4}{9}+\frac{2}{3}-1}{\frac{4}{3}+1} \Rightarrow x_2 = \dfrac{2}{3} - \dfrac{4+6-9}{12+9} = \dfrac{2}{3} - \dfrac{1}{21} = \dfrac{13}{21} \approx .61905$; $x_0 = -1 \Rightarrow x_1 = 1 - \dfrac{1-1-1}{-2+1} = -2$

$\Rightarrow x_2 = -2 - \dfrac{4-2-1}{-4+1} = -\dfrac{5}{3} \approx -1.66667$

3. $y = x^4 + x - 3 \Rightarrow y' = 4x^3 + 1 \Rightarrow x_{n+1} = x_n - \dfrac{x_n^4 + x_n - 3}{4x_n^3 + 1}$; $x_0 = 1 \Rightarrow x_1 = 1 - \dfrac{1+1-3}{4+1} = \dfrac{6}{5}$

$\Rightarrow x_2 = \dfrac{6}{5} - \dfrac{\frac{1296}{625}+\frac{6}{5}-3}{\frac{864}{125}+1} = \dfrac{6}{5} - \dfrac{1296+750-1875}{4320+625} = \dfrac{6}{5} - \dfrac{171}{4945} = \dfrac{5763}{4945} \approx 1.16542$; $x_0 = -1 \Rightarrow x_1 = -1 - \dfrac{1-1-3}{-4+1}$

$= -2 \Rightarrow x_2 = -2 - \dfrac{16-2-3}{-32+1} = -2 + \dfrac{11}{31} = -\dfrac{51}{31} \approx -1.64516$

5. $y = x^4 - 2 \Rightarrow y' = 4x^3 \Rightarrow x_{n+1} = x_n - \dfrac{x_n^4 - 2}{4x_n^3}$; $x_0 = 1 \Rightarrow x_1 = 1 - \dfrac{1-2}{4} = \dfrac{5}{4} \Rightarrow x_2 = \dfrac{5}{4} - \dfrac{\frac{625}{256}-2}{\frac{125}{16}} = \dfrac{5}{4} - \dfrac{625-512}{2000}$

$= \dfrac{5}{4} - \dfrac{113}{2000} = \dfrac{2500-113}{2000} = \dfrac{2387}{2000} \approx 1.1935$

7. From the graph we let $x_0 = 0.5$ and $f(x) = \cos x - 2x$

$\Rightarrow x_{n+1} = x_n - \dfrac{\cos(x_n) - 2x_n}{-\sin(x_n) - 2} \Rightarrow x_1 = .45063$

$\Rightarrow x_2 = .45018 \Rightarrow$ at $x \approx 0.45$ we have $\cos x = 2x$.

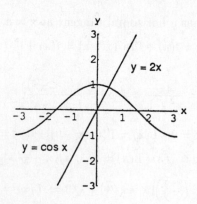

9. If $f(x) = x^3 + 2x - 4$, then $f(1) = -1 < 0$ and $f(2) = 8 > 0 \Rightarrow$ by the Intermediate Value Theorem the equation

$x^3 + 2x - 4 = 0$ has a solution between 1 and 2. Consequently, $f'(x) = 3x^2 + 2$ and $x_{n+1} = x_n - \dfrac{x_n^3 + 2x_n - 4}{3x_n^2 + 2}$.

Then $x_0 = 1 \Rightarrow x_1 = 1.2 \Rightarrow x_2 = 1.17975 \Rightarrow x_3 = 1.179509 \Rightarrow x_4 = 1.1795090 \Rightarrow$ the root is approximately 1.17951.

11. $f(x_0) = 0$ and $f'(x_0) \neq 0 \Rightarrow x_{n+1} = x_n - \dfrac{f(x_n)}{f'(x_n)}$ gives $x_1 = x_0 \Rightarrow x_2 = x_0 \Rightarrow x_n = x_0$ for all $n \geq 0$. That is, all of

the approximations in Newton's method will be the root of $f(x) = 0$ as well as x_0.

13. If $x_0 = h > 0 \Rightarrow x_1 = x_0 - \dfrac{f(x_0)}{f'(x_0)} = h - \dfrac{f(h)}{f'(h)}$

$= h - \dfrac{\sqrt{h}}{\left(\dfrac{1}{2\sqrt{h}}\right)} = h - (\sqrt{h})(2\sqrt{h}) = -h;$

if $x_0 = -h < 0 \Rightarrow x_1 = x_0 - \dfrac{f(x_0)}{f'(x_0)} = -h - \dfrac{f(-h)}{f'(-h)}$

$= -h - \dfrac{\sqrt{h}}{\left(\dfrac{-1}{2\sqrt{h}}\right)} = -h + (\sqrt{h})(2\sqrt{h}) = h.$

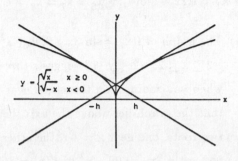

15. (a) The points of intersection of $y = x^3$ and $y = 3x + 1$, or of $y = x^3 - 3x$ and $y = 1$, have the same x-values as

the roots of $f(x) = x^3 - 3x - 1$ or the solutions of $g'(x) = 0$.

(b) $f(x) = x^3 - 3x - 1 \Rightarrow f'(x) = 3x^2 - 3 \Rightarrow x_{n+1} = x_n - \dfrac{x_n^3 - 3x_n - 1}{3x_n^2 - 3} \Rightarrow$ the two negative zeros are -1.53209

and -0.34730

(c) The estimated solutions of $x^3 - 3x - 1 = 0$ are

$-1.53209, -0.34730, 1.87939.$

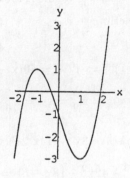

(d) The estimated x values where $g(x) = 0.25x^4 - 1.5x^2 - x + 5$

has horizontal tangents are the roots of $g'(x) = x^3 - 3x - 1$,

and these are $-1.53209, -0.34730, 1.87939.$

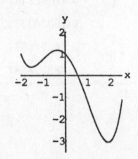

17. $f(x) = x^3 + 3.6x^2 - 36.4 \Rightarrow f'(x) = 3x^2 + 7.2x \Rightarrow x_{n+1} = x_n - \dfrac{x_n^3 + 3.6x_n^2 - 36.4}{3x_n^2 + 7.2x_n}$; $x_0 = 2 \Rightarrow x_1 = 2.53\overline{03}$

$\Rightarrow x_2 = 2.45418225 \Rightarrow x_3 = 2.45238021 \Rightarrow x_4 = 2.45237921$ which is 2.45 to two decimal places. Recall that

$x = 10^4[H_3O^+] \Rightarrow [H_3O^+] = (x)(10^{-4}) = (2.45)(10^{-4}) = 0.000245$

19. $f(x) = \tan x - 2x \Rightarrow f'(x) = \sec^2 x - 2 \Rightarrow x_{n+1} = x_n - \dfrac{\tan(x_n) - 2x_n}{\sec^2(x_n)}$; $x_0 = 1 \Rightarrow x_1 = 1.31047803$

$\Rightarrow x_2 = 1.223929097 \Rightarrow x_6 = x_7 = x_8 = 1.16556119$

21. (a) The graph of $f(x) = \sin 3x - 0.99 + x^2$ in the window

$-2 \le x \le 2$, $-2 \le y \le 3$ suggests three roots. However,

when you zoom in on the x-axis near $x = 1.2$, you can see

that the graph lies above the axis there. There are only

two roots, one near $x = -1$, the other near $x = 0.4$.

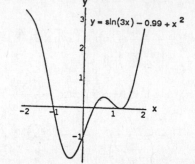

(b) $f(x) = \sin 3x - 0.99 + x^2 \Rightarrow f'(x) = 3\cos 3x + 2x$

$\Rightarrow x_{n+1} = x_n - \dfrac{\sin(3x_n) - 0.99 + x_n^2}{3\cos(3x_n) + 2x_n}$ and the solutions

are approximately 0.35003501505249 and -1.0261731615301

23. $f(x) = 2x^4 - 4x^2 + 1 \Rightarrow f'(x) = 8x^3 - 8x \Rightarrow x_{n+1} = x_n - \dfrac{2x_n^4 - 4x_n^2 + 1}{8x_n^3 - 8x_n}$; if $x_0 = -2$, then $x_6 = -1.30656296$; if

$x_0 = -0.5$, then $x_3 = -0.5411961$; the roots are approximately ± 0.5411961 and ± 1.30656296 because $f(x)$ is

an even function.

25. $f(x) = (x-1)^{40} \Rightarrow f'(x) = 40(x-1)^{39} \Rightarrow x_{n+1} = x_n - \dfrac{(x_n - 1)^{40}}{40(x_n - 1)^{39}} = \dfrac{39x_n + 1}{40}$. With $x_0 = 2$, our computer

gave $x_{87} = x_{88} = x_{89} = \cdots = x_{200} = 1.11051$, coming within 0.11051 of the root $x = 1$.

27. We wish to solve $8x^4 - 14x^3 - 9x^2 + 11x - 1 = 0$. Let $f(x) = 8x^4 - 14x^3 - 9x^2 + 11x - 1$, then

$f'(x) = 32x^3 - 42x^2 - 18x + 11 \Rightarrow x_{n+1} = x_n - \dfrac{8x_n^4 - 14x_n^3 - 9x_n^2 + 11x_n - 1}{32x_n^3 - 42x_n^2 - 18x_n + 11}$.

x_0	approximation of corresponding root
-1.0	-0.976823589
0.1	0.100363332
0.6	0.642746671
2.0	1.983713587

CHAPTER 3 PRACTICE EXERCISES

1. No, since $f(x) = x^3 + 2x + \tan x \Rightarrow f'(x) = 3x^2 + 2 + \sec^2 x > 0 \Rightarrow f(x)$ is always increasing on its domain

3. No absolute minimum because $\lim\limits_{x \to \infty} (7 + x)(11 - 3x)^{1/3} = -\infty$. Next $f'(x) =$

   $(11 - 3x)^{1/3} - (7 + x)(11 - 3x)^{-2/3} = \dfrac{(11 - 3x) - (7 + x)}{(11 - 3x)^{2/3}} = \dfrac{4(1 - x)}{(11 - 3x)^{2/3}} \Rightarrow x = 1$ and $x = \dfrac{11}{3}$ are critical points.

   Since $f' > 0$ if $x < 1$ and $f' < 0$ if $x > 1$, $f(1) = 16$ is the absolute maximum.

5. Yes, because at each point of $[0, 1]$ except $x = 0$, the function's value is a local minimum value as well as a local maximum value. At $x = 0$ the function's value, 0, is not a local minimum value because each open interval around $x = 0$ on the x-axis contains points to the left of 0 where f equals -1.

7. No, because the interval $0 < x < 1$ fails to be closed. The Max-Min Theorem says that if the function is continuous throughout a finite closed interval $a \leq x \leq b$ then the existence of absolute extrema is guaranteed on that interval.

9. (a) There appear to be local minima at $x = -1.75$ and 1.8. Points of inflection are indicated at $x = 0$ and $x = \pm 1$.

   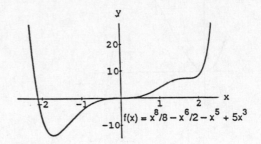

   (b) $f'(x) = x^7 - 3x^5 - 5x^4 + 15x^2 = x^2(x^2 - 3)(x^3 - 5)$. The pattern $y' = ---\ |\ +++\ |\ +++\ |\ ---\ |\ +++$

   $\qquad\qquad\qquad\qquad\qquad\qquad\qquad\qquad\qquad\qquad\qquad -\sqrt{3} \quad 0 \quad \sqrt[3]{5} \quad \sqrt{3}$

   indicates a local maximum at $x = \sqrt[3]{5}$ and local minima at $x = \pm\sqrt{3}$.

   (c)

   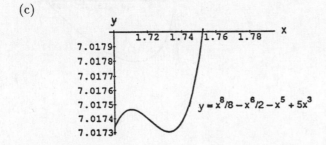

11. (a) $g(t) = \sin^2 t - 3t \Rightarrow g'(t) = 2 \sin t \cos t - 3 = \sin(2t) - 3 \Rightarrow g' < 0 \Rightarrow g(t)$ is always falling and hence must decrease on every interval in its domain.

    (b) One, since $\sin^2 t - 3t - 5 = 0$ and $\sin^2 t - 3t = 5$ have the same solutions: $f(t) = \sin^2 t - 3t - 5$ has the same derivative as $g(t)$ in part (a) and is always decreasing with $f(-3) > 0$ and $f(0) < 0$. The Intermediate Value Theorem guarantees the continuous function f has a root in $[-3, 0]$.

13. (a) $f(x) = x^4 + 2x^2 - 2 \Rightarrow f'(x) = 4x^3 + 4x$. Since $f(0) = -2 < 0$, $f(1) = 1 > 0$ and $f'(x) \geq 0$ for $0 \leq x \leq 1$, we may conclude from the Intermediate Value Theorem that $f(x)$ has exactly one solution when $0 \leq x \leq 1$.

    (b) $x^2 = \dfrac{-2 \pm \sqrt{4 + 8}}{2} > 0 \Rightarrow x^2 = \sqrt{3} - 1$ and $x \geq 0 \Rightarrow x \approx \sqrt{.7320508076} \approx .8555996772$

15. Let $V(t)$ represent the volume of the water in the reservoir at time t, in minutes, let $V(0) = a_0$ be the initial amount and $V(1440) = a_0 + (1400)(43{,}560)(7.48)$ gallons be the amount of water contained in the reservoir after the rain, where 24 hr = 1440 min. Assume that $V(t)$ is continuous on $[0, 1440]$ and differentiable on $(0, 1440)$. The Mean Value Theorem says that for some t_0 in $(0, 1440)$ we have $V'(t_0) = \dfrac{V(1440) - V(0)}{1440 - 0}$
$= \dfrac{a_0 + (1400)(43{,}560)(7.48) - a_0}{1440} = \dfrac{456{,}160{,}320 \text{ gal}}{1440 \text{ min}} = 316{,}778$ gal/min. Therefore at t_0 the reservoir's volume was increasing at a rate in excess of 225,000 gal/min.

17. No, $\dfrac{x}{x+1} = 1 + \dfrac{-1}{x+1} \Rightarrow \dfrac{x}{x+1}$ differs from $\dfrac{-1}{x+1}$ by the constant 1. Both functions have the same derivative
$\dfrac{d}{dx}\left(\dfrac{x}{x+1}\right) = \dfrac{(x+1) - x(1)}{(x+1)^2} = \dfrac{1}{(x+1)^2} = \dfrac{d}{dx}\left(\dfrac{-1}{x+1}\right).$

19.

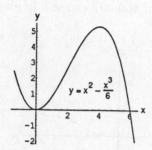

21.

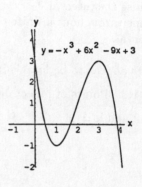

23.

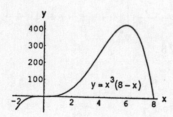

25.

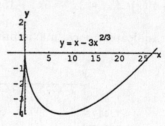

27.

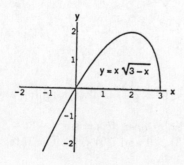

29. (a) $y' = 16 - x^2 \Rightarrow y' = ---\,|\,+++\,|\,--- \Rightarrow$ the curve is rising on $(-4, 4)$, falling on $(-\infty, -4)$ and $(4, \infty)$

$\Rightarrow$ a local maximum at $x = 4$ and a local minimum at $x = -4$; $y'' = -2x \Rightarrow y'' = +++\,|\,--- \Rightarrow$ the curve is concave up on $(-\infty, 0)$, concave down on $(0, \infty) \Rightarrow$ a point of inflection at $x = 0$

(b)

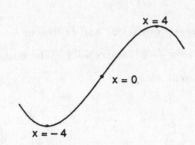

31. (a) $y' = 6x(x+1)(x-2) = 6x^3 - 6x^2 - 12x \Rightarrow y' = ---|_{-1} +++|_{0} ---|_{2} +++ \Rightarrow$ the graph is rising on $(-1,0)$

and $(2,\infty)$, falling on $(-\infty,-1)$ and $(0,2) \Rightarrow$ a local maximum at $x = 0$, local minima at $x = -1$ and

$x = 2$; $y'' = 18x^2 - 12x - 12 = 6(3x^2 - 2x - 2) = 6\left(x - \dfrac{1-\sqrt{7}}{3}\right)\left(x - \dfrac{1+\sqrt{7}}{3}\right) \Rightarrow$

$y'' = +++ \left|_{\frac{1-\sqrt{7}}{3}} --- \right|_{\frac{1+\sqrt{7}}{3}} +++ \Rightarrow$ the curve is concave up on $\left(-\infty, \dfrac{1-\sqrt{7}}{3}\right)$ and $\left(\dfrac{1+\sqrt{7}}{3}, \infty\right)$, concave down

on $\left(\dfrac{1-\sqrt{7}}{3}, \dfrac{1+\sqrt{7}}{3}\right) \Rightarrow$ points of inflection at $x = \dfrac{1 \pm \sqrt{7}}{3}$

(b)

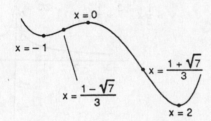

33. (a) $y' = x^4 - 2x^2 = x^2(x^2 - 2) \Rightarrow y' = +++ \left|_{-\sqrt{2}} --- \right|_{0} --- \left|_{\sqrt{2}} +++ \Rightarrow\right.$ the curve is rising on $\left(-\infty, -\sqrt{2}\right)$ and

$\left(\sqrt{2}, \infty\right)$, falling on $\left(-\sqrt{2}, \sqrt{2}\right) \Rightarrow$ a local maximum at $x = -\sqrt{2}$ and a local minimum at $x = \sqrt{2}$;

$y'' = 4x^3 - 4x = 4x(x-1)(x+1) \Rightarrow y'' = ---|_{-1} +++|_{0} ---|_{1} +++ \Rightarrow$ concave up on $(-1,0)$ and $(1,\infty)$,

concave down on $(-\infty,-1)$ and $(0,1) \Rightarrow$ points of inflection at $x = 0$ and $x = \pm 1$

(b)

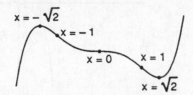

35. The values of the first derivative indicate that the curve is rising on $(0, \infty)$ and falling on $(-\infty, 0)$. The slope of the curve approaches $-\infty$ as $x \to 0^-$, and approaches ∞ as $x \to 0^+$ and $x \to 1$. The curve should therefore have a cusp and local minimum at $x = 0$, and a vertical tangent at $x = 1$.

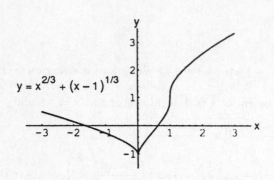

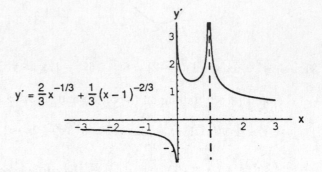

37. The values of the first derivative indicate that the curve is always rising. The slope of the curve approaches ∞ as $x \to 0$ and as $x \to 1$, indicating vertical tangents at both $x = 0$ and $x = 1$.

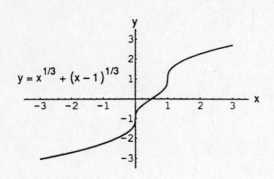

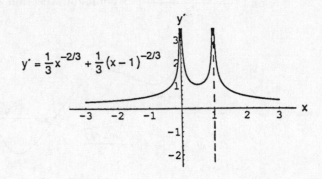

39. $y = \dfrac{x+1}{x-3} = 1 + \dfrac{4}{x-3}$

41. $y = \dfrac{x^2+1}{x} = x + \dfrac{1}{x}$

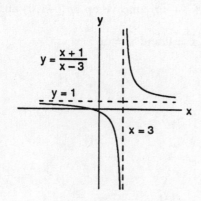

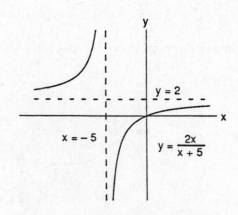

43. $y = \dfrac{x^3 + 2}{2x} = \dfrac{x^2}{2} + \dfrac{1}{x}$

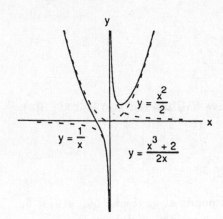

45. $y = \dfrac{x^2 - 4}{x^2 - 3} = 1 - \dfrac{1}{x^2 - 3}$

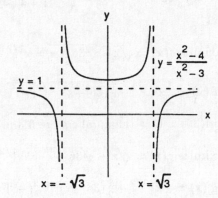

47. $y = \csc x - \dfrac{1}{x^2}$

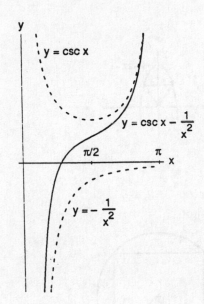

49. (a) $t = 0, 6, 12$ (b) $t = 3, 9$ (c) $6 < t < 12$ (d) $0 < t < 6,\ 12 < t < 14$

51. $\displaystyle \lim_{x \to \infty} \frac{2x + 3}{5x + 7} = \lim_{x \to \infty} \frac{2 + \frac{3}{x}}{5 + \frac{7}{x}} = \frac{2 + 0}{5 + 0} = \frac{2}{5}$

53. $\displaystyle \lim_{x \to \infty} \frac{x^2 - 4x + 8}{3x^3} = \lim_{x \to \infty} \left(\frac{1}{3x} - \frac{4}{3x^2} + \frac{8}{3x^3} \right) = 0 - 0 + 0 = 0$

55. $\displaystyle \lim_{x \to -\infty} \frac{x^2 - 7x}{x + 1} = \lim_{x \to -\infty} \left(\frac{x - 7}{1 + \frac{1}{x}} \right) = -\infty$

57. $\displaystyle \lim_{x \to \infty} \frac{|\sin x|}{\lfloor x \rfloor} \le \lim_{x \to \infty} \frac{1}{\lfloor x \rfloor} = 0$ since $\lfloor x \rfloor \to \infty$ as $x \to \infty$

59. $\lim\limits_{x\to\infty} \dfrac{x + \sin x + 2\sqrt{x}}{x + \sin x} = \lim\limits_{x\to\infty}\left(\dfrac{1 + \frac{\sin x}{x} + \frac{2}{\sqrt{x}}}{1 + \frac{\sin x}{x}}\right) = \dfrac{1 + 0 + 0}{1 + 0} = 1$

61. (a) Maximize $f(x) = \sqrt{x} - \sqrt{36 - x} = x^{1/2} - (36 - x)^{1/2}$ where $0 \le x \le 36$

$\Rightarrow f'(x) = \frac{1}{2}x^{-1/2} - \frac{1}{2}(36 - x)^{-1/2}(-1) = \dfrac{\sqrt{36 - x} + \sqrt{x}}{2\sqrt{x}\sqrt{36 - x}} \Rightarrow$ derivative fails to exist at 0 and 36; $f(0) = -6$,

and $f(36) = 6 \Rightarrow$ the numbers are 0 and 36

(b) Maximize $g(x) = \sqrt{x} + \sqrt{36 - x} = x^{1/2} + (36 - x)^{1/2}$ where $0 \le x \le 36$

$\Rightarrow g'(x) = \frac{1}{2}x^{-1/2} + \frac{1}{2}(36 - x)^{-1/2}(-1) = \dfrac{\sqrt{36 - x} - \sqrt{x}}{2\sqrt{x}\sqrt{36 - x}} \Rightarrow$ critical points at 0, 18 and 36; $g(0) = 6$,

$g(18) = 2\sqrt{18} = 6\sqrt{2}$ and $g(36) = 6 \Rightarrow$ the numbers are 18 and 18

63. $A(x) = \frac{1}{2}(2x)(27 - x^2)$ for $0 \le x \le \sqrt{27}$

$\Rightarrow A'(x) = 3(3 + x)(3 - x)$ and $A''(x) = -6x$.

The critical points are -3 and 3, but -3 is not in the

domain. Since $A''(3) = -18 < 0$ and $A(\sqrt{27}) = 0$,

the maximum occurs at $x = 3 \Rightarrow$ the largest area is

$A(3) = 54$ sq units.

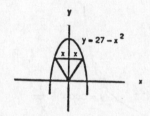

65. From the diagram we have $\left(\frac{h}{2}\right)^2 + r^2 = \left(\sqrt{3}\right)^2$

$\Rightarrow r^2 = \dfrac{12 - h^2}{4}$. The volume of the cylinder is

$V = \pi r^2 h = \pi\left(\dfrac{12 - h^2}{4}\right)h = \frac{\pi}{4}(12h - h^3)$, where

$0 \le h \le 2\sqrt{3}$. Then $V'(h) = \frac{3\pi}{4}(2 + h)(2 - h)$

$\Rightarrow$ the critical points are -2 and 2, but -2 is not in

the domain. At $h = 2$ there is a maximum since

$V''(2) = -3\pi < 0$. The dimensions of the largest

cylinder are radius $= \sqrt{2}$ and height $= 2$.

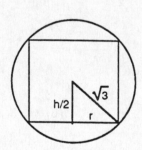

67. From the diagram in the text we have the cost $c = (50{,}000)x + (30{,}000)(20 - y)$ and $x = \sqrt{y^2 + 144}$, where

$0 \le y \le 20$. Thus $c(y) = 50{,}000\sqrt{y^2 + 144} + 30{,}000(20 - y) \Rightarrow c'(y) = \dfrac{10{,}000\left(5y - 3\sqrt{y^2 + 144}\right)}{\sqrt{y^2 + 144}}$. The critical

points are -9 and 9, but -9 is not in the domain. Also $c(0) = \$1{,}200{,}000$, $c(9) = \$1{,}080{,}000$, and

$c(20) = \$1{,}166{,}190$. The minimum cost occurs when $x = 15$ miles and $y = 9$ miles.

69. The profit $P = 2px + py = 2px + p\left(\dfrac{40 - 10x}{5 - x}\right)$, where p is the profit on grade B tires and $0 \leq x \leq 4$. Thus

$P'(x) = \dfrac{2p}{(5 - x)^2}\left(x^2 - 10x + 20\right) \Rightarrow$ the critical points are $\left(5 - \sqrt{5}\right)$, 5, and $\left(5 + \sqrt{5}\right)$, but only $\left(5 - \sqrt{5}\right)$ is in

the domain. Now $P'(x) > 0$ for $0 < x < \left(5 - \sqrt{5}\right)$ and $P'(x) < 0$ for $\left(5 - \sqrt{5}\right) < x < 4 \Rightarrow$ at $x = \left(5 - \sqrt{5}\right)$ there

is a local maximum. Also $P(0) = 8p$, $P\left(5 - \sqrt{5}\right) = 4p\left(5 - \sqrt{5}\right) \approx 11p$, and $P(4) = 8p \Rightarrow$ at $x = \left(5 - \sqrt{5}\right)$ there

is an absolute maximum. The maximum occurs when $x = \left(5 - \sqrt{5}\right)$ and $y = 2\left(5 - \sqrt{5}\right)$, the units are

hundreds of tires, i.e., $x \approx 276$ tires and $y \approx 553$ tires.

71. (a) If $f(x) = \tan x$ and $x = -\dfrac{\pi}{4}$, then $f'(x) = \sec^2 x$,

$f\left(-\dfrac{\pi}{4}\right) = -1$ and $f'\left(-\dfrac{\pi}{4}\right) = 2$. The linearization of

$f(x)$ is $L(x) = 2\left(x + \dfrac{\pi}{4}\right) + (-1) = 2x + \dfrac{\pi - 2}{2}$.

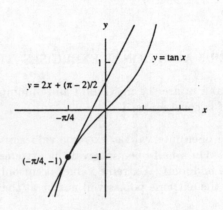

(b) If $f(x) = \sec x$ and $x = -\dfrac{\pi}{4}$, then $f'(x) = \sec x \tan x$,

$f\left(-\dfrac{\pi}{4}\right) = \sqrt{2}$ and $f'\left(-\dfrac{\pi}{4}\right) = -\sqrt{2}$. The linearization of

$f(x)$ is $L(x) = -\sqrt{2}\left(x + \dfrac{\pi}{4}\right) + \sqrt{2} = -\sqrt{2}x + \dfrac{\sqrt{2}(4 - \pi)}{4}$.

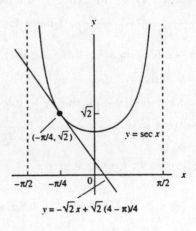

73. $f(x) = \sqrt{x + 1} + \sin x - 0.5 = (x + 1)^{1/2} + \sin x - 0.5 \Rightarrow f'(x) = \left(\dfrac{1}{2}\right)(x + 1)^{-1/2} + \cos x$

$\Rightarrow L(x) = f'(0)(x - 0) + f(0) = 1.5(x - 0) + 0.5 \Rightarrow L(x) = 1.5x + 0.5$, the linearization of $f(x)$.

75. When the volume is $V = \dfrac{1}{3}\pi r^2 h$, then $dV = \dfrac{2}{3}\pi r_0 h\, dr$ estimates the change in the volume for fixed h.

77. (a) $S = 6r^2 \Rightarrow dS = 12r\, dr$. We want $|dS| \leq (2\%)\,S \Rightarrow |12r\, dr| \leq \dfrac{12r^2}{100} \Rightarrow |dr| \leq \dfrac{r}{100}$. The measurement of the

edge r must have an error less than 1%.

(b) When $V = r^3$, then $dV = 3r^2\, dr$. The accuracy of the volume is $\left(\dfrac{dV}{V}\right)(100\%) = \left(\dfrac{3r^2\, dr}{r^3}\right)(100\%)$

$= \left(\dfrac{3}{r}\right)(dr)(100\%) = \left(\dfrac{3}{r}\right)\left(\dfrac{r}{100}\right)(100\%) = 3\%$

79. With h representing the height of the tree, we have $h = 100 \tan \theta$, with θ being the angle of elevation of the tree top. Recall that $1° = \frac{\pi}{180}$ radians. Now $dh = 100 \sec^2 \theta \, d\theta \Rightarrow dh = \left[100 \sec^2 \left(\frac{\pi}{6} \right) \right] \left(\frac{\pi}{180} \right) = \frac{20\pi}{27}$. The error is $\pm \frac{20\pi}{27} \approx \pm 2.3271$ ft.

81. $g(x) = 3x - x^3 + 4 \Rightarrow g(2) = 2 > 0$ and $g(3) = -14 < 0 \Rightarrow g(x) = 0$ in the interval $[2,3]$ by the Intermediate Value Theorem. Then $g'(x) = 3 - 3x^2 \Rightarrow x_{n+1} = x_n - \frac{3x_n - x_n^3 + 4}{3 - 3x_n^2}$; $x_0 = 2 \Rightarrow x_1 = 2.\overline{22} \Rightarrow x_2 = 2.196215$, and so forth to $x_5 = 2.195823345$.

CHAPTER 3 ADDITIONAL EXERCISES–THEORY, EXAMPLES, APPLICATIONS

1. If M and m are the maximum and minimum values, respectively, then $m \leq f(x) \leq M$ for all $x \in I$. If $m = M$ then f is constant on I.

3. On an open interval the extreme values of a continuous function (if any) must occur at an interior critical point. On a half-open interval the extreme values of a continuous function may be at a critical point or at the closed endpoint. Extreme values occur only where $f' = 0$, f' does not exist, or at the endpoints of the interval. Thus the extreme points will not be at the ends of an open interval.

5. (a) If $y' = 6(x+1)(x-2)^2$, then $y' < 0$ for $x < -1$ and $y' > 0$ for $x > -1$. The sign pattern is

    $f' = --- | +++ | +++ \Rightarrow$ f has a local minimum at $x = -1$. Also $y'' = 6(x-2)^2 + 12(x+1)(x-2)$
    $ {}_{-1} {}_{2}$

    $= 6(x-2)(3x) \Rightarrow y'' > 0$ for $x < 0$ or $x > 2$, while $y'' < 0$ for $0 < x < 2$. Therefore f has points of inflection at $x = 0$ and $x = 2$.

   (b) If $y' = 6x(x+1)(x-2)$, then $y' < 0$ for $x < -1$ and $0 < x < 2$; $y' > 0$ for $-1 < x < 0$ and $x > 2$. The sign sign pattern is $y' = --- | +++ | --- | +++$. Therefore f has a local maximum at $x = 0$ and
    $ {}_{-1} {}_{0} {}_{2}$

    local minima at $x = -1$ and $x = 2$. Also, $y'' = 6 \left[x - \left(\frac{1 - \sqrt{7}}{3} \right) \right] \left[x - \left(\frac{1 + \sqrt{7}}{3} \right) \right]$, so $y'' < 0$ for

    $\frac{1 - \sqrt{7}}{3} < x < \frac{1 + \sqrt{7}}{3}$ and $y'' > 0$ for all other $x \Rightarrow$ f has points of inflection at $x = \frac{1 \pm \sqrt{7}}{3}$.

7. If f is continuous on $[a,c)$ and $f'(x) \leq 0$ on $[a,c)$, then by the Mean Value Theorem for all $x \in [a,c)$ we have $\frac{f(c) - f(x)}{c - x} \leq 0 \Rightarrow f(c) - f(x) \leq 0 \Rightarrow f(x) \geq f(c)$. Also if f is continuous on $(c,b]$ and $f'(x) \geq 0$ on $(c,b]$, then for all $x \in (c,b]$ we have $\frac{f(x) - f(c)}{x - c} \geq 0 \Rightarrow f(x) - f(c) \geq 0 \Rightarrow f(x) \geq f(c)$. Therefore $f(x) \geq f(c)$ for all $x \in [a,b]$.

9. No. Corollary 1 requires that $f'(x) = 0$ for <u>all</u> x in some interval I, not $f'(x) = 0$ at a single point in I.

11. From (ii), $f(-1) = \frac{-1+a}{b-c+2} = 0 \Rightarrow a = 1$; from (iii), $1 = \lim_{x \to \infty} f(x) = \lim_{x \to \infty} \frac{x+1}{bx^2 + cx + 2} = \lim_{x \to \infty} \frac{1 + \frac{1}{x}}{bx + c + \frac{2}{x}}$

   $\Rightarrow b = 0$ (because $b = 1 \Rightarrow \lim_{x \to \infty} f(x) = 0$). Also, if $c = 0$ then $\lim_{x \to \infty} f(x) = \infty$ so we must have $c = 1$. In summary, $a = 1$, $b = 0$, and $c = 1$.

13. The area of the $\triangle ABC$ is $A(x) = \frac{1}{2}(2)\sqrt{1-x^2} = \left(1-x^2\right)^{1/2}$,

where $0 \le x \le 1$. Thus $A'(x) = \dfrac{-x}{\sqrt{1-x^2}} \Rightarrow 0$ and ± 1 are

critical points. Also $A(\pm 1) = 0$ so $A(0) = 1$ is the

maximum. When $x = 0$ the $\triangle ABC$ is isosceles since

$AC = BC = \sqrt{2}$.

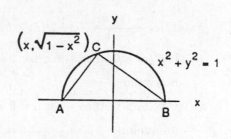

15. The time it would take the water to hit the ground from height y is $\sqrt{\dfrac{2y}{g}}$, where g is the acceleration of

gravity. The product of time and exit velocity (rate) yields the distance the water travels:

$D(y) = \sqrt{\dfrac{2y}{g}}\sqrt{64(h-y)} = 8\sqrt{\dfrac{2}{g}}\left(hy - y^2\right)^{1/2},\ 0 \le y \le h \Rightarrow D'(y) = -4\sqrt{\dfrac{2}{g}}\left(hy - y^2\right)^{-1/2}(h - 2y) \Rightarrow 0, \dfrac{h}{2}$ and h

are critical points. Now $D(0) = 0$, $D\left(\dfrac{h}{2}\right) = \dfrac{8h}{\sqrt{g}}$ and $D(h) = 0 \Rightarrow$ the best place to drill the hole is at $y = \dfrac{h}{2}$.

17. The surface area of the cylinder is $S = 2\pi r^2 + 2\pi rh$. From

the diagram we have $\dfrac{r}{R} = \dfrac{H-h}{H} \Rightarrow h = \dfrac{RH - rH}{R}$ and

$S(r) = 2\pi r(r + h) = 2\pi r\left(r + H - r\dfrac{H}{R}\right) = 2\pi\left(1 - \dfrac{H}{R}\right)r^2 + 2\pi Hr$,

where $0 \le r \le R$.

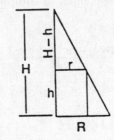

Case 1: $H < R \Rightarrow S(r)$ is a quadratic equation containing the

origin and concave upward $\Rightarrow S(r)$ is maximum at $r = R$.

Case 2: $H = R \Rightarrow S(r)$ is a linear equation containing the origin with a positive slope $\Rightarrow S(r)$ is maximum at

r = R.

Case 3: $H > R \Rightarrow S(r)$ is a quadratic equation containing the origin and concave downward. Then

$\dfrac{dS}{dr} = 4\pi\left(1 - \dfrac{H}{R}\right)r + 2\pi H$ and $\dfrac{dS}{dr} = 0 \Rightarrow 4\pi\left(1 - \dfrac{H}{R}\right)r + 2\pi H = 0 \Rightarrow r = \dfrac{RH}{2(H-R)}$. For simplification

we let $r^* = \dfrac{RH}{2(H-R)}$.

(a) If $R < H < 2R$, then $0 \ge H - 2R \Rightarrow H \ge 2(H - R) \Rightarrow \dfrac{RH}{2(H-R)} \ge R$ which is impossible.

(b) If $H = 2R$, then $r^* = \dfrac{2R^2}{2R} = R \Rightarrow S(r)$ is maximum at $r = R$.

(c) If $H > 2R$, then $2R + H \le 2H \Rightarrow H \le 2(H - R) \Rightarrow \dfrac{H}{2(H-R)} \le 1 \Rightarrow \dfrac{RH}{2(H-R)} \le R \Rightarrow r^* \le R$. Therefore,

$S(r)$ is a maximum at $r = r^* = \dfrac{RH}{2(H-R)}$.

Conclusion: If $H \in (0, R]$ or $H = 2R$, then the maximum surface area is at $r = R$. If $H \in (R, 2R)$, then $r > R$

which is not possible. If $H \in (2R, \infty)$, then the maximum is at $r = r^* = \dfrac{RH}{2(H-R)}$.

19. $\displaystyle\lim_{h \to 0} \dfrac{f'(c+h) - f'(c)}{h} = f''(c) \Leftrightarrow$ for $\epsilon = \frac{1}{2}\left|f''(c)\right| > 0$ there exists a $\delta > 0$ such that $0 < |h| < \delta$

$\Rightarrow \left|\dfrac{f'(c+h) - f'(c)}{h} - f''(c)\right| < \frac{1}{2}\left|f''(c)\right|$. Then $f'(c) = 0 \Rightarrow -\frac{1}{2}\left|f''(c)\right| < \dfrac{f'(c+h)}{h} - f''(c) < \frac{1}{2}\left|f''(c)\right|$

$\Rightarrow f''(c) - \frac{1}{2}\left|f''(c)\right| < \dfrac{f'(c+h)}{h} < f''(c) + \frac{1}{2}\left|f''(c)\right|$. If $f''(c) < 0$, then $\left|f''(c)\right| = -f''(c)$

$\Rightarrow \frac{3}{2}f''(c) < \frac{f'(c+h)}{h} < \frac{1}{2}f''(c) < 0$; likewise if $f''(c) > 0$, then $0 < \frac{1}{2}f''(c) < \frac{f'(c+h)}{h} < \frac{3}{2}f''(c)$.

(a) If $f''(c) < 0$, then $-\delta < h < 0 \Rightarrow f'(c+h) > 0$ and $0 < h < \delta \Rightarrow f'(c+h) < 0$. Therefore, f(c) is a local maximum.

(b) If $f''(c) > 0$, then $-\delta < h < 0 \Rightarrow f'(c+h) < 0$ and $0 < h < \delta \Rightarrow f'(c+h) > 0$. Therefore, f(c) is a local minimum.

21. (a) $(1)^2 = \frac{4\pi^2 L}{32.2} \Rightarrow L \approx 0.8156$ ft

(b) $2T\,dT = \frac{4\pi^2}{g}\,dL \Rightarrow dT = \frac{2\pi^2}{Tg}\,dL = \frac{2\pi^2}{\left(\frac{2\pi\sqrt{L}}{\sqrt{g}}\right)g}\,dL = \frac{\pi}{\sqrt{gL}}\,dL \approx \left(\frac{\pi}{\sqrt{32.2}\,\sqrt{0.8156}}\right)(0.01) = 0.00613$ sec.

(c) The original clock completes 1 swing every second or $(24)(60)(60) = 86{,}400$ swings per day. The new clock completes 1 swing every 1.00613 seconds. Therefore it takes $(86{,}400)(1.00613) = 86{,}929.632$ seconds for the new clock to complete the same number of swings. Thus the new clock loses $\frac{529.632}{60} \approx 8.83$ min/day.

23. $\displaystyle\lim_{x \to \pm\infty} \frac{3x^4 - 2x^3 + 5x + 1}{3x^4} = \lim_{x \to \pm\infty} \frac{3 - \frac{2}{x} + \frac{5}{x^3} + \frac{1}{x^4}}{3} = \frac{3 - 0 + 0 + 0}{3} = 1 \Rightarrow y = 3x^4$ is an end behavior model for $f(x) = 3x^4 - 2x^3 + 5x + 1$.

NOTES.

CHAPTER 4 INTEGRATION

4.1 INDEFINITE INTEGRALS

1. (a) x^2 (b) $\frac{x^3}{3}$ (c) $\frac{x^3}{3} - x^2 + x$

3. (a) x^{-3} (b) $-\frac{x^{-3}}{3}$ (c) $-\frac{x^{-3}}{3} + x^2 + 3x$

5. (a) $\frac{-1}{x}$ (b) $\frac{-5}{x}$ (c) $2x + \frac{5}{x}$

7. (a) $\sqrt{x^3}$ (b) $\sqrt{x}$ (c) $\frac{2}{3}\sqrt{x^3} + 2\sqrt{x}$

9. (a) $x^{2/3}$ (b) $x^{1/3}$ (c) $x^{-1/3}$

11. (a) $\cos(\pi x)$ (b) $-3\cos x$ (c) $\frac{-\cos(\pi x)}{\pi} + \cos(3x)$

13. (a) $\tan x$ (b) $2\tan\left(\frac{x}{3}\right)$ (c) $-\frac{2}{3}\tan\left(\frac{3x}{2}\right)$

15. (a) $-\csc x$ (b) $\frac{1}{5}\csc(5x)$ (c) $2\csc\left(\frac{\pi x}{2}\right)$

17. $(\sin x - \cos x)^2 = \sin^2 x - 2\sin x\cos x + \cos^2 x = \sin^2 x + \cos^2 x - \sin 2x = 1 - \sin 2x.$ An antiderivative is
$x + \frac{\cos 2x}{2}$.

19. $\int (x+1)\,dx = \frac{x^2}{2} + x + C$ 21. $\int \left(3t^2 + \frac{t}{2}\right)dt = t^3 + \frac{t^2}{4} + C$

23. $\int \left(2x^3 - 5x + 7\right)dx = \frac{1}{2}x^4 - \frac{5}{2}x^2 + 7x + C$

25. $\int \left(\frac{1}{x^2} - x^2 - \frac{1}{3}\right)dx = \int \left(x^{-2} - x^2 - \frac{1}{3}\right)dx = \frac{x^{-1}}{-1} - \frac{x^3}{3} - \frac{1}{3}x + C = -\frac{1}{x} - \frac{x^3}{3} - \frac{x}{3} + C$

27. $\int x^{-1/3}\,dx = \frac{x^{2/3}}{\frac{2}{3}} + C = \frac{3}{2}x^{2/3} + C$

29. $\int \left(\sqrt{x} + \sqrt[3]{x}\right)dx = \int \left(x^{1/2} + x^{1/3}\right)dx = \frac{x^{3/2}}{\frac{3}{2}} + \frac{x^{4/3}}{\frac{4}{3}} + C = \frac{2}{3}x^{3/2} + \frac{3}{4}x^{4/3} + C$

31. $\int \left(8y - \frac{2}{y^{1/4}}\right)dy = \int \left(8y - 2y^{-1/4}\right)dy = \frac{8y^2}{2} - 2\left(\frac{y^{3/4}}{\frac{3}{4}}\right) + C = 4y^2 - \frac{8}{3}y^{3/4} + C$

33. $\int 2x\left(1-x^{-3}\right)dx = \int\left(2x-2x^{-2}\right)dx = \frac{2x^2}{2}-2\left(\frac{x^{-1}}{-1}\right)+C = x^2+\frac{2}{x}+C$

35. $\int\frac{t\sqrt{t}+\sqrt{t}}{t^2}\,dt = \int\left(\frac{t^{3/2}}{t^2}+\frac{t^{1/2}}{t^2}\right)dt = \int\left(t^{-1/2}+t^{-3/2}\right)dt = \frac{t^{1/2}}{\frac{1}{2}}+\left(\frac{t^{-1/2}}{-\frac{1}{2}}\right)+C = 2\sqrt{t}-\frac{2}{\sqrt{t}}+C$

37. $\int -2\cos t\,dt = -2\sin t+C$

39. $\int 7\sin\frac{\theta}{3}\,d\theta = -21\cos\frac{\theta}{3}+C$

41. $\int -3\csc^2 x\,dx = 3\cot x+C$

43. $\int\frac{\csc\theta\cot\theta}{2}\,d\theta = -\frac{1}{2}\csc\theta+C$

45. $\int\left(4\sec x\tan x-2\sec^2 x\right)dx = 4\sec x-2\tan x+C$

47. $\int\left(\sin 2x-\csc^2 x\right)dx = -\frac{1}{2}\cos 2x+\cot x+C$

49. $\int 4\sin^2 y\,dy = \int 4\left(\frac{1}{2}-\frac{1}{2}\cos 2y\right)dy = \int\left(2-2\cos 2y\right)dy = 2y-\sin 2y+C$

51. $\int\frac{1+\cos 4t}{2}\,dt = \int\left(\frac{1}{2}+\frac{1}{2}\cos 4t\right)dt = \frac{1}{2}t+\frac{1}{2}\left(\frac{\sin 4t}{4}\right)+C = \frac{t}{2}+\frac{\sin 4t}{8}+C$

53. $\int\left(1+\tan^2\theta\right)d\theta = \int\sec^2\theta\,d\theta = \tan\theta+C$

55. $\int\cot^2 x\,dx = \int\left(\csc^2 x-1\right)dx = -\cot x-x+C$

57. $\int\cos\theta\left(\tan\theta+\sec\theta\right)d\theta = \int\left(\sin\theta+1\right)d\theta = -\cos\theta+\theta+C$

59. $\frac{d}{dx}\left(\frac{(7x-2)^4}{28}+C\right) = \frac{4(7x-2)^3(7)}{28} = (7x-2)^3$

61. $\frac{d}{dx}\left(\frac{1}{5}\tan(5x-1)+C\right) = \frac{1}{5}\left(\sec^2(5x-1)\right)(5) = \sec^2(5x-1)$

63. $\frac{d}{dx}\left(\frac{-1}{x+1}+C\right) = (-1)(-1)(x+1)^{-2} = \frac{1}{(x+1)^2}$

65. (a) Wrong: $\frac{d}{dx}\left(\frac{x^2}{2}\sin x+C\right) = \frac{2x}{2}\sin x+\frac{x^2}{2}\cos x = x\sin x+\frac{x^2}{2}\cos x$

(b) Wrong: $\frac{d}{dx}\left(-x\cos x+C\right) = -\cos x+x\sin x$

(c) Right: $\frac{d}{dx}\left(-x\cos x+\sin x+C\right) = -\cos x+x\sin x+\cos x = x\sin x$

67. (a) Wrong: $\dfrac{d}{dx}\left(\dfrac{(2x+1)^3}{3}+C\right)=\dfrac{3(2x+1)^2(2)}{3}=2(2x+1)^2$

    (b) Wrong: $\dfrac{d}{dx}\left((2x+1)^3+C\right)=3(2x+1)^2(2)=6(2x+1)^2$

    (c) Right: $\dfrac{d}{dx}\left((2x+1)^3+C\right)=6(2x+1)^2$

69. (a) $\displaystyle\int f(x)\,dx=1-\sqrt{x}+C_1=-\sqrt{x}+C$ (b) $\displaystyle\int g(x)\,dx=x+2+C_1=x+C$

    (c) $\displaystyle\int -f(x)\,dx=-(1-\sqrt{x})+C_1=\sqrt{x}+C$ (d) $\displaystyle\int -g(x)\,dx=-(x+2)+C_1=-x+C$

    (e) $\displaystyle\int [f(x)+g(x)]\,dx=(1-\sqrt{x})+(x+2)+C_1=x-\sqrt{x}+C$

    (f) $\displaystyle\int [f(x)-g(x)]\,dx=(1-\sqrt{x})-(x+2)+C_1=-x-\sqrt{x}+C$

    (g) $\displaystyle\int [x+f(x)]\,dx=\dfrac{x^2}{2}+(1-\sqrt{x})+C_1=\dfrac{x^2}{2}-\sqrt{x}+C$

    (h) $\displaystyle\int [g(x)-4]\,dx=(x+2)-4x+C_1=-3x+C$

4.2 DIFFERENTIAL EQUATIONS, INITIAL VALUE PROBLEMS, AND MATHEMATICAL MODELING

1. Graph (b), because $\dfrac{dy}{dx}=2x\Rightarrow y=x^2+C$. Then $y(1)=4\Rightarrow C=3$.

3. $\dfrac{dy}{dx}=2x-7\Rightarrow y=x^2-7x+C$; at $x=2$ and $y=0$ we have $0=2^2-7(2)+C\Rightarrow C=10\Rightarrow y=x^2-7x+10$

5. $\dfrac{dy}{dx}=\dfrac{1}{x^2}+x=x^{-2}+x\Rightarrow y=-x^{-1}+\dfrac{x^2}{2}+C$; at $x=2$ and $y=1$ we have $1=-2^{-1}+\dfrac{2^2}{2}+C\Rightarrow C=-\dfrac{1}{2}$

   $\Rightarrow y=-x^{-1}+\dfrac{x^2}{2}-\dfrac{1}{2}$ or $y=-\dfrac{1}{x}+\dfrac{x^2}{2}-\dfrac{1}{2}$

7. $\dfrac{dy}{dx}=3\sqrt{x}=3x^{1/2}\Rightarrow y=2x^{3/2}+C$; at $x=9$ and $y=4$ we have $4=2(9)^{3/2}+C\Rightarrow C=-50\Rightarrow y=2x^{3/2}-50$

9. $\dfrac{ds}{dt}=1+\cos t\Rightarrow s=t+\sin t+C$; at $t=0$ and $s=4$ we have $4=0+\sin 0+C\Rightarrow C=4\Rightarrow s=t+\sin t+4$

11. $\dfrac{dr}{d\theta}=-\pi\sin\pi\theta\Rightarrow r=\cos(\pi\theta)+C$; at $r=0$ and $\theta=0$ we have $0=\cos(\pi 0)+C\Rightarrow C=-1\Rightarrow r=\cos(\pi\theta)-1$

13. $\dfrac{dv}{dt}=\dfrac{1}{2}\sec t\tan t\Rightarrow v=\dfrac{1}{2}\sec t+C$; at $v=1$ and $t=0$ we have $1=\dfrac{1}{2}\sec(0)+C\Rightarrow C=\dfrac{1}{2}\Rightarrow v=\dfrac{1}{2}\sec t+\dfrac{1}{2}$

15. $\dfrac{d^2y}{dx^2}=2-6x\Rightarrow\dfrac{dy}{dx}=2x-3x^2+C_1$; at $\dfrac{dy}{dx}=4$ and $x=0$ we have $4=2(0)-3(0)^2+C_1\Rightarrow C_1=4$

    $\Rightarrow\dfrac{dy}{dx}=2x-3x^2+4\Rightarrow y=x^2-x^3+4x+C_2$; at $y=1$ and $x=0$ we have $1=0^2-0^3+4(0)+C_2\Rightarrow C_2=1$

    $\Rightarrow y=x^2-x^3+4x+1$

17. $\frac{d^2r}{dt^2} = \frac{2}{t^3} = 2t^{-3} \Rightarrow \frac{dr}{dt} = -t^{-2} + C_1$; at $\frac{dr}{dt} = 1$ and $t = 1$ we have $1 = -(1)^{-2} + C_1 \Rightarrow C_1 = 2 \Rightarrow \frac{dr}{dt} = -t^{-2} + 2$

$\Rightarrow r = t^{-1} + 2t + C_2$; at $r = 1$ and $t = 1$ we have $1 = 1^{-1} + 2(1) + C_2 \Rightarrow C_2 = -2 \Rightarrow r = t^{-1} + 2t - 2$ or

$r = \frac{1}{t} + 2t - 2$

19. $\frac{d^3y}{dx^3} = 6 \Rightarrow \frac{d^2y}{dx^2} = 6x + C_1$; at $\frac{d^2y}{dx^2} = -8$ and $x = 0$ we have $-8 = 6(0) + C_1 \Rightarrow C_1 = -8 \Rightarrow \frac{d^2y}{dx^2} = 6x - 8$

$\Rightarrow \frac{dy}{dx} = 3x^2 - 8x + C_2$; at $\frac{dy}{dx} = 0$ and $x = 0$ we have $0 = 3(0)^2 - 8(0) + C_2 \Rightarrow C_2 = 0 \Rightarrow \frac{dy}{dx} = 3x^2 - 8x$

$\Rightarrow y = x^3 - 4x^2 + C_3$; at $y = 5$ and $x = 0$ we have $5 = 0^3 - 4(0)^2 + C_3 \Rightarrow C_3 = 5 \Rightarrow y = x^3 - 4x^2 + 5$

21. $y^{(4)} = -\sin t + \cos t \Rightarrow y''' = \cos t + \sin t + C_1$; at $y''' = 7$ and $t = 0$ we have $7 = \cos(0) + \sin(0) + C_1$

$\Rightarrow C_1 = 6 \Rightarrow y''' = \cos t + \sin t + 6 \Rightarrow y'' = \sin t - \cos t + 6t + C_2$; at $y'' = -1$ and $t = 0$ we have

$-1 = \sin(0) - \cos(0) + 6(0) + C_2 \Rightarrow C_2 = 0 \Rightarrow y'' = \sin t - \cos t + 6t \Rightarrow y' = -\cos t - \sin t + 3t^2 + C_3$;

at $y' = -1$ and $t = 0$ we have $-1 = -\cos(0) - \sin(0) + 3(0)^2 + C_3 \Rightarrow C_3 = 0 \Rightarrow y' = -\cos t - \sin t + 3t^2$

$\Rightarrow y = -\sin t + \cos t + t^3 + C_4$; at $y = 0$ and $t = 0$ we have $0 = -\sin(0) + \cos(0) + 0^3 + C_4 \Rightarrow C_4 = -1$

$\Rightarrow y = -\sin t + \cos t + t^3 - 1$

23. $v = \frac{ds}{dt} = 9.8t + 5 \Rightarrow s = 4.9t^2 + 5t + C$; at $s = 10$ and $t = 0$ we have $C = 10 \Rightarrow s = 4.9t^2 + 5t + 10$

25. $v = \frac{ds}{dt} = \sin(\pi t) \Rightarrow s = -\frac{1}{\pi}\cos(\pi t) + C$; at $s = 0$ and $t = 0$ we have $C = \frac{1}{\pi} \Rightarrow s = \frac{1 - \cos(\pi t)}{\pi}$

27. $a = 32 \Rightarrow v = 32t + C_1$; at $v = 20$ and $t = 0$ we have $C_1 = 20 \Rightarrow v = 32t + 20 \Rightarrow s = 16t^2 + 20t + C_2$; at $s = 5$

and $t = 0$ we have $C_2 = 5 \Rightarrow s = 16t^2 + 20t + 5$

29. $a = -4\sin(2t) \Rightarrow v = 2\cos(2t) + C_1$; at $v = 2$ and $t = 0$ we have $C_1 = 0 \Rightarrow v = 2\cos(2t)$

$\Rightarrow s = \sin(2t) + C_2$; at $s = -3$ and $t = 0$ we have $C_2 = -3 \Rightarrow s = \sin(2t) - 3$

31. $m = y' = 3\sqrt{x} = 3x^{1/2} \Rightarrow y = 2x^{3/2} + C$; at $(9, 4)$ we have $4 = 2(9)^{3/2} + C \Rightarrow C = -50 \Rightarrow y = 2x^{3/2} - 50$

33. $\frac{dy}{dx} = 1 - \frac{4}{3}x^{1/3} \Rightarrow y = \int \left(1 - \frac{4}{3}x^{1/3}\right) dx = x - x^{4/3} + C$; at $(1, 0.5)$ on the curve we have $0.5 = 1 - 1^{4/3} + C$

$\Rightarrow C = 0.5 \Rightarrow y = x - x^{4/3} + \frac{1}{2}$

35. $\frac{dy}{dx} = \sin x - \cos x \Rightarrow y = \int (\sin x - \cos x)\, dx = -\cos x - \sin x + C$; at $(-\pi, -1)$ on the curve we have

$-1 = -\cos(-\pi) - \sin(-\pi) + C \Rightarrow C = -2 \Rightarrow y = -\cos x - \sin x - 2$

37. $\frac{dy}{dx} = 2x \Rightarrow y' = \underset{0}{--- \,|\, +++} \Rightarrow$ the curve is rising

on $(0, \infty)$ and falling on $(-\infty, 0) \Rightarrow$ a local minimum

at $x = 0$; $\frac{d^2y}{dx^2} = 2 \Rightarrow y'' = ++++++ \Rightarrow$ the curve

is always concave up; $\frac{dy}{dx} = 2x \Rightarrow y = x^2 + C$, which

is a curve having the properties described above

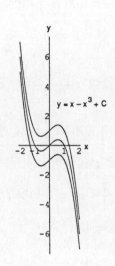

39. $\frac{dy}{dx} = 1 - 3x^2 \Rightarrow y' = \underset{-\sqrt{\frac{1}{3}}}{---} \,|\, \underset{\sqrt{\frac{1}{3}}}{+++} \,|\, --- \Rightarrow$ the curve

is rising on $\left(-\sqrt{\frac{1}{3}}, \sqrt{\frac{1}{3}}\right)$, falling on $\left(-\infty, -\sqrt{\frac{1}{3}}\right)$ and

$\left(\sqrt{\frac{1}{3}}, \infty\right) \Rightarrow$ a local maximum at $x = \sqrt{\frac{1}{3}}$ and a local

minimum at $x = -\sqrt{\frac{1}{3}}$; $\frac{d^2y}{dx^2} = -6x \Rightarrow y'' = \underset{0}{+++ \,|\, ---}$

$\Rightarrow$ the curve is concave up on $(-\infty, 0)$, concave down on

$(0, \infty) \Rightarrow$ a point of inflection at $x = 0$; $\frac{dy}{dx} = 1 - 3x^2$

$\Rightarrow y = x - x^3 + C$, which is a curve having the properties

described above

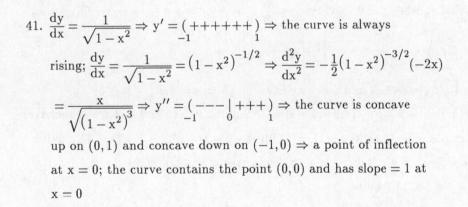

41. $\frac{dy}{dx} = \frac{1}{\sqrt{1-x^2}} \Rightarrow y' = \underset{-1 \qquad 1}{(++++++)} \Rightarrow$ the curve is always

rising; $\frac{dy}{dx} = \frac{1}{\sqrt{1-x^2}} = \left(1 - x^2\right)^{-1/2} \Rightarrow \frac{d^2y}{dx^2} = -\frac{1}{2}\left(1 - x^2\right)^{-3/2}(-2x)$

$= \frac{x}{\sqrt{\left(1-x^2\right)^3}} \Rightarrow y'' = \underset{-1 \quad 0 \quad 1}{(--- \,|\, +++)} \Rightarrow$ the curve is concave

up on $(0, 1)$ and concave down on $(-1, 0) \Rightarrow$ a point of inflection

at $x = 0$; the curve contains the point $(0, 0)$ and has slope $= 1$ at

$x = 0$

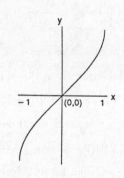

43. $\frac{dy}{dx} = \frac{1}{x^2 + 1} - 1 = \frac{-x^2}{x^2 + 1} \Rightarrow y' = ------- \Rightarrow$ the curve is

always falling; $\frac{dy}{dx} = \frac{1}{x^2 + 1} - 1 = \left(x^2 + 1\right)^{-1} - 1$

$\Rightarrow \frac{d^2y}{dx^2} = -\left(x^2 + 1\right)^{-2}(2x) = \frac{-2x}{\left(x^2 + 1\right)^2} \Rightarrow y'' = \underset{0}{+++ \,|\, ---}$

$\Rightarrow$ the curve is concave up on $(-\infty, 0)$ and concave down on

$(0, \infty) \Rightarrow$ a point of inflection a $x = 0$; the curve contains the

point $(0, 1)$ and has slope $= 0$ at $x = 0$

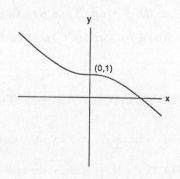

45. $a(t) = v'(t) = 1.6 \Rightarrow v(t) = 1.6t + C$; at $(0,0)$ we have $C = 0 \Rightarrow v(t) = 1.6t$. When $t = 30$, then $v(30) = 48$ m/sec.

47. $a(t) = v'(t) = 9.8 \Rightarrow v(t) = 9.8t + C_1$; at $(0,0)$ we have $C_1 = 0 \Rightarrow s'(t) = v(t) = 9.8t \Rightarrow s(t) = 4.9t^2 + C_2$; at $(0,0)$ we have $C_2 = 0 \Rightarrow s(t) = 4.9t^2$. Then $s(t) = 10 \Rightarrow t^2 = \frac{10}{4.9} \Rightarrow t = \sqrt{\frac{10}{4.9}}$, and $v\left(\sqrt{\frac{10}{4.9}}\right) = 9.8 \sqrt{\frac{10}{4.9}}$

$$= \frac{2(4.9)\sqrt{10}}{\sqrt{4.9}} = (2)\sqrt{4.9}\sqrt{10} = 14 \text{ m/sec}.$$

49. Step 1: $\frac{d^2s}{dt^2} = -k \Rightarrow \frac{ds}{dt} = -kt + C_1$; at $\frac{ds}{dt} = 88$ and $t = 0$ we have $C_1 = 88 \Rightarrow \frac{ds}{dt} = -kt + 88 \Rightarrow$

$s = -k\left(\frac{t^2}{2}\right) + 88t + C_2$; at $s = 0$ and $t = 0$ we have $C_2 = 0 \Rightarrow s = -\frac{kt^2}{2} + 88t$

Step 2: $\frac{ds}{dt} = 0 \Rightarrow 0 = -kt + 88 \Rightarrow t = \frac{88}{k}$

Step 3: $242 = \frac{-k\left(\frac{88}{k}\right)^2}{2} + 88\left(\frac{88}{k}\right) \Rightarrow 242 = -\frac{(88)^2}{2k} + \frac{(88)^2}{k} \Rightarrow 242 = \frac{(88)^2}{2k} \Rightarrow k = 16$

51. (a) $v = \int a \, dt = \int \left(15t^{1/2} - 3t^{-1/2}\right) dt = 10t^{3/2} - 6t^{1/2} + C$; $\frac{ds}{dt}(1) = 4 \Rightarrow 4 = 10(1)^{3/2} - 6(1)^{1/2} + C \Rightarrow C = 0$

$\Rightarrow v = 10t^{3/2} - 6t^{1/2}$

(b) $s = \int v \, dt = \int \left(10t^{3/2} - 6t^{1/2}\right) dt = 4t^{5/2} - 4t^{3/2} + C$; $s(1) = 0 \Rightarrow 0 = 4(1)^{5/2} - 4(1)^{3/2} + C \Rightarrow C = 0$

$\Rightarrow s = 4t^{5/2} - 4t^{3/2}$

53. $\frac{d^2s}{dt^2} = a \Rightarrow \frac{ds}{dt} = \int a \, dt = at + C$; $\frac{ds}{dt} = v_0$ when $t = 0 \Rightarrow C = v_0 \Rightarrow \frac{ds}{dt} = at + v_0 \Rightarrow s = \frac{at^2}{2} + v_0 t + C_1$; $s = s_0$

when $t = 0 \Rightarrow s_0 = \frac{a(0)^2}{2} + v_0(0) + C_1 \Rightarrow C_1 = s_0 \Rightarrow s = \frac{at^2}{2} + v_0 t + s_0$

55. (a) $\frac{ds}{dt} = 9.8t - 3 \Rightarrow s = 4.9t^2 - 3t + C$; 1) at $s = 5$ and $t = 0$ we have $C = 5 \Rightarrow s = 4.9t^2 - 3t + 5$;

displacement $= s(3) - s(1) = ((4.9)(9) - 9 + 5) - (4.9 - 3 + 5) = 33.2$ units; 2) at $s = 2$ and $t = 0$ we have

$C = -2 \Rightarrow s = 4.9t^2 - 3t - 2$; displacement $= s(3) - s(1) = ((4.9)(9) - 9 - 2) - (4.9 - 3 - 2) = 33.2$ units;

3) at $s = s_0$ and $t = 0$ we have $C = s_0 \Rightarrow s = 4.9t^2 - 3t + s_0$; displacement $= s(3) - s(1)$

$= ((4.9)(9) - 9 + s_0) - (4.9 - 3 + s_0) = 33.2$ units

(b) True. Given an antiderivative f(t) of the velocity function, we know that the body's position function is

s = f(t) + C for some constant C. Therefore, the displacement from t = a to t = b is (f(b) + C) − (f(a) + C)

= f(b) − f(a). Thus we can find the displacement from any antiderivative f as the numerical difference

f(b) − f(a) without knowing the exact values of C and s.

4.3   INTEGRATION BY SUBSTITUTION

1. Let $u = 3x \Rightarrow du = 3 \, dx \Rightarrow \frac{1}{3} du = dx$

$$\int \sin 3x \, dx = \int \frac{1}{3} \sin u \, du = -\frac{1}{3} \cos u + C = -\frac{1}{3} \cos 3x + C$$

3. Let $u = 2t \Rightarrow du = 2 \, dt \Rightarrow \frac{1}{2} \, du = dt$

$$\int \sec 2t \, \tan 2t \, dt = \int \frac{1}{2} \sec u \, \tan u \, du = \frac{1}{2} \sec u + C = \frac{1}{2} \sec 2t + C$$

5. Let $u = 7x - 2 \Rightarrow du = 7 \, dx \Rightarrow \frac{1}{7} \, du = dx$

$$\int 28(7x - 2)^{-5} \, dx = \int \frac{1}{7}(28)u^{-5} \, du = \int 4u^{-5} \, du = -u^{-4} + C = -(7x - 2)^{-4} + C$$

7. Let $u = 1 - r^3 \Rightarrow du = -3r^2 \, dr \Rightarrow -3 \, du = 9r^2 \, dr$

$$\int \frac{9r^2 \, dr}{\sqrt{1 - r^3}} = \int -3u^{-1/2} \, du = -3(2)u^{1/2} + C = -6(1 - r^3)^{1/2} + C$$

9. Let $u = x^{3/2} - 1 \Rightarrow du = \frac{3}{2}x^{1/2} \, dx \Rightarrow \frac{2}{3} \, du = \sqrt{x} \, dx$

$$\int \sqrt{x} \, \sin^2(x^{3/2} - 1) \, dx = \int \frac{2}{3} \sin^2 u \, du = \frac{2}{3}\left(\frac{u}{2} - \frac{1}{4} \sin 2u\right) + C = \frac{1}{3}(x^{3/2} - 1) - \frac{1}{6} \sin(2x^{3/2} - 2) + C$$

11. (a) Let $u = \cot 2\theta \Rightarrow du = -2 \csc^2 2\theta \, d\theta \Rightarrow -\frac{1}{2} \, du = \csc^2 2\theta \, d\theta$

$$\int \csc^2 2\theta \, \cot 2\theta \, d\theta = -\int \frac{1}{2} u \, du = -\frac{1}{2}\left(\frac{u^2}{2}\right) + C = -\frac{u^2}{4} + C = -\frac{1}{4} \cot^2 2\theta + C$$

(b) Let $u = \csc 2\theta \Rightarrow du = -2 \csc 2\theta \, \cot 2\theta \, d\theta \Rightarrow -\frac{1}{2} \, du = \csc 2\theta \, \cot 2\theta \, d\theta$

$$\int \csc^2 2\theta \, \cot 2\theta \, d\theta = \int -\frac{1}{2} u \, du = -\frac{1}{2}\left(\frac{u^2}{2}\right) + C = -\frac{u^2}{4} + C = -\frac{1}{4} \csc^2 2\theta + C$$

13. Let $u = 3 - 2s \Rightarrow du = -2 \, ds \Rightarrow -\frac{1}{2} \, du = ds$

$$\int \sqrt{3 - 2s} \, ds = \int \sqrt{u}\left(-\frac{1}{2} \, du\right) = -\frac{1}{2} \int u^{1/2} \, du = \left(-\frac{1}{2}\right)\left(\frac{2}{3}u^{3/2}\right) + C = -\frac{1}{3}(3 - 2s)^{3/2} + C$$

15. Let $u = 5s + 4 \Rightarrow du = 5 \, ds \Rightarrow \frac{1}{5} \, du = ds$

$$\int \frac{1}{\sqrt{5s + 4}} \, ds = \int \frac{1}{\sqrt{u}}\left(\frac{1}{5} \, du\right) = \frac{1}{5} \int u^{-1/2} \, du = \left(\frac{1}{5}\right)\left(2u^{1/2}\right) + C = \frac{2}{5}\sqrt{5s + 4} + C$$

17. Let $u = 1 - \theta^2 \Rightarrow du = -2\theta \, d\theta \Rightarrow -\frac{1}{2} \, du = \theta \, d\theta$

$$\int \theta \sqrt[4]{1 - \theta^2} \, d\theta = \int \sqrt[4]{u}\left(-\frac{1}{2} \, du\right) = -\frac{1}{2} \int u^{1/4} \, du = \left(-\frac{1}{2}\right)\left(\frac{4}{5}u^{5/4}\right) + C = -\frac{2}{5}(1 - \theta^2)^{5/4} + C$$

19. Let $u = 7 - 3y^2 \Rightarrow du = -6y \, dy \Rightarrow -\frac{1}{2} \, du = 3y \, dy$

$$\int 3y\sqrt{7 - 3y^2} \, dy = \int \sqrt{u}\left(-\frac{1}{2} \, du\right) = -\frac{1}{2} \int u^{1/2} \, du = \left(-\frac{1}{2}\right)\left(\frac{2}{3}u^{3/2}\right) + C = -\frac{1}{3}(7 - 3y^2)^{3/2} + C$$

21. Let $u = 1 + \sqrt{x} \Rightarrow du = \frac{1}{2\sqrt{x}} \, dx \Rightarrow 2 \, du = \frac{1}{\sqrt{x}} \, dx$

$$\int \frac{1}{\sqrt{x}(1 + \sqrt{x})^2} \, dx = \int \frac{2 \, du}{u^2} = -\frac{2}{u} + C = \frac{-2}{1 + \sqrt{x}} + C$$

23. Let $u = 3z + 4 \Rightarrow du = 3\ dz \Rightarrow \frac{1}{3} du = dz$

$$\int \cos(3z+4)\ dz = \int (\cos u)\left(\frac{1}{3} du\right) = \frac{1}{3} \int \cos u\ du = \frac{1}{3} \sin u + C = \frac{1}{3} \sin(3z+4) + C$$

25. Let $u = 3x + 2 \Rightarrow du = 3\ dx \Rightarrow \frac{1}{3} du = dx$

$$\int \sec^2(3x+2)\ dx = \int (\sec^2 u)\left(\frac{1}{3} du\right) = \frac{1}{3} \int \sec^2 u\ du = \frac{1}{3} \tan u + C = \frac{1}{3} \tan(3x+2) + C$$

27. Let $u = \sin\left(\frac{x}{3}\right) \Rightarrow du = \frac{1}{3} \cos\left(\frac{x}{3}\right) dx \Rightarrow 3\ du = \cos\left(\frac{x}{3}\right) dx$

$$\int \sin^5\left(\frac{x}{3}\right) \cos\left(\frac{x}{3}\right) dx = \int u^5 (3\ du) = 3\left(\frac{1}{6} u^6\right) + C = \frac{1}{2} \sin^6\left(\frac{x}{3}\right) + C$$

29. Let $u = \frac{r^3}{18} - 1 \Rightarrow du = \frac{r^2}{6}\ dr \Rightarrow 6\ du = r^2\ dr$

$$\int r^2 \left(\frac{r^3}{18} - 1\right)^5 dr = \int u^5 (6\ du) = 6\int u^5\ du = 6\left(\frac{u^6}{6}\right) + C = \left(\frac{r^3}{18} - 1\right)^6 + C$$

31. Let $u = x^{3/2} + 1 \Rightarrow du = \frac{3}{2} x^{1/2}\ dx \Rightarrow \frac{2}{3} du = x^{1/2}\ dx$

$$\int x^{1/2} \sin\left(x^{3/2} + 1\right) dx = \int (\sin u)\left(\frac{2}{3} du\right) = \frac{2}{3} \int \sin u\ du = \frac{2}{3}(-\cos u) + C = -\frac{2}{3} \cos\left(x^{3/2} + 1\right) + C$$

33. Let $u = \sec\left(v + \frac{\pi}{2}\right) \Rightarrow du = \sec\left(v + \frac{\pi}{2}\right) \tan\left(v + \frac{\pi}{2}\right) dv$

$$\int \sec\left(v + \frac{\pi}{2}\right) \tan\left(v + \frac{\pi}{2}\right) dv = \int du = u + C = \sec\left(v + \frac{\pi}{2}\right) + C$$

35. Let $u = \cos(2t+1) \Rightarrow du = -2 \sin(2t+1) \Rightarrow -\frac{1}{2} du = \sin(2t+1)$

$$\int \frac{\sin(2t+1)}{\cos^2(2t+1)}\ dt = \int -\frac{1}{2} \frac{du}{u^2} = \frac{1}{2u} + C = \frac{1}{2 \cos(2t+1)} + C$$

37. Let $u = \cot y \Rightarrow du = -\csc^2 y\ dy \Rightarrow -du = \csc^2 y\ dy$

$$\int \sqrt{\cot y}\ \csc^2 y\ dy = \int \sqrt{u}\ (-du) = -\int u^{1/2}\ du = -\frac{2}{3} u^{3/2} + C = -\frac{2}{3}(\cot y)^{3/2} + C = -\frac{2}{3}\left(\cot^3 y\right)^{1/2} + C$$

39. Let $u = \frac{1}{t} - 1 = t^{-1} - 1 \Rightarrow du = -t^{-2}\ dt \Rightarrow -du = \frac{1}{t^2}\ dt$

$$\int \frac{1}{t^2} \cos\left(\frac{1}{t} - 1\right) dt = \int (\cos u)(-du) = -\int \cos u\ du = -\sin u + C = -\sin\left(\frac{1}{t} - 1\right) + C$$

41. Let $u = \sin\frac{1}{\theta} \Rightarrow du = \left(\cos \frac{1}{\theta}\right)\left(-\frac{1}{\theta^2}\right) d\theta \Rightarrow -du = \frac{1}{\theta^2} \cos \frac{1}{\theta}\ d\theta$

$$\int \frac{1}{\theta^2} \sin \frac{1}{\theta} \cos \frac{1}{\theta}\ d\theta = \int -u\ du = -\frac{1}{2} u^2 + C = -\frac{1}{2} \sin^2 \frac{1}{\theta} + C$$

43. Let $u = s^3 + 2s^2 - 5s + 5 \Rightarrow du = (3s^2 + 4s - 5)\, ds$

$$\int (s^3 + 2s^2 - 5s + 5)(3s^2 + 4s - 5)\, ds = \int u\, du = \frac{u^2}{2} + C = \frac{(s^3 + 2s^2 - 5s + 5)^2}{2} + C$$

45. Let $u = 1 + t^4 \Rightarrow du = 4t^3\, dt \Rightarrow \frac{1}{4}\, du = t^3\, dt$

$$\int t^3 (1 + t^4)^3\, dt = \int u^3 \left(\tfrac{1}{4}\, du\right) = \tfrac{1}{4}\left(\tfrac{1}{4} u^4\right) + C = \tfrac{1}{16}(1 + t^4)^4 + C$$

47. (a) Let $u = \tan x \Rightarrow du = \sec^2 x\, dx$; $v = u^3 \Rightarrow dv = 3u^2\, du \Rightarrow 6\, dv = 18u^2\, du$; $w = 2 + v \Rightarrow dw = dv$

$$\int \frac{18 \tan^2 x \sec^2 x}{(2 + \tan^3 x)^2}\, dx = \int \frac{18u^2}{(2 + u^3)^2}\, du = \int \frac{6\, dv}{(2 + v)^2} = \int \frac{6\, dw}{w^2} = 6 \int w^{-2}\, dw = -6w^{-1} + C = -\frac{6}{2 + v} + C$$

$$= -\frac{6}{2 + u^3} + C = -\frac{6}{2 + \tan^3 x} + C$$

(b) Let $u = \tan^3 x \Rightarrow du = 3 \tan^2 x \sec^2 x\, dx \Rightarrow 6\, du = 18 \tan^2 x \sec^2 x\, dx$; $v = 2 + u \Rightarrow dv = du$

$$\int \frac{18 \tan^2 x \sec^2 x}{(2 + \tan^3 x)^2}\, dx = \int \frac{6\, du}{(2 + u)^2} = \int \frac{6\, dv}{v^2} = -\frac{6}{v} + C = -\frac{6}{2 + u} + C = -\frac{6}{2 + \tan^3 x} + C$$

(c) Let $u = 2 + \tan^3 x \Rightarrow du = 3 \tan^2 x \sec^2 x\, dx \Rightarrow 6\, du = 18 \tan^2 x \sec^2 x\, dx$

$$\int \frac{18 \tan^2 x \sec^2 x}{(2 + \tan^3 x)^2}\, dx = \int \frac{6\, du}{u^2} = -\frac{6}{u} + C = -\frac{6}{2 + \tan^3 x} + C$$

49. Let $u = 3(2r - 1)^2 + 6 \Rightarrow du = 6(2r - 1)(2)\, dr \Rightarrow \frac{1}{12}\, du = (2r - 1)\, dr$; $v = \sqrt{u} \Rightarrow dv = \frac{1}{2\sqrt{u}}\, du \Rightarrow \frac{1}{6}\, dv$

$$= \frac{1}{12\sqrt{u}}\, du$$

$$\int \frac{(2r - 1) \cos \sqrt{3(2r - 1)^2 + 6}}{\sqrt{3(2r - 1)^2 + 6}}\, dr = \int \left(\frac{\cos \sqrt{u}}{\sqrt{u}}\right)\left(\tfrac{1}{12}\, du\right) = \int (\cos v)\left(\tfrac{1}{6}\, dv\right) = \tfrac{1}{6} \sin v + C = \tfrac{1}{6} \sin \sqrt{u} + C$$

$$= \tfrac{1}{6} \sin \sqrt{3(2r - 1)^2 + 6} + C$$

51. Let $u = 3t^2 - 1 \Rightarrow du = 6t\, dt \Rightarrow 2\, du = 12t\, dt$

$$s = \int 12t(3t^2 - 1)^3\, dt = \int u^3 (2\, du) = 2\left(\tfrac{1}{4} u^4\right) + C = \tfrac{1}{2} u^4 + C = \tfrac{1}{2}(3t^2 - 1)^4 + C;$$

$$s = 3 \text{ when } t = 1 \Rightarrow 3 = \tfrac{1}{2}(3 - 1)^4 + C \Rightarrow 3 = 8 + C \Rightarrow C = -5 \Rightarrow s = \tfrac{1}{2}(3t^2 - 1)^4 - 5$$

53. Let $u = t + \frac{\pi}{12} \Rightarrow du = dt$

$$s = \int 8 \sin^2\left(t + \tfrac{\pi}{12}\right) dt = \int 8 \sin^2 u\, du = 8\left(\tfrac{u}{2} - \tfrac{1}{4} \sin 2u\right) + C = 4\left(t + \tfrac{\pi}{12}\right) - 2 \sin\left(2t + \tfrac{\pi}{6}\right) + C;$$

$$s = 8 \text{ when } t = 0 \Rightarrow 8 = 4\left(\tfrac{\pi}{12}\right) - 2 \sin\left(\tfrac{\pi}{6}\right) + C \Rightarrow C = 8 - \tfrac{\pi}{3} + 1 = 9 - \tfrac{\pi}{3} \Rightarrow s = 4t - 2 \sin\left(2t + \tfrac{\pi}{6}\right) + 9$$

55. Let $u = 2t - \frac{\pi}{2} \Rightarrow du = 2\ dt \Rightarrow -2\ du = -4\ dt$

$\frac{ds}{dt} = \int -4 \sin\left(2t - \frac{\pi}{2}\right) dt = \int (\sin u)(-2\ du) = 2 \cos u + C_1 = 2 \cos\left(2t - \frac{\pi}{2}\right) + C_1;$

at $t = 0$ and $\frac{ds}{dt} = 100$ we have $100 = 2 \cos\left(-\frac{\pi}{2}\right) + C_1 \Rightarrow C_1 = 100 \Rightarrow \frac{ds}{dt} = 2 \cos\left(2t - \frac{\pi}{2}\right) + 100$

$\Rightarrow s = \int \left(2 \cos\left(2t - \frac{\pi}{2}\right) + 100\right) dt = \int (\cos u + 50)\ du = \sin u + 50u + C_2 = \sin\left(2t - \frac{\pi}{2}\right) + 50\left(2t - \frac{\pi}{2}\right) + C_2;$

at $t = 0$ and $s = 0$ we have $0 = \sin\left(-\frac{\pi}{2}\right) + 50\left(-\frac{\pi}{2}\right) + C_2 \Rightarrow C_2 = 1 + 25\pi$

$\Rightarrow s = \sin\left(2t - \frac{\pi}{2}\right) + 100t - 25\pi + (1 + 25\pi) \Rightarrow s = \sin\left(2t - \frac{\pi}{2}\right) + 100t + 1$

57. Let $u = 2t \Rightarrow du = 2\ dt \Rightarrow 3\ du = 6\ dt$

$s = \int 6 \sin 2t\ dt = \int (\sin u)(3\ du) = -3 \cos u + C = -3 \cos 2t + C;$

at $t = 0$ and $s = 0$ we have $0 = -3 \cos 0 + C \Rightarrow C = 3 \Rightarrow s = 3 - 3 \cos 2t \Rightarrow s\left(\frac{\pi}{2}\right) = 3 - 3 \cos(\pi) = 6$ m

59. All three integrations are correct. In each case, the derivative of the function on the right is the integrand on the left, and each formula has an arbitrary constant for generating the remaining antiderivatives. Moreover, $\sin^2 x + C_1 = 1 - \cos^2 x + C_1 \Rightarrow C_2 = 1 + C_1;$ also $-\cos^2 x + C_2 = -\frac{\cos 2x}{2} - \frac{1}{2} + C_2 \Rightarrow C_3 = C_2 - \frac{1}{2} = C_1 + \frac{1}{2}.$

4.4 ESTIMATING WITH FINITE SUMS

1. Using midpoints of the intervals, Area $\approx (0.25)(2) + (1.0)(2) + (2.0)(2) + (3.25)(2) + (4.0)(2) + (4.0)(2)$
   $+ (3.35)(2) + (2.25)(2) + (1.3)(2) + (0.75)(2) + (0.25)(2) = 44.8$ mg $\cdot$ sec/L. Cardiac output
   $= \frac{\text{amount of dye}}{\text{area under curve}} \times 60 \approx \frac{5\ \text{mg}}{44.5\ \text{mg} \cdot \text{sec/L}} \times 60\ \frac{\text{sec}}{\text{min}} \approx 6.7$ L/min.

3. (a) $D \approx (0)(1) + (12)(1) + (22)(1) + (10)(1) + (5)(1) + (13)(1) + (11)(1) + (6)(1) + (2)(1) + (6)(1) = 87$ inches
   (b) $D \approx (12)(1) + (22)(1) + (10)(1) + (5)(1) + (13)(1) + (11)(1) + (6)(1) + (2)(1) + (6)(1) + (0)(1) = 87$ inches

5. (a) $D \approx (0)(10) + (44)(10) + (15)(10) + (35)(10) + (30)(10) + (44)(10) + (35)(10) + (15)(10) + (22)(10)$
   $+ (35)(10) + (44)(10) + (30)(10) = 3490$ feet ≈ 0.73 miles
   (b) $D \approx (44)(10) + (15)(10) + (35)(10) + (30)(10) + (44)(10) + (35)(10) + (15)(10) + (22)(10) + (35)(10)$
   $+ (44)(10) + (30)(10) + (35)(10) = 3840$ feet ≈ 0.66 miles

7. (a) $S_2 = 4\left(9 - (-2)^2\right)(2) + 4\left(9 - 0^2\right)(2) = 8(9 - 4) + 8(9) = 112$

   (b) $\frac{|V - S_2|}{V} = \frac{\left(\frac{368}{3}\right) - 112}{368} = \frac{32}{368} \approx 9\%$

9. (a) $S_4 = \pi\left[\sqrt{16-(-2)^2}\,\right]^2(2) + \pi\left[\sqrt{16-0^2}\,\right]^2(2) + \pi\left[\sqrt{16-(2)^2}\,\right]^2(2) = \pi[(16-4)+(16-0)+(16-4)](2)$

   $= 80\pi$

   (b) $\dfrac{|V-S_4|}{V} = \dfrac{\left(\frac{256}{3}\right)\pi - 80\pi}{\left(\frac{256}{3}\right)\pi} = \dfrac{16}{256} \approx 6\%$

11. (a) $S_8 = \pi\left[\left(16-0^2\right) + \left(16-\left(\frac{1}{2}\right)^2\right) + \left(16-(1)^2\right) + \left(16-\left(\frac{3}{2}\right)^2\right) + \left(16-(2)^2\right) + \left(16-\left(\frac{5}{2}\right)^2\right)\right.$

   $\left. + \left(16-(3)^2\right) + \left(16-\left(\frac{7}{2}\right)\right)^2\right]\left(\frac{1}{2}\right) = \frac{\pi}{2}\left[128 - \frac{1}{4} - 1 - \frac{9}{4} - 4 - \frac{25}{4} - 9 - \frac{49}{4}\right] = \frac{372\pi}{8} = \frac{93\pi}{2}$, overestimates

   (b) $V = \frac{2}{3}\pi r^3 = \frac{128\pi}{3} \Rightarrow \dfrac{|V-S_8|}{V} = \dfrac{\left(\frac{93}{2}\right)\pi - \left(\frac{128}{3}\right)\pi}{\left(\frac{128}{3}\right)\pi} = \dfrac{23}{256} \approx 9\%$

13. (a) $S_4 = (2f(1))^2\,\Delta x + (2f(2))^2\,\Delta x + (2f(3))^2\,\Delta x + (2f(4))^2\,\Delta x = 4(1)(1) + 4(2)(1) + 4(3)(1) + 4(4)(1) = 40$

   (b) $\dfrac{|V-S_4|}{V} = \dfrac{40-32}{32} = \dfrac{8}{32} = 25\%$

   (c) $S_8 = 4\left[\left(f\left(\frac{1}{2}\right)\right)^2 + (f(1))^2 + \left(f\left(\frac{3}{2}\right)\right)^2 + (f(2))^2 + \left(f\left(\frac{5}{2}\right)\right)^2 + (f(3))^2 + \left(f\left(\frac{7}{2}\right)\right)^2 + (f(4))^2\right]\left(\frac{1}{2}\right)$

   $= 2\left(\frac{1}{2} + 1 + \frac{3}{2} + 2 + \frac{5}{2} + 3 + \frac{7}{2} + 4\right) = 2\left(\frac{36}{2}\right) = 36;\ \dfrac{|V-S_8|}{V} = \dfrac{36-32}{32} = \dfrac{4}{32} = 12.5\%$

15. To have the same orientation as Exercise 11, tip the bowl sideways (assume the water is ice). The water covers the interval $[4,8]$. The function which will give us the values of the radii of the approximating cylinders is the equation of the upper semicircle formed by intersecting the hemisphere with the xy-plane,

   $f(x) = \sqrt{64 - x^2}$. Using $x = \frac{1}{2}$ and left-endpoints for

   each interval $\Rightarrow S_8 = \pi\left[\left(64-(4)^2\right) + \left(64-\left(\frac{9}{2}\right)^2\right)\right.$

   $+ \left(64-(5)^2\right) + \left(64-\left(\frac{11}{2}\right)^2\right) + \left(64-(6)^2\right)$

   $\left. + \left(64-\left(\frac{13}{2}\right)^2\right) + \left(64-(7)^2\right) + \left(64-\left(\frac{15}{2}\right)^2\right)\right]\left(\frac{1}{2}\right) = \frac{\pi}{2}\left(512 - 16 - \frac{81}{4} - 25 - \frac{121}{4} - 36 - \frac{169}{4} - 49 - \frac{225}{4}\right)$

   $= \frac{\pi}{2}\left(386 - \frac{596}{4}\right) = \frac{\pi}{8}(1544 - 596) = \frac{948}{8}\pi = 118.5\pi;\ \dfrac{|V-S_8|}{V} = \dfrac{\left|\left(\frac{320}{3}\right)\pi - \left(\frac{948}{8}\right)\pi\right|}{\left(\frac{320}{3}\right)\pi} = \dfrac{2844-2560}{2560} \approx 11\%$

17. (a) $S_5 = \pi\left[\left(\sqrt{0}\right)^2 + \left(\sqrt{1}\right)^2 + \left(\sqrt{2}\right)^2 + \left(\sqrt{3}\right)^2 + \left(\sqrt{4}\right)^2\right](1) = 10\pi$, underestimates

   (b) $\dfrac{|V-S_5|}{V} = \dfrac{\left(\frac{25}{2}\right)\pi - 10\pi}{\left(\frac{25}{2}\right)\pi} = \dfrac{5}{25} = 20\%$

19. Partition $[0,2]$ into the four subintervals $[0,0.5]$, $[0.5,1]$, $[1,1.5]$, and $[1.5,2]$. The midpoints of these subintervals are $m_1 = 0.25$, $m_2 = 0.75$, $m_3 = 1.25$, and $m_4 = 1.75$. The heights of the four approximating rectangles are $f(m_1) = (0.25)^3 = \frac{1}{64}$, $f(m_2) = (0.75)^3 = \frac{27}{64}$, $f(m_3) = (1.25)^3 = \frac{125}{64}$, and $f(m_4) = (1.75)^3 = \frac{343}{64}$

$\Rightarrow$ Average value $\approx \dfrac{\frac{1}{64} + \frac{27}{64} + \frac{125}{64} + \frac{343}{64}}{4} = \dfrac{1 + 27 + 125 + 343}{4 \cdot 64} = \dfrac{496}{256} = \dfrac{31}{16}$. Notice that the average value is

approximated by $\dfrac{\left(\frac{1}{4}\right)^3 + \left(\frac{3}{4}\right)^3 + \left(\frac{5}{4}\right)^3 + \left(\frac{7}{4}\right)^3}{4} = \frac{1}{2}\left[\left(\frac{1}{4}\right)^3\left(\frac{1}{2}\right) + \left(\frac{3}{4}\right)^3\left(\frac{1}{2}\right) + \left(\frac{5}{4}\right)^3\left(\frac{1}{2}\right) + \left(\frac{7}{4}\right)^3\left(\frac{1}{2}\right)\right]$

$= \dfrac{1}{\text{length of } [0,2]} \cdot \begin{bmatrix}\text{approximate area under}\\\text{curve } f(x) = x^3\end{bmatrix}$. We use this observation in solving the next several exercises.

21. Partition $[0,2]$ into the four subintervals $[0,0.5]$, $[0.5,1]$, $[1,1.5]$, and $[1.5,2]$. The midpoints of the subintervals are $m_1 = 0.25$, $m_2 = 0.75$, $m_3 = 1.25$, and $m_4 = 1.75$. The heights of the four approximating rectangles are $f(m_1) = \frac{1}{2} + \sin^2 \frac{\pi}{4} = \frac{1}{2} + \frac{1}{2} = 1$, $f(m_2) = \frac{1}{2} + \sin^2 \frac{3\pi}{4} = \frac{1}{2} + \frac{1}{2} = 1$, $f(m_3) = \frac{1}{2} + \sin^2 \frac{5\pi}{4} = \frac{1}{2} + \left(-\frac{1}{\sqrt{2}}\right)^2$

$= \frac{1}{2} + \frac{1}{2} = 1$, and $f(m_4) = \frac{1}{2} + \sin^2 \frac{7\pi}{4} = \frac{1}{2} + \left(-\frac{1}{\sqrt{2}}\right)^2 = 1$. The width of each rectangle is $\Delta x = \frac{1}{2}$. Thus,

Area $\approx (1 + 1 + 1 + 1)\left(\frac{1}{2}\right) = 2 \Rightarrow$ average value $\approx \dfrac{\text{area}}{\text{length of } [0,2]} = \dfrac{2}{2} = 1$.

23. (a) Because the acceleration is decreasing, an upper estimate is obtained using left end-points in summing acceleration $\cdot \Delta t$. Thus, $\Delta t = 1$ and speed $\approx [32.00 + 19.41 + 11.77 + 7.14 + 4.33](1) = 74.65$ ft/sec

(b) Using right end-points we obtain a lower estimate: speed $\approx [19.41 + 11.77 + 7.14 + 4.33 + 2.63](1)$

  $= 45.28$ ft/sec

(c) Upper estimates for the speed at each second are:

t	0	1	2	3	4	5
v	0	32.00	51.41	63.18	70.32	74.65

Thus, the distance fallen when $t = 3$ seconds is $s \approx [32.00 + 51.41 + 63.18](1) = 146.59$ ft.

25. Since the leakage is increasing, an upper estimate uses right end-points and a lower estimate uses left end-points:

(a) upper estimate $= (70)(1) + (97)(1) + (136)(1) + (190)(1) + (265)(1) = 758$ gal,

  lower estimate $= (50)(1) + (70)(1) + (97)(1) + (136)(1) + (190)(1) = 543$ gal.

(b) upper estimate $= (70 + 97 + 136 + 190 + 265 + 369 + 516 + 720) = 2363$ gal,

  lower estimate $= (50 + 70 + 97 + 136 + 190 + 265 + 369 + 516) = 1693$ gal.

(c) worst case: $2363 + 720t = 25,000 \Rightarrow t \approx 31.4$ hrs;

  best case: $1693 + 720t = 25,000 \Rightarrow t \approx 32.4$ hrs

4.5 RIEMANN SUMS AND DEFINITE INTEGRALS

1. $\displaystyle\sum_{k=1}^{2} \frac{6k}{k+1} = \frac{6(1)}{1+1} + \frac{6(2)}{2+1} = \frac{6}{2} + \frac{12}{3} = 7$

3. $\displaystyle\sum_{k=1}^{4} \cos k\pi = \cos(1\pi) + \cos(2\pi) + \cos(3\pi) + \cos(4\pi) = -1 + 1 - 1 + 1 = 0$

5. $\displaystyle\sum_{k=1}^{3} (-1)^{k+1} \sin \frac{\pi}{k} = (-1)^{1+1} \sin \frac{\pi}{1} + (-1)^{2+1} \sin \frac{\pi}{2} + (-1)^{3+1} \sin \frac{\pi}{3} = 0 - 1 + \frac{\sqrt{3}}{2} = \frac{\sqrt{3}-2}{2}$

7. (a) $\displaystyle\sum_{k=1}^{6} 2^{k-1} = 2^{1-1} + 2^{2-1} + 2^{3-1} + 2^{4-1} + 2^{5-1} + 2^{6-1} = 1 + 2 + 4 + 8 + 16 + 32$

   (b) $\displaystyle\sum_{k=0}^{5} 2^{k} = 2^{0} + 2^{1} + 2^{2} + 2^{3} + 2^{4} + 2^{5} = 1 + 2 + 4 + 8 + 16 + 32$

   (c) $\displaystyle\sum_{k=-1}^{4} 2^{k+1} = 2^{-1+1} + 2^{0+1} + 2^{1+1} + 2^{2+1} + 2^{3+1} + 2^{4+1} = 1 + 2 + 4 + 8 + 16 + 32$

   All of them represent $1 + 2 + 4 + 8 + 16 + 32$

9. (a) $\displaystyle\sum_{k=2}^{4} \frac{(-1)^{k-1}}{k-1} = \frac{(-1)^{2-1}}{2-1} + \frac{(-1)^{3-1}}{3-1} + \frac{(-1)^{4-1}}{4-1} = -1 + \frac{1}{2} - \frac{1}{3}$

   (b) $\displaystyle\sum_{k=0}^{2} \frac{(-1)^{k}}{k+1} = \frac{(-1)^{0}}{0+1} + \frac{(-1)^{1}}{1+1} + \frac{(-1)^{2}}{2+1} = 1 - \frac{1}{2} + \frac{1}{3}$

   (c) $\displaystyle\sum_{k=-1}^{1} \frac{(-1)^{k}}{k+2} = \frac{(-1)^{-1}}{-1+2} + \frac{(-1)^{0}}{0+2} + \frac{(-1)^{1}}{1+2} = -1 + \frac{1}{2} - \frac{1}{3}$

   (a) and (c) are equivalent; (b) is not equivalent to the other two.

11. $\displaystyle\sum_{k=1}^{6} k$ 13. $\displaystyle\sum_{k=1}^{4} \frac{1}{2^{k}}$ 15. $\displaystyle\sum_{k=1}^{5} (-1)^{k+1} \frac{1}{k}$

17. (a) $\displaystyle\sum_{k=1}^{n} 3a_{k} = 3 \sum_{k=1}^{n} a_{k} = 3(-5) = -15$

   (b) $\displaystyle\sum_{k=1}^{n} \frac{b_{k}}{6} = \frac{1}{6} \sum_{k=1}^{n} b_{k} = \frac{1}{6}(6) = 1$

   (c) $\displaystyle\sum_{k=1}^{n} (a_{k} + b_{k}) = \sum_{k=1}^{n} a_{k} + \sum_{k=1}^{n} b_{k} = -5 + 6 = 1$

   (d) $\displaystyle\sum_{k=1}^{n} (a_{k} - b_{k}) = \sum_{k=1}^{n} a_{k} - \sum_{k=1}^{n} b_{k} = -5 - 6 = -11$

   (e) $\displaystyle\sum_{k=1}^{n} (b_{k} - 2a_{k}) = \sum_{k=1}^{n} b_{k} - 2 \sum_{k=1}^{n} a_{k} = 6 - 2(-5) = 16$

19. (a) $\displaystyle\sum_{k=1}^{10} k = \frac{10(10+1)}{2} = 55$ (b) $\displaystyle\sum_{k=1}^{10} k^2 = \frac{10(10+1)(2(10)+1)}{6} = 385$

(c) $\displaystyle\sum_{k=1}^{10} k^3 = \left[\frac{10(10+1)}{2}\right]^2 = 55^2 = 3025$

21. $\displaystyle\sum_{k=1}^{7} -2k = -2\sum_{k=1}^{7} k = -2\left(\frac{7(7+1)}{2}\right) = -56$

23. $\displaystyle\sum_{k=1}^{6} \left(3 - k^2\right) = \sum_{k=1}^{6} 3 - \sum_{k=1}^{6} k^2 = 3(6) - \frac{6(6+1)(2(6)+1)}{6} = -73$

25. $\displaystyle\sum_{k=1}^{5} k(3k+5) = \sum_{k=1}^{5} \left(3k^2 + 5k\right) = 3\sum_{k=1}^{5} k^2 + 5\sum_{k=1}^{5} k = 3\left(\frac{5(5+1)(2(5)+1)}{6}\right) + 5\left(\frac{5(5+1)}{2}\right) = 240$

27. $\displaystyle\sum_{k=1}^{5} \frac{k^3}{225} + \left(\sum_{k=1}^{5} k\right)^3 = \frac{1}{225}\sum_{k=1}^{5} k^3 + \left(\sum_{k=1}^{5} k\right)^3 = \frac{1}{225}\left(\frac{5(5+1)}{2}\right)^2 + \left(\frac{5(5+1)}{2}\right)^3 = 3376$

29. (a)   (b)   (c)

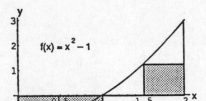 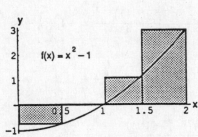 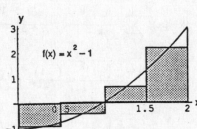

31. (a)   (b)   (c)

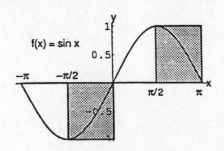 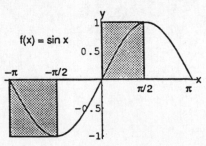 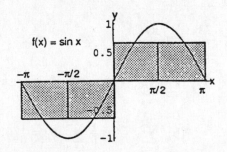

33. $|x_1 - x_0| = |1.2 - 0| = 1.2$, $|x_2 - x_1| = |1.5 - 1.2| = 0.3$, $|x_3 - x_2| = |2.3 - 1.5| = 0.8$, $|x_4 - x_3| = |2.6 - 2.3| = 0.3$, and $|x_5 - x_4| = |3 - 2.6| = 0.4$; the largest is $\|P\| = 1.2$.

35. $\displaystyle\int_{0}^{2} x^2 \, dx$ 37. $\displaystyle\int_{-7}^{5} \left(x^2 - 3x\right) dx$ 39. $\displaystyle\int_{2}^{3} \frac{1}{1-x} \, dx$

41. $\displaystyle\int_{-\pi/4}^{0} (\sec x) \, dx$ 43. $\displaystyle\int_{-2}^{1} 5 \, dx = 5(1 - (-2)) = 15$

45. $\displaystyle\int_0^3 (-160)\, dx = (-160)(3 - 0) = -480$

47. $\displaystyle\int_{-2.1}^{3.4} 0.5\, dx = 0.5(3.4 - (-2.1)) = 2.75$

49. The area of the trapezoid is $A = \frac{1}{2}(B + b)h$

$= \frac{1}{2}(5 + 2)(6) = 21 \Rightarrow \displaystyle\int_{-2}^{4} \left(\frac{x}{2} + 3\right) dx$

$= 21$ square units

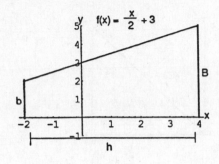

51. The area of the semicircle is $A = \frac{1}{2}\pi r^2 = \frac{1}{2}\pi(3)^2$

$= \frac{9}{2}\pi \Rightarrow \displaystyle\int_{-5}^{3} \sqrt{9 - x^2}\, dx = \frac{9}{2}\pi$ square units

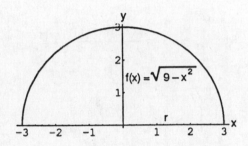

53. The area of the triangle on the left is $A = \frac{1}{2}bh = \frac{1}{2}(2)(2)$

$= 2$. The area of the triangle on the right is $A = \frac{1}{2}bh$

$= \frac{1}{2}(1)(1) = \frac{1}{2}$. Then, the total area is $2.5 \Rightarrow \displaystyle\int_{-2}^{1} |x|\, dx$

$= 2.5$ square units

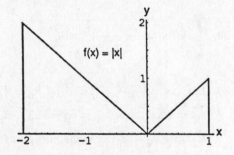

55. The area of the triangular peak is $A = \frac{1}{2}bh = \frac{1}{2}(2)(1) = 1$.

The area of the rectangular base is $S = \ell w = (2)(1) = 2$.

Then the total area is $3 \Rightarrow \displaystyle\int_{-1}^{1} \left(2 - |x|\right) dx = 3$ square units

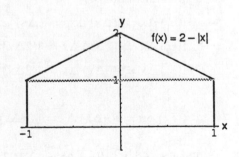

57. $\displaystyle\int_0^b x\,dx = \frac{1}{2}(b)(b) = \frac{b^2}{2}$

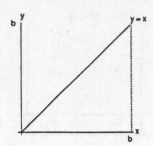

59. $\displaystyle\int_a^b 2s\,ds = \frac{1}{2}b(2b) - \frac{1}{2}a(2a) = b^2 - a^2$

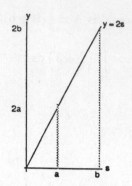

61. $\displaystyle\int_1^{\sqrt{2}} x\,dx = \frac{\left(\sqrt{2}\right)^2}{2} - \frac{(1)^2}{2} = \frac{1}{2}$

63. $\displaystyle\int_\pi^{2\pi} \theta\,d\theta = \frac{(2\pi)^2}{2} - \frac{\pi^2}{2} = \frac{3\pi^2}{2}$

65. $\displaystyle\int_0^{\sqrt[3]{7}} x^2\,dx = \frac{\left(\sqrt[3]{7}\right)^3}{3} = \frac{7}{3}$

67. $\displaystyle\int_0^{1/2} t^2\,dt = \frac{\left(\frac{1}{2}\right)^3}{3} = \frac{1}{24}$

69. $\displaystyle\int_a^{2a} x\,dx = \frac{(2a)^2}{2} - \frac{a^2}{2} = \frac{3a^2}{2}$

71. $\displaystyle\int_0^{\sqrt[3]{b}} x^2\,dx = \frac{\left(\sqrt[3]{b}\right)^3}{3} = \frac{b}{3}$

73. Let $\Delta x = \dfrac{b-0}{n} = \dfrac{b}{n}$ and let $x_0 = 0$, $x_1 = \Delta x$,

$x_2 = 2\Delta x, \ldots, x_{n-1} = (n-1)\Delta x$, $x_n = n\Delta x = b$.

Let the c_k's be the right end-points of the subintervals

$\Rightarrow c_1 = x_1$, $c_2 = x_2$, and so on.   The rectangles

defined have areas:

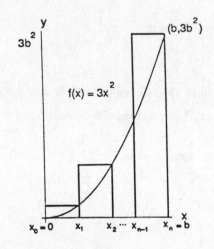

$f(c_1)\,\Delta x = f(\Delta x)\,\Delta x = 3(\Delta x)^2\,\Delta x = 3(\Delta x)^3$

$f(c_2)\,\Delta x = f(2\Delta x)\,\Delta x = 3(2\Delta x)^2\,\Delta x = 3(2)^2(\Delta x)^3$

$f(c_3)\,\Delta x = f(3\Delta x)\,\Delta x = 3(3\Delta x)^2\,\Delta x = 3(3)^2(\Delta x)^3$

$\vdots$

$f(c_n)\,\Delta x = f(n\Delta x)\,\Delta x = 3(n\Delta x)^2\,\Delta x = 3(n)^2(\Delta x)^3$

Then $\displaystyle S_n = \sum_{k=1}^n f(c_k)\,\Delta x = \sum_{k=1}^n 3k^2(\Delta x)^3 = 3(\Delta x)^3 \sum_{k=1}^n k^2 = 3\left(\frac{b^3}{n^3}\right)\left(\frac{n(n+1)(2n+1)}{6}\right) = \frac{b^3}{2}\left(2 + \frac{3}{n} + \frac{1}{n^2}\right)$

$\Rightarrow \displaystyle\int_0^b 3x^2\,dx = \lim_{n\to\infty} \frac{b^3}{2}\left(2 + \frac{3}{n} + \frac{1}{n^2}\right) = b^3.$

75. Let $\Delta x = \frac{b-0}{n} = \frac{b}{n}$ and let $x_0 = 0$, $x_1 = \Delta x$,

$x_2 = 2\Delta x, \ldots, x_{n-1} = (n-1)\Delta x$, $x_n = n\Delta x = b$.

Let the c_k's be the right end-points of the subintervals

$\Rightarrow c_1 = x_1$, $c_2 = x_2$, and so on. The rectangles

defined have areas:

$f(c_1)\,\Delta x = f(\Delta x)\,\Delta x = 2(\Delta x)(\Delta x) = 2(\Delta x)^2$

$f(c_2)\,\Delta x = f(2\Delta x)\,\Delta x = 2(2\Delta x)(\Delta x) = 2(2)(\Delta x)^2$

$f(c_3)\,\Delta x = f(3\Delta x)\,\Delta x = 2(3\Delta x)(\Delta x) = 2(3)(\Delta x)^2$

$\vdots$

$f(c_n)\,\Delta x = f(n\Delta x)\,\Delta x = 2(n\Delta x)(\Delta x) = 2(n)(\Delta x)^2$

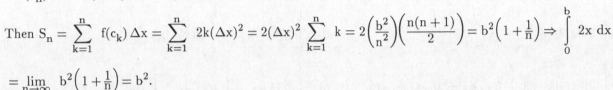

Then $S_n = \sum_{k=1}^{n} f(c_k)\,\Delta x = \sum_{k=1}^{n} 2k(\Delta x)^2 = 2(\Delta x)^2 \sum_{k=1}^{n} k = 2\left(\frac{b^2}{n^2}\right)\left(\frac{n(n+1)}{2}\right) = b^2\left(1+\frac{1}{n}\right) \Rightarrow \int_0^b 2x\,dx$

$= \lim_{n\to\infty} b^2\left(1+\frac{1}{n}\right) = b^2.$

77. To find where $x - x^2 \geq 0$, let $x - x^2 = 0 \Rightarrow x(1-x) = 0 \Rightarrow x = 0$ or $x = 1$. If $0 < x < 1$, then $x^2 < x$

$\Rightarrow 0 < x - x^2 \Rightarrow a = 0$ and $b = 1$ maximize the integral.

79. (a) $U = \max_1 \Delta x + \max_2 \Delta x + \ldots + \max_n \Delta x$ where $\max_1 = f(x_1)$, $\max_2 = f(x_2)$, ..., $\max_n = f(x_n)$ since f is

increasing on $[a,b]$; $L = \min_1 \Delta x + \min_2 \Delta x + \ldots + \min_n \Delta x$ where $\min_1 = f(x_0)$, $\min_2 = f(x_1), \ldots,$

$\min_n = f(x_{n-1})$ since f is increasing on $[a,b]$. Therefore

$U - L = (\max_1 - \min_1)\,\Delta x + (\max_2 - \min_2)\,\Delta x + \ldots + (\max_n - \min_n)\,\Delta x$

$= (f(x_1) - f(x_0))\,\Delta x + (f(x_2) - f(x_1))\Delta x + \ldots + (f(x_n) - f(x_{n-1}))\,\Delta x = (f(x_n) - f(x_0))\,\Delta x = (f(b) - f(a))\,\Delta x.$

(b) $U = \max_1 \Delta x_1 + \max_2 \Delta x_2 + \ldots + \max_n \Delta x_n$ where $\max_1 = f(x_1)$, $\max_2 = f(x_2)$, ..., $\max_n = f(x_n)$ since f

is increasing on $[a,b]$; $L = \min_1 \Delta x_1 + \min_2 \Delta x_2 + \ldots + \min_n \Delta x_n$ where

$\min_1 = f(x_0)$, $\min_2 = f(x_1), \ldots, \min_n = f(x_{n-1})$ since f is increasing on $[a,b]$. Therefore

$U - L = (\max_1 - \min_1)\,\Delta x_1 + (\max_2 - \min_2)\,\Delta x_2 + \ldots + (\max_n - \min_n)\,\Delta x_n$

$= (f(x_1) - f(x_0))\,\Delta x_1 + (f(x_2) - f(x_1))\Delta x_2 + \ldots + (f(x_n) - f(x_{n-1}))\,\Delta x_n$

$\leq (f(x_1) - f(x_0))\,\Delta x_{max} + (f(x_2) - f(x_1))\,\Delta x_{max} + \ldots + (f(x_n) - f(x_{n-1}))\,\Delta x_{max} = \|P\|.$ Then

$U - L \leq (f(x_n) - f(x_0))\,\Delta x_{max} = (f(b) - f(a))\,\Delta x_{max} = |f(b) - f(a)|\,\Delta x_{max}$ since $f(b) \geq f(a)$. Thus

$\lim_{\|P\|\to 0} (U - L) = \lim_{\|P\|\to 0} (f(b) - f(a))\,\Delta x_{max} = 0$, since $\Delta x_{max} = \|P\|.$

81. Partition the interval $[0,b]$ into n subintervals, each of length $\Delta x = \frac{b}{n}$ with the points

$x_0 = 0$, $x_1 = \Delta x$, $x_2 = 2\Delta x$, ..., $x_{n-1} = (n-1)\Delta x$, $x_n = n\Delta x = b$. The inscribed rectangles so determined have

areas: $f(x_0)\,\Delta x = (0)\,\Delta x$, $f(x_1)\,\Delta x = (\Delta x)^2\,\Delta x$, $f(x_2)\,\Delta x = (2\Delta x)^2\,\Delta x$, ..., $f(x_{n-1}) = ((n-1)\Delta x)^2\,\Delta x$. The sum

of these areas is $S_n = \left(0 + 1^2 + 2^2 + \ldots + (n-1)^2\right)\Delta x^3 = \frac{(n-1)(n)(2(n-1)+1)}{6}\left(\frac{b}{n}\right)^3 = \frac{b^3}{6}\left(\frac{n}{n}\right)\left(\frac{n-1}{n}\right)\left(\frac{2n-1}{n}\right)$

$\Rightarrow \int_0^b x^2\,dx = \lim_{n\to\infty} S_n = \lim_{n\to\infty}\left(\frac{b^3}{6}\left(\frac{n}{n}\right)\left(\frac{n-1}{n}\right)\left(\frac{2n-1}{n}\right)\right) = \left(\frac{b^3}{6}\right)(1)(1)(2) = \frac{b^3}{3}.$

83. Partition $[0,1]$ into n subintervals, each of length $\Delta x = \frac{1}{n}$ with the points $x_0 = 0$, $x_1 = \frac{1}{n}$, $x_2 = \frac{2}{n}$, ..., $x_n = \frac{n}{n}$ $= 1$. The inscribed rectangles so determined have areas

$f(x_0)\,\Delta x = (0)^2\,\Delta x$, $f(x_1)\,\Delta x = \left(\frac{1}{n}\right)^2\Delta x$, $f(x_2)\,\Delta x = \left(\frac{2}{n}\right)^2\Delta x$, ..., $f(x_{n-1}) = \left(\frac{n-1}{n}\right)^2\Delta x$. The sum of these areas

is $S_n = \left(0^2 + \left(\frac{1}{n}\right)^2 + \left(\frac{2}{n}\right)^2 + \cdots + \left(\frac{n-1}{n}\right)^2\right)\Delta x = \left(\frac{1^2}{n^2} + \frac{2^2}{n^2} + \cdots + \frac{(n-1)^2}{n^2}\right)\frac{1}{n} = \frac{1^2}{n^3} + \frac{2^2}{n^3} + \cdots + \frac{(n-1)^2}{n^3}$. Then

$$\lim_{n\to\infty} S_n = \lim_{n\to\infty}\left(\frac{1^2}{n^3} + \frac{2^2}{n^3} + \cdots + \frac{(n-1)^2}{n^3}\right) = \int_0^1 x^2\,dx = \frac{1^3}{3} = \frac{1}{3}.$$

4.6 PROPERTIES, AREA, AND THE MEAN VALUE THEOREM

1. (a) $\displaystyle\int_2^2 g(x)\,dx = 0$

   (b) $\displaystyle\int_5^1 g(x)\,dx = -\int_1^5 g(x)\,dx = -8$

   (c) $\displaystyle\int_1^2 3f(x)\,dx = 3\int_1^2 f(x)\,dx = 3(-4) = -12$

   (d) $\displaystyle\int_2^5 f(x)\,dx = \int_1^5 f(x)\,dx - \int_1^2 f(x)\,dx = 6 - (-4) = 10$

   (e) $\displaystyle\int_1^5 [f(x) - g(x)]\,dx = \int_1^5 f(x)\,dx - \int_1^5 g(x)\,dx = 6 - 8 = -2$

   (f) $\displaystyle\int_1^5 [4f(x) - g(x)]\,dx = 4\int_1^5 f(x)\,dx - \int_1^5 g(x)\,dx = 4(6) - 8 = 16$

3. (a) $\displaystyle\int_1^2 f(u)\,du = \int_1^2 f(x)\,dx = 5$

   (b) $\displaystyle\int_1^2 \sqrt{3}\,f(z)\,dz = \sqrt{3}\int_1^2 f(z)\,dz = 5\sqrt{3}$

   (c) $\displaystyle\int_2^1 f(t)\,dt = -\int_1^2 f(t)\,dt = -5$

   (d) $\displaystyle\int_1^2 [-f(x)]\,dx = -\int_1^2 f(x)\,dx = -5$

5. (a) $\displaystyle\int_3^4 f(z)\,dz = \int_0^4 f(z)\,dz - \int_0^3 f(z)\,dz = 7 - 3 = 4$

   (b) $\displaystyle\int_4^3 f(t)\,dt = -\int_3^4 f(t)\,dt = -4$

7. $\displaystyle\int_3^1 7\,dx = 7(1 - 3) = -14$

9. $\displaystyle\int_0^2 5x\,dx = 5\int_0^2 x\,dx = 5\left[\frac{2^2}{2} - \frac{0^2}{2}\right] = 10$

11. $\displaystyle\int_0^2 (2t - 3)\,dt = 2\int_1^1 t\,dt - \int_0^2 3\,dt = 2\left[\frac{2^2}{2} - \frac{0^2}{2}\right] - 3(2 - 0) = 4 - 6 = -2$

13. $\int_2^1 \left(1+\frac{z}{2}\right) dz = \int_2^1 1\, dz + \int_2^1 \frac{z}{2}\, dz = \int_2^1 1\, dz - \frac{1}{2} \int_1^2 z\, dz = 1[1-2] - \frac{1}{2}\left[\frac{2^2}{2} - \frac{1^2}{2}\right] = -1 - \frac{1}{2}\left(\frac{3}{2}\right) = -\frac{7}{4}$

15. $\int_1^2 3u^2\, du = 3 \int_1^2 u^2\, du = 3\left[\int_0^2 u^2\, du - \int_0^1 u^2\, du\right] = 3\left(\left[\frac{2^3}{3} - \frac{0^3}{3}\right] - \left[\frac{1^3}{3} - \frac{0^3}{3}\right]\right) = 3\left[\frac{2^3}{3} - \frac{1^3}{3}\right] = 3\left(\frac{7}{3}\right) = 7$

17. $\int_0^2 \left(3x^2 + x - 5\right) dx = 3 \int_0^2 x^2\, dx + \int_0^2 x\, dx - \int_0^2 5\, dx = 3\left[\frac{2^3}{3} - \frac{0^3}{3}\right] + \left[\frac{2^2}{2} - \frac{0^2}{2}\right] - 5[2-0] = (8+2) - 10 = 0$

19. The area of R_1 is $\int_0^1 3\, dx = 3(1-0)$ (the area of the

rectangle). The area of R_2 is the area of the rectangle

bounded by the lines $x=1$, $x=2$, $y=0$, and $y=3$

minus the area under the curve over $[1,2] \Rightarrow$ Area of R_2

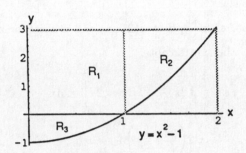

$= (3)(1) - \int_1^2 \left(x^2-1\right) dx = 3 - \int_1^2 x^2\, dx - \int_1^2 (-1)\, dx$

$= 3 - \left[\int_0^2 x^2\, dx - \int_0^1 x^2\, dx\right] + \int_1^2 1\, dx = 3 - \left[\frac{2^3}{3} - \frac{1^3}{3}\right] + 1(2-1) = \frac{5}{3}.$ The area of R_3

$= -\int_0^1 \left(x^2-1\right) dx = -\int_0^1 x^2\, dx + \int_0^1 1\, dx = -\left(\frac{1^3}{3} - \frac{0^3}{3}\right) + 1(1-0) = \frac{2}{3}.$ Therefore the total area is

$R_1 + R_2 + R_3 = 3 + \frac{5}{3} + \frac{2}{3} = \frac{16}{3}$ or $5\frac{1}{3}.$

21. The shaded area is $\int_0^3 \left(3x - x^2\right) dx - \int_3^4 \left(3x - x^2\right) dx = 3 \int_0^3 x\, dx - \int_0^3 x^2\, dx - 3 \int_3^4 x\, dx + \int_3^4 x^2\, dx$

$= 3\left(\frac{3^2}{2} - \frac{0^2}{2}\right) - \frac{3^3}{3} - 3\left(\frac{4^2}{2} - \frac{3^2}{2}\right) + \int_0^4 x^2\, dx - \int_0^3 x^2\, dx = \left(\frac{27}{2} - 9 - \frac{21}{2}\right) + \left(\frac{4^3}{3} - \frac{3^3}{3}\right) = -6 + \frac{37}{3} = \frac{19}{3}$

23. $x^2 - 6x + 8 = 0 \Rightarrow (x-4)(x-2) = 0 \Rightarrow x = 4$ or

$x = 2$, the x-intercepts.

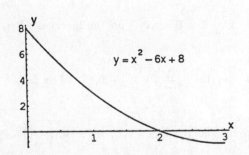

(a) $\int_0^3 \left(x^2 - 6x + 8\right) dx = \int_0^3 x^2\, dx - 6 \int_0^3 x\, dx + \int_0^3 8\, dx$

$= \frac{3^3}{3} - 6\left(\frac{3^2}{2} - \frac{0^2}{2}\right) + 8(3-0) = 6$

(b) Area $= \int\limits_0^2 (x^2 - 6x + 8)\,dx + \left(- \int\limits_2^3 (x^2 - 6x + 8)\,dx\right)$

$= \left(\int\limits_0^2 x^2\,dx - 6\int\limits_0^2 x\,dx + \int\limits_0^2 8\,dx\right) - \left(\int\limits_2^3 x^2\,dx - 6\int\limits_2^3 x\,dx + \int\limits_2^3 8\,dx\right)$

$= \left[\frac{2^3}{3} - 6\left(\frac{2^2}{2} - \frac{0^2}{2}\right) + 8(2-0)\right] - \left(\int\limits_0^3 x^2\,dx - \int\limits_0^2 x^2\,dx - 6\left(\frac{3^2}{2} - \frac{2^2}{2}\right) + 8(3-2)\right)$

$= \left(\frac{8}{3} - 12 + 16\right) - \left(\frac{3^3}{3} - \frac{2^3}{3} - 15 + 8\right) = \frac{22}{3} = 7\frac{1}{3}$

25. $2x - x^2 = 0 \Rightarrow x(2 - x) = 0 \Rightarrow x = 0$ or $x = 2$,

the x-intercepts.

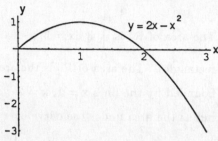

$y = 2x - x^2$

(a) $\int\limits_0^3 (2x - x^2)\,dx = 2\int\limits_0^3 x\,dx - \int\limits_0^3 x^2\,dx$

$= 2\left(\frac{3^2}{2} - \frac{0^2}{2}\right) - \frac{3^3}{3} = 0$

(b) Area $= \int\limits_0^2 (2x - x^2)\,dx - \int\limits_2^3 (2x - x^2)\,dx = 2\int\limits_0^2 x\,dx - \int\limits_0^2 x^2\,dx - \left(2\int\limits_2^3 x\,dx - \int\limits_2^3 x^2\,dx\right)$

$= 2\left(\frac{2^2}{2} - \frac{0^2}{2}\right) - \frac{2^3}{3} - 2\left(\frac{3^2}{2} - \frac{2^2}{2}\right) + \left(\int\limits_0^3 x^2\,dx - \int\limits_0^2 x^2\,dx\right) = 4 - \frac{8}{3} - 5 + \frac{3^3}{3} - \frac{2^3}{3} = \frac{8}{3}$

27. $\text{av}(f) = \left(\frac{1}{\sqrt{3} - 0}\right)\int\limits_0^{\sqrt{3}} (x^2 - 1)\,dx$

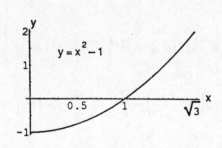

$y = x^2 - 1$

$= \frac{1}{\sqrt{3}}\int\limits_0^{\sqrt{3}} x^2\,dx - \frac{1}{\sqrt{3}}\int\limits_0^{\sqrt{3}} 1\,dx$

$= \frac{1}{\sqrt{3}}\left(\frac{(\sqrt{3})^3}{3}\right) - \frac{1}{\sqrt{3}}(\sqrt{3} - 0) = 1 - 1 = 0;$

$x^2 - 1 = 0 = \text{av}(f)$ on the interval when $x = 1$.

29. $\text{av}(f) = \left(\frac{1}{1 - 0}\right)\int\limits_0^1 (-3x^2 - 1)\,dx =$

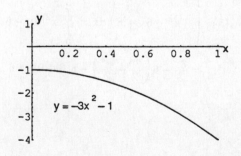

$= -3\int\limits_0^1 x^2\,dx - \int\limits_0^1 1\,dx = -3\left(\frac{1^3}{3}\right) - (1 - 0)$

$y = -3x^2 - 1$

$= -2; \; -3x^2 - 1 = -2 = \text{av}(f) \Rightarrow -3x^2 = -1$

$\Rightarrow x^2 = \frac{1}{3} \Rightarrow x = \pm\frac{1}{\sqrt{3}} \Rightarrow x = \frac{\sqrt{3}}{3}$ on the interval.

31. $av(f) = \left(\frac{1}{3-0}\right) \int_0^3 (t-1)^2 \, dt$

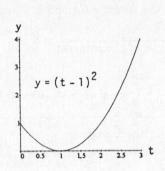

$y = (t-1)^2$

$= \frac{1}{3} \int_0^3 t^2 \, dt - \frac{2}{3} \int_0^3 t \, dt + \frac{1}{3} \int_0^3 1 \, dt$

$= \frac{1}{3}\left(\frac{3^3}{3}\right) - \frac{2}{3}\left(\frac{3^2}{2} - \frac{0^2}{2}\right) + \frac{1}{3}(3-0) = 1;$

$(t-1)^2 = 1 = av(f) \Rightarrow t-1 = \pm 1 \Rightarrow$

$t = 0$ and $t = 2$ on the interval.

33. (a) $av(g) = \left(\frac{1}{1-(-1)}\right) \int_{-1}^1 \left(|x|-1\right) dx$

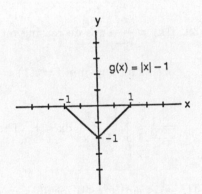

$g(x) = |x| - 1$

$= \frac{1}{2} \int_{-1}^0 (-x-1) \, dx + \frac{1}{2} \int_0^1 (x-1) \, dx$

$= -\frac{1}{2} \int_{-1}^0 x \, dx - \frac{1}{2} \int_{-1}^0 1 \, dx + \frac{1}{2} \int_0^1 x \, dx - \frac{1}{2} \int_0^1 1 \, dx$

$= -\frac{1}{2}\left(\frac{0^2}{2} - \frac{(-1)^2}{2}\right) - \frac{1}{2}(0-(-1)) + \frac{1}{2}\left(\frac{1^2}{2} - \frac{0^2}{2}\right) - \frac{1}{2}(1-0)$

$= -\frac{1}{2}; |x|-1 = -\frac{1}{2} = av(g) \Rightarrow |x| = \frac{1}{2} \Rightarrow x = \pm\frac{1}{2}$ on the interval.

(b) $av(g) = \left(\frac{1}{3-1}\right) \int_1^3 \left(|x|-1\right) dx = \frac{1}{2} \int_1^3 (x-1) \, dx$

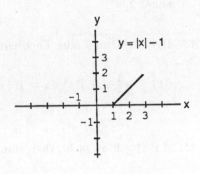

$y = |x| - 1$

$= \frac{1}{2} \int_1^3 x \, dx - \frac{1}{2} \int_1^3 1 \, dx = \frac{1}{2}\left(\frac{3^2}{2} - \frac{1^2}{2}\right) - \frac{1}{2}(3-1)$

$= 1; |x|-1 = 1 = av(g) \Rightarrow |x| = 2 \Rightarrow x = 2$ on

the interval.

(c) $av(g) = \left(\frac{1}{3-(-1)}\right) \int_{-1}^3 \left(|x|-1\right) dx$

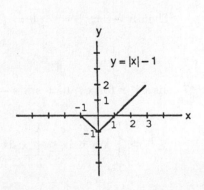

$y = |x| - 1$

$= \frac{1}{4} \int_{-1}^1 \left(|x|-1\right) dx + \frac{1}{4} \int_1^3 \left(|x|-1\right) dx$

$= \frac{1}{4}(-1+2) = \frac{1}{4}$ (see parts (a) and (b) above);

$|x|-1 = \frac{1}{4} = av(g) \Rightarrow |x| = \frac{5}{4} \Rightarrow x = \frac{5}{4}$ on the

interval.

35. The height of the triangle formed is 3 and the base is $6 \left(|2 - (-4)| \right) \Rightarrow$ the area is $\frac{1}{2} bh = \frac{1}{2}(6)(3) = 9$

$\Rightarrow av(f) = \dfrac{\text{area}}{\text{length of the interval}} = \dfrac{9}{6} = \dfrac{3}{2}.$

37. $av(f) = \left(\dfrac{1}{2\pi - 0} \right) \displaystyle\int_{0}^{2\pi} \sin t \, dt = \dfrac{1}{2\pi}(0) = 0$ because of

symmetry about the x-axis.

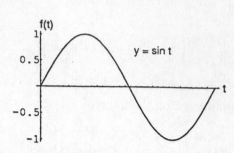

39. $f(x) = \dfrac{1}{1 + x^2}$ is decreasing on $[0, 1] \Rightarrow$ maximum value of f occurs at $0 \Rightarrow \max f = f(0) = 1$; minimum value of f

occurs at $1 \Rightarrow \min f = f(1) = \dfrac{1}{1 + 1^2} = \dfrac{1}{2}$. Therefore, $(1 - 0) \min f \le \displaystyle\int_{0}^{1} \dfrac{1}{1 + x^2} \, dx \le (1 - 0) \max f$

$\Rightarrow \dfrac{1}{2} \le \displaystyle\int_{0}^{1} \dfrac{1}{1 + x^2} \, dx \le 1.$ That is, an upper bound $= 1$ and a lower bound $= \dfrac{1}{2}$.

41. $-1 \le \sin(x^2) \le 1$ for all $x \Rightarrow (1 - 0)(-1) \le \displaystyle\int_{0}^{1} \sin(x^2) \, dx \le (1 - 0)(1)$ or $\displaystyle\int_{0}^{1} \sin x^2 \, dx \le 1 \Rightarrow \displaystyle\int_{0}^{1} \sin x^2 \, dx$ cannot

equal 2.

43. By the Mean Value Theorem for Integrals, since f is continuous, then at some point c in $[1, 2]$,

$f(c) = \dfrac{1}{2 - 1} \displaystyle\int_{1}^{2} f(x) \, dx = 1(4) = 4.$

45. If $f(x) \ge 0$ on $[a, b]$, then $\min f \ge 0$ and $\max f \ge 0$ on $[a, b]$. Now, $(b - a) \min f \le \displaystyle\int_{a}^{b} f(x) \, dx \le (b - a) \max f.$

Then $b \ge a \Rightarrow b - a \ge 0 \Rightarrow (b - a) \min f \ge 0 \Rightarrow \displaystyle\int_{a}^{b} f(x) \, dx \ge 0.$

47. $\sin x \le x$ for $x \ge 0 \Rightarrow \sin x - x \le 0$ for $x \ge 0 \Rightarrow \displaystyle\int_{0}^{1} (\sin x - x) \, dx \le 0$ (see Exercise 46) $\Rightarrow \displaystyle\int_{0}^{1} \sin x \, dx - \displaystyle\int_{0}^{1} x \, dx$

$\le 0 \Rightarrow \displaystyle\int_{0}^{1} \sin x \, dx \le \displaystyle\int_{0}^{1} x \, dx \Rightarrow \displaystyle\int_{0}^{1} \sin x \, dx \le \left(\dfrac{1^2}{2} - \dfrac{0^2}{2} \right) \Rightarrow \displaystyle\int_{0}^{1} \sin x \, dx \le \dfrac{1}{2}.$ Thus an upper bound is $\dfrac{1}{2}$.

49. Yes, for the following reasons: $\text{av}(f) = \frac{1}{b-a} \int_a^b f(x)\, dx$ is a constant K. Thus $\int_a^b \text{av}(f)\, dx = \int_a^b K\, dx$

$$= K(b-a) \Rightarrow \int_a^b \text{av}(f)\, dx = (b-a)K = (b-a)\cdot\frac{1}{b-a}\int_a^b f(x)\, dx = \int_a^b f(x)\, dx.$$

51. The car drove the first 150 miles in 5 hours and the second 150 miles in 3 hours, which means it drove 300 miles in 8 hours, for an average of $\frac{300}{8}$ mi/hr $= 37.5$ mi/hr. In terms of average values of functions, the function whose average value we seek is

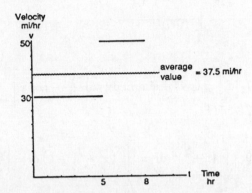

$$v(t) = \begin{cases} 30, & 0 \le t \le 5 \\ 50, & 5 < 1 \le 8 \end{cases}, \text{ and the average value is}$$

$\frac{(30)(5) + (50)(3)}{8} = 37.5$. It does not help to consider

$$v(s) = \begin{cases} 30, & 0 \le s \le 150 \\ 50, & 150 < s \le 300 \end{cases} \text{ whose average value is } \frac{(30)(150) + (50)(150)}{300} = 40 \text{ (mi/hr)/mi because we want}$$

the average speed with respect to time, not distance.

4.7 THE FUNDAMENTAL THEOREM

1. $\int_{-2}^{0} (2x + 5)\, dx = \left[x^2 + 5x\right]_{-2}^{0} = \left(0^2 + 5(0)\right) - \left((-2)^2 + 5(-2)\right) = 6$

3. $\int_{0}^{4} \left(3x - \frac{x^3}{4}\right) dx = \left[\frac{3x^2}{2} - \frac{x^4}{16}\right]_{0}^{4} = \left(\frac{3(4)^2}{2} - \frac{4^4}{16}\right) - \left(\frac{3(0)^2}{2} - \frac{(0)^4}{16}\right) = 8$

5. $\int_{0}^{1} \left(x^2 + \sqrt{x}\right) dx = \left[\frac{x^3}{3} + \frac{2}{3}x^{3/2}\right]_{0}^{1} = \left(\frac{1}{3} + \frac{2}{3}\right) - 0 = 1$

7. $\int_{1}^{32} x^{-6/5}\, dx = \left[-5x^{-1/5}\right]_{1}^{32} = \left(-\frac{5}{2}\right) - (-5) = \frac{5}{2}$

9. $\int_{0}^{\pi} \sin x\, dx = \left[-\cos x\right]_{0}^{\pi} = (-\cos \pi) - (-\cos 0) = -(-1) - (-1) = 2$

11. $\int_{0}^{\pi/3} 2\sec^2 x\, dx = \left[2\tan x\right]_{0}^{\pi/3} = \left(2\tan\left(\frac{\pi}{3}\right)\right) - (2\tan 0) = 2\sqrt{3} - 0 = 2\sqrt{3}$

13. $\displaystyle\int_{\pi/4}^{3\pi/4} \csc x \cot x \, dx = [-\csc x]_{\pi/4}^{3\pi/4} = \left(-\csc\left(\frac{3\pi}{4}\right)\right) - \left(-\csc\left(\frac{\pi}{4}\right)\right) = -\sqrt{2} - \left(-\sqrt{2}\right) = 0$

15. $\displaystyle\int_{\pi/2}^{0} \frac{1+\cos 2t}{2} \, dt = \int_{\pi/2}^{0} \left(\frac{1}{2} + \frac{1}{2}\cos 2t\right) dt = \left[\frac{1}{2}t + \frac{1}{4}\sin 2t\right]_{\pi/2}^{0} = \left(\frac{1}{2}(0) + \frac{1}{4}\sin 2(0)\right) - \left(\frac{1}{2}\left(\frac{\pi}{2}\right) + \frac{1}{4}\sin 2\left(\frac{\pi}{2}\right)\right)$

$= -\frac{\pi}{4}$

17. $\displaystyle\int_{-\pi/2}^{\pi/2} \left(8y^2 + \sin y\right) dy = \left[\frac{8y^3}{3} - \cos y\right]_{-\pi/2}^{\pi/2} = \left(\frac{8\left(\frac{\pi}{2}\right)^3}{3} - \cos\frac{\pi}{2}\right) - \left(\frac{8\left(-\frac{\pi}{2}\right)^3}{3} - \cos\left(-\frac{\pi}{2}\right)\right) = \frac{2\pi^3}{3}$

19. $\displaystyle\int_{1}^{-1} (r+1)^2 \, dr = \int_{1}^{-1} \left(r^2 + 2r + 1\right) dr = \left[\frac{r^3}{3} + r^2 + r\right]_{1}^{-1} = \left(\frac{(-1)^3}{3} + (-1)^2 + (-1)\right) - \left(\frac{1^3}{3} + 1^2 + 1\right) = -\frac{8}{3}$

21. $\displaystyle\int_{\sqrt{2}}^{1} \left(\frac{u^7}{2} - \frac{1}{u^5}\right) du = \int_{\sqrt{2}}^{1} \left(\frac{u^7}{2} - u^{-5}\right) du = \left[\frac{u^8}{16} + \frac{1}{4u^4}\right]_{\sqrt{2}}^{1} = \left(\frac{1^8}{16} + \frac{1}{4(1)^4}\right) - \left(\frac{(\sqrt{2})^8}{16} + \frac{1}{4(\sqrt{2})^4}\right) = -\frac{3}{4}$

23. $\displaystyle\int_{1}^{\sqrt{2}} \frac{s^2 + \sqrt{s}}{s^2} \, ds = \int_{1}^{\sqrt{2}} \left(1 + s^{-3/2}\right) ds = \left[s - \frac{2}{\sqrt{s}}\right]_{1}^{\sqrt{2}} = \left(\sqrt{2} - \frac{2}{\sqrt{\sqrt{2}}}\right) - \left(1 - \frac{2}{\sqrt{1}}\right) = \sqrt{2} - 2^{3/4} + 1$

$= \sqrt{2} - \sqrt[4]{8} + 1$

25. $\displaystyle\int_{-4}^{4} |x| \, dx = \int_{-4}^{0} |x| \, dx = \int_{0}^{4} |x| \, dx = -\int_{4}^{0} x \, dx + \int_{0}^{4} x \, dx = \left[-\frac{x^2}{2}\right]_{-4}^{0} + \left[\frac{x^2}{2}\right]_{0}^{4} = \left(-\frac{0^2}{2} + \frac{(-4)^2}{2}\right) + \left(\frac{4^2}{2} - \frac{0^2}{2}\right)$

$= 16$

27. Let $u = 1 - 2x \Rightarrow du = -2 \, dx$

$\displaystyle\int (1-2x)^3 \, dx = \int -\frac{1}{2}u^3 \, du = -\frac{1}{8}u^4 + C \Rightarrow \int_{0}^{1} (1-2x)^3 \, dx = \left[-\frac{1}{8}(1-2x)^4\right]_{0}^{1} = -\frac{1}{8}(-1)^4 - \left(-\frac{1}{8}\right)(1)^4 = 0$

29. Let $u = t^2 + 1 \Rightarrow du = 2t \, dt$

$\displaystyle\int t\sqrt{t^2+1} \, dt = \int \frac{1}{2}u^{1/2} \, du = \frac{1}{3}u^{3/2} + C \Rightarrow \int_{0}^{1} t\sqrt{t^2+1} \, dt = \left[\frac{1}{3}\left(t^2+1\right)^{3/2}\right]_{0}^{1} = \frac{1}{3}(2)^{3/2} - \frac{1}{3}(1)^{3/2}$

$= \frac{1}{3}\left(2\sqrt{2} - 1\right)$

31. Let $u = 1 + \frac{\theta}{2} \Rightarrow du = \frac{1}{2} \, d\theta$

$\displaystyle\int \sin^2\left(1 + \frac{\theta}{2}\right) d\theta = \int 2\sin^2 u \, du = 2\left(\frac{u}{2} - \frac{1}{4}\sin 2u\right) + C \Rightarrow \int_{0}^{\pi} \sin^2\left(1 + \frac{\theta}{2}\right) d\theta = \left[\left(1 + \frac{\theta}{2}\right) - \frac{1}{2}\sin(2+\theta)\right]_{0}^{\pi}$

$= \left[\left(1 + \frac{\pi}{2}\right) - \frac{1}{2}\sin(2+\pi)\right] - \left(1 - \frac{1}{2}\sin 2\right) = \frac{\pi}{2} + \sin 2$

33. Let $u = \sin\frac{x}{4} \Rightarrow du = \frac{1}{4}\cos\frac{x}{4}\,dx$

$\int \sin^2\frac{x}{4}\cos\frac{x}{4}\,dx = \int 4u^2\,du = \frac{4}{3}u^3 + C \Rightarrow \int\limits_0^\pi \sin^2\frac{x}{4}\cos\frac{x}{4}\,dx = \left[\frac{4}{3}\sin^3\frac{x}{4}\right]_0^\pi = \frac{4}{3}\sin^3\frac{\pi}{4} - \frac{4}{3}\sin 0$

$= \frac{4}{3}\left(\frac{\sqrt{2}}{2}\right)^3 = \frac{2\sqrt{2}}{6} = \frac{\sqrt{2}}{3}$

35. $-x^2 - 2x = 0 \Rightarrow -x(x+2) = 0 \Rightarrow x = 0$ or $x = -2$; Area

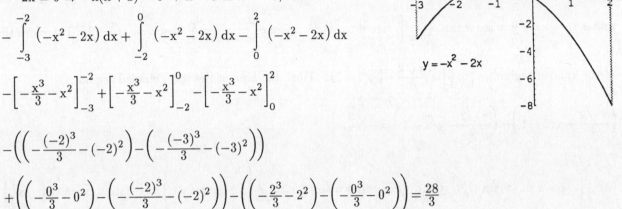

$= -\int\limits_{-3}^{-2} \left(-x^2 - 2x\right)dx + \int\limits_{-2}^0 \left(-x^2 - 2x\right)dx - \int\limits_0^2 \left(-x^2 - 2x\right)dx$

$= -\left[-\frac{x^3}{3} - x^2\right]_{-3}^{-2} + \left[-\frac{x^3}{3} - x^2\right]_{-2}^0 - \left[-\frac{x^3}{3} - x^2\right]_0^2$

$= -\left(\left(-\frac{(-2)^3}{3} - (-2)^2\right) - \left(-\frac{(-3)^3}{3} - (-3)^2\right)\right)$

$\quad + \left(\left(-\frac{0^3}{3} - 0^2\right) - \left(-\frac{(-2)^3}{3} - (-2)^2\right)\right) - \left(\left(-\frac{2^3}{3} - 2^2\right) - \left(-\frac{0^3}{3} - 0^2\right)\right) = \frac{28}{3}$

37. $x^3 - 3x^2 + 2x = 0 \Rightarrow x(x^2 - 3x + 2) = 0$

$\Rightarrow x(x-2)(x-1) = 0,\ x = 0,\ 1,$ or 2;

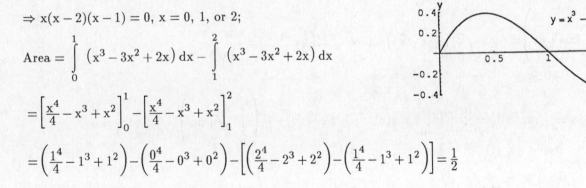

Area $= \int\limits_0^1 \left(x^3 - 3x^2 + 2x\right)dx - \int\limits_1^2 \left(x^3 - 3x^2 + 2x\right)dx$

$= \left[\frac{x^4}{4} - x^3 + x^2\right]_0^1 - \left[\frac{x^4}{4} - x^3 + x^2\right]_1^2$

$= \left(\frac{1^4}{4} - 1^3 + 1^2\right) - \left(\frac{0^4}{4} - 0^3 + 0^2\right) - \left[\left(\frac{2^4}{4} - 2^3 + 2^2\right) - \left(\frac{1^4}{4} - 1^3 + 1^2\right)\right] = \frac{1}{2}$

39. $x^{1/3} = 0 \Rightarrow x = 0$; Area $= -\int\limits_{-1}^0 x^{1/3}\,dx + \int\limits_0^8 x^{1/3}\,dx$

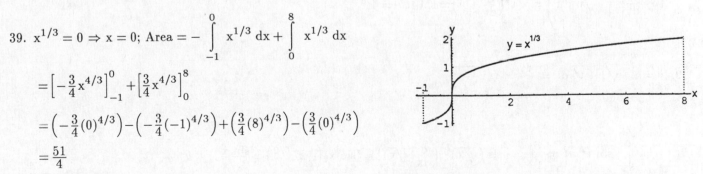

$= \left[-\frac{3}{4}x^{4/3}\right]_{-1}^0 + \left[\frac{3}{4}x^{4/3}\right]_0^8$

$= \left(-\frac{3}{4}(0)^{4/3}\right) - \left(-\frac{3}{4}(-1)^{4/3}\right) + \left(\frac{3}{4}(8)^{4/3}\right) - \left(\frac{3}{4}(0)^{4/3}\right)$

$= \frac{51}{4}$

41. The area of the rectangle bounded by the lines $y = 2$, $y = 0$, $x = \pi$, and $x = 0$ is 2π. The area under the curve $y = 1 + \cos x$ on $[0, \pi]$ is $\int\limits_0^\pi (1 + \cos x)\,dx = [x + \sin x]_0^\pi = (\pi + \sin\pi) - (0 + \sin 0) = \pi$. Therefore the area of the shaded region is $2\pi - \pi = \pi$.

43. On $\left[-\frac{\pi}{4}, 0\right]$: The area of the rectangle bounded by the lines $y = \sqrt{2}$, $y = 0$, $\theta = 0$, and $\theta = -\frac{\pi}{4}$ is $\sqrt{2}\left(\frac{\pi}{4}\right)$

$= \frac{\pi\sqrt{2}}{4}$. The area between the curve $y = \sec\theta\tan\theta$ and $\theta = 0$ is $-\int_{-\pi/4}^{0} \sec\theta\tan\theta \, d\theta = [-\sec\theta]_{-\pi/4}^{0}$

$= (-\sec 0) - \left(-\sec\left(-\frac{\pi}{4}\right)\right) = \sqrt{2} - 1$. Therefore the area of the shaded region on $\left[-\frac{\pi}{4}, 0\right]$ is $\frac{\pi\sqrt{2}}{4} + (\sqrt{2} - 1)$.

On $\left[0, \frac{\pi}{4}\right]$: The area of the rectangle bounded by $\theta = \frac{\pi}{4}$, $\theta = 0$, $y = \sqrt{2}$, and $y = 0$ is $\sqrt{2}\left(\frac{\pi}{4}\right) = \frac{\pi\sqrt{2}}{4}$. The area

under the curve $y = \sec\theta\tan\theta$ is $\int_{0}^{\pi/4} \sec\theta\tan\theta \, d\theta = [\sec\theta]_{0}^{\pi/4} = \sec\frac{\pi}{4} - \sec 0 = \sqrt{2} - 1$. Therefore the area

of the shaded region on $\left[0, \frac{\pi}{4}\right]$ is $\frac{\pi\sqrt{2}}{4} - (\sqrt{2} - 1)$. Thus, the area of the total shaded region is

$\left(\frac{\pi\sqrt{2}}{4} + \sqrt{2} - 1\right) + \left(\frac{\pi\sqrt{2}}{4} - \sqrt{2} + 1\right) = \frac{\pi\sqrt{2}}{2}$.

45. (a) $\int_{0}^{\sqrt{x}} \cos t \, dt = [\sin t]_{0}^{\sqrt{x}} = \sin\sqrt{x} - \sin 0 = \sin\sqrt{x} \Rightarrow \frac{d}{dx}\left(\int_{0}^{\sqrt{x}} \cos t \, dt\right) = \frac{d}{dx}(\sin\sqrt{x}) = \cos\sqrt{x}\left(\frac{1}{2}x^{-1/2}\right)$

$= \frac{\cos\sqrt{x}}{2\sqrt{x}}$

(b) $\frac{d}{dx}\left(\int_{0}^{\sqrt{x}} \cos t \, dt\right) = (\cos\sqrt{x})\left(\frac{d}{dx}(\sqrt{x})\right) = (\cos\sqrt{x})\left(\frac{1}{2}x^{-1/2}\right) = \frac{\cos\sqrt{x}}{2\sqrt{x}}$

47. (a) $\int_{0}^{t^4} \sqrt{u} \, du = \int_{0}^{t^4} u^{1/2} \, du = \left[\frac{2}{3}u^{3/2}\right]_{0}^{t^4} = \frac{2}{3}(t^4)^{3/2} - 0 = \frac{2}{3}t^6 \Rightarrow \frac{d}{dt}\left(\int_{0}^{t^4} \sqrt{u} \, du\right) = \frac{d}{dt}\left(\frac{2}{3}t^6\right) = 4t^5$

(b) $\frac{d}{dt}\left(\int_{0}^{t^4} \sqrt{u} \, du\right) = \sqrt{t^4}\left(\frac{d}{dt}(t^4)\right) = t^2(4t^3) = 4t^5$

49. $y = \int_{0}^{x} \sqrt{1 + t^2} \, dt \Rightarrow \frac{dy}{dx} = \sqrt{1 + x^2}$

51. $y = \int_{0}^{\sqrt{x}} \sin t^2 \, dt \Rightarrow \frac{dy}{dx} = \left(\sin(\sqrt{x})^2\right)\left(\frac{d}{dx}(\sqrt{x})\right) = (\sin x)\left(\frac{1}{2}x^{-1/2}\right) = \frac{\sin x}{2\sqrt{x}}$

53. $y = \int_{0}^{\sin x} \frac{dt}{\sqrt{1 - t^2}}, |x| < \frac{\pi}{2} \Rightarrow \frac{dy}{dx} = \frac{1}{\sqrt{1 - \sin^2 x}}\left(\frac{d}{dx}(\sin x)\right) = \frac{1}{\sqrt{\cos^2 x}}(\cos x) = \frac{\cos x}{|\cos x|} = \frac{\cos x}{\cos x} = 1$ since $|x| < \frac{\pi}{2}$

55. $y = \int_{\pi}^{x} \frac{1}{t} \, dt - 3 \Rightarrow \frac{dy}{dx} = \frac{1}{x}$ and $y(\pi) = \int_{\pi}^{\pi} \frac{1}{t} \, dt - 3 = 0 - 3 = -3 \Rightarrow$ (d) is a solution to this problem.

57. $y = \int_{0}^{x} \sec t \, dt + 4 \Rightarrow \frac{dy}{dx} = \sec x$ and $y(0) = \int_{0}^{0} \sec t \, dt + 4 = 0 + 4 = 4 \Rightarrow$ (b) is a solution to this problem.

59. $y = \int_{2}^{x} \sec t \, dt + 3$

61. $s = \int_{t_0}^{t} f(x) \, dx + s_0$

63. (a) $6 - x - x^2 = 0 \Rightarrow x^2 + x - 6 = 0$

$\Rightarrow (x+3)(x-2) = 0 \Rightarrow x = -3$ or $x = 2$;

$\text{Area} = \int_{-3}^{2} (6 - x - x^2) \, dx = \left[6x - \frac{x^2}{2} - \frac{x^3}{2} \right]_{-3}^{2}$

$= \left(6(2) - \frac{2^2}{2} - \frac{2^3}{3} \right) - \left(6(-3) - \frac{(-3)^2}{2} - \frac{(-3)^3}{3} \right)$

$= \frac{125}{6}$

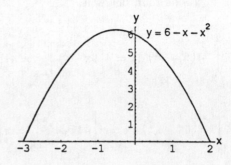

(b) $y' = -1 - 2x = 0 \Rightarrow x = -\frac{1}{2}$; $y' > 0$ for $x < -\frac{1}{2}$ and $y' < 0$ for $x > -\frac{1}{2} \Rightarrow x = -\frac{1}{2}$ yields a local maximum

$\Rightarrow \text{height} = y\left(-\frac{1}{2} \right) = \frac{25}{4}$

(c) $\text{Base} = 2 - (-3) = 5$, $\text{height} = y\left(-\frac{1}{2} \right) = \frac{25}{4} \Rightarrow \text{Area} = \frac{2}{3}(\text{Base})(\text{Height}) = \frac{2}{3}(5)\left(\frac{25}{4} \right) = \frac{125}{6}$

(d) $\text{Area} = \int_{-b/2}^{b/2} \left(h - \left(\frac{4h}{b^2} \right)x^2 \right) dx = \left[hx - \frac{4hx^3}{3b^2} \right]_{-b/2}^{b/2}$

$= \left(h\left(\frac{b}{2} \right) - \frac{4h\left(\frac{b}{2} \right)^3}{3b^2} \right) - \left(h\left(-\frac{b}{2} \right) - \frac{4h\left(-\frac{b}{2} \right)^3}{3b^2} \right)$

$= \left(\frac{bh}{2} - \frac{bh}{6} \right) - \left(-\frac{bh}{2} + \frac{bh}{6} \right) = bh - \frac{bh}{3} = \frac{2}{3}bh$

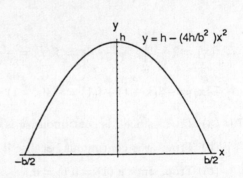

65. $\frac{dc}{dx} = \frac{1}{2\sqrt{x}} = \frac{1}{2}x^{-1/2} \Rightarrow c = \int_{0}^{x} \frac{1}{2}t^{-1/2} dt = \left[t^{1/2} \right]_{0}^{x} = \sqrt{x}$

(a) $c(100) - c(1) = \sqrt{100} - \sqrt{1} = \$9.00$

(b) $c(400) - c(100) = \sqrt{400} - \sqrt{100} = \$10.00$

67. (a) $v = \frac{ds}{dt} = \frac{d}{dt} \int_{0}^{t} f(x) \, dx = f(t) \Rightarrow v(5) = f(5) = 2$ m/sec

(b) $a = \frac{df}{dt}$ is negative since the slope of the tangent line at $t = 5$ is negative

(c) $s = \int_{0}^{3} f(x)\, dx = \frac{1}{2}(3)(3) = \frac{9}{2}$ m since the integral is the area of the triangle formed by $y = f(x)$, the x-axis,

and $x = 3$

(d) $t = 6$ since after $t = 6$ to $t = 9$, the region lies below the x-axis

(e) At $t = 4$ and $t = 7$, since there are horizontal tangents there

(f) Toward the origin between $t = 6$ and $t = 9$ since the velocity is negative on this interval. Away from the origin between $t = 0$ and $t = 6$ since the velocity is positive there.

(g) Right or positive side, because the integral of f from 0 to 9 is positive, there being more area above the x-axis than below it.

69. $\int_{-2}^{2} 4(9 - x^2)\, dx = 4\left[9x - \frac{x^3}{3}\right]_{-2}^{2} = 4\left[\left(18 - \frac{8}{3}\right) - \left(-18 - \frac{(-2)^3}{3}\right)\right] = 4\left(36 - \frac{16}{3}\right) = \frac{368}{3}$

71. $\int_{4}^{8} \pi(64 - x^2)\, dx = \pi\left[64x - \frac{x^3}{3}\right]_{4}^{8} = \pi\left[\left(512 - \frac{512}{3}\right) - \left(256 - \frac{64}{3}\right)\right] = \pi\left(256 - \frac{448}{3}\right) = \frac{320\pi}{3}$

73. $k > 0 \Rightarrow$ one arch of $y = \sin kx$ will occur over the interval $\left[0, \frac{\pi}{k}\right] \Rightarrow$ the area $= \int_{0}^{\pi/k} \sin kx\, dx = \left[-\frac{1}{k}\cos kx\right]_{0}^{\pi/k}$

$= -\frac{1}{k}\cos\left(k\left(\frac{\pi}{k}\right)\right) - \left(-\frac{1}{k}\cos(0)\right) = \frac{2}{k}$

75. $\int_{1}^{x} f(t)\, dt = x^2 - 2x + 1 \Rightarrow f(x) = \frac{d}{dx}\int_{1}^{x} f(t)\, dt = \frac{d}{dx}(x^2 - 2x + 1) = 2x - 2$

77. $f(x) = 2 - \int_{2}^{x+1} \frac{9}{1+t}\, dt \Rightarrow f'(x) = -\frac{9}{1+(x+1)} = \frac{-9}{x+2} \Rightarrow f'(1) = -3; f(1) = 2 - \int_{2}^{1+1} \frac{9}{1+t}\, dt = 2 - 0 = 2;$

$L(x) = -3(x - 1) + f(1) = -3(x - 1) + 2 = -3x + 5$

79. (a) True: since f is continuous, g is differentiable by Part 1 of the Fundamental Theorem of Calculus.

(b) True: g is continuous because it is differentiable.

(c) True, since $g'(1) = f(1) = 0$.

(d) False, since $g''(1) = f'(1) > 0$.

(e) True, since $g'(1) = 0$ and $g''(1) = f'(1) > 0$.

(f) False: $g''(x) = f'(x) > 0$, so g'' never changes sign.

(g) True, since $g'(1) = f(1) = 0$ and $g'(x) = f(x)$ is an increasing function of x (because $f'(x) > 0$).

81.

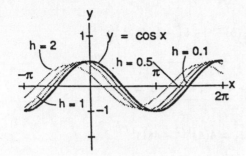

4.8 SUBSTITUTION IN DEFINITE INTEGRALS

1. (a) Let $u = y + 1 \Rightarrow du = dy$; $y = 0 \Rightarrow u = 1$, $y = 3 \Rightarrow u = 4$

$$\int_0^3 \sqrt{y+1}\ dy = \int_1^4 u^{1/2}\ du = \left[\tfrac{2}{3}u^{3/2}\right]_1^4 = \left(\tfrac{2}{3}\right)(4)^{3/2} - \left(\tfrac{2}{3}\right)(1)^{3/2} = \left(\tfrac{2}{3}\right)(8) - \left(\tfrac{2}{3}\right)(1) = \tfrac{14}{3}$$

(b) Use the same substitution for u as in part (a); $y = -1 \Rightarrow u = 0$, $y = 0 \Rightarrow u = 1$

$$\int_{-1}^0 \sqrt{y+1}\ dy = \int_0^1 u^{1/2}\ du = \left[\tfrac{2}{3}u^{3/2}\right]_0^1 = \left(\tfrac{2}{3}\right)(1)^{3/2} - 0 = \tfrac{2}{3}$$

3. (a) Let $u = \tan x \Rightarrow du = \sec^2 x\ dx$; $x = 0 \Rightarrow u = 0$, $x = \tfrac{\pi}{4} \Rightarrow u = 1$

$$\int_0^{\pi/4} \tan x \sec^2 x\ dx = \int_0^1 u\ du = \left[\tfrac{u^2}{2}\right]_0^1 = \tfrac{1^2}{2} - 0 = \tfrac{1}{2}$$

(b) Use the same substitution as in part (a); $x = -\tfrac{\pi}{4} \Rightarrow u = -1$, $x = 0 \Rightarrow u = 0$

$$\int_{-\pi/4}^0 \tan x \sec^2 x\ dx = \int_{-1}^0 u\ du = \left[\tfrac{u^2}{2}\right]_{-1}^0 = 0 - \tfrac{1}{2} = -\tfrac{1}{2}$$

5. (a) $u = 1 + t^4 \Rightarrow du = 4t^3\ dt \Rightarrow \tfrac{1}{4}\ du = t^3\ dt$; $t = 0 \Rightarrow u = 1$, $t = 1 \Rightarrow u = 2$

$$\int_0^1 t^3\left(1 + t^4\right)^3\ dt = \int_1^2 \tfrac{1}{4}u^3\ du = \left[\tfrac{u^4}{16}\right]_1^2 = \tfrac{2^4}{16} - \tfrac{1^4}{16} = \tfrac{15}{16}$$

(b) Use the same substitution as in part (a); $t = -1 \Rightarrow u = 2$, $t = 1 \Rightarrow u = 2$

$$\int_{-1}^1 t^3\left(1 + t^4\right)^3\ dt = \int_2^2 \tfrac{1}{4}u^3\ du = 0$$

7. (a) Let $u = 4 + r^2 \Rightarrow du = 2r\ dr \Rightarrow \tfrac{1}{2}\ du = r\ dr$; $r = -1 \Rightarrow u = 5$, $r = 1 \Rightarrow u = 5$

$$\int_{-1}^1 \frac{5r}{\left(4 + r^2\right)^2}\ dr = 5\int_5^5 \tfrac{1}{2}u^{-2}\ du = 0$$

(b) Use the same substitution as in part (a); $r = 0 \Rightarrow u = 4$, $r = 1 \Rightarrow u = 5$

$$\int_0^1 \frac{5r}{\left(4+r^2\right)^2}\, dr = 5 \int_4^5 \tfrac{1}{2} u^{-2}\, du = 5\left[-\tfrac{1}{2} u^{-1}\right]_4^5 = 5\left(-\tfrac{1}{2}(5)^{-1}\right) - 5\left(-\tfrac{1}{2}(4)^{-1}\right) = \tfrac{1}{8}$$

9. (a) Let $u = x^2 + 1 \Rightarrow du = 2x\, dx \Rightarrow 2\, du = 4x\, dx$; $x = 0 \Rightarrow u = 1$, $x = \sqrt{3} \Rightarrow u = 4$

$$\int_0^{\sqrt{3}} \frac{4x}{\sqrt{x^2+1}}\, dx = \int_1^4 \frac{2}{\sqrt{u}}\, du = \int_1^4 2u^{-1/2}\, du = \left[4u^{1/2}\right]_1^4 = 4(4)^{1/2} - 4(1)^{1/2} = 4$$

(b) Use the same substitution as in part (a); $x = -\sqrt{3} \Rightarrow u = 4$, $x = \sqrt{3} \Rightarrow u = 4$

$$\int_{-\sqrt{3}}^{\sqrt{3}} \frac{4x}{\sqrt{x^2+1}}\, dx = \int_4^4 \frac{2}{\sqrt{u}}\, du = 0$$

11. (a) Let $u = 1 - \cos 3t \Rightarrow du = 3\sin 3t\, dt \Rightarrow \tfrac{1}{3} du = \sin 3t\, dt$; $t = 0 \Rightarrow u = 0$, $t = \tfrac{\pi}{6} \Rightarrow u = 1 - \cos \tfrac{\pi}{2} = 1$

$$\int_0^{\pi/6} (1 - \cos 3t) \sin 3t\, dt = \int_0^1 \tfrac{1}{3} u\, du = \left[\tfrac{1}{3}\left(\tfrac{u^2}{2}\right)\right]_0^1 = \tfrac{1}{6}(1)^2 - \tfrac{1}{6}(0)^2 = \tfrac{1}{6}$$

(b) Use the same substitution as in part (a); $t = \tfrac{\pi}{6} \Rightarrow u = 1$, $t = \tfrac{\pi}{3} \Rightarrow u = 1 - \cos \pi = 2$

$$\int_{\pi/6}^{\pi/3} (1 - \cos 3t) \sin 3t\, dt = \int_1^2 \tfrac{1}{3} u\, du = \left[\tfrac{1}{3}\left(\tfrac{u^2}{2}\right)\right]_1^2 = \tfrac{1}{6}(2)^2 - \tfrac{1}{6}(1)^2 = \tfrac{1}{2}$$

13. (a) Let $u = 4 + 3\sin z \Rightarrow du = 3\cos z\, dz \Rightarrow \tfrac{1}{3} du = \cos z\, dz$; $z = 0 \Rightarrow u = 4$, $z = 2\pi \Rightarrow u = 4$

$$\int_9^{2\pi} \frac{\cos z}{\sqrt{4 + 3\sin z}}\, dz = \int_4^4 \frac{1}{\sqrt{u}}\left(\tfrac{1}{3}\, du\right) = 0$$

(b) Use the same substitution as in part (a); $z = -\pi \Rightarrow u = 4 + 3\sin(-\pi) = 4$, $z = \pi \Rightarrow u = 4$

$$\int_{-\pi}^{\pi} \frac{\cos z}{\sqrt{4 + 3\sin z}}\, dz = \int_4^4 \frac{1}{\sqrt{u}}\left(\tfrac{1}{3}\, du\right) = 0$$

15. Let $u = t^5 + 2t \Rightarrow du = \left(5t^4 + 2\right) dt$; $t = 0 \Rightarrow u = 0$, $t = 1 \Rightarrow u = 3$

$$\int_0^1 \sqrt{t^5 + 2t}\left(5t^4 + 2\right) dt = \int_0^3 u^{1/2}\, du = \left[\tfrac{2}{3} u^{3/2}\right]_0^3 = \tfrac{2}{3}(3)^{3/2} - \tfrac{2}{3}(0)^{3/2} = 2\sqrt{3}$$

17. Let $u = \cos 2\theta \Rightarrow du = -2\sin 2\theta\, d\theta \Rightarrow -\tfrac{1}{2} du = \sin 2\theta\, d\theta$; $\theta = 0 \Rightarrow u = 1$, $\theta = \tfrac{\pi}{6} \Rightarrow u = \cos 2\left(\tfrac{\pi}{6}\right) = \tfrac{1}{2}$

$$\int_0^{\pi/6} \cos^{-3} 2\theta \sin 2\theta\, d\theta = \int_1^{1/2} u^{-3}\left(-\tfrac{1}{2}\, du\right) = -\tfrac{1}{2} \int_1^{1/2} u^{-3}\, du = \left[-\tfrac{1}{2}\left(\tfrac{u^{-2}}{-2}\right)\right]_1^{1/2} = \frac{1}{4\left(\tfrac{1}{2}\right)^2} - \frac{1}{4(1)^2} = \tfrac{3}{4}$$

19. Let $u = 5 - 4 \cos t \Rightarrow du = 4 \sin t \, dt \Rightarrow \frac{1}{4} du = \sin t \, dt$; $t = 0 \Rightarrow u = 5 - 4 \cos 0 = 1$, $t = \pi \Rightarrow$
$u = 5 - 4 \cos \pi = 9$

$$\int_0^\pi 5 (5 - 4 \cos t)^{1/4} \sin t \, dt = \int_1^9 5u^{1/4} \left(\frac{1}{4} du\right) = \frac{5}{4} \int_1^9 u^{1/4} \, du = \left[\frac{5}{4}\left(\frac{4}{5} u^{5/4}\right)\right]_1^9 = 9^{5/4} - 1$$

21. Let $u = 4y - y^2 + 4y^3 + 1 \Rightarrow du = \left(4 - 2y + 12y^2\right) dy$; $y = 0 \Rightarrow u = 1$, $y = 1 \Rightarrow u = 4(1) - (1)^2 + 4(1)^3 + 1 = 8$

$$\int_0^1 \left(4y - y^2 + 4y^3 + 1\right)^{-2/3} \left(12y^2 - 2y + 4\right) dy = \int_1^8 u^{-2/3} \, du = \left[3u^{1/3}\right]_1^8 = 3(8)^{1/3} - 3(1)^{1/3} = 3$$

23. Let $u = \theta^{3/2} \Rightarrow du = \frac{3}{2} \theta^{1/2} \, d\theta \Rightarrow \frac{2}{3} du = \sqrt{\theta} \, d\theta$; $\theta = 0 \Rightarrow u = 0$, $\theta = \sqrt[3]{\pi^2} \Rightarrow u = \pi$

$$\int_0^{\sqrt[3]{\pi^2}} \sqrt{\theta} \cos^2 \left(\theta^{3/2}\right) d\theta = \int_0^\pi \cos^2 u \left(\frac{2}{3} du\right) = \left[\frac{2}{3}\left(\frac{u}{2} + \frac{1}{4} \sin 2u\right)\right]_0^\pi = \frac{2}{3}\left(\frac{\pi}{2} + \frac{1}{4} \sin 2\pi\right) - \frac{2}{3}(0) = \frac{\pi}{3}$$

25. Let $u = 4 - x^2 \Rightarrow du = -2x \, dx \Rightarrow -\frac{1}{2} du = dx$; $x = -2 \Rightarrow u = 0$, $x = 0 \Rightarrow u = 4$, $x = 2 \Rightarrow u = 0$

$$A = - \int_{-2}^0 x\sqrt{4 - x^2} \, dx + \int_0^2 x\sqrt{4 - x^2} \, dx = - \int_0^4 -\frac{1}{2} u^{1/2} \, du + \int_4^0 -\frac{1}{2} u^{1/2} \, du = 2 \int_0^4 \frac{1}{2} u^{1/2} \, du = \int_0^4 u^{1/2} \, du$$

$$= \left[\frac{2}{3} u^{3/2}\right]_0^4 = \frac{2}{3}(4)^{3/2} - \frac{2}{3}(0)^{3/2} = \frac{16}{3}$$

27. Let $u = 1 + \cos x \Rightarrow du = -\sin x \, dx \Rightarrow -du = \sin x \, dx$; $x = -\pi \Rightarrow u = 1 + \cos(-\pi) = 0$, $x = 0$
$\Rightarrow u = 1 + \cos 0 = 2$

$$A = - \int_{-\pi}^0 3 (\sin x) \sqrt{1 + \cos x} \, dx = - \int_0^2 3u^{1/2} (-du) = 3 \int_0^2 u^{1/2} \, du = \left[2u^{3/2}\right]_0^2 = 2(2)^{3/2} - 2(0)^{3/2} = 2^{5/2}$$

29. Let $u = 2x \Rightarrow du = 2 \, dx \Rightarrow \frac{1}{2} du = dx$; $x = 1 \Rightarrow u = 2$, $x = 3 \Rightarrow u = 6$

$$\int_1^3 \frac{\sin 2x}{x} \, dx = \int_2^6 \frac{\sin u}{\left(\frac{u}{2}\right)} \left(\frac{1}{2} du\right) = \int_2^6 \frac{\sin u}{u} \, du = \left[F(u)\right]_2^6 = F(6) - F(2)$$

31. (a) Let $u = -x \Rightarrow du = -dx$; $x = -1 \Rightarrow u = 1$, $x = 0 \Rightarrow u = 0$

f odd $\Rightarrow f(-x) = -f(x)$. Then $\displaystyle\int_{-1}^0 f(x) \, dx = \int_1^0 f(-u)(-du) = \int_1^0 -f(u)(-du) = \int_1^0 f(u) \, du = - \int_0^1 f(u) \, du$

$= -3$

(b) Let $u = -x \Rightarrow du = -dx$; $x = -1 \Rightarrow u = 1$, $x = 0 \Rightarrow u = 0$

f even $\Rightarrow f(-x) = f(x)$. Then $\displaystyle\int_{-1}^0 f(x) \, dx = \int_1^0 f(-u)(-du) = - \int_1^0 f(u) \, du = \int_0^1 f(u) \, du = 3$

33. Let $u = a - x \Rightarrow du = -dx$; $x = 0 \Rightarrow u = a$, $x = a \Rightarrow u = 0$

$$I = \int_0^a \frac{f(x)\,dx}{f(x) + f(a-x)} = \int_a^0 \frac{f(a-u)}{f(a-u) + f(u)}(-du) = \int_0^a \frac{f(a-u)\,du}{f(u) + f(a-u)} = \int_0^a \frac{f(a-x)\,dx}{f(x) + f(a-x)}$$

$$\Rightarrow I + I = \int_0^a \frac{f(x)\,dx}{f(x) + f(a-x)} + \int_0^a \frac{f(a-x)\,dx}{f(x) + f(a-x)} = \int_0^a \frac{f(x) + f(a-x)}{f(x) + f(a-x)}\,dx = \int_0^a dx = [x]_0^a = a - 0 = a.$$

Therefore, $2I = a \Rightarrow I = \frac{a}{2}$.

35. Let $u = x + c \Rightarrow du = dx$; $x = a - c \Rightarrow u = a$, $x = b - c \Rightarrow u = b$

$$\int_{a-c}^{b-c} f(x+c)\,dx = \int_a^b f(u)\,du = \int_a^b f(x)\,dx$$

4.9 NUMERICAL INTEGRATION

1. $\int_1^2 x\,dx$

  I. (a) For $n = 4$, $h = \frac{b-a}{n} = \frac{2-1}{4} = \frac{1}{4} \Rightarrow \frac{h}{2} = \frac{1}{8}$;

	x_i	$f(x_i)$	m	$mf(x_i)$
x_0	1	1	1	1
x_1	5/4	5/4	2	5/2
x_2	3/2	3/2	2	3
x_3	7/4	7/4	2	7/2
x_4	2	2	1	2

$$\sum mf(x_i) = 12 \Rightarrow T = \frac{1}{8}(12) = \frac{3}{2};$$

$$f(x) = x \Rightarrow f'(x) = 1 \Rightarrow f'' = 0 \Rightarrow M = 0$$
$$\Rightarrow |E_T| = 0$$

  (b) $\int_1^2 x\,dx = \left[\frac{x^2}{2}\right]_1^2 = 2 - \frac{1}{2} = \frac{3}{2} \Rightarrow |E_T| = \int_1^2 x\,dx - T = 0$

  (c) $\frac{|E_T|}{\text{True Value}} \times 100 = 0\%$

 II. (a) For $n = 4$, $h = \frac{b-a}{n} = \frac{2-1}{4} = \frac{1}{4} \Rightarrow \frac{h}{3} = \frac{1}{12}$;

	x_i	$f(x_i)$	m	$mf(x_i)$
x_0	1	1	1	1
x_1	5/4	5/4	4	5
x_2	3/2	3/2	2	3
x_3	7/4	7/4	4	7
x_4	2	2	1	2

$$\sum mf(x_i) = 18 \Rightarrow S = \frac{1}{12}(18) = \frac{3}{2};$$

$$f^{(4)}(x) = 0 \Rightarrow M = 0 \Rightarrow |E_S| = 0$$

  (b) $\int_1^2 x\,dx = \frac{3}{2} \Rightarrow |E_S| = \int_1^2 x\,dx - S = \frac{3}{2} - \frac{3}{2} = 0$

  (c) $\frac{|E_S|}{\text{True Value}} \times 100 = 0\%$

3. $\displaystyle\int_{-1}^{1} (x^2 + 1)\, dx$

I.   (a) For $n = 4$, $h = \dfrac{b-a}{n} = \dfrac{1-(-1)}{4} = \dfrac{2}{4} = \dfrac{1}{2} \Rightarrow \dfrac{h}{2} = \dfrac{1}{4}$;

	x_i	$f(x_i)$	m	$mf(x_i)$
x_0	-1	2	1	2
x_1	$-1/2$	$5/4$	2	$5/2$
x_2	0	1	2	2
x_3	$1/2$	$5/4$	2	$5/2$
x_4	1	2	1	2

$$\sum mf(x_i) = 11 \Rightarrow T = \tfrac{1}{4}(11) = 2.75;$$

$$f(x) = x^2 + 1 \Rightarrow f'(x) = 2x \Rightarrow f''(x) = 2 \Rightarrow M = 2$$

$$\Rightarrow |E_T| \le \frac{1-(-1)}{12}\left(\frac{1}{2}\right)^2 (2) = \frac{1}{12} \text{ or } 0.08333$$

(b) $\displaystyle\int_{-1}^{1} (x^2 + 1)\, dx = \left[\frac{x^3}{3} + x\right]_{-1}^{1} = \left(\frac{1}{3} + 1\right) - \left(-\frac{1}{3} - 1\right) = \frac{8}{3} \Rightarrow E_T = \int_{-1}^{1} (x^2 + 1)\, dx - T = \frac{8}{3} - \frac{11}{4} = -\frac{1}{12}$

$$\Rightarrow |E_T| = \left|-\frac{1}{12}\right| \approx 0.08333$$

(c) $\dfrac{|E_T|}{\text{True Value}} \times 100 = \left(\dfrac{\frac{1}{12}}{\frac{8}{3}}\right) \times 100 \approx 3\%$

II. (a) For $n = 4$, $h = \dfrac{b-a}{n} = \dfrac{3-1}{4} = \dfrac{2}{4} = \dfrac{1}{2} \Rightarrow \dfrac{h}{3} = \dfrac{1}{6}$;

	x_i	$f(x_i)$	m	$mf(x_i)$
x_0	-1	2	1	2
x_1	$-1/2$	$5/4$	4	5
x_2	0	1	2	2
x_3	$1/2$	$5/4$	4	5
x_4	1	2	1	2

$$\sum mf(x_i) = 16 \Rightarrow S = \tfrac{1}{6}(16) = \tfrac{8}{3} = 2.66667;$$

$$f^{(3)}(x) = 0 \Rightarrow f^{(4)}(x) = 0 \Rightarrow M = 0 \Rightarrow |E_S| = 0$$

(b) $\displaystyle\int_{-1}^{1} (x^2 + 1)\, dx = \left[\frac{x^3}{3} + x\right]_{-1}^{1} = \frac{8}{3} \Rightarrow |E_S| = \int_{-1}^{1} (x^2 + 1)\, dx - S = \frac{8}{3} - \frac{8}{3} = 0$

(c) $\dfrac{|E_S|}{\text{True Value}} \times 100 = 0\%$

5. $\displaystyle\int_{0}^{2} (t^3 + t)\, dt$

I.   (a) For $n = 4$, $h = \dfrac{b-a}{n} = \dfrac{2-0}{4} = \dfrac{2}{4} = \dfrac{1}{2} \Rightarrow \dfrac{h}{2} = \dfrac{1}{4}$;

	t_i	$f(t_i)$	m	$mf(t_i)$
t_0	0	0	1	0
t_1	$1/2$	$5/8$	2	$5/4$
t_2	1	2	2	4
t_3	$3/2$	$39/8$	2	$39/4$
t_4	2	10	1	10

$$\sum mf(t_i) = 25 \Rightarrow T = \tfrac{1}{4}(25) = \tfrac{25}{4};$$

$$f(t) = t^3 + t \Rightarrow f'(t) = 3t^2 + 1 \Rightarrow f''(t) = 6t \Rightarrow M = 12$$

$$= f''(2) \Rightarrow |E_T| \le \frac{2-0}{12}\left(\frac{1}{2}\right)^2(12) = \frac{1}{2}$$

(b) $\displaystyle\int_0^2 \left(t^3 + t\right) dt = \left[\frac{t^4}{4} + \frac{t^2}{2}\right]_0^2 = \left(\frac{2^4}{4} + \frac{2^2}{2}\right) - 0 = 6 \Rightarrow |E_T| = \int_0^2 \left(t^3 + t\right) dt - T = 6 - \frac{25}{4} = -\frac{1}{4} \Rightarrow |E_T| = \frac{1}{4}$

(c) $\displaystyle\frac{|E_T|}{\text{True Value}} \times 100 = \frac{\left|-\frac{1}{4}\right|}{6} \times 100 \approx 4\%$

II. (a) For $n = 4$, $h = \frac{b-a}{n} = \frac{2-0}{4} = \frac{2}{4} = \frac{1}{2} \Rightarrow \frac{h}{3} = \frac{1}{6}$;

$$\sum mf(t_i) = 36 \Rightarrow S = \frac{1}{6}(36) = 6\,;$$

$$f^{(3)}(t) = 6 \Rightarrow f^{(4)}(t) = 0 \Rightarrow M = 0 \Rightarrow |E_S| = 0$$

	t_i	$f(t_i)$	m	$mf(t_i)$
t_0	0	0	1	0
t_1	1/2	5/8	4	5/2
t_2	1	2	2	4
t_3	3/2	39/8	4	39/2
t_4	2	10	1	10

(b) $\displaystyle\int_0^2 \left(t^3 + t\right) dt = 6 \Rightarrow |E_S| = \int_0^2 \left(t^3 + t\right) dt - S = 6 - 6 = 0$

(c) $\displaystyle\frac{|E_S|}{\text{True Value}} \times 100 = 0\%$

7. $\displaystyle\int_1^2 \frac{1}{s^2}\, ds$

I. (a) For $n = 4$, $h = \frac{b-a}{n} = \frac{2-1}{4} = \frac{1}{4} \Rightarrow \frac{h}{2} = \frac{1}{8}$;

$$\sum mf(s_i) = \frac{179,573}{44,100} \Rightarrow T = \frac{1}{8}\left(\frac{179,573}{44,100}\right) = \frac{179,573}{352,800}$$

$$\approx 0.50899; \ f(s) = \frac{1}{s^2} \Rightarrow f'(s) = -\frac{2}{s^3} \Rightarrow f''(s) = \frac{6}{s^4}$$

$$\Rightarrow M = 6 = f''(1) \Rightarrow |E_T| \le \frac{2-1}{12}\left(\frac{1}{4}\right)^2(6) = \frac{1}{32} = 0.03125$$

	s_i	$f(s_i)$	m	$mf(s_i)$
s_0	1	1	1	1
s_1	5/4	16/25	2	32/25
s_2	3/2	4/9	2	8/9
s_3	7/4	16/49	2	32/49
s_4	2	1/4	1	1/4

(b) $\displaystyle\int_1^2 \frac{1}{s^2}\, ds = \int_1^2 s^{-2}\, ds = \left[-\frac{1}{s}\right]_1^2 = -\frac{1}{2} - \left(-\frac{1}{1}\right) = \frac{1}{2} \Rightarrow E_T = \int_1^2 \frac{1}{s^2}\, ds - T = \frac{1}{2} - 0.50899 = -0.00899$

$\Rightarrow |E_T| = 0.00899$

(c) $\displaystyle\frac{|E_T|}{\text{True Value}} \times 100 = \frac{0.00899}{0.5} \times 100 \approx 2\%$

II. (a) For $n = 4$, $h = \frac{b-a}{n} = \frac{2-1}{4} = \frac{1}{4} \Rightarrow \frac{h}{3} = \frac{1}{12}$;

	s_i	$f(s_i)$	m	$mf(s_i)$
s_0	1	1	1	1
s_1	5/4	16/25	4	64/25
s_2	3/2	4/9	2	8/9
s_3	7/4	16/49	4	64/49
s_4	2	1/4	1	1/4

$$\sum mf(s_i) = \frac{264{,}821}{44{,}100} \Rightarrow S = \frac{1}{12}\left(\frac{264{,}821}{44{,}100}\right) = \frac{264{,}821}{529{,}200}$$

$$\approx 0.50041; \ f^{(3)}(s) = -\frac{24}{s^5} \Rightarrow f^{(4)}(s) = \frac{120}{s^6}$$

$$\Rightarrow M = 120 \Rightarrow |E_S| \le \left|\frac{2-1}{180}\right|\left(\frac{1}{4}\right)^4 (120) = \frac{1}{384} \approx 0.00260$$

(b) $\int_1^2 \frac{1}{s^2}\, ds = \frac{1}{2} \Rightarrow E_S = \int_1^2 \frac{1}{s^2}\, ds - S = \frac{1}{2} - 0.050041 = -0.00041 \Rightarrow |E_S| = 0.00041$

(c) $\dfrac{|E_S|}{\text{True Value}} \times 100 = \dfrac{0.0004}{0.5} \times 100 \approx 0.08\%$

9. $\displaystyle\int_0^\pi \sin t \, dt$

I. (a) For $n = 4$, $h = \frac{b-a}{n} = \frac{\pi - 0}{4} = \frac{\pi}{4} \Rightarrow \frac{h}{2} = \frac{\pi}{8}$;

	t_i	$f(t_i)$	m	$mf(t_i)$
t_0	0	0	1	0
t_1	$\pi/4$	$\sqrt{2}/2$	2	$\sqrt{2}$
t_2	$\pi/2$	1	2	2
t_3	$3\pi/4$	$\sqrt{2}/2$	2	$\sqrt{2}$
t_4	π	0	1	0

$$\sum mf(t_i) = 2 + 2\sqrt{2} \approx 4.9294 \Rightarrow T = \frac{\pi}{8}\left(2 + 2\sqrt{2}\right)$$

$$\approx 1.89612; \ f(t) = \sin t \Rightarrow f'(t) = \cos t \Rightarrow f''(t) = -\sin t$$

$$\Rightarrow M = 1 \Rightarrow |E_T| \le \frac{\pi - 0}{12}\left(\frac{\pi}{4}\right)^2 (1) = \frac{\pi^3}{192} \approx 0.16149$$

(b) $\displaystyle\int_0^\pi \sin t \, dt = [-\cos t]_0^\pi = (-\cos \pi) - (-\cos 0) = 2 \Rightarrow |E_T| = \int_0^\pi \sin t \, dt - T \approx 2 - 1.89612 = 0.10388$

(c) $\dfrac{|E_T|}{\text{True Value}} \times 100 = \dfrac{0.10388}{2} \times 100 \approx 5\%$

II. (a) For $n = 4$, $h = \frac{b-a}{n} = \frac{\pi - 0}{4} = \frac{\pi}{4} \Rightarrow \frac{h}{3} = \frac{\pi}{12}$;

	t_i	$f(t_i)$	m	$mf(t_i)$
t_0	0	0	1	0
t_1	$\pi/4$	$\sqrt{2}/2$	4	$2\sqrt{2}$
t_2	$\pi/2$	1	2	2
t_3	$3\pi/4$	$\sqrt{2}/2$	4	$2\sqrt{2}$
t_4	π	0	1	0

$$\sum mf(t_i) = 2 + 4\sqrt{2} \approx 7.6569 \Rightarrow S = \frac{\pi}{12}\left(2 + 4\sqrt{2}\right)$$

$$\approx 2.00456; \ f^{(3)}(t) = -\cos t \Rightarrow f^{(4)}(t) = \sin t$$

$$\Rightarrow M = 1 \Rightarrow |E_S| \le \frac{\pi - 0}{180}\left(\frac{\pi}{4}\right)^4 (1) \approx 0.00664$$

(b) $\displaystyle\int_0^\pi \sin t\ dt \Rightarrow E_S = \int_0^\pi \sin t\ dt - S \approx 2 - 2.00456 = -0.00456 \Rightarrow |E_S| \approx 0.00456$

(c) $\displaystyle \frac{|E_S|}{\text{True Value}} \times 100 = \frac{0.00456}{2} \times 100 \approx 0\%$

11. (a) $n = 8 \Rightarrow h = \frac{1}{8} \Rightarrow \frac{h}{2} = \frac{1}{16}$;

$\displaystyle\sum mf(x_i) = 1(0.0) + 2(0.12402) + 2(0.24206) + 2(0.34763) + 2(0.43301) + 2(0.48789) + 2(0.49608)$
$+ 2(0.42361) + 1(0) = 5.1086 \Rightarrow T = \frac{1}{16}(5.1086) = 0.31929$

(b) $n = 8 \Rightarrow h = \frac{1}{8} \Rightarrow \frac{h}{3} = \frac{1}{24}$;

$\displaystyle\sum mf(x_i) = 1(0.0) + 4(0.12402) + 2(0.24206) + 4(0.34763) + 2(0.43301) + 4(0.48789) + 2(0.49608)$
$+ 4(0.42361) + 1(0) = 7.8749 \Rightarrow S = \frac{1}{24}(7.8749) = 0.32812$

(c) Let $u = 1 - x^2 \Rightarrow du = -2x\ dx \Rightarrow -\frac{1}{2}du = x\ dx$; $x = 0 \Rightarrow u = 1$, $x = 1 \Rightarrow u = 0$

$\displaystyle\int_0^1 x\sqrt{1-x^2}\ dx = \int_1^0 \sqrt{u}\left(-\frac{1}{2}du\right) = \frac{1}{2}\int_0^1 u^{1/2}\ du = \left[\frac{1}{2}\left(\frac{u^{3/2}}{\frac{3}{2}}\right)\right]_0^1 = \left[\frac{1}{3}u^{3/2}\right]_0^1 = \frac{1}{3}(\sqrt{1})^3 - \frac{1}{3}(\sqrt{0})^3 = \frac{1}{3}$;

$\displaystyle E_T = \int_0^1 x\sqrt{1-x^2}\ dx - T \approx \frac{1}{3} - 0.31929 = 0.01404$; $\displaystyle E_S = \int_0^1 x\sqrt{1-x^2}\ dx - S \approx \frac{1}{3} - 0.32812 = 0.00521$

13. (a) $n = 8 \Rightarrow h = \frac{\pi}{8} \Rightarrow \frac{h}{2} = \frac{\pi}{16}$;

$\displaystyle\sum mf(t_i) = 1(0.0) + 2(1.99138) + 2(1.26906) + 2(1.05961) + 2(0.75) + 2(0.48821) + 2(0.28946) + 2(0.13429)$
$+ 1(0) = 9.96402 \Rightarrow T = \frac{\pi}{16}(9.96402) \approx 1.95643$

(b) $n = 8 \Rightarrow h = \frac{\pi}{8} \Rightarrow \frac{h}{3} = \frac{\pi}{24}$;

$\displaystyle\sum mf(t_i) = 1(0.0) + 4(0.99138) + 2(1.26906) + 4(1.05961) + 2(0.75) + 4(0.48821) + 2(0.28946) + 4(0.13429)$
$+ 1(0) = 15.311 \Rightarrow S \approx \frac{\pi}{24}(15.311) \approx 2.00421$

(c) Let $u = 2 + \sin t \Rightarrow du = \cos t\ dt$; $t = -\frac{\pi}{2} \Rightarrow u = 2 + \sin\left(-\frac{\pi}{2}\right) = 1$, $t = \frac{\pi}{2} \Rightarrow u = 2 + \sin\frac{\pi}{2} = 3$

$\displaystyle\int_{-\pi/2}^{\pi/2} \frac{3\cos t}{(2+\sin t)^2}\ dt = \int_1^3 \frac{3}{u^2}\ du = 3\int_1^3 u^{-2}\ du = \left[3\left(\frac{u^{-1}}{-1}\right)\right]_1^3 = 3\left(-\frac{1}{3}\right) - 3\left(-\frac{1}{1}\right) = 2$;

$\displaystyle E_T = \int_{-\pi/2}^{\pi/2} \frac{3\cos t}{(2+\sin t)^2}\ dt - T \approx 2 - 1.95643 = 0.04357$; $\displaystyle E_S = \int_{-\pi/2}^{\pi/2} \frac{3\cos t}{(2+\sin t)^2}\ dt - S$

$\approx 2 - 2.00421 = -0.00421$

15. (a) $M = 0$ (see Exercise 1): Then $n = 1 \Rightarrow h = 1 \Rightarrow |E_T| = \frac{1}{12}(1)^2(0) = 0 < 10^{-4}$

(b) $M = 0$ (see Exercise 1): Then $n = 2$ (n must be even) $\Rightarrow h = \frac{1}{2} \Rightarrow |E_S| = \frac{1}{180}\left(\frac{1}{2}\right)^4(0) = 0 < 10^{-4}$

17. (a) $M = 2$ (see Exercise 3): Then $h = \frac{2}{n} \Rightarrow |E_T| \leq \frac{2}{12}\left(\frac{2}{n}\right)^2 (2) = \frac{4}{3n^2} < 10^{-4} \Rightarrow n^2 > \frac{4}{3}(10^4) \Rightarrow n > \sqrt{\frac{4}{3}(10^4)}$

   $\Rightarrow n > 115.4$, so let $n = 116$

(b) $M = 0$ (see Exercise 3): Then $n = 2$ (n must be even) $\Rightarrow h = 1 \Rightarrow |E_S| = \frac{2}{180}(1)^4(0) = 0 < 10^{-4}$

19. (a) $M = 12$ (see Exercise 5): Then $h = \frac{2}{n} \Rightarrow |E_T| \leq \frac{2}{12}\left(\frac{2}{n}\right)^2 (12) = \frac{8}{n^2} < 10^{-4} \Rightarrow n^2 > 8(10^4) \Rightarrow n > \sqrt{8(10^4)}$

   $\Rightarrow n > 282.8$, so let $n = 283$

(b) $M = 0$ (see Exercise 5): Then $n = 2$ (n must be even) $\Rightarrow h = 1 \Rightarrow |E_S| = \frac{2}{180}(1)^4(0) = 0 < 10^{-4}$

21. (a) $M = 6$ (see Exercise 7): Then $h = \frac{1}{n} \Rightarrow |E_T| \leq \frac{1}{12}\left(\frac{1}{n}\right)^2 (6) = \frac{1}{2n^2} < 10^{-4} \Rightarrow n^2 > \frac{1}{2}(10^4) \Rightarrow n > \sqrt{\frac{1}{2}(10^4)}$

   $\Rightarrow n > 70.7$, so let $n = 71$

(b) $M = 120$ (see Exercise 7): Then $h = \frac{1}{n} \Rightarrow |E_S| = \frac{1}{180}\left(\frac{1}{n}\right)^4 (120) = \frac{2}{3n^4} < 10^{-4} \Rightarrow n^4 > \frac{2}{3}(10^4)$

   $\Rightarrow n > \sqrt[4]{\frac{2}{3}(10^4)} \Rightarrow n > 9.04$, so let $n = 10$ (n must be even)

23. (a) $f(x) = \sqrt{x+1} \Rightarrow f'(x) = \frac{1}{2}(x+1)^{-1/2} \Rightarrow f''(x) = -\frac{1}{4}(x+1)^{-3/2} = -\frac{1}{4(\sqrt{x+1})^3} \Rightarrow M = \frac{1}{4(\sqrt{1})^3} = \frac{1}{4}$.

   Then $h = \frac{3}{n} \Rightarrow |E_T| \leq \frac{3}{12}\left(\frac{3}{n}\right)^2\left(\frac{1}{4}\right) = \frac{9}{16n^2} < 10^{-4} \Rightarrow n^2 > \frac{9}{16}(10^4) \Rightarrow n > \sqrt{\frac{9}{16}(10^4)} \Rightarrow n > 75$, so let $n = 76$

(b) $f^{(3)}(x) = \frac{3}{8}(x+1)^{-5/2} \Rightarrow f^{(4)}(x) = -\frac{15}{16}(x+1)^{-7/2} = -\frac{15}{16(\sqrt{x+1})^7} \Rightarrow M = \frac{15}{16(\sqrt{1})^7} = \frac{15}{16}$. Then $h = \frac{3}{n}$

   $\Rightarrow |E_S| \leq \frac{3}{180}\left(\frac{3}{n}\right)^4\left(\frac{15}{16}\right) = \frac{3^5(15)}{16(180)n^4} < 10^{-4} \Rightarrow n^4 > \frac{3^5(15)(10^4)}{16(180)} \Rightarrow n > \sqrt[4]{\frac{3^5(15)(10^4)}{16(180)}} \Rightarrow n > 10.6$, so let

   $n = 12$ (n must be even)

25. (a) $f(x) = \sin(x+1) \Rightarrow f'(x) = \cos(x+1) \Rightarrow f''(x) = -\sin(x+1) \Rightarrow M = 1$. Then $h = \frac{2}{n} \Rightarrow |E_T| \leq \frac{2}{12}\left(\frac{2}{n}\right)^2 (1)$

   $= \frac{8}{12n^2} < 10^{-4} \Rightarrow n^2 > \frac{8(10^4)}{12} \Rightarrow n > \sqrt{\frac{8(10^4)}{12}} \Rightarrow n > 81.6$, so let $n = 82$

(b) $f^{(3)}(x) = -\cos(x+1) \Rightarrow f^{(4)}(x) = \sin(x+1) \Rightarrow M = 1$. Then $h = \frac{2}{n} \Rightarrow |E_S| \leq \frac{2}{180}\left(\frac{2}{n}\right)^4 (1) = \frac{32}{180n^4} < 10^{-4}$

   $\Rightarrow n^4 > \frac{32(10^4)}{180} \Rightarrow n > \sqrt[4]{\frac{32(10^4)}{180}} \Rightarrow n > 6.49$, so let $n = 8$ (n must be even)

27. Using Simpson's Rule, $h = 200 \Rightarrow \frac{h}{3} = \frac{200}{3}$;

$$\sum mf(x_i) = 20{,}260 \Rightarrow \text{Area} \approx \frac{200}{3}(20{,}260)$$

$= 1{,}350{,}666.667 \text{ ft}^2$. Since the average depth $= 20$ ft

we obtain Volume $\approx 20\,(\text{Area}) \approx 27{,}013{,}333.33 \text{ ft}^3$.

Now, Number of fish $= \dfrac{\text{Volume}}{1000} = 27{,}013$ (to the nearest

fish) $\Rightarrow$ Maximum to be caught $= 75\%$ of $27{,}013 = 20{,}260$

$\Rightarrow$ Number of licenses $= \dfrac{20{,}260}{20} = 1013$

	x_i	$f(x_i)$	m	$mf(x_i)$
x_0	0	0	1	0
x_1	200	520	4	2080
x_2	400	800	2	1600
x_3	600	1000	4	4000
x_4	800	1140	2	2280
x_4	1000	1160	4	4640
x_6	1200	1110	2	2220
x_7	1400	860	4	3440
x_8	1600	0	1	0

29. Using Simpson's Rule, $h = \dfrac{b-a}{n} = \dfrac{24-0}{6} = \dfrac{24}{6} = 4$;

$$\sum my_i = 350 \Rightarrow S = \frac{4}{3}(350) = \frac{1400}{3} \approx 466.7 \text{ in.}^2$$

	x_i	y_i	m	my_i
x_0	0	0	1	0
x_1	4	18.75	4	75
x_2	8	24	2	48
x_3	12	26	4	104
x_4	16	24	2	48
x_5	20	18.75	4	75
x_6	24	0	1	0

31. $n = 2 \Rightarrow h = \dfrac{2-0}{2} = 1 \Rightarrow \dfrac{h}{3} = \dfrac{1}{3}$;

$$\sum mf(x_i) = 12 \Rightarrow S = \frac{1}{3}(12) = 4;$$

$$\int_0^2 x^3\,dx = \left[\frac{x^4}{4}\right]_0^2 = \frac{2^4}{4} - \frac{0^4}{4} = 4$$

	x_i	$f(x_i)$	m	$mf(x_i)$
x_0	0	0	1	0
x_1	1	1	4	4
x_2	2	8	1	8

33. (a) $n = 10 \Rightarrow h = \dfrac{\pi - 0}{10} = \dfrac{\pi}{10} \Rightarrow \dfrac{h}{2} = \dfrac{\pi}{20}$;

$$\sum mf(x_i) = 1(0) + 2(0.09708) + 2(0.36932) + 2(0.76248) + 2(1.19513) + 2(1.57080) + 2(1.79270)$$

$$+ 2(1.77912) + 2(1.47727) + 2(0.87372) + 1(0) = 19.83523 \Rightarrow T = \frac{\pi}{20}(19.83523) = 3.11571$$

(b) $\pi - 3.11571 \approx 0.02588$

(c) With $M = 3.11$, we get $|E_T| \le \dfrac{\pi}{12}\left(\dfrac{\pi}{10}\right)^2(3.11) = \dfrac{\pi^3}{1200}(3.11) < 0.08036$

35. The fourth derivative of $f(x) = x^{3/2}$ is $f^{(4)} = \frac{9}{16}x^{-5/2}$, which approaches ∞ as $x \to 0^+$. There is, therefore, no finite upper bound M for the values of $\left|f^{(4)}\right|$ on $[0, 4]$, the interval of integration, and Equation (7) gives no estimate for $\left|E_S\right|$ no matter how small you take the step size h.

This problem cannot be avoided by changing to the trapezoidal rule. The values of the integrand's second derivative $f''(x) = \frac{3}{4}x^{-1/2}$ are also unbounded on $[0, 4]$, and Equation (3) provides no estimate for $\left|E_T\right|$ no matter what you use for h.

Exercises 37-39 were done using a graphing calculator with $n = 50$

37. 1.08943   39. 0.82812

CHAPTER 4 PRACTICE EXERCISES

1. (a) Each time subinterval is of length $\Delta t = 0.4$ sec. The distance traveled over each subinterval, using the midpoint rule, is $\Delta h = \frac{1}{2}(v_i + v_{i+1})\Delta t$, where v_i is the velocity at the left, and v_{i+1} the velocity at the right, endpoint of the subinterval. We then add Δh to the height attained so far at the left endpoint v_i to arrive at the height associated with velocity v_{i+1} at the right endpoint. Using this methodology we build the following table based on the figure in the text:

t (sec)	0	0.4	0.8	1.2	1.6	2.0	2.4	2.8	3.2	3.6	4.0	4.4	4.8	5.2	5.6	6.0
v (fps)	0	10	25	55	100	190	180	170	155	140	130	120	105	90	80	65
h (ft)	0	2	9	25	56	114	188	258	323	382	436	486	531	570	604	633

t (sec)	6.4	6.8	7.2	7.6	8.0
v (fps)	52	40	30	15	0
h (ft)	656	674	688	697	700

NOTE: Your table values may vary slightly from ours depending on the v-values you read from the graph. Remember that some shifting of the graph occurs in the printing process.

The total height attained is about 700 ft.

(b) The graph is based on the table in part (a).

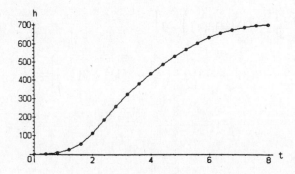

3. (a) $\displaystyle\sum_{k=1}^{10} \frac{a_k}{4} = \frac{1}{4} \sum_{k=1}^{10} a_k = \frac{1}{4}(-2) = -\frac{1}{2}$ (b) $\displaystyle\sum_{k=1}^{10} (b_k - 3a_k) = \sum_{k=1}^{10} b_k - 3 \sum_{k=1}^{10} a_k = 25 - 3(-2) = 31$

(c) $\displaystyle\sum_{k=1}^{10} (a_k + b_k - 1) = \sum_{k=1}^{10} a_k + \sum_{k=1}^{10} b_k - \sum_{k=1}^{10} 1 = -2 + 25 - (1)(10) = 13$

(d) $\displaystyle\sum_{k=1}^{10} \left(\frac{5}{2} - b_k\right) = \sum_{k=1}^{10} \frac{5}{2} - \sum_{k=1}^{10} b_k = \frac{5}{2}(10) - 25 = 0$

5. Let $u = 2x - 1 \Rightarrow du = 2\,dx \Rightarrow \frac{1}{2}\,du = dx;\ x = 1 \Rightarrow u = 1,\ x = 5 \Rightarrow u = 9$

$$\int_1^5 (2x - 1)^{-1/2}\,dx = \int_1^9 u^{-1/2}\left(\frac{1}{2}\,du\right) = \left[u^{1/2}\right]_1^9 = 3 - 1 = 2$$

7. Let $u = \frac{x}{2} \Rightarrow 2\,du = dx;\ x = -\pi \Rightarrow u = -\frac{\pi}{2},\ x = 0 \Rightarrow u = 0$

$$\int_{-\pi}^0 \cos\left(\frac{x}{2}\right)dx = \int_{-\pi/2}^0 (\cos u)(2\,du) = [2\sin u]_{-\pi/2}^0 = 2\sin 0 - 2\sin\left(-\frac{\pi}{2}\right) = 2(0 - (-1)) = 2$$

9. (a) $\displaystyle\int_{-2}^2 f(x)\,dx = \frac{1}{3}\int_{-2}^2 3f(x)\,dx = \frac{1}{3}(12) = 4$ (b) $\displaystyle\int_2^5 f(x)\,dx = \int_{-2}^5 f(x)\,dx - \int_{-2}^2 f(x)\,dx = 6 - 4 = 2$

(c) $\displaystyle\int_5^{-2} g(x)\,dx = -\int_{-2}^5 g(x)\,dx = -2$ (d) $\displaystyle\int_{-2}^5 (-\pi g(x))\,dx = -\pi\int_{-2}^5 g(x)\,dx = -\pi(2) = -2\pi$

(e) $\displaystyle\int_{-2}^5 \left(\frac{f(x) + g(x)}{5}\right)dx = \frac{1}{5}\int_{-2}^5 f(x)\,dx + \frac{1}{5}\int_{-2}^5 g(x)\,dx = \frac{1}{5}(6) + \frac{1}{5}(2) = \frac{8}{5}$

11. $x^2 - 4x + 3 = 0 \Rightarrow (x - 3)(x - 1) = 0 \Rightarrow x = 3\ \text{ or } x = 1;$

Area $= \displaystyle\int_0^1 (x^2 - 4x + 3)\,dx - \int_1^3 (x^2 - 4x + 3)\,dx$

$= \left[\dfrac{x^3}{3} - 2x^2 + 3x\right]_0^1 - \left[\dfrac{x^3}{3} - 2x^2 + 3x\right]_1^3$

$= \left[\left(\dfrac{1^3}{3} - 2(1)^2 + 3(1)\right) - 0\right]$

$- \left[\left(\dfrac{3^3}{3} - 2(3)^2 + 3(3)\right) - \left(\dfrac{1^3}{3} - 2(1)^2 + 3(1)\right)\right] = \left(\dfrac{1}{3} + 1\right) - \left[0 - \left(\dfrac{1}{3} + 1\right)\right] = \dfrac{8}{3}$

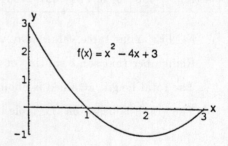

$f(x) = x^2 - 4x + 3$

13. $5 - 5x^{2/3} = 0 \Rightarrow 1 - x^{2/3} = 0 \Rightarrow x = \pm 1$;

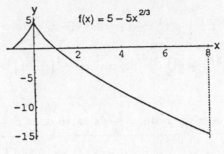

$$\text{Area} = \int_{-1}^{1} \left(5 - 5x^{2/3}\right) dx - \int_{1}^{8} \left(5 - 5x^{2/3}\right) dx$$

$$= \left[5x - 3x^{5/3}\right]_{-1}^{1} - \left[5x - 3x^{5/3}\right]_{1}^{8}$$

$$= \left[\left(5(1) - 3(1)^{5/3}\right) - \left(5(-1) - 3(-1)^{5/3}\right)\right]$$

$$- \left[\left(5(8) - 3(8)^{5/3}\right) - \left(5(1) - 3(1)^{5/3}\right)\right] = [2 - (-2)] - [(40 - 96) - 2] = 62$$

15. $y = \int \dfrac{x^2 + 1}{x^2} \, dx = \int \left(1 + x^{-2}\right) dx = x - x^{-1} + C = x - \dfrac{1}{x} + C$; $y = -1$ when $x = 1 \Rightarrow 1 - \dfrac{1}{1} + C = -1$

$\Rightarrow C = -1 \Rightarrow y = x - \dfrac{1}{x} - 1$

17. $\dfrac{dr}{dt} = \int \left(15\sqrt{t} + \dfrac{3}{\sqrt{t}}\right) dt = \int \left(15t^{1/2} + 3t^{-1/2}\right) dt = 10t^{3/2} + 6t^{1/2} + C$; $\dfrac{dr}{dt} = 8$ when $t = 1$

$\Rightarrow 10(1)^{3/2} + 6(1)^{1/2} + C = 8 \Rightarrow C = -8$. Thus $\dfrac{dr}{dt} = 10t^{3/2} + 6t^{1/2} - 8 \Rightarrow r = \int \left(10t^{3/2} + 6t^{1/2} - 8\right) dt$

$= 4t^{5/2} + 4t^{3/2} - 8t + C$; $r = 0$ when $t = 1 \Rightarrow 4(1)^{5/2} + 4(1)^{3/2} - 8(1) + C_1 = 0 \Rightarrow C_1 = 0$. Therefore,

$r = 4t^{5/2} + 4t^{3/2} - 8t$

19. $y = x^2 + \int_{1}^{x} \dfrac{1}{t} \, dt \Rightarrow \dfrac{dy}{dx} = 2x + \dfrac{1}{x} \Rightarrow \dfrac{d^2y}{dx^2} = 2 - \dfrac{1}{x^2}$; $y(1) = 1 + \int_{1}^{1} \dfrac{1}{t} \, dt = 1$ and $y'(1) = 2 + 1 = 3$

21. $y = \int_{5}^{x} \dfrac{\sin t}{t} \, dt - 3 \Rightarrow \dfrac{dy}{dx} = \dfrac{\sin x}{x}$; $x = 5 \Rightarrow y = \int_{5}^{5} \dfrac{\sin t}{t} \, dt - 3 = -3$

23. $\int \left(x^3 + 5x - 7\right) dx = \dfrac{x^4}{4} + \dfrac{5x^2}{2} - 7x + C$

25. $\int \left(3\sqrt{t} + \dfrac{4}{t^2}\right) dt = \int \left(3t^{1/2} + 4t^{-2}\right) dt = \dfrac{3t^{3/2}}{\left(\frac{3}{2}\right)} + \dfrac{4t^{-1}}{-1} + C = 2t^{3/2} - \dfrac{4}{t} + C$

27. Let $u = r^2 + 5 \Rightarrow du = 2r \, dr \Rightarrow \dfrac{1}{2} \, du = r \, dr$

$\int \dfrac{r \, dr}{\left(r^2 + 5\right)^2} = \int \dfrac{\left(\frac{1}{2}\right) du}{u^2} = \dfrac{1}{2} \int u^{-2} \, du = \dfrac{1}{2}\left(\dfrac{u^{-1}}{-1}\right) + C = -\dfrac{1}{2}u^{-1} + C = -\dfrac{1}{2\left(r^2 + 5\right)} + C$

29. Let $u = 2 - \theta^2 \Rightarrow du = -2\theta \, d\theta \Rightarrow -\dfrac{1}{2} \, du = \theta \, d\theta$

$\int 3\theta\sqrt{2 - \theta^2} \, d\theta = \int \sqrt{u}\left(-\dfrac{3}{2} \, du\right) = -\dfrac{3}{2} \int u^{1/2} \, du = -\dfrac{3}{2}\left(\dfrac{u^{3/2}}{\frac{3}{2}}\right) + C = -u^{3/2} + C = -\left(2 - \theta^2\right)^{3/2} + C$

31. Let $u = 1 + x^4 \Rightarrow du = 4x^3 \, dx \Rightarrow \frac{1}{4} \, du = x^3 \, dx$

$$\int x^3 \left(1 + x^4\right)^{-1/4} dx = \int u^{-1/4}\left(\frac{1}{4} \, du\right) = \frac{1}{4}\int u^{-1/4} \, du = \frac{1}{4}\left(\frac{u^{3/4}}{\frac{3}{4}}\right) + C = \frac{1}{3}u^{3/4} + C = \frac{1}{3}\left(1 + x^4\right)^{3/4} + C$$

33. Let $u = \frac{s}{10} \Rightarrow du = \frac{1}{10} \, ds \Rightarrow 10 \, du = ds$

$$\int \sec^2 \frac{s}{10} \, ds = \int \left(\sec^2 u\right)(10 \, du) = 10 \int \sec^2 u \, du = 10 \tan u + C = 10 \tan \frac{s}{10} + C$$

35. Let $u = \sqrt{2}\,\theta \Rightarrow du = \sqrt{2} \, d\theta \Rightarrow \frac{1}{\sqrt{2}} \, du = d\theta$

$$\int \csc \sqrt{2}\theta \cot \sqrt{2}\theta \, d\theta = \int (\csc u \cot u)\left(\frac{1}{\sqrt{2}} \, du\right) = \frac{1}{\sqrt{2}}(-\csc u) + C = -\frac{1}{\sqrt{2}} \csc \sqrt{2}\theta + C$$

37. Let $u = \frac{x}{4} \Rightarrow du = \frac{1}{4} \, dx \Rightarrow 4 \, du = dx$

$$\int \sin^2 \frac{x}{4} \, dx = \int \left(\sin^2 u\right)(4 \, du) = \int 4\left(\frac{1 - \cos 2u}{2}\right) du = 2\int (1 - \cos 2u) \, du = 2\left(u - \frac{\sin 2u}{2}\right) + C$$

$$= 2u - \sin 2u + C = 2\left(\frac{x}{4}\right) - \sin 2\left(\frac{x}{4}\right) + C = \frac{x}{2} - \sin \frac{x}{2} + C$$

39. Let $u = \cos x \Rightarrow du = -\sin x \, dx \Rightarrow -du = \sin x \, dx$

$$\int 2(\cos x)^{-1/2} \sin x \, dx = \int 2u^{-1/2}(-du) = -2\int u^{-1/2} \, du = -2\left(\frac{u^{1/2}}{\frac{1}{2}}\right) + C = -4u^{1/2} + C$$

$$= -4(\cos x)^{1/2} + C$$

41. Let $u = 2\theta + 1 \Rightarrow du = 2 \, d\theta \Rightarrow \frac{1}{2} \, du = d\theta$

$$\int [2\theta + 1 + 2\cos(2\theta + 1)] \, d\theta = \int (u + 2\cos u)\left(\frac{1}{2} \, du\right) = \frac{u^2}{4} + \sin u + C_1 = \frac{(2\theta + 1)^2}{4} + \sin(2\theta + 1) + C_1$$

$$= \theta^2 + \theta + \sin(2\theta + 1) + C, \text{ where } C = C_1 + \frac{1}{4} \text{ is still an arbitrary constant}$$

43. $\int \left(t - \frac{2}{t}\right)\left(t + \frac{2}{t}\right) dt = \int \left(t^2 - \frac{4}{t^2}\right) dt = \int \left(t^2 - 4t^{-2}\right) dt = \frac{t^3}{3} - 4\left(\frac{t^{-1}}{-1}\right) + C = \frac{t^3}{3} + \frac{4}{t} + C$

45. $\displaystyle\int_{-1}^{1} \left(3x^2 - 4x + 7\right) dx = \left[x^3 - 2x^2 + 7x\right]_{-1}^{1} = \left[1^3 - 2(1)^2 + 7(1)\right] - \left[(-1)^3 - 2(-1)^2 + 7(-1)\right] = 6 - (-10) = 16$

47. $\displaystyle\int_{1}^{2} \frac{4}{x^2} \, dx = \int_{1}^{2} 4x^{-2} \, dx = \left[-4x^{-1}\right]_{1}^{2} = \left(\frac{-4}{2}\right) - \left(\frac{-4}{1}\right) = 2$

49. $\displaystyle\int_{1}^{4} \frac{dt}{t\sqrt{t}} = \int_{1}^{4} \frac{dt}{t^{3/2}} = \int_{1}^{4} t^{-3/2} \, dt = \left[-2t^{-1/2}\right]_{1}^{4} = \frac{-2}{\sqrt{4}} - \frac{(-2)}{\sqrt{1}} = 1$

51. Let $u = 2x + 1 \Rightarrow du = 2\,dx \Rightarrow 18\,du = 36\,dx$; $x = 0 \Rightarrow u = 1$, $x = 1 \Rightarrow u = 3$

$$\int_0^1 \frac{36\,dx}{(2x+1)^3} = \int_1^3 18u^{-3}\,du = \left[\frac{18u^{-2}}{-2}\right]_1^3 = \left[\frac{-9}{u^2}\right]_1^3 = \left(\frac{-9}{3^2}\right) - \left(\frac{-9}{1^2}\right) = 8$$

53. Let $u = 1 - x^{2/3} \Rightarrow du = -\frac{2}{3}x^{-1/3}\,dx \Rightarrow -\frac{3}{2}\,du = x^{-1/3}\,dx$; $x = \frac{1}{8} \Rightarrow u = 1 - \left(\frac{1}{8}\right)^{2/3} = \frac{3}{4}$,

$x = 1 \Rightarrow u = 1 - 1^{2/3} = 0$

$$\int_{1/8}^1 x^{-1/3}\left(1 - x^{2/3}\right)^{3/2}\,dx = \int_{3/4}^0 u^{3/2}\left(-\frac{3}{2}\,du\right) = \left[\left(-\frac{3}{2}\right)\left(\frac{u^{5/2}}{\frac{5}{2}}\right)\right]_{3/4}^0 = \left[-\frac{3}{5}u^{5/2}\right]_{3/4}^0 = -\frac{3}{5}(0)^{5/2} - \left(-\frac{3}{5}\right)\left(\frac{3}{4}\right)^{5/2}$$

$$= \frac{27\sqrt{3}}{160}$$

55. Let $u = 5r \Rightarrow du = 5\,dr \Rightarrow \frac{1}{5}\,du = dr$; $r = 0 \Rightarrow u = 0$, $r = \pi \Rightarrow u = 5\pi$

$$\int_0^\pi \sin^2 5r\,dr = \int_0^{5\pi} (\sin^2 u)\left(\frac{1}{5}\,du\right) = \frac{1}{5}\left[\frac{u}{2} - \frac{\sin 2u}{4}\right]_0^{5\pi} = \left(\frac{\pi}{2} - \frac{\sin 10\pi}{20}\right) - \left(0 - \frac{\sin 0}{20}\right) = \frac{\pi}{2}$$

57. $\displaystyle\int_0^{\pi/3} \sec^2\theta\,d\theta = [\tan\theta]_0^{\pi/3} = \tan\frac{\pi}{3} - \tan 0 = \sqrt{3}$

59. Let $u = \frac{x}{6} \Rightarrow du = \frac{1}{6}\,dx \Rightarrow 6\,du = dx$; $x = \pi \Rightarrow u = \frac{\pi}{6}$, $x = 3\pi \Rightarrow u = \frac{\pi}{2}$

$$\int_\pi^{3\pi} \cot^2\frac{x}{6}\,dx = \int_{\pi/6}^{\pi/2} 6\cot^2 u\,du = 6\int_{\pi/6}^{\pi/2}\left(\csc^2 u - 1\right)du = [6(-\cot u - u)]_{\pi/6}^{\pi/2} = 6\left(-\cot\frac{\pi}{2} - \frac{\pi}{2}\right) - 6\left(-\cot\frac{\pi}{6} - \frac{\pi}{6}\right)$$

$$= 6\sqrt{3} - 2\pi$$

61. $\displaystyle\int_{-\pi/3}^0 \sec x \tan x\,dx = [\sec x]_{-\pi/3}^0 = \sec 0 - \sec\left(-\frac{\pi}{3}\right) = 1 - 2 = -1$

63. Let $u = \sin x \Rightarrow du = \cos x\,dx$; $x = 0 \Rightarrow u = 0$, $x = \frac{\pi}{2} \Rightarrow u = 1$

$$\int_0^{\pi/2} 5(\sin x)^{3/2}\cos x\,dx = \int_0^1 5u^{3/2}\,du = \left[5\left(\frac{2}{5}\right)u^{5/2}\right]_0^1 = \left[2u^{5/2}\right]_0^1 = 2(1)^{5/2} - 2(0)^{5/2} = 2$$

65. Let $u = \sin 3x \Rightarrow du = 3\cos 3x\,dx \Rightarrow \frac{1}{3}\,du = \cos 3x\,dx$; $x = -\frac{\pi}{2} \Rightarrow u = \sin\left(-\frac{3\pi}{2}\right) = 1$, $x = \frac{\pi}{2} \Rightarrow u = \sin\left(\frac{3\pi}{2}\right)$

$= -1$

$$\int_{-\pi/2}^{\pi/2} 15\sin^4 3x\cos 3x\,dx = \int_1^{-1} 15u^4\left(\frac{1}{3}\,du\right) = \int_1^{-1} 5u^4\,du = \left[u^5\right]_1^{-1} = (-1)^5 - (1)^5 = -2$$

67. Let $u = 1 + 3 \sin^2 x \Rightarrow du = 6 \sin x \cos x \, dx \Rightarrow \frac{1}{2} du = 3 \sin x \cos x \, dx; \ x = 0 \Rightarrow u = 1, \ x = \frac{\pi}{2}$

$\Rightarrow u = 1 + 3 \sin^2 \frac{\pi}{2} = 4$

$$\int_0^{\pi/2} \frac{3 \sin x \cos x}{\sqrt{1 + 3 \sin^2 x}} \, dx = \int_1^4 \frac{1}{\sqrt{u}}\left(\frac{1}{2} du\right) = \int_1^4 \frac{1}{2} u^{-1/2} \, du = \left[\frac{1}{2}\left(\frac{u^{1/2}}{\frac{1}{2}}\right)\right]_1^4 = \left[u^{1/2}\right]_1^4 = 4^{1/2} - 1^{1/2} = 1$$

69. Let $u = \sec \theta \Rightarrow du = \sec \theta \tan \theta \, d\theta; \ \theta = 0 \Rightarrow u = \sec 0 = 1, \ \theta = \frac{\pi}{3} \Rightarrow u = \sec \frac{\pi}{3} = 2$

$$\int_0^{\pi/3} \frac{\tan \theta}{\sqrt{2 \sec \theta}} \, d\theta = \int_0^{\pi/3} \frac{\sec \theta \tan \theta}{\sec \theta \sqrt{2 \sec \theta}} \, d\theta = \int_0^{\pi/3} \frac{\sec \theta \tan \theta}{\sqrt{2}(\sec \theta)^{3/2}} \, d\theta = \int_1^2 \frac{1}{\sqrt{2} \, u^{3/2}} \, du = \frac{1}{\sqrt{2}} \int_1^2 u^{-3/2} \, du$$

$$= \frac{1}{\sqrt{2}}\left[\frac{u^{-1/2}}{\left(-\frac{1}{2}\right)}\right]_1^2 = \left[-\frac{2}{\sqrt{2u}}\right]_1^2 = -\frac{2}{\sqrt{2(2)}} - \left(-\frac{2}{\sqrt{2(1)}}\right) = \sqrt{2} - 1$$

71. (a) $\text{av}(f) = \frac{1}{1-(-1)} \int_{-1}^1 (mx + b) \, dx = \frac{1}{2}\left[\frac{mx^2}{2} + bx\right]_{-1}^1 = \frac{1}{2}\left[\left(\frac{m(1)^2}{2} + b(1)\right) - \left(\frac{m(-1)^2}{2} + b(-1)\right)\right] = \frac{1}{2}(2b) = b$

(b) $\text{av}(f) = \frac{1}{k-(-k)} \int_{-k}^k (mx + b) \, dx = \frac{1}{2k}\left[\frac{mx^2}{2} + bx\right]_{-k}^k = \frac{1}{2k}\left[\left(\frac{m(k)^2}{2} + b(k)\right) - \left(\frac{m(-k)^2}{2} + b(-k)\right)\right]$

$= \frac{1}{2k}(2bk) = b$

73. $f'_{av} = \frac{1}{b-a} \int_a^b f'(x) \, dx = \frac{1}{b-a}[f(x)]_a^b = \frac{1}{b-a}[f(b) - f(a)] = \frac{f(b) - f(a)}{b-a}$ so the average value of f' over $[a, b]$ is the

slope of the secant line joining the points $(a, f(a))$ and $(b, f(b))$.

75. $|E_s| \leq \frac{3-1}{180}(h)^4 M$ where $h = \frac{3-1}{n} = \frac{2}{n}$; $f(x) = \frac{1}{x} = x^{-1} \Rightarrow f'(x) = -x^{-2} \Rightarrow f''(x) = 2x^{-3} \Rightarrow f'''(x) = -6x^{-4}$

$\Rightarrow f^{(4)}(x) = 24x^{-5}$ which is decreasing on $[1, 3] \Rightarrow$ maximum of $f^{(4)}(x)$ on $[1, 3]$ is $f^{(4)}(1) = 24 \Rightarrow M = 24$. Then

$|E_s| \leq 0.0001 \Rightarrow \left(\frac{3-1}{180}\right)\left(\frac{2}{n}\right)^4 (24) \leq 0.0001 \Rightarrow \left(\frac{768}{180}\right)\left(\frac{1}{n^4}\right) \leq 0.0001 \Rightarrow \frac{1}{n^4} \leq (0.0001)\left(\frac{180}{768}\right) \Rightarrow n^4 \geq 10{,}000\left(\frac{768}{180}\right)$

$\Rightarrow n \geq 14.37 \Rightarrow n \geq 16$ (n must be even)

77. $h = \frac{b-a}{n} = \frac{\pi - 0}{6} = \frac{\pi}{6} \Rightarrow \frac{h}{2} = \frac{\pi}{12}$;

$\sum_{i=0}^6 mf(x_i) = 12 \Rightarrow T = \left(\frac{\pi}{12}\right)(12) = \pi$;

	x_i	$f(x_i)$	m	$mf(x_i)$
x_0	0	0	1	0
x_1	$\pi/6$	1/2	2	1
x_2	$\pi/3$	3/2	2	3
x_3	$\pi/2$	2	2	4
x_4	$2\pi/3$	3/2	2	3
x_5	$5\pi/6$	1/2	2	1
x_6	π	0	1	0

	x_i	$f(x_i)$	m	$mf(x_i)$
x_0	0	0	1	0
x_1	$\pi/6$	1/2	4	2
x_2	$\pi/3$	3/2	2	3
x_3	$\pi/2$	2	4	8
x_4	$2\pi/3$	3/2	2	3
x_5	$5\pi/6$	1/2	4	2
x_6	π	0	1	0

$$\sum_{i=0}^{6} mf(x_i) = 18 \text{ and } \frac{h}{3} = \frac{\pi}{18} \Rightarrow S = \left(\frac{\pi}{18}\right)(18) = \pi.$$

79. $y_{av} = \dfrac{1}{365-0} \displaystyle\int_0^{365} \left[37\sin\left(\dfrac{2\pi}{365}(x-101)\right) + 25\right] dx = \dfrac{1}{365}\left[-37\left(\dfrac{365}{2\pi}\right)\cos\left(\dfrac{2\pi}{365}(x-101)\right) + 25x\right]_0^{365}$

$= \dfrac{1}{365}\left[\left(-37\left(\dfrac{365}{2\pi}\right)\cos\left[\dfrac{2\pi}{365}(365-101)\right] + 25(365)\right) - \left(-37\left(\dfrac{365}{2\pi}\right)\cos\left[\dfrac{2\pi}{365}(0-101)\right] + 25(0)\right)\right]$

$= -\dfrac{37}{2\pi}\cos\left(\dfrac{2\pi}{365}(264)\right) + 25 + \dfrac{37}{2\pi}\cos\left(\dfrac{2\pi}{365}(-101)\right) = -\dfrac{37}{2\pi}\left(\cos\left(\dfrac{2\pi}{365}(264)\right) - \cos\left(\dfrac{2\pi}{365}(-101)\right)\right) + 25$

$\approx -\dfrac{37}{2\pi}(0.16705 - 0.16705) + 25 = 25°\,\text{F}$

81. Yes. The function f, being differentiable on [a, b], is then continuous on [a, b]. The Fundamental Theorem of Calculus says that every continuous function on [a, b] is the derivative of a function on [a, b].

83. $y = \displaystyle\int_x^1 \sqrt{1+t^2}\,dt = -\int_1^x \sqrt{1+t^2}\,dt \Rightarrow \dfrac{dy}{dx} = \dfrac{d}{dx}\left[-\int_1^x \sqrt{1+t^2}\,dt\right] = -\dfrac{d}{dx}\left[\int_1^x \sqrt{1+t^2}\,dt\right] = -\sqrt{1+x^2}$

85. Using the trapezoidal rule, $h = 15 \Rightarrow \dfrac{h}{2} = 7.5$;

$\sum mf(x_i) = 794.8 \Rightarrow \text{Area} \approx (794.8)(7.5) = 5961 \text{ ft}^2;$

The cost is $\text{Area} \cdot (\$2.10/\text{ft}^2) \approx \left(5961 \text{ ft}^2\right)\left(\$2.10/\text{ft}^2\right)$

$= \$12{,}518.10 \Rightarrow$ the job cannot be done for $11,000.

	x_i	$f(x_i)$	m	$mf(x_i)$
x_0	0	0	1	0
x_1	15	36	2	72
x_2	30	54	2	108
x_3	45	51	2	102
x_4	60	49.5	2	99
x_5	75	54	2	108
x_6	90	64.4	2	128.8
x_7	105	67.5	2	135
x_8	120	42	1	42

87. $\text{av}(I) = \dfrac{1}{30} \displaystyle\int_0^{30} (1200 - 40t)\,dt = \dfrac{1}{30}\left[1200t - 20t^2\right]_0^{30} = \dfrac{1}{30}\left[\left((1200(30) - 20(30)^2\right) - \left(1200(0) - 20(0)^2\right)\right]$

$= \dfrac{1}{30}(18{,}000) = 600;$ Average Daily Holding Cost $= (600)(\$0.03) = \$18$

89. $\text{av}(I) = \frac{1}{30} \int_0^{30} \left(450 - \frac{t^2}{2}\right) dt = \frac{1}{30}\left[450t - \frac{t^3}{6}\right]_0^{30} = \frac{1}{30}\left[450(30) - \frac{30^3}{6} - 0\right] = 300;$ Average Daily Holding Cost

$= (300)(\$0.02) = \$6$

CHAPTER 4 ADDITIONAL EXERCISES–THEORY, EXAMPLES, APPLICATIONS

1. (a) Yes, because $\int_0^1 f(x)\, dx = \frac{1}{7}\int_0^1 7f(x)\, dx = \frac{1}{7}(7) = 1$

   (b) No. For example, $\int_0^1 8x\, dx = \left[4x^2\right]_0^1 = 4,$ but $\int_0^1 \sqrt{8x}\, dx = \left[2\sqrt{2}\left(\frac{x^{3/2}}{\frac{3}{2}}\right)\right]_0^1 = \frac{4\sqrt{2}}{3}(1^{3/2} - 0^{3/2})$

   $= \frac{4\sqrt{2}}{3} \neq \sqrt{4}$

3. $y = \frac{1}{a}\int_0^x f(t)\sin a(x-t)\, dt = \frac{1}{a}\int_0^x f(t)\sin ax \cos at\, dt - \frac{1}{a}\int_0^x f(t)\cos ax \sin at\, dt$

   $= \frac{\sin ax}{a}\int_0^x f(t)\cos at\, dt - \frac{\cos ax}{a}\int_0^x f(t)\sin at\, dt \Rightarrow \frac{dy}{dx} = \cos ax \int_0^x f(t)\cos at\, dt$

   $+ \frac{\sin ax}{a}\left(\frac{d}{dx}\int_0^x f(t)\cos at\, dt\right) + \sin ax \int_0^x f(t)\sin at\, dt - \frac{\cos ax}{a}\left(\frac{d}{dx}\int_0^x f(t)\sin at\, dt\right)$

   $= \cos ax \int_0^x f(t)\cos at\, dt + \frac{\sin ax}{a}(f(x)\cos ax) + \sin ax \int_0^x f(t)\sin at\, dt - \frac{\cos ax}{a}(f(x)\sin ax)$

   $\Rightarrow \frac{dy}{dx} = \cos ax \int_0^x f(t)\cos at\, dt + \sin ax \int_0^x f(t)\sin at\, dt.$ Next,

   $\frac{d^2y}{dx^2} = -a\sin ax \int_0^x f(t)\cos at\, dt + (\cos ax)\left(\frac{d}{dx}\int_0^x f(t)\cos at\, dt\right) + a\cos ax \int_0^x f(t)\sin at\, dt$

   $+ (\sin ax)\left(\frac{d}{dx}\int_0^x f(t)\sin at\, dt\right) = -a\sin ax \int_0^x f(t)\cos at\, dt + (\cos ax)f(x)\cos ax$

   $+ a\cos ax \int_0^x f(t)\sin at\, dt + (\sin ax)f(x)\sin ax = -a\sin ax \int_0^x f(t)\cos at\, dt + a\cos ax \int_0^x f(t)\sin at\, dt + f(x).$

   Therefore, $y'' + a^2 y = a\cos ax \int_0^x f(t)\sin at\, dt - a\sin ax \int_0^x f(t)\cos at\, dt + f(x)$

   $+ a^2\left(\frac{\sin ax}{a}\int_0^x f(t)\cos at\, dt - \frac{\cos ax}{a}\int_0^x f(t)\sin at\, dt\right) = f(x).$ Note also that $y'(0) = y(0) = 0.$

5. (a) $\displaystyle\int_0^{x^2} f(t)\ dt = x\cos\pi x \Rightarrow \frac{d}{dx}\int_0^{x^2} f(t)\ dt = \cos\pi x - \pi x\sin\pi x \Rightarrow f(x^2)(2x) = \cos\pi x - \pi x\sin\pi x$

$\Rightarrow f(x^2) = \dfrac{\cos\pi x - \pi x\sin\pi x}{2x}$. Thus, $x = 2 \Rightarrow f(4) = \dfrac{\cos 2\pi - 2\pi\sin 2\pi}{4} = \dfrac{1}{4}$

(b) $\displaystyle\int_0^{f(x)} t^2\ dt = \left[\frac{t^3}{3}\right]_0^{f(x)} = \frac{1}{3}(f(x))^3 \Rightarrow \frac{1}{3}(f(x))^3 = x\cos\pi x \Rightarrow (f(x))^3 = 3x\cos\pi x \Rightarrow f(x) = \sqrt[3]{3x\cos\pi x}$

$\Rightarrow f(4) = \sqrt[3]{3(4)\cos 4\pi} = \sqrt[3]{12}$

7. $\displaystyle\int_1^b f(x)\ dx = \sqrt{b^2+1} - \sqrt{2} \Rightarrow f(b) = \frac{d}{db}\int_1^b f(x)\ dx = \frac{1}{2}(b^2+1)^{-1/2}(2b) = \frac{b}{\sqrt{b^2+1}} \Rightarrow f(x) = \frac{x}{\sqrt{x^2+1}}$

9. $\dfrac{dy}{dx} = 3x^2 + 2 \Rightarrow y = \displaystyle\int(3x^2+2)\ dx = x^3 + 2x + C$. Then $(1,-1)$ on the curve $\Rightarrow 1^3 + 2(1) + C = -1 \Rightarrow C = -4$

$\Rightarrow y = x^3 + 2x - 4$

11. $\displaystyle\int_{-8}^3 f(x)\ dx = \int_{-8}^0 x^{2/3}\ dx + \int_0^3 -4\ dx$

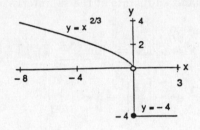

$= \left[\frac{3}{5}x^{5/3}\right]_{-8}^0 + [-4x]_0^3$

$= \left(0 - \frac{3}{5}(-8)^{5/3}\right) + (-4(3) - 0) = \frac{96}{5} - 12$

$= \frac{36}{5}$

13. $\displaystyle\int_0^2 f(t)\ dt = \int_0^1 t\ dt + \int_1^2 \sin\pi t\ dt$

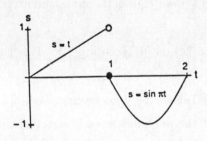

$= \left[\frac{t^2}{2}\right]_0^1 + \left[-\frac{1}{\pi}\cos\pi t\right]_1^2$

$= \left(\frac{1}{2} - 0\right) + \left[-\frac{1}{\pi}\cos 2\pi - \left(-\frac{1}{\pi}\cos\pi\right)\right]$

$= \frac{1}{2} - \frac{2}{\pi}$

15. $\displaystyle\int_{-2}^2 f(x)\ dx = \int_{-2}^{-1} dx + \int_{-1}^1 (1-x^2)\ dx + \int_1^2 2\ dx$

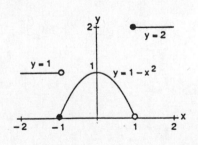

$= [x]_{-2}^{-1} + \left[x - \frac{x^3}{3}\right]_{-1}^1 + [2x]_1^2$

$= -1 - (-2) + \left(1 - \frac{1^3}{3}\right) - \left(-1 - \frac{(-1)^3}{3}\right) + 2(2) - 2(1)$

$= 1 + \frac{2}{3} - \left(-\frac{2}{3}\right) + 4 - 2 = \frac{13}{3}$

17. Ave. value $= \dfrac{1}{b-a} \displaystyle\int_a^b f(x)\,dx = \dfrac{1}{2-0} \displaystyle\int_0^2 f(x)\,dx = \dfrac{1}{2}\left[\displaystyle\int_0^1 x\,dx + \displaystyle\int_1^2 (x-1)\,dx \right] = \dfrac{1}{2}\left[\dfrac{x^2}{2} \right]_0^1 + \dfrac{1}{2}\left[\dfrac{x^2}{2} - x \right]_1^2$

$= \dfrac{1}{2}\left[\dfrac{1^2}{2} - 0 + \left(\dfrac{2^2}{2} - 2 \right) - \left(\dfrac{1^2}{2} - 1 \right) \right] = \dfrac{1}{2}$

19. $f(x) = \displaystyle\int_{1/x}^{x} \dfrac{1}{t}\,dt \Rightarrow f'(x) = \dfrac{1}{x}\left(\dfrac{dx}{dx} \right) - \left(\dfrac{1}{\frac{1}{x}} \right)\left(\dfrac{d}{dx}\left(\dfrac{1}{x} \right) \right) = \dfrac{1}{x} - x\left(-\dfrac{1}{x^2} \right) = \dfrac{1}{x} + \dfrac{1}{x} = \dfrac{2}{x}$

21. $g(y) = \displaystyle\int_{\sqrt{y}}^{2\sqrt{y}} \sin t^2\,dt \Rightarrow g'(y) = \left(\sin\left(2\sqrt{y}\right)^2 \right)\left(\dfrac{d}{dy}\left(2\sqrt{y}\right) \right) - \left(\sin\left(\sqrt{y}\right)^2 \right)\left(\dfrac{d}{dy}\left(\sqrt{y}\right) \right) = \dfrac{\sin 4y}{\sqrt{y}} - \dfrac{\sin y}{2\sqrt{y}}$

23. Let $f(x) = x^5$ on $[0,1]$. Partition $[0,1]$ into n subintervals with $\Delta x = \dfrac{1-0}{n} = \dfrac{1}{n}$. Then $\dfrac{1}{n}, \dfrac{2}{n}, \ldots, \dfrac{n}{n}$ are the

right-hand endpoints of the subintervals. Since f is increasing on $[0,1]$, $U = \displaystyle\sum_{j=1}^{\infty} \left(\dfrac{j}{n} \right)^5 \left(\dfrac{1}{n} \right)$ is the upper sum for

$f(x) = x^5$ on $[0,1] \Rightarrow \displaystyle\lim_{n\to\infty} \sum_{j=1}^{\infty} \left(\dfrac{j}{n} \right)^5 \left(\dfrac{1}{n} \right) = \lim_{n\to\infty} \dfrac{1}{n}\left[\left(\dfrac{1}{n} \right)^5 + \left(\dfrac{2}{n} \right)^5 + \ldots + \left(\dfrac{n}{n} \right)^5 \right] = \lim_{n\to\infty} \left[\dfrac{1^5 + 2^5 + \ldots + n^5}{n^6} \right]$

$= \displaystyle\int_0^1 x^5\,dx = \left[\dfrac{x^6}{6} \right]_0^1 = \dfrac{1}{6}$

25. Let $y = f(x)$ on $[0,1]$. Partition $[0,1]$ into n subintervals with $\Delta x = \dfrac{1-0}{n} = \dfrac{1}{n}$. Then $\dfrac{1}{n}, \dfrac{2}{n}, \ldots, \dfrac{n}{n}$ are the

right-hand endpoints of the subintervals. Since f is continuous on $[0,1]$, $\displaystyle\sum_{j=1}^{\infty} f\left(\dfrac{j}{n} \right)\left(\dfrac{1}{n} \right)$ is a Riemann sum of

$y = f(x)$ on $[0,1] \Rightarrow \displaystyle\lim_{n\to\infty} \sum_{j=1}^{\infty} f\left(\dfrac{j}{n} \right)\left(\dfrac{1}{n} \right) = \lim_{n\to\infty} \dfrac{1}{n}\left[f\left(\dfrac{1}{n} \right) + f\left(\dfrac{2}{n} \right) + \ldots + f\left(\dfrac{n}{n} \right) \right] = \displaystyle\int_0^1 f(x)\,dx$

27. (a) Let the polygon be inscribed in a circle of radius r. If we draw a radius from the center of the circle (and

the polygon) to each vertex of the polygon, we have n isosceles triangles formed (the equal sides are equal

to r, the radius of the circle) and a vertex angle of θ_n where $\theta_n = \dfrac{2\pi}{n}$. The area of each triangle is

$A_n = \dfrac{1}{2} r^2 \sin \theta_n \Rightarrow$ the area of the polygon is $A = n A_n = \dfrac{nr^2}{2} \sin \theta_n = \dfrac{nr^2}{2} \sin \dfrac{2\pi}{n}$.

(b) $\displaystyle\lim_{n\to\infty} A = \lim_{n\to\infty} \dfrac{nr^2}{2} \sin \dfrac{2\pi}{n} = \lim_{n\to\infty} \dfrac{n\pi r^2}{2\pi} \sin \dfrac{2\pi}{n} = \lim_{n\to\infty} \left(\pi r^2 \right) \dfrac{\sin\left(\dfrac{2\pi}{n} \right)}{\left(\dfrac{2\pi}{n} \right)} = \left(\pi r^2 \right) \lim_{2\pi/n \to 0} \dfrac{\sin\left(\dfrac{2\pi}{n} \right)}{\left(\dfrac{2\pi}{n} \right)} = \pi r^2$

NOTES:

CHAPTER 5 APPLICATIONS OF INTEGRALS

5.1 AREAS BETWEEN CURVES

1. For the sketch given, $a = 0$, $b = \pi$; $f(x) - g(x) = 1 - \cos^2 x = \sin^2 x = \dfrac{1 - \cos 2x}{2}$;

$$A = \int_0^\pi \frac{(1 - \cos 2x)}{2}\, dx = \frac{1}{2} \int_0^\pi (1 - \cos 2x)\, dx = \frac{1}{2}\left[x - \frac{\sin 2x}{2}\right]_0^\pi = \frac{1}{2}\left[(\pi - 0) - (0 - 0)\right] = \frac{\pi}{2}$$

3. For the sketch given, $c = 0$, $d = 1$; $f(y) - g(y) = y^2 - y^3$;

$$A = \int_0^1 \left(y^2 - y^3\right) dy = \int_0^1 y^2\, dy - \int_0^1 y^3\, dy = \left[\frac{y^3}{3}\right]_0^1 - \left[\frac{y^4}{4}\right]_0^1 = \frac{(1 - 0)}{3} - \frac{(1 - 0)}{4} = \frac{1}{3} - \frac{1}{4} = \frac{1}{12}$$

5. For the sketch given, $a = -2$, $b = 2$; $f(x) - g(x) = 2x^2 - \left(x^4 - 2x^2\right) = 4x^2 - x^4$;

$$A = \int_{-2}^2 \left(4x^2 - x^4\right) dx = \left[\frac{4x^3}{3} - \frac{x^5}{5}\right]_{-2}^2 = \left(\frac{32}{3} - \frac{32}{5}\right) - \left[-\frac{32}{3} - \left(-\frac{32}{5}\right)\right] = \frac{64}{3} - \frac{64}{5} = \frac{320 - 192}{15} = \frac{128}{15}$$

7. We want the area between the line $y = 1$, $0 \le x \le 2$, and the curve $y = \dfrac{x^2}{4}$, *minus* the area of a triangle

   (formed by $y = x$ and $y = 1$) with base 1 and height 1. Thus, $A = \displaystyle\int_0^2 \left(1 - \frac{x^2}{4}\right) dx - \frac{1}{2}(1)(1) = \left[x - \frac{x^3}{12}\right]_0^2 - \frac{1}{2}$

$$= \left(2 - \frac{8}{12}\right) - \frac{1}{2} = 2 - \frac{2}{3} - \frac{1}{2} = \frac{5}{6}$$

9. AREA $= A1 + A2$

   A1: For the sketch given, $a = -3$ and we find b by solving the equations $y = x^2 - 4$ and $y = -x^2 - 2x$

   simultaneously for x: $x^2 - 4 = -x^2 - 2x \Rightarrow 2x^2 + 2x - 4 = 0 \Rightarrow 2(x + 2)(x - 1) \Rightarrow x = -2$ or $x = 1$ so

   $b = -2$: $f(x) - g(x) = \left(x^2 - 4\right) - \left(-x^2 - 2x\right) = 2x^2 + 2x - 4 \Rightarrow A1 = \displaystyle\int_{-3}^{-2} \left(2x^2 + 2x - 4\right) dx$

$$= \left[\frac{2x^3}{3} + \frac{2x^2}{2} - 4x\right]_{-3}^{-2} = \left(-\frac{16}{3} + 4 + 8\right) - (-18 + 9 + 12) = 9 - \frac{16}{3} = \frac{11}{3};$$

   A2: For the sketch given, $a = -2$ and $b = 1$: $f(x) - g(x) = \left(-x^2 - 2x\right) - \left(x^2 - 4\right) = -2x^2 - 2x + 4$

$$\Rightarrow A2 = -\int_{-2}^1 \left(2x^2 + 2x - 4\right) dx = -\left[\frac{2x^3}{3} + x^2 - 4x\right]_{-2}^1 = -\left(\frac{2}{3} + 1 - 4\right) + \left(-\frac{16}{3} + 4 + 8\right)$$

$$= -\frac{2}{3} - 1 + 4 - \frac{16}{3} + 4 + 8 = 9;$$

   Therefore, AREA $= A1 + A2 = \dfrac{11}{3} + 9 = \dfrac{38}{3}$

11. $AREA = A1 + A2 + A3$

A1:   For the sketch given, $a = -2$ and $b = -1$: $f(x) - g(x) = (-x + 2) - (4 - x^2) = x^2 - x - 2$

$$\Rightarrow A1 = \int_{-2}^{-1} (x^2 - x - 2)\, dx = \left[\frac{x^3}{3} - \frac{x^2}{2} - 2x\right]_{-2}^{-1} = \left(-\frac{1}{3} - \frac{1}{2} + 2\right) - \left(-\frac{8}{3} - \frac{4}{2} + 4\right) = \frac{7}{3} - \frac{1}{2} = \frac{14 - 3}{6} = \frac{11}{6};$$

A2:   For the sketch given, $a = -1$ and $b = 2$: $f(x) - g(x) = (4 - x^2) - (-x + 2) = -(x^2 - x - 2)$

$$\Rightarrow A2 = -\int_{-1}^{2} (x^2 - x - 2)\, dx = -\left[\frac{x^3}{3} - \frac{x^2}{2} - 2x\right]_{-1}^{2} = -\left(\frac{8}{3} - \frac{4}{2} - 4\right) + \left(-\frac{1}{3} - \frac{1}{2} + 2\right) = -3 + 8 - \frac{1}{2} = \frac{9}{2};$$

A3:   For the sketch given, $a = 2$ and $b = 3$: $f(x) - g(x) = (-x + 2) - (4 - x^2) = x^2 - x - 2$

$$\Rightarrow A3 = \int_{2}^{3} (x^2 - x - 2)\, dx = \left[\frac{x^3}{3} - \frac{x^2}{2} - 2x\right]_{2}^{3} = \left(\frac{27}{3} - \frac{9}{2} - 6\right) - \left(\frac{8}{3} - \frac{4}{2} - 4\right) = 9 - \frac{9}{2} - \frac{8}{3};$$

Therefore, $AREA = A1 + A2 + A3 = \frac{11}{6} + \frac{9}{2} + \left(9 - \frac{9}{2} - \frac{8}{3}\right) = 9 - \frac{5}{6} = \frac{49}{6}$

13. $a = -2$, $b = 2$;

$f(x) - g(x) = 2 - (x^2 - 2) = 4 - x^2$

$$\Rightarrow A = \int_{-2}^{2} (4 - x^2)\, dx = \left[4x - \frac{x^3}{3}\right]_{-2}^{2} = \left(8 - \frac{8}{3}\right) - \left(-8 + \frac{8}{3}\right)$$

$$= 2 \cdot \left(\frac{24}{3} - \frac{8}{3}\right) = \frac{32}{3}$$

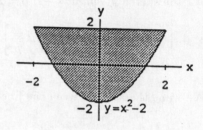

15. $a = 0$, $b = 2$;

$$f(x) - g(x) = 8x - x^4 \Rightarrow A = \int_{0}^{2} (8x - x^4)\, dx$$

$$= \left[\frac{8x^2}{2} - \frac{x^5}{5}\right]_{0}^{2} = 16 - \frac{32}{5} = \frac{80 - 32}{5} = \frac{48}{5}$$

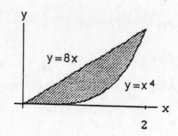

17. Limits of integration: $x^2 = -x^2 + 4x \Rightarrow 2x^2 - 4x = 0$

$\Rightarrow 2x(x - 2) = 0 \Rightarrow a = 0$ and $b = 2$;

$f(x) - g(x) = (-x^2 + 4x) - x^2 = -2x^2 + 4x$

$$\Rightarrow A = \int_{0}^{2} (-2x^2 + 4x)\, dx = \left[\frac{-2x^3}{3} + \frac{4x^2}{2}\right]_{0}^{2}$$

$$= -\frac{16}{3} + \frac{16}{2} = \frac{-32 + 48}{6} = \frac{8}{3}$$

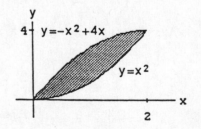

19. Limits of integration: $x^4 - 4x^2 + 4 = x^2$

$\Rightarrow x^4 - 5x^2 + 4 = 0 \Rightarrow (x^2 - 4)(x^2 - 1) = 0$

$\Rightarrow (x + 2)(x - 2)(x + 1)(x - 1) = 0 \Rightarrow x = -2, -1, 1, 2;$

$f(x) - g(x) = (x^4 - 4x^2 + 4) - x^2 = x^4 - 5x^2 + 4$ and

$g(x) - f(x) = x^2 - (x^4 - 4x^2 + 4) = -x^4 + 5x^2 - 4$

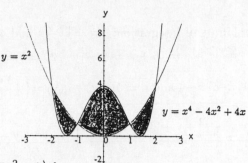

$\Rightarrow A = \int_{-2}^{-1} \left(-x^4 + 5x^2 - 4\right) dx + \int_{-1}^{1} \left(x^4 - 5x^2 + 4\right) dx + \int_{1}^{2} \left(-x^4 + 5x^2 - 4\right) dx$

$= \left[-\dfrac{x^5}{5} + \dfrac{5x^3}{3} - 4x\right]_{-2}^{-1} + \left[\dfrac{x^5}{5} - \dfrac{5x^3}{3} + 4x\right]_{-1}^{1} + \left[\dfrac{-x^5}{5} + \dfrac{5x^3}{3} - 4x\right]_{1}^{2} = \left(\dfrac{1}{5} - \dfrac{5}{3} + 4\right) - \left(\dfrac{32}{5} - \dfrac{40}{3} + 8\right) + \left(\dfrac{1}{5} - \dfrac{5}{3} + 4\right)$

$-\left(-\dfrac{1}{5} + \dfrac{5}{3} - 4\right) + \left(-\dfrac{32}{5} + \dfrac{40}{3} - 8\right) - \left(-\dfrac{1}{5} + \dfrac{5}{3} - 4\right) = -\dfrac{60}{5} + \dfrac{60}{3} = \dfrac{300 - 180}{15} = 8$

21. Limits of integration: $y = \sqrt{|x|} = \begin{cases} \sqrt{-x}, & x \le 0 \\ \sqrt{x}, & x \ge 0 \end{cases}$ and

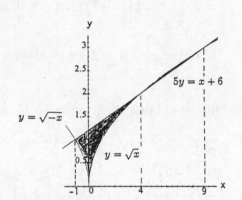

$5y = x + 6$ or $y = \dfrac{x}{5} + \dfrac{6}{5}$; for $x \le 0$: $\sqrt{-x} = \dfrac{x}{5} + \dfrac{6}{5}$

$\Rightarrow 5\sqrt{-x} = x + 6 \Rightarrow 25(-x) = x^2 + 12x + 36$

$\Rightarrow x^2 + 37x + 36 = 0 \Rightarrow (x + 1)(x + 36) = 0$

$\Rightarrow x = -1, -36$ (but $x = -36$ is not a solution);

for $x \ge 0$: $5\sqrt{x} = x + 6 \Rightarrow 25x = x^2 + 12x + 36$

$\Rightarrow x^2 - 13x + 36 = 0 \Rightarrow (x - 4)(x - 9) = 0$

$\Rightarrow x = 4, 9$; there are three intersection points and

$A = \int_{-1}^{0} \left(\dfrac{x + 6}{5} - \sqrt{-x}\right) dx + \int_{0}^{4} \left(\dfrac{x + 6}{5} - \sqrt{x}\right) dx + \int_{4}^{9} \left(\sqrt{x} - \dfrac{x + 6}{5}\right) dx$

$= \left[\dfrac{(x + 6)^2}{10} + \dfrac{2}{3}(-x)^{3/2}\right]_{-1}^{0} + \left[\dfrac{(x + 6)^2}{10} - \dfrac{2}{3}x^{3/2}\right]_{0}^{4} + \left[\dfrac{2}{3}x^{3/2} - \dfrac{(x + 6)^2}{10}\right]_{4}^{9}$

$= \left(\dfrac{36}{10} - \dfrac{25}{10} - \dfrac{2}{3}\right) + \left(\dfrac{100}{10} - \dfrac{2}{3}\cdot 4^{3/2} - \dfrac{36}{10} + 0\right) + \left(\dfrac{2}{3}\cdot 9^{3/2} - \dfrac{225}{10} - \dfrac{2}{3}\cdot 4^{3/2} + \dfrac{100}{10}\right) = -\dfrac{50}{10} + \dfrac{20}{3} = \dfrac{5}{3}$

23. Limits of integration: $c = 0$ and $d = 3$;

$f(y) - g(y) = 2y^2 - 0 = 2y^2$

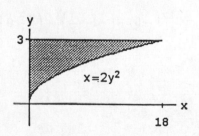

$\Rightarrow A = \int_{0}^{3} 2y^2\, dy = \left[\dfrac{2y^3}{3}\right]_{0}^{3} = 2\cdot 9 = 18$

18

25. Limits of integration: $4x = y^2 - 4$ and $4x = 16 + y$

$\Rightarrow y^2 - 4 = 16 + y \Rightarrow y^2 - y - 20 = 0 \Rightarrow (y - 5)(y + 4) = 0$

$\Rightarrow c = -4$ and $d = 5$; $f(y) - g(y) = \left(\dfrac{16 + y}{4}\right) - \left(\dfrac{y^2 - 4}{4}\right)$

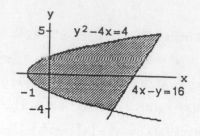

$= \dfrac{-y^2 + y + 20}{4} \Rightarrow A = \dfrac{1}{4} \displaystyle\int_{-4}^{5} (-y^2 + y + 20)\, dy$

$= \dfrac{1}{4}\left[-\dfrac{y^3}{3} + \dfrac{y^2}{2} + 20y\right]_{-4}^{5} = \dfrac{1}{4}\left(-\dfrac{125}{3} + \dfrac{25}{2} + 100\right) - \dfrac{1}{4}\left(\dfrac{64}{3} + \dfrac{16}{2} - 80\right) = \dfrac{1}{4}\left(-\dfrac{189}{3} + \dfrac{9}{2} + 180\right) = \dfrac{243}{8}$

27. Limits of integration: $x = -y^2$ and $x = 2 - 3y^2$

$\Rightarrow -y^2 = 2 - 3y^2 \Rightarrow 2y^2 - 2 = 0 \Rightarrow 2(y - 1)(y + 1) = 0$

$\Rightarrow c = -1$ and $d = 1$; $f(y) - g(y) = (2 - 3y^2) - (-y^2)$

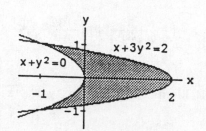

$= 2 - 2y^2 = 2(1 - y^2) \Rightarrow A = 2 \displaystyle\int_{-1}^{1} (1 - y^2)\, dy$

$= 2\left[y - \dfrac{y^3}{3}\right]_{-1}^{1} = 2\left(1 - \dfrac{1}{3}\right) - 2\left(-1 + \dfrac{1}{3}\right) = 4\left(\dfrac{2}{3}\right) = \dfrac{8}{3}$

29. Limits of integration: $x = y^2 - 1$ and $x = |y|\sqrt{1 - y^2}$

$\Rightarrow y^2 - 1 = |y|\sqrt{1 - y^2} \Rightarrow y^4 - 2y^2 + 1 = y^2(1 - y^2)$

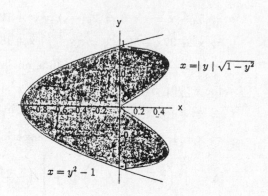

$\Rightarrow y^4 - 2y^2 + 1 = y^2 - y^4 \Rightarrow 2y^4 - 3y^2 + 1 = 0$

$\Rightarrow (2y^2 - 1)(y^2 - 1) = 0 \Rightarrow 2y^2 - 1 = 0$ or $y^2 - 1 = 0$

$\Rightarrow y^2 = \dfrac{1}{2}$ or $y^2 = 1 \Rightarrow y = \pm\dfrac{\sqrt{2}}{2}$ or $y = \pm 1.$

Substitution shows that $\dfrac{\pm\sqrt{2}}{2}$ are not solutions $\Rightarrow y = \pm 1$;

for $-1 \le y \le 0$, $f(x) - g(x) = -y\sqrt{1 - y^2} - (y^2 - 1)$

$= 1 - y^2 - y(1 - y^2)^{1/2}$, and by symmetry of the graph, $A = 2 \displaystyle\int_{-1}^{0} \left[1 - y^2 - y(1 - y^2)^{1/2}\right] dy$

$= 2 \displaystyle\int_{-1}^{0} (1 - y^2)\, dy - 2 \displaystyle\int_{-1}^{0} y(1 - y^2)^{1/2}\, dy = 2\left[y - \dfrac{y^3}{3}\right]_{-1}^{0} + 2\left(\dfrac{1}{2}\right)\left[\dfrac{2(1 - y^2)^{3/2}}{3}\right]_{-1}^{0}$

$= 2\left[(0 - 0) - \left(-1 + \dfrac{1}{3}\right)\right] + \left(\dfrac{2}{3} - 0\right) = 2$

31. Limits of integration: $y = -4x^2 + 4$ and $y = x^4 - 1$

$\Rightarrow x^4 - 1 = -4x^2 + 4 \Rightarrow x^4 + 4x^2 - 5 = 0$

$\Rightarrow (x^2 + 5)(x - 1)(x + 1) = 0 \Rightarrow a = -1$ and $b = 1$;

$f(x) - g(x) = -4x^2 + 4 - x^4 + 1 = -4x^2 - x^4 + 5$

$\Rightarrow A = \int_{-1}^{1} (-4x^2 - x^4 + 5)\, dx = \left[-\frac{4x^3}{3} - \frac{x^5}{5} + 5x \right]_{-1}^{1}$

$= \left(-\frac{4}{3} - \frac{1}{5} + 5 \right) - \left(\frac{4}{3} + \frac{1}{5} - 5 \right) = 2\left(-\frac{4}{3} - \frac{1}{5} + 5 \right) = \frac{104}{15}$

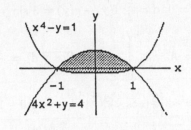

33. Limits of integration: $x = 4 - 4y^2$ and $x = 1 - y^4$

$\Rightarrow 4 - 4y^2 = 1 - y^4 \Rightarrow y^4 - 4y^2 + 3 = 0$

$\Rightarrow (y - \sqrt{3})(y + \sqrt{3})(y - 1)(y + 1) = 0 \Rightarrow c = -1$

and $d = 1$ since $x \geq 0$; $f(y) - g(y) = (4 - 4y^2) - (1 - y^4)$

$= 3 - 4y^2 + y^4 \Rightarrow A = \int_{-1}^{1} (3 - 4y^2 + y^4)\, dy$

$= \left[3y - \frac{4y^3}{3} + \frac{y^5}{5} \right]_{-1}^{1} = 2\left(3 - \frac{4}{3} + \frac{1}{5} \right) = \frac{56}{15}$

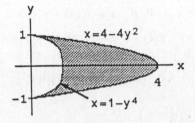

35. $a = 0$, $b = \pi$; $f(x) - g(x) = 2\sin x - \sin 2x$

$\Rightarrow A = \int_{0}^{\pi} (2\sin x - \sin 2x)\, dx = \left[-2\cos x + \frac{\cos 2x}{2} \right]_{0}^{\pi}$

$= \left[-2(-1) + \frac{1}{2} \right] - \left(-2 \cdot 1 + \frac{1}{2} \right) = 4$

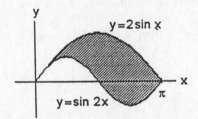

37. $a = -1$, $b = 1$; $f(x) - g(x) = (1 - x^2) - \cos\left(\frac{\pi x}{2} \right)$

$\Rightarrow A = \int_{-1}^{1} \left[1 - x^2 - \cos\left(\frac{\pi x}{2} \right) \right] dx = \left[x - \frac{x^3}{3} - \frac{2}{\pi} \sin\left(\frac{\pi x}{2} \right) \right]_{-1}^{1}$

$= \left(1 - \frac{1}{3} - \frac{2}{\pi} \right) - \left(-1 + \frac{1}{3} + \frac{2}{\pi} \right) = 2\left(\frac{2}{3} - \frac{2}{\pi} \right) = \frac{4}{3} - \frac{4}{\pi}$

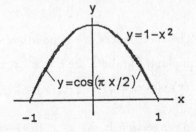

39. $a = -\frac{\pi}{4}$, $b = \frac{\pi}{4}$; $f(x) - g(x) = \sec^2 x - \tan^2 x$

$\Rightarrow A = \int_{-\pi/4}^{\pi/4} (\sec^2 x - \tan^2 x)\, dx = \int_{-\pi/4}^{\pi/4} \left[\sec^2 x - (\sec^2 x - 1) \right] dx$

$= \int_{-\pi/4}^{\pi/4} 1 \cdot dx = [x]_{-\pi/4}^{\pi/4} = \frac{\pi}{4} - \left(-\frac{\pi}{4} \right) = \frac{\pi}{2}$

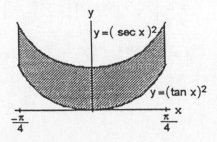

41. $c = 0$, $d = \frac{\pi}{2}$; $f(y) - g(y) = 3 \sin y \sqrt{\cos y} - 0 = 3 \sin y \sqrt{\cos y}$

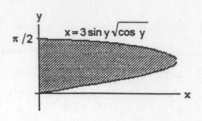

$$\Rightarrow A = 3 \int_0^{\pi/2} \sin y \sqrt{\cos y} \, dy = -3 \int_0^{\pi/2} \sqrt{\cos y} \, d(\cos y)$$

$$= -3\left[\frac{2}{3}(\cos y)^{3/2}\right]_0^{\pi/2} = -2(0 - 1) = 2$$

43. $A = A_1 + A_2$

Limits of integration: $x = y^3$ and $x = y \Rightarrow y = y^3$

$\Rightarrow y^3 - y = 0 \Rightarrow y(y - 1)(y + 1) = 0 \Rightarrow c_1 = -1$, $d_1 = 0$

and $c_2 = 0$, $d_2 = 1$; $f_1(y) - g_1(y) = y^3 - y$ and

$f_2(y) - g_2(y) = y - y^3 \Rightarrow$ by symmetry about the origin,

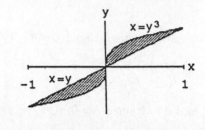

$$A_1 + A_2 = 2A_2 \Rightarrow A = 2 \int_0^1 (y - y^3) \, dy = 2\left[\frac{y^2}{2} - \frac{y^4}{4}\right]_0^1$$

$$= 2\left(\frac{1}{2} - \frac{1}{4}\right) = \frac{1}{2}$$

45. $A = A_1 + A_2$

Limits of integration: $y = x$ and $y = \frac{1}{x^2} \Rightarrow x = \frac{1}{x^2}$, $x \neq 0$

$\Rightarrow x^3 = 1 \Rightarrow x = 1$, $f_1(x) - g_1(x) = x - 0 = x$

$$\Rightarrow A_1 = \int_0^1 x \, dx = \left[\frac{x^2}{2}\right]_0^1 = \frac{1}{2}; \ f_2(x) - g_2(x) = \frac{1}{x^2} - 0$$

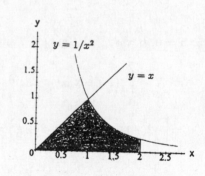

$$= x^{-2} \Rightarrow A_2 = \int_1^2 x^{-2} \, dx = \left[\frac{-1}{x}\right]_1^2 = -\frac{1}{2} + 1 = \frac{1}{2};$$

$$A = A_1 + A_2 = \frac{1}{2} + \frac{1}{2} = 1$$

47. (a) The coordinates of the points of intersection of the

line and parabola are $c = x^2 \Rightarrow x = \pm\sqrt{c}$ and $y = c$

(b) $f(y) - g(y) = \sqrt{y} - (-\sqrt{y}) = 2\sqrt{y} \Rightarrow$ the area of the

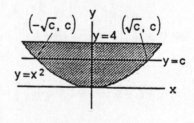

lower section is, $A_L = \int_0^c [f(y) - g(y)] \, dy$

$$= 2 \int_0^c \sqrt{y} \, dy = 2\left[\frac{2}{3}y^{3/2}\right]_0^c = \frac{4}{3}c^{3/2}. \text{ The area of the entire shaded region can be found by setting } c = 4:$$

$$A = \left(\frac{4}{3}\right)4^{3/2} = \frac{4 \cdot 8}{3} = \frac{32}{3}. \text{ Since c divides the region into subsections of equal area we have } A = 2A_L$$

$$\Rightarrow \frac{32}{3} = 2\left(\frac{4}{3}c^{3/2}\right) \Rightarrow c = 4^{2/3}$$

(c) $f(x) - g(x) = c - x^2 \Rightarrow A_L = \int\limits_{-\sqrt{c}}^{\sqrt{c}} [f(x) - g(x)]\, dx = \int\limits_{-\sqrt{c}}^{\sqrt{c}} (c - x^2)\, dx = \left[cx - \frac{x^3}{3} \right]_{-\sqrt{c}}^{\sqrt{c}} = 2\left[c^{3/2} - \frac{c^{3/2}}{3} \right]$

$= \frac{4}{3} c^{3/2}$. Again, the area of the whole shaded region can be found by setting $c = 4 \Rightarrow A = \frac{32}{3}$. From the

condition $A = 2A_L$, we get $\frac{4}{3} c^{3/2} = \frac{32}{3} \Rightarrow c = 4^{2/3}$ as in part (b).

49. Limits of integration: $y = 1 + \sqrt{x}$ and $y = \frac{2}{\sqrt{x}}$

$\Rightarrow 1 + \sqrt{x} = \frac{2}{\sqrt{x}}, x \neq 0 \Rightarrow \sqrt{x} + x = 2 \Rightarrow x = (2 - x)^2$

$\Rightarrow x = 4 - 4x + x^2 \Rightarrow x^2 - 5x + 4 = 0$

$\Rightarrow (x - 4)(x - 1) = 0 \Rightarrow x = 1, 4$ (but $x = 4$ does not

satisfy the equation); $y = \frac{2}{\sqrt{x}}$ and $y = \frac{x}{4} \Rightarrow \frac{2}{\sqrt{x}} = \frac{x}{4}$

$\Rightarrow 8 = x\sqrt{x} \Rightarrow 64 = x^3 \Rightarrow x = 4$ since $x > 0$;

Therefore, AREA $= A_1 + A_2$: $f_1(x) - g_1(x) = \left(1 + x^{1/2} \right) - \frac{x}{4}$

$\Rightarrow A_1 = \int\limits_0^1 \left(1 + x^{1/2} - \frac{x}{4} \right) dx = \left[x + \frac{2}{3} x^{3/2} - \frac{x^2}{8} \right]_0^1 = \left(1 + \frac{2}{3} - \frac{1}{8} \right) - 0 = \frac{37}{24}$; $f_2(x) - g_2(x) = 2x^{-1/2} - \frac{x}{4}$

$\Rightarrow A_2 = \int\limits_1^4 \left(2x^{-1/2} - \frac{x}{4} \right) dx = \left[4x^{1/2} - \frac{x^2}{8} \right]_1^4 = \left(4 \cdot 2 - \frac{16}{8} \right) - \left(4 - \frac{1}{8} \right) = 4 - \frac{15}{8} = \frac{17}{8}$;

Therefore, AREA $= A_1 + A_2 = \frac{37}{24} + \frac{17}{8} = \frac{37 + 51}{24} = \frac{88}{24} = \frac{11}{3}$

51. Area between parabola and $y = a^2$: $A = 2 \int\limits_0^a (a^2 - x^2)\, dx = 2\left[a^2 x - \frac{1}{3} x^3 \right]_0^a = 2\left(a^3 - \frac{a^3}{3} \right) - 0 = \frac{4a^3}{3}$;

Area of triangle AOC: $\frac{1}{2}(2a)(a^2) = a^3$; limit of ratio $= \lim\limits_{a \to 0^+} \dfrac{a^3}{\left(\frac{4a^3}{3} \right)} = \frac{3}{4}$ which is independent of a

53. Neither one; they are both zero. Neither integral takes into account the changes in the formulas for the region's upper and lower bounding curves at $x = 0$. The area of the shaded region is actually

$A = \int\limits_{-1}^0 [-x - (x)]\, dx + \int\limits_0^1 [x - (-x)]\, dx = \int\limits_{-1}^0 -2x\, dx + \int\limits_0^1 2x\, dx$.

5.2 FINDING VOLUMES BY SLICING

1. (a) $A = \pi(\text{radius})^2$ and radius $= \sqrt{1 - x^2} \Rightarrow A(x) = \pi(1 - x^2)$

   (b) $A = \text{width} \cdot \text{height}$, width $= \text{height} = 2\sqrt{1 - x^2} \Rightarrow A(x) = 4(1 - x^2)$

   (c) $A = (\text{side})^2$ and diagonal $= \sqrt{2}(\text{side}) \Rightarrow A = \frac{(\text{diagonal})^2}{2}$; diagonal $= 2\sqrt{1 - x^2} \Rightarrow A(x) = 2(1 - x^2)$

   (d) $A = \frac{\sqrt{3}}{4}(\text{side})^2$ and side $= 2\sqrt{1 - x^2} \Rightarrow A(x) = \sqrt{3}(1 - x^2)$

3. $A(x) = \dfrac{(\text{diagonal})^2}{2} = \dfrac{\left(\sqrt{x}-(-\sqrt{x})\right)^2}{2} = 2x$ (see Exercise 1c); $a = 0,\ b = 4$;

$$V = \int\limits_a^b A(x)\,dx = \int\limits_0^4 2x\,dx = \left[x^2\right]_0^4 = 16$$

5. $A(x) = (\text{edge})^2 = \left[\sqrt{1-x^2}-\left(-\sqrt{1-x^2}\right)\right]^2 = \left(2\sqrt{1-x^2}\right)^2 = 4(1-x^2)$; $a = -1,\ b = 1$;

$$V = \int\limits_a^b A(x)\,dx = \int\limits_{-1}^1 4(1-x^2)\,dx = 4\left[x-\frac{x^3}{3}\right]_{-1}^1 = 8\left(1-\tfrac{1}{3}\right) = \frac{16}{3}$$

7. (a) STEP 1) $A(x) = \frac{1}{2}(\text{side})\cdot(\text{side})\cdot\left(\sin\frac{\pi}{3}\right) = \frac{1}{2}\cdot\left(2\sqrt{\sin x}\right)\cdot\left(2\sqrt{\sin x}\right)\left(\sin\frac{\pi}{3}\right) = \sqrt{3}\sin x$

STEP 2) $a = 0,\ b = \pi$

STEP 3) $V = \int\limits_a^b A(x)\,dx = \sqrt{3}\int\limits_0^\pi \sin x\,dx = \left[-\sqrt{3}\cos x\right]_0^\pi = \sqrt{3}(1+1) = 2\sqrt{3}$

(b) STEP 1) $A(x) = (\text{side})^2 = \left(2\sqrt{\sin x}\right)\left(2\sqrt{\sin x}\right) = 4\sin x$

STEP 2) $a = 0,\ b = \pi$

STEP 3) $V = \int\limits_a^b A(x)\,dx = \int\limits_0^\pi 4\sin x\,dx = [-4\cos x]_0^\pi = 8$

9. $A(y) = \frac{\pi}{4}(\text{diameter})^2 = \frac{\pi}{4}\left(\sqrt{5}y^2-0\right)^2 = \frac{5\pi}{4}y^4$;

$c = 0,\ d = 2;\ V = \int\limits_c^d A(y)\,dy = \int\limits_0^2 \frac{5\pi}{4}y^4\,dy$

$= \left[\left(\frac{5\pi}{4}\right)\left(\frac{y^5}{5}\right)\right]_0^2 = \frac{\pi}{4}\left(2^5-0\right) = 8\pi$

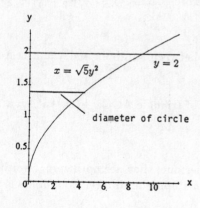

11. (a) It follows from Cavalieri's Theorem that the volume of a column is the same as the volume of a right prism with a square base of side length s and altitude h. Thus, STEP 1) $A(x) = (\text{side length})^2 = s^2$;

STEP 2) $a = 0,\ b = h$; STEP 3) $V = \int\limits_a^b A(x)\,dx = \int\limits_0^h s^2\,dx = s^2 h$

(b) From Cavalieri's Theorem we conclude that the volume of the column is the same as the volume of the prism described above, regardless of the number of turns $\Rightarrow V = s^2 h$

13. We are given that

   (a) $f(x) - g(x) = s(x) - t(x)$ at any x in the interval $[a, b]$

   (see accompanying figure). The area of region A is

   (b) $A = \int_a^b [f(x) - g(x)] \, dx$; using part (a) we get:

   (c) $A = \int_a^b [s(x) - t(x)] \, dx$; but the area of region B is

   (d) $B = \int_a^b [s(x) - t(x)] \, dx$

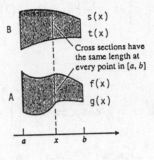

From parts (c) and (d) we conclude that $A = B$.

5.3 VOLUMES OF SOLIDS OF REVOLUTION–DISKS AND WASHERS

1. $R(x) = y = 1 - \frac{x}{2} \Rightarrow V = \int_0^2 \pi [R(x)]^2 \, dx = \pi \int_0^2 \left(1 - \frac{x}{2}\right)^2 dx = \pi \int_0^2 \left(1 - x + \frac{x^2}{4}\right) dx = \pi \left[x - \frac{x^2}{2} + \frac{x^3}{12}\right]_0^2$

   $= \pi \left(2 - \frac{4}{2} + \frac{8}{12}\right) = \frac{2\pi}{3}$

3. $R(x) = \tan\left(\frac{\pi}{4}y\right);\ u = \frac{\pi}{4}y \Rightarrow du = \frac{\pi}{4} dy \Rightarrow 4\, du = \pi\, dy;\ y = 0 \Rightarrow u = 0,\ y = 1 \Rightarrow u = \frac{\pi}{4};$

   $V = \int_0^1 \pi [R(y)]^2 \, dy = \pi \int_0^1 \left[\tan\left(\frac{\pi}{4}y\right)\right]^2 dy = 4 \int_0^{\pi/4} \tan^2 u\, du = 4 \int_0^{\pi/4} \left(-1 + \sec^2 u\right) du = 4[-u + \tan u]_0^{\pi/4}$

   $= 4\left(-\frac{\pi}{4} + 1 - 0\right) = 4 - \pi$

5. $R(x) = x^2 \Rightarrow V = \int_0^2 \pi [R(x)]^2 \, dx = \pi \int_0^2 \left(x^2\right)^2 dx$

   $= \pi \int_0^2 x^4 \, dx = \pi \left[\frac{x^5}{5}\right]_0^2 = \frac{32\pi}{5}$

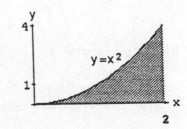

7. $R(x) = \sqrt{9-x^2} \Rightarrow V = \int\limits_{-3}^{3} \pi[R(x)]^2 \, dx = \pi \int\limits_{-3}^{3} (9-x^2) \, dx$

$$= \pi\left[9x - \frac{x^3}{3}\right]_{-3}^{3} = 2\pi\left[9(3) - \frac{27}{3}\right] = 2 \cdot \pi \cdot 18 = 36\pi$$

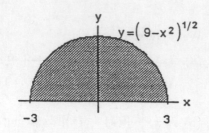

9. $R(x) = \sqrt{\cos x} \Rightarrow V = \int\limits_{0}^{\pi/2} \pi[R(x)]^2 \, dx = \pi \int\limits_{0}^{\pi/2} \cos x \, dx$

$$= \pi[\sin x]_0^{\pi/2} = \pi(1-0) = \pi$$

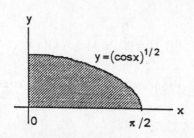

11. $R(x) = \sqrt{2} - \sec x \tan x \Rightarrow V = \int\limits_{0}^{\pi/4} \pi[R(x)]^2 \, dx$

$$= \pi \int\limits_{0}^{\pi/4} \left(\sqrt{2} - \sec x \tan x\right)^2 \, dx$$

$$= \pi \int\limits_{0}^{\pi/4} \left(2 - 2\sqrt{2} \sec x \tan x + \sec^2 x \tan^2 x\right) \, dx$$

$$= \pi\left(\int\limits_{0}^{\pi/4} 2 \, dx - 2\sqrt{2} \int\limits_{0}^{\pi/4} \sec x \tan x \, dx + \int\limits_{0}^{\pi/4} (\tan x)^2 \sec^2 x \, dx\right)$$

$$= \pi\left([2x]_0^{\pi/4} - 2\sqrt{2}[\sec x]_0^{\pi/4} + \left[\frac{\tan^3 x}{3}\right]_0^{\pi/4}\right) = \pi\left[\left(\frac{\pi}{2} - 0\right) - 2\sqrt{2}(\sqrt{2}-1) + \frac{1}{3}(1^3 - 0)\right] = \pi\left(\frac{\pi}{2} + 2\sqrt{2} - \frac{11}{3}\right)$$

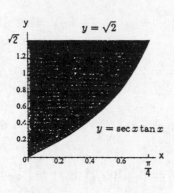

13. $R(y) = \sqrt{5} \cdot y^2 \Rightarrow V = \int\limits_{-1}^{1} \pi[R(y)]^2 \, dy = \pi \int\limits_{-1}^{1} 5y^4 \, dy$

$$= \pi[y^5]_{-1}^{1} = \pi[1 - (-1)] = 2\pi$$

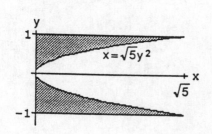

15. $R(y) = \sqrt{2 \sin 2y} \Rightarrow V = \displaystyle\int_0^{\pi/2} \pi [R(y)]^2 \, dy = \pi \int_0^{\pi/2} 2 \sin 2y \, dy$

$= \pi \left[-\cos 2y \right]_0^{\pi/2} = \pi [1 - (-1)] = 2\pi$

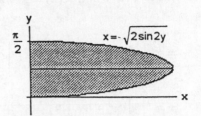
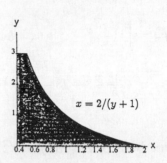

17. $R(y) = \dfrac{2}{y+1} \Rightarrow V = \displaystyle\int_0^3 \pi [R(y)]^2 \, dy = 4\pi \int_0^3 \dfrac{1}{(y+1)^2} \, dy$

$= 4\pi \left[\dfrac{-1}{y+1} \right]_0^3 = 4\pi \left[-\dfrac{1}{4} - (-1) \right] = 3\pi$

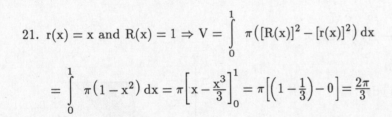

19. For the sketch given, $a = -\dfrac{\pi}{2}$, $b = \dfrac{\pi}{2}$; $R(x) = 1$, $r(x) = \sqrt{\cos x}$; $V = \displaystyle\int_a^b \pi \left([R(x)]^2 - [r(x)]^2 \right) dx$

$= \displaystyle\int_{-\pi/2}^{\pi/2} \pi (1 - \cos x) \, dx = 2\pi \int_0^{\pi/2} (1 - \cos x) \, dx = 2\pi [x - \sin x]_0^{\pi/2} = 2\pi \left(\dfrac{\pi}{2} - 1 \right) = \pi^2 - 2\pi$

21. $r(x) = x$ and $R(x) = 1 \Rightarrow V = \displaystyle\int_0^1 \pi \left([R(x)]^2 - [r(x)]^2 \right) dx$

$= \displaystyle\int_0^1 \pi \left(1 - x^2 \right) dx = \pi \left[x - \dfrac{x^3}{3} \right]_0^1 = \pi \left[\left(1 - \dfrac{1}{3} \right) - 0 \right] = \dfrac{2\pi}{3}$

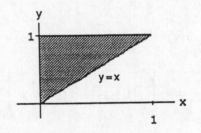

23. $r(x) = 2\sqrt{x}$ and $R(x) = 2 \Rightarrow V = \displaystyle\int_0^1 \pi \left([R(x)]^2 - [r(x)]^2 \right) dx$

$= \pi \displaystyle\int_0^1 (4 - 4x) \, dx = 4\pi \left[x - \dfrac{x^2}{2} \right]_0^1 = 4\pi \left(1 - \dfrac{1}{2} \right) = 2\pi$

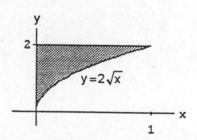

25. $r(x) = x^2 + 1$ and $R(x) = x + 3 \Rightarrow V = \int_{-1}^{2} \pi\left([R(x)]^2 - [r(x)]^2\right) dx$

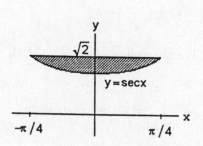

$= \pi \int_{-1}^{2} \left[(x+3)^2 - (x^2+1)^2\right] dx$

$= \pi \int_{-1}^{2} \left[(x^2 + 6x + 9) - (x^4 + 2x^2 + 1)\right] dx$

$= \pi \int_{-1}^{2} \left(-x^4 - x^2 + 6x + 8\right) dx = \pi\left[-\frac{x^5}{5} - \frac{x^3}{3} + \frac{6x^2}{2} + 8x\right]_{-1}^{2} = \pi\left[\left(-\frac{32}{5} - \frac{8}{3} + \frac{24}{2} + 16\right) - \left(\frac{1}{5} + \frac{1}{3} + \frac{6}{2} - 8\right)\right]$

$= \pi\left(-\frac{33}{5} - 3 + 28 - 3 + 8\right) = \pi\left(\frac{5\cdot 30 - 33}{5}\right) = \frac{117\pi}{5}$

27. $r(x) = \sec x$ and $R(x) = \sqrt{2} \Rightarrow V = \int_{-\pi/4}^{\pi/4} \pi\left([R(x)]^2 - [r(x)]^2\right) dx$

$= \pi \int_{-\pi/4}^{\pi/4} \left(2 - \sec^2 x\right) dx = \pi[2x - \tan x]_{-\pi/4}^{\pi/4}$

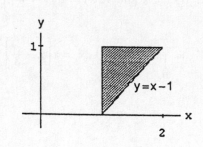

$= \pi\left[\left(\frac{\pi}{2} - 1\right) - \left(-\frac{\pi}{2} + 1\right)\right] = \pi(\pi - 2)$

29. $r(y) = 1$ and $R(y) = 1 + y \Rightarrow V = \int_{0}^{1} \pi\left([R(y)]^2 - [r(y)]^2\right) dy$

$= \pi \int_{0}^{1} \left[(1+y)^2 - 1\right] dy = \pi \int_{0}^{1} \left(1 + 2y + y^2 - 1\right) dy$

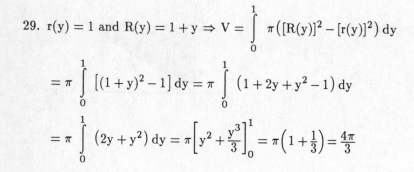

$= \pi \int_{0}^{1} \left(2y + y^2\right) dy = \pi\left[y^2 + \frac{y^3}{3}\right]_{0}^{1} = \pi\left(1 + \frac{1}{3}\right) = \frac{4\pi}{3}$

31. $R(y) = 2$ and $r(y) = \sqrt{y} \Rightarrow V = \int_{0}^{4} \pi\left([R(y)]^2 - [r(y)]^2\right) dy$

$= \pi \int_{0}^{4} (4 - y) \, dy = \pi\left[4y - \frac{y^2}{2}\right]_{0}^{4} = \pi(16 - 8) = 8\pi$

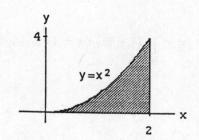

33. $R(y) = \sqrt{3}$ and $r(y) = \sqrt{3-y^2} \Rightarrow V = \int_0^{\sqrt{3}} \pi\left([R(y)]^2 - [r(y)]^2\right) dy$

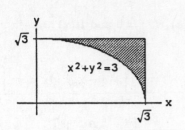

$= \pi \int_0^{\sqrt{3}} [3-(3-y^2)] \, dy = \pi \int_0^{\sqrt{3}} y^2 \, dy = \pi\left[\frac{y^3}{3}\right]_0^{\sqrt{3}} = \pi\sqrt{3}$

35. $R(y) = 2$ and $r(y) = 1 + \sqrt{y} \Rightarrow V = \int_0^1 \pi\left([R(y)]^2 - [r(y)]^2\right) dy$

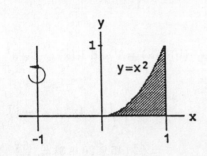

$= \pi \int_0^1 \left[4 - \left(1 + \sqrt{y}\right)^2\right] dy = \pi \int_0^1 \left(4 - 1 - 2\sqrt{y} - y\right) dy$

$= \pi \int_0^1 \left(3 - 2\sqrt{y} - y\right) dy = \pi\left[3y - \frac{4}{3}y^{3/2} - \frac{y^2}{2}\right]_0^1$

$= \pi\left(3 - \frac{4}{3} - \frac{1}{2}\right) = \pi\left(\frac{18 - 8 - 3}{6}\right) = \frac{7\pi}{6}$

37. (a) $r(x) = \sqrt{x}$ and $R(x) = 2 \Rightarrow V = \int_0^4 \pi\left([R(x)]^2 - [r(x)]^2\right) dx$

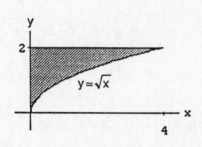

$= \pi \int_0^4 (4 - x) \, dx = \pi\left[4x - \frac{x^2}{2}\right]_0^4 = \pi(16 - 8) = 8\pi$

(b) $r(y) = 0$ and $R(y) = y^2 \Rightarrow V = \int_0^2 \pi\left([R(y)]^2 - [r(y)]^2\right) dy$

$= \pi \int_0^2 y^4 \, dy = \pi\left[\frac{y^5}{5}\right]_0^2 = \frac{32\pi}{5}$

(c) $r(x) = 0$ and $R(x) = 2 - \sqrt{x} \Rightarrow V = \int_0^4 \pi\left([R(x)]^2 - [r(x)]^2\right) dx = \pi \int_0^4 \left(2 - \sqrt{x}\right)^2 dx$

$= \pi \int_0^4 \left(4 - 4\sqrt{x} + x\right) dx = \pi\left[4x - \frac{8x^{3/2}}{3} + \frac{x^2}{2}\right]_0^4 = \pi\left(16 - \frac{64}{3} + \frac{16}{2}\right) = \frac{8\pi}{3}$

(d) $r(y) = 4 - y^2$ and $R(y) = 4 \Rightarrow V = \int_0^2 \pi\left([R(y)]^2 - [r(y)]^2\right) dy = \pi \int_0^2 \left[16 - \left(4 - y^2\right)^2\right] dy$

$= \pi \int_0^2 \left(16 - 16 + 8y^2 - y^4\right) dy = \pi \int_0^2 \left(8y^2 - y^4\right) dy = \pi\left[\frac{8}{3}y^3 - \frac{y^5}{5}\right]_0^2 = \pi\left(\frac{64}{3} - \frac{32}{5}\right) = \frac{224\pi}{15}$

39. (a) $r(x) = 0$ and $R(x) = 1 - x^2 \Rightarrow V = \int_{-1}^{1} \pi \left([R(x)]^2 - [r(x)]^2 \right) dx$

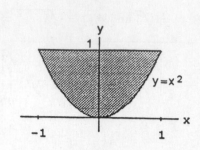

$$= \pi \int_{-1}^{1} \left(1 - x^2 \right)^2 dx = \pi \int_{-1}^{1} \left(1 - 2x^2 + x^4 \right) dx$$

$$= \pi \left[x - \frac{2x^3}{3} + \frac{x^5}{5} \right]_{-1}^{1} = 2\pi \left(1 - \frac{2}{3} + \frac{1}{5} \right) = 2\pi \left(\frac{15 - 10 + 3}{15} \right)$$

$$= \frac{16\pi}{15}$$

(b) $r(x) = 1$ and $R(x) = 2 - x^2 \Rightarrow V = \int_{-1}^{1} \pi \left([R(x)]^2 - [r(x)]^2 \right) dx = \pi \int_{-1}^{1} \left[\left(2 - x^2 \right)^2 - 1 \right] dx$

$$= \pi \int_{-1}^{1} \left(4 - 4x^2 + x^4 - 1 \right) dx = \pi \int_{-1}^{1} \left(3 - 4x^2 + x^4 \right) dx = \pi \left[3x - \frac{4}{3}x^3 + \frac{x^5}{5} \right]_{-1}^{1} = 2\pi \left(3 - \frac{4}{3} + \frac{1}{5} \right)$$

$$= \frac{2\pi}{15}(45 - 20 + 3) = \frac{56\pi}{15}$$

(c) $r(x) = 1 + x^2$ and $R(x) = 2 \Rightarrow V = \int_{-1}^{1} \pi \left([R(x)]^2 - [r(x)]^2 \right) dx = \pi \int_{-1}^{1} \left[4 - \left(1 + x^2 \right)^2 \right] dx$

$$= \pi \int_{-1}^{1} \left(4 - 1 - 2x^2 - x^4 \right) dx = \pi \int_{-1}^{1} \left(3 - 2x^2 - x^4 \right) dx = \pi \left[3x - \frac{2}{3}x^3 - \frac{x^5}{5} \right]_{-1}^{1} = 2\pi \left(3 - \frac{2}{3} - \frac{1}{5} \right)$$

$$= \frac{2\pi}{15}(45 - 10 - 3) = \frac{64\pi}{15}$$

41. $R(y) = \sqrt{256 - y^2} \Rightarrow V = \int_{-16}^{-7} \pi [R(y)]^2 \, dy = \pi \int_{-16}^{-7} \left(256 - y^2 \right) dy = \pi \left[256y - \frac{y^3}{3} \right]_{-16}^{-7}$

$$= \pi \left[(256)(-7) + \frac{7^3}{3} - \left((256)(-16) + \frac{16^3}{3} \right) \right] = \pi \left(\frac{7^3}{3} + 256(16 - 7) - \frac{16^3}{3} \right) = 1053\pi \ \text{cm}^3$$

43. (a) $R(x) = |c - \sin x|$, so $V = \pi \int_{0}^{\pi} [R(x)]^2 \, dx = \pi \int_{0}^{\pi} (c - \sin x)^2 \, dx = \pi \int_{0}^{\pi} \left(c^2 - 2c \sin x + \sin^2 x \right) dx$

$$= \pi \int_{0}^{\pi} \left(c^2 - 2c \sin x + \frac{1 - \cos 2x}{2} \right) dx = \pi \int_{0}^{\pi} \left(c^2 + \frac{1}{2} - 2c \sin x - \frac{\cos 2x}{2} \right) dx$$

$$= \pi \left[\left(c^2 + \frac{1}{2} \right)x + 2c \cos x - \frac{\sin 2x}{4} \right]_{0}^{\pi} = \pi \left[\left(c^2\pi + \frac{\pi}{2} - 2c - 0 \right) - (0 + 2c - 0) \right] = \pi \left(c^2\pi + \frac{\pi}{2} - 4c \right). \text{ Let }$$

$V(c) = \pi \left(c^2\pi + \frac{\pi}{2} - 4c \right)$. We find the extreme values of $V(c)$: $\frac{dV}{dc} = \pi(2c\pi - 4) = 0 \Rightarrow c = \frac{2}{\pi}$ is a critical

point, and $V\left(\frac{2}{\pi} \right) = \pi \left(\frac{4}{\pi} + \frac{\pi}{2} - \frac{8}{\pi} \right) = \pi \left(\frac{\pi}{2} - \frac{4}{\pi} \right) = \frac{\pi^2}{2} - 4$; Evaluate V at the endpoints: $V(0) = \frac{\pi^2}{2}$ and

$V(1) = \pi \left(\frac{3}{2}\pi - 4 \right) = \frac{\pi^2}{2} - (4 - \pi)\pi$. Now we see that the function's absolute minimum value is $\frac{\pi^2}{2} - 4$,

taken on at the critical point $c = \frac{2}{\pi}$. (See also the accompanying graph.)

(b) From the discussion in part (a) we conclude that the function's absolute maximum value is $\frac{\pi^2}{2}$, taken on at the endpoint $c = 0$.

(c) The graph of the solid's volume as a function of c for $0 \leq c \leq 1$ is given at the right. As c moves away from $[0, 1]$ the volume of the solid increases without bound. If we approximate the solid as a set of solid disks, we can see that the radius of a typical disk increases without bounds as c moves away from $[0, 1]$.

45. $R(y) = b + \sqrt{a^2 - y^2}$ and $r(y) = b - \sqrt{a^2 - y^2}$

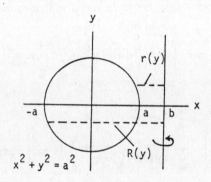

$$\Rightarrow V = \int_{-a}^{a} \pi \left([R(y)]^2 - [r(y)]^2\right) dy$$

$$= \pi \int_{-a}^{a} \left[\left(b + \sqrt{a^2 - y^2}\right)^2 - \left(b - \sqrt{a^2 - y^2}\right)^2\right] dy$$

$$= \pi \int_{-a}^{a} 4b\sqrt{a^2 - y^2}\, dy = 4b\pi \int_{-a}^{a} \sqrt{a^2 - y^2}\, dy$$

$$= 4b\pi \cdot \text{area of semicircle of radius } a = 4b\pi \cdot \frac{\pi a^2}{2} = 2a^2 b\pi^2$$

47. (a) $R(x) = \sqrt{a^2 - x^2} \Rightarrow V = \int_{-a}^{a} \pi [R(x)]^2\, dx = \pi \int_{-a}^{a} \left(a^2 - x^2\right) dx = \pi \left[a^2 x - \frac{x^3}{3}\right]_{-a}^{a}$

$$= \pi \left[\left(a^3 - \frac{a^3}{3}\right) - \left(-a^3 + \frac{a^3}{3}\right)\right] = 2\pi \left(\frac{2a^3}{3}\right) = \frac{4}{3}\pi a^3, \text{ the volume of a sphere of radius } a$$

(b) $R(x) = \frac{rx}{h} \Rightarrow V = \int_{0}^{h} \pi [R(x)]^2\, dx = \pi \int_{0}^{h} \frac{r^2 x^2}{h^2}\, dx$

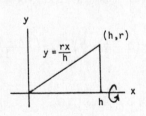

$$= \frac{\pi r^2}{h^2} \left[\frac{x^3}{3}\right]_{0}^{h} = \left(\frac{\pi r^2}{h^2}\right)\left(\frac{h^3}{3}\right) = \frac{1}{3}\pi r^2 h, \text{ the volume of}$$

a cone of radius r and height h

5.4 CYLINDRICAL SHELLS

1. For the sketch given, $a = 0$, $b = 2$;

$$V = \int_{a}^{b} 2\pi \binom{\text{shell}}{\text{radius}}\binom{\text{shell}}{\text{height}} dx = \int_{0}^{2} 2\pi x\left(1 + \frac{x^2}{4}\right) dx = 2\pi \int_{0}^{2} \left(x + \frac{x^3}{4}\right) dx = 2\pi \left[\frac{x^2}{2} + \frac{x^4}{16}\right]_{0}^{2} = 2\pi \left(\frac{4}{2} + \frac{16}{16}\right)$$

$$= 2\pi \cdot 3 = 6\pi$$

3. For the sketch given, c = 0, d = $\sqrt{2}$;

$$V = \int_c^d 2\pi \left(\begin{array}{c}\text{shell} \\ \text{radius}\end{array}\right)\left(\begin{array}{c}\text{shell} \\ \text{height}\end{array}\right) dy = \int_0^{\sqrt{2}} 2\pi y \cdot (y^2) dy = 2\pi \int_0^{\sqrt{2}} y^3\, dy = 2\pi \left[\frac{y^4}{4}\right]_0^{\sqrt{2}} = 2\pi$$

5. For the sketch given, a = 0, b = $\sqrt{3}$;

$$V = \int_a^b 2\pi \left(\begin{array}{c}\text{shell} \\ \text{radius}\end{array}\right)\left(\begin{array}{c}\text{shell} \\ \text{height}\end{array}\right) dx = \int_0^{\sqrt{3}} 2\pi x \cdot \left(\sqrt{x^2+1}\right) dx;$$

$$\left[u = x^2 + 1 \Rightarrow du = 2x\, dx; x = 0 \Rightarrow u = 1, x = \sqrt{3} \Rightarrow u = 4\right]$$

$$\rightarrow V = \pi \int_1^4 u^{1/2}\, du = \pi \left[\frac{2}{3}u^{3/2}\right]_1^4 = \frac{2\pi}{3}\left(4^{3/2} - 1\right) = \left(\frac{2\pi}{3}\right)(8-1) = \frac{14\pi}{3}$$

7. a = 0, b = 2;

$$V = \int_a^b 2\pi \left(\begin{array}{c}\text{shell} \\ \text{radius}\end{array}\right)\left(\begin{array}{c}\text{shell} \\ \text{height}\end{array}\right) dx = \int_0^2 2\pi x \left[x - \left(-\frac{x}{2}\right)\right] dx$$

$$= \int_0^2 2\pi x^2 \cdot \frac{3}{2}\, dx = \pi \int_0^2 3x^2\, dx = \pi \left[x^3\right]_0^2 = 8\pi$$

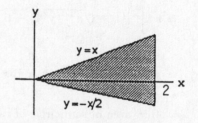

9. a = 0, b = 1;

$$V = \int_a^b 2\pi \left(\begin{array}{c}\text{shell} \\ \text{radius}\end{array}\right)\left(\begin{array}{c}\text{shell} \\ \text{height}\end{array}\right) dx = \int_0^1 2\pi x \left[(2-x) - x^2\right] dx$$

$$= 2\pi \int_0^1 \left(2x - x^2 - x^3\right) dx = 2\pi \left[x^2 - \frac{x^3}{3} - \frac{x^4}{4}\right]_0^1$$

$$= 2\pi \left(1 - \frac{1}{3} - \frac{1}{4}\right) = 2\pi \left(\frac{12 - 4 - 3}{12}\right) = \frac{10\pi}{12} = \frac{5\pi}{6}$$

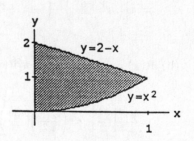

11. a = 0, b = 4;

$$V = \int_a^b 2\pi \left(\begin{array}{c}\text{shell} \\ \text{radius}\end{array}\right)\left(\begin{array}{c}\text{shell} \\ \text{height}\end{array}\right) dx = \int_0^4 2\pi x \left(\sqrt{x}\right) dx$$

$$= 2\pi \int_0^4 x^{3/2}\, dx = 2\pi \left[\frac{2}{5}x^{5/2}\right]_0^4 = 2\pi \left(\frac{2}{5}\right)(2^5) = \frac{128\pi}{5}$$

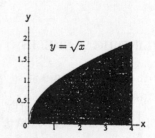

13. $a = \frac{1}{2}$, $b = 2$;

$$V = \int_a^b 2\pi \binom{\text{shell}}{\text{radius}} \binom{\text{shell}}{\text{height}} dx = \int_{1/2}^2 2\pi x \left(\frac{1}{x}\right) dx$$

$$= 2\pi \int_{1/2}^2 dx = 2\pi [x]_{1/2}^2 = 2\pi \left(2 - \frac{1}{2}\right) = 3\pi$$

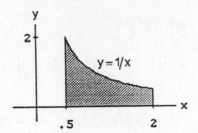

15. $c = 0$, $d = 2$;

$$V = \int_c^d 2\pi \binom{\text{shell}}{\text{radius}} \binom{\text{shell}}{\text{height}} dy = \int_0^2 2\pi y \left[\sqrt{y} - (-y)\right] dy$$

$$= 2\pi \int_0^2 \left(y^{3/2} + y^2\right) dy = 2\pi \left[\frac{2y^{5/2}}{5} + \frac{y^3}{3}\right]_0^2$$

$$= 2\pi \left[\frac{2}{5}(\sqrt{2})^5 + \frac{2^3}{3}\right] = 2\pi \left(\frac{8\sqrt{2}}{5} + \frac{8}{3}\right) = 16\pi \left(\frac{\sqrt{2}}{5} + \frac{1}{3}\right)$$

$$= \frac{16\pi}{15}(3\sqrt{2} + 5)$$

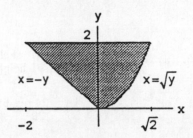

17. $c = 0$, $d = 2$;

$$V = \int_c^d 2\pi \binom{\text{shell}}{\text{radius}} \binom{\text{shell}}{\text{height}} dy = \int_0^2 2\pi y \left(2y - y^2\right) dy$$

$$= 2\pi \int_0^2 \left(2y^2 - y^3\right) dy = 2\pi \left[\frac{2y^3}{3} - \frac{y^4}{4}\right]_0^2 = 2\pi \left(\frac{16}{3} - \frac{16}{4}\right)$$

$$= 32\pi \left(\frac{1}{3} - \frac{1}{4}\right) = \frac{32\pi}{12} = \frac{8\pi}{3}$$

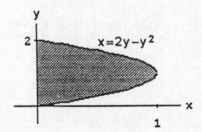

19. $c = 0$, $d = 1$;

$$V = \int_c^d 2\pi \binom{\text{shell}}{\text{radius}} \binom{\text{shell}}{\text{height}} dy = 2\pi \int_0^1 y[y - (-y)] dy$$

$$= 2\pi \int_0^1 2y^2 dy = \frac{4\pi}{3}[y^3]_0^1 = \frac{4\pi}{3}$$

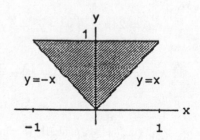

21. $c = 0$, $d = 2$;

$$V = \int_c^d 2\pi \binom{\text{shell}}{\text{radius}}\binom{\text{shell}}{\text{height}} dy = \int_0^2 2\pi y\left[(2+y) - y^2\right] dy$$

$$= 2\pi \int_0^2 \left(2y + y^2 - y^3\right) dy = 2\pi\left[y^2 + \frac{y^3}{3} - \frac{y^4}{4}\right]_0^2$$

$$= 2\pi\left(4 + \frac{8}{3} - \frac{16}{4}\right) = \frac{\pi}{6}(48 + 32 - 48) = \frac{16\pi}{3}$$

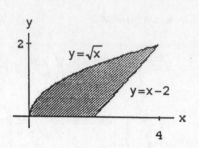

23. (a) $V = \int_c^d 2\pi \binom{\text{shell}}{\text{radius}}\binom{\text{shell}}{\text{height}} dy = \int_0^1 2\pi y \cdot 12\left(y^2 - y^3\right) dy = 24\pi \int_0^1 \left(y^3 - y^4\right) dy = 24\pi\left[\frac{y^4}{4} - \frac{y^5}{5}\right]_0^1$

$$= 24\pi\left(\frac{1}{4} - \frac{1}{5}\right) = \frac{24\pi}{20} = \frac{6\pi}{5}$$

(b) $V = \int_c^d 2\pi \binom{\text{shell}}{\text{radius}}\binom{\text{shell}}{\text{height}} dy = \int_0^1 2\pi(1-y)\left[12\left(y^2 - y^3\right)\right] dy = 24\pi \int_0^1 (1-y)\left(y^2 - y^3\right) dy$

$$= 24\pi \int_0^1 \left(y^2 - 2y^3 + y^4\right) dy = 24\pi\left[\frac{y^3}{3} - \frac{y^4}{2} + \frac{y^5}{5}\right]_0^1 = 24\pi\left(\frac{1}{3} - \frac{1}{2} + \frac{1}{5}\right) = 24\pi\left(\frac{1}{30}\right) = \frac{4\pi}{5}$$

(c) $V = \int_c^d 2\pi \binom{\text{shell}}{\text{radius}}\binom{\text{shell}}{\text{height}} dy = \int_0^1 2\pi\left(\frac{8}{5} - y\right)\left[12\left(y^2 - y^3\right)\right] dy = 24\pi \int_0^1 \left(\frac{8}{5} - y\right)\left(y^2 - y^3\right) dy$

$$= 24\pi \int_0^1 \left(\frac{8}{5}y^2 - \frac{13}{5}y^3 + y^4\right) dy = 24\pi\left[\frac{8}{15}y^3 - \frac{13}{20}y^4 + \frac{y^5}{5}\right]_0^1 = 24\pi\left(\frac{8}{15} - \frac{13}{20} + \frac{1}{5}\right) = \frac{24\pi}{60}(32 - 39 + 12)$$

$$= \frac{24\pi}{12} = 2\pi$$

(d) $V = \int_c^d 2\pi \binom{\text{shell}}{\text{radius}}\binom{\text{shell}}{\text{height}} dy = \int_0^1 2\pi\left(y + \frac{2}{5}\right)\left[12\left(y^2 - y^3\right)\right] dy = 24\pi \int_0^1 \left(y + \frac{2}{5}\right)\left(y^2 - y^3\right) dy$

$$= 24\pi \int_0^1 \left(y^3 - y^4 + \frac{2}{5}y^2 - \frac{2}{5}y^3\right) dy = 24\pi \int_0^1 \left(\frac{2}{5}y^2 + \frac{3}{5}y^3 - y^4\right) dy = 24\pi\left[\frac{2}{15}y^3 + \frac{3}{20}y^4 - \frac{y^5}{5}\right]_0^1$$

$$= 24\pi\left(\frac{2}{15} + \frac{3}{20} - \frac{1}{5}\right) = \frac{24\pi}{60}(8 + 9 - 12) = \frac{24\pi}{12} = 2\pi$$

25. (a) $V = \int_c^d 2\pi \binom{\text{shell}}{\text{radius}}\binom{\text{shell}}{\text{height}} dy = \int_1^2 2\pi y(y - 1) dy$

$$= 2\pi \int_1^2 \left(y^2 - y\right) dy = 2\pi\left[\frac{y^3}{3} - \frac{y^2}{2}\right]_1^2 = 2\pi\left[\left(\frac{8}{3} - \frac{4}{2}\right) - \left(\frac{1}{3} - \frac{1}{2}\right)\right]$$

$$= 2\pi\left(\frac{7}{3} - 2 + \frac{1}{2}\right) = \frac{\pi}{3}(14 - 12 + 3) = \frac{5\pi}{3}$$

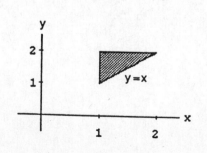

(b) $V = \int_a^b 2\pi \left(\begin{smallmatrix}\text{shell}\\\text{radius}\end{smallmatrix}\right)\left(\begin{smallmatrix}\text{shell}\\\text{height}\end{smallmatrix}\right) dx = \int_1^2 2\pi x(2-x)\, dx = 2\pi \int_1^2 (2x - x^2)\, dx = 2\pi \left[x^2 - \frac{x^3}{3} \right]_1^2$

$= 2\pi\left[\left(4 - \frac{8}{3}\right) - \left(1 - \frac{1}{3}\right)\right] = 2\pi\left[\left(\frac{12-8}{3}\right) - \left(\frac{3-1}{3}\right)\right] = 2\pi\left(\frac{4}{3} - \frac{2}{3}\right) = \frac{4\pi}{3}$

(c) $V = \int_a^b 2\pi \left(\begin{smallmatrix}\text{shell}\\\text{radius}\end{smallmatrix}\right)\left(\begin{smallmatrix}\text{shell}\\\text{height}\end{smallmatrix}\right) dx = \int_1^2 2\pi \left(\frac{10}{3} - x\right)(2-x)\, dx = 2\pi \int_1^2 \left(\frac{20}{3} - \frac{16}{3}x + x^2\right) dx$

$= 2\pi \left[\frac{20}{3}x - \frac{8}{3}x^2 + \frac{1}{3}x^3\right]_1^2 = 2\pi\left[\left(\frac{40}{3} - \frac{32}{3} + \frac{8}{3}\right) - \left(\frac{20}{3} - \frac{8}{3} + \frac{1}{3}\right)\right] = 2\pi\left(\frac{3}{3}\right) = 2\pi$

(d) $V = \int_c^d 2\pi \left(\begin{smallmatrix}\text{shell}\\\text{radius}\end{smallmatrix}\right)\left(\begin{smallmatrix}\text{shell}\\\text{height}\end{smallmatrix}\right) dy = \int_1^2 2\pi (y-1)(y-1)\, dy = 2\pi \int_1^2 (y-1)^2 = 2\pi\left[\frac{(y-1)^3}{3}\right]_1^2 = \frac{2\pi}{3}$

27. (a) $V = \int_c^d 2\pi \left(\begin{smallmatrix}\text{shell}\\\text{radius}\end{smallmatrix}\right)\left(\begin{smallmatrix}\text{shell}\\\text{height}\end{smallmatrix}\right) dy = \int_0^1 2\pi y\left[1 - (y - y^3)\right] dy$

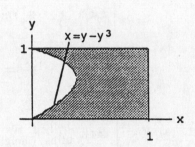

$= 2\pi \int_0^1 (y - y^2 + y^4)\, dy = 2\pi\left[\frac{y^2}{2} - \frac{y^3}{3} + \frac{y^5}{5}\right]_0^1$

$= 2\pi\left(\frac{1}{2} - \frac{1}{3} + \frac{1}{5}\right) = \frac{2\pi}{30}(15 - 10 + 6)$

$= \frac{11\pi}{15}$

(b) Use the washer method:

$V = \int_c^d \pi\left[R^2(y) - r^2(y)\right] dy = \int_0^1 \pi\left[1^2 - (y - y^3)^2\right] dy = \pi \int_0^1 (1 - y^2 - y^6 + 2y^4)\, dy = \pi\left[y - \frac{y^3}{3} - \frac{y^7}{7} + \frac{2y^5}{5}\right]_0^1$

$= \pi\left(1 - \frac{1}{3} - \frac{1}{7} + \frac{2}{5}\right) = \frac{\pi}{105}(105 - 35 - 15 + 42) = \frac{97\pi}{105}$

(c) Use the washer method:

$V = \int_c^d \pi\left[R^2(y) - r^2(y)\right] dy = \int_0^1 \pi\left[\left[1 - (y - y^3)\right]^2 - 0\right] dy = \pi \int_0^1 \left[1 - 2(y - y^3) + (y - y^3)^2\right] dy$

$= \pi \int_0^1 (1 + y^2 + y^6 - 2y + 2y^3 - 2y^4)\, dy = \pi\left[y + \frac{y^3}{3} + \frac{y^7}{7} - y^2 + \frac{y^4}{2} - \frac{2y^5}{5}\right]_0^1 = \pi\left(1 + \frac{1}{3} + \frac{1}{7} - 1 + \frac{1}{2} - \frac{2}{5}\right)$

$= \frac{\pi}{210}(70 + 30 + 105 - 2\cdot 42) = \frac{121\pi}{210}$

(d) $V = \int_c^d 2\pi \left(\begin{smallmatrix}\text{shell}\\\text{radius}\end{smallmatrix}\right)\left(\begin{smallmatrix}\text{shell}\\\text{height}\end{smallmatrix}\right) dy = \int_0^1 2\pi (1-y)\left[1 - (y - y^3)\right] dy = 2\pi \int_0^1 (1-y)(1 - y + y^3)\, dy$

$= 2\pi \int_0^1 (1 - y + y^3 - y + y^2 - y^4)\, dy = 2\pi \int_0^1 (1 - 2y + y^2 + y^3 - y^4)\, dy = 2\pi\left[y - y^2 + \frac{y^3}{3} + \frac{y^4}{4} - \frac{y^5}{5}\right]_0^1$

$= 2\pi\left(1 - 1 + \frac{1}{3} + \frac{1}{4} - \frac{1}{5}\right) = \frac{2\pi}{60}(20 + 15 - 12) = \frac{23\pi}{30}$

29. (a) $V = \int_c^d 2\pi \left(\begin{matrix}\text{shell} \\ \text{radius}\end{matrix}\right)\left(\begin{matrix}\text{shell} \\ \text{height}\end{matrix}\right) dy = \int_0^8 2\pi y \left(y^{1/3} - \frac{y}{4}\right) dy$

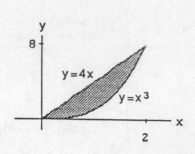

$= 2\pi \int_0^8 \left(y^{4/3} - \frac{y^2}{4}\right) dy = 2\pi \left[\frac{3}{7} y^{7/3} - \frac{y^3}{12}\right]_0^8$

$= 2\pi \left(\frac{3}{7} \cdot 2^7 - \frac{8^3}{12}\right) = 2\pi \left(\frac{3}{7} \cdot 2^7 - \frac{2^9}{12}\right) = 2\pi \cdot 2^7 \left(\frac{3}{7} - \frac{1}{3}\right)$

$= \frac{\pi \cdot 2^8}{21}(9 - 7) = \frac{\pi \cdot 2^9}{21} = \frac{512\pi}{21}$

(b) $V = \int_c^d 2\pi \left(\begin{matrix}\text{shell} \\ \text{radius}\end{matrix}\right)\left(\begin{matrix}\text{shell} \\ \text{height}\end{matrix}\right) dy = \int_0^8 2\pi(8 - y)\left(y^{1/3} - \frac{y}{4}\right) dy = 2\pi \int_0^8 \left(8y^{1/3} - 2y - y^{4/3} + \frac{y^2}{4}\right) dy$

$= 2\pi \left[6y^{4/3} - y^2 - \frac{3}{7} y^{7/3} + \frac{y^3}{12}\right]_0^8 = 2\pi \left[(6)(16) - 64 - \left(\frac{3}{7}\right)(2^7) + \frac{8 \cdot 64}{12}\right] = 2\pi \left(96 - 64 - \frac{384}{7} + \frac{128}{3}\right)$

$= 2\pi \left(32 - \frac{384}{7} + \frac{128}{3}\right) = \frac{2\pi}{21}(672 - 1152 + 896) = \frac{2\pi}{21}(416) = \frac{832\pi}{21}$

31. (a) $V = \int_a^b 2\pi \left(\begin{matrix}\text{shell} \\ \text{radius}\end{matrix}\right)\left(\begin{matrix}\text{shell} \\ \text{height}\end{matrix}\right) dx = \int_0^1 2\pi x\left[(2x - x^2) - x\right] dx$

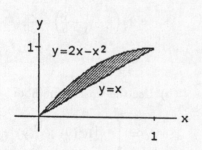

$= 2\pi \int_0^1 x(x - x^2) dx = 2\pi \int_0^1 (x^2 - x^3) dx$

$= 2\pi \left[\frac{x^3}{3} - \frac{x^4}{4}\right]_0^1 = 2\pi \left(\frac{1}{3} - \frac{1}{4}\right) = \frac{\pi}{6}$

(b) $V = \int_a^b 2\pi \left(\begin{matrix}\text{shell} \\ \text{radius}\end{matrix}\right)\left(\begin{matrix}\text{shell} \\ \text{height}\end{matrix}\right) dx = \int_0^1 2\pi(1 - x)\left[(2x - x^2) - x\right] dx = 2\pi \int_0^1 (1 - x)(x - x^2) dx$

$= 2\pi \int_0^1 (x - 2x^2 + x^3) dx = 2\pi \left[\frac{x^2}{2} - \frac{2}{3} x^3 + \frac{x^4}{4}\right]_0^1 = 2\pi \left(\frac{1}{2} - \frac{2}{3} + \frac{1}{4}\right) = \frac{2\pi}{12}(6 - 8 + 3) = \frac{\pi}{6}$

33. (a) $V = \int_a^b \pi \left[R^2(x) - r^2(x)\right] dx = \pi \int_{1/16}^1 \left(x^{-1/2} - 1\right) dx$

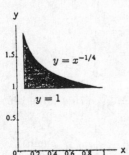

$= \pi \left[2x^{1/2} - x\right]_{1/16}^1 = \pi \left[(2 - 1) - \left(2 \cdot \frac{1}{4} - \frac{1}{16}\right)\right]$

$= \pi \left(1 - \frac{7}{16}\right) = \frac{9\pi}{16}$

(b) $V = \int_a^b 2\pi \left(\begin{matrix}\text{shell} \\ \text{radius}\end{matrix}\right)\left(\begin{matrix}\text{shell} \\ \text{height}\end{matrix}\right) dy = \int_1^2 2\pi y \left(\frac{1}{y^4} - \frac{1}{16}\right) dy = 2\pi \int_1^2 \left(y^{-3} - \frac{y}{16}\right) dy = 2\pi \left[-\frac{1}{2} y^{-2} - \frac{y^2}{32}\right]_1^2$

$= 2\pi \left[\left(-\frac{1}{8} - \frac{1}{8}\right) - \left(-\frac{1}{2} - \frac{1}{32}\right)\right] = 2\pi \left(\frac{1}{4} + \frac{1}{32}\right) = \frac{2\pi}{32}(8 + 1) = \frac{9\pi}{16}$

35. (a) $xf(x) = \begin{cases} x \cdot \frac{\sin x}{x}, & 0 < x \le \pi \\ x, & x = 0 \end{cases} \Rightarrow xf(x) = \begin{cases} \sin x, & 0 < x \le \pi \\ 0, & x = 0 \end{cases}$; since $\sin 0 = 0$ we have

$xf(x) = \begin{cases} \sin x, & 0 < x \le \pi \\ \sin x, & x = 0 \end{cases} \Rightarrow xf(x) = \sin x, \; 0 \le x \le \pi$

(b) $V = \int\limits_a^b 2\pi \left(\begin{array}{c} \text{shell} \\ \text{radius} \end{array}\right)\left(\begin{array}{c} \text{shell} \\ \text{height} \end{array}\right) dx = \int\limits_0^\pi 2\pi x \cdot f(x) \, dx$ and $x \cdot f(x) = \sin x, \; 0 \le x \le \pi$ by part (a)

$\Rightarrow V = 2\pi \int\limits_0^\pi \sin x \, dx = 2\pi[-\cos x]_0^\pi = 2\pi(-\cos \pi + \cos 0) = 4\pi$

37. (a) *Disc:* $V = V_1 - V_2$

$V_1 = \int\limits_{a_1}^{b_1} \pi[R_1(x)]^2 \, dx$ and $V_2 = \int\limits_{a_2}^{b_2} \pi[R_2(x)]^2$ with $R_1(x) = \sqrt{\dfrac{x+2}{3}}$ and $R_2(x) = \sqrt{x}$,

$a_1 = -2, \; b_1 = 1; \; a_2 = 0, \; b_2 = 1 \Rightarrow$ two integrals are required

(b) *Washer:* $V = V_1 - V_2$

$V_1 = \int\limits_{a_1}^{b_1} \pi\left([R_1(x)]^2 - [r_1(x)]^2\right) dx$ with $R_1(x) = \sqrt{\dfrac{x+2}{3}}$ and $r_1(x) = 0$; $a_1 = -2$ and $b_1 = 0$;

$V_2 = \int\limits_{a_2}^{b_2} \pi\left([R_2(x)]^2 - [r_2(x)]^2\right) dx$ with $R_2(x) = \sqrt{\dfrac{x+2}{3}}$ and $r_2(x) = \sqrt{x}$; $a_2 = 0$ and $b_2 = 1$

$\Rightarrow$ two integrals are required

(c) *Shell:* $V = \int\limits_c^d 2\pi \left(\begin{array}{c} \text{shell} \\ \text{radius} \end{array}\right)\left(\begin{array}{c} \text{shell} \\ \text{height} \end{array}\right) dy = \int\limits_c^d 2\pi y \left(\begin{array}{c} \text{shell} \\ \text{height} \end{array}\right) dy$ where shell height $- y^2 - (3y^2 - 2) = 2 - 2y^2$;

$c = 0$ and $d = 1$. Only *one* integral is required. It is, therefore preferable to use the *shell* method. However, whichever method you use, you will get $V = \pi$.

39. By the shell method we have $2\pi b^3 = 2\pi \int\limits_0^b xf(x) \, dx \Rightarrow x^3 = \int\limits_0^x tf(t) \, dt$, where $x > 0$. By the Fundamental Theorem of Calculus we have $3x^2 = xf(x) \Rightarrow f(x) = 3x$.

5.5 LENGTHS OF PLANE CURVES

1. (a) $\dfrac{dy}{dx} = 2x \Rightarrow \left(\dfrac{dy}{dx}\right)^2 = 4x^2 \Rightarrow L = \displaystyle\int_{-1}^{2} \sqrt{1 + \left(\dfrac{dy}{dx}\right)^2}\, dx$ (b)

$$= \int_{-1}^{2} \sqrt{1 + 4x^2}\, dx$$

(c) $L \approx 6.13$

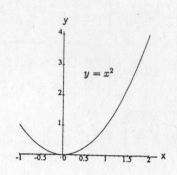

3. (a) $\dfrac{dx}{dy} = \cos y \Rightarrow \left(\dfrac{dx}{dy}\right)^2 = \cos^2 y$ (b)

$$\Rightarrow L = \int_{0}^{\pi} \sqrt{1 + \cos^2 y}\, dy$$

(c) $L \approx 3.82$

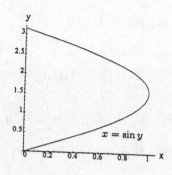

5. (a) $2y + 2 = 2\dfrac{dx}{dy} \Rightarrow \left(\dfrac{dx}{dy}\right)^2 = (y+1)^2$ (b)

$$\Rightarrow L = \int_{-1}^{3} \sqrt{1 + (y+1)^2}\, dy$$

(c) $L \approx 9.29$

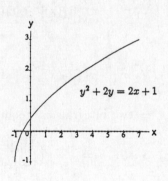

7. (a) $\dfrac{dy}{dx} = \tan x \Rightarrow \left(\dfrac{dy}{dx}\right)^2 = \tan^2 x$ (b)

$$\Rightarrow L = \int_{0}^{\pi/6} \sqrt{1 + \tan^2 x}\, dx = \int_{0}^{\pi/6} \sqrt{\dfrac{\sin^2 x + \cos^2 x}{\cos^2 x}}\, dx$$

$$= \int_{0}^{\pi/6} \dfrac{dx}{\cos x} = \int_{0}^{\pi/6} \sec x\, dx$$

(c) $L \approx 0.55$

9. $\dfrac{dy}{dx} = \dfrac{1}{3} \cdot \dfrac{3}{2}(x^2+2)^{1/2} \cdot 2x = \sqrt{(x^2+2)} \cdot x$

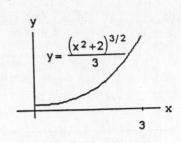

$\Rightarrow L = \displaystyle\int_0^3 \sqrt{1+(x^2+2)x^2}\ dx = \int_0^3 \sqrt{1+2x^2+x^4}\ dx$

$= \displaystyle\int_0^3 \sqrt{(1+x^2)^2}\ dx = \int_0^3 (1+x^2)\ dx = \left[x+\dfrac{x^3}{3}\right]_0^3$

$= 3 + \dfrac{27}{3} = 12$

11. $\dfrac{dx}{dy} = y^2 - \dfrac{1}{4y^2} \Rightarrow \left(\dfrac{dx}{dy}\right)^2 = y^4 - \dfrac{1}{2} + \dfrac{1}{16y^4}$

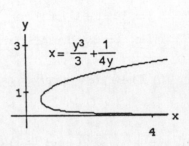

$\Rightarrow L = \displaystyle\int_1^3 \sqrt{1+y^4-\dfrac{1}{2}+\dfrac{1}{16y^4}}\ dy = \int_1^3 \sqrt{y^4+\dfrac{1}{2}+\dfrac{1}{16y^4}}\ dy$

$= \displaystyle\int_1^3 \sqrt{\left(y^2+\dfrac{1}{4y^2}\right)^2}\ dy = \int_1^3 \left(y^2+\dfrac{1}{4y^2}\right) dy = \left[\dfrac{y^3}{3}-\dfrac{y^{-1}}{4}\right]_1^3$

$= \left(\dfrac{27}{3}-\dfrac{1}{12}\right)-\left(\dfrac{1}{3}-\dfrac{1}{4}\right) = 9 - \dfrac{1}{12} - \dfrac{1}{3} + \dfrac{1}{4} = 9 + \dfrac{(-1-4+3)}{12} = 9 + \dfrac{(-2)}{12} = \dfrac{53}{6}$

13. $\dfrac{dx}{dy} = y^3 - \dfrac{1}{4y^3} \Rightarrow \left(\dfrac{dx}{dy}\right)^2 = y^6 - \dfrac{1}{2} + \dfrac{1}{16y^6}$

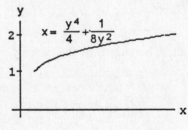

$\Rightarrow L = \displaystyle\int_1^2 \sqrt{1+y^6-\dfrac{1}{2}+\dfrac{1}{16y^6}}\ dy = \int_1^2 \sqrt{y^6+\dfrac{1}{2}+\dfrac{1}{16y^6}}\ dy$

$= \displaystyle\int_1^2 \sqrt{\left(y^3+\dfrac{y^{-3}}{4}\right)^2}\ dy = \int_1^2 \left(y^3+\dfrac{y^{-3}}{4}\right) dy$

$= \left[\dfrac{y^4}{4}-\dfrac{y^{-2}}{8}\right]_1^2 = \left(\dfrac{16}{4}-\dfrac{1}{(16)(2)}\right)-\left(\dfrac{1}{4}-\dfrac{1}{8}\right) = 4 - \dfrac{1}{32} - \dfrac{1}{4} + \dfrac{1}{8} = \dfrac{128-1-8+4}{32} = \dfrac{123}{32}$

15. $\dfrac{dy}{dx} = x^{1/3} - \dfrac{1}{4}x^{-1/3} \Rightarrow \left(\dfrac{dy}{dx}\right)^2 = x^{2/3} - \dfrac{1}{2} + \dfrac{x^{-2/3}}{16}$

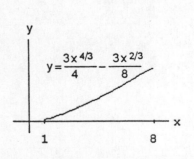

$\Rightarrow L = \displaystyle\int_1^8 \sqrt{1+x^{2/3}-\dfrac{1}{2}+\dfrac{x^{-2/3}}{16}}\ dx = \int_1^8 \sqrt{x^{2/3}+\dfrac{1}{2}+\dfrac{x^{-2/3}}{16}}\ dx$

$= \displaystyle\int_1^8 \sqrt{\left(x^{1/3}+\dfrac{1}{4}x^{-1/3}\right)^2}\ dx = \int_1^8 \left(x^{1/3}+\dfrac{1}{4}x^{-1/3}\right) dx$

$= \left[\dfrac{3}{4}x^{4/3}+\dfrac{3}{8}x^{2/3}\right]_1^8 = \dfrac{3}{8}\left[2x^{4/3}+x^{2/3}\right]_1^8 = \dfrac{3}{8}\left[(2\cdot2^4+2^2)-(2+1)\right] = \dfrac{3}{8}(32+4-3) = \dfrac{99}{8}$

17. $\frac{dx}{dy} = \sqrt{\sec^4 y - 1} \Rightarrow \left(\frac{dx}{dy}\right)^2 = \sec^4 y - 1$

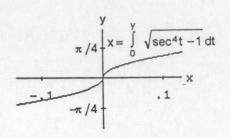

$\Rightarrow L = \int_{-\pi/4}^{\pi/4} \sqrt{1 + (\sec^4 y - 1)} \, dy = \int_{-\pi/4}^{\pi/4} \sec^2 y \, dy$

$= \left[\tan y\right]_{-\pi/4}^{\pi/4} = 1 - (-1) = 2$

19. (a) $\left(\frac{dy}{dx}\right)^2 = \frac{1}{4x} \Rightarrow \frac{dy}{dx} = \frac{1}{2\sqrt{x}}$ or $\frac{dy}{dx} = -\frac{1}{2\sqrt{x}} \Rightarrow y = x^{1/2} + C$ or $y = -x^{1/2} + C$; $(1,1)$ on the curve

$\Rightarrow 1 = 1 + C$ for $y = x^{1/2} + C \Rightarrow C = 0$ and $y = \sqrt{x}$; $(1,1)$ on the curve $\Rightarrow 1 = -1 + C$ for $y = -x^{1/2} + C$

$\Rightarrow C = 2$ and $y = -\sqrt{x} + 2$; in summary, $y = \sqrt{x}$ or $y = -\sqrt{x} + 2$

(b) There are two curves, as derived in part (a), because $\left(\frac{dy}{dx}\right)^2 = \frac{1}{4x}$ gives two derivative functions for $\frac{dy}{dx}$.

21. $y = \int_0^x \sqrt{\cos 2t} \, dt \Rightarrow \frac{dy}{dx} = \sqrt{\cos 2x} \Rightarrow \left(\frac{dy}{dx}\right)^2 = \cos 2x \Rightarrow L = \int_0^{\pi/4} \sqrt{1 + \cos 2x} \, dx = \int_0^{\pi/4} \sqrt{2 \cos^2 x} \, dx$

$= \sqrt{2} \int_0^{\pi/4} |\cos x| \, dx = \sqrt{2} \int_0^{\pi/4} \cos x \, dx = \sqrt{2} \left[\sin x\right]_0^{x/4} = \sqrt{2} \left(\frac{1}{\sqrt{2}} - 0\right) = 1$

23. The length of the curve $y = \sin\left(\frac{3\pi}{20} x\right)$ from 0 to 20 is: $L = \int_0^{20} \sqrt{1 + \left(\frac{dy}{dx}\right)^2} \, dx$; $\frac{dy}{dx} = \frac{3\pi}{20} \cos\left(\frac{3\pi}{20} x\right) \Rightarrow \left(\frac{dy}{dx}\right)^2$

$= \frac{9\pi^2}{400} \cos^2\left(\frac{3\pi}{20} x\right) \Rightarrow L = \int_0^{20} \sqrt{1 + \frac{9\pi^2}{400} \cos^2\left(\frac{3\pi}{20} x\right)} \, dx$. Using numerical integration we find $L \approx 21.07$ in

25. $\sqrt{2} x = \int_0^x \sqrt{1 + \left(\frac{dy}{dt}\right)^2} \, dt, x \geq 0 \Rightarrow \sqrt{2} = \sqrt{1 + \left(\frac{dy}{dx}\right)^2} \Rightarrow \frac{dy}{dx} = \pm 1 \Rightarrow y = f(x) = \pm x + C$ where C is any

real number.

5.6 AREAS OF SURFACES OF REVOLUTION

1. (a) $\dfrac{dy}{dx} = \sec^2 x \Rightarrow \left(\dfrac{dy}{dx}\right)^2 = \sec^4 x$ (b)

$$\Rightarrow S = 2\pi \int_0^{\pi/4} (\tan x) \sqrt{1 + \sec^4 x}\; dx$$

(c) $S \approx 3.84$

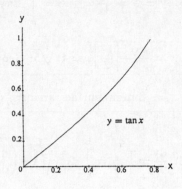

$y = \tan x$

3. (a) $xy = 1 \Rightarrow x = \dfrac{1}{y} \Rightarrow \dfrac{dx}{dy} = -\dfrac{1}{y^2} \Rightarrow \left(\dfrac{dx}{dy}\right)^2 = \dfrac{1}{y^4}$ (b)

$$\Rightarrow S = 2\pi \int_1^2 \frac{1}{y}\sqrt{1 + y^{-4}}\; dy$$

(c) $S \approx 5.02$

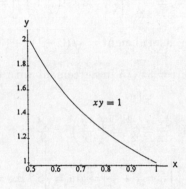

$xy = 1$

5. (a) $x^{1/2} + y^{1/2} = 3 \Rightarrow y = \left(3 - x^{1/2}\right)^2$ (b)

$$\Rightarrow \frac{dy}{dx} = 2\left(3 - x^{1/2}\right)\left(-\tfrac{1}{2}x^{-1/2}\right)$$

$$\Rightarrow \left(\frac{dy}{dx}\right)^2 = \left(1 - 3x^{-1/2}\right)^2$$

$$\Rightarrow S = 2\pi \int_1^4 \left(3 - x^{1/2}\right)^2 \sqrt{1 + \left(1 - 3x^{-1/2}\right)^2}\; dx$$

(c) $S \approx 63.37$

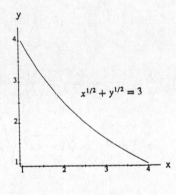

$x^{1/2} + y^{1/2} = 3$

7. (a) $\dfrac{dx}{dy} = \tan y \Rightarrow \left(\dfrac{dx}{dy}\right)^2 = \tan^2 y$ (b)

$$\Rightarrow S = 2\pi \int_0^{\pi/3} \left(\int_0^y \tan t\; dt\right) \sqrt{1 + \tan^2 y}\; dy$$

$$= 2\pi \int_0^{\pi/3} \left(\int_0^y \tan t\; dt\right) \sec y\; dy$$

(c) $S \approx 2.08$

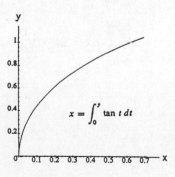

$x = \displaystyle\int_0^y \tan t\; dt$

9. $y = \frac{x}{2} \Rightarrow \frac{dy}{dx} = \frac{1}{2}$; $S = \int\limits_{a}^{b} 2\pi y \sqrt{1 + \left(\frac{dy}{dx}\right)^2}\, dx \Rightarrow S = \int\limits_{0}^{4} 2\pi \left(\frac{x}{2}\right)\sqrt{1 + \frac{1}{4}}\, dx = \frac{\pi\sqrt{5}}{2} \int\limits_{0}^{4} x\, dx$

$= \frac{\pi\sqrt{5}}{2}\left[\frac{x^2}{2}\right]_0^4 = 4\pi\sqrt{5}$; Geometry formula: base circumference $= 2\pi(2)$, slant height $= \sqrt{4^2 + 2^2} = 2\sqrt{5}$

$\Rightarrow$ Lateral surface area $= \frac{1}{2}(4\pi)(2\sqrt{5}) = 4\pi\sqrt{5}$ in agreement with the integral value

11. $\frac{dy}{dx} = \frac{1}{2}$; $S = \int\limits_{a}^{b} 2\pi y \sqrt{1 + \left(\frac{dy}{dx}\right)^2}\, dx = \int\limits_{1}^{3} 2\pi \frac{(x+1)}{2}\sqrt{1 + \left(\frac{1}{2}\right)^2}\, dx = \frac{\pi\sqrt{5}}{2}\int\limits_{1}^{3}(x+1)\, dx = \frac{\pi\sqrt{5}}{2}\left[\frac{x^2}{2} + x\right]_1^3$

$= \frac{\pi\sqrt{5}}{2}\left[\left(\frac{9}{2} + 3\right) - \left(\frac{1}{2} + 1\right)\right] = \frac{\pi\sqrt{5}}{2}(4 + 2) = 3\pi\sqrt{5}$; Geometry formula: $r_1 = \frac{1}{2} + \frac{1}{2} = 1$, $r_2 = \frac{3}{2} + \frac{1}{2} = 2$,

slant height $= \sqrt{(2-1)^2 + (3-1)^2} = \sqrt{5} \Rightarrow$ Frustum surface area $= \pi(r_1 + r_2) \times$ slant height $= \pi(1+2)\sqrt{5}$

$= 3\pi\sqrt{5}$ in agreement with the integral value

13. $\frac{dy}{dx} = \frac{x^2}{3} \Rightarrow \left(\frac{dy}{dx}\right)^2 = \frac{x^4}{9} \Rightarrow S = \int\limits_{0}^{2} \frac{2\pi x^3}{9}\sqrt{1 + \frac{x^4}{9}}\, dx$;

$\left[u = 1 + \frac{x^4}{9} \Rightarrow du = \frac{4}{9}x^3\, dx \Rightarrow \frac{1}{4}\, du = \frac{x^3}{9}\, dx\right.$;

$x = 0 \Rightarrow u = 1$, $x = 2 \Rightarrow u = \frac{25}{9}\bigg] \to S = 2\pi \int\limits_{1}^{25/9} u^{1/2}\cdot\frac{1}{4}\, du$

$= \frac{\pi}{2}\left[\frac{2}{3}u^{3/2}\right]_1^{25/9} = \frac{\pi}{3}\left(\frac{125}{27} - 1\right) = \frac{\pi}{3}\left(\frac{125 - 27}{27}\right) = \frac{98\pi}{81}$

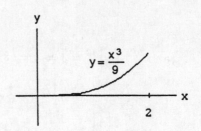

15. $\frac{dy}{dx} = \frac{1}{2}\frac{(2 - 2x)}{\sqrt{2x - x^2}} = \frac{1 - x}{\sqrt{2x - x^2}} \Rightarrow \left(\frac{dy}{dx}\right)^2 = \frac{(1 - x)^2}{2x - x^2}$

$\Rightarrow S = \int\limits_{0.5}^{1.5} 2\pi\sqrt{2x - x^2}\sqrt{1 + \frac{(1 - x)^2}{2x - x^2}}\, dx$

$= 2\pi \int\limits_{0.5}^{1.5} \sqrt{2x - x^2}\,\frac{\sqrt{2x - x^2 + 1 - 2x + x^2}}{\sqrt{2x - x^2}}\, dx$

$= 2\pi \int\limits_{0.5}^{1.5} dx = 2\pi[x]_{0.5}^{1.5} = 2\pi$

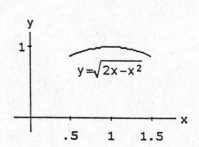

17. $\frac{dx}{dy} = y^2 \Rightarrow \left(\frac{dx}{dy}\right)^2 = y^4 \Rightarrow S = \int\limits_0^1 \frac{2\pi y^3}{3} \sqrt{1 + y^4}\ dy;$

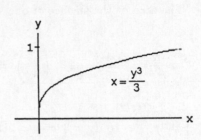

$\left[u = 1 + y^4 \Rightarrow du = 4y^3\ dy \Rightarrow \frac14\ du = y^3\ dy;\ y = 0\right.$

$\left. \Rightarrow u = 1,\ y = 1 \Rightarrow u = 2\right] \to S = \int\limits_1^2 2\pi\left(\tfrac13\right)u^{1/2}\left(\tfrac14\ du\right)$

$= \frac{\pi}{6} \int\limits_1^2 u^{1/2}\ du = \frac{\pi}{6}\left[\tfrac23 u^{3/2}\right]_1^2 = \frac{\pi}{9}\left(\sqrt{8} - 1\right)$

19. $\frac{dx}{dy} = \frac{-1}{\sqrt{4 - y}} \Rightarrow \left(\frac{dx}{dy}\right)^2 = \frac{1}{4 - y} \Rightarrow S = \int\limits_0^{15/4} 2\pi \cdot 2\sqrt{4 - y}\ \sqrt{1 + \frac{1}{4 - y}}\ dy = 4\pi \int\limits_0^{15/4} \sqrt{(4 - y) + 1}\ dy$

$= 4\pi \int\limits_0^{15/4} \sqrt{5 - y}\ dy = -4\pi\left[\tfrac23(5 - y)^{3/2}\right]_0^{15/4} = -\frac{8\pi}{3}\left[\left(5 - \tfrac{15}{4}\right)^{3/2} - 5^{3/2}\right] = -\frac{8\pi}{3}\left[\left(\tfrac54\right)^{3/2} - 5^{3/2}\right]$

$= \frac{8\pi}{3}\left(5\sqrt{5} - \frac{5\sqrt{5}}{8}\right) = \frac{8\pi}{3}\left(\frac{40\sqrt{5} - 5\sqrt{5}}{8}\right) = \frac{35\pi\sqrt{5}}{3}$

21. $ds = \sqrt{dx^2 + dy^2} = \sqrt{\left(y^3 - \frac{1}{4y^3}\right)^2 + 1}\ dy = \sqrt{\left(y^6 - \frac12 + \frac{1}{16y^6}\right) + 1}\ dy = \sqrt{\left(y^6 + \frac12 + \frac{1}{16y^6}\right)}\ dy$

$= \sqrt{\left(y^3 + \frac{1}{4y^3}\right)^2}\ dy = \left(y^3 + \frac{1}{4y^3}\right) dy;\ S = \int\limits_1^2 2\pi y\ ds = 2\pi \int\limits_1^2 y\left(y^3 + \frac{1}{4y^3}\right) dy = 2\pi \int\limits_1^2 \left(y^4 + \tfrac14 y^{-2}\right) dy$

$= 2\pi\left[\frac{y^5}{5} - \tfrac14 y^{-1}\right]_1^2 = 2\pi\left[\left(\tfrac{32}{5} - \tfrac18\right) - \left(\tfrac15 - \tfrac14\right)\right] = 2\pi\left(\tfrac{31}{5} + \tfrac18\right) = \frac{2\pi}{40}(8 \cdot 31 + 5) = \frac{253\pi}{20}$

23. $y = \sqrt{a^2 - x^2} \Rightarrow \frac{dy}{dx} = \frac12\left(a^2 - x^2\right)^{-1/2}(-2x) = \frac{-x}{\sqrt{a^2 - x^2}} \Rightarrow \left(\frac{dy}{dx}\right)^2 = \frac{x^2}{\left(a^2 - x^2\right)}$

$\Rightarrow S = 2\pi \int\limits_{-a}^a \sqrt{a^2 - x^2}\ \sqrt{1 + \frac{x^2}{\left(a^2 - x^2\right)}}\ dx = 2\pi \int\limits_{-a}^a \sqrt{\left(a^2 - x^2\right) + x^2}\ dx = 2\pi \int\limits_{-a}^a a\ dx = 2\pi a[x]_{-a}^a$

$= 2\pi a[a - (-a)] = (2\pi a)(2a) = 4\pi a^2$

25. (a) $y = \cos x \Rightarrow \frac{dy}{dx} = -\sin x \Rightarrow \left(\frac{dy}{dx}\right)^2 = \sin^2 x \Rightarrow S = 2\pi \int\limits_{-\pi/2}^{\pi/2} (\cos x) \sqrt{1 + \sin^2 x}\ dx$

(b) $S \approx 14.4236$

27. The area of the surface of one wok is $S = \displaystyle\int_c^d 2\pi x \sqrt{1 + \left(\dfrac{dx}{dy}\right)^2}\, dy$. Now, $x^2 + y^2 = 16^2 \Rightarrow x = \sqrt{16^2 - y^2}$

$$\Rightarrow \dfrac{dx}{dy} = \dfrac{-y}{\sqrt{16^2 - y^2}} \Rightarrow \left(\dfrac{dx}{dy}\right)^2 = \dfrac{y^2}{16^2 - y^2}; \; S = \int_{-16}^{-7} 2\pi\sqrt{16^2 - y^2}\sqrt{1 + \dfrac{y^2}{16^2 - y^2}}\, dy = 2\pi \int_{-16}^{-7} \sqrt{(16^2 - y^2) + y^2}\, dy$$

$$= 2\pi \int_{-16}^{-7} 16\, dy = 32\pi \cdot 9 = 288\pi \approx 904.78 \text{ cm}^2.$$ The enamel needed to cover one surface of one wok is

$V = S \cdot 0.5 \text{ mm} = S \cdot 0.05 \text{ cm} = (904.78)(0.05) \text{ cm}^3 = 45.24 \text{ cm}^3.$ For 5000 woks, we need

$5000 \cdot V = 5000 \cdot 45.24 \text{ cm}^3 = (5)(45.24)\text{L} = 226.2\text{L} \Rightarrow 226.2$ liters of each color are needed.

29. $y = \sqrt{R^2 - x^2} \Rightarrow \dfrac{dy}{dx} = -\dfrac{1}{2}\dfrac{2x}{\sqrt{R^2 - x^2}} = \dfrac{-x}{\sqrt{R^2 - x^2}} \Rightarrow \left(\dfrac{dx}{dy}\right)^2 = \dfrac{x^2}{R^2 - x^2}; \; S = 2\pi \int_a^{a+h} \sqrt{R^2 - x^2}\sqrt{1 + \dfrac{x^2}{R^2 - x^2}}\, dx$

$$= 2\pi \int_a^{a+h} \sqrt{(R^2 - x^2) + x^2}\, dx = 2\pi R \int_a^{a+h} dx = 2\pi R h$$

31. $y = x \Rightarrow \left(\dfrac{dy}{dx}\right) = 1 \Rightarrow \left(\dfrac{dy}{dx}\right)^2 = 1 \Rightarrow S = 2\pi \int_{-1}^{2} |x|\sqrt{1+1}\, dx = 2\pi \int_{-1}^{0} (-x)\sqrt{2}\, dx + 2\pi \int_{0}^{2} x\sqrt{2}\, dx$

$$= -2\sqrt{2}\pi\left[\dfrac{x^2}{2}\right]_{-1}^{0} + 2\sqrt{2}\pi\left[\dfrac{x^2}{2}\right]_{0}^{2} = -2\sqrt{2}\pi\left(0 - \dfrac{1}{2}\right) + 2\sqrt{2}\pi(2 - 0) = 5\sqrt{2}\pi$$

33. $y = \sin x \Rightarrow \dfrac{dy}{dx} = \cos x \Rightarrow \left(\dfrac{dy}{dx}\right)^2 = \cos^2 x \Rightarrow S = \int_0^{\pi} 2\pi(\sin x)\sqrt{1 + \cos^2 x}\, dx$; a numerical integration gives

$S \approx 14.4$

35. $y = x + \sin 2x \Rightarrow \dfrac{dy}{dx} = 1 + 2\cos 2x \Rightarrow \left(\dfrac{dy}{dx}\right)^2 = (1 + 2\cos 2x)^2$; by symmetry of the graph we have that

$S = 2 \displaystyle\int_0^{2\pi/3} 2\pi(x + \sin 2x)\sqrt{1 + (1 + 2\cos 2x)^2}\, dx$; a numerical integration gives $S \approx 54.9$

37. (a) An equation of the tangent line segment is

(see figure) $y = f(m_k) + f'(m_k)(x - m_k)$.

When $x = x_{k-1}$ we have

$r_1 = f(m_k) + f'(m_k)(x_{k-1} - m_k)$

$= f(m_k) + f'(m_k)\left(-\dfrac{\Delta x_k}{2}\right) = f(m_k) - f'(m_k)\dfrac{\Delta x_k}{2}$;

when $x = x_k$ we have

$r_2 = f(m_k) + f'(m_k)(x_k - m_k)$

$= f(m_k) + f'(m_k)\dfrac{\Delta x_k}{2}$;

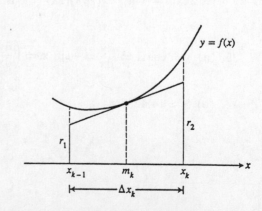

(b) $\ell^2 = (\Delta x_k)^2 + (r_2 - r_1)^2 = (\Delta x_k)^2 + \left[f'(m_k) \frac{\Delta x_k}{2} - \left(-f'(m_k) \frac{\Delta x_k}{2} \right) \right]^2 = (\Delta x_k)^2 + [f'(m_k)\Delta x_k]^2$

$\Rightarrow \ell = \sqrt{(\Delta x_k)^2 + [f'(m_k)\Delta x_k]^2}$, as claimed

(c) From geometry it is a fact that the lateral surface area of the frustrum obtained by revolving the tangent

line segment about the x-axis is given by $\Delta S_k = \pi(r_1 + r_2)\ell = \pi[2f(m_k)] \sqrt{(\Delta x_k)^2 + [f'(m_k)\Delta x_k]^2}$

using parts (a) and (b) above. Thus, $\Delta S_k = 2\pi f(m_k) \sqrt{1 + [f'(m_k)]^2} \, \Delta x_k$.

(d) $S = \lim\limits_{n \to \infty} \sum\limits_{k=1}^{n} \Delta S_k = \lim\limits_{n \to \infty} \sum\limits_{k=1}^{n} 2\pi f(m_k) \sqrt{1 + [f'(m_k)]^2} \, \Delta x_k = \int\limits_{a}^{b} 2\pi f(x) \sqrt{1 + [f'(x)]^2} \, dx$

5.7 MOMENTS AND CENTERS OF MASS

1. Because the children are balanced, the moment of the system about the origin must be equal to zero:
$5 \cdot 80 = x \cdot 100 \Rightarrow x = 4$ ft, the distance of the 100-lb child from the fulcrum.

3. The center of mass of each rod is in its center (see Example 1). The rod system is equivalent to two point

masses located at the centers of the rods at coordinates $\left(\frac{L}{2}, 0 \right)$ and $\left(0, \frac{L}{2} \right)$. Therefore $\bar{x} = \frac{m_y}{m}$

$= \frac{x_1 m_1 + x_2 m_2}{m_1 + m_2} = \frac{\frac{L}{2} \cdot m + 0}{m + m} = \frac{L}{4}$ and $\bar{y} = \frac{m_x}{m} = \frac{y_1 m_2 + y_2 m_2}{m_1 + m_2} = \frac{0 + \frac{L}{2} \cdot m}{m + m} = \frac{L}{4} \Rightarrow \left(\frac{L}{4}, \frac{L}{4} \right)$ is the center of

mass location.

5. $M_0 = \int\limits_{0}^{2} x \cdot 4 \, dx = \left[4 \frac{x^2}{2} \right]_0^2 = 4 \cdot \frac{4}{2} = 8; \; M = \int\limits_{0}^{2} 4 \, dx = [4x]_0^2 = 4 \cdot 2 = 8 \Rightarrow \bar{x} = \frac{M_0}{M} = 1$

7. $M_0 = \int\limits_{0}^{3} x \left(1 + \frac{x}{3} \right) dx = \int\limits_{0}^{3} \left(x + \frac{x^2}{3} \right) dx = \left[\frac{x^2}{2} + \frac{x^3}{9} \right]_0^3 = \left(\frac{9}{2} + \frac{27}{9} \right) = \frac{15}{2}; \; M = \int\limits_{0}^{3} \left(1 + \frac{x}{3} \right) dx = \left[x + \frac{x^2}{6} \right]_0^3$

$= 3 + \frac{9}{6} = \frac{9}{2} \Rightarrow \bar{x} = \frac{M_0}{M} = \frac{\left(\frac{15}{2} \right)}{\left(\frac{9}{2} \right)} = \frac{15}{9} = \frac{5}{3}$

9. $M_0 = \int\limits_{1}^{4} x \left(1 + \frac{1}{\sqrt{x}} \right) dx = \int\limits_{1}^{4} \left(x + x^{1/2} \right) dx = \left[\frac{x^2}{2} + \frac{2x^{3/2}}{3} \right]_1^4 = \left(8 + \frac{16}{3} \right) - \left(\frac{1}{2} + \frac{2}{3} \right) = \frac{15}{2} + \frac{14}{3} = \frac{45 + 28}{6} = \frac{73}{6};$

$M = \int\limits_{1}^{4} \left(1 + x^{-1/2} \right) dx = \left[x + 2x^{1/2} \right]_1^4 = (4 + 4) - (1 + 2) = 5 \Rightarrow \bar{x} = \frac{M_0}{M} = \frac{\left(\frac{73}{6} \right)}{5} = \frac{73}{30}$

11. $M_0 = \int\limits_0^1 x(2-x)\,dx + \int\limits_1^2 x \cdot x\,dx = \int\limits_0^1 (2x - x^2)\,dx + \int\limits_1^2 x^2\,dx = \left[\frac{2x^2}{2} - \frac{x^3}{3}\right]_0^1 + \left[\frac{x^3}{3}\right]_1^2 = \left(1 - \frac{1}{3}\right) + \left(\frac{8}{3} - \frac{1}{3}\right)$

$= \frac{9}{3} = 3; \ M = \int\limits_0^1 (2-x)\,dx + \int\limits_1^2 x\,dx = \left[2x - \frac{x^2}{2}\right]_0^1 + \left[\frac{x^2}{2}\right]_1^2 = \left(2 - \frac{1}{2}\right) + \left(\frac{4}{2} - \frac{1}{2}\right) = 3 \Rightarrow \bar{x} = \frac{M_0}{M} = 1$

13. Since the plate is symmetric about the y-axis and its density is constant, the distribution of mass is symmetric about the y-axis and the center of mass lies on the y-axis. This means that $\bar{x} = 0$.

It remains to find $\bar{y} = \dfrac{M_x}{M}$. We model the distribution of mass with *vertical* strips. The typical strip has center of mass:

$(\tilde{x}, \tilde{y}) = \left(x, \dfrac{x^2 + 4}{2}\right)$, length: $4 - x^2$, width: dx, area:

$dA = (4 - x^2)\,dx$, mass: $dm = \delta\,dA = \delta(4 - x^2)\,dx$. The moment

of the strip about the x-axis is $\tilde{y}\,dm = \left(\dfrac{x^2 + 4}{2}\right)\delta(4 - x^2)\,dx = \dfrac{\delta}{2}(16 - x^4)\,dx$. The moment of the plate about

the x-axis is $M_x = \int \tilde{y}\,dm = \int\limits_{-2}^2 \dfrac{\delta}{2}(16 - x^4)\,dx = \dfrac{\delta}{2}\left[16x - \dfrac{x^5}{5}\right]_{-2}^2 = \dfrac{\delta}{2}\left[\left(16 \cdot 2 - \dfrac{2^5}{5}\right) - \left(-16 \cdot 2 + \dfrac{2^5}{5}\right)\right]$

$= \dfrac{\delta \cdot 2}{2}\left(32 - \dfrac{32}{5}\right) = \dfrac{128\delta}{5}$. The mass of the plate is $M = \int \delta(4 - x^2)\,dx = \delta\left[4x - \dfrac{x^3}{3}\right]_{-2}^2 = 2\delta\left(8 - \dfrac{8}{3}\right) = \dfrac{32\delta}{3}$.

Therefore $\bar{y} = \dfrac{M_x}{M} = \dfrac{\left(\dfrac{128\delta}{5}\right)}{\left(\dfrac{32\delta}{3}\right)} = \dfrac{12}{5}$. The plate's center of mass is the point $(\bar{x}, \bar{y}) = \left(0, \dfrac{12}{5}\right)$.

15. Intersection points: $x - x^2 = -x \Rightarrow 2x - x^2 = 0$

$\Rightarrow x(2 - x) = 0 \Rightarrow x = 0$ or $x = 2$. The typical *vertical*

strip has center of mass: $(\tilde{x}, \tilde{y}) = \left(x, \dfrac{(x - x^2) + (-x)}{2}\right)$

$= \left(x, -\dfrac{x^2}{2}\right)$, length: $(x - x^2) - (-x) = 2x - x^2$, width: dx,

area: $dA = (2x - x^2)\,dx$, mass: $dm = \delta\,dA = \delta(2x - x^2)\,dx$.

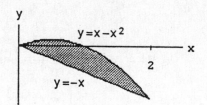

The moment of the strip about the x-axis is $\tilde{y}\,dm = \left(-\dfrac{x^2}{2}\right)\delta(2x - x^2)\,dx$; about the y-axis it is

$\tilde{x}\,dm = x \cdot \delta(2x - x^2)\,dx$. Thus, $M_x = \int \tilde{y}\,dm = -\int\limits_0^2 \left(\dfrac{\delta}{2}x^2\right)(2x - x^2)\,dx = -\dfrac{\delta}{2}\int\limits_0^2 (2x^3 - x^4)\,dx$

$= -\dfrac{\delta}{2}\left[\dfrac{x^4}{2} - \dfrac{x^5}{5}\right]_0^2 = -\dfrac{\delta}{2}\left(2^3 - \dfrac{2^5}{5}\right) = -\dfrac{\delta}{2} \cdot 2^3\left(1 - \dfrac{4}{5}\right) = -\dfrac{4\delta}{5}; \ M_y = \int \tilde{x}\,dm = \int\limits_0^2 x \cdot \delta(2x - x^2)\,dx$

$= \delta\int\limits_0^2 (2x^2 - x^3) = \delta\left[\dfrac{2}{3}x^3 - \dfrac{x^4}{4}\right]_0^2 = \delta\left(2 \cdot \dfrac{2^3}{3} - \dfrac{2^4}{4}\right) = \dfrac{\delta \cdot 2^4}{12} = \dfrac{4\delta}{3}; \ M = \int dm = \int\limits_0^2 \delta(2x - x^2)\,dx$

$$= \delta \int_0^2 (2x - x^2)\, dx = \delta \left[x^2 - \frac{x^3}{3} \right]_0^2 = \delta \left(4 - \frac{8}{3} \right) = \frac{4\delta}{3}. \text{ Therefore, } \bar{x} = \frac{M_y}{M} = \left(\frac{4\delta}{3} \right)\left(\frac{3}{4\delta} \right) = 1 \text{ and } \bar{y} = \frac{M_x}{M}$$

$$= \left(-\frac{4\delta}{5} \right)\left(\frac{3}{4\delta} \right) = -\frac{3}{5} \Rightarrow (\bar{x}, \bar{y}) = \left(1, -\frac{3}{5} \right) \text{ is the center of mass.}$$

17. The typical *horizontal* strip has center of mass:

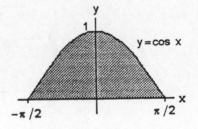

$$(\tilde{x}, \tilde{y}) = \left(\frac{y - y^3}{2}, y \right), \text{ length: } y - y^3, \text{ width: } dy,$$

area: $dA = (y - y^3)\, dy$, mass: $dm = \delta\, dA = \delta(y - y^3)\, dy$.

The moment of the strip about the y-axis is

$$\tilde{x}\, dm = \delta \left(\frac{y - y^3}{2} \right)(y - y^3)\, dy = \frac{\delta}{2}(y - y^3)^2\, dy$$

$$= \frac{\delta}{2}(y^2 - 2y^4 + y^6)\, dy; \text{ the moment about the x-axis is } \tilde{y}\, dm = \delta y(y - y^3)\, dy = \delta(y^2 - y^4)\, dy. \text{ Thus,}$$

$$M_x = \int \tilde{y}\, dm = \delta \int_0^1 (y^2 - y^4)\, dy = \delta \left[\frac{y^3}{3} - \frac{y^5}{5} \right]_0^1 = \delta \left(\frac{1}{3} - \frac{1}{5} \right) = \frac{2\delta}{15}; \quad M_y = \int \tilde{x}\, dm = \frac{\delta}{2} \int_0^1 (y^2 - 2y^4 + y^6)\, dy$$

$$= \frac{\delta}{2} \left[\frac{y^3}{3} - \frac{2y^5}{5} + \frac{y^7}{7} \right]_0^1 = \frac{\delta}{2} \left(\frac{1}{3} - \frac{2}{5} + \frac{1}{7} \right) = \frac{\delta}{2}\left(\frac{35 - 42 + 15}{3 \cdot 5 \cdot 7} \right) = \frac{4\delta}{105}; \quad M = \int dm = \delta \int_0^1 (y - y^3)\, dy$$

$$= \delta \left[\frac{y^2}{2} - \frac{y^4}{4} \right]_0^1 = \delta \left(\frac{1}{2} - \frac{1}{4} \right) = \frac{\delta}{4}. \text{ Therefore, } \bar{x} = \frac{M_y}{M} = \left(\frac{4\delta}{105} \right)\left(\frac{4}{\delta} \right) = \frac{16}{105} \text{ and } \bar{y} = \frac{M_x}{M} = \left(\frac{2\delta}{15} \right)\left(\frac{4}{\delta} \right) = \frac{8}{15}$$

$$\Rightarrow (\bar{x}, \bar{y}) = \left(\frac{16}{105}, \frac{8}{15} \right) \text{ is the center of mass.}$$

19. Applying the symmetry argument analogous to the one used
in Exercise 13, we find $\bar{x} = 0$. The typical *vertical* strip has
center of mass: $(\tilde{x}, \tilde{y}) = \left(x, \frac{\cos x}{2} \right)$, length: $\cos x$, width: dx,
area: $dA = \cos x\, dx$, mass: $dm = \delta\, dA = \delta \cos x\, dx$. The
moment of the strip about the x-axis is $\tilde{y}\, dm = \delta \cdot \frac{\cos x}{2} \cdot \cos x\, dx$

$$= \frac{\delta}{2} \cos^2 x\, dx = \frac{\delta}{2}\left(\frac{1 + \cos 2x}{2} \right) dx = \frac{\delta}{4}(1 + \cos 2x)\, dx; \text{ thus,}$$

$$M_x = \int \tilde{y}\, dm = \int_{-\pi/2}^{\pi/2} \frac{\delta}{4}(1 + \cos 2x)\, dx = \frac{\delta}{4}\left[x + \frac{\sin 2x}{2} \right]_{-\pi/2}^{\pi/2} = \frac{\delta}{4}\left[\left(\frac{\pi}{2} + 0 \right) - \left(-\frac{\pi}{2} \right) \right] = \frac{\delta\pi}{4}; \quad M = \int dm$$

$$= \delta \int_{-\pi/2}^{\pi/2} \cos x\, dx = \delta[\sin x]_{-\pi/2}^{\pi/2} = 2\delta. \text{ Therefore, } \bar{y} = \frac{M_x}{M} = \frac{\delta\pi}{4 \cdot 2\delta} = \frac{\pi}{8} \Rightarrow (\bar{x}, \bar{y}) = \left(0, \frac{\pi}{8} \right) \text{ is the center of mass.}$$

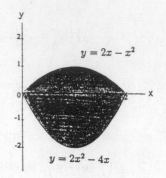

21. Since the plate is symmetric about the line x = 1 and its density is constant, the distribution of mass is symmetric about this line and the center of mass lies on it. This means that $\bar{x} = 1$. The typical *vertical* strip has center of mass:

$$(\tilde{x}, \tilde{y}) = \left(x, \frac{(2x - x^2) + (2x^2 - 4x)}{2}\right) = \left(x, \frac{x^2 - 2x}{2}\right),$$

length: $(2x - x^2) - (2x^2 - 4x) = -3x^2 + 6x = 3(2x - x^2)$,

width: dx, area: $dA = 3(2x - x^2)\,dx$, mass: $dm = \delta\,dA$

$= 3\delta(2x - x^2)\,dx$. The moment about the x-axis is

$\tilde{y}\,dm = \frac{3}{2}\delta(x^2 - 2x)(2x - x^2)\,dx = -\frac{3}{2}\delta(x^2 - 2x)^2\,dx = -\frac{3}{2}\delta(x^4 - 4x^3 + 4x^2)\,dx$. Thus, $M_x = \int \tilde{y}\,dm$

$= -\int_0^2 \frac{3}{2}\delta(x^4 - 4x^3 + 4x^2)\,dx = -\frac{3}{2}\delta\left[\frac{x^5}{5} - x^4 + \frac{4}{3}x^3\right]_0^2 = -\frac{3}{2}\delta\left(\frac{2^5}{5} - 2^4 + \frac{4}{3}\cdot 2^3\right) = -\frac{3}{2}\delta \cdot 2^4\left(\frac{2}{5} - 1 + \frac{2}{3}\right)$

$= -\frac{3}{2}\delta \cdot 2^4\left(\frac{6 - 15 + 10}{15}\right) = -\frac{8\delta}{5}$; $M = \int dm = \int_0^2 3\delta(2x - x^2)\,dx = 3\delta\left[x^2 - \frac{x^3}{3}\right]_0^2 = 3\delta\left(4 - \frac{8}{3}\right) = 4\delta.$

Therefore, $\bar{y} = \frac{M_x}{M} = \left(-\frac{8\delta}{5}\right)\left(\frac{1}{4\delta}\right) = -\frac{2}{5} \Rightarrow (\bar{x}, \bar{y}) = \left(1, -\frac{2}{5}\right)$ is the center of mass.

23. Since the plate is symmetric about the line x = y and its density is constant, the distribution of mass is symmetric about this line. This means that $\bar{x} = \bar{y}$. The typical *vertical* strip has

center of mass: $(\tilde{x}, \tilde{y}) = \left(x, \frac{3 + \sqrt{9 - x^2}}{2}\right),$

length: $3 - \sqrt{9 - x^2}$, width: dx, area: $dA = (3 - \sqrt{9 - x^2})\,dx,$

mass: $dm = \delta\,dA = \delta(3 - \sqrt{9 - x^2})\,dx$. The moment about the

x-axis is $\tilde{y}\,dm = \delta\frac{(3 + \sqrt{9 - x^2})(3 - \sqrt{9 - x^2})}{2}\,dx$

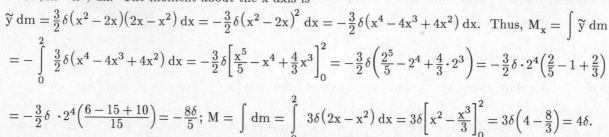

$= \frac{\delta}{2}[9 - (9 - x^2)]\,dx = \frac{\delta x^2}{2}\,dx$. Thus, $M_x = \int_0^3 \frac{\delta x^2}{2}\,dx = \frac{\delta}{6}[x^3]_0^3 = \frac{9\delta}{2}$. The area equals the area of a square

with side length 3 minus one quarter the area of a disk with radius $3 \Rightarrow A = 3^2 - \frac{\pi 9}{4} = \frac{9}{4}(4 - \pi) \Rightarrow M = \delta A$

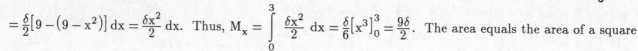

$= \frac{9\delta}{4}(4 - \pi)$. Therefore, $\bar{y} = \frac{M_x}{M} = \left(\frac{9\delta}{2}\right)\left[\frac{4}{9\delta(4 - \pi)}\right] = \frac{2}{4 - \pi} \Rightarrow (\bar{x}, \bar{y}) = \left(\frac{2}{4 - \pi}, \frac{2}{4 - \pi}\right)$ is the center of mass.

25. $M_x = \int \tilde{y}\, dm = \int\limits_1^2 \dfrac{\left(\dfrac{2}{x^2}\right)}{2} \cdot \delta \cdot \left(\dfrac{2}{x^2}\right) dx$

$= \int\limits_1^2 \left(\dfrac{1}{x^2}\right)(x^2)\left(\dfrac{2}{x^2}\right) dx = \int\limits_1^2 \dfrac{2}{x^2}\, dx = 2\int\limits_1^2 x^{-2}\, dx$

$= 2\left[-x^{-1}\right]_1^2 = 2\left[\left(-\tfrac{1}{2}\right) - (-1)\right] = 2\left(\tfrac{1}{2}\right) = 1;$

$M_y = \int \tilde{x}\, dm = \int\limits_1^2 x \cdot \delta \cdot \left(\dfrac{2}{x^2}\right) dx$

$= \int\limits_1^2 x(x^2)\left(\dfrac{2}{x^2}\right) dx = 2\int\limits_1^2 x\, dx = 2\left[\dfrac{x^2}{2}\right]_1^2 = 2\left(2 - \tfrac{1}{2}\right) = 4 - 1 = 3;\ M = \int dm = \int\limits_1^2 \delta\left(\dfrac{2}{x^2}\right) dx$

$= \int\limits_1^2 x^2\left(\dfrac{2}{x^2}\right) dx = 2\int\limits_1^2 dx = 2[x]_1^2 = 2(2-1) = 2.$ So $\bar{x} = \dfrac{M_y}{M} = \dfrac{3}{2}$ and $\bar{y} = \dfrac{M_x}{M} = \dfrac{1}{2} \Rightarrow (\bar{x}, \bar{y}) = \left(\dfrac{3}{2}, \dfrac{1}{2}\right)$ is the

center of mass.

27. (a) We use the shell method:

$$V = \int\limits_a^b 2\pi \left(\genfrac{}{}{0pt}{}{\text{shell}}{\text{radius}}\right)\left(\genfrac{}{}{0pt}{}{\text{shell}}{\text{height}}\right) dx = \int\limits_1^4 2\pi x\left[\dfrac{4}{\sqrt{x}} - \left(-\dfrac{4}{\sqrt{x}}\right)\right] dx$$

$= 16\pi \int\limits_1^4 \dfrac{x}{\sqrt{x}}\, dx = 16\pi \int\limits_1^4 x^{1/2}\, dx = 16\pi\left[\dfrac{2}{3}x^{3/2}\right]_1^4$

$= 16\pi\left(\dfrac{2}{3}\cdot 8 - \dfrac{2}{3}\right) = \dfrac{32\pi}{3}(8-1) = \dfrac{224\pi}{3}$

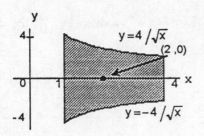

(b) Since the plate is symmetric about the x-axis and its density $\delta(x) = \dfrac{1}{x}$ is a function of x alone, the

distribution of its mass is symmetric about the x-axis. This means that $\bar{y} = 0$. We use the vertical strip

approach to find $\bar{x}$: $M_y = \int \tilde{x}\, dm = \int\limits_1^4 x \cdot \left[\dfrac{4}{\sqrt{x}} - \left(-\dfrac{4}{\sqrt{x}}\right)\right] \cdot \delta\, dx = \int\limits_1^4 x \cdot \dfrac{8}{\sqrt{x}} \cdot \dfrac{1}{x}\, dx = 8\int\limits_1^4 x^{-1/2}\, dx$

$= 8\left[2x^{1/2}\right]_1^4 = 8(2\cdot 2 - 2) = 16;\ M = \int dm = \int\limits_1^4 \left[\dfrac{4}{\sqrt{x}} - \left(\dfrac{-4}{\sqrt{x}}\right)\right] \cdot \delta\, dx = 8\int\limits_1^4 \left(\dfrac{1}{\sqrt{x}}\right)\left(\dfrac{1}{x}\right) dx = 8\int\limits_1^4 x^{-3/2}\, dx$

$= 8\left[-2x^{-1/2}\right]_1^4 = 8[-1 - (-2)] = 8.$ So $\bar{x} = \dfrac{M_y}{M} = \dfrac{16}{8} = 2 \Rightarrow (\bar{x}, \bar{y}) = (2, 0)$ is the center of mass.

29. The mass of a horizontal strip is $dm = \delta\, dA = \delta L$, where L is the width of the triangle at a distance of y above

its base on the x-axis as shown in the figure in the text. Also, by similar triangles we have $\dfrac{L}{b} = \dfrac{h - y}{h}$

$\Rightarrow L = \dfrac{b}{h}(h - y).$ Thus, $M_x = \int \tilde{y}\, dm = \int\limits_0^h \delta y\left(\dfrac{b}{h}\right)(h - y)\, dy = \dfrac{\delta b}{h}\int\limits_0^h \left(hy - y^2\right) dy = \dfrac{\delta b}{h}\left[\dfrac{hy^2}{2} - \dfrac{y^3}{3}\right]_0^h$

$$= \frac{\delta b}{h}\left(\frac{h^3}{2} - \frac{h^3}{3}\right) = \delta b h^2\left(\frac{1}{2} - \frac{1}{3}\right) = \frac{\delta b h^2}{6}; \quad M = \int dm = \int_0^h \delta\left(\frac{b}{h}\right)(h-y)\,dy = \frac{\delta b}{h}\int_0^h (h-y)\,dy = \frac{\delta b}{h}\left[hy - \frac{y^2}{2}\right]_0^h$$

$$= \frac{\delta b}{h}\left(h^2 - \frac{h^2}{2}\right) = \frac{\delta b h}{2}. \text{ So } \overline{y} = \frac{M_x}{M} = \left(\frac{\delta b h^2}{6}\right)\left(\frac{2}{\delta b h}\right) = \frac{h}{3} \Rightarrow \text{ the center of mass lies above the base of the}$$

triangle one-third of the way toward the opposite vertex. Similarly the other two sides of the triangle can be placed on the x-axis and the same results will occur. Therefore the centroid does lie at the intersection of the medians, as claimed.

31. From the symmetry about the line $x = y$ it follows that

$\overline{x} = \overline{y}$. It also follows that the line through the points $(0,0)$

and $\left(\frac{1}{2}, \frac{1}{2}\right)$ is a median $\Rightarrow \overline{y} = \overline{x} = \frac{2}{3}\cdot\left(\frac{1}{2} - 0\right) = \frac{1}{3}$

$\Rightarrow (\overline{x}, \overline{y}) = \left(\frac{1}{3}, \frac{1}{3}\right).$

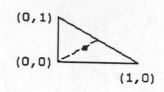

33. The point of intersection of the median from the vertex $(0, b)$

to the opposite side has coordinates $\left(0, \frac{a}{2}\right) \Rightarrow \overline{y} = (b - 0)\cdot\frac{1}{3}$

$= \frac{b}{3}$ and $\overline{x} = \left(\frac{a}{2} - 0\right)\cdot\frac{2}{3} = \frac{a}{3} \Rightarrow (\overline{x}, \overline{y}) = \left(\frac{a}{3}, \frac{b}{3}\right).$

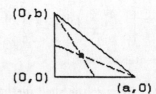

35. $y = x^{1/2} \Rightarrow dy = \frac{1}{2}x^{-1/2}\,dx \Rightarrow ds = \sqrt{(dx)^2 + (dy)^2} = \sqrt{1 + \frac{1}{4x}}\,dx; \quad M_x = \delta\int_0^2 \sqrt{x}\sqrt{1 + \frac{1}{4x}}\,dx$

$$= \delta\int_0^2 \sqrt{x + \frac{1}{4}}\,dx = \frac{2\delta}{3}\left[\left(x + \frac{1}{4}\right)^{3/2}\right]_0^2 = \frac{2\delta}{3}\left[\left(2 + \frac{1}{4}\right)^{3/2} - \left(\frac{1}{4}\right)^{3/2}\right] = \frac{2\delta}{3}\left[\left(\frac{9}{4}\right)^{3/2} - \left(\frac{1}{4}\right)^{3/2}\right] = \frac{2\delta}{3}\left(\frac{27}{8} - \frac{1}{8}\right) = \frac{13\delta}{6}$$

37. From Example 6 we have $M_x = \int_0^\pi a(a\sin\theta)(k\sin\theta)\,d\theta = a^2 k\int_0^\pi \sin^2\theta\,d\theta = \frac{a^2 k}{2}\int_0^\pi (1 - \cos 2\theta)\,d\theta$

$$= \frac{a^2 k}{2}\left[\theta - \frac{\sin 2\theta}{2}\right]_0^\pi = \frac{a^2 k\pi}{2}; \quad M_y = \int_0^\pi a(a\cos\theta)(k\sin\theta)\,d\theta = a^2 k\int_0^\pi \sin\theta\cos\theta\,d\theta = \frac{a^2 k}{2}\left[\sin^2\theta\right]_0^\pi = 0;$$

$$M = \int_0^\pi ak\sin\theta\,d\theta = ak[-\cos\theta]_0^\pi = 2ak. \text{ Therefore, } \overline{x} = \frac{M_y}{M} = 0 \text{ and } \overline{y} = \frac{M_x}{M} = \left(\frac{a^2 k\pi}{2}\right)\left(\frac{1}{2ak}\right) = \frac{a\pi}{4} \Rightarrow \left(0, \frac{a\pi}{4}\right)$$

is the center of mass.

39. Consider the curve as an infinite number of line segments joined together. From the derivation of arc

length we have that the length of a particular segment is $ds = \sqrt{(dx)^2 + (dy)^2}$. This implies that

$$M_x = \int \delta y\,ds, \quad M_y = \int \delta x\,ds \text{ and } M = \int \delta\,ds. \text{ If } \delta \text{ is constant, then } \overline{x} = \frac{M_y}{M} = \frac{\int x\,ds}{\int ds} = \frac{\int x\,ds}{\text{length}} \text{ and}$$

$$\overline{y} = \frac{M_x}{M} = \frac{\int y\,ds}{\int ds} = \frac{\int y\,ds}{\text{length}}.$$

41. A generalization of Example 6 yields $M_x = \int \tilde{y} \, dm = \displaystyle\int_{\pi/2 - \alpha}^{\pi/2 + \alpha} a^2 \sin\theta \, d\theta = a^2[-\cos\theta]_{\pi/2-\alpha}^{\pi/2+\alpha}$

$= a^2\left[-\cos\left(\dfrac{\pi}{2}+\alpha\right)+\cos\left(\dfrac{\pi}{2}-\alpha\right)\right] = a^2(\sin\alpha + \sin\alpha) = 2a^2 \sin\alpha; \ M = \int dm = \displaystyle\int_{\pi/2-\alpha}^{\pi/2+\alpha} a \, d\theta = a[\theta]_{\pi/2-\alpha}^{\pi/2+\alpha}$

$= a\left[\left(\dfrac{\pi}{2}+\alpha\right)-\left(\dfrac{\pi}{2}-\alpha\right)\right] = 2a\alpha$. Thus, $\overline{y} = \dfrac{M_x}{M} = \dfrac{2a^2 \sin\alpha}{2a\alpha} = \dfrac{a \sin\alpha}{\alpha}$. Now $s = a(2\alpha)$ and $a \sin\alpha = \dfrac{c}{2}$

$\Rightarrow c = 2a \sin\alpha$. Then $\overline{y} = \dfrac{a(2a \sin\alpha)}{2a\alpha} = \dfrac{ac}{s}$, as claimed.

5.8 WORK

1. The force required to lift the water is equal to the water's weight, which varies steadily from 40 lb to 0 lb over

   the 20-ft lift. When the bucket is x ft off the ground, the water weighs: $F(x) = 40\left(\dfrac{20-x}{20}\right) = 40\left(1-\dfrac{x}{20}\right)$

   $= 40 - 2x$ lb. The work done is: $W = \displaystyle\int_a^b F(x) \, dx = \int_0^{20} (40-2x) \, dx = \left[40x - x^2\right]_0^{20} = (40)(20) - 20^2$

   $= 800 - 400 = 400$ ft $\cdot$ lb

3. The force required to haul up the rope is equal to the rope's weight, which varies steadily and is proportional to

   x, the length of the rope still hanging: $F(x) = 0.624x$. The work done is: $W = \displaystyle\int_0^{50} F(x) \, dx = \int_0^{50} 0.624x \, dx$

   $= 0.624\left[\dfrac{x^2}{2}\right]_0^{50} = 780$ J

5. The force required to lift the cable is equal to the weight of the cable paid out: $F(x) = (4.5)(180 - x)$ where x

   is the position of the car off the first floor. The work done is: $W = \displaystyle\int_0^{180} F(x) \, dx = 4.5 \int_0^{180} (180-x) \, dx$

   $= 4.5\left[180x - \dfrac{x^2}{2}\right]_0^{180} = 4.5\left(180^2 - \dfrac{180^2}{2}\right) = \dfrac{4.5 \cdot 180^2}{2} = 72{,}900$ ft $\cdot$ lb

7. The force against the piston is $F = pA$. If $V = Ax$, where x is the height of the cylinder, then $dV = A \, dx$

   $\Rightarrow \text{Work} = \displaystyle\int F \, dx = \int pA \, dx = \int_{(p_1, V_1)}^{(p_2, V_2)} p \, dV.$

9. The force required to stretch the spring from its natural length of 2 m to a length of 5 m is $F(x) = kx$. The

   work done by F is $W = \displaystyle\int_0^3 F(x) \, dx = k \int_0^3 x \, dx = \dfrac{k}{2}[x^2]_0^3 = \dfrac{9k}{2}$. This work is equal to 1800 J $\Rightarrow \dfrac{9}{2}k = 1800$

   $\Rightarrow k = 400$ N/m

11. We find the force constant from Hooke's law: $F = kx$. A force of 2 N stretches the spring to 0.02 m

$\Rightarrow 2 = k \cdot (0.02) \Rightarrow k = 100\,\frac{N}{m}$. The force of 4 N will stretch the rubber band y m, where $F = ky \Rightarrow y = \frac{F}{k}$

$\Rightarrow y = \dfrac{4N}{100\frac{N}{m}} \Rightarrow y = 0.04$ m $= 4$ cm. The work done to stretch the rubber band 0.04 m is $W = \displaystyle\int_{0}^{0.04} kx\,dx$

$= 100 \displaystyle\int_{0}^{0.04} x\,dx = 100\left[\frac{x^2}{2}\right]_{0}^{0.04} = \frac{(100)(0.04)^2}{2} = 0.08$ J

13. (a) We find the spring's constant from Hooke's law: $F = kx \Rightarrow k = \frac{F}{x} = \frac{21,714}{8-5} = \frac{21,714}{3} \Rightarrow k = 7238\,\frac{lb}{in}$

   (b) The work done to compress the assembly the first half inch is $W = \displaystyle\int_{0}^{0.5} kx\,dx = 7238 \displaystyle\int_{0}^{0.5} x\,dx$

   $= 7238\left[\frac{x^2}{2}\right]_{0}^{0.5} = (7238)\frac{(0.5)^2}{2} = \frac{(7238)(0.25)}{2} \approx 905$ in·lb. The work done to compress the assembly the

   second half inch is: $W = \displaystyle\int_{0.5}^{1.0} kx\,dx = 7238 \displaystyle\int_{0.5}^{1.0} x\,dx = 7238\left[\frac{x^2}{2}\right]_{0.5}^{1.0} = \frac{7238}{2}\left[1 - (0.5)^2\right] = \frac{(7238)(0.75)}{2}$

   ≈ 2714 in·lb

15. We will use the coordinate system given.

   (a) The typical slab between the planes at y and $y + \Delta y$ has a volume of $\Delta V = (10)(12)\,\Delta y = 120\,\Delta y$ ft^3. The force F required to lift the slab is equal to its weight: $F = 62.4\,\Delta V = 62.4 \cdot 120\,\Delta y$ lb. The distance through which F must act is about y ft, so the work done lifting the slab is about $\Delta W = $ force × distance $= 62.4 \cdot 120 \cdot y \cdot \Delta y$ ft·lb. The work it takes to lift all

   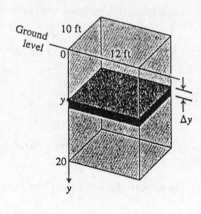

   the water is approximately $W \approx \displaystyle\sum_{0}^{20} \Delta W$

   $= \displaystyle\sum_{0}^{20} 62.4 \cdot 120y \cdot \Delta y$ ft·lb. This is a Riemann sum for

   the function $62.4 \cdot 120y$ over the interval $0 \le y \le 20$.

   The work of pumping the tank empty is the limit of these sums: $W = \displaystyle\int_{0}^{20} 62.4 \cdot 120y\,dy$

   $= (62.4)(120)\left[\frac{y^2}{2}\right]_{0}^{20} = (62.4)(120)\left(\frac{400}{2}\right) = (62.4)(120)(200) = 1{,}497{,}600$ ft·lb

   (b) The time t it takes to empty the full tank with $\left(\frac{5}{11}\right)$-hp motor is $t = \dfrac{W}{250\,\frac{ft \cdot lb}{sec}} = \dfrac{1{,}497{,}600\ ft \cdot lb}{250\,\frac{ft \cdot lb}{sec}}$

   $= 5990.4$ sec $= 1.664$ hr $\Rightarrow t \approx 1$ hr and 40 min

   (c) Following all the steps of part (a), we find that the work it takes to lower the water level 10 ft is

   $W = \displaystyle\int_{0}^{10} 62.4 \cdot 120y\,dy = (62.4)(120)\left[\frac{y^2}{2}\right]_{0}^{10} = (62.4)(120)\left(\frac{100}{2}\right) = 374{,}400$ ft·lb and the time is

$$t \approx \frac{W}{250 \; \frac{ft \cdot lb}{sec}} = 1497.6 \; \text{sec} = 0.416 \; \text{hr} \approx 25 \; \text{min}$$

(d) In a location where water weighs $62.26 \; \frac{lb}{ft^3}$:

    a) $W = (62.26)(24,000) = 1,494,240 \; ft \cdot lb.$

    b) $t = \dfrac{1,494,240}{250} = 5976.96 \; \text{sec} \approx 1.660 \; \text{hr} \Rightarrow t \approx 1 \; \text{hr and } 40 \; \text{min}$

  In a location where water weighs $62.59 \; \frac{lb}{ft^3}$

    a) $W = (62.59)(24,000) = 1,502,160 \; ft \cdot lb$

    b) $t = \dfrac{1,502,160}{250} = 6008.64 \; \text{sec} \approx 1.669 \; \text{hr} \Rightarrow t \approx 1 \; \text{hr and } 40.1 \; \text{min}$

17. Using exactly the same procedure as done in Example 6 we change only the distance through which F must act:

distance $\approx (10 - y)$ m. Then $\Delta W = 245,000\pi(10 - y)\,\Delta y$ J $\Rightarrow W \approx \sum\limits_{0}^{10} \Delta W = \sum\limits_{0}^{10} 245,000\pi(10 - y)\,\Delta y$

$$\Rightarrow W = \int\limits_{0}^{10} 245,000\pi(10 - y)\,dy = 245,000\pi \int\limits_{0}^{10} (10 - y)\,dy = 245,000\pi\left[10y - \frac{y^2}{2}\right]_0^{10} = 245,000\pi\left(100 - \frac{100}{2}\right)$$

$$\approx (245,000\pi)(50) \approx 38,484,510 \; \text{J}$$

19. The typical slab between the planes at y and and $y + \Delta y$ has a volume of $\Delta V = \pi(\text{radius})^2(\text{thickness})$

$= \pi\left(\dfrac{20}{2}\right)^2 \Delta y = \pi \cdot 100 \; \Delta y \; ft^3$. The force F required to lift the slab is equal to its weight:

$F = 51.2\,\Delta V = 51.2 \cdot 100\pi\,\Delta y \; lb \Rightarrow F = 5120\pi\,\Delta y \; lb$. The distance through which F must act is about

$(30 - y)$ ft. The work it takes to lift all the kerosene is approximately $W \approx \sum\limits_{0}^{30} \Delta W$

$= \sum\limits_{0}^{30} 5120\pi(30 - y)\,\Delta y \; ft \cdot lb$ which is a Riemann sum. The work to pump the tank dry is the limit of

these sums: $W = \int\limits_{0}^{30} 5120\pi(30 - y)\,dy = 5120\pi\left[30y - \frac{y^2}{2}\right]_0^{30} = 5120\pi\left(\frac{900}{2}\right) = (5120)(450\pi)$

$$\approx 7,238,229.48 \; ft \cdot lb$$

21. We imagine the milkshake divided into thin slabs by planes perpendicular to the y-axis at the points of a partition of the interval $[0,7]$. The typical slab between the planes at y and $y + \Delta y$ has a volume of about

$\Delta V = \pi(\text{radius})^2(\text{thickness}) = \pi\left(\dfrac{y + 17.5}{14}\right)^2 \Delta y \; in^3$. The force $F(y)$ required to lift this slab is equal to its

weight: $F(y) = \dfrac{4}{9}\,\Delta V = \dfrac{4\pi}{9}\left(\dfrac{y + 17.5}{14}\right)^2 \Delta y \; oz$. The distance through which $F(y)$ must act to lift this slab to

the level of 1 inch above the top is about $(8 - y)$ in. The work done lifting the slab is about

$\Delta W = \left(\dfrac{4\pi}{9}\right)\dfrac{(y + 17.5)^2}{14^2}(8 - y)\,\Delta y \; in \cdot oz$. The work done lifting all the slabs from $y = 0$ to $y = 7$ is

approximately $W = \sum\limits_{0}^{7} \dfrac{4\pi}{9 \cdot 14^2}(y + 17.5)^2(8 - y)\,\Delta y \; in \cdot oz$ which is a Riemann sum. The work is the limit of

these sums as the norm of the partition goes to zero: $W = \int\limits_{0}^{7} \dfrac{4\pi}{9 \cdot 14^2}(y + 17.5)^2(8 - y)\,dy$

$$= \frac{4\pi}{9 \cdot 14^2} \int_0^7 \left(2450 - 26.25y - 27y^2 - y^3\right) dy = \frac{4\pi}{9 \cdot 14^2}\left[-\frac{y^4}{4} - 9y^3 - \frac{26.25}{2}y^2 + 2450y\right]_0^7$$

$$= \frac{4\pi}{9 \cdot 14^2}\left[-\frac{7^4}{4} - 9 \cdot 7^3 - \frac{26.25}{2} \cdot 7^2 + 2450 \cdot 7\right] \approx 91.32 \text{ in} \cdot \text{oz}$$

23. The typical slab between the planes of y and $y + \Delta y$ has a volume of about $\Delta V = \pi(\text{radius})^2(\text{thickness})$
$= \pi\left(\sqrt{y}\right)^2 \Delta y = xy \, \Delta y \text{ m}^3$. The force $F(y)$ is equal to the slab's weight: $F(y) = 10{,}000 \, \frac{N}{m^3} \cdot \Delta V$
$= \pi 10{,}000y \, \Delta y$ N. The height of the tank is $4^2 = 16$ m. The distance through which $F(y)$ must act to lift
the slab to the level of the top of the tank is about $(16 - y)$ m, so the work done lifting the slab is about
$\Delta W = 10{,}000\pi y(16 - y)\Delta y \text{ N} \cdot \text{m}$. The work done lifting all the slabs from $y = 0$ to $y = 16$ to the top is
approximately $W \approx \sum_0^{16} 10{,}000\pi y(16 - y)\Delta y$. Taking the limit of these Riemann sums, we get

$$W = \int_0^{16} 10{,}000\pi y(16 - y) \, dy = 10{,}000\pi \int_0^{16}\left(16y - y^2\right) dy = 10{,}000\pi\left[\frac{16y^2}{2} - \frac{y^3}{3}\right]_0^{16} = 10{,}000\pi\left(\frac{16^3}{2} - \frac{16^3}{3}\right)$$

$$= \frac{10{,}000 \cdot \pi \cdot 16^3}{6} \approx 21{,}446{,}605.9 \text{ J}$$

25. The typical slab between the planes at y and $y + \Delta y$ has a volume of about $\Delta V = \pi(\text{radius})^2(\text{thickness})$
$= \pi\left(\sqrt{100 - y^2}\right)^2 \Delta y = \pi\left(100 - y^2\right)\Delta y \text{ ft}^3$. The force is $F(y) = \frac{56 \text{ lb}}{\text{ft}^3} \cdot \Delta V = 56\pi\left(100 - y^2\right)\Delta y$ lb. The
distance through which $F(y)$ must act to lift the slab to the level of 2 ft above the top of the tank is about
$(12 - y)$ ft, so the work done is $\Delta W \approx 56\pi\left(100 - y^2\right)(12 - y)\Delta y \text{ lb} \cdot \text{ft}$. The work done lifting all the slabs
from $y = 0$ ft to $y = 10$ ft is approximately $W \approx \sum_0^{10} 56\pi\left(100 - y^2\right)(12 - y)\Delta y \text{ lb} \cdot \text{ft}$. Taking the limit of these

Riemann sums, we get $W = \int_0^{10} 56\pi\left(100 - y^2\right)(12 - y) \, dy = 56\pi \int_0^{10}\left(100 - y^2\right)(12 - y) \, dy$

$$= 56\pi \int_0^{10}\left(1200 - 100y - 12y^2 + y^3\right) dy = 56\pi\left[1200y - \frac{100y^2}{2} - \frac{12y^3}{3} + \frac{y^4}{4}\right]_0^{10}$$

$$= 56\pi\left(12{,}000 - \frac{10{,}000}{2} - 4 \cdot 1000 + \frac{10{,}000}{4}\right) = (56\pi)\left(12 - 5 - 4 + \frac{5}{2}\right)(1000) \approx 967{,}611 \text{ ft} \cdot \text{lb}.$$

It would cost $(0.5)(967.611) = 483{,}805$¢ $= \$4838.05$. Yes, we can afford to hire the firm.

27. Work $= \displaystyle\int_{6{,}370{,}000}^{35{,}780{,}000} \frac{1000\,\text{MG}}{r^2} \, dr = 1000\,\text{MG} \int_{6{,}370{,}000}^{35{,}780{,}000} \frac{dr}{r^2} = 1000\,\text{MG}\left[-\frac{1}{r}\right]_{6{,}370{,}000}^{35{,}780{,}000}$

$$= (1000)\left(5.975 \cdot 10^{24}\right)\left(6.672 \cdot 10^{-11}\right)\left(\frac{1}{6{,}370{,}000} - \frac{1}{35{,}780{,}000}\right) \approx 5.144 \times 10^{10} \text{ J}$$

29. $F = m\frac{dv}{dt} = mv\frac{dv}{dx}$ by the chain rule $\Rightarrow W = \int_{x_1}^{x_2} mv\frac{dv}{dx}\,dx = m\int_{x_1}^{x^2}\left(v\frac{dv}{dx}\right)dx = m\left[\frac{1}{2}v^2(x)\right]_{x_1}^{x_2}$

$= \frac{1}{2}m\left[v^2(x_2) - v^2(x_1)\right] = \frac{1}{2}mv_2^2 - \frac{1}{2}mv_1^2$, as claimed.

31. $90\text{ mph} = \frac{90\text{ mi}}{1\text{ hr}}\cdot\frac{1\text{ hr}}{60\text{ min}}\cdot\frac{1\text{ min}}{60\text{ sec}}\cdot\frac{5280\text{ ft}}{1\text{ mi}} = 132\text{ ft/sec}$; $m = \frac{0.3125\text{ lb}}{32\text{ ft/sec}^2} = \frac{0.3125}{32}$ slugs;

$W = \left(\frac{1}{2}\right)\left(\frac{0.3125\text{ lb}}{32\text{ ft/sec}^2}\right)(132\text{ ft/sec})^2 \approx 85.1\text{ ft}\cdot\text{lb}$

33. weight $= 2\text{ oz} = \frac{1}{8}\text{ lb} \Rightarrow m = \frac{\frac{1}{8}}{32}$ slugs $= \frac{1}{256}$ slugs; $124\text{ mph} = \frac{(124)(5280)}{(60)(60)} \approx 181.87\text{ ft/sec}$;

$W = \left(\frac{1}{2}\right)\left(\frac{1}{256}\text{ slugs}\right)(181.87\text{ ft/sec})^2 \approx 64.6\text{ ft}\cdot\text{lb}$

35. weight $= 6.5\text{ oz} = \frac{6.5}{16}\text{ lb} \Rightarrow m = \frac{6.5}{(16)(32)}$ slugs; $W = \left(\frac{1}{2}\right)\left(\frac{6.5}{(16)(32)}\text{ slugs}\right)(132\text{ ft/sec})^2 \approx 110.6\text{ ft}\cdot\text{lb}$

5.9 FLUID PRESSURES AND FORCES

1. Using the coordinate system of Example 4, we find for a full tank that the typical horizontal strip at level y

   has strip depth: $(90 - y)$ and strip length: $\pi \times$ tank diameter $= 90\pi$. The total fluid force against the

   cylindrical inside wall of a full tank is $F_{\text{Full}} = \int_0^{90} w(\text{depth})(\text{length})\,dy = \int_0^{90} 100(90 - y)(90\pi)\,dy$

   $= 9000\pi\int_0^{90}(90 - y)\,dy = 9000\pi\left[90y - \frac{y^2}{2}\right]_0^{90} = 9000\pi\left(90\cdot90 - \frac{90^2}{2}\right) = \frac{9000\pi\cdot8100}{2} \approx 114{,}511{,}052\text{ lb}$. When

   the tank is half full we have: strip depth: $(45 - y) \Rightarrow F_{\text{Hf}} = \int_0^{45} w(\text{depth})(\text{length})\,dy = \int_0^{45} 100(45 - y)(90\pi)\,dy$

   $= 9000\pi\left[45y - \frac{y^2}{2}\right]_0^{45} = 9000\pi\left(45\cdot45 - \frac{45^2}{2}\right) = \frac{9000\pi\cdot2025}{2} \approx 28{,}627{,}763\text{ lb}$

3. To find the width of the plate at a typical depth y, we first find an equation for the line of the plate's
   right-hand edge: $y = x - 5$. If we let x denote the width of the right-hand half of the triangle at depth y, then
   $x = 5 + y$ and the total width is $L(y) = 2x = 2(5 + y)$. The depth of the strip is $(-y)$. The force exerted by the

   water against one side of the plate is therefore $F = \int_{-5}^{-2} w(-y)\cdot L(y)\,dy = \int_{-5}^{-2} 62.4\cdot(-y)\cdot2(5 + y)\,dy$

   $= 124.8\int_{-5}^{-2}(-5y - y^2)\,dy = 124.8\left[-\frac{5}{2}y^2 - \frac{1}{3}y^3\right]_{-5}^{-2} = 124.8\left[\left(-\frac{5}{2}\cdot4 + \frac{1}{3}\cdot8\right) - \left(-\frac{5}{2}\cdot25 + \frac{1}{3}\cdot125\right)\right]$

   $= (124.8)\left(\frac{105}{2} - \frac{117}{3}\right) = (124.8)\left(\frac{315 - 234}{6}\right) = 1684.8\text{ lb}$

5. Using the coordinate system of Exercise 4, we find the equation for the line of the plate's right-hand edge is $y = x - 3 \Rightarrow x = y + 3$. Thus the total width is $L(y) = 2x = 2(y + 3)$. The depth of the strip changes to $(4 - y)$

$$\Rightarrow F = \int_{-3}^{0} w(4 - y)L(y)\, dy = \int_{-3}^{0} 62.4 \cdot (4 - y) \cdot 2(y + 3)\, dy = 124.8 \int_{-3}^{0} \left(12 + y - y^2\right) dy$$

$$= 124.8 \left[12y + \frac{y^2}{2} - \frac{y^3}{3}\right]_{-3}^{0} = (-124.8)\left(-36 + \frac{9}{2} + 9\right) = (-124.8)\left(-\frac{45}{2}\right) = 2808 \text{ lb}$$

7. Using the coordinate system of Exercise 4, we find the equation for the line of the plate's right-hand edge to be $y = 2x - 4 \Rightarrow x = \frac{y + 4}{2}$ and $L(y) = 2x = y + 4$. The depth of the strip is $(1 - y)$.

(a) $F = \int_{-4}^{0} w(1 - y)L(y)\, dy = \int_{-4}^{0} 62.4 \cdot (1 - y)(y + 4)\, dy = 62.4 \int_{-4}^{0} \left(4 - 3y - y^2\right) dy = 62.4 \left[4y - \frac{3y^2}{2} - \frac{y^3}{3}\right]_{-4}^{0}$

$$= (-62.4)\left[(-4)(4) - \frac{(3)(16)}{2} + \frac{64}{3}\right] = (-62.4)\left(-16 - 24 + \frac{64}{3}\right) = \frac{(-62.4)(-120 + 64)}{3} = 1164.8 \text{ lb}$$

(b) $F = (-64.0)\left[(-4)(4) - \frac{(3)(16)}{2} + \frac{64}{3}\right] = \frac{(-64.0)(-120 + 64)}{3} \approx 1194.7 \text{ lb}$

9. (a) An equation of the right-hand edge is $y = \frac{3}{2}x \Rightarrow x = \frac{2}{3}y$ and $L(y) = 2x = \frac{4y}{3}$. The depth of the strip

is $(3 - y) \Rightarrow F = \int_{0}^{3} w(3 - y)L(y)\, dy = \int_{0}^{3} (62.4)(3 - y)\left(\frac{4}{3}y\right) dy = (62.4) \cdot \left(\frac{4}{3}\right) \int_{0}^{3} \left(3y - y^2\right) dy$

$$= (62.4)\left(\frac{4}{3}\right)\left[\frac{3}{2}y^2 - \frac{y^3}{3}\right]_{0}^{3} = (62.4)\left(\frac{4}{3}\right)\left[\frac{27}{2} - \frac{27}{3}\right] = (62.4)\left(\frac{4}{3}\right)\left(\frac{27}{6}\right) = 374.4 \text{ lb}$$

(b) We want to find a new water level Y such that $F_Y = \frac{1}{2}(374.4) = 187.2$ lb. The new depth of the strip is $(Y - y)$, and Y is the new upper limit of integration. Thus, $F_Y = \int_{0}^{Y} w(Y - y)L(y)\, dy$

$$= 62.4 \int_{0}^{Y} (Y - y)\left(\frac{4}{3}y\right) dy = (62.4)\left(\frac{4}{3}\right) \int_{0}^{Y} \left(Yy - y^2\right) dy = (62.4)\left(\frac{4}{3}\right)\left[Y \cdot \frac{y^2}{2} - \frac{y^3}{3}\right]_{0}^{Y} = (62.4)\left(\frac{4}{3}\right)\left(\frac{Y^3}{2} - \frac{Y^3}{3}\right)$$

$$= (62.4)\left(\frac{2}{9}\right)Y^3. \text{ Therefore } Y^3 = \frac{9F_Y}{2 \cdot (62.4)} = \frac{(9)(187.2)}{124.8} \Rightarrow Y = \sqrt[3]{\frac{(9)(187.2)}{124.8}} = \sqrt[3]{13.5} \approx 2.3811 \text{ ft. So,}$$

$\Delta Y = 3 - Y \approx 3 - 2.3811 \approx 0.6189 \text{ ft} \approx 7.5 \text{ in to the nearest half inch}$

(c) No, it does not matter how long the trough is.

11. Using the coordinate system given in the accompanying figure, we see that the total width is $L(y) = 63$ and the depth of the

strip is $(33.5 - y) \Rightarrow F = \int_{0}^{33} w(33.5 - y)L(y)\, dy$

$$= \int_{0}^{33} \frac{64}{12^3} \cdot (33.5 - y) \cdot 63\, dy = \left(\frac{64}{12^3}\right)(63) \int_{0}^{33} (33.5 - y)\, dy$$

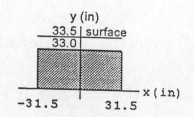

$$= \left(\frac{64}{12^3}\right)(63)\left[33.5y - \frac{y^2}{2}\right]_0^{33} = \left(\frac{64 \cdot 63}{12^3}\right)\left[(33.5)(33) - \frac{33^2}{2}\right] = \frac{(64)(63)(33)(67 - 33)}{(2)(12^3)} = 1309 \text{ lb}$$

13. Use the same coordinate system as in Exercise 10 with $L(y) = 3.75$ and the depth of a typical strip being

$(7.75 - y)$. Then $F = \displaystyle\int_0^{7.75} w(7.75 - y)L(y)\, dy = \left(\frac{64.5}{12^3}\right)(3.75) \int_0^{7.75} (7.75 - y)\, dy = \left(\frac{64.5}{12^3}\right)(3.75)\left[7.75y - \frac{y^2}{2}\right]_0^{7.75}$

$$= \left(\frac{64.5}{12^3}\right)(3.75)\frac{(7.75)^2}{2} \approx 4.2 \text{ lb}$$

15. Using the coordinate system given in the accompanying
figure, we see that the right-hand edge is $x = \sqrt{1 - y^2}$
so the total width is $L(y) = 2x = 2\sqrt{1 - y^2}$ and the depth
of the strip is $(-y)$. The force exerted by the water is

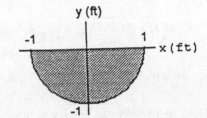

therefore $F = \displaystyle\int_{-1}^0 w \cdot (-y) \cdot 2\sqrt{1 - y^2}\, dy$

$$= 62.4 \int_{-1}^0 \sqrt{1 - y^2}\, d(1 - y^2) = 62.4\left[\frac{2}{3}(1 - y^2)^{3/2}\right]_{-1}^0 = (62.4)\left(\frac{2}{3}\right)(1 - 0) = 416 \text{ lb}$$

17. The coordinate system is given in the text. The right-hand edge is $x = \sqrt{y}$ and the total width is $L(y) = 2x$
$= 2\sqrt{y}$.

(a) The depth of the strip is $(2 - y)$ so the force exerted by the liquid on the gate is $F = \displaystyle\int_0^1 w(2 - y)L(y)\, dy$

$$= \int_0^1 50(2 - y) \cdot 2\sqrt{y}\, dy = 100 \int_0^1 (2 - y)\sqrt{y}\, dy = 100 \int_0^1 \left(2y^{1/2} - y^{3/2}\right) dy = 100\left[\frac{4}{3}y^{3/2} - \frac{2}{5}y^{5/2}\right]_0^1$$

$$= 100\left(\frac{4}{3} - \frac{2}{5}\right) = \left(\frac{100}{15}\right)(20 - 6) = 93.33 \text{ lb}$$

(b) Suppose that H is the maximum height to which the container can be filled without exceeding its design

limitation. The depth of a typical strip is $(H - y)$ and the force is $F = \displaystyle\int_0^1 w(H - y)L(y)\, dy = F_{max}$, where

$$F_{max} = 160 \text{ lb. Therefore, } F_{max} = w \int_0^1 (H - y) \cdot 2\sqrt{y}\, dy = 100 \int_0^1 (H - y)\sqrt{y}\, dy$$

$$= 100 \int_0^1 \left(Hy^{1/2} - y^{3/2}\right) dy = 100\left[\frac{2}{3}Hy^{3/2} - \frac{2}{5}y^{5/2}\right]_0^1 = 100\left(\frac{2H}{3} - \frac{2}{5}\right) = \left(\frac{100}{15}\right)(10H - 6). \text{ When}$$

$$F_{max} = 160 \text{ lb we have } 160 = \left(\frac{100}{15}\right)(10H - 6) \Rightarrow 10H - 6 = 24 \Rightarrow H = 3 \text{ ft}$$

19. Suppose that h is the maximum height. Using the coordinate system given in the text, we find an equation for the line of the end plate's right-hand edge is $y = \frac{5}{2}x \Rightarrow x = \frac{2}{5}y$. The total width is $L(y) = 2x = \frac{4}{5}y$ and the

depth of the typical horizontal strip at level y is $(h - y)$. Then the force is $F = \int_0^h w(h-y)L(y)\, dy = F_{max}$,

where $F_{max} = 6667$ lb. Hence, $F_{max} = w \int_0^h (h-y) \cdot \frac{4}{5}y\, dy = (62.4)\left(\frac{4}{5}\right)\int_0^h (hy - y^2)\, dy$

$= (62.4)\left(\frac{4}{5}\right)\left[\frac{hy^2}{2} - \frac{y^3}{3}\right]_0^h = (62.4)\left(\frac{4}{5}\right)\left(\frac{h^3}{2} - \frac{h^3}{3}\right) = (62.4)\left(\frac{4}{5}\right)\left(\frac{1}{6}\right)h^3 = (10.4)\left(\frac{4}{5}\right)h^3 \Rightarrow h = \sqrt[3]{\left(\frac{5}{4}\right)\left(\frac{F_{max}}{10.4}\right)}$

$= \sqrt[3]{\left(\frac{5}{4}\right)\left(\frac{6667}{10.4}\right)} \approx 9.288$ ft. The volume of water which the tank can hold is $V = \frac{1}{2}(\text{Base})(\text{Height}) \cdot 30$, where

Height $= h$ and $\frac{1}{2}(\text{Base}) = \frac{2}{5}h \Rightarrow V = \left(\frac{2}{5}h^2\right)(30) = 12h^2 \approx 12(9.288)^2 \approx 1035$ ft^3.

21. The pressure at level y is $p(y) = w \cdot y \Rightarrow$ the average

pressure is $\overline{p} = \frac{1}{b}\int_0^b p(y)\, dy = \frac{1}{b}\int_0^b w \cdot y\, dy = \frac{1}{b}w\left[\frac{y^2}{2}\right]_0^b$

$= \left(\frac{w}{b}\right)\left(\frac{b^2}{2}\right) = \frac{wb}{2}$. This is the pressure at level $\frac{b}{2}$, which

is the pressure at the middle of the plate.

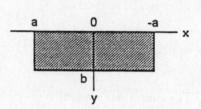

23. When the water reaches the top of the tank the force on the movable side is $\int_{-2}^{0} (62.4)\left(2\sqrt{4 - y^2}\right)(-y)\, dy$

$= (62.4)\int_{-2}^{0} \left(4 - y^2\right)^{1/2}(-2y)\, dy = (62.4)\left[\frac{2}{3}\left(4 - y^2\right)^{3/2}\right]_{-2}^{0} = (62.4)\left(\frac{2}{3}\right)\left(4^{3/2}\right) = 332.8$ ft $\cdot$ lb. The force

compressing the spring is $F = 100x$, so when the tank is full we have $332.8 = 100x \Rightarrow x \approx 3.33$ ft. Therefore the movable end does not reach the required 5 ft to allow drainage $\Rightarrow$ the tank will overflow.

5.10 THE BASIC PATTERN AND OTHER MODELING APPLICATIONS

1. (b) Total distance $= \int_a^b |v(t)|\, dt = \int_0^{2\pi} 5|\cos t|\, dt$

$= 5\left(\int_0^{\pi/2} \cos t\, dt - \int_{\pi/2}^{3\pi/2} \cos t\, dt + \int_{3\pi/2}^{2\pi} \cos t\, dt\right)$

$= 5\left([\sin t]_0^{\pi/2} - [\sin t]_{\pi/2}^{3\pi/2} + [\sin t]_{3\pi/2}^{2\pi}\right)$

$= 5[1 - (-1 - 1) - (-1)] = 20$ m

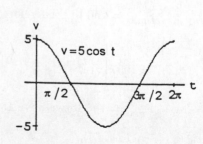

(c) Displacement $= \int_a^b v(t)\,dt = 5\int_0^{2\pi}\cos t\,dt = 5[\sin t]_0^{2\pi} = 0$ m

3. (b) Total distance $= \int_a^b |v(t)|\,dt = \int_0^{\pi/2}|6\sin 3t|\,dt$

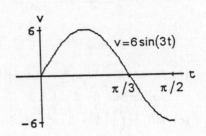

$$= 6\left(\int_0^{\pi/3}\sin 3t\,dt - \int_{\pi/3}^{\pi/2}\sin 3t\,dt\right)$$

$$= \frac{6}{3}\left([-\cos 3t]_0^{\pi/3} - [-\cos 3t]_{\pi/3}^{\pi/2}\right)$$

$$= 2[(-(-1)-(-1))+(0-(-1))] = 6 \text{ m}$$

(c) Displacement $= \int_a^b v(t)\,dt = 6\int_0^{\pi/2}\sin 3t\,dt = [-2\cos t]_0^{\pi/2} = -2(0-1) = 2$ m

5. (b) Total distance $= \int_a^b |v(t)|\,dt = \int_0^{10}|49-9.8t|\,dt$

$$= \int_0^5 (49-9.8t)\,dt - \int_5^{10}(49-9.8t)\,dt$$

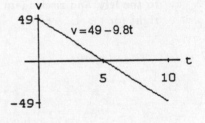

$$= \left[49t - \frac{9.8t^2}{2}\right]_0^5 - \left[49t - \frac{9.8t^2}{2}\right]_5^{10}$$

$$= \left[49\cdot5 - \frac{(9.8)\cdot5^2}{2}\right] - \left[\left(49\cdot10 - \frac{(9.8)\cdot10^2}{2}\right) - \left(49\cdot5 - \frac{(9.8)\cdot5^2}{2}\right)\right]$$

$$= (49\cdot10) - (9.8)\cdot25 - (49\cdot10) + (9.8)\cdot50 = (9.8)\cdot25 = 245 \text{ m}$$

(c) Displacement $= \int_a^b v(t)\,dt = \int_0^{10}(49-9.8t)\,dt = \left[49t - \frac{(9.8)t^2}{2}\right]_0^{10} = 490 - \frac{980}{2} = 0$ m

7. (b) Total distance $= \int_a^b |v(t)|\,dt = \int_0^2 |6t^2 - 18t + 12|\,dt$

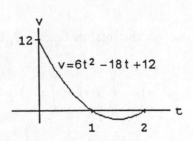

$$= \int_0^1 (6t^2 - 18t + 12)\,dt - \int_1^2 (6t^2 - 18t + 12)\,dt$$

$$= \left[2t^3 - 9t^2 + 12t\right]_0^1 - \left[2t^3 - 9t^2 + 12t\right]_1^2$$

$$= (2-9+12) - [(2\cdot8 - 9\cdot4 + 24) - (2-9+12)]$$

$$= 10 - 16 + 36 - 24 = 6 \text{ m}$$

(c) Displacement $= \int_a^b v(t)\,dt = \int_0^2 (6t^2 - 18t + 12)\,dt = \left[2t^3 - 9t^2 + 12t\right]_0^2 = 16 - 36 + 24 = 4$ m

9. (a) $v(t) = \frac{ds}{dt} = 3\left(\frac{1}{3}\right)t^2 - 6t + 8 = t^2 - 6t + 8$

$\Rightarrow v(0) = 8 > 0 \Rightarrow$ the body is moving to the right

at $t = 0$

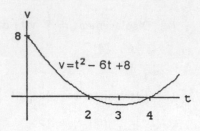

(b) $v(t) = t^2 - 6t + 8 = (t - 4)(t - 2)$ is sketched at the right $\Rightarrow v(t) < 0$ when $2 < t < 4 \Rightarrow$ the particle moves

to the left for $2 < t < 4$

(c) $s(3) = \left(\frac{1}{3}\right)3^3 - 3 \cdot 3^2 + 8 \cdot 3 = 9 - 27 + 24 = 6$ m

(d) The total distance is $\int_0^3 |v(t)| \, dt = \int_0^2 \left(t^2 - 6t + 8\right) dt - \int_2^3 \left(t^2 - 6t + 8\right) dt$

$= \left[\frac{t^3}{3} - 3t^2 + 8t\right]_0^2 - \left[\frac{t^3}{3} - 3t^2 + 8t\right]_2^3 = [s(2) - s(0)] - [s(3) - s(2)] = 2s(2) - s(0) - s(3)$

$= 2\left(\frac{8}{3} - 12 + 16\right) - 0 - 6 = 2\left(\frac{20}{3}\right) - 6 = \frac{22}{3}$ m

(e) The graph rises from $t = 0$ to $t = 2$ as the body moves
to the right, falls from $t = 2$ to $t = 4$ as the body moves
to the left, and rises again as the particle moves to the
right for $t > 4$.

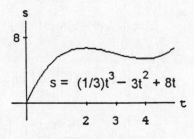

11. (a) The total distance is $\int_a^b |v(t)| \, dt = -\int_0^1 v(t) \, dt + \int_1^5 v(t) \, dt - \int_5^7 v(t) \, dt$. All the integrals above can be

computed using formulas from geometry (the integrals consist of triangles and rectangles):

$-\int_0^1 v(t) \, dt = \frac{1}{2}(1)(2) = 1, \int_1^5 v(t) \, dt = \frac{1}{2}(1)(2) + (1)(2) + \frac{1}{2}(2)(2) = 5, -\int_5^7 v(t) \, dt = \frac{1}{2}(2)(1) = 1$

$\Rightarrow$ the total distance $= 7$; Displacement $= \int_a^b v(t) \, dt = \int_0^7 v(t) \, dt = -\left(-\int_0^1 v(t) \, dt\right) + \int_1^5 v(t) \, dt$

$-\left(-\int_5^7 v(t) \, dt\right) = -1 + 5 - 1 = 3$

(b) The total distance is $\int_a^b |v(t)| \, dt = \int_0^2 v(t) \, dt - \int_2^7 v(t) \, dt + \int_7^{10} v(t) \, dt$. Using formulas from geometry for

areas of triangles and rectangles: $\int_0^2 v(t) \, dt = \frac{1}{2}(2)(3) = 3, -\int_2^7 v(t) \, dt = \frac{1}{2}(1)(3) + (3)(3) + \frac{1}{2}(1)(3) = 12,$

$$\int_{7}^{10} v(t)\, dt = \tfrac{1}{2}(3)(3) = \tfrac{9}{2} \Rightarrow \text{the total distance} = 19.5; \ \text{Displacement} = \int_{a}^{b} v(t)\, dt = \int_{0}^{10} v(t)\, dt$$

$$= \int_{0}^{2} v(t)\, dt - \left(-\int_{2}^{7} v(t)\, dt\right) + \int_{7}^{10} v(t)\, dt = 3 - 12 + \tfrac{9}{2} = -4.5$$

13.

row	1	2	3	4	5	6	7	8	9	10	11	12	13	14	15
number of shrimp-colored squares	17	18	18	15	16	18	15	19	24	25	26	22	20	21	19

Take the sum of these numbers and divide the result by $30 \times 15 = 450$. Thus, the proportion is $\frac{293}{450} \approx 0.65$ or about 65%

15. $S = \displaystyle\int_{a}^{b} 2\pi f(x)\, dx = \int_{0}^{\sqrt{3}} 2\pi \cdot \frac{x}{\sqrt{3}}\, dx = \frac{\pi}{\sqrt{3}}\big[x^2\big]_{0}^{\sqrt{3}} = \frac{3\pi}{\sqrt{3}} = \sqrt{3}\,\pi$

17. (a) Displacement Volume $V \approx \frac{h}{3}(y_0 + 4y_1 + 2y_2 + 4y_3 + \ldots + 2y_{n-2} + 4y_{n-1} + y_n)$, $x_0 = 0$, $x_n = 10 - h$,

$h = 2.54$, $n = 10 \Rightarrow V = \displaystyle\int_{x_0}^{x_n} A(x)\, dx \approx \frac{2.54}{3}[0 + 4(1.07) + 2(3.84) + 4(7.82) + 2(12.20) + 4(15.18)$

$+ 2(16.14) + 4(14.00) + 2(9.21) + 4(3.24) + 0] = \frac{2.54}{3}(4.28 + 7.68 + 31.28 + 24.4 + 60.72 + 32.28$

$+ 56 + 18.42 + 12.96) = \frac{2.54}{3}(248.02) = 209.99 \approx 210 \text{ ft}^3$

(b) The weight of water displaced is approximately $64 \cdot 210 = 13{,}440$ lb

19. The centroid of the square is located at $(2,2)$. The volume is $V = (2\pi)(\overline{y})(A) = (2\pi)(2)(8) = 32\pi$ and the surface area is $S = (2\pi)(\overline{y})(L) = (2\pi)(2)(4\sqrt{8}) = 32\sqrt{2}\,\pi$ (where $\sqrt{8}$ is the length of a side).

21. The centroid is located at $(2,0) \Rightarrow V = (2\pi)(\overline{x})(A) = (2\pi)(2)(\pi) = 4\pi^2$

23. $S = 2\pi\,\overline{y}\,L \Rightarrow 4\pi a^2 = (2\pi\overline{y})(\pi a) \Rightarrow \overline{y} = \frac{2a}{\pi}$, and by symmetry $\overline{x} = 0$

25. $V = 2\pi\,\overline{y}A \Rightarrow \frac{4}{3}\pi ab^2 = (2\pi\overline{y})\left(\frac{\pi ab}{2}\right) \Rightarrow \overline{y} = \frac{4b}{3\pi}$ and by symmetry $\overline{x} = 0$

27. $V = 2\pi\rho\,A = (2\pi)(\text{area of the region}) \cdot (\text{distance from the centroid to the line } y = x - a)$. We must find the distance from $\left(0, \frac{4a}{3\pi}\right)$ to $y = x - a$. The line containing the centroid and perpendicular to $y = x - a$ has slope -1 and contains the point $\left(0, \frac{4a}{3\pi}\right)$. This line is $y = -x + \frac{4a}{3\pi}$. The intersection of $y = x - a$ and $y = -x + \frac{4a}{3\pi}$ is the point $\left(\frac{4a + 3a\pi}{6\pi}, \frac{4a - 3a\pi}{6\pi}\right)$. Thus, the distance from the centroid to the line $y = x - a$ is

$$\sqrt{\left(\frac{4a + 3a\pi}{6\pi}\right)^2 + \left(\frac{4a}{3\pi} - \frac{4a}{6\pi} + \frac{3a\pi}{6\pi}\right)^2} = \frac{\sqrt{2}\,(4a + 3a\pi)}{6\pi} \Rightarrow V = (2\pi)\left(\frac{\sqrt{2}\,(4a + 3a\pi)}{6\pi}\right)\left(\frac{\pi a^2}{2}\right) = \frac{\sqrt{2}\,\pi a^3(4 + 3\pi)}{6}$$

29. From Example 4 and Pappus's Theorem for Volumes we have the moment about the x-axis is $M_x = \bar{y} M$

$$= \left(\frac{4a}{3\pi}\right)\left(\frac{\pi a^2}{2}\right) = \frac{2a^3}{3}.$$

CHAPTER 5 PRACTICE EXERCISES

1. $f(x) = x$, $g(x) = \frac{1}{x^2}$, $a = 1$, $b = 2 \Rightarrow A = \int_a^b [f(x) - g(x)]\, dx$

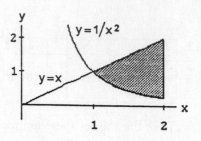

$$= \int_1^2 \left(x - \frac{1}{x^2}\right) dx = \left[\frac{x^2}{2} + \frac{1}{x}\right]_1^2 = \left(\frac{4}{2} + \frac{1}{2}\right) - \left(\frac{1}{2} + 1\right) = 1$$

3. $f(x) = \left(1 - \sqrt{x}\right)^2$, $g(x) = 0$, $a = 0$, $b = 1 \Rightarrow A = \int_a^b [f(x) - g(x)]\, dx = \int_0^1 \left(1 - \sqrt{x}\right)^2 dx = \int_0^1 \left(1 - 2\sqrt{x} + x\right) dx$

$$= \int_0^1 \left(1 - 2x^{1/2} + x\right) dx = \left[x - \frac{4}{3}x^{3/2} + \frac{x^2}{2}\right]_0^1 = 1 - \frac{4}{3} + \frac{1}{2} = \frac{1}{6}(6 - 8 + 3) = \frac{1}{6}$$

5. $f(y) = 2y^2$, $g(y) = 0$, $c = 0$, $d = 3$

$$\Rightarrow A = \int_c^d [f(y) - g(y)]\, dy = \int_0^3 \left(2y^2 - 0\right) dy$$

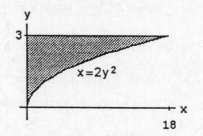

$$= 2\int_0^3 y^2\, dy = \frac{2}{3}\left[y^3\right]_0^3 = 18$$

7. Let us find the intersection points: $\dfrac{y^2}{4} = \dfrac{y+2}{4}$

$$\Rightarrow y^2 - y - 2 = 0 \Rightarrow (y - 2)(y + 1) = 0 \Rightarrow y = -1$$

or $y = 2 \Rightarrow c = -1$, $d = 2$; $f(y) = \dfrac{y+2}{4}$, $g(y) = \dfrac{y^2}{4}$

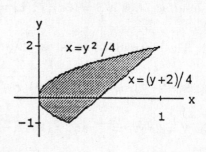

$$\Rightarrow A = \int_c^d [f(y) - g(y)]\, dy = \int_{-1}^2 \left(\frac{y+2}{4} - \frac{y^2}{4}\right) dy$$

$$= \frac{1}{4}\int_{-1}^2 \left(y + 2 - y^2\right) dy = \frac{1}{4}\left[\frac{y^2}{2} + 2y - \frac{y^3}{3}\right]_{-1}^2 = \frac{1}{4}\left[\left(\frac{4}{2} + 4 - \frac{8}{3}\right) - \left(\frac{1}{2} - 2 + \frac{1}{3}\right)\right] = \frac{9}{8}$$

9. $f(x) = x$, $g(x) = \sin x$, $a = 0$, $b = \frac{\pi}{4}$

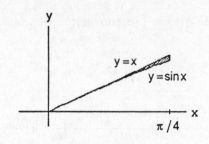

$$\Rightarrow A = \int_a^b [f(x) - g(x)] \, dx = \int_0^{\pi/4} (x - \sin x) \, dx$$

$$= \left[\frac{x^2}{2} + \cos x\right]_0^{\pi/4} = \left(\frac{\pi^2}{32} + \frac{\sqrt{2}}{2}\right) - 1$$

11. $a = 0$, $b = \pi$, $f(x) - g(x) = 2\sin x - \sin 2x$

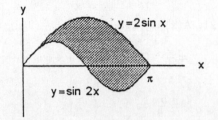

$$\Rightarrow A = \int_0^\pi (2\sin x - \sin 2x) \, dx = \left[-2\cos x + \frac{\cos 2x}{2}\right]_0^\pi$$

$$= \left[-2 \cdot (-1) + \frac{1}{2}\right] - \left(-2 \cdot 1 + \frac{1}{2}\right) = 4$$

13. $f(y) = \sqrt{y}$, $g(y) = 2 - y$, $c = 1$, $d = 2$

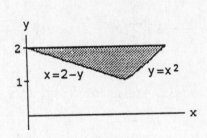

$$\Rightarrow A = \int_c^d [f(y) - g(y)] \, dy = \int_1^2 \left[\sqrt{y} - (2 - y)\right] dy$$

$$= \int_1^2 (\sqrt{y} - 2 + y) \, dy = \left[\frac{2}{3}y^{3/2} - 2y + \frac{y^2}{2}\right]_1^2$$

$$= \left(\frac{4}{3}\sqrt{2} - 4 + 2\right) - \left(\frac{2}{3} - 2 + \frac{1}{2}\right) = \frac{4}{3}\sqrt{2} - \frac{7}{6} = \frac{8\sqrt{2} - 7}{6}$$

15. $f(x) = x^3 - 3x^2 = x^2(x - 3) \Rightarrow f'(x) = 3x^2 - 6x = 3x(x - 2) \Rightarrow f' = {+}{+}{+} \underset{0}{|} {-}{-}{-}{-} \underset{2}{|} {+}{+}{+}$

$\Rightarrow f(0) = 0$ is a maximum and $f(2) = -4$ is a minimum. Then $A = -\int_0^3 (x^3 - 3x^2) \, dx = -\left[\frac{x^4}{4} - x^3\right]_0^3$

$$= -\left(\frac{81}{4} - 27\right) = \frac{27}{4}$$

17. The area above the x-axis is $A_1 = \int_0^1 \left(y^{2/3} - y\right) dy$

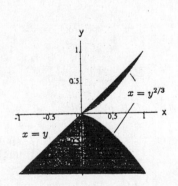

$$= \left[\frac{3y^{5/3}}{5} - \frac{y^2}{2}\right]_0^1 = \frac{1}{10}; \text{ the area below the x-axis is}$$

$$A_2 = \int_{-1}^0 \left(y^{2/3} - y\right) dy = \left[\frac{3y^{5/3}}{5} - \frac{y^2}{2}\right]_{-1}^0 = \frac{11}{10}$$

$\Rightarrow$ the total area is $A_1 + A_2 = \frac{6}{5}$

19. $A(x) = \frac{\pi}{4}(\text{diameter})^2 = \frac{\pi}{4}\left(\sqrt{x} - x^2\right)^2 = \frac{\pi}{4}\left(x - 2\sqrt{x}\cdot x^2 + x^4\right);$

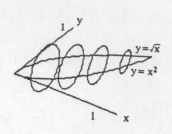

$a = 0, b = 1 \Rightarrow V = \int_a^b A(x)\,dx = \frac{\pi}{4}\int_0^1 \left(x - 2x^{5/2} + x^4\right)dx$

$= \frac{\pi}{4}\left[\frac{x^2}{2} - \frac{4}{7}x^{7/2} + \frac{x^5}{5}\right]_0^1 = \frac{\pi}{4}\left(\frac{1}{2} - \frac{4}{7} + \frac{1}{5}\right) = \frac{\pi}{4\cdot 70}(35 - 40 + 14)$

$= \frac{9\pi}{280}$

21. $A(x) = \frac{\pi}{4}(\text{diameter})^2 = \frac{\pi}{4}(2\sin x - 2\cos x)^2$

$= \frac{\pi}{4}\cdot 4\left(\sin^2 x - 2\sin x\cos x + \cos^2 x\right) = \pi(1 - \sin 2x);\ a = \frac{\pi}{4},$

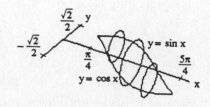

$b = \frac{5\pi}{4} \Rightarrow V = \int_a^b A(x)\,dx = \pi\int_{\pi/4}^{5\pi/4}(1 - \sin 2x)\,dx$

$= \pi\left[x + \frac{\cos 2x}{2}\right]_{\pi/4}^{5\pi/4} = \pi\left[\left(\frac{5\pi}{4} + \frac{\cos\frac{5\pi}{2}}{2}\right) - \left(\frac{\pi}{4} - \frac{\cos\frac{\pi}{2}}{2}\right)\right] = \pi^2$

23. $A(x) = \frac{\pi}{4}(\text{diameter})^2 = \frac{\pi}{4}\left(2\sqrt{x} - \frac{x^2}{4}\right)^2 = \frac{\pi}{4}\left(4x - x^{5/2} + \frac{x^4}{16}\right);$

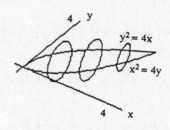

$a = 0, b = 4 \Rightarrow V = \int_a^b A(x)\,dx = \frac{\pi}{4}\int_0^4\left(4x - x^{5/2} + \frac{x^4}{16}\right)dx$

$= \frac{\pi}{4}\left[2x^2 - \frac{2}{7}x^{7/2} + \frac{x^5}{5\cdot 16}\right]_0^4 = \frac{\pi}{4}\left(32 - 32\cdot\frac{8}{7} + \frac{2}{5}\cdot 32\right)$

$= \frac{32\pi}{4}\left(1 - \frac{8}{7} + \frac{2}{5}\right) = \frac{8\pi}{35}(35 - 40 + 14) = \frac{72\pi}{35}$

25. (a) *disk method*:

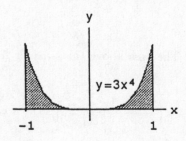

$$V = \int_a^b \pi R^2(x)\,dx = \int_{-1}^1 \pi\left(3x^4\right)^2 dx = \pi\int_{-1}^1 9x^8\,dx$$

$$= \pi\left[x^9\right]_{-1}^1 = 2\pi$$

(b) *shell method*:

$$V = \int_a^b 2\pi\binom{\text{shell}}{\text{radius}}\binom{\text{shell}}{\text{height}}dx = \int_0^1 2\pi x\left(3x^4\right)dx = 2\pi\cdot 3\int_0^1 x^5\,dx = 2\pi\cdot 3\left[\frac{x^6}{6}\right]_0^1 = \pi$$

Note: The lower limit of integration is 0 rather than -1.

(c) *shell method:*

$$V = \int_a^b 2\pi \left(\begin{array}{c}\text{shell}\\\text{radius}\end{array}\right)\left(\begin{array}{c}\text{shell}\\\text{height}\end{array}\right) dx = 2\pi \int_{-1}^1 (1-x)(3x^4)\, dx = 2\pi\left[\frac{3x^5}{5} - \frac{x^6}{2}\right]_{-1}^1 = 2\pi\left[\left(\frac{3}{5} - \frac{1}{2}\right) - \left(-\frac{3}{5} - \frac{1}{2}\right)\right] = \frac{12\pi}{5}$$

(d) *washer method:*

$$R(x) = 3,\ r(x) = 3 - 3x^4 = 3\left(1 - x^4\right) \Rightarrow V = \int_a^b \pi\left[R^2(x) - r^2(x)\right] dx = \int_{-1}^1 \pi\left[9 - 9\left(1 - x^4\right)^2\right] dx$$

$$= 9\pi \int_{-1}^1 \left[1 - \left(1 - 2x^4 + x^8\right)\right] dx = 9\pi \int_{-1}^1 \left(2x^4 - x^8\right) dx = 9\pi\left[\frac{2x^5}{5} - \frac{x^9}{9}\right]_{-1}^1 = 18\pi\left[\frac{2}{5} - \frac{1}{9}\right] = \frac{2\pi \cdot 13}{5} = \frac{26\pi}{5}$$

27. (a) *disk method:*

$$V = \pi \int_1^5 \left(\sqrt{x-1}\right)^2 dx = \int_1^5 (x-1)\, dx = \pi\left[\frac{x^2}{2} - x\right]_1^5$$

$$= \pi\left[\left(\frac{25}{2} - 5\right) - \left(\frac{1}{2} - 1\right)\right] = \pi\left(\frac{24}{2} - 4\right) = 8\pi$$

(b) *washer method:*

$$R(y) = 5,\ r(y) = y^2 + 1 \Rightarrow V = \int_c^d \pi\left[R^2(y) - r^2(y)\right] dy = \pi \int_{-2}^2 \left[25 - \left(y^2 + 1\right)^2\right] dy$$

$$= \pi \int_{-2}^2 \left(25 - y^4 - 2y^2 - 1\right) dy = \pi \int_{-2}^2 \left(24 - y^4 - 2y^2\right) dy = \pi\left[24y - \frac{y^5}{5} - \frac{2}{3}y^3\right]_{-2}^2 = 2\pi\left(24 \cdot 2 - \frac{32}{5} - \frac{2}{3} \cdot 8\right)$$

$$= 32\pi\left(3 - \frac{2}{5} - \frac{1}{3}\right) = \frac{32\pi}{15}(45 - 6 - 5) = \frac{1088\pi}{15}$$

(c) *disk method:*

$$R(y) = 5 - \left(y^2 + 1\right) = 4 - y^2 \Rightarrow V = \int_c^d \pi R^2(y)\, dy = \int_{-2}^2 \pi\left(4 - y^2\right)^2 dy = \pi \int_{-2}^2 \left(16 - 8y^2 + y^4\right) dy$$

$$= \pi\left[16y - \frac{8y^3}{3} + \frac{y^5}{5}\right]_{-2}^2 = 2\pi\left(32 - \frac{64}{3} + \frac{32}{5}\right) = 64\pi\left(1 - \frac{2}{3} + \frac{1}{5}\right) = \frac{64\pi}{15}(15 - 10 + 3) = \frac{512\pi}{15}$$

29. *disk method:*

$$R(x) = \tan x,\ a = 0,\ b = \frac{\pi}{3} \Rightarrow V = \pi \int_0^{\pi/3} \tan^2 x\, dx = \pi \int_0^{\pi/3} \left(\sec^2 x - 1\right) dx = \pi[\tan x - x]_0^{\pi/3} = \frac{\pi\left(3\sqrt{3} - \pi\right)}{3}$$

31. (a) *disk method:*

$$V = \pi \int_0^2 \left(x^2 - 2x\right)^2 dx = \pi \int_0^2 \left(x^4 - 4x^3 + 4x^2\right) dx = \pi\left[\frac{x^5}{5} - x^4 + \frac{4}{3}x^3\right]_0^2 = \pi\left(\frac{32}{5} - 16 + \frac{32}{3}\right)$$

$$= \frac{16\pi}{15}(6 - 15 + 10) = \frac{16\pi}{15}$$

(b) *disk method:*

$$V = 2\pi - \pi \int_0^2 \left[1 + (x^2 - 2x)\right]^2 dx = 2\pi - \pi \int_0^2 \left[1 + 2(x^2 - 2x) + (x^2 - 2x)^2\right] dx$$

$$= 2\pi - \pi \int_0^2 \left(1 + 2x^2 - 4x + x^4 - 4x^3 + 4x^2\right) dx = 2\pi - \pi \int_0^2 \left(x^4 - 4x^3 + 6x^2 - 4x + 1\right) dx$$

$$= 2\pi - \pi \left[\frac{x^5}{5} - x^4 + 2x^3 - 2x^2 + x\right]_0^2 = 2\pi - \pi \left(\frac{32}{5} - 16 + 16 - 8 + 2\right) = 2\pi - \frac{\pi}{5}(32 - 30) = 2\pi - \frac{2\pi}{5} = \frac{8\pi}{5}$$

(c) *shell method:*

$$V = \int_a^b 2\pi \binom{\text{shell}}{\text{radius}}\binom{\text{shell}}{\text{height}} dx = 2\pi \int_0^2 (2-x)\left[-(x^2 - 2x)\right] dx = 2\pi \int_0^2 (2-x)(2x - x^2) dx$$

$$= 2\pi \int_0^2 \left(4x - 2x^2 - 2x^2 + x^3\right) dx = 2\pi \int_0^2 \left(x^3 - 4x^2 + 4x\right) dx = 2\pi\left[\frac{x^4}{4} - \frac{4}{3}x^3 + 2x^2\right]_0^2 = 2\pi\left(4 - \frac{32}{3} + 8\right)$$

$$= \frac{2\pi}{3}(36 - 32) = \frac{8\pi}{3}$$

(d) *disk method:*

$$V = \pi \int_0^2 \left[2 - (x^2 - 2x)\right]^2 dx - \pi \int_0^2 2^2 dx = \pi \int_0^2 \left[4 - 4(x^2 - 2x) + (x^2 - 2x)^2\right] dx - 8\pi$$

$$= \pi \int_0^2 \left(4 - 4x^2 + 8x + x^4 - 4x^3 + 4x^2\right) dx - 8\pi = \pi \int_0^2 \left(x^4 - 4x^3 + 8x + 4\right) dx - 8\pi$$

$$= \pi\left[\frac{x^5}{5} - x^4 + 4x^2 + 4x\right]_0^2 - 8\pi = \pi\left(\frac{32}{5} - 16 + 16 + 8\right) - 8\pi = \frac{\pi}{5}(32 + 40) - 8\pi = \frac{72\pi}{5} - \frac{40\pi}{5} = \frac{32\pi}{5}$$

33. The volume cut out is equivalent to the volume of the solid generated by revolving the region shown here about the x-axis. Using the *shell* method:

$$V = \int_c^d 2\pi \binom{\text{shell}}{\text{radius}}\binom{\text{shell}}{\text{height}} dy = \int_0^{\sqrt{3}} 2\pi y\left[\sqrt{4 - y^2} - (-\sqrt{4 - y^2})\right] dy$$

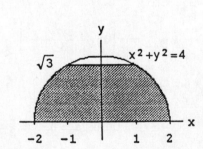

$$= 2\pi \int_0^{\sqrt{3}} 2y\sqrt{4 - y^2} \, dy = -2\pi \int_0^{\sqrt{3}} \sqrt{4 - y^2} \, d(4 - y^2)$$

$$= (-2\pi)\left(\frac{2}{3}\right)\left[(4 - y^2)^{3/2}\right]_0^{\sqrt{3}} = -\frac{4\pi}{3}(1 - 8) = \frac{28\pi}{3}$$

35. $y = x^{1/2} - \frac{x^{3/2}}{3} \Rightarrow \frac{dy}{dx} = \frac{1}{2}x^{-1/2} - \frac{1}{2}x^{1/2} \Rightarrow \left(\frac{dy}{dx}\right)^2 = \frac{1}{4}\left(\frac{1}{x} - 2 + x\right) \Rightarrow L = \int_1^4 \sqrt{1 + \frac{1}{4}\left(\frac{1}{x} - 2 + x\right)}\, dx$

$\Rightarrow L = \int_1^4 \sqrt{\frac{1}{4}\left(\frac{1}{x} + 2 + x\right)}\, dx = \int_1^4 \sqrt{\frac{1}{4}\left(x^{-1/2} + x^{1/2}\right)^2}\, dx = \int_1^4 \frac{1}{2}\left(x^{-1/2} + x^{1/2}\right) dx = \frac{1}{2}\left[2x^{1/2} + \frac{2}{3}x^{3/2}\right]_1^4$

$= \frac{1}{2}\left[\left(4 + \frac{2}{3}\cdot 8\right) - \left(2 + \frac{2}{3}\right)\right] = \frac{1}{2}\left(2 + \frac{14}{3}\right) = \frac{10}{3}$

37. $y = \frac{5}{12}x^{6/5} - \frac{5}{8}x^{4/5} \Rightarrow \frac{dy}{dx} = \frac{1}{2}x^{1/5} - \frac{1}{2}x^{-1/5} \Rightarrow \left(\frac{dy}{dx}\right)^2 = \frac{1}{4}\left(x^{2/5} - 2 + x^{-2/5}\right)$

$\Rightarrow L = \int_1^{32} \sqrt{1 + \frac{1}{4}\left(x^{2/5} - 2 + x^{-2/5}\right)}\, dx \Rightarrow L = \int_1^{32} \sqrt{\frac{1}{4}\left(x^{2/5} + 2 + x^{-2/5}\right)}\, dx = \int_1^{32} \sqrt{\frac{1}{4}\left(x^{1/5} + x^{-1/5}\right)^2}\, dx$

$= \int_1^{32} \frac{1}{2}\left(x^{1/5} + x^{-1/5}\right) dx = \frac{1}{2}\left[\frac{5}{6}x^{6/5} + \frac{5}{4}x^{4/5}\right]_1^{32} = \frac{1}{2}\left[\left(\frac{5}{6}\cdot 2^6 + \frac{5}{4}\cdot 2^4\right) - \left(\frac{5}{6} + \frac{5}{4}\right)\right] = \frac{1}{2}\left(\frac{315}{6} + \frac{75}{4}\right)$

$= \frac{1}{48}(1260 + 450) = \frac{1710}{48} = \frac{285}{8}$

39. $S = \int_a^b 2\pi y \sqrt{1 + \left(\frac{dy}{dx}\right)^2}\, dx;\ \frac{dy}{dx} = \frac{1}{\sqrt{2x+1}} \Rightarrow \left(\frac{dy}{dx}\right)^2 = \frac{1}{2x+1} \Rightarrow S = \int_0^3 2\pi\sqrt{2x+1}\sqrt{1 + \frac{1}{2x+1}}\, dx$

$= 2\pi \int_0^3 \sqrt{2x+1}\sqrt{\frac{2x+2}{2x+1}}\, dx = 2\sqrt{2}\pi \int_0^3 \sqrt{x+1}\, dx = 2\sqrt{2}\pi\left[\frac{2}{3}(x+1)^{3/2}\right]_0^3 = 2\sqrt{2}\pi \cdot \frac{2}{3}(8 - 1) = \frac{28\pi\sqrt{2}}{3}$

41. $S = \int_c^d 2\pi x \sqrt{1 + \left(\frac{dx}{dy}\right)^2}\, dy;\ \frac{dx}{dy} = \frac{\left(\frac{1}{2}\right)(4-2y)}{\sqrt{4y-y^2}} = \frac{2-y}{\sqrt{4y-y^2}} \Rightarrow 1 + \left(\frac{dx}{dy}\right)^2 = \frac{4y - y^2 + 4 - 4y + y^2}{4y - y^2} = \frac{4}{4y-y^2}$

$\Rightarrow S = \int_1^2 2\pi\sqrt{4y-y^2}\sqrt{\frac{4}{4y-y^2}}\, dy = 4\pi \int_1^2 dx = 4\pi$

43. Intersection points: $3 - x^2 = 2x^2 \Rightarrow 3x^2 - 3 = 0$
$\Rightarrow 3(x-1)(x+1) = 0 \Rightarrow x = -1$ or $x = 1$. Applying the
symmetry argument analogous to the one used in Exercise
5.7.13, we find that $\bar{x} = 0$. The typical *vertical* strip has

center of mass: $(\tilde{x}, \tilde{y}) = \left(x, \frac{2x^2 + (3-x^2)}{2}\right) = \left(x, \frac{x^2+3}{2}\right)$,

length: $(3 - x^2) - 2x^2 = 3(1 - x^2)$, width: dx,

area: $dA = 3(1 - x^2)\, dx$, and mass: $dm = \delta \cdot dA = 3\delta(1 - x^2)\, dx$

$\Rightarrow$ the moment about the x-axis is $\tilde{y}\, dm = \frac{3}{2}\delta(x^2 + 3)(1 - x^2)\, dx = \frac{3}{2}\delta(-x^4 - 2x^2 + 3)\, dx$

$\Rightarrow M_x = \int \tilde{y}\, dm = \frac{3}{2}\delta \int_{-1}^1 (-x^4 - 2x^2 + 3)\, dx = \frac{3}{2}\delta\left[-\frac{x^5}{5} - \frac{2x^3}{3} + 3x\right]_{-1}^1 = 3\delta\left(-\frac{1}{5} - \frac{2}{3} + 3\right)$

$= \frac{3\delta}{15}(-3 - 10 + 45) = \frac{32\delta}{5}$; $M = \int dm = 3\delta \int\limits_{-1}^{1} \left(1 - x^2\right) dx = 3\delta\left[x - \frac{x^3}{3}\right]_{-1}^{1} = 6\delta\left(1 - \frac{1}{3}\right) = 4\delta$

$\Rightarrow \bar{y} = \frac{M_x}{M} = \frac{32\delta}{5 \cdot 4\delta} = \frac{8}{5}$.   Therefore, the centroid is $(\bar{x}, \bar{y}) = \left(0, \frac{8}{5}\right)$.

45. The typical *vertical* strip has: center of mass: $(\tilde{x}, \tilde{y}) = \left(x, \dfrac{4 + \frac{x^2}{4}}{2}\right)$,

length: $4 - \frac{x^2}{4}$, width: dx, area: $dA = \left(4 - \frac{x^2}{4}\right) dx$,

mass: $dm = \delta \cdot dA = \delta\left(4 - \frac{x^2}{4}\right) dx \Rightarrow$ the moment about the x-axis is

$y = (1/4)x^2$

$\tilde{y}\, dm = \delta \cdot \dfrac{\left(4 + \frac{x^2}{4}\right)}{2}\left(4 - \frac{x^2}{4}\right) dx = \frac{\delta}{2}\left(16 - \frac{x^4}{16}\right) dx$; the moment about

the y-axis is $\tilde{x}\, dm = \delta\left(4 - \frac{x^2}{4}\right) \cdot x\, dx = \delta\left(4x - \frac{x^3}{4}\right) dx$. Thus, $M_x = \int \tilde{y}\, dm = \frac{\delta}{2}\int\limits_{0}^{4}\left(16 - \frac{x^4}{16}\right) dx$

$= \frac{\delta}{2}\left[16x - \frac{x^5}{5 \cdot 16}\right]_{0}^{4} = \frac{\delta}{2}\left[64 - \frac{64}{5}\right] = \frac{128\delta}{5}$; $M_y = \int \tilde{x}\, dm = \delta\int\limits_{0}^{4}\left(4x - \frac{x^3}{4}\right) dx = \delta\left[2x^2 - \frac{x^4}{16}\right]_{0}^{4}$

$= \delta(32 - 16) = 16\delta$; $M = \int dm = \delta\int\limits_{-0}^{4}\left(4 - \frac{x^2}{4}\right) dx = \delta\left[4x - \frac{x^3}{12}\right]_{0}^{4} = \delta\left(16 - \frac{64}{12}\right) = \frac{32\delta}{3}$

$\Rightarrow \bar{x} = \frac{M_y}{M} = \frac{16 \cdot \delta \cdot 3}{32 \cdot \delta} = \frac{3}{2}$ and $\bar{y} = \frac{M_x}{M} = \frac{128 \cdot \delta \cdot 3}{5 \cdot 32 \cdot \delta} = \frac{12}{5}$. Therefore, the centroid is $(\bar{x}, \bar{y}) = \left(\frac{3}{2}, \frac{12}{5}\right)$.

47. A typical horizontal strip has: center of mass: $(\tilde{x}, \tilde{y}) = \left(\dfrac{y^2 + 2y}{2}, y\right)$,

length: $2y - y^2$, width: dy, area: $dA = \left(2y - y^2\right) dy$,

mass: $dm = \delta \cdot dA = (1 + y)\left(2y - y^2\right) dy \Rightarrow$ the moment about the

x-axis is $\tilde{y}\, dm = y(1 + y)\left(2y - y^2\right) dy = \left(2y^2 + 2y^3 - y^3 - y^4\right) dy$

$= \left(2y^2 + y^3 - y^4\right) dy$; the moment about the y-axis is

$x = y^2$

$x = 2y$

$\tilde{x}\, dm = \left(\dfrac{y^2 + 2y}{2}\right)(1 + y)\left(2y - y^2\right) dy = \frac{1}{2}\left(4y^2 - y^4\right)(1 + y)\, dy$

$= \frac{1}{2}\left(4y^2 + 4y^3 - y^4 - y^5\right) dy \Rightarrow M_x = \int \tilde{y}\, dm = \int\limits_{0}^{2}\left(2y^2 + y^3 - y^4\right) dy = \left[\frac{2}{3}y^3 + \frac{y^4}{4} - \frac{y^5}{5}\right]_{0}^{2}$

$= \left(\frac{16}{3} + \frac{16}{4} - \frac{32}{5}\right) = 16\left(\frac{1}{3} + \frac{1}{4} - \frac{2}{5}\right) = \frac{16}{60}(20 + 15 - 24) = \frac{4}{15}(11) = \frac{44}{15}$; $M_y = \int \tilde{x}\, dm$

$= \int\limits_{0}^{2} \frac{1}{2}\left(4y^2 + 4y^3 - y^4 - y^5\right) dy = \frac{1}{2}\left[\frac{4}{3}y^3 + y^4 - \frac{y^5}{5} - \frac{y^6}{6}\right]_{0}^{2} = \frac{1}{2}\left(\frac{4 \cdot 2^3}{3} + 2^4 - \frac{2^5}{5} - \frac{2^6}{6}\right)$

$= 4\left(\frac{4}{3} + 2 - \frac{4}{5} - \frac{8}{6}\right) = 4\left(2 - \frac{4}{5}\right) = \frac{24}{5}$; $M = \int dm = \int\limits_{0}^{2}(1 + y)\left(2y - y^2\right) dy = \int\limits_{0}^{2}\left(2y + y^2 - y^3\right) dy$

$$= \left[y^2 + \frac{y^3}{3} - \frac{y^4}{4} \right]_0^2 = \left(4 + \frac{8}{3} - \frac{16}{4} \right) = \frac{8}{3} \Rightarrow \overline{x} = \frac{M_y}{M} = \left(\frac{24}{5} \right) \left(\frac{3}{8} \right) = \frac{9}{5} \text{ and } \overline{y} = \frac{M_x}{M} = \left(\frac{44}{15} \right) \left(\frac{3}{8} \right) = \frac{44}{40} = \frac{11}{10}. \text{ Therefore,}$$

the center of mass is $(\overline{x}, \overline{y}) = \left(\frac{9}{5}, \frac{11}{10} \right)$.

49. The equipment alone: the force required to lift the equipment is equal to its weight $\Rightarrow F_1(x) = 100$ N.

The work done is $W_1 = \int_a^b F_1(x)\, dx = \int_0^{40} 100\, dx = [100x]_0^{40} = 4000$ J; the rope alone: the force required

to lift the rope is equal to the weight of the rope paid out at elevation $x \Rightarrow F_2(x) = 0.8(40 - x)$. The work

done is $W_2 = \int_a^b F_2(x)\, dx = \int_0^{40} 0.8(40 - x)\, dx = 0.8 \left[40x - \frac{x^2}{2} \right]_0^{40} = 0.8 \left(40^2 - \frac{40^2}{2} \right) = \frac{(0.8)(1600)}{2} = 640$ J;

the total work is $W = W_1 + W_2 = 4000 + 640 = 4640$ J

51. Force constant: $F = kx \Rightarrow 20 = k \cdot 1 \Rightarrow k = 20$ lb/ft; the work to stretch the spring 1 ft is

$$W = \int_0^1 kx\, dx = k \int_0^1 x\, dx = \left[20 \frac{x^2}{2} \right]_0^1 = 10 \text{ ft} \cdot \text{lb; the work to stretch the spring an additional foot is}$$

$$W = \int_1^2 kx\, dx = k \int_1^2 x\, dx = 20 \left[\frac{x^2}{2} \right]_1^2 = 20 \left(\frac{4}{2} - \frac{1}{2} \right) = 20 \left(\frac{3}{2} \right) = 30 \text{ ft} \cdot \text{lb}$$

53. We imagine the water divided into thin slabs by planes perpendicular to the y-axis at the points of a partition of the interval $[0, 8]$. The typical slab between the planes at y and

$y + \Delta y$ has a volume of about $\Delta V = \pi (\text{radius})^2 (\text{thickness})$

$= \pi \left(\frac{5}{4} y \right)^2 = \frac{25\pi}{16} y^2 \Delta y$ ft^3. The force $F(y)$ required to lift

this slab is equal to its weight: $F(y) = 62.4 \Delta V$

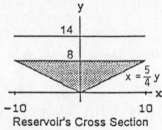

Reservoir's Cross Section

$= \frac{(62.4)(25)}{16} \pi y^2 \Delta y$ lb. The distance through which $F(y)$

must act to lift this slab to the level 6 ft above the top is about $(6 + 8 - y)$ ft, so the work done lifting the slab

is about $\Delta W = \frac{(62.4)(25)}{16} \pi y^2 (14 - y) \Delta y$ ft $\cdot$ lb. The work done lifting all the slabs from $y = 0$ to $y = 8$ to the

level 6 ft above the top is approximately $W \approx \sum_0^8 \frac{(62.4)(25)}{16} \pi y^2 (14 - y) \Delta y$ ft $\cdot$ lb so the work to pump the

water is the limit of these Riemann sums as the norm of the partition goes to zero:

$$W = \int_0^8 \frac{(62.4)(25)}{(16)} \pi y^2 (14 - y)\, dy = \frac{(62.4)(25)\pi}{16} \int_0^8 (14y^2 - y^3)\, dy = (62.4) \left(\frac{25\pi}{16} \right) \left[\frac{14}{3} y^3 - \frac{y^4}{4} \right]_0^8$$

$$= (62.4) \left(\frac{25\pi}{16} \right) \left(\frac{14}{3} \cdot 8^3 - \frac{8^4}{4} \right) \approx 418{,}208.81 \text{ ft} \cdot \text{lb}$$

55. The tank's cross section looks like the figure in Exercise 53 with right edge given by $x = \frac{5}{10} y = \frac{y}{2}$. A typical

horizontal slab has volume $\Delta V = \pi (\text{radius})^2 (\text{thickness}) = \pi \left(\frac{y}{2} \right)^2 \Delta y = \frac{\pi}{4} y^2 \Delta y$. The force required to lift this

slab is its weight: $F(y) = 60 \cdot \frac{\pi}{4} y^2 \Delta y$. The distance through which $F(y)$ must act is $(2 + 10 - y)$ ft, so the work to pump the liquid is $W = 60 \int\limits_{0}^{10} \pi(12-y)\left(\dfrac{y^2}{4}\right) dy = 15\pi\left[\dfrac{12y^3}{3} - \dfrac{y^4}{4}\right]_0^{10} = 22{,}500\pi$ ft · lb; the time needed

to empty the tank is $\dfrac{22{,}500 \text{ ft} \cdot \text{lb}}{275 \text{ ft} \cdot \text{lb/sec}} \approx 257$ sec

57. $F = \int\limits_{a}^{b} W \cdot \left(\begin{array}{c}\text{strip}\\\text{depth}\end{array}\right) \cdot L(y)\, dy \Rightarrow F = 2 \int\limits_{0}^{2} (62.4)(2-y)(2y)\, dy = 249.6 \int\limits_{0}^{2} \left(2y - y^2\right) dy = 249.6\left[y^2 - \dfrac{y^3}{3}\right]_0^2$

$= (249.6)\left(4 - \dfrac{8}{3}\right) = (249.6)\left(\dfrac{4}{3}\right) = 332.8$ lb

59. $F = \int\limits_{a}^{b} W \cdot \left(\begin{array}{c}\text{strip}\\\text{depth}\end{array}\right) \cdot L(y)\, dy \Rightarrow F = 62.4 \int\limits_{0}^{4} (9-y)\left(2 \cdot \dfrac{\sqrt{y}}{2}\right) dy = 62.4 \int\limits_{0}^{4} \left(9y^{1/2} - 3y^{3/2}\right) dy$

$= 62.4\left[6y^{3/2} - \dfrac{2}{5}y^{5/2}\right]_0^4 = (62.4)\left(6 \cdot 8 - \dfrac{2}{5} \cdot 32\right) = \left(\dfrac{62.4}{5}\right)(48 \cdot 5 - 64) = \dfrac{(62.4)(176)}{5} = 2196.48$ lb

61. $F = \omega_1 \int\limits_{0}^{6} (8-y)(2)(6-y)\, dy + \omega_2 \int\limits_{-6}^{0} (8-y)(2)(y+6)\, dy = 2\omega_1 \int\limits_{0}^{6} \left(48 - 14y + y^2\right) dy + 2\omega_2 \int\limits_{6}^{0} \left(48 + 2y - y^2\right) dy$

$= 2\omega_1\left[48y - 7y^2 + \dfrac{y^3}{3}\right]_0^6 + 2\omega_2\left[48y + y^2 - \dfrac{y^3}{3}\right]_{-6}^0 = 216\omega_1 + 360\omega_2$

63. Total distance $= \int\limits_{a}^{b} |v|\, dt = \int\limits_{0}^{2} v\, dt + \left(-\int\limits_{2}^{6} v\, dt\right)$

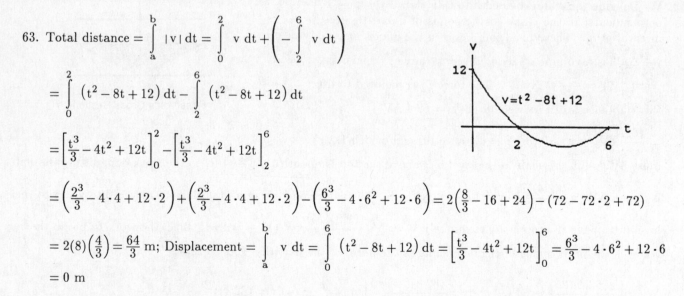

$= \int\limits_{0}^{2} \left(t^2 - 8t + 12\right) dt - \int\limits_{2}^{6} \left(t^2 - 8t + 12\right) dt$

$= \left[\dfrac{t^3}{3} - 4t^2 + 12t\right]_0^2 - \left[\dfrac{t^3}{3} - 4t^2 + 12t\right]_2^6$

$= \left(\dfrac{2^3}{3} - 4 \cdot 4 + 12 \cdot 2\right) + \left(\dfrac{2^3}{3} - 4 \cdot 4 + 12 \cdot 2\right) - \left(\dfrac{6^3}{3} - 4 \cdot 6^2 + 12 \cdot 6\right) = 2\left(\dfrac{8}{3} - 16 + 24\right) - (72 - 72 \cdot 2 + 72)$

$= 2(8)\left(\dfrac{4}{3}\right) = \dfrac{64}{3}$ m; Displacement $= \int\limits_{a}^{b} v\, dt = \int\limits_{0}^{6} \left(t^2 - 8t + 12\right) dt = \left[\dfrac{t^3}{3} - 4t^2 + 12t\right]_0^6 = \dfrac{6^3}{3} - 4 \cdot 6^2 + 12 \cdot 6$

$= 0$ m

65. Total distance $= \int\limits_{0}^{3\pi/2} |v|\, dt = 5 \int\limits_{0}^{\pi/2} \cos t\, dt - 5 \int\limits_{\pi/2}^{3\pi/2} \cos t\, dt$

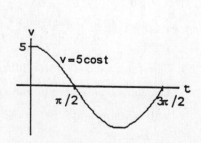

$= 5[\sin t]_0^{\pi/2} - 5[\sin t]_{\pi/2}^{3\pi/2} = 5[(1-0) - (-1-1)] = 15$ m;

Displacement $= 5 \int\limits_{0}^{3\pi/2} \cos t\, dt = 5[\sin t]_0^{3\pi/2} = 5[-1-0] = -5$ m

CHAPTER 5   ADDITIONAL EXERCISES–THEORY, EXAMPLES, APPLICATIONS

1. $V = \pi \displaystyle\int_a^b [f(x)]^2 \, dx = b^2 - ab \Rightarrow \pi \displaystyle\int_a^x [f(t)]^2 \, dt = x^2 - ax$ for all $x > a \Rightarrow \pi [f(x)]^2 = 2x - a \Rightarrow f(x) = \sqrt{\dfrac{2x-a}{\pi}}$

3. $s(x) = Cx \Rightarrow \displaystyle\int_0^x \sqrt{1 + [f'(t)]^2} \, dt = Cx \Rightarrow \sqrt{1 + [f'(x)]^2} = C \Rightarrow f'(x) = \sqrt{C^2 - 1}$ for $C \geq 1$

$\Rightarrow f(x) = \displaystyle\int_0^x \sqrt{C^2 - 1} \, dt + k.$ Then $f(0) = a \Rightarrow a = 0 + k \Rightarrow f(x) = \displaystyle\int_0^x \sqrt{C^2 - 1} \, dt + a \Rightarrow f(x) = x\sqrt{C^2 - 1} + a,$

where $C \geq 1.$

5. From the symmetry of $y = 1 - x^n$, n even, about the y-axis for $-1 \leq x \leq 1$, we have $\overline{x} = 0$. To find $\overline{y} = \dfrac{M_x}{M}$, we

use the vertical strips technique. The typical strip has center of mass: $(\widetilde{x}, \widetilde{y}) = \left(x, \dfrac{1 - x^n}{2}\right)$, length: $1 - x^n$,

width: dx, area: $dA = (1 - x^n) \, dx$, mass: $dm = 1 \cdot dA = (1 - x^n) \, dx$. The moment of the strip about the

x-axis is $\widetilde{y} \, dm = \dfrac{(1 - x^n)^2}{2} \, dx \Rightarrow M_x = \displaystyle\int_{-1}^1 \dfrac{(1 - x^n)^2}{2} \, dx = 2 \displaystyle\int_0^1 \dfrac{1}{2}(1 - 2x^n + x^{2n}) \, dx = \left[x - \dfrac{2x^{n+1}}{n+1} + \dfrac{x^{2n+1}}{2n+1}\right]_0^1$

$= 1 - \dfrac{2}{n+1} + \dfrac{1}{2n+1} = \dfrac{(n+1)(2n+1) - 2(2n+1) + (n+1)}{(n+1)(2n+1)} = \dfrac{2n^2 + 3n + 1 - 4n - 2 + n + 1}{(n+1)(2n+1)} = \dfrac{2n^2}{(n+1)(2n+1)}.$

Also, $M = \displaystyle\int_{-1}^1 dA = \displaystyle\int_{-1}^1 (1 - x^n) \, dx = 2 \displaystyle\int_0^1 (1 - x^n) \, dx = 2\left[x - \dfrac{x^{n+1}}{n+1}\right]_0^1 = 2\left(1 - \dfrac{1}{n+1}\right) = \dfrac{2n}{n+1}.$ Therefore,

$\overline{y} = \dfrac{M_x}{M} = \dfrac{2n^2}{(n+1)(2n+1)} \cdot \dfrac{(n+1)}{2n} = \dfrac{n}{2n+1} \Rightarrow \left(0, \dfrac{n}{2n+1}\right)$ is the location of the centroid. As $n \to \infty$, $\overline{y} \to \dfrac{1}{2}$ so

the limiting position of the centroid is $\left(0, \dfrac{1}{2}\right)$.

7. (a) Consider a single vertical strip with center of mass $(\widetilde{x}, \widetilde{y})$. If the plate lies to the right of the line, then

   the moment of this strip about the line $x = b$ is $(\widetilde{x} - b) \, dm = (\widetilde{x} - b) \delta \, dA \Rightarrow$ the plate's first moment

   about $x = b$ is the integral $\displaystyle\int (x - b)\delta \, dA = \displaystyle\int \delta x \, dA - \displaystyle\int \delta b \, dA = M_y - b \, \delta A.$

   (b) If the plate lies to the left of the line, the moment of a vertical strip about the line $x = b$ is

   $(b - \widetilde{x}) \, dm = (b - \widetilde{x})\delta \, dA \Rightarrow$ the plate's first moment about $x = b$ is $\displaystyle\int (b - x)\delta \, dA = \displaystyle\int b\delta \, dA - \displaystyle\int \delta x \, dA$

   $= b\delta A - M_y.$

9. (a) On $[0, a]$ a typical *vertical* strip has center of mass: $(\widetilde{x}, \widetilde{y}) = \left(x, \dfrac{\sqrt{b^2 - x^2} + \sqrt{a^2 - x^2}}{2}\right),$

   length: $\sqrt{b^2 - x^2} - \sqrt{a^2 - x^2}$, width: dx, area: $dA = \left(\sqrt{b^2 - x^2} - \sqrt{a^2 - x^2}\right) dx$, mass: $dm = \delta \, dA$

$= \delta\left(\sqrt{b^2 - x^2} - \sqrt{a^2 - x^2}\right) dx$. On $[a, b]$ a typical *vertical* strip has center of mass:

$(\tilde{x}, \tilde{y}) = \left(x, \dfrac{\sqrt{b^2 - x^2}}{2}\right)$, length: $\sqrt{b^2 - x^2}$, width: dx, area: $dA = \sqrt{b^2 - x^2}\ dx$,

mass: $dm = \delta\ dA = \delta\sqrt{b^2 - x^2}\ dx$. Thus, $M_x = \displaystyle\int \tilde{y}\ dm$

$$= \int_0^a \tfrac{1}{2}\left(\sqrt{b^2 - x^2} + \sqrt{a^2 - x^2}\right)\delta\left(\sqrt{b^2 - x^2} - \sqrt{a^2 - x^2}\right) dx + \int_a^b \tfrac{1}{2}\sqrt{b^2 - x^2}\,\delta\sqrt{b^2 - x^2}\ dx$$

$$= \frac{\delta}{2}\int_0^a \left[(b^2 - x^2) - (a^2 - x^2)\right] dx + \frac{\delta}{2}\int_a^b (b^2 - x^2)\ dx = \frac{\delta}{2}\int_0^a (b^2 - a^2)\ dx + \frac{\delta}{2}\int_a^b (b^2 - x^2)\ dx$$

$$= \frac{\delta}{2}\left[(b^2 - a^2)x\right]_0^a + \frac{\delta}{2}\left[b^2 x - \frac{x^3}{3}\right]_a^b = \frac{\delta}{2}\left[(b^2 - a^2)a\right] + \frac{\delta}{2}\left[\left(b^3 - \frac{b^3}{3}\right) - \left(b^2 a - \frac{a^3}{3}\right)\right]$$

$$= \frac{\delta}{2}(ab^2 - a^3) + \frac{\delta}{2}\left(\frac{2}{3}b^3 - ab^2 + \frac{a^3}{3}\right) = \frac{\delta b^3}{3} - \frac{\delta a^3}{3} = \delta\left(\frac{b^3 - a^3}{3}\right);\ M_y = \int \tilde{x}\ dm$$

$$= \int_0^a x\delta\left(\sqrt{b^2 - x^2} - \sqrt{a^2 - x^2}\right) dx + \int_a^b x\delta\sqrt{b^2 - x^2}\ dx$$

$$= \delta\int_0^a x(b^2 - x^2)^{1/2}\ dx - \delta\int_0^a x(a^2 - x^2)^{1/2}\ dx + \delta\int_a^b x(b^2 - x^2)^{1/2}\ dx$$

$$= \frac{-\delta}{2}\left[\frac{2(b^2 - x^2)^{3/2}}{3}\right]_0^a + \frac{\delta}{2}\left[\frac{2(a^2 - x^2)^{3/2}}{3}\right]_0^a - \frac{\delta}{2}\left[\frac{2(b^2 - x^2)^{3/2}}{3}\right]_a^b$$

$$= -\frac{\delta}{3}\left[(b^2 - a^2)^{3/2} - (b^2)^{3/2}\right] + \frac{\delta}{3}\left[0 - (a^2)^{3/2}\right] - \frac{\delta}{3}\left[0 - (b^2 - a^2)^{3/2}\right] = \frac{\delta b^3}{3} - \frac{\delta a^3}{3} = \frac{\delta(b^3 - a^3)}{3} = M_x;$$

We calculate the mass geometrically: $M = \delta A = \delta\left(\dfrac{\pi b^2}{4}\right) - \delta\left(\dfrac{\pi a^2}{4}\right) = \dfrac{\delta\pi}{4}(b^2 - a^2)$. Thus, $\bar{x} = \dfrac{M_y}{M}$

$$= \frac{\delta(b^3 - a^3)}{3} \cdot \frac{4}{\delta\pi(b^2 - a^2)} = \frac{4}{3\pi}\left(\frac{b^3 - a^3}{b^2 - a^2}\right) = \frac{4}{3\pi}\frac{(b - a)(a^2 + ab + b^2)}{(b - a)(b + a)} = \frac{4(a^2 + ab + b^2)}{3\pi(a + b)};\ \text{likewise}$$

$$\bar{y} = \frac{M_x}{M} = \frac{4(a^2 + ab + b^2)}{3\pi(a + b)}.$$

(b) $\displaystyle\lim_{b \to a} \frac{4}{3\pi}\left(\frac{a^2 + ab + b^2}{a + b}\right) = \left(\frac{4}{3\pi}\right)\left(\frac{a^2 + a^2 + a^2}{a + a}\right) = \left(\frac{4}{3\pi}\right)\left(\frac{3a^2}{2a}\right) = \frac{2a}{\pi} \Rightarrow (\bar{x}, \bar{y}) = \left(\frac{2a}{\pi}, \frac{2a}{\pi}\right)$ is the limiting

position of the centroid as $b \to a$. This is the centroid of a circle of radius a (and we note the two circles coincide when $b = a$).

11. $y = 2\sqrt{x} \Rightarrow ds = \sqrt{\dfrac{1}{x} + 1}\ dx \Rightarrow A = \displaystyle\int_0^3 2\sqrt{x}\sqrt{\dfrac{1}{x} + 1}\ dx = \dfrac{4}{3}\left[(1 + x)^{3/2}\right]_0^3 = \dfrac{28}{3}$

13. $F = ma = t^2 \Rightarrow \frac{d^2x}{dt^2} = a = \frac{t^2}{m} \Rightarrow v = \frac{dx}{dt} = \frac{t^3}{3m} + C$; $v = 0$ when $t = 0 \Rightarrow C = 0 \Rightarrow \frac{dx}{dt} = \frac{t^3}{3m} \Rightarrow x = \frac{t^4}{12m} + C_1$;

$x = 0$ when $t = 0 \Rightarrow C_1 = 0 \Rightarrow x = \frac{t^4}{12m}$. Then $x = h \Rightarrow t = (12mh)^{1/4}$. The work done is

$$W = \int F\,dx = \int_0^{(12mh)^{1/4}} F(t) \cdot \frac{dx}{dt}\,dt = \int_0^{(12mh)^{1/4}} t^2 \cdot \frac{t^3}{3m}\,dt = \frac{1}{3m}\left[\frac{t^6}{6}\right]_0^{(12mh)^{1/4}} = \left(\frac{1}{18m}\right)(12mh)^{6/4}$$

$$= \frac{(12mh)^{3/2}}{18m} = \frac{12mh \cdot \sqrt{12mh}}{18m} = \frac{2h}{3} \cdot 2\sqrt{3mh} = \frac{4h}{3}\sqrt{3mh}$$

15. The submerged triangular plate is depicted in the figure
    at the right. The hypotenuse of the triangle has slope -1
    $\Rightarrow y - (-2) = -(x - 0) \Rightarrow x = -(y + 2)$ is an equation of
    the hypotenuse. Using a typical horizontal strip, the fluid

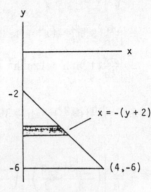

pressure is $F = \int (62.4) \cdot \left(\begin{array}{c}\text{strip} \\ \text{depth}\end{array}\right) \cdot \left(\begin{array}{c}\text{strip} \\ \text{length}\end{array}\right) dy$

$$= \int_{-6}^{-2} (62.4)(-y)[-(y+2)]\,dy = 62.4 \int_{-6}^{-2} (y^2 + 2y)\,dy$$

$$= 62.4\left[\frac{y^3}{3} + y^2\right]_{-6}^{-2} = (62.4)\left[\left(-\frac{8}{3} + 4\right) - \left(-\frac{216}{3} + 36\right)\right] = (62.4)\left(\frac{208}{3} - 32\right) = \frac{(62.4)(112)}{3} \approx 2329.6 \text{ lb}$$

17. (a) We establish a coordinate system as shown. A typical
    horizontal strip has: center of pressure: $(\tilde{x}, \tilde{y}) = \left(\frac{b}{2}, y\right)$,
    length: $L(y) = b$, width: dy, area: $dA = b\,dy$,

    pressure: $dp = \omega|y|\,dA = \omega b|y|\,dy \Rightarrow F_x = \int \tilde{y}\,dp$

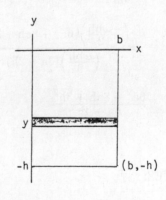

$$= \int_{-h}^0 y \cdot \omega b|y|\,dy = -\omega b \int_{-h}^0 y^2\,dy = -\omega b\left[\frac{y^3}{3}\right]_{-h}^0$$

$$= -\omega b\left[0 - \left(\frac{-h^3}{3}\right)\right] = \frac{-\pi b h^3}{3}; \quad F = \int dp = \int_a^b \omega|y|\,L(y)\,dy$$

$$= -\omega b \int_{-h}^0 y\,dy = -\omega b\left[\frac{y^2}{2}\right]_{-h}^0 = -\omega b\left[0 - \frac{h^2}{2}\right] = \frac{\omega b h^2}{2}. \text{ Thus, } \bar{y} = \frac{F_x}{F} = \frac{\left(\frac{-\omega b h^3}{3}\right)}{\left(\frac{\omega b h^2}{2}\right)} = \frac{-2h}{3} \Rightarrow \text{the distance}$$

below the surface is $\frac{2}{3}h$.

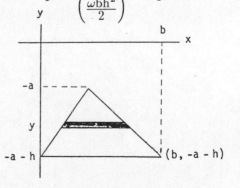

(b) A typical horizontal strip has length $L(y)$. By similar
    triangles from the figure at the right, $\frac{L(y)}{b} = \frac{-y - a}{h}$

    $\Rightarrow L(y) = -\frac{b}{h}(y + a)$. Thus, a typical strip has center

    of pressure: $(\tilde{x}, \tilde{y}) = (\tilde{x}, y)$, length: $L(y) = -\frac{b}{h}(y + a)$, width: dy, area: $dA = -\frac{b}{h}(y + a)\, dy$, pressure:

$$dp = \omega\,|y|\,dA = \omega(-y)\left(-\frac{b}{h}\right)(y + a) = \frac{\omega b}{h}(y^2 + ay)$$

$$\Rightarrow F_x = \int \tilde{y}\, dp = \int_{-(a+h)}^{-a} y \cdot \frac{\omega b}{h}(y^2 + ay)\, dy = \frac{\omega b}{h}\int_{-(a+h)}^{-a} (y^3 + ay^2)\, dy = \frac{\omega b}{h}\left[\frac{y^4}{4} + \frac{ay^3}{3}\right]_{-(a+h)}^{-a}$$

$$= \frac{\omega b}{h}\left[\left(\frac{a^4}{4} - \frac{a^4}{3}\right) - \left(\frac{(a+h)^4}{4} - \frac{a(a+h)^3}{3}\right)\right] = \frac{\omega b}{h}\left[\frac{a^4 - (a+h)^4}{4} - \frac{a^4 - a(a+h)^3}{3}\right]$$

$$= \frac{\omega b}{12h}\left[3\left(a^4 - \left(a^4 + 4a^3h + 6a^2h^2 + 4ah^3 + h^4\right)\right) - 4\left(a^4 - a\left(a^3 + 3a^2h + 3ah^2 + h^3\right)\right)\right]$$

$$= \frac{\omega b}{12h}\left(12a^3h + 12a^2h^2 + 4ah^3 - 12a^3h - 18a^2h^2 - 12ah^3 - 3h^4\right) = \frac{\omega b}{12h}\left(-6a^2h^2 - 8ah^3 - 3h^4\right)$$

$$= \frac{-\omega bh}{12}\left(6a^2 + 8ah + 3h^2\right);\ F = \int dp = \int \omega\,|y|\,L(y)\, dy = \frac{\omega b}{h}\int_{-(a+h)}^{-a} (y^2 + ay)\, dy = \frac{\omega b}{h}\left[\frac{y^3}{3} + \frac{ay^2}{2}\right]_{-(a+h)}^{-a}$$

$$= \frac{\omega b}{h}\left[\left(\frac{-a^3}{3} + \frac{a^3}{2}\right) - \left(\frac{-(a+h)^3}{3} + \frac{a(a+h)^2}{2}\right)\right] = \frac{\omega b}{h}\left[\frac{(a+h)^3 - a^3}{3} + \frac{a^3 - a(a+h)^2}{2}\right]$$

$$= \frac{\omega b}{h}\left[\frac{a^3 + 3a^2h + 3ah^2 + h^3 - a^3}{3} + \frac{a^3 - \left(a^3 + 2a^2h + ah^2\right)}{2}\right] = \frac{\omega b}{6h}\left[2\left(3a^2h + 3ah^2 + h^3\right) - 3\left(2a^2h + ah^2\right)\right]$$

$$= \frac{\omega b}{6h}\left(6a^2h + 6ah^2 + 2h^3 - 6a^2h - 3ah^2\right) = \frac{\omega b}{6h}\left(3ah^2 + 2h^3\right) = \frac{\omega bh}{6}(3a + 2h).\ \text{ Thus, } \overline{y} = \frac{F_x}{F}$$

$$= \frac{\left(\frac{-\omega bh}{12}\right)\left(6a^2 + 8ah + 3h^2\right)}{\left(\frac{\omega bh}{6}\right)(3a + 2h)} = \left(\frac{-1}{2}\right)\left(\frac{6a^2 + 8ah + 3h^2}{3a + 2h}\right) \Rightarrow \text{ the distance below the surface is}$$

$$\frac{6a^2 + 8ah + 3h^2}{6a + 4h}\ .$$

NOTES:

CHAPTER 6 TRANSCENDENTAL FUNCTIONS

6.1 INVERSE FUNCTIONS AND THEIR DERIVATIVES

1. Yes one-to-one, the graph passes the horizontal test

3. Not one-to-one, the graph fails the horizontal test

5. Yes one-to-one, the graph passes the horizontal test

7. Domain: $0 < x \leq 1$, Range: $0 \leq y$

9. Domain: $-1 \leq x \leq 1$, Range: $-\frac{\pi}{2} \leq y \leq \frac{\pi}{2}$

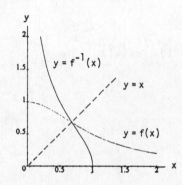

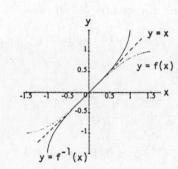

11. The graph is symmetric about $y = x$.

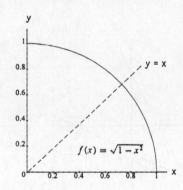

   (b) $y = \sqrt{1 - x^2} \Rightarrow y^2 = 1 - x^2 \Rightarrow x^2 = 1 - y^2 \Rightarrow x = \sqrt{1 - y^2} \Rightarrow y = \sqrt{1 - x^2} = f^{-1}(x)$

13. Step 1: $y = x^2 + 1 \Rightarrow x^2 = y - 1 \Rightarrow x = \sqrt{y - 1}$

    Step 2: $y = \sqrt{x - 1} = f^{-1}(x)$

15. Step 1: $y = x^3 - 1 \Rightarrow x^3 = y + 1 \Rightarrow x = (y + 1)^{1/3}$

    Step 2: $y = \sqrt[3]{x + 1} = f^{-1}(x)$

17. Step 1: $y = (x+1)^2 \Rightarrow \sqrt{y} = x+1 \Rightarrow x = \sqrt{y}-1$

 Step 2: $y = \sqrt{x}-1 = f^{-1}(x)$

19. Step 1: $y = x^5 \Rightarrow x = y^{1/5}$

 Step 2: $y = \sqrt[5]{x} = f^{-1}(x)$;

 Domain and Range of f^{-1}: all reals;

 $f\left(f^{-1}(x)\right) = \left(x^{1/5}\right)^5 = x$ and $f^{-1}(f(x)) = \left(x^5\right)^{1/5} = x$

21. Step 1: $y = x^3 + 1 \Rightarrow x^3 = y-1 \Rightarrow x = (y-1)^{1/3}$

 Step 2: $y = \sqrt[3]{x-1} = f^{-1}(x)$;

 Domain and Range of f^{-1}: all reals;

 $f\left(f^{-1}(x)\right) = \left((x-1)^{1/3}\right)^3 + 1 = (x-1)+1 = x$ and $f^{-1}(f(x)) = \left((x^3+1)-1\right)^{1/3} = \left(x^3\right)^{1/3} = x$

23. Step 1: $y = \dfrac{1}{x^2} \Rightarrow x^2 = \dfrac{1}{y} \Rightarrow x = \dfrac{1}{\sqrt{y}}$

 Step 2: $y = \dfrac{1}{\sqrt{x}} = f^{-1}(x)$

 Domain of f^{-1}: $x > 0$, Range of f^{-1}: $y > 0$;

 $f\left(f^{-1}(x)\right) = \dfrac{1}{\left(\frac{1}{\sqrt{x}}\right)^2} = \dfrac{1}{\left(\frac{1}{x}\right)} = x$ and $f^{-1}(f(x)) = \dfrac{1}{\sqrt{\frac{1}{x^2}}} = \dfrac{1}{\left(\frac{1}{x}\right)} = x$ since $x > 0$

25. (a) $y = 2x+3 \Rightarrow 2x = y-3$

  $\Rightarrow x = \dfrac{y}{2} - \dfrac{3}{2} \Rightarrow f^{-1}(x) = \dfrac{x}{2} - \dfrac{3}{2}$

 (c) $\left.\dfrac{df}{dx}\right|_{x=-1} = 2$, $\left.\dfrac{df^{-1}}{dx}\right|_{x=1} = \dfrac{1}{2}$

          (b)

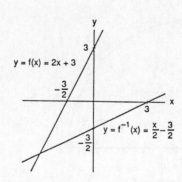

27. (a) $y = \dfrac{1}{5}x + 7 \Rightarrow \dfrac{1}{5}x = y-7$

  $\Rightarrow x = 5y - 35 \Rightarrow f^{-1}(x) = 5x - 35$

 (c) $\left.\dfrac{df}{dx}\right|_{x=-1} = \dfrac{1}{5}$, $\left.\dfrac{df^{-1}}{dx}\right|_{x=34/5} = 5$

          (b)

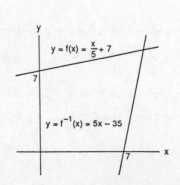

29. (a) $f(g(x)) = \left(\sqrt[3]{x}\right)^3 = x$, $g(f(x)) = \sqrt[3]{x^3} = x$ (b)

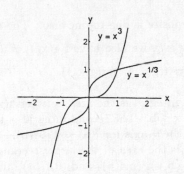

(c) $f'(x) = 3x^2 \Rightarrow f'(1) = 3$, $f'(-1) = 3$;

$\quad g'(x) = \frac{1}{3}x^{-2/3} \Rightarrow g'(1) = \frac{1}{3}$, $g'(-1) = \frac{1}{3}$

(d) The line $y = 0$ is tangent to $f(x) = x^3$ at $(0,0)$;

$\quad$ the line $x = 0$ is tangent to $g(x) = \sqrt[3]{x}$ at $(0,0)$

31. $\dfrac{df}{dx} = 3x^2 - 6x \Rightarrow \dfrac{df^{-1}}{dx}\bigg|_{x=f(3)} = \dfrac{1}{\frac{df}{dx}}\bigg|_{x=3} = \dfrac{1}{9}$ 33. $\dfrac{df^{-1}}{dx}\bigg|_{x=4} = \dfrac{df^{-1}}{dx}\bigg|_{x=f(2)} = \dfrac{1}{\frac{df}{dx}}\bigg|_{x=2} = \dfrac{1}{\left(\frac{1}{3}\right)} = 3$

35. (a) $y = mx \Rightarrow x = \frac{1}{m}y \Rightarrow f^{-1}(x) = \frac{1}{m}x$

(b) The graph of $y = f^{-1}(x)$ is a line through the origin with slope $\frac{1}{m}$.

37. (a) $y = x + 1 \Rightarrow x = y - 1 \Rightarrow f^{-1}(x) = x - 1$

(b) $y = x + b \Rightarrow x = y - b \Rightarrow f^{-1}(x) = x - b$

(c) Their graphs will be parallel to one another and lie on opposite sides of the line $y = x$ equidistant from that line.

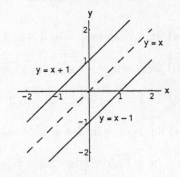

39. Let $x_1 \neq x_2$ be two numbers in the domain of an increasing function f. Then, either $x_1 < x_2$ or $x_1 > x_2$ which implies $f(x_1) < f(x_2)$ or $f(x_1) > f(x_2)$, since $f(x)$ is increasing. In either case, $f(x_1) \neq f(x_2)$ and f is one-to-one. Similar arguments hold if f is decreasing.

41. $f(x)$ is increasing since $x_2 > x_1 \Rightarrow 27x_2^3 > 27x_1^3$; $y = 27x^3 \Rightarrow x = \frac{1}{3}y^{1/3} \Rightarrow f^{-1}(x) = \frac{1}{3}x^{1/3}$;

$\dfrac{df}{dx} = 81x^2 \Rightarrow \dfrac{df^{-1}}{dx} = \dfrac{1}{81x^2}\bigg|_{\frac{1}{3}x^{1/3}} = \dfrac{1}{9x^{2/3}} = \dfrac{1}{9}x^{-2/3}$

43. $f(x)$ is decreasing since $x_2 > x_1 \Rightarrow (1-x_2)^3 < (1-x_1)^3$; $y = (1-x)^3 \Rightarrow x = 1 - y^{1/3} \Rightarrow f^{-1}(x) = 1 - x^{1/3}$;

$\dfrac{df}{dx} = -3(1-x)^2 \Rightarrow \dfrac{df^{-1}}{dx} = \dfrac{1}{-3(1-x)^2}\bigg|_{1-x^{1/3}} = \dfrac{-1}{3x^{2/3}} = -\dfrac{1}{3}x^{-2/3}$

45. The function $g(x)$ is also one-to-one. The reasoning: $f(x)$ is one-to-one means that if $x_1 \neq x_2$ then $f(x_1) \neq f(x_2)$, so $-f(x_1) \neq -f(x_2)$ and therefore $g(x_1) \neq g(x_2)$. Therefore $g(x)$ is one-to-one as well.

47. The composite is one-to-one also. The reasoning: If $x_1 \neq x_2$ then $g(x_1) \neq g(x_2)$ because g is one-to-one. Since $g(x_1) \neq g(x_2)$, we also have $f(g(x_1)) \neq f(g(x_2))$ because f is one-to-one. Thus, $f \circ g$ is one-to-one because $x_1 \neq x_2 \Rightarrow f(g(x_1)) \neq f(g(x_2))$.

49. The first integral is the area between $f(x)$ and the x-axis over $a \leq x \leq b$. The second integral is the area between $f(x)$ and the y-axis for $f(a) \leq y \leq f(b)$. The sum of the integrals is the area of the larger rectangle with corners at $(0,0)$, $(b,0)$, $(b,f(b))$ and $(0,f(b))$ minus the area of the smaller rectangle with vertices at $(0,0)$, $(a,0)$, $(a,f(a))$ and $(0,f(a))$. That is, the sum of the integrals is $bf(b) - af(a)$.

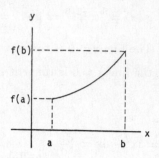

51. $(g \circ f)(x) = x \Rightarrow g(f(x)) = x \Rightarrow g'(f(x))f'(x) = 1$

6.2 NATURAL LOGARITHMS

1. (a) $\ln 0.75 = \ln \frac{3}{4} = \ln 3 - \ln 4 = \ln 3 - \ln 2^2 = \ln 3 - 2\ln 2$

    (b) $\ln \frac{4}{9} = \ln 4 - \ln 9 = \ln 2^2 - \ln 3^2 = 2\ln 2 - 2\ln 3$

    (c) $\ln \frac{1}{2} = \ln 1 - \ln 2 = -\ln 2$ (d) $\ln \sqrt[3]{9} = \frac{1}{3}\ln 9 = \frac{1}{3}\ln 3^2 = \frac{2}{3}\ln 3$

    (e) $\ln 3\sqrt{2} = \ln 3 + \ln 2^{1/2} = \ln 3 + \frac{1}{2}\ln 2$

    (f) $\ln \sqrt{13.5} = \frac{1}{2}\ln 13.5 = \frac{1}{2}\ln \frac{27}{2} = \frac{1}{2}(\ln 3^3 - \ln 2) = \frac{1}{2}(3\ln 3 - \ln 2)$

3. (a) $\ln \sin \theta - \ln\left(\frac{\sin \theta}{5}\right) = \ln\left(\frac{\sin \theta}{\left(\frac{\sin \theta}{5}\right)}\right) = \ln 5$ (b) $\ln(3x^2 - 9x) + \ln\left(\frac{1}{3x}\right) = \ln\left(\frac{3x^2 - 9x}{3x}\right) = \ln(x - 3)$

    (c) $\frac{1}{2}\ln(4t^4) - \ln 2 = \ln \sqrt{4t^4} - \ln 2 = \ln 2t^2 - \ln 2 = \ln\left(\frac{2t^2}{2}\right) = \ln(t^2)$

5. $y = \ln 3x \Rightarrow y' = \left(\frac{1}{3x}\right)(3) = \frac{1}{x}$ 7. $y = \ln(t^2) \Rightarrow \frac{dy}{dt} = \left(\frac{1}{t^2}\right)(2t) = \frac{2}{t}$

9. $y = \ln \frac{3}{x} = \ln 3x^{-1} \Rightarrow \frac{dy}{dx} = \left(\frac{1}{3x^{-1}}\right)(-3x^{-2}) = -\frac{1}{x}$

11. $y = \ln(\theta + 1) \Rightarrow \frac{dy}{d\theta} = \left(\frac{1}{\theta + 1}\right)(1) = \frac{1}{\theta + 1}$ 13. $y = \ln x^3 \Rightarrow \frac{dy}{dx} = \left(\frac{1}{x^3}\right)(3x^2) = \frac{3}{x}$

15. $y = t(\ln t)^2 \Rightarrow \frac{dy}{dt} = (\ln t)^2 + 2t(\ln t) \cdot \frac{d}{dt}(\ln t) = (\ln t)^2 + \frac{2t \ln t}{t} = (\ln t)^2 + 2\ln t$

17. $y = \frac{x^4}{4}\ln x - \frac{x^4}{16} \Rightarrow \frac{dy}{dx} = x^3 \ln x + \frac{x^4}{4} \cdot \frac{1}{x} - \frac{4x^3}{16} = x^3 \ln x$

19. $y = \dfrac{\ln t}{t} \Rightarrow \dfrac{dy}{dt} = \dfrac{t\left(\frac{1}{t}\right) - (\ln t)(1)}{t^2} = \dfrac{1 - \ln t}{t^2}$

21. $y = \dfrac{\ln x}{1 + \ln x} \Rightarrow y' = \dfrac{(1 + \ln x)\left(\frac{1}{x}\right) - (\ln x)\left(\frac{1}{x}\right)}{(1 + \ln x)^2} = \dfrac{\frac{1}{x} + \frac{\ln x}{x} - \frac{\ln x}{x}}{(1 + \ln x)^2} = \dfrac{1}{x(1 + \ln x)^2}$

23. $y = \ln(\ln x) \Rightarrow y' = \left(\dfrac{1}{\ln x}\right)\left(\dfrac{1}{x}\right) = \dfrac{1}{x \ln x}$

25. $y = \theta[\sin(\ln \theta) + \cos(\ln \theta)] \Rightarrow \dfrac{dy}{d\theta} = [\sin(\ln \theta) + \cos(\ln \theta)] + \theta\left[\cos(\ln \theta)\cdot\frac{1}{\theta} - \sin(\ln \theta)\cdot\frac{1}{\theta}\right]$

$= \sin(\ln \theta) + \cos(\ln \theta) + \cos(\ln \theta) - \sin(\ln \theta) = 2\cos(\ln \theta)$

27. $y = \ln \dfrac{1}{x\sqrt{x+1}} = -\ln x - \dfrac{1}{2}\ln(x+1) \Rightarrow y' = -\dfrac{1}{x} - \dfrac{1}{2}\left(\dfrac{1}{x+1}\right) = -\dfrac{2(x+1)+x}{2x(x+1)} = -\dfrac{3x+2}{2x(x+1)}$

29. $y = \dfrac{1 + \ln t}{1 - \ln t} \Rightarrow \dfrac{dy}{dt} = \dfrac{(1 - \ln t)\left(\frac{1}{t}\right) - (1 + \ln t)\left(\frac{-1}{t}\right)}{(1 - \ln t)^2} - \dfrac{\frac{1}{t} - \frac{\ln t}{t} + \frac{1}{t} + \frac{\ln t}{t}}{(1 - \ln t)^2} = \dfrac{2}{t(1 - \ln t)^2}$

31. $y = \ln(\sec(\ln \theta)) \Rightarrow \dfrac{dy}{d\theta} = \dfrac{1}{\sec(\ln \theta)}\cdot\dfrac{d}{d\theta}(\sec(\ln \theta)) = \dfrac{\sec(\ln \theta)\tan(\ln \theta)}{\sec(\ln \theta)}\cdot\dfrac{d}{d\theta}(\ln \theta) = \dfrac{\tan(\ln \theta)}{\theta}$

33. $y = \ln\left(\dfrac{(x^2+1)^5}{\sqrt{1-x}}\right) = 5\ln(x^2+1) - \dfrac{1}{2}\ln(1-x) \Rightarrow y' = \dfrac{5 \cdot 2x}{x^2+1} - \dfrac{1}{2}\left(\dfrac{1}{1-x}\right)(-1) = \dfrac{10x}{x^2+1} + \dfrac{1}{2(1-x)}$

35. $y = \displaystyle\int_{x^2/2}^{x^2} \ln\sqrt{t}\, dt \Rightarrow \dfrac{dy}{dx} = (\ln\sqrt{x^2})\cdot\dfrac{d}{dx}(x^2) - \left(\ln\sqrt{\dfrac{x^2}{2}}\right)\cdot\dfrac{d}{dx}\left(\dfrac{x^2}{2}\right) = 2x \ln|x| - x \ln\dfrac{|x|}{\sqrt{2}}$

37. $y = \sqrt{x(x+1)} = (x(x+1))^{1/2} \Rightarrow \ln y = \dfrac{1}{2}\ln(x(x+1)) \Rightarrow 2\ln y = \ln x + \ln(x+1) \Rightarrow \dfrac{2y'}{y} = \dfrac{1}{x} + \dfrac{1}{x+1}$

$\Rightarrow y' = \left(\dfrac{1}{2}\right)\sqrt{x(x+1)}\left(\dfrac{1}{x} + \dfrac{1}{x+1}\right) = \dfrac{\sqrt{x(x+1)}\,(2x+1)}{2x(x+1)} = \dfrac{2x+1}{2\sqrt{x(x+1)}}$

39. $y = \sqrt{\dfrac{t}{t+1}} = \left(\dfrac{t}{t+1}\right)^{1/2} \Rightarrow \ln y = \dfrac{1}{2}[\ln t - \ln(t+1)] \Rightarrow \dfrac{1}{y}\dfrac{dy}{dt} = \dfrac{1}{2}\left(\dfrac{1}{t} - \dfrac{1}{t+1}\right)$

$\Rightarrow \dfrac{dy}{dt} = \dfrac{1}{2}\sqrt{\dfrac{t}{t+1}}\left(\dfrac{1}{t} - \dfrac{1}{t+1}\right) = \dfrac{1}{2}\sqrt{\dfrac{t}{t+1}}\left[\dfrac{1}{t(t+1)}\right] = \dfrac{1}{2\sqrt{t}\,(t+1)^{3/2}}$

41. $y = \sqrt{\theta+3}\,(\sin\theta) = (\theta+3)^{1/2}\sin\theta \Rightarrow \ln y = \dfrac{1}{2}\ln(\theta+3) + \ln(\sin\theta) \Rightarrow \dfrac{1}{y}\dfrac{dy}{d\theta} = \dfrac{1}{2(\theta+3)} + \dfrac{\cos\theta}{\sin\theta}$

$\Rightarrow \dfrac{dy}{d\theta} = \sqrt{\theta+3}\,(\sin\theta)\left[\dfrac{1}{2(\theta+3)} + \cot\theta\right]$

43. $y = t(t+1)(t+2) \Rightarrow \ln y = \ln t + \ln(t+1) + \ln(t+2) \Rightarrow \frac{1}{y}\frac{dy}{dt} = \frac{1}{t} + \frac{1}{t+1} + \frac{1}{t+2}$

$\Rightarrow \frac{dy}{dt} = t(t+1)(t+2)\left[\frac{1}{t} + \frac{1}{t+1} + \frac{1}{t+2}\right] = t(t+1)(t+2)\left[\frac{(t+1)(t+2) + t(t+2) + t(t+1)}{t(t+1)(t+2)}\right] = 3t^2 + 6t + 2$

45. $y = \frac{\theta + 5}{\theta \cos \theta} \Rightarrow \ln y = \ln(\theta+5) - \ln\theta - \ln(\cos\theta) \Rightarrow \frac{1}{y}\frac{dy}{d\theta} = \frac{1}{\theta+5} - \frac{1}{\theta} + \frac{\sin\theta}{\cos\theta}$

$\Rightarrow \frac{dy}{d\theta} = \left(\frac{\theta+5}{\theta\cos\theta}\right)\left(\frac{1}{\theta+5} - \frac{1}{\theta} + \tan\theta\right)$

47. $y = \frac{x\sqrt{x^2+1}}{(x+1)^{2/3}} \Rightarrow \ln y = \ln x + \frac{1}{2}\ln(x^2+1) - \frac{2}{3}\ln(x+1) \Rightarrow \frac{y'}{y} = \frac{1}{x} + \frac{x}{x^2+1} - \frac{2}{3(x+1)}$

$\Rightarrow y' = \frac{x\sqrt{x^2+1}}{(x+1)^{2/3}}\left(\frac{1}{x} + \frac{x}{x^2+1} - \frac{2}{3(x+1)}\right)$

49. $y = \sqrt[3]{\frac{x(x-2)}{x^2+1}} \Rightarrow \ln y = \frac{1}{3}\left[\ln x + \ln(x-2) - \ln(x^2+1)\right] \Rightarrow \frac{y'}{y} = \frac{1}{3}\left(\frac{1}{x} + \frac{1}{x-2} - \frac{2x}{x^2+1}\right)$

$\Rightarrow y' = \frac{1}{3}\sqrt[3]{\frac{x(x-2)}{x^2+1}}\left(\frac{1}{x} + \frac{1}{x-2} - \frac{2x}{x^2+1}\right)$

51. $\int_{-3}^{-2} \frac{1}{x}\,dx = \left[\ln|x|\right]_{-3}^{-2} = \ln 2 - \ln 3 = \ln\frac{2}{3}$ 53. $\int \frac{2y}{y^2-25}\,dy = \ln\left|y^2-25\right| + C$

55. $\int_0^\pi \frac{\sin t}{2-\cos t}\,dt = \left[\ln|2-\cos t|\right]_0^\pi = \ln 3 - \ln 1 = \ln 3$; or let $u = 2 - \cos t \Rightarrow du = \sin t\,dt$ with $t = 0$

$\Rightarrow u = 1$ and $t = \pi \Rightarrow u = 3 \Rightarrow \int_0^\pi \frac{\sin t}{2-\cos t}\,dt = \int_1^3 \frac{1}{u}\,du = \left[\ln|u|\right]_1^3 = \ln 3 - \ln 1 = \ln 3$

57. Let $u = \ln x \Rightarrow du = \frac{1}{x}\,dx$; $x = 1 \Rightarrow u = 0$ and $x = 2 \Rightarrow u = \ln 2$;

$\int_1^2 \frac{2\ln x}{x}\,dx = \int_0^{\ln 2} 2u\,du = \left[u^2\right]_0^{\ln 2} = (\ln 2)^2$

59. Let $u = \ln x \Rightarrow du = \frac{1}{x}\,dx$; $x = 2 \Rightarrow u = \ln 2$ and $x = 4 \Rightarrow u = \ln 4$;

$\int_2^4 \frac{dx}{x(\ln x)^2} = \int_{\ln 2}^{\ln 4} u^{-2}\,du = \left[-\frac{1}{u}\right]_{\ln 2}^{\ln 4} = -\frac{1}{\ln 4} + \frac{1}{\ln 2} = -\frac{1}{\ln 2^2} + \frac{1}{\ln 2} = -\frac{1}{2\ln 2} + \frac{1}{\ln 2} = \frac{1}{2\ln 2} = \frac{1}{\ln 4}$

61. Let $u = 6 + 3\tan t \Rightarrow du = 3\sec^2 t\,dt$;

$\int \frac{3\sec^2 t}{6+3\tan t}\,dt = \int \frac{du}{u} = \ln|u| + C = \ln|6+3\tan t| + C$

63. Let $u = \cos \frac{x}{2} \Rightarrow du = -\frac{1}{2} \sin \frac{x}{2} dx \Rightarrow -2\, du = \sin \frac{x}{2} dx;\ x = 0 \Rightarrow u = 1$ and $x = \frac{\pi}{2} \Rightarrow u = \frac{1}{\sqrt{2}}$;

$$\int_0^{\pi/2} \tan \frac{x}{2}\, dx = \int_0^{\pi/2} \frac{\sin \frac{x}{2}}{\cos \frac{x}{2}}\, dx = -2 \int_1^{1/\sqrt{2}} \frac{du}{u} = \left[-2 \ln |u|\right]_1^{1/\sqrt{2}} = -2 \ln \frac{1}{\sqrt{2}} = 2 \ln \sqrt{2} = \ln 2$$

65. Let $u = \sin \frac{\theta}{3} \Rightarrow du = \frac{1}{3} \cos \frac{\theta}{3}\, d\theta \Rightarrow 6\, du = 2 \cos \frac{\theta}{3}\, d\theta;\ \theta = \frac{\pi}{2} \Rightarrow u = \frac{1}{2}$ and $\theta = \pi \Rightarrow u = \frac{\sqrt{3}}{2}$;

$$\int_{\pi/2}^{\pi} 2 \cot \frac{\theta}{3}\, d\theta = \int_{\pi/2}^{\pi} \frac{2 \cos \frac{\theta}{3}}{\sin \frac{\theta}{3}}\, d\theta = 6 \int_{1/2}^{\sqrt{3}/2} \frac{du}{u} = 6\left[\ln |u|\right]_{1/2}^{\sqrt{3}/2} = 6\left(\ln \frac{\sqrt{3}}{2} - \ln \frac{1}{2}\right) = 6 \ln \sqrt{3} = \ln 27$$

67. $\int \frac{dx}{2\sqrt{x} + 2x} = \int \frac{dx}{2\sqrt{x}(1 + \sqrt{x})}$; let $u = 1 + \sqrt{x} \Rightarrow du = \frac{1}{2\sqrt{x}}\, dx;\ \int \frac{dx}{2\sqrt{x}(1 + \sqrt{x})} = \int \frac{du}{u} = \ln |u| + C$

$= \ln \left|1 + \sqrt{x}\right| + C = \ln (1 + \sqrt{x}) + C$

69. (a) $f(x) = \ln(\cos x) \Rightarrow f'(x) = -\frac{\sin x}{\cos x} = -\tan x = 0 \Rightarrow x = 0;\ f'(x) > 0$ for $-\frac{\pi}{4} \leq x < 0$ and $f'(x) < 0$ for

$0 < x \leq \frac{\pi}{3} \Rightarrow$ there is a relative maximum at $x = 0$ with $f(0) = \ln(\cos 0) = \ln 1 = 0;\ f\left(-\frac{\pi}{4}\right) = \ln\left(\cos\left(-\frac{\pi}{4}\right)\right)$

$= \ln\left(\frac{1}{\sqrt{2}}\right) = -\frac{1}{2} \ln 2$ and $f\left(\frac{\pi}{3}\right) = \ln\left(\cos\left(\frac{\pi}{3}\right)\right) = \ln \frac{1}{2} = -\ln 2$. Therefore, the absolute minimum occurs at

$x = \frac{\pi}{3}$ with $f\left(\frac{\pi}{3}\right) = -\ln 2$ and the absolute maximum occurs at $x = 0$ with $f(0) = 0$.

(b) $f(x) = \cos(\ln x) \Rightarrow f'(x) = \frac{-\sin(\ln x)}{x} = 0 \Rightarrow x = 1;\ f'(x) > 0$ for $\frac{1}{2} \leq x < 1$ and $f'(x) < 0$ for $1 < x \leq 2$

$\Rightarrow$ there is a relative maximum at $x = 1$ with $f(1) = \cos(\ln 1) = \cos 0 = 1;\ f\left(\frac{1}{2}\right) = \cos\left(\ln\left(\frac{1}{2}\right)\right)$

$= \cos(-\ln 2) = \cos(\ln 2)$ and $f(2) = \cos(\ln 2)$. Therefore, the absolute minimum occurs at $x = \frac{1}{2}$ and

$x = 2$ with $f\left(\frac{1}{2}\right) = f(2) = \cos(\ln 2)$, and the absolute maximum occurs at $x = 1$ with $f(1) = 1$.

71. $\int_1^5 (\ln 2x - \ln x)\, dx = \int_1^5 (-\ln x + \ln 2 + \ln x)\, dx = (\ln 2) \int_1^5 dx = (\ln 2)(5 - 1) = \ln 2^4 = \ln 16$

73. $V = \pi \int_0^3 \left(\frac{2}{\sqrt{y+1}}\right)^2 dy = 4\pi \int_0^3 \frac{1}{y+1}\, dy = 4\pi \left[\ln |y+1|\right]_0^3 = 4\pi(\ln 4 - \ln 1) = 4\pi \ln 4$

75. $V = 2\pi \int_{1/2}^2 x\left(\frac{1}{x^2}\right) dx = 2\pi \int_{1/2}^2 \frac{1}{x}\, dx = 2\pi \left[\ln |x|\right]_{1/2}^2 = 2\pi\left(\ln 2 - \ln \frac{1}{2}\right) = 2\pi(2 \ln 2) = \pi \ln 2^4 = \pi \ln 16$

77. (a) $y = \frac{x^2}{8} - \ln x \Rightarrow 1 + (y')^2 = 1 + \left(\frac{x}{4} - \frac{1}{x}\right)^2 = 1 + \left(\frac{x^2 - 4}{4x}\right)^2 = \left(\frac{x^2 + 4}{4x}\right)^2 \Rightarrow L = \int\limits_{4}^{8} \sqrt{1 + (y')^2} \, dx$

$= \int\limits_{4}^{8} \frac{x^2 + 4}{4x} \, dx = \int\limits_{4}^{8} \left(\frac{x}{4} + \frac{1}{x}\right) dx = \left[\frac{x^2}{8} + \ln |x|\right]_{4}^{8} = (8 + \ln 8) - (2 + \ln 4) = 6 + \ln 2$

(b) $x = \left(\frac{y}{4}\right)^2 - 2 \ln \left(\frac{y}{4}\right) \Rightarrow \frac{dx}{dy} = \frac{y}{8} - \frac{2}{y} \Rightarrow 1 + \left(\frac{dx}{dy}\right)^2 = 1 + \left(\frac{y}{8} - \frac{2}{y}\right)^2 = 1 + \left(\frac{y^2 - 16}{8y}\right)^2 = \left(\frac{y^2 + 16}{8y}\right)^2$

$\Rightarrow L = \int\limits_{4}^{12} \sqrt{1 + \left(\frac{dx}{dy}\right)^2} \, dy = \int\limits_{4}^{12} \frac{y^2 + 16}{8y} \, dy = \int\limits_{4}^{12} \left(\frac{y}{8} + \frac{2}{y}\right) dy = \left[\frac{y^2}{16} + 2 \ln y\right]_{4}^{12} = (9 + 2 \ln 12) - (1 + 2 \ln 4)$

$= 8 + 2 \ln 3 = 8 + \ln 9$

79. (a) $M_y = \int\limits_{1}^{2} x\left(\frac{1}{x}\right) dx = 1; \; M_x = \int\limits_{1}^{2} \left(\frac{1}{2x}\right)\left(\frac{1}{x}\right) dx = \frac{1}{2} \int\limits_{1}^{2} \frac{1}{x^2} \, dx = \left[-\frac{1}{2x}\right]_{1}^{2} = \frac{1}{4}, \; M = \int\limits_{1}^{2} \frac{1}{x} \, dx = \left[\ln |x|\right]_{1}^{2} = \ln 2$

$\Rightarrow \bar{x} = \frac{M_x}{M} = \frac{1}{\ln 2} \approx 1.44 \text{ and } \bar{y} = \frac{M_y}{M} = \frac{\left(\frac{1}{4}\right)}{\ln 2} \approx 0.36$

(b)

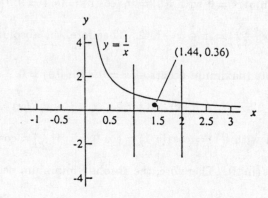

81. $\frac{dy}{dx} = 1 + \frac{1}{x}$ at $(1, 3) \Rightarrow y = x + \ln |x| + C; \; y = 3$ at $x = 1 \Rightarrow C = 2 \Rightarrow y = x + \ln |x| + 2$

83. (a) $L(x) = f(0) + f'(0) \cdot x$, and $f(x) = \ln (1 + x) \Rightarrow f'(x)\big|_{x=0} = \frac{1}{1+x}\big|_{x=0} = 1 \Rightarrow L(x) = \ln 1 + 1 \cdot x \Rightarrow L(x) = x$

(b) $|E(x)| \le \frac{1}{2} M(x - a)^2$, where $a = 0$ and M is an upper bound for max $|f''(x)|$ over $0 \le x \le 0.1$. Now,

$f''(x) = -\frac{1}{(x+1)^2} \Rightarrow |f''(x)| \le 1$ on $0 \le x \le 0.1 \Rightarrow |E(x)| \le \frac{1}{2}(1)(x)^2 \le \frac{1}{2}(1)(0.1)^2 \le 0.005$

(c) The approximation $y = x$ for $\ln (1 + x)$ is best for smaller positive values of x; in particular for $0 \le x \le 0.1$ in the graph. As x increases, so does the error $x - \ln (1 + x)$. From the graph an upper bound for the error is $0.5 - \ln (1 + 0.5) \approx 0.095$; i.e., $|E(x)| \le 0.095$ for $0 \le x \le 0.5$. Note from the graph that $0.1 - \ln (1 + 0.1) \approx 0.00469$ estimates the error in replacing $\ln (1 + x)$ by x over $0 \le x \le 0.1$. This is consistent with the estimate given in part (b) above.

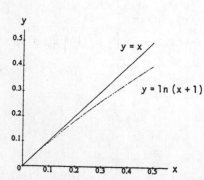

85. $\lim\limits_{x\to\infty} \dfrac{\ln x^2}{\ln x} = \lim\limits_{x\to\infty} \dfrac{2\ln x}{\ln x} = 2$; in general, $\lim\limits_{x\to\infty} \dfrac{\ln x^n}{\ln x^m} = \lim\limits_{x\to\infty} \dfrac{n\ln x}{m\ln x} = \dfrac{n}{m}$

87. $y = \ln kx \Rightarrow y = \ln x + \ln k$; thus the graph of
$y = \ln kx$ is the graph of $y = \ln x$ shifted vertically
by $\ln k$, $k > 0$.

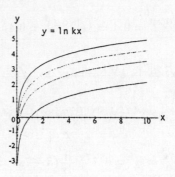

89. (b) As a increases, the value of $a + \sin x$ gets closer
and closer to $a \pm 1$. Thus, $\ln(a + \sin x)$ looks more
and more like the constant value $\ln a$ for larger
and larger values of a $\Rightarrow$ the curves flatten as a
increases.

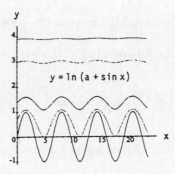

6.3 THE EXPONENTIAL FUNCTION

1. (a) $e^{\ln 7.2} = 7.2$ (b) $e^{-\ln x^2} = \dfrac{1}{e^{\ln x^2}} = \dfrac{1}{x^2}$ (c) $e^{\ln x - \ln y} = e^{\ln(x/y)} = \dfrac{x}{y}$

3. (a) $2\ln\sqrt{e} = 2\ln e^{1/2} = (2)\left(\dfrac{1}{2}\right)\ln e = 1$ (b) $\ln(\ln e^e) = \ln(e\ln e) = \ln e = 1$

   (c) $\ln e^{\left(-x^2 - y^2\right)} = \left(-x^2 - y^2\right)\ln e = -x^2 - y^2$

5. $\ln y = 2t + 4 \Rightarrow e^{\ln y} = e^{2t+4} \Rightarrow y = e^{2t+4}$

7. $\ln(y - 40) = 5t \Rightarrow e^{\ln(y-40)} = e^{5t} \Rightarrow y - 40 = e^{5t} \Rightarrow y = e^{5t} + 40$

9. $\ln(y - 1) - \ln 2 = x + \ln x \Rightarrow \ln(y - 1) - \ln 2 - \ln x = x \Rightarrow \ln\left(\dfrac{y-1}{2x}\right) = x \Rightarrow e^{\ln\left(\frac{y-1}{2x}\right)} = e^x \Rightarrow \dfrac{y-1}{2x} = e^x$
   $\Rightarrow y - 1 = 2xe^x \Rightarrow y = 2xe^x + 1$

11. (a) $e^{2k} = 4 \Rightarrow \ln e^{2k} = \ln 4 \Rightarrow 2k\ln e = \ln 2^2 \Rightarrow 2k = 2\ln 2 \Rightarrow k = \ln 2$

    (b) $100e^{10k} = 200 \Rightarrow e^{10k} = 2 \Rightarrow \ln e^{10k} = \ln 2 \Rightarrow 10k\ln e = \ln 2 \Rightarrow 10k = \ln 2 \Rightarrow k = \dfrac{\ln 2}{10}$

    (c) $e^{k/1000} = a \Rightarrow \ln e^{k/1000} = \ln a \Rightarrow \dfrac{k}{1000}\ln e = \ln a \Rightarrow \dfrac{k}{1000} = \ln a \Rightarrow k = 1000\ln a$

13. (a) $e^{-0.3t} = 27 \Rightarrow \ln e^{-0.3t} = \ln 3^3 \Rightarrow (-0.3t) \ln e = 3 \ln 3 \Rightarrow -0.3t = 3 \ln 3 \Rightarrow t = -10 \ln 3$

(b) $e^{kt} = \frac{1}{2} \Rightarrow \ln e^{kt} = \ln 2^{-1} = kt \ln e = -\ln 2 \Rightarrow t = -\frac{\ln 2}{k}$

(c) $e^{(\ln 0.2)t} = 0.4 \Rightarrow \left(e^{\ln 0.2}\right)^t = 0.4 \Rightarrow 0.2^t = 0.4 \Rightarrow \ln 0.2^t = \ln 0.4 \Rightarrow t \ln 0.2 = \ln 0.4 \Rightarrow t = \frac{\ln 0.4}{\ln 0.2}$

15. $e^{\sqrt{t}} = x^2 \Rightarrow \ln e^{\sqrt{t}} = \ln x^2 \Rightarrow \sqrt{t} = 2 \ln x \Rightarrow t = 4(\ln x)^2$

17. $y = e^{-5x} \Rightarrow y' = e^{-5x} \frac{d}{dx}(-5x) \Rightarrow y' = -5e^{-5x}$

19. $y = e^{5-7x} \Rightarrow y' = e^{5-7x} \frac{d}{dx}(5-7x) \Rightarrow y' = -7e^{5-7x}$

21. $y = xe^x - e^x \Rightarrow y' = (e^x + xe^x) - e^x = xe^x$

23. $y = (x^2 - 2x + 2)e^x \Rightarrow y' = (2x-2)e^x + (x^2 - 2x + 2)e^x = x^2 e^x$

25. $y = e^\theta(\sin\theta + \cos\theta) \Rightarrow y' = e^\theta(\sin\theta + \cos\theta) + e^\theta(\cos\theta - \sin\theta) = 2e^\theta \cos\theta$

27. $y = \cos\left(e^{-\theta^2}\right) \Rightarrow \frac{dy}{d\theta} = -\sin\left(e^{-\theta^2}\right) \frac{d}{d\theta}\left(e^{-\theta^2}\right) = \left(-\sin\left(e^{-\theta^2}\right)\right)\left(e^{-\theta^2}\right)\frac{d}{d\theta}(-\theta^2) = 2\theta e^{-\theta^2} \sin\left(e^{-\theta^2}\right)$

29. $y = \ln\left(3te^{-t}\right) = \ln 3 + \ln t + \ln e^{-t} = \ln 3 + \ln t - t \Rightarrow \frac{dy}{dt} = \frac{1}{t} - 1 = \frac{1-t}{t}$

31. $y = \ln \frac{e^\theta}{1 + e^\theta} = \ln e^\theta - \ln(1 + e^\theta) = \theta - \ln(1 + e^\theta) \Rightarrow \frac{dy}{d\theta} = 1 - \left(\frac{1}{1+e^\theta}\right)\frac{d}{d\theta}(1 + e^\theta) = 1 - \frac{e^\theta}{1 + e^\theta} = \frac{1}{1 + e^\theta}$

33. $y = e^{(\cos t + \ln t)} = e^{\cos t} e^{\ln t} = te^{\cos t} \Rightarrow \frac{dy}{dt} = e^{\cos t} + te^{\cos t}\frac{d}{dt}(\cos t) = (1 - t\sin t)e^{\cos t}$

35. $\displaystyle\int_0^{\ln x} \sin e^t \, dt \Rightarrow y' = (\sin e^{\ln x}) \cdot \frac{d}{dx}(\ln x) = \frac{\sin x}{x}$

37. $\ln y = e^y \sin x \Rightarrow \left(\frac{1}{y}\right)y' = (y'e^y)(\sin x) + e^y \cos x \Rightarrow y'\left(\frac{1}{y} - e^y \sin x\right) = e^y \cos x$

$\Rightarrow y'\left(\frac{1 - ye^y \sin x}{y}\right) = e^y \cos x \Rightarrow y' = \frac{ye^y \cos x}{1 - ye^y \sin x}$

39. $e^{2x} = \sin(x + 3y) \Rightarrow 2e^{2x} = (1 + 3y')\cos(x + 3y) \Rightarrow 1 + 3y' = \frac{2e^{2x}}{\cos(x + 3y)} \Rightarrow 3y' = \frac{2e^{2x}}{\cos(x + 3y)} - 1$

$\Rightarrow y' = \frac{2e^{2x} - \cos(x + 3y)}{3\cos(x + 3y)}$

41. $\displaystyle\int \left(e^{3x} + 5e^{-x}\right) dx = \frac{e^{3x}}{3} - 5e^{-x} + C$

43. $\displaystyle\int_{\ln 2}^{\ln 3} e^x \, dx = \left[e^x\right]_{\ln 2}^{\ln 3} = e^{\ln 3} - e^{\ln 2} = 3 - 2 = 1$

45. $\displaystyle\int 8e^{(x+1)} \, dx = 8e^{(x+1)} + C$

47. $\int\limits_{\ln 4}^{\ln 9} e^{x/2}\,dx = \left[2e^{x/2}\right]_{\ln 4}^{\ln 9} = 2\left[e^{(\ln 9)/2} - e^{(\ln 4)/2}\right] = 2\left(e^{\ln 3} - e^{\ln 2}\right) = 2(3-2) = 2$

49. Let $u = r^{1/2} \Rightarrow du = \frac{1}{2}r^{-1/2}\,dr \Rightarrow 2\,du = r^{-1/2}\,dr$;

$\int \dfrac{e^{\sqrt{r}}}{\sqrt{r}}\,dr = \int e^{r^{1/2}} \cdot r^{-1/2}\,dr = 2\int e^u\,du = 2e^u + C = 2e^{r^{1/2}} + C = 2e^{\sqrt{r}} + C$

51. Let $u = -t^2 \Rightarrow du = -2t\,dt \Rightarrow -du = 2t\,dt$;

$\int 2te^{-t^2}\,dt = -\int e^u\,du = -e^u + C = -e^{-t^2} + C$

53. Let $u = \frac{1}{x} \Rightarrow du = -\frac{1}{x^2}\,dx \Rightarrow -du = \frac{1}{x^2}\,dx$;

$\int \dfrac{e^{1/x}}{x^2}\,dx = \int -e^u\,du = -e^u + C = -e^{1/x} + C$

55. Let $u = \tan\theta \Rightarrow du = \sec^2\theta\,d\theta$; $\theta = 0 \Rightarrow u = 0$, $\theta = \frac{\pi}{4} \Rightarrow u = 1$;

$\int\limits_{0}^{\pi/4} \left(1 + e^{\tan\theta}\right)\sec^2\theta\,d\theta = \int\limits_{0}^{\pi/4} \sec^2\theta\,d\theta + \int\limits_{0}^{1} e^u\,du = [\tan\theta]_0^{\pi/4} + [e^u]_0^1 = \left[\tan\left(\frac{\pi}{4}\right) - \tan(0)\right] + \left(e^1 - e^0\right)$

$= (1-0) + (e-1) = e$

57. Let $u = \sec\pi t \Rightarrow du = \frac{1}{\pi}\sec\pi t \tan\pi t\,dt \Rightarrow \pi\,du = \sec\pi t \tan\pi t\,dt$;

$\int e^{\sec(\pi t)}\sec(\pi t)\tan(\pi t)\,dt = \frac{1}{\pi}\int e^u\,du = \frac{e^u}{\pi} + C = \frac{e^{\sec(\pi t)}}{\pi} + C$

59. Let $u = e^v \Rightarrow du = e^v\,dv \Rightarrow 2\,du = 2e^v\,dv$; $v = \ln\frac{\pi}{6} \Rightarrow u = \frac{\pi}{6}$, $v = \ln\frac{\pi}{2} \Rightarrow u = \frac{\pi}{2}$;

$\int\limits_{\ln(\pi/6)}^{\ln(\pi/2)} 2e^v\cos e^v\,dv = 2\int\limits_{\pi/6}^{\pi/2} \cos u\,du = [2\sin u]_{\pi/6}^{\pi/2} = 2\left[\sin\left(\frac{\pi}{2}\right) - \sin\left(\frac{\pi}{6}\right)\right] = 2\left(1 - \frac{1}{2}\right) = 1$

61. Let $u = 1 + e^r \Rightarrow du = e^r\,dr$;

$\int \dfrac{e^r}{1 + e^r}\,dr = \int \frac{1}{u}\,du = \ln|u| + C = \ln\left(1 + e^r\right) + C$

63. $\dfrac{dy}{dt} = e^t\sin\left(e^t - 2\right) \Rightarrow y = \int e^t\sin\left(e^t - 2\right)\,dt$;

let $u = e^t - 2 \Rightarrow du = e^t\,dt \Rightarrow y = \int \sin u\,du = -\cos u + C = -\cos\left(e^t - 2\right) + C$; $y(\ln 2) = 0$

$\Rightarrow -\cos\left(e^{\ln 2} - 2\right) + C = 0 \Rightarrow -\cos(2-2) + C = 0 \Rightarrow C = \cos 0 = 1$; thus, $y = 1 - \cos\left(e^t - 2\right)$

65. $\frac{d^2y}{dx^2} = 2e^{-x} \Rightarrow \frac{dy}{dx} = -2e^{-x} + C$; $x = 0$ and $\frac{dy}{dx} = 0 \Rightarrow 0 = -2e^0 + C \Rightarrow C = 2$; thus $\frac{dy}{dx} = -2e^{-x} + 2$

$\Rightarrow y = 2e^{-x} + 2x + C_1$; $x = 0$ and $y = 1 \Rightarrow 1 = 2e^0 + C_1 \Rightarrow C_1 = -1 \Rightarrow y = 2e^{-x} + 2x - 1 = 2(e^{-x} + x) - 1$

67. $f(x) = e^x - 2x \Rightarrow f'(x) = e^x - 2$; $f'(x) = 0 \Rightarrow e^x = 2 \Rightarrow x = \ln 2$; $f(0) = 1$, the absolute maximum;

$f(\ln 2) = 2 - 2\ln 2 \approx 0.613706$, the absolute minimum; $f(1) = e - 2 \approx 0.71828$, a relative or local maximum

since $f''(x) = e^x$ is always positive

69. $f(x) = x^2 \ln \frac{1}{x} \Rightarrow f'(x) = 2x \ln \frac{1}{x} + x^2\left(\frac{1}{\frac{1}{x}}\right)(-x^{-2}) = 2x \ln \frac{1}{x} - x = -x(2 \ln x + 1)$; $f'(x) = 0 \Rightarrow x = 0$ or

$\ln x = -\frac{1}{2}$. Since $x = 0$ is not in the domain of f, $x = e^{-1/2} = \frac{1}{\sqrt{e}}$. Also, $f'(x) > 0$ for $0 < x < \frac{1}{\sqrt{e}}$ and

$f'(x) < 0$ for $x > \frac{1}{\sqrt{e}}$. Therefore, $f\left(\frac{1}{\sqrt{e}}\right) = \frac{1}{e} \ln \sqrt{e} = \frac{1}{e} \ln e^{1/2} = \frac{1}{2e} \ln e = \frac{1}{2e}$ is the absolute maximum value

of f assumed at $x = \frac{1}{\sqrt{e}}$.

71. $\int_0^{\ln 3} (e^{2x} - e^x)\, dx = \left[\frac{e^{2x}}{2} - e^x\right]_0^{\ln 3} = \left(\frac{e^{2\ln 3}}{2} - e^{\ln 3}\right) - \left(\frac{e^0}{2} - e^0\right) = \left(\frac{9}{2} - 3\right) - \left(\frac{1}{2} - 1\right) = \frac{8}{2} - 2 = 2$

73. $L = \int_0^1 \sqrt{1 + \frac{e^x}{4}}\, dx \Rightarrow \frac{dy}{dx} = \frac{e^{x/2}}{2} \Rightarrow y = e^{x/2} + C$; $y(0) = 0 \Rightarrow 0 = e^0 + C \Rightarrow C = -1 \Rightarrow y = e^{x/2} - 1$

75. (a) $\frac{d}{dx}(x \ln x - x + C) = x \cdot \frac{1}{x} + \ln x - 1 + 0 = \ln x$

(b) average value $= \frac{1}{e-1} \int_1^e \ln x\, dx = \frac{1}{e-1}[x \ln x - x]_1^e = \frac{1}{e-1}[(e \ln e - e) - (1 \ln 1 - 1)]$

$= \frac{1}{e-1}(e - e + 1) = \frac{1}{e-1}$

77. (a) $f(x) = e^x \Rightarrow f'(x) = e^x$; $L(x) = f(0) + f'(0)(x - 0) \Rightarrow L(x) = 1 + x$

(b) $f(0) = 1$ and $L(0) = 1 \Rightarrow$ error $= 0$; $f(0.2) = e^{0.2} \approx 1.22140$ and $L(0.2) = 1.2 \Rightarrow$ error ≈ 0.02140

(c) Since $y'' = e^x > 0$, the tangent line linear

approximation always lies below the curve $y = e^x$.

Thus $L(x) = x + 1$ always underestimates e^x.

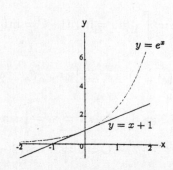

79. $f(x) = \ln(x) - 1 \Rightarrow f'(x) = \frac{1}{x} \Rightarrow x_{n+1} = x_n - \dfrac{\ln(x_n) - 1}{\left(\frac{1}{x_n}\right)} \Rightarrow x_{n+1} = x_n[2 - \ln(x_n)]$. Then $x_1 = 2$

$\Rightarrow x_2 = 2.61370564$, $x_3 = 2.71624393$ and $x_5 = 2.71828183$.

81. Note that $y = \ln x$ and $e^y = x$ are the same curve; $\displaystyle\int_1^a \ln x \, dx$ = area under the curve between 1 and a;

$\displaystyle\int_0^{\ln a} e^y \, dy$ = area to the left of the curve. The sum of these areas is equal to the area of the rectangle

$\Rightarrow \displaystyle\int_1^a \ln x \, dx + \int_0^{\ln a} e^y \, dy = a \ln a.$

6.4 a^x and $\log_a x$

1. (a) $5^{\log_5 7} = 7$
   (b) $8^{\log_8 \sqrt{2}} = \sqrt{2}$
   (c) $1.3^{\log_{1.3} 75} = 75$

   (d) $\log_4 16 = \log_4 4^2 = 2 \log_4 4 = 2 \cdot 1 = 2$
   (e) $\log_3 \sqrt{3} = \log_3 3^{1/2} = \frac{1}{2} \log_3 3 = \frac{1}{2} \cdot 1 = \frac{1}{2} = 0.5$

   (f) $\log_4 \left(\frac{1}{4}\right) = \log_4 4^{-1} = -1 \log_4 4 = -1 \cdot 1 = -1$

3. (a) Let $z = \log_4 x \Rightarrow 4^z = x \Rightarrow 2^{2z} = x \Rightarrow \left(2^z\right)^2 = x \Rightarrow 2^z = \sqrt{x}$

   (b) Let $z = \log_3 x \Rightarrow 3^z = x \Rightarrow \left(3^z\right)^2 = x^2 \Rightarrow 3^{2z} = x^2 \Rightarrow 9^z = x^2$

   (c) $\log_2\left(e^{(\ln 2)\sin x}\right) = \log_2 2^{\sin x} = \sin x$

5. (a) $\dfrac{\log_2 x}{\log_3 x} = \dfrac{\ln x}{\ln 2} \div \dfrac{\ln x}{\ln 3} = \dfrac{\ln x}{\ln 2} \cdot \dfrac{\ln 3}{\ln x} = \dfrac{\ln 3}{\ln 2}$
   (b) $\dfrac{\log_2 x}{\log_8 x} = \dfrac{\ln x}{\ln 2} \div \dfrac{\ln x}{\ln 8} = \dfrac{\ln x}{\ln 2} \cdot \dfrac{\ln 8}{\ln x} = \dfrac{3 \ln 2}{\ln 2} = 3$

   (c) $\dfrac{\log_x a}{\log_{x^2} a} = \dfrac{\ln a}{\ln x} \div \dfrac{\ln a}{\ln x^2} = \dfrac{\ln a}{\ln x} \cdot \dfrac{\ln x^2}{\ln a} = \dfrac{2 \ln x}{\ln x} = 2$

7. $3^{\log_3 (7)} + 2^{\log_2 (5)} = 5^{\log_5 (x)} = 7 + 5 = x \Rightarrow x = 12$

9. $3^{\log_3\left(x^2\right)} = 5 e^{\ln x} - 3 \cdot 10^{\log_{10} (2)} \Rightarrow x^2 = 5x - 6 \Rightarrow x^2 - 5x + 6 = 0 \Rightarrow (x-2)(x-3) = 0 \Rightarrow x = 2$ or $x = 3$

11. $y = 2^x \Rightarrow y' = 2^x \ln 2$

13. $y = 5^{\sqrt{s}} \Rightarrow \dfrac{dy}{ds} = 5^{\sqrt{s}} (\ln 5)\left(\frac{1}{2} s^{-1/2}\right) = \left(\dfrac{\ln 5}{2\sqrt{s}}\right) 5^{\sqrt{s}}$

15. $y = x^\pi \Rightarrow y' = \pi x^{(\pi - 1)}$

17. $y = (\cos \theta)^{\sqrt{2}} \Rightarrow \frac{dy}{d\theta} = -\sqrt{2}\,(\cos \theta)^{\left(\sqrt{2}-1\right)}(\sin \theta)$

19. $y = 7^{\sec \theta}\ln 7 \Rightarrow \frac{dy}{d\theta} = \left(7^{\sec \theta}\ln 7\right)(\ln 7)(\sec \theta \tan \theta) = 7^{\sec \theta}(\ln 7)^2(\sec \theta \tan \theta)$

21. $y = 2^{\sin 3t} \Rightarrow \frac{dy}{dt} = \left(2^{\sin 3t}\ln 2\right)(\cos 3t)(3) = (3\cos 3t)\left(2^{\sin 3t}\right)(\ln 2)$

23. $y = \log_2 5\theta = \frac{\ln 5\theta}{\ln 2} \Rightarrow \frac{dy}{d\theta} = \left(\frac{1}{\ln 2}\right)\left(\frac{1}{5\theta}\right)(5) = \frac{1}{\theta \ln 2}$

25. $y = \frac{\ln x}{\ln 4} + \frac{\ln x^2}{\ln 4} = \frac{\ln x}{\ln 4} + 2\frac{\ln x}{\ln 4} = 3\frac{\ln x}{\ln 4} \Rightarrow y' = \frac{3}{x \ln 4}$

27. $y = \log_2 r \cdot \log_4 r = \left(\frac{\ln r}{\ln 2}\right)\left(\frac{\ln r}{\ln 4}\right) = \frac{\ln^2 r}{(\ln 2)(\ln 4)} \Rightarrow \frac{dy}{dr} = \left[\frac{1}{(\ln 2)(\ln 4)}\right](2 \ln r)\left(\frac{1}{r}\right) = \frac{2 \ln r}{r(\ln 2)(\ln 4)}$

29. $y = \log_3 \left(\left(\frac{x+1}{x-1}\right)^{\ln 3}\right) = \frac{\ln\left(\frac{x+1}{x-1}\right)^{\ln 3}}{\ln 3} = \frac{(\ln 3)\ln\left(\frac{x+1}{x-1}\right)}{\ln 3} = \ln\left(\frac{x+1}{x-1}\right) = \ln(x+1) - \ln(x-1)$

$\Rightarrow \frac{dy}{dx} = \frac{1}{x+1} - \frac{1}{x-1} = \frac{-2}{(x+1)(x-1)}$

31. $y = \theta \sin(\log_7 \theta) = \theta \sin\left(\frac{\ln \theta}{\ln 7}\right) \Rightarrow \frac{dy}{d\theta} = \sin\left(\frac{\ln \theta}{\ln 7}\right) + \theta\left[\cos\left(\frac{\ln \theta}{\ln 7}\right)\right]\left(\frac{1}{\theta \ln 7}\right) = \sin(\log_7 \theta) + \left(\frac{1}{\ln 7}\right)\cos(\log_7 \theta)$

33. $y = \log_5 e^x = \frac{\ln e^x}{\ln 5} = \frac{x}{\ln 5} \Rightarrow y' = \frac{1}{\ln 5}$

35. $y = 3^{\log_2 t} = 3^{(\ln t)/(\ln 2)} \Rightarrow \frac{dy}{dt} = \left[3^{(\ln t)/(\ln 2)}(\ln 3)\right]\left(\frac{1}{t \ln 2}\right) = \frac{1}{t}\left(\log_2 3\right)3^{\log_2 t}$

37. $y = \log_2\left(8t^{\ln 2}\right) = \frac{\ln 8 + \ln\left(t^{\ln 2}\right)}{\ln 2} = \frac{3 \ln 2 + (\ln 2)(\ln t)}{\ln 2} = 3 + \ln t \Rightarrow \frac{dy}{dt} = \frac{1}{t}$

39. $y = (x+1)^x \Rightarrow \ln y = \ln(x+1)^x = x \ln(x+1) \Rightarrow \frac{y'}{y} = \ln(x+1) + x \cdot \frac{1}{(x+1)} \Rightarrow y' = (x+1)^x\left[\frac{x}{x+1} + \ln(x+1)\right]$

41. $y = \left(\sqrt{t}\right)^t = \left(t^{1/2}\right)^t = t^{t/2} \Rightarrow \ln y = \ln t^{t/2} = \left(\frac{t}{2}\right)\ln t \Rightarrow \frac{1}{y}\frac{dy}{dt} = \left(\frac{1}{2}\right)(\ln t) + \left(\frac{t}{2}\right)\left(\frac{1}{t}\right) = \frac{\ln t}{2} + \frac{1}{2}$

$\Rightarrow \frac{dy}{dt} = \left(\sqrt{t}\right)^t\left(\frac{\ln t}{2} + \frac{1}{2}\right)$

43. $y = (\sin x)^x \Rightarrow \ln y = \ln(\sin x)^x = x \ln(\sin x) \Rightarrow \frac{y'}{y} = \ln(\sin x) + x\left(\frac{\cos x}{\sin x}\right) \Rightarrow y' = (\sin x)^x[\ln(\sin x) + x \cot x]$

45. $y = x^{\ln x},\ x > 0 \Rightarrow \ln y = (\ln x)^2 \Rightarrow \frac{y'}{y} = 2(\ln x)\left(\frac{1}{x}\right) \Rightarrow y' = \left(x^{\ln x}\right)\left(\frac{\ln x^2}{x}\right)$

47. $\displaystyle\int 5^x\,dx = \frac{5^x}{\ln 5} + C$

49. $\displaystyle\int_0^1 2^{-\theta}\, d\theta = \int_0^1 \left(\tfrac{1}{2}\right)^\theta d\theta = \left[\frac{\left(\tfrac{1}{2}\right)^\theta}{\ln\left(\tfrac{1}{2}\right)}\right]_0^1 = \frac{\tfrac{1}{2}}{\ln\left(\tfrac{1}{2}\right)} - \frac{1}{\ln\left(\tfrac{1}{2}\right)} = -\frac{\tfrac{1}{2}}{\ln\left(\tfrac{1}{2}\right)} = \frac{-1}{2(\ln 1 - \ln 2)} = \frac{1}{2\ln 2}$

51. Let $u = x^2 \Rightarrow du = 2x\, dx \Rightarrow \tfrac{1}{2}\, du = x\, dx$; $x = 1 \Rightarrow u = 1$, $x = \sqrt{2} \Rightarrow u = 2$;

$\displaystyle\int_1^{\sqrt{2}} x 2^{\left(x^2\right)}\, dx = \int_1^2 \left(\tfrac{1}{2}\right) 2^u\, du = \tfrac{1}{2}\left[\frac{2^u}{\ln 2}\right]_1^2 = \left(\frac{1}{2\ln 2}\right)\left(2^2 - 2^1\right) = \frac{1}{\ln 2}$

53. Let $u = \cos t \Rightarrow du = -\sin t\, dt \Rightarrow -du = \sin t\, dt$; $t = 0 \Rightarrow u = 1$, $t = \tfrac{\pi}{2} \Rightarrow u = 0$;

$\displaystyle\int_0^{\pi/2} 7^{\cos t} \sin t\, dt = -\int_1^0 7^u\, du = \left[-\frac{7^u}{\ln 7}\right]_1^0 = \left(\frac{-1}{\ln 7}\right)\left(7^0 - 7\right) = \frac{6}{\ln 7}$

55. Let $u = x^{2x} \Rightarrow \ln u = 2x \ln x \Rightarrow \tfrac{1}{u}\frac{du}{dx} = 2\ln x + (2x)\left(\tfrac{1}{x}\right) \Rightarrow \frac{du}{dx} = 2u(\ln x + 1) \Rightarrow \tfrac{1}{2}\, du = x^{2x}(1 + \ln x)\, dx$;

$x = 2 \Rightarrow u = 2^4 = 16$, $x = 4 \Rightarrow u = 4^8 = 65{,}536$;

$\displaystyle\int_2^4 x^{2x}(1 + \ln x)\, dx = \tfrac{1}{2}\int_{16}^{65{,}536} du = \tfrac{1}{2}[u]_{16}^{65{,}536} = \tfrac{1}{2}(65{,}536 - 16) = \frac{65{,}520}{2} = 32{,}760$

57. $\displaystyle\int 3x^{\sqrt{3}}\, dx = \frac{3x^{\left(\sqrt{3}+1\right)}}{\sqrt{3}+1} + C$

59. $\displaystyle\int_0^3 \left(\sqrt{2}+1\right)x^{\sqrt{2}}\, dx = \left[x^{\left(\sqrt{2}+1\right)}\right]_0^3 = 3^{\left(\sqrt{2}+1\right)}$

61. $\displaystyle\int \frac{\log_{10} x}{x}\, dx = \int \left(\frac{\ln x}{\ln 10}\right)\left(\tfrac{1}{x}\right) dx$; $\left[u = \ln x \Rightarrow du = \tfrac{1}{x}\, dx\right]$

$\displaystyle\rightarrow \int \left(\frac{\ln x}{\ln 10}\right)\left(\tfrac{1}{x}\right) dx = \frac{1}{\ln 10}\int u\, du = \left(\frac{1}{\ln 10}\right)\left(\tfrac{1}{2}u^2\right) + C = \frac{(\ln x)^2}{2\ln 10} + C$

63. $\displaystyle\int_1^4 \frac{\ln 2 \log_2 x}{x}\, dx = \int_1^4 \left(\frac{\ln 2}{x}\right)\left(\frac{\ln x}{\ln 2}\right) dx = \int_1^4 \frac{\ln x}{x}\, dx = \left[\tfrac{1}{2}(\ln x)^2\right]_1^4 = \tfrac{1}{2}\left[(\ln 4)^2 - (\ln 1)^2\right] = \tfrac{1}{2}(\ln 4)^2$

$= \tfrac{1}{2}(2\ln 2)^2 = 2(\ln 2)^2$

65. $\displaystyle\int_0^2 \frac{\log_2 (x+2)}{x+2}\, dx = \frac{1}{\ln 2}\int_0^2 [\ln(x+2)]\left(\frac{1}{x+2}\right) dx = \left(\frac{1}{\ln 2}\right)\left[\frac{(\ln(x+2))^2}{2}\right]_0^2 = \left(\frac{1}{\ln 2}\right)\left[\frac{(\ln 4)^2}{2} - \frac{(\ln 2)^2}{2}\right]$

$= \left(\frac{1}{\ln 2}\right)\left[\frac{4(\ln 2)^2}{2} - \frac{(\ln 2)^2}{2}\right] = \tfrac{3}{2}\ln 2$

67. $\displaystyle\int_0^9 \frac{2\log_{10}(x+1)}{x+1}\,dx = \frac{2}{\ln 10}\int_0^9 \ln(x+1)\left(\frac{1}{x+1}\right)dx = \left(\frac{2}{\ln 10}\right)\left[\frac{(\ln(x+1))^2}{2}\right]_0^9 = \left(\frac{2}{\ln 10}\right)\left[\frac{(\ln 10)^2}{2} - \frac{(\ln 1)^2}{2}\right]$

    $= \ln 10$

69. $\displaystyle\int \frac{dx}{x\log_{10} x} = \int \left(\frac{\ln 10}{\ln x}\right)\left(\frac{1}{x}\right)dx = (\ln 10)\int \left(\frac{1}{\ln x}\right)\left(\frac{1}{x}\right)dx; \left[u = \ln x \Rightarrow du = \frac{1}{x}\,dx\right]$

    $\rightarrow (\ln 10)\displaystyle\int \left(\frac{1}{\ln x}\right)\left(\frac{1}{x}\right)dx = (\ln 10)\int \frac{1}{u}\,du = (\ln 10)\ln|u| + C = (\ln 10)\ln|\ln x| + C$

71. $\displaystyle\int_1^{\ln x} \frac{1}{t}\,dt = \left[\ln|t|\right]_1^{\ln x} = \ln|\ln x| - \ln 1 = \ln(\ln x),\ x > 1$

73. $\displaystyle\int_1^{1/x} \frac{1}{t}\,dt = \left[\ln|t|\right]_1^{1/x} = \ln\left|\frac{1}{x}\right| - \ln 1 = (\ln 1 - \ln|x|) - \ln 1 = -\ln x,\ x > 0$

75. $A = \displaystyle\int_{-2}^2 \frac{2x}{1+x^2}\,dx = 2\int_0^2 \frac{2x}{1+x^2}\,dx; \left[u = 1 + x^2 \Rightarrow du = 2x\,dx;\ x = 0 \Rightarrow u = 1,\ x = 2 \Rightarrow u = 5\right]$

    $\rightarrow A = 2\displaystyle\int_1^5 \frac{1}{u}\,du = 2\left[\ln|u|\right]_1^5 = 2(\ln 5 - \ln 1) = 2\ln 5$

77. Let $\left[H_3O^+\right] = x$ and solve the equations $7.37 = -\log_{10} x$ and $7.44 = -\log_{10} x$. The solutions of these equations are $10^{-7.37}$ and $10^{-7.44}$. Consequently, the bounds for $\left[H_3O^+\right]$ are $\left[10^{-7.44}, 10^{-7.37}\right]$.

79. Let $O = $ original sound level $= 10\log_{10}\left(I \times 10^{12}\right)$ db from Equation (8) in the text. Solving

    $O + 10 = 10\log_{10}\left(kI \times 10^{12}\right)$ for $k \Rightarrow 10\log_{10}\left(I \times 10^{12}\right) + 10 = 10\log_{10}\left(kI \times 10^{12}\right) \Rightarrow \log_{10}\left(I \times 10^{12}\right) + 1$

    $= \log_{10}\left(kI \times 10^{12}\right) \Rightarrow \log_{10}\left(I \times 10^{12}\right) + 1 = \log_{10} k + \log_{10}\left(I \times 10^{12}\right) \Rightarrow 1 = \log_{10} k \Rightarrow 1 = \frac{\ln k}{\ln 10}$

    $\Rightarrow \ln k = \ln 10 \Rightarrow k = 10$

81. (a) If $x = \left[H_3O^+\right]$ and $S - x = [OH^-]$, then $x(S - x) = 10^{-14} \Rightarrow S = x + \frac{10^{-14}}{x} \Rightarrow \frac{dS}{dx} = 1 - \frac{10^{-14}}{x^2}$

    and $\dfrac{d^2S}{dx^2} = \dfrac{2 \cdot 10^{-14}}{x^3} > 0 \Rightarrow$ a minimum exists at $x = 10^{-7}$

    (b) $\text{pH} = -\log_{10}\left(10^{-7}\right) = 7$

    (c) $\dfrac{[OH^-]}{\left[H_3O^+\right]} = \dfrac{S-x}{x} = \dfrac{\left(x + \frac{10^{-14}}{x}\right) - x}{x} = \dfrac{10^{-14}}{x^2} \Rightarrow$ the ratio $\dfrac{[OH^-]}{\left[H_3O^+\right]}$ equals 1 at $x = 10^{-7}$

83. From zooming in on the graph at the right, we estimate
the third root to be $x \approx -0.76666$

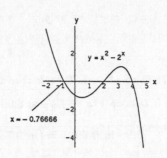

85. (a) $f(x) = 2^x \Rightarrow f'(x) = 2^x \ln 2$; $L(x) = \left(2^0 \ln 2\right)x + 2^0 = x \ln 2 + 1 \approx 0.69x + 1$

(b)

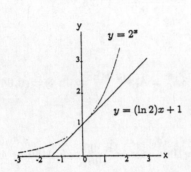

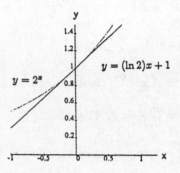

87. (a) $\log_3 8 = \dfrac{\ln 8}{\ln 3} \approx 1.89279$ (b) $\log_7 0.5 = \dfrac{\ln 0.5}{\ln 7} \approx -0.35621$

(c) $\log_{20} 17 = \dfrac{\ln 17}{\ln 20} \approx 0.94575$ (d) $\log_{0.5} 7 = \dfrac{\ln 7}{\ln 0.5} \approx -2.80735$

(e) $\ln x = (\log_{10} x)(\ln 10) = 2.3 \ln 10 \approx 5.29595$ (f) $\ln x = (\log_2 x)(\ln 2) = 1.4 \ln 2 \approx 0.97041$

(g) $\ln x = (\log_2 x)(\ln 2) = -1.5 \ln 2 \approx -1.03972$ (h) $\ln x = (\log_{10} x)(\ln 10) = -0.7 \ln 10 \approx -1.61181$

6.5 GROWTH AND DECAY

1. (a) $y = y_0 e^{kt} \Rightarrow 0.99 y_0 = y_0 e^{1000k} \Rightarrow k = \dfrac{\ln 0.99}{1000} \approx -0.00001$

(b) $0.9 = e^{(-0.00001)t} \Rightarrow (-0.00001)t = \ln (0.9) \Rightarrow t = \dfrac{\ln (0.9)}{-0.00001} \approx 10{,}536$ years

(c) $y = y_0 e^{(20,000)k} = y_0 e^{-0.2} \approx y_0(0.82) \Rightarrow 82\%$

3. $\dfrac{dy}{dt} = -0.6y \Rightarrow y = y_0 e^{-0.6t}$; $y_0 = 100 \Rightarrow y = 100 e^{-0.6t} \Rightarrow y = 100 e^{-0.6} \approx 54.88$ grams when $t = 1$ hr

5. $L(x) = L_0 e^{-kx} \Rightarrow \dfrac{L_0}{2} = L_0 e^{-18k} \Rightarrow \ln \frac{1}{2} = -18k \Rightarrow k = \dfrac{\ln 2}{18} \approx 0.0385 \Rightarrow L(x) = L_0 e^{-0.0385x}$; when the intensity

is one-tenth of the surface value, $\dfrac{L_0}{10} = L_0 e^{-0.0385x} \Rightarrow \ln 10 = 0.0385x \Rightarrow x \approx 59.8$ ft

7. $y = y_0 e^{kt}$ and $y_0 = 1 \Rightarrow y = e^{kt} \Rightarrow$ at $y = 2$ and $t = 0.5$ we have $2 = e^{0.5k} \Rightarrow \ln 2 = 0.5k \Rightarrow k = \frac{\ln 2}{0.5} = \ln 4$.
   Therefore, $y = e^{(\ln 4)t} \Rightarrow y = e^{24 \ln 4} = 4^{24} = 2.8147497 \times 10^{14}$ at the end of 24 hrs

9. (a) $10{,}000 e^{k(1)} = 7500 \Rightarrow e^k = 0.75 \Rightarrow k = \ln 0.75$ and $y = 10{,}000 e^{(\ln 0.75)t}$. Now $1000 = 10{,}000 e^{(\ln 0.75)t}$
   $\Rightarrow \ln 0.1 = (\ln 0.75)t \Rightarrow t = \frac{\ln 0.1}{\ln 0.75} \approx 8.00$ years (to the nearest hundredth of a year)

   (b) $1 = 10{,}000 e^{(\ln 0.75)t} \Rightarrow \ln 0.0001 = (\ln 0.75)t \Rightarrow t = \frac{\ln 0.0001}{\ln 0.75} \approx 32.02$ years (to the nearest hundredth of a
   year)

11. $0.9 P_0 = P_0 e^k \Rightarrow k = \ln 0.9$; when the well's output falls to one-fifth of its present value $P = 0.2 P_0$
   $\Rightarrow 0.2 P_0 = P_0 e^{(\ln 0.9)t} \Rightarrow 0.2 = e^{(\ln 0.9)t} \Rightarrow \ln (0.2) = (\ln 0.9)t \Rightarrow t = \frac{\ln 0.2}{\ln 0.9} \approx 15.28$ yr

13. (a) $A_0 e^{(0.04)5} = A_0 e^{0.2}$

   (b) $2A_0 = A_0 e^{(0.04)t} \Rightarrow \ln 2 = (0.04)t \Rightarrow t = \frac{\ln 2}{0.04} \approx 17.33$ years; $3A_0 = A_0 e^{(0.04)t} \Rightarrow \ln 3 = (0.04)t$
   $\Rightarrow t = \frac{\ln 3}{0.04} \approx 27.47$ years

15. $A(100) = 90{,}000 \Rightarrow 90{,}000 = 1000 e^{r(100)} \Rightarrow 90 = e^{100r} \Rightarrow \ln 90 = 100r \Rightarrow r = \frac{\ln 90}{100} \approx 0.0450$ or 4.50%

17. $y = y_0 e^{-0.18t}$ represents the decay equation; solving $(0.9)y_0 = y_0 e^{-0.18t} \Rightarrow t = \frac{\ln (0.9)}{-0.18} \approx 0.585$ days

19. $y = y_0 e^{-kt} = y_0 e^{-(k)(3/k)} = y_0 e^{-3} = \frac{y_0}{e^3} < \frac{y_0}{20} = (0.05)(y_0) \Rightarrow$ after three mean lifetimes less than 5% remains

21. $T - T_s = (T_0 - T_s) e^{-kt}$, $T_0 = 90°C$, $T_s = 20°C$, $T = 60°C \Rightarrow 60 - 20 = 70 e^{-10k} \Rightarrow \frac{4}{7} = e^{-10k}$

   $\Rightarrow k = \frac{\ln\left(\frac{7}{4}\right)}{10} \approx 0.05596$

   (a) $35 - 20 = 70 e^{-0.05596t} \Rightarrow t \approx 27.5$ min is the total time $\Rightarrow$ it will take $27.5 - 10 = 17.5$ to reach $35°C$
   (b) $T - T_s = (T_0 - T_s) e^{-kt}$, $T_0 = 90°C$, $T_s = -15°C \Rightarrow 35 + 15 = 105 e^{-0.05596t} \Rightarrow t \approx 13.26$ min

23. $T - T_s = (T_0 - T_s) e^{-kt} \Rightarrow 39 - T_s = (46 - T_s) e^{-10k}$ and $33 - T_s = (46 - T_s) e^{-20k} \Rightarrow \frac{39 - T_s}{46 - T_s} = e^{-10k}$ and

   $\frac{33 - T_s}{46 - T_s} = e^{-20k} = \left(e^{-10k}\right)^2 \Rightarrow \frac{33 - T_s}{46 - T_s} = \left(\frac{39 - T_s}{46 - T_s}\right)^2 \Rightarrow (33 - T_s)(46 - T_s) = (39 - T_s)^2 \Rightarrow 1518 - 79 T_s + T_s^2$

   $= 1521 - 78 T_s + T_s^2 \Rightarrow -T_s = 3 \Rightarrow T_s = -3°C$

25. From Example 5, the half-life of carbon-14 is 5700 yr $\Rightarrow \frac{1}{2} c_0 = c_0 e^{-k(5700)} \Rightarrow k = \frac{\ln 2}{5700} \approx 0.0001216$

   $\Rightarrow c = c_0 e^{-0.0001216t} \Rightarrow (0.445)c_0 = c_0 e^{-0.0001216t} \Rightarrow t = \frac{\ln (0.445)}{-0.0001216} \approx 6658$ years

27. From Exercise 25, $k \approx 0.0001216$ for carbon-14. Thus, $c = c_0 e^{-0.0001216t} \Rightarrow (0.995)c_0 = c_0 e^{-0.0001216t}$

$\Rightarrow t = \dfrac{\ln{(0.995)}}{-0.0001216} \approx 41$ years old

6.6 L'HÔPITAL'S RULE

1. $\displaystyle\lim_{x\to 2}\ \frac{x-2}{x^2-4} = \lim_{x\to 2}\ \frac{1}{2x} = \frac{1}{4}$

3. $\displaystyle\lim_{t\to -3}\ \frac{t^3-4t+15}{t^2-t-12} = \lim_{t\to -3}\ \frac{3t^2-4}{2t-1} = \frac{3(-3)^2-4}{2(-3)-1} = -\frac{23}{7}$

5. $\displaystyle\lim_{x\to\infty}\ \frac{5x^2-3x}{7x^2+1} = \lim_{x\to\infty}\ \frac{10x-3}{14x} = \lim_{x\to\infty}\ \frac{10}{14} = \frac{5}{7}$

7. $\displaystyle\lim_{t\to 0}\ \frac{\sin t^2}{t} = \lim_{t\to 0}\ \frac{\left(\cos t^2\right)(2t)}{1} = 0$

9. $\displaystyle\lim_{x\to 0}\ \frac{8x^2}{\cos x-1} = \lim_{x\to 0}\ \frac{16x}{-\sin x} = \lim_{x\to 0}\ \frac{16}{-\cos x} = \frac{16}{-1} = -16$

11. $\displaystyle\lim_{\theta\to\pi/2}\ \frac{2\theta-\pi}{\cos{(2\pi-\theta)}} = \lim_{\theta\to\pi/2}\ \frac{2}{\sin{(2\pi-\theta)}} = \frac{2}{\sin\left(\frac{3\pi}{2}\right)} = -2$

13. $\displaystyle\lim_{\theta\to\pi/2}\ \frac{1-\sin\theta}{1+\cos 2\theta} = \lim_{\theta\to\pi/2}\ \frac{-\cos\theta}{-2\sin 2\theta} = \lim_{\theta\to\pi/2}\ \frac{\sin\theta}{-4\cos 2\theta} = \frac{1}{(-4)(-1)} = \frac{1}{4}$

15. $\displaystyle\lim_{x\to 0}\ \frac{x^2}{\ln{(\sec x)}} = \lim_{x\to 0}\ \frac{2x}{\left(\dfrac{\sec x \tan x}{\sec x}\right)} = \lim_{x\to 0}\ \frac{2x}{\tan x} = \lim_{x\to 0}\ \frac{2}{\sec^2 x} = \frac{2}{1^2} = 2$

17. $\displaystyle\lim_{t\to 0}\ \frac{t(1-\cos t)}{t-\sin t} = \lim_{t\to 0}\ \frac{(1-\cos t)+t(\sin t)}{1-\cos t} = \lim_{t\to 0}\ \frac{\sin t+(\sin t+t\cos t)}{\sin t}$

$= \displaystyle\lim_{t\to 0}\ \frac{\cos t+\cos t+\cos t-t\sin t}{\cos t} = \frac{1+1+1-0}{1} = 3$

19. $\displaystyle\lim_{x\to(\pi/2)^-}\ \left(x-\frac{\pi}{2}\right)\sec x = \lim_{x\to(\pi/2)^-}\ \frac{\left(x-\dfrac{\pi}{2}\right)}{\cos x} = \lim_{x\to(\pi/2)^-}\ \left(\frac{1}{-\sin x}\right) = \frac{1}{-1} = -1$

21. $\displaystyle\lim_{\theta\to 0}\ \frac{3^{\sin\theta}-1}{\theta} = \lim_{\theta\to 0}\ \frac{3^{\sin\theta}(\ln 3)(\cos\theta)}{1} = \frac{\left(3^0\right)(\ln 3)(1)}{1} = \ln 3$

23. $\displaystyle\lim_{x\to 0}\ \frac{x\,2^x}{2^x-1} = \lim_{x\to 0}\ \frac{(1)\left(2^x\right)+(x)(\ln 2)\left(2^x\right)}{(\ln 2)\left(2^x\right)} = \frac{1\cdot 2^0+0}{(\ln 2)\cdot 2^0} = \frac{1}{\ln 2}$

25. $\displaystyle\lim_{x\to\infty}\ \frac{\ln{(x+1)}}{\log_2 x} = \lim_{x\to\infty}\ \frac{\ln{(x+1)}}{\left(\dfrac{\ln x}{\ln 2}\right)} = (\ln 2)\lim_{x\to\infty}\ \frac{\left(\dfrac{1}{x+1}\right)}{\left(\dfrac{1}{x}\right)} = (\ln 2)\lim_{x\to\infty}\ \frac{x}{x+1} = (\ln 2)\lim_{x\to\infty}\ \frac{1}{1} = \ln 2$

27. $\lim\limits_{x\to 0^+} \dfrac{\ln\left(x^2+2x\right)}{\ln x} = \lim\limits_{x\to 0^+} \dfrac{\left(\frac{2x+2}{x^2+2x}\right)}{\left(\frac{1}{x}\right)} = \lim\limits_{x\to 0^+} \dfrac{2x^2+2x}{x^2+2x} = \lim\limits_{x\to 0^+} \dfrac{4x+2}{2x+2} = \lim\limits_{x\to 0^+} \dfrac{2}{2} = 1$

29. $\lim\limits_{y\to 0} \dfrac{\sqrt{5y+25}-5}{y} = \lim\limits_{y\to 0} \dfrac{(5y+25)^{1/2}-5}{y} = \lim\limits_{y\to 0} \dfrac{\left(\frac{1}{2}\right)(5y+25)^{-1/2}(5)}{1} = \lim\limits_{y\to 0} \dfrac{5}{2\sqrt{5y+25}} = \dfrac{1}{2}$

31. $\lim\limits_{x\to\infty} \left[\ln 2x - \ln(x+1)\right] = \lim\limits_{x\to\infty} \ln\left(\dfrac{2x}{x+1}\right) = \ln\left(\lim\limits_{x\to\infty} \dfrac{2x}{x+1}\right) = \ln\left(\lim\limits_{x\to\infty} \dfrac{2}{1}\right) = \ln 2$

33. $\lim\limits_{x\to 0^+} \left(\dfrac{1}{x} - \dfrac{1}{\sin x}\right) = \lim\limits_{x\to 0^+} \left(\dfrac{\sin x - x}{x\sin x}\right) = \lim\limits_{x\to 0^+} \dfrac{\cos x - 1}{\sin x + x\cos x} = \lim\limits_{x\to 0^+} \dfrac{-\sin x}{\cos x + \cos x - x\sin x} = \dfrac{0}{2} = 0$

35. $\lim\limits_{x\to 1^+} \left(\dfrac{1}{x-1} - \dfrac{1}{\ln x}\right) = \lim\limits_{x\to 1^+} \left(\dfrac{\ln x - (x-1)}{(x-1)(\ln x)}\right) = \lim\limits_{x\to 1^+} \left(\dfrac{\frac{1}{x}-1}{(\ln x)+(x-1)\left(\frac{1}{x}\right)}\right) = \lim\limits_{x\to 1^+} \left(\dfrac{1-x}{(x\ln x)+x-1}\right)$

$= \lim\limits_{x\to 1^+} \left(\dfrac{-1}{(\ln x + 1)+1}\right) = \dfrac{-1}{(0+1)+1} = -\dfrac{1}{2}$

37. $\lim\limits_{x\to\infty} \displaystyle\int_x^{2x} \dfrac{1}{t}\,dt = \lim\limits_{x\to\infty} \left[\ln|t|\right]_x^{2x} = \lim\limits_{x\to\infty} \ln\left(\dfrac{2x}{x}\right) = \ln 2$

39. $\lim\limits_{\theta\to 0} \dfrac{\cos\theta - 1}{e^\theta - \theta - 1} = \lim\limits_{\theta\to 0} \dfrac{-\sin\theta}{e^\theta - 1} = \lim\limits_{\theta\to 0} \dfrac{-\cos\theta}{e^\theta} = -1$

41. $\lim\limits_{t\to\infty} \dfrac{e^t + t^2}{e^t - 1} = \lim\limits_{t\to\infty} \dfrac{e^t + 2t}{e^t} = \lim\limits_{t\to\infty} \dfrac{e^t + 2}{e^t} = \lim\limits_{t\to\infty} \dfrac{e^t}{e^t} = 1$

43. The limit leads to the indeterminate form 1^∞. Let $f(x) = x^{1/(1-x)} \Rightarrow \ln f(x) = \ln\left(x^{1/(1-x)}\right) = \dfrac{\ln x}{1-x}$. Now

$\lim\limits_{x\to 1^+} \ln f(x) = \lim\limits_{x\to 1^+} \dfrac{\ln x}{1-x} = \lim\limits_{x\to 1^+} \dfrac{\left(\frac{1}{x}\right)}{-1} = -1$. Therefore $\lim\limits_{x\to 1^+} x^{1/(1-x)} = \lim\limits_{x\to 1^+} f(x) = \lim\limits_{x\to 1^+} e^{\ln f(x)} = e^{-1} = \dfrac{1}{e}$

45. The limit leads to the indeterminate form ∞^0. Let $f(x) = (\ln x)^{1/x} \Rightarrow \ln f(x) = \ln(\ln x)^{1/x} = \dfrac{\ln(\ln x)}{x}$. Now

$\lim\limits_{x\to\infty} \ln f(x) = \lim\limits_{x\to\infty} \dfrac{\ln(\ln x)}{x} = \lim\limits_{x\to\infty} \dfrac{\left(\frac{1}{x\ln x}\right)}{1} = 0$. Therefore $\lim\limits_{x\to\infty} (\ln x)^{1/x} = \lim\limits_{x\to\infty} f(x)$

$= \lim\limits_{x\to\infty} e^{\ln f(x)} = e^0 = 1$

47. The limit leads to the indeterminate form 0^0. Let $f(x) = x^{-1/\ln x} \Rightarrow \ln f(x) = -\dfrac{\ln x}{\ln x} = -1$. Therefore

$\lim\limits_{x\to 0^+} x^{-1/\ln x} = \lim\limits_{x\to 0^+} f(x) = \lim\limits_{x\to 0^+} e^{\ln f(x)} = e^{-1} = \dfrac{1}{e}$

49. The limit leads to the indeterminate form ∞^0. Let $f(x) = (1 + 2x)^{1/(2 \ln x)} \Rightarrow \ln f(x) = \dfrac{\ln (1 + 2x)}{2 \ln x}$

$\Rightarrow \lim\limits_{x \to \infty} \ln f(x) = \lim\limits_{x \to \infty} \dfrac{\ln (1 + 2x)}{2 \ln x} = \lim\limits_{x \to \infty} \dfrac{x}{1 + 2x} = \lim\limits_{x \to \infty} \dfrac{1}{2} = \dfrac{1}{2}$. Therefore $\lim\limits_{x \to \infty} (1 + 2x)^{1/(2 \ln x)}$

$= \lim\limits_{x \to \infty} f(x) = \lim\limits_{x \to \infty} e^{\ln f(x)} = e^{1/2}$

51. The limit leads to the indeterminate form 0^0. Let $f(x) = x^x \Rightarrow \ln f(x) = x \ln x \Rightarrow \ln f(x) = \dfrac{\ln x}{\left(\frac{1}{x}\right)}$

$= \lim\limits_{x \to 0^+} \ln f(x) = \lim\limits_{x \to 0^+} \dfrac{\ln x}{\left(\frac{1}{x}\right)} = \lim\limits_{x \to 0^+} \dfrac{\left(\frac{1}{x}\right)}{\left(-\frac{1}{x^2}\right)} = \lim\limits_{x \to 0^+} (-x) = 0$. Therefore $\lim\limits_{x \to 0^+} x^x = \lim\limits_{x \to 0^+} f(x)$

$= \lim\limits_{x \to 0^+} e^{\ln f(x)} = e^0 = 1$

53. $\lim\limits_{x \to \infty} \dfrac{\sqrt{9x + 1}}{\sqrt{x + 1}} = \sqrt{\lim\limits_{x \to \infty} \dfrac{9x + 1}{x + 1}} = \sqrt{\lim\limits_{x \to \infty} \dfrac{9}{1}} = \sqrt{9} = 3$

55. $\lim\limits_{x \to \pi/2^-} \dfrac{\sec x}{\tan x} = \lim\limits_{x \to \pi/2^-} \left(\dfrac{1}{\cos x}\right)\left(\dfrac{\cos x}{\sin x}\right) = \lim\limits_{x \to \pi/2^-} \dfrac{1}{\sin x} = 1$

57. Part (b) is correct because part (a) is neither in the $\frac{0}{0}$ nor $\frac{\infty}{\infty}$ form and so l'Hôpital's rule may not be used.

59. Part (d) is correct, the other parts are indeterminate forms and cannot be calculated by the incorrect arithmetic

61. If $f(x)$ is to be continuous at $x = 0$, then $\lim\limits_{x \to 0} f(x) = f(0) \Rightarrow c = f(0) = \lim\limits_{x \to 0} \dfrac{9x - 3 \sin 3x}{5x^3} = \lim\limits_{x \to 0} \dfrac{9 - 9 \cos 3x}{15x^2}$

$= \lim\limits_{x \to 0} \dfrac{27 \sin 3x}{30x} = \lim\limits_{x \to 0} \dfrac{81 \cos 3x}{30} = \dfrac{27}{10}$.

63. Let $f(k) = \left(1 + \dfrac{r}{k}\right)^k \Rightarrow \ln f(k) = \dfrac{\ln\left(1 + rk^{-1}\right)}{k^{-1}} \Rightarrow \lim\limits_{k \to \infty} \dfrac{\ln\left(1 + rk^{-1}\right)}{k^{-1}} = \lim\limits_{k \to \infty} \dfrac{\left(\dfrac{-rk^{-2}}{1 + rk^{-1}}\right)}{-k^{-2}} = \lim\limits_{k \to \infty} \dfrac{r}{1 + rk^{-1}}$

$= \lim\limits_{k \to \infty} \dfrac{rk}{k + r} = \lim\limits_{k \to \infty} \dfrac{r}{1} = r$. Therefore $\lim\limits_{k \to \infty} \left(1 + \dfrac{r}{k}\right)^k = \lim\limits_{k \to \infty} f(k) = \lim\limits_{k \to \infty} e^{\ln f(k)} = e^r$.

65. (a) The limit leads to the indeterminate form 1^∞. Let $f(x) = \left(1 + \dfrac{1}{x}\right)^x \Rightarrow \ln f(x) = x \ln\left(1 + \dfrac{1}{x}\right) \Rightarrow \lim\limits_{x \to \infty} \ln f(x)$

$= \lim\limits_{x \to \infty} \dfrac{\ln\left(1 + \frac{1}{x}\right)}{\left(\frac{1}{x}\right)} = \lim\limits_{x \to \infty} \dfrac{\ln\left(1 + x^{-1}\right)}{x^{-1}} = \lim\limits_{x \to \infty} \dfrac{\left(\dfrac{-x^{-2}}{1 + x^{-1}}\right)}{-x^{-2}} = \lim\limits_{x \to \infty} \dfrac{1}{1 + \left(\frac{1}{x}\right)} = \dfrac{1}{1 + 0} = 1$

$\Rightarrow \lim\limits_{x \to \infty} \left(1 + \dfrac{1}{x}\right)^x = \lim\limits_{x \to \infty} f(x) = \lim\limits_{x \to \infty} e^{\ln f(x)} = e^1 = e$

(b)

x	$\left(1 + \frac{1}{x}\right)^x$
10	2.5937424601
100	2.70481382942
1000	2.71692393224
10,000	2.71814592683
100,000	2.71826823717

(c)

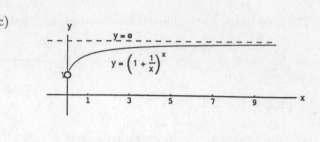

67. (b) The limit leads to the indeterminate form $\infty - \infty$:

$$\lim_{x \to \infty} \left(x - \sqrt{x^2 + x}\right) = \lim_{x \to \infty} \left(x - \sqrt{x^2 + x}\right)\left(\frac{x + \sqrt{x^2 + x}}{x + \sqrt{x^2 + x}}\right)$$

$$= \lim_{x \to \infty} \left(\frac{x^2 - \left(x^2 + x\right)}{x + \sqrt{x^2 + x}}\right) = \lim_{x \to \infty} \left(\frac{-x}{x + \sqrt{x^2 + x}}\right)$$

$$= \lim_{x \to \infty} \left(\frac{-1}{1 + \sqrt{1 + \frac{1}{x}}}\right) = \frac{-1}{1 + \sqrt{1 + 0}} = -\frac{1}{2}$$

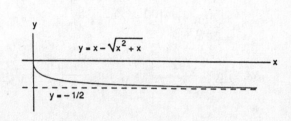

69. The graph indicates a limit near -1. The limit leads to the indeterminate form $\frac{0}{0}$: $\lim_{x \to 1} \dfrac{2x^2 - (3x + 1)\sqrt{x} + 2}{x - 1}$

$$= \lim_{x \to 1} \frac{2x^2 - 3x^{3/2} - x^{1/2} + 2}{x - 1} = \lim_{x \to 1} \frac{4x - \frac{9}{2}x^{1/2} - \frac{1}{2}x^{-1/2}}{1}$$

$$= \frac{4 - \frac{9}{2} - \frac{1}{2}}{1} = \frac{4 - 5}{1} = -1$$

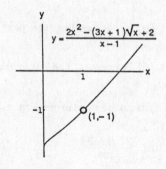

71. (a) We should assign the value 1 to $f(x) = (\sin x)^x$ to make it continuous at $x = 0$.

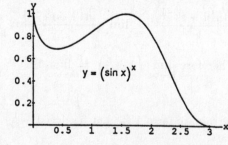

(b) $\ln f(x) = x \ln(\sin x) = \dfrac{\ln(\sin x)}{\left(\frac{1}{x}\right)} \Rightarrow \lim_{x \to 0^+} \ln f(x) = \lim_{x \to 0^+} \dfrac{\ln(\sin x)}{\left(\frac{1}{x}\right)} = \lim_{x \to 0^+} \dfrac{\left(\frac{1}{\sin x}\right)(\cos x)}{\left(-\frac{1}{x^2}\right)}$

$$= \lim_{x \to 0} \frac{-x^2}{\tan x} = \lim_{x \to 0} \frac{-2x}{\sec^2 x} = 0 \Rightarrow \lim_{x \to 0} f(x) = e^0 = 1$$

(c) The maximum value of $f(x)$ is close to 2 near the point $x \approx 1.55$ (see the graph in part (a)).

(d) The root in question is near 1.57.

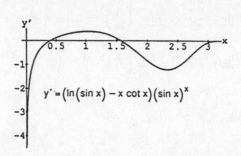

(e) $y' = 0 \Rightarrow (\ln(\sin x) - x \cot x)(\sin x)^x = 0 \Rightarrow \ln(\sin x) - x \cot x = 0.$ Let $g(x) = \ln(\sin x) - x \cot x$

$\Rightarrow g'(x) = \cot x - \cot x + x \csc^2 x = x \csc^2 x.$ Using Newton's method, $g(x) = 0 \Rightarrow x_{n+1} = x_n - \dfrac{g(x_n)}{g'(x_n)}$

$= x_n - \dfrac{\ln(\sin x_n) - x_n \cot x_n}{x_n \csc^2 x_n}.$ Then $x_1 = 1.55 \Rightarrow x_2 = 1.57093 \Rightarrow x_3 = 1.57080 \Rightarrow x_4 = 1.57080$

$\Rightarrow x_k = 1.57080,\ k \geq 3.$

(f)

x	1.55	1.57	1.57080
$(\sin x)^x$	0.999664854	0.999999502	1

73. (b) $\displaystyle\lim_{k \to 0} \frac{x^k - 1}{k} = \lim_{k \to 0} \frac{x^k \ln x}{1} = \ln x$

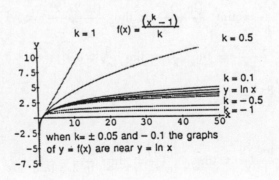

6.7 RELATIVE RATES OF GROWTH

1. (a) slower, $\displaystyle\lim_{x \to \infty} \frac{x+3}{e^x} = \lim_{x \to \infty} \frac{1}{e^x} = 0$

(b) slower, $\displaystyle\lim_{x \to \infty} \frac{x^3 + \sin^2 x}{e^x} = \lim_{x \to \infty} \frac{3x^2 + 2 \sin x \cos x}{e^x} = \lim_{x \to \infty} \frac{6x + 2 \cos 2x}{e^x} = \lim_{x \to \infty} \frac{6 - 4 \sin 2x}{e^x} = 0$ by the

Sandwich Theorem because $\dfrac{2}{e^x} \leq \dfrac{6 - 4 \sin 2x}{e^x} \leq \dfrac{10}{e^x}$ for all reals and $\displaystyle\lim_{x \to \infty} \frac{2}{e^x} = 0 = \lim_{x \to \infty} \frac{10}{e^x}$

(c) slower, $\displaystyle\lim_{x \to \infty} \frac{\sqrt{x}}{e^x} = \lim_{x \to \infty} \frac{x^{1/2}}{e^x} = \lim_{x \to \infty} \frac{\left(\frac{1}{2}\right) x^{-1/2}}{e^x} = \lim_{x \to \infty} \frac{1}{2\sqrt{x}\, e^x} = 0$

(d) faster, $\displaystyle\lim_{x \to \infty} \frac{4^x}{e^x} = \lim_{x \to \infty} \left(\frac{4}{e}\right)^x = \infty$ since $\frac{4}{e} > 1$

(e) slower, $\displaystyle\lim_{x\to\infty} \frac{\left(\frac{3}{2}\right)^x}{e^x} = \lim_{x\to\infty} \left(\frac{3}{2e}\right)^x = 0$ since $\frac{3}{2e} < 1$

(f) slower, $\displaystyle\lim_{x\to\infty} \frac{e^{x/2}}{e^x} = \lim_{x\to\infty} \frac{1}{e^{x/2}} = 0$

(g) same, $\displaystyle\lim_{x\to\infty} \frac{\left(\frac{e^x}{2}\right)}{e^x} = \lim_{x\to\infty} \frac{1}{2} = \frac{1}{2}$

(h) slower, $\displaystyle\lim_{x\to\infty} \frac{\log_{10} x}{e^x} = \lim_{x\to\infty} \frac{\ln x}{(\ln 10)\,e^x} = \lim_{x\to\infty} \frac{\left(\frac{1}{x}\right)}{(\ln 10)\,e^x} = \lim_{x\to\infty} \frac{1}{(\ln 10)x e^x} = 0$

3. (a) same, $\displaystyle\lim_{x\to\infty} \frac{x^2 + 4x}{x^2} = \lim_{x\to\infty} \frac{2x + 4}{2x} = \lim_{x\to\infty} \frac{2}{2} = 1$

(b) faster, $\displaystyle\lim_{x\to\infty} \frac{x^5 - x^2}{x^2} = \lim_{x\to\infty} (x^3 - 1) = \infty$

(c) same, $\displaystyle\lim_{x\to\infty} \frac{\sqrt{x^4 + x^3}}{x^2} = \sqrt{\lim_{x\to\infty} \frac{x^4 + x^3}{x^4}} = \sqrt{\lim_{x\to\infty} \left(1 + \frac{1}{x}\right)} = \sqrt{1} = 1$

(d) same, $\displaystyle\lim_{x\to\infty} \frac{(x+3)^2}{x^2} = \lim_{x\to\infty} \frac{2(x+3)}{2x} = \lim_{x\to\infty} \frac{2}{2} = 1$

(e) slower, $\displaystyle\lim_{x\to\infty} \frac{x \ln x}{x^2} = \lim_{x\to\infty} \frac{\ln x}{x} = \lim_{x\to\infty} \frac{\left(\frac{1}{x}\right)}{1} = 0$

(f) faster, $\displaystyle\lim_{x\to\infty} \frac{2^x}{x^2} = \lim_{x\to\infty} \frac{(\ln 2)\,2^x}{2x} = \lim_{x\to\infty} \frac{(\ln 2)^2\,2^x}{2} = \infty$

(g) slower, $\displaystyle\lim_{x\to\infty} \frac{x^3 e^{-x}}{x^2} = \lim_{x\to\infty} \frac{x}{e^x} = \lim_{x\to\infty} \frac{1}{e^x} = 0$

(h) same, $\displaystyle\lim_{x\to\infty} \frac{8x^2}{x^2} = \lim_{x\to\infty} 8 = 8$

5. (a) same, $\displaystyle\lim_{x\to\infty} \frac{\log_3 x}{\ln x} = \lim_{x\to\infty} \frac{\left(\frac{\ln x}{\ln 3}\right)}{\ln x} = \lim_{x\to\infty} \frac{1}{\ln 3} = \frac{1}{\ln 3}$

(b) same, $\displaystyle\lim_{x\to\infty} \frac{\ln 2x}{\ln x} = \lim_{x\to\infty} \frac{\left(\frac{2}{2x}\right)}{\left(\frac{1}{x}\right)} = 1$

(c) same, $\displaystyle\lim_{x\to\infty} \frac{\ln \sqrt{x}}{\ln x} = \lim_{x\to\infty} \frac{\left(\frac{1}{2}\right)\ln x}{\ln x} = \lim_{x\to\infty} \frac{1}{2} = \frac{1}{2}$

(d) faster, $\displaystyle\lim_{x\to\infty} \frac{\sqrt{x}}{\ln x} = \lim_{x\to\infty} \frac{x^{1/2}}{\ln x} = \lim_{x\to\infty} \frac{\left(\frac{1}{2}\right)x^{-1/2}}{\left(\frac{1}{x}\right)} = \lim_{x\to\infty} \frac{x}{2\sqrt{x}} = \lim_{x\to\infty} \frac{\sqrt{x}}{2} = \infty$

(e) faster, $\displaystyle\lim_{x\to\infty} \frac{x}{\ln x} = \lim_{x\to\infty} \frac{1}{\left(\frac{1}{x}\right)} = \lim_{x\to\infty} x = \infty$

(f) same, $\lim\limits_{x\to\infty} \dfrac{5 \ln x}{\ln x} = \lim\limits_{x\to\infty} 5 = 5$

(g) slower, $\lim\limits_{x\to\infty} \dfrac{\left(\frac{1}{x}\right)}{\ln x} = \lim\limits_{x\to\infty} \dfrac{1}{x \ln x} = 0$

(h) faster, $\lim\limits_{x\to\infty} \dfrac{e^x}{\ln x} = \lim\limits_{x\to\infty} \dfrac{e^x}{\left(\frac{1}{x}\right)} = \lim\limits_{x\to\infty} x e^x = \infty$

7. $\lim\limits_{x\to\infty} \dfrac{e^x}{e^{x/2}} = \lim\limits_{x\to\infty} e^{x/2} = \infty \Rightarrow e^x$ is faster than $e^{x/2}$; since for $x > e^e$ we have $\ln x > e$ and $\lim\limits_{x\to\infty} \dfrac{(\ln x)^x}{e^x}$

$= \lim\limits_{x\to\infty} \left(\dfrac{\ln x}{e}\right)^x = \infty \Rightarrow (\ln x)^x$ is faster than e^x; since $x > \ln x$ for all $x > 0$ and $\lim\limits_{x\to\infty} \dfrac{x^x}{(\ln x)^x} = \lim\limits_{x\to\infty} \left(\dfrac{x}{\ln x}\right)^x$

$= \infty \Rightarrow x^x$ is faster than $(\ln x)^x$. Therefore, slowest to fastest are: $e^{x/2}$, e^x, $(\ln x)^x$, x^x so the order is d, a, c, b

9. (a) false; $\lim\limits_{x\to\infty} \dfrac{x}{x} = 1$

(b) false; $\lim\limits_{x\to\infty} \dfrac{x}{x+5} = \dfrac{1}{1} = 1$

(c) true; $x < x + 5 \Rightarrow \dfrac{x}{x+5} < 1$ if $x > 1$ (or sufficiently large)

(d) true; $x < 2x \Rightarrow \dfrac{x}{2x} < 1$ if $x > 1$ (or sufficiently large)

(e) true; $\lim\limits_{x\to\infty} \dfrac{e^x}{e^{2x}} = \lim\limits_{x\to 0} \dfrac{1}{e^x} = 0$

(f) true; $\dfrac{x + \ln x}{x} = 1 + \dfrac{\ln x}{x} < 1 + \dfrac{\sqrt{x}}{x} = 1 + \dfrac{1}{\sqrt{x}} < 2$ if $x > 1$ (or sufficiently large)

(g) false; $\lim\limits_{x\to\infty} \dfrac{\ln x}{\ln 2x} = \lim\limits_{x\to\infty} \dfrac{\left(\frac{1}{x}\right)}{\left(\frac{2}{2x}\right)} = \lim\limits_{x\to\infty} 1 = 1$

(h) true; $\dfrac{\sqrt{x^2+5}}{x} < \dfrac{\sqrt{(x+5)^2}}{x} < \dfrac{x+5}{x} = 1 + \dfrac{5}{x} < 6$ if $x > 1$ (or sufficiently large)

11. If $f(x)$ and $g(x)$ grow at the same rate, then $\lim\limits_{x\to\infty} \dfrac{f(x)}{g(x)} = L \neq 0 \Rightarrow \lim\limits_{x\to\infty} \dfrac{g(x)}{f(x)} = \dfrac{1}{L} \neq 0$. Then

$\left|\dfrac{f(x)}{g(x)} - L\right| < 1$ if x is sufficiently large $\Rightarrow L - 1 < \dfrac{f(x)}{g(x)} < L + 1 \Rightarrow \dfrac{f(x)}{g(x)} \leq |L| + 1$ if x is sufficiently large

$\Rightarrow f = O(g)$. Similarly, $\dfrac{g(x)}{f(x)} \leq \left|\dfrac{1}{L}\right| + 1 \Rightarrow g = O(f)$.

13. When the degree of f is less than or equal to the degree of g since $\lim\limits_{x\to\infty} \dfrac{f(x)}{g(x)} = 0$ when the degree of f is smaller than the degree of g, and $\lim\limits_{x\to\infty} \dfrac{f(x)}{g(x)} = \dfrac{a}{b}$ (the ratio of the leading coefficients) when the degrees are the same.

15. Polynomials of a greater degree grow at a greater rate than polynomials of a lesser degree. Polynomials of the same degree grow at the same rate.

17. $\lim\limits_{x\to\infty} \dfrac{\sqrt{10x+1}}{\sqrt{x}} = \sqrt{\lim\limits_{x\to\infty}\dfrac{10x+1}{x}} = \sqrt{10}$ and $\lim\limits_{x\to\infty}\dfrac{\sqrt{x+1}}{\sqrt{x}} = \sqrt{\lim\limits_{x\to\infty}\dfrac{x+1}{x}} = \sqrt{1} = 1$. Since the growth rate
is transitive, we conclude that $\sqrt{10x+1}$ and $\sqrt{x+1}$ have the same growth rate $\left(\text{that of } \sqrt{x}\right)$.

19. $\lim\limits_{x\to\infty}\dfrac{x^n}{e^x} = \lim\limits_{x\to\infty}\dfrac{nx^{n-1}}{e^x} = \ldots = \lim\limits_{x\to\infty}\dfrac{n!}{e^x} = 0 \Rightarrow x^n = o\left(e^x\right)$ for any non-negative integer n

21. (a) $\lim\limits_{x\to\infty}\dfrac{x^{1/n}}{\ln x} = \lim\limits_{x\to\infty}\dfrac{x^{(1-n)/n}}{n\left(\frac{1}{x}\right)} = \left(\dfrac{1}{n}\right)\lim\limits_{x\to\infty} x^{1/n} = \infty \Rightarrow \ln x = o\left(x^{1/n}\right)$ for any positive integer n

(b) $\ln\left(e^{17{,}000{,}000}\right) = 17{,}000{,}000 < \left(e^{17\times 10^6}\right)^{1/10^6} = e^{17} \approx 24{,}154{,}952.75$

(c) $x \approx 3.430631121 \times 10^{15}$

(d) In the interval $\left[3.41\times 10^{15}, 3.45\times 10^{15}\right]$ we have
$\ln x = 10\ln\left(\ln x\right)$.   The graphs cross at about
3.4306311×10^{15}.

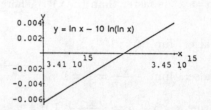

23. (a) $\lim\limits_{n\to\infty}\dfrac{n\log_2 n}{n\left(\log_2 n\right)^2} = \lim\limits_{n\to\infty}\dfrac{1}{\log_2 n} = 0 \Rightarrow n\log_2 n$ grows

slower than $n\left(\log_2 n\right)^2$; $\lim\limits_{n\to\infty}\dfrac{n\log_2 n}{n^{3/2}} = \lim\limits_{n\to\infty}\dfrac{\left(\frac{\ln n}{\ln 2}\right)}{n^{1/2}}$

$= \dfrac{1}{\ln 2}\lim\limits_{n\to\infty}\dfrac{\left(\frac{1}{n}\right)}{\left(\frac{1}{2}\right)n^{-1/2}} = \dfrac{2}{\ln 2}\lim\limits_{n\to\infty}\dfrac{1}{n^{1/2}} = 0 \Rightarrow n\log_2 n$

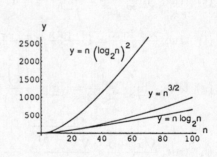

grows slower than $n^{3/2}$. Therefore, $n\log_2 n$ grows at the
slowest rate $\Rightarrow$ the algorithm that takes $O(n\log_2 n)$ steps
is the most efficient in the long run.

25. It could take one million steps for a sequential search, but at most 20 steps for a binary search because
$2^{19} = 524{,}288 < 1{,}000{,}000 < 1{,}048{,}576 = 2^{20}$.

6.8 INVERSE TRIGONOMETRIC FUNCTIONS

1. (a) $\dfrac{\pi}{4}$ (b) $-\dfrac{\pi}{3}$ (c) $\dfrac{\pi}{6}$ 3. (a) $-\dfrac{\pi}{6}$ (b) $\dfrac{\pi}{4}$ (c) $-\dfrac{\pi}{3}$

5. (a) $\dfrac{\pi}{3}$ (b) $\dfrac{3\pi}{4}$ (c) $\dfrac{\pi}{6}$ 7. (a) $\dfrac{3\pi}{4}$ (b) $\dfrac{\pi}{6}$ (c) $\dfrac{2\pi}{3}$

9. (a) $\dfrac{\pi}{4}$ (b) $-\dfrac{\pi}{3}$ (c) $\dfrac{\pi}{6}$ 11. (a) $\dfrac{3\pi}{4}$ (b) $\dfrac{\pi}{6}$ (c) $\dfrac{2\pi}{3}$

13. $\alpha = \sin^{-1}\left(\dfrac{5}{13}\right) \Rightarrow \cos\alpha = \dfrac{12}{13}$, $\tan\alpha = \dfrac{5}{12}$, $\sec\alpha = \dfrac{13}{12}$, $\csc\alpha = \dfrac{13}{5}$, and $\cot\alpha = \dfrac{12}{5}$

15. $\alpha = \sec^{-1}\left(-\sqrt{5}\right) \Rightarrow \sin\alpha = \dfrac{2}{\sqrt{5}}$, $\cos\alpha = -\dfrac{1}{\sqrt{5}}$, $\tan\alpha = -2$, $\csc\alpha = \dfrac{\sqrt{5}}{2}$, and $\cot\alpha = -\dfrac{1}{2}$

17. $\sin\left(\cos^{-1}\dfrac{\sqrt{2}}{2}\right) = \sin\left(\dfrac{\pi}{4}\right) = \dfrac{1}{\sqrt{2}}$

19. $\tan\left(\sin^{-1}\left(-\dfrac{1}{2}\right)\right) = \tan\left(-\dfrac{\pi}{6}\right) = -\dfrac{1}{\sqrt{3}}$

21. $\csc\left(\sec^{-1}2\right) + \cos\left(\tan^{-1}(-\sqrt{3})\right) = \csc\left(\cos^{-1}\left(\dfrac{1}{2}\right)\right) + \cos\left(-\dfrac{\pi}{3}\right) = \csc\left(\dfrac{\pi}{3}\right) + \cos\left(-\dfrac{\pi}{3}\right) = \dfrac{2}{\sqrt{3}} + \dfrac{1}{2} = \dfrac{4+\sqrt{3}}{2\sqrt{3}}$

23. $\sin\left(\sin^{-1}\left(-\dfrac{1}{2}\right) + \cos^{-1}\left(-\dfrac{1}{2}\right)\right) = \sin\left(-\dfrac{\pi}{6} + \dfrac{2\pi}{3}\right) = \sin\left(\dfrac{\pi}{2}\right) = 1$

25. $\sec\left(\tan^{-1}1 + \csc^{-1}1\right) = \sec\left(\dfrac{\pi}{4} + \sin^{-1}\dfrac{1}{1}\right) = \sec\left(\dfrac{\pi}{4} + \dfrac{\pi}{2}\right) = \sec\left(\dfrac{3\pi}{4}\right) = -\sqrt{2}$

27. $\sec^{-1}\left(\sec\left(-\dfrac{\pi}{6}\right)\right) = \sec^{-1}\left(\dfrac{2}{\sqrt{3}}\right) = \cos^{-1}\left(\dfrac{\sqrt{3}}{2}\right) = \dfrac{\pi}{6}$

29. $\alpha = \tan^{-1}\dfrac{x}{2}$ indicates the diagram $\qquad$ 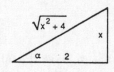 $\qquad \Rightarrow \sec\left(\tan^{-1}\dfrac{x}{2}\right) = \sec\alpha = \dfrac{\sqrt{x^2+4}}{2}$

31. $\alpha = \sec^{-1}3y$ indicates the diagram $\qquad$ 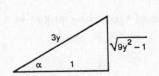 $\qquad \Rightarrow \tan\left(\sec^{-1}3y\right) = \tan\alpha = \sqrt{9y^2-1}$

33. $\alpha = \sin^{-1}x$ indicates the diagram $\qquad$ 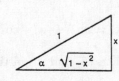 $\qquad \Rightarrow \cos\left(\sin^{-1}x\right) = \cos\alpha = \sqrt{1-x^2}$

35. $\alpha = \tan^{-1}\sqrt{x^2-2x}$ indicates the diagram $\qquad$ 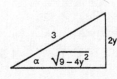 $\qquad \Rightarrow \sin\left(\tan^{-1}\sqrt{x^2-2x}\right)$

$\quad = \sin\alpha = \dfrac{\sqrt{x^2-2x}}{x-1}$

37. $\alpha = \sin^{-1}\dfrac{2y}{3}$ indicates the diagram $\qquad$ 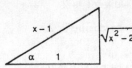 $\qquad \Rightarrow \cos\left(\sin^{-1}\dfrac{2y}{3}\right) = \cos\alpha = \dfrac{\sqrt{9-4y^2}}{3}$

39. $\alpha = \sec^{-1}\dfrac{x}{4}$ indicates the diagram $\qquad$  $\qquad \Rightarrow \sin\left(\sec^{-1}\dfrac{x}{4}\right) = \sin\alpha = \dfrac{\sqrt{x^2-16}}{x}$

41. $\displaystyle\lim_{x\to 1^-} \sin^{-1} x = \frac{\pi}{2}$

43. $\displaystyle\lim_{x\to\infty} \tan^{-1} x = \frac{\pi}{2}$

45. $\displaystyle\lim_{x\to\infty} \sec^{-1} x = \frac{\pi}{2}$

47. $\displaystyle\lim_{x\to\infty} \csc^{-1} x = \lim_{x\to\infty} \sin^{-1}\left(\frac{1}{x}\right) = 0$

49. The angle α is the large angle between the wall and the right end of the blackboard minus the small angle between the left end of the blackboard and the wall $\Rightarrow \alpha = \cot^{-1}\left(\frac{x}{15}\right) - \cot^{-1}\left(\frac{x}{3}\right)$.

51. $V = \left(\frac{1}{3}\right)\pi r^2 h = \left(\frac{1}{3}\right)\pi(3\sin\theta)^2(3\cos\theta) = 9\pi\left(\cos\theta - \cos^3\theta\right)$, where $0 \le \theta \le \frac{\pi}{2} \Rightarrow \frac{dV}{d\theta} = -9\pi(\sin\theta)\left(1 - 3\cos^2\theta\right)$

$= 0 \Rightarrow \sin\theta = 0$ or $\cos\theta = \pm\frac{1}{\sqrt{3}} \Rightarrow$ the critical points are: 0, $\cos^{-1}\left(\frac{1}{\sqrt{3}}\right)$, and $\cos^{-1}\left(-\frac{1}{\sqrt{3}}\right)$; but

$\cos^{-1}\left(-\frac{1}{\sqrt{3}}\right)$ is not in the domain. When $\theta = 0$, we have a minimum and when $\theta = \cos^{-1}\left(\frac{1}{\sqrt{3}}\right) \approx 54.7°$, we

have a maximum volume.

53. Take each square as a unit square. From the diagram we have the following: the smallest angle α has a tangent of $1 \Rightarrow \alpha = \tan^{-1} 1$; the middle angle β has a tangent of $2 \Rightarrow \beta = \tan^{-1} 2$; and the largest angle γ has a tangent of $3 \Rightarrow \gamma = \tan^{-1} 3$. The sum of these three angles is $\pi \Rightarrow \alpha + \beta + \gamma = \pi$ $\Rightarrow \tan^{-1} 1 + \tan^{-1} 2 + \tan^{-1} 3 = \pi$.

55. $\sin^{-1}(1) + \cos^{-1}(1) = \frac{\pi}{2} + 0 = \frac{\pi}{2}$; $\sin^{-1}(0) + \cos^{-1}(0) = 0 + \frac{\pi}{2} = \frac{\pi}{2}$; and $\sin^{-1}(-1) + \cos^{-1}(-1) = -\frac{\pi}{2} + \pi = \frac{\pi}{2}$.

If $x \in (-1, 0)$ and $x = -a$, then $\sin^{-1}(x) + \cos^{-1}(x) = \sin^{-1}(-a) + \cos^{-1}(-a) = -\sin^{-1} a + \left(\pi - \cos^{-1} a\right)$

$= \pi - \left(\sin^{-1} a + \cos^{-1} a\right) = \pi - \frac{\pi}{2} = \frac{\pi}{2}$ from Equations (5) and (6) in the text.

57. (a) Defined; there is an angle whose tangent is 2.

    (b) Not defined; there is no angle whose cosine is 2.

59. (a) Not defined; there is no angle whose secant is 0.

    (b) Not defined; there is no angle whose sine is $\sqrt{2}$.

61. (a) $\sec^{-1} 1.5 = \cos^{-1}\frac{1}{1.5} \approx 0.84107$

    (b) $\csc^{-1}(-1.5) = \sin^{-1}\left(-\frac{1}{1.5}\right) \approx -0.72973$

    (c) $\cot^{-1} 2 = \frac{\pi}{2} - \tan^{-1} 2 \approx 0.46365$

63. (a) Domain: all real numbers except those having

the form $\frac{\pi}{2} + k\pi$ where k is an integer.

Range: $-\frac{\pi}{2} < y < \frac{\pi}{2}$

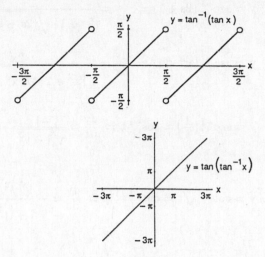

(b) Domain: $-\infty < x < \infty$; Range: $-\infty < y < \infty$

The graph of $y = \tan^{-1}(\tan x)$ is periodic, the

graph of $y = \tan(\tan^{-1}x) = x$ for $-\infty \le x < \infty$.

65. (a) Domain: $-\infty < x < \infty$; Range: $0 \le y \le \pi$

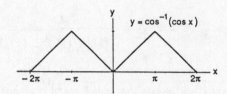

(b) Domain: $-1 \le x \le 1$; Range: $-1 \le y \le 1$

The graph of $y = \cos^{-1}(\cos x)$ is periodic; the

graph of $y = \cos(\cos^{-1}x) = x$ for $-1 \le x \le 1$.

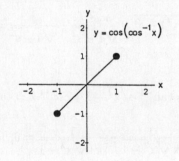

67. The graphs are identical for $y = 2\sin(2\tan^{-1}x)$

$= 4\left[\sin(\tan^{-1}x)\right]\left[\cos(\tan^{-1}x)\right] = 4\left(\dfrac{x}{\sqrt{x^2+1}}\right)\left(\dfrac{1}{\sqrt{x^2+1}}\right)$

$= \dfrac{4x}{x^2+1}$ from the triangle

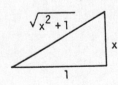

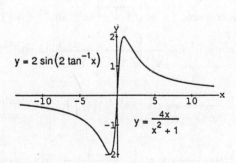

6.9 DERIVATIVES OF INVERSE TRIGONOMETRIC FUNCTIONS; INTEGRALS

1. $y = \cos^{-1}(x^2) \Rightarrow \dfrac{dy}{dx} = -\dfrac{2x}{\sqrt{1-(x^2)^2}} = \dfrac{-2x}{\sqrt{1-x^4}}$

3. $y = \sin^{-1}\sqrt{2}t \Rightarrow \dfrac{dy}{dt} = \dfrac{\sqrt{2}}{\sqrt{1-(\sqrt{2}t)^2}} = \dfrac{\sqrt{2}}{\sqrt{1-2t^2}}$

5. $y = \sec^{-1}(2s+1) \Rightarrow \dfrac{dy}{ds} = \dfrac{2}{|2s+1|\sqrt{(2s+1)^2-1}} = \dfrac{2}{|2s+1|\sqrt{4s^2+4s}} = \dfrac{1}{|2s+1|\sqrt{s^2+s}}$

7. $y = \csc^{-1}(x^2+1) \Rightarrow \dfrac{dy}{dx} = -\dfrac{2x}{|x^2+1|\sqrt{(x^2+1)^2-1}} = \dfrac{-2x}{(x^2+1)\sqrt{x^4+2x^2}}$

9. $y = \sec^{-1}\left(\dfrac{1}{t}\right) = \cos^{-1}t \Rightarrow \dfrac{dy}{dt} = \dfrac{-1}{\sqrt{1-t^2}}$

11. $y = \cot^{-1}\sqrt{t} = \cot^{-1}t^{1/2} \Rightarrow \dfrac{dy}{dt} = -\dfrac{\left(\frac{1}{2}\right)t^{-1/2}}{1+\left(t^{1/2}\right)^2} = \dfrac{-1}{2\sqrt{t}(1+t)}$

13. $y = \ln\left(\tan^{-1}x\right) \Rightarrow \dfrac{dy}{dx} = \dfrac{\left(\dfrac{1}{1+x^2}\right)}{\tan^{-1}x} = \dfrac{1}{\left(\tan^{-1}x\right)\left(1+x^2\right)}$

15. $y = \csc^{-1}\left(e^t\right) \Rightarrow \dfrac{dy}{dt} = -\dfrac{e^t}{|e^t|\sqrt{(e^t)^2-1}} = \dfrac{-1}{\sqrt{e^{2t}-1}}$

17. $y = s\sqrt{1-s^2} + \cos^{-1}s = s\left(1-s^2\right)^{1/2} + \cos^{-1}s \Rightarrow \dfrac{dy}{ds} = \left(1-s^2\right)^{1/2} + s\left(\dfrac{1}{2}\right)\left(1-s^2\right)^{-1/2}(-2s) - \dfrac{1}{\sqrt{1-s^2}}$

$= \sqrt{1-s^2} - \dfrac{s^2}{\sqrt{1-s^2}} - \dfrac{1}{\sqrt{1-s^2}} = \sqrt{1-s^2} - \dfrac{s^2+1}{\sqrt{1-s^2}} = \dfrac{1-s^2-s^2-1}{\sqrt{1-s^2}} = \dfrac{-2s^2}{\sqrt{1-s^2}}$

19. $y = \tan^{-1}\sqrt{x^2-1} + \csc^{-1}x = \tan^{-1}\left(x^2-1\right)^{1/2} + \csc^{-1}x \Rightarrow \dfrac{dy}{dx} = \dfrac{\left(\frac{1}{2}\right)\left(x^2-1\right)^{-1/2}(2x)}{1+\left[\left(x^2-1\right)^{1/2}\right]^2} - \dfrac{1}{|x|\sqrt{x^2-1}}$

$= \dfrac{1}{x\sqrt{x^2-1}} - \dfrac{1}{|x|\sqrt{x^2-1}} = 0, \text{ for } x > 1$

21. $y = x\sin^{-1}x + \sqrt{1-x^2} = x\sin^{-1}x + \left(1-x^2\right)^{1/2} \Rightarrow \dfrac{dy}{dx} = \sin^{-1}x + x\left(\dfrac{1}{\sqrt{1-x^2}}\right) + \left(\dfrac{1}{2}\right)\left(1-x^2\right)^{-1/2}(-2x)$

$= \sin^{-1}x + \dfrac{x}{\sqrt{1-x^2}} - \dfrac{x}{\sqrt{1-x^2}} = \sin^{-1}x$

23. $\displaystyle\int \dfrac{1}{\sqrt{49-x^2}}\,dx = \sin^{-1}\left(\dfrac{x}{7}\right) + C$

25. $\displaystyle\int \dfrac{1}{17+x^2}\,dx = \int \dfrac{1}{\left(\sqrt{17}\right)^2+x^2}\,dx = \dfrac{1}{\sqrt{17}}\tan^{-1}\dfrac{x}{\sqrt{17}} + C$

27. $\int \dfrac{dx}{x\sqrt{25x^2 - 2}} = \int \dfrac{du}{u\sqrt{u^2 - 2}}$, where $u = 5x$ and $du = 5\,dx$

$\qquad = \dfrac{1}{\sqrt{2}}\sec^{-1}\left|\dfrac{u}{\sqrt{2}}\right| + C = \dfrac{1}{\sqrt{2}}\sec^{-1}\left|\dfrac{5x}{\sqrt{2}}\right| + C$

29. $\displaystyle\int_0^1 \dfrac{4\,ds}{\sqrt{4 - s^2}} = \left[4\sin^{-1}\dfrac{s}{2}\right]_0^1 = 4\left(\sin^{-1}\dfrac{1}{2} - \sin^{-1}0\right) = 4\left(\dfrac{\pi}{6} - 0\right) = \dfrac{2\pi}{3}$

31. $\displaystyle\int_0^2 \dfrac{dt}{8 + 2t^2} = \dfrac{1}{\sqrt{2}}\int_0^{2\sqrt{2}} \dfrac{du}{8 + u^2}$, where $u = \sqrt{2}t$ and $du = \sqrt{2}\,dt$; $t = 0 \Rightarrow u = 0$, $t = 2 \Rightarrow u = 2\sqrt{2}$

$\qquad = \left[\dfrac{1}{\sqrt{2}}\cdot\dfrac{1}{\sqrt{8}}\tan^{-1}\dfrac{u}{\sqrt{8}}\right]_0^{2\sqrt{2}} = \dfrac{1}{4}\left(\tan^{-1}\dfrac{2\sqrt{2}}{\sqrt{8}} - \tan^{-1}0\right) = \dfrac{1}{4}(\tan^{-1}1 - \tan^{-1}0) = \dfrac{1}{4}\left(\dfrac{\pi}{4} - 0\right) = \dfrac{\pi}{16}$

33. $\displaystyle\int_{-1}^{-\sqrt{2}/2} \dfrac{dy}{y\sqrt{4y^2 - 1}} = \int_{-2}^{-\sqrt{2}} \dfrac{du}{u\sqrt{u^2 - 1}}$, where $u = 2y$ and $du = 2\,dy$; $y = -1 \Rightarrow u = -2$, $y = -\dfrac{\sqrt{2}}{2} \Rightarrow u = -\sqrt{2}$

$\qquad = \left[\sec^{-1}|u|\right]_{-2}^{-\sqrt{2}} = \sec^{-1}\left|-\sqrt{2}\right| - \sec^{-1}|-2| = \dfrac{\pi}{4} - \dfrac{\pi}{3} = -\dfrac{\pi}{12}$

35. $\int \dfrac{3\,dr}{\sqrt{1 - 4(r-1)^2}} = \dfrac{3}{2}\int \dfrac{du}{\sqrt{1 - u^2}}$, where $u = 2(r-1)$ and $du = 2\,dr$

$\qquad = \dfrac{3}{2}\sin^{-1}u + C = \dfrac{3}{2}\sin^{-1}2(r-1) + C$

37. $\int \dfrac{dx}{2 + (x-1)^2} = \int \dfrac{du}{2 + u^2}$, where $u = x - 1$ and $du = dx$

$\qquad = \dfrac{1}{\sqrt{2}}\tan^{-1}\dfrac{u}{\sqrt{2}} + C = \dfrac{1}{\sqrt{2}}\tan^{-1}\left(\dfrac{x-1}{\sqrt{2}}\right) + C$

39. $\int \dfrac{dx}{(2x-1)\sqrt{(2x-1)^2 - 4}} = \dfrac{1}{2}\int \dfrac{du}{u\sqrt{u^2 - 4}}$, where $u = 2x - 1$ and $du = 2\,dx$

$\qquad = \dfrac{1}{2}\cdot\dfrac{1}{2}\sec^{-1}\left|\dfrac{u}{2}\right| + C = \dfrac{1}{4}\sec^{-1}\left|\dfrac{2x-1}{2}\right| + C$

41. $\displaystyle\int_{-\pi/2}^{\pi/2} \dfrac{2\cos\theta\,d\theta}{1 + (\sin\theta)^2} = 2\int_{-1}^{1} \dfrac{du}{1 + u^2}$, where $u = \sin\theta$ and $du = \cos\theta\,d\theta$

$\qquad = \left[2\tan^{-1}u\right]_{-1}^{1} = 2\left(\tan^{-1}1 - \tan^{-1}(-1)\right) = 2\left[\dfrac{\pi}{4} - \left(-\dfrac{\pi}{4}\right)\right] = \pi$

43. $\displaystyle\int_0^{\ln\sqrt{3}} \frac{e^x\,dx}{1+e^{2x}} = \int_1^{\sqrt{3}} \frac{du}{1+u^2}$, where $u = e^x$ and $du = e^x\,dx$; $x = 0 \Rightarrow u = 1$, $x = \ln\sqrt{3} \Rightarrow u = \sqrt{3}$

$\displaystyle = \left[\tan^{-1} u\right]_1^{\sqrt{3}} = \tan^{-1}\sqrt{3} - \tan^{-1}1 = \frac{\pi}{3} - \frac{\pi}{4} = \frac{\pi}{12}$

45. $\displaystyle\int \frac{y\,dy}{\sqrt{1-y^4}} = \frac{1}{2}\int \frac{du}{\sqrt{1-u^2}}$, where $u = y^2$ and $du = 2y\,dy$

$\displaystyle = \frac{1}{2}\sin^{-1}u + C = \frac{1}{2}\sin^{-1}y^2 + C$

47. $\displaystyle\int \frac{dx}{\sqrt{-x^2+4x-3}} = \int \frac{dx}{\sqrt{1-(x^2-4x+4)}} = \int \frac{dx}{\sqrt{1-(x-2)^2}} = \sin^{-1}(x-2) + C$

49. $\displaystyle\int_{-1}^0 \frac{6\,dt}{\sqrt{3-2t-t^2}} = 6\int_{-1}^0 \frac{dt}{\sqrt{4-(t^2+2t+1)}} = 6\int_{-1}^0 \frac{dt}{\sqrt{2^2-(t+1)^2}} = 6\left[\sin^{-1}\left(\frac{t+1}{2}\right)\right]_{-1}^0$

$\displaystyle = 6\left[\sin^{-1}\left(\frac{1}{2}\right) - \sin^{-1}0\right] = 6\left(\frac{\pi}{6} - 0\right) = \pi$

51. $\displaystyle\int \frac{dy}{y^2-2y+5} = \int \frac{dy}{4+y^2-2y+1} = \int \frac{dy}{2^2+(y-1)^2} = \frac{1}{2}\tan^{-1}\left(\frac{y-1}{2}\right) + C$

53. $\displaystyle\int_1^2 \frac{8\,dx}{x^2-2x+2} = 8\int_1^2 \frac{dx}{1+(x^2-2x+1)} = 8\int_1^2 \frac{dx}{1+(x-1)^2} = 8\left[\tan^{-1}(x-1)\right]_1^2$

$\displaystyle = 8\left(\tan^{-1}1 - \tan^{-1}0\right) = 8\left(\frac{\pi}{4} - 0\right) = 2\pi$

55. $\displaystyle\int \frac{dx}{(x+1)\sqrt{x^2+2x}} = \int \frac{dx}{(x+1)\sqrt{x^2+2x+1-1}} = \int \frac{dx}{(x+1)\sqrt{(x+1)^2-1}}$

$\displaystyle = \int \frac{du}{u\sqrt{u^2-1}}$, where $u = x+1$ and $du = dx$

$\displaystyle = \sec^{-1}|u| + C = \sec^{-1}|x+1| + C$

57. $\displaystyle\int \frac{e^{\sin^{-1}x}}{\sqrt{1-x^2}}\,dx = \int e^u\,du$, where $u = \sin^{-1}x$ and $du = \frac{dx}{\sqrt{1-x^2}}$

$\displaystyle = e^u + C = e^{\sin^{-1}x} + C$

59. $\displaystyle\int \frac{\left(\sin^{-1}x\right)^2}{\sqrt{1-x^2}}\,dx = \int u^2\,du$, where $u = \sin^{-1}x$ and $du = \frac{dx}{\sqrt{1-x^2}}$

$\displaystyle = \frac{u^3}{3} + C = \frac{\left(\sin^{-1}x\right)^3}{3}$

61. $\int \frac{1}{(\tan^{-1}y)(1+y^2)}\,dy = \int \frac{\left(\frac{1}{1+y^2}\right)}{\tan^{-1}y}\,dy = \int \frac{1}{u}\,du$, where $u = \tan^{-1}y$ and $du = \frac{dy}{1+y^2}$

$= \ln|u| + C = \ln\left|\tan^{-1}y\right| + C$

63. $\int_{\sqrt{2}}^{2} \frac{\sec^2(\sec^{-1}x)}{x\sqrt{x^2-1}}\,dx = \int_{\pi/4}^{\pi/3} \sec^2 u\,du$, where $u = \sec^{-1}x$ and $du = \frac{dx}{x\sqrt{x^2-1}}$; $x = \sqrt{2} \Rightarrow u = \frac{\pi}{4}$, $x = 2 \Rightarrow u = \frac{\pi}{3}$

$= [\tan u]_{\pi/4}^{\pi/3} = \tan\frac{\pi}{3} - \tan\frac{\pi}{4} = \sqrt{3} - 1$

65. $\lim_{x\to 0} \frac{\sin^{-1}5x}{x} = \lim_{x\to 0} \frac{\left(\frac{5}{\sqrt{1-25x^2}}\right)}{1} = 5$

67. $\lim_{x\to\infty} x\tan^{-1}\left(\frac{2}{x}\right) = \lim_{x\to\infty} \frac{\tan^{-1}(2x^{-1})}{x^{-1}} = \lim_{x\to\infty} \frac{\left(\frac{-2x^{-2}}{1+4x^{-2}}\right)}{-x^{-2}} = \lim_{x\to\infty} \frac{2}{1+4x^{-2}} = 2$

69. If $y = \ln x - \frac{1}{2}\ln(1+x^2) - \frac{\tan^{-1}x}{x} + C$, then $dy = \left[\frac{1}{x} - \frac{x}{1+x^2} - \frac{\left(\frac{x}{1+x^2}\right) - \tan^{-1}x}{x^2}\right]dx$

$= \left(\frac{1}{x} - \frac{x}{1+x^2} - \frac{1}{x(1+x^2)} + \frac{\tan^{-1}x}{x^2}\right)dx = \frac{x(1+x^2) - x^3 - x + (\tan^{-1}x)(1+x^2)}{x^2(1+x^2)}\,dx = \frac{\tan^{-1}x}{x^2}\,dx$,

which verifies the formula

71. If $y = x(\sin^{-1}x)^2 - 2x + 2\sqrt{1-x^2}\,\sin^{-1}x + C$, then

$dy = \left[(\sin^{-1}x)^2 + \frac{2x(\sin^{-1}x)}{\sqrt{1-x^2}} - 2 + \frac{-2x}{\sqrt{1-x^2}}\sin^{-1}x + 2\sqrt{1-x^2}\left(\frac{1}{\sqrt{1-x^2}}\right)\right]dx = (\sin^{-1}x)^2\,dx$, which verifies

the formula

73. $\frac{dy}{dx} = \frac{1}{\sqrt{1-x^2}} \Rightarrow dy = \frac{dx}{\sqrt{1-x^2}} \Rightarrow y = \sin^{-1}x + C$; $x = 0$ and $y = 0 \Rightarrow 0 = \sin^{-1}0 + C \Rightarrow C = 0 \Rightarrow y = \sin^{-1}x$

75. $\frac{dy}{dx} = \frac{1}{x\sqrt{x^2-1}} \Rightarrow dy = \frac{dx}{x\sqrt{x^2-1}} \Rightarrow y = \sec^{-1}|x| + C$; $x = 2$ and $y = \pi \Rightarrow \pi = \sec^{-1}2 + C \Rightarrow C = \pi - \sec^{-1}2$

$= \pi - \frac{\pi}{3} = \frac{2\pi}{3} \Rightarrow y = \sec^{-1}(x) + \frac{2\pi}{3}$, $x > 1$

77. $\alpha(x) = \cot^{-1}\left(\frac{x}{15}\right) - \cot^{-1}\left(\frac{x}{3}\right)$, $x > 0 \Rightarrow \alpha'(x) = \frac{-15}{225+x^2} + \frac{3}{9+x^2} = \frac{-15(9+x^2) + 3(225+x^2)}{(225+x^2)(9+x^2)}$; solving

$\alpha'(x) = 0 \Rightarrow -135 - 15x^2 + 675 + 3x^2 = 0 \Rightarrow x = 3\sqrt{5}$; $\alpha'(x) > 0$ when $0 < x < 3\sqrt{5}$ and $\alpha'(x) < 0$ for

$x > 3\sqrt{5} \Rightarrow$ there is a maximum at $3\sqrt{5}$ ft from the front of the room

79. Yes, $\sin^{-1} x$ and $\cos^{-1} x$ differ by the constant $\frac{\pi}{2}$

81. $\cos^{-1} u = \frac{\pi}{2} - \sin^{-1} u \Rightarrow \frac{d}{dx}\left(\cos^{-1} u\right) = \frac{d}{dx}\left(\frac{\pi}{2} - \sin^{-1} u\right) = 0 - \dfrac{\dfrac{du}{dx}}{\sqrt{1 - u^2}}, \ |u| < 1$

83. $\csc^{-1} u = \frac{\pi}{2} - \sec^{-1} u \Rightarrow \frac{d}{dx}\left(\csc^{-1} u\right) = \frac{d}{dx}\left(\frac{\pi}{2} - \sec^{-1} u\right) = 0 - \dfrac{\dfrac{du}{dx}}{|u|\sqrt{u^2 - 1}} = -\dfrac{\dfrac{du}{dx}}{|u|\sqrt{u^2 - 1}}, \ |u| > 1$

85. $f(x) = \sin x \Rightarrow f'(x) = \cos x \Rightarrow \dfrac{df^{-1}}{dx} = \dfrac{1}{(\cos x)_{\sin^{-1} x}} \Rightarrow \dfrac{df^{-1}}{dx} = \dfrac{1}{\cos\left(\sin^{-1} x\right)} = \dfrac{1}{\sqrt{1 - \sin^2\left(\sin^{-1} x\right)}} = \dfrac{1}{\sqrt{1 - x^2}}$

87. The functions f and g have the same derivative (for $x \geq 0$), namely $\dfrac{1}{\sqrt{x}\,(x + 1)}$. The functions therefore differ

by a constant. To identify the constant we can set x equal to 0 in the equation $f(x) = g(x) + C$, obtaining

$\sin^{-1}(-1) = 2\tan^{-1}(0) + C \Rightarrow -\frac{\pi}{2} = 0 + C \Rightarrow C = -\frac{\pi}{2}$. For $x \geq 0$, we have $\sin^{-1}\left(\frac{x - 1}{x + 1}\right) = 2\tan^{-1}\sqrt{x} - \frac{\pi}{2}$.

89. $V = \pi \displaystyle\int_{-\sqrt{3}/3}^{\sqrt{3}} \left(\frac{1}{\sqrt{1 + x^2}}\right)^2 dx = \pi \displaystyle\int_{-\sqrt{3}/3}^{\sqrt{3}} \frac{1}{1 + x^2}\, dx = \pi\left[\tan^{-1} x\right]_{-\sqrt{3}/3}^{\sqrt{3}} = \pi\left[\tan^{-1}\sqrt{3} - \tan^{-1}\left(-\frac{\sqrt{3}}{3}\right)\right]$

$= \pi\left[\frac{\pi}{3} - \left(-\frac{\pi}{6}\right)\right] = \frac{\pi^2}{2}$

91. (a) $A(x) = \frac{\pi}{4}(\text{diameter})^2 = \frac{\pi}{4}\left[\frac{1}{\sqrt{1 + x^2}} - \left(-\frac{1}{\sqrt{1 + x^2}}\right)\right]^2 = \frac{\pi}{1 + x^2} \Rightarrow V = \displaystyle\int_a^b A(x)\, dx = \displaystyle\int_{-1}^1 \frac{\pi\, dx}{1 + x^2}$

$= \pi\left[\tan^{-1} x\right]_{-1}^1 = (\pi)(2)\left(\frac{\pi}{4}\right) = \frac{\pi^2}{2}$

(b) $A(x) = (\text{edge})^2 = \left[\frac{1}{\sqrt{1 + x^2}} - \left(-\frac{1}{\sqrt{1 + x^2}}\right)\right]^2 = \frac{\pi}{1 + x^2} \Rightarrow V = \displaystyle\int_a^b A(x)\, dx = \displaystyle\int_{-1}^1 \frac{4\, dx}{1 + x^2}$

$= 4\left[\tan^{-1} x\right]_{-1}^1 = 4\left[\tan^{-1}(1) - \tan^{-1}(-1)\right] = 4\left[\frac{\pi}{4} - \left(-\frac{\pi}{4}\right)\right] = 2\pi$

93. A calculator or computer numerical integrator yields $\sin^{-1} 0.6 \approx 0.643517104$.

95. The values of f increase over the interval $[-1, 1]$ because
$f' > 0$, and the graph of f steepens as the values of f'
increase towards the ends of the interval. The graph of f
is concave down to the left of the origin where $f'' < 0$,
and concave up to the right of the origin where $f'' > 0$.
There is an inflection point at $x = 0$ where $f'' = 0$ and
f' has a local minimum value.

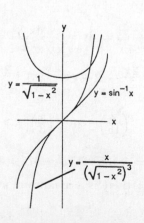

6.10 HYPERBOLIC FUNCTIONS

1. $\sinh x = -\frac{3}{4} \Rightarrow \cosh x = \sqrt{1 + \sinh^2 x} = \sqrt{1 + \left(-\frac{3}{4}\right)^2} = \sqrt{1 + \frac{9}{16}} = \sqrt{\frac{25}{16}} = \frac{5}{4}$, $\tanh x = \frac{\sinh x}{\cosh x} = \frac{\left(-\frac{3}{4}\right)}{\left(\frac{5}{4}\right)} = -\frac{3}{5}$,

$\coth x = \frac{1}{\tanh x} = -\frac{5}{3}$, $\operatorname{sech} x = \frac{1}{\cosh x} = \frac{4}{5}$, and $\operatorname{csch} x = \frac{1}{\sin x} = -\frac{4}{3}$

3. $\cosh x = \frac{17}{15}$, $x > 0 \Rightarrow \sinh x = \sqrt{\cosh^2 x - 1} = \sqrt{\left(\frac{17}{15}\right)^2 - 1} = \sqrt{\frac{289}{225} - 1} = \sqrt{\frac{64}{225}} = \frac{8}{15}$, $\tanh x = \frac{\sinh x}{\cosh x} = \frac{\left(\frac{8}{15}\right)}{\left(\frac{17}{15}\right)}$

$= \frac{8}{17}$, $\coth x = \frac{1}{\tanh x} = \frac{17}{8}$, $\operatorname{sech} x = \frac{1}{\cosh x} = \frac{15}{17}$, and $\operatorname{csch} x = \frac{1}{\sinh x} = \frac{15}{8}$

5. $2\cosh(\ln x) = 2\left(\frac{e^{\ln x} + e^{-\ln x}}{2}\right) = e^{\ln x} + \frac{1}{e^{\ln x}} = x + \frac{1}{x}$

7. $\cosh 5x + \sinh 5x = \frac{e^{5x} + e^{-5x}}{2} + \frac{e^{5x} - e^{-5x}}{2} = e^{5x}$

9. $(\sinh x + \cosh x)^4 = \left(\frac{e^x - e^{-x}}{2} + \frac{e^x + e^{-x}}{2}\right)^4 = \left(e^x\right)^4 = e^{4x}$

11. (a) $\sinh 2x = \sinh(x + x) = \sinh x \cosh x + \cosh x \sinh x = 2 \sinh x \cosh x$

(b) $\cosh 2x = \cosh(x + x) = \cosh x \cosh x + \sinh x \sin x = \cosh^2 x + \sinh^2 x$

13. $y = 6 \sinh \frac{x}{3} \Rightarrow \frac{dy}{dx} = 6\left(\cosh \frac{x}{3}\right)\left(\frac{1}{3}\right) = 2 \cosh \frac{x}{3}$

15. $y = 2\sqrt{t} \tanh \sqrt{t} = 2t^{1/2} \tanh t^{1/2} \Rightarrow \frac{dy}{dt} = \left[\operatorname{sech}^2\left(t^{1/2}\right)\right]\left(\frac{1}{2} t^{-1/2}\right)\left(2t^{1/2}\right) + \left(\tanh t^{1/2}\right)\left(t^{-1/2}\right)$

$= \operatorname{sech}^2 \sqrt{t} + \frac{\tanh \sqrt{t}}{\sqrt{t}}$

17. $y = \ln(\sinh z) \Rightarrow \frac{dy}{dz} = \frac{\cosh z}{\sinh z} = \coth z$

19. $y = (\operatorname{sech} \theta)(1 - \ln \operatorname{sech} \theta) \Rightarrow \frac{dy}{d\theta} = \left(-\frac{-\operatorname{sech} \theta \tanh \theta}{\operatorname{sech} \theta}\right)(\operatorname{sech} \theta) + (-\operatorname{sech} \theta \tanh \theta)(1 - \ln \operatorname{sech} \theta)$

$= \operatorname{sech} \theta \tanh \theta - (\operatorname{sech} \theta \tanh \theta)(1 - \ln \operatorname{sech} \theta) = (\operatorname{sech} \theta \tanh \theta)[1 - (1 - \ln \operatorname{sech} \theta)]$

$= (\operatorname{sech} \theta \tanh \theta)(\ln \operatorname{sech} \theta)$

21. $y = \ln \cosh v - \frac{1}{2} \tanh^2 v \Rightarrow \frac{dy}{dv} = \frac{\sinh v}{\cosh v} - \left(\frac{1}{2}\right)(2 \tanh v)\left(\operatorname{sech}^2 v\right) = \tanh v - (\tanh v)\left(\operatorname{sech}^2 v\right)$

$= (\tanh v)\left(1 - \operatorname{sech}^2 v\right) = (\tanh v)\left(\tanh^2 v\right) = \tanh^3 v$

23. $y = \left(x^2 + 1\right) \operatorname{sech}(\ln x) = \left(x^2 + 1\right)\left(\frac{2}{e^{\ln x} + e^{-\ln x}}\right) = \left(x^2 + 1\right)\left(\frac{2}{x + x^{-1}}\right) = \left(x^2 + 1\right)\left(\frac{2x}{x^2 + 1}\right) = 2x \Rightarrow \frac{dy}{dx} = 2$

25. $y = \sinh^{-1}\sqrt{x} = \sinh^{-1}(x^{1/2}) \Rightarrow \dfrac{dy}{dx} = \dfrac{\left(\frac{1}{2}\right)x^{-1/2}}{\sqrt{1+\left(x^{1/2}\right)^2}} = \dfrac{1}{2\sqrt{x}\sqrt{1+x}} = \dfrac{1}{2\sqrt{x(1+x)}}$

27. $y = (1-\theta)\tanh^{-1}\theta \Rightarrow \dfrac{dy}{d\theta} = (1-\theta)\left(\dfrac{1}{1-\theta^2}\right) + (-1)\tanh^{-1}\theta = \dfrac{1}{1+\theta} - \tanh^{-1}\theta$

29. $y = (1-t)\coth^{-1}\sqrt{t} = (1-t)\coth^{-1}(t^{1/2}) \Rightarrow \dfrac{dy}{dt} = (1-t)\left[\dfrac{\left(\frac{1}{2}\right)t^{-1/2}}{1-\left(t^{1/2}\right)^2}\right] + (-1)\coth^{-1}(t^{1/2}) = \dfrac{1}{2\sqrt{t}} - \coth^{-1}\sqrt{t}$

31. $y = \cos^{-1}x - x\,\mathrm{sech}^{-1}x \Rightarrow \dfrac{dy}{dx} = \dfrac{-1}{\sqrt{1-x^2}} - \left[x\left(\dfrac{-1}{x\sqrt{1-x^2}}\right) + (1)\,\mathrm{sech}^{-1}x\right] = \dfrac{-1}{\sqrt{1-x^2}} + \dfrac{1}{\sqrt{1-x^2}} - \mathrm{sech}^{-1}x$

$= -\,\mathrm{sech}^{-1}x$

33. $y = \mathrm{csch}^{-1}\left(\frac{1}{2}\right)^{\theta} \Rightarrow \dfrac{dy}{d\theta} = -\dfrac{\left[\ln\left(\frac{1}{2}\right)\right]\left(\frac{1}{2}\right)^{\theta}}{\left(\frac{1}{2}\right)^{\theta}\sqrt{1+\left[\left(\frac{1}{2}\right)^{\theta}\right]^2}} = -\dfrac{\ln(1)-\ln(2)}{\sqrt{1+\left(\frac{1}{2}\right)^{2\theta}}} = \dfrac{\ln 2}{\sqrt{1+\left(\frac{1}{2}\right)^{2\theta}}}$

35. $y = \sinh^{-1}(\tan x) \Rightarrow \dfrac{dy}{dx} = \dfrac{\sec^2 x}{\sqrt{1+(\tan x)^2}} = \dfrac{\sec^2 x}{\sqrt{\sec^2 x}} = \dfrac{\sec^2 x}{|\sec x|} = \dfrac{|\sec x||\sec x|}{|\sec x|} = |\sec x|$

37. (a) If $y = \tan^{-1}(\sinh x) + C$, then $\dfrac{dy}{dx} = \dfrac{\cosh x}{1+\sinh^2 x} = \dfrac{\cosh x}{\cosh^2 x} = \mathrm{sech}\,x$, which verifies the formula

(b) If $y = \sin^{-1}(\tanh x) + C$, then $\dfrac{dy}{dx} = \dfrac{\mathrm{sech}^2 x}{\sqrt{1-\tanh^2 x}} = \dfrac{\mathrm{sech}^2 x}{\mathrm{sech}\,x} = \mathrm{sech}\,x$, which verifies the formula

39. If $y = \dfrac{x^2-1}{2}\coth^{-1}x + \dfrac{x}{2} + C$, then $\dfrac{dy}{dx} = x\coth^{-1}x + \left(\dfrac{x^2-1}{2}\right)\left(\dfrac{1}{1-x^2}\right) + \dfrac{1}{2} = x\coth^{-1}x$, which verifies
the formula

41. $\displaystyle\int \sinh 2x\,dx = \dfrac{1}{2}\int \sinh u\,du$, where $u = 2x$ and $du = 2\,dx$

$= \dfrac{\cosh u}{2} + C = \dfrac{\cosh 2x}{2} + C$

43. $\displaystyle\int 6\cosh\left(\dfrac{x}{2} - \ln 3\right)dx = 12\int \cosh u\,du$, where $u = \dfrac{x}{2} - \ln 3$ and $du - \dfrac{1}{2}\,dx$

$= 12\sinh u + C = 12\sinh\left(\dfrac{x}{2} - \ln 3\right) + C$

45. $\displaystyle\int \tanh\dfrac{x}{7}\,dx = 7\int \dfrac{\sinh u}{\cosh u}\,du$, where $u = \dfrac{x}{7}$ and $du = \dfrac{1}{7}\,dx$

$= 7\ln|\cosh u| + C_1 = 7\ln\left|\cosh\dfrac{x}{7}\right| + C_1 = 7\ln\left|\dfrac{e^{x/7}+e^{-x/7}}{2}\right| + C_1 = 7\ln\left|e^{x/7}+e^{-x/7}\right| - 7\ln 2 + C_1$

$= 7\ln\left|e^{x/7}+e^{-x/7}\right| + C$

47. $\displaystyle\int \operatorname{sech}^2\left(x-\tfrac{1}{2}\right)dx = \int \sec^2 u\, du$, where $u = \left(x-\tfrac{1}{2}\right)$ and $du = dx$

$\qquad = \tanh u + C = \tan\left(x-\tfrac{1}{2}\right)+C$

49. $\displaystyle\int \frac{\operatorname{sech}\sqrt{t}\,\tanh\sqrt{t}}{\sqrt{t}}\,dt = 2\int \operatorname{sech} u\,\tanh u\, du$, where $u = \sqrt{t} = t^{1/2}$ and $du = \dfrac{dt}{2\sqrt{t}}$

$\qquad = 2(-\operatorname{sech} u) + C = -2\operatorname{sech}\sqrt{t} + C$

51. $\displaystyle\int_{\ln 2}^{\ln 4} \coth x\, dx = \int_{\ln 2}^{\ln 4} \frac{\cosh x}{\sinh x}\, dx = \int_{3/4}^{15/8} \frac{1}{u}\, du = \Big[\ln|u|\Big]_{3/4}^{15/8} = \ln\left|\frac{15}{8}\right| - \ln\left|\frac{3}{4}\right| = \ln\left|\frac{15}{8}\cdot\frac{4}{3}\right| = \ln\frac{5}{2},$

where $u = \sinh x$, $du = \cosh x\, dx$, the lower limit is $\sinh(\ln 2) = \dfrac{e^{\ln 2} - e^{-\ln 2}}{2} = \dfrac{2 - \left(\frac{1}{2}\right)}{2} = \dfrac{3}{4}$ and the upper

limit is $\sinh(\ln 4) = \dfrac{e^{\ln 4} - e^{-\ln 4}}{2} = \dfrac{4 - \left(\frac{1}{4}\right)}{2} = \dfrac{15}{8}$

53. $\displaystyle\int_{-\ln 4}^{-\ln 2} 2e^\theta \cosh\theta\, d\theta = \int_{-\ln 4}^{-\ln 2} 2e^\theta\left(\frac{e^\theta + e^{-\theta}}{2}\right)d\theta = \int_{-\ln 4}^{-\ln 2}\left(e^{2\theta} + 1\right)d\theta = \left[\frac{e^{2\theta}}{2} + \theta\right]_{-\ln 4}^{-\ln 2}$

$\qquad = \left(\dfrac{e^{-2\ln 2}}{2} - \ln 2\right) - \left(\dfrac{e^{-2\ln 4}}{2} - \ln 4\right) = \left(\dfrac{1}{8} - \ln 2\right) - \left(\dfrac{1}{32} - \ln 4\right) = \dfrac{3}{32} - \ln 2 + 2\ln 2 = \dfrac{3}{32} + \ln 2$

55. $\displaystyle\int_{-\pi/4}^{\pi/4} \cosh(\tan\theta)\sec^2\theta\, d\theta = \int_{-1}^{1} \cosh u\, du = [\sinh u]_{-1}^{1} = \sinh(1) - \sinh(-1) = \left(\dfrac{e^1 - e^{-1}}{2}\right) - \left(\dfrac{e^{-1} - e^1}{2}\right)$

$\qquad = \dfrac{e - e^{-1} - e^{-1} + e}{2} = e - e^{-1}$, where $u = \tan\theta$, $du = \sec^2\theta\, d\theta$, the lower limit is $\tan\left(-\dfrac{\pi}{4}\right) = -1$ and the upper

limit is $\tan\left(\dfrac{\pi}{4}\right) = 1$

57. $\displaystyle\int_{1}^{2} \frac{\cosh(\ln t)}{t}\, dt = \int_{0}^{\ln 2} \cosh u\, du = [\sinh u]_{0}^{\ln 2} = \sinh(\ln 2) - \sinh(0) = \dfrac{e^{\ln 2} - e^{-\ln 2}}{2} - 0 = \dfrac{2 - \frac{1}{2}}{2} = \dfrac{3}{4}$, where

$u = \ln t$, $du = \dfrac{1}{t}\, dt$, the lower limit is $\ln 1 = 0$ and the upper limit is $\ln 2$

59. $\displaystyle\int_{-\ln 2}^{0} \cosh^2\left(\frac{x}{2}\right)dx = \int_{-\ln 2}^{0} \frac{\cosh x + 1}{2}\, dx = \frac{1}{2}\int_{-\ln 2}^{0}(\cosh x + 1)\, dx = \frac{1}{2}[\sinh x + x]_{-\ln 2}^{0}$

$\qquad = \dfrac{1}{2}[(\sinh 0 + 0) - (\sinh(-\ln 2) - \ln 2)] = \dfrac{1}{2}\left[(0+0) - \left(\dfrac{e^{-\ln 2} - e^{\ln 2}}{2} - \ln 2\right)\right] = \dfrac{1}{2}\left[-\dfrac{\left(\frac{1}{2}\right) - 2}{2} + \ln 2\right]$

$\qquad = \dfrac{1}{2}\left(1 - \dfrac{1}{4} + \ln 2\right) = \dfrac{3}{8} + \dfrac{1}{2}\ln 2 = \dfrac{3}{8} + \ln\sqrt{2}$

61. $\sinh^{-1}\left(\frac{-5}{12}\right) = \ln\left(-\frac{5}{12} + \sqrt{\frac{25}{144}+1}\right) = \ln\left(\frac{2}{3}\right)$

63. $\tanh^{-1}\left(-\frac{1}{2}\right) = \frac{1}{2}\ln\left(\frac{1-(1/2)}{1+(1/2)}\right) = -\frac{\ln 3}{2}$

65. $\text{sech}^{-1}\left(\frac{3}{5}\right) = \ln\left(\frac{1+\sqrt{1-(9/25)}}{(3/5)}\right) = \ln 3$

67. (a) $\displaystyle\int_0^{2\sqrt{3}} \frac{dx}{\sqrt{4+x^2}} = \left[\sinh^{-1}\frac{x}{2}\right]_0^{2\sqrt{3}} = \sinh^{-1}\sqrt{3} - \sinh 0 = \sinh^{-1}\sqrt{3}$

(b) $\sinh^{-1}\sqrt{3} = \ln\left(\sqrt{3} + \sqrt{3+1}\right) = \ln\left(\sqrt{3}+2\right)$

69. (a) $\displaystyle\int_{5/4}^2 \frac{1}{1-x^2}\,dx = \left[\coth^{-1}x\right]_{5/4}^2 = \coth^{-1}2 - \coth^{-1}\frac{5}{4}$

(b) $\coth^{-1}2 - \coth^{-1}\frac{5}{4} = \frac{1}{2}\left[\ln 3 - \ln\left(\frac{9/4}{1/4}\right)\right] = \frac{1}{2}\ln\frac{1}{3}$

71. (a) $\displaystyle\int_{1/5}^{3/13} \frac{dx}{x\sqrt{1-16x^2}} = \int_{4/5}^{12/13} \frac{du}{u\sqrt{a^2-u^2}}$, where $u = 4x$, $du = 4\,dx$, $a = 1$

$= \left[-\text{sech}^{-1}u\right]_{4/5}^{12/13} = -\text{sech}^{-1}\frac{12}{13} + \text{sech}^{-1}\frac{4}{5}$

(b) $-\text{sech}^{-1}\frac{12}{13} + \text{sech}^{-1}\frac{4}{5} = -\ln\left(\frac{1+\sqrt{1-(12/13)^2}}{(12/13)}\right) + \ln\left(\frac{1+\sqrt{1-(4/5)^2}}{(4/5)}\right)$

$= -\ln\left(\frac{13+\sqrt{169-144}}{12}\right) + \ln\left(\frac{5+\sqrt{25-16}}{4}\right) = \ln\left(\frac{5+3}{4}\right) - \ln\left(\frac{13+5}{12}\right) = \ln 2 - \ln\frac{3}{2}$

$= = \ln\left(2\cdot\frac{2}{3}\right) = \ln\frac{4}{3}$

73. (a) $\displaystyle\int_0^\pi \frac{\cos x}{\sqrt{1+\sin^2 x}}\,dx = \int_0^0 \frac{1}{\sqrt{1+u^2}}\,du = \left[\sinh^{-1}u\right]_0^0 = \sinh^{-1}0 - \sinh^{-1}0 = 0$, where $u = \sin x$, $du = \cos x\,dx$

(b) $\sinh^{-1}0 - \sinh^{-1}0 = \ln\left(0 + \sqrt{0+1}\right) - \ln\left(0 + \sqrt{0+1}\right) = 0$

75. (a) Let $E(x) = \dfrac{f(x)+f(-x)}{2}$ and $O(x) = \dfrac{f(x)-f(-x)}{2}$. Then $E(x) + O(x) = \dfrac{f(x)+f(-x)}{2} + \dfrac{f(x)-f(-x)}{2}$

$= \dfrac{2f(x)}{2} = f(x)$. Also, $E(-x) = \dfrac{f(-x)+f(-(-x))}{2} = \dfrac{f(x)+f(-x)}{2} = E(x) \Rightarrow E(x)$ is even, and

$O(-x) = \dfrac{f(-x)-f(-(-x))}{2} = -\dfrac{f(x)-f(-x)}{2} = -O(x) \Rightarrow O(x)$ is odd. Consequently, $f(x)$ can be written as a sum of an even and an odd function.

(b) $f(x) = \dfrac{f(x) + f(-x)}{2}$ because $\dfrac{f(x) - f(-x)}{2} = 0$ and $f(x) = \dfrac{f(x) - f(-x)}{2}$ because $\dfrac{f(x) + f(-x)}{2} = 0$; thus

$f(x) = \dfrac{2f(x)}{2} + 0$ and $f(x) = 0 + \dfrac{2f(x)}{2}$

77. (a) $m\dfrac{dv}{dt} = mg - kv^2 \Rightarrow \dfrac{m\dfrac{dv}{dt}}{mg - kv^2} = 1 \Rightarrow \dfrac{\dfrac{1}{g}\dfrac{dv}{dt}}{1 - \dfrac{kv^2}{mg}} = 1 \Rightarrow \dfrac{\sqrt{\dfrac{k}{mg}}\,dv}{1 - \sqrt{v\left(\dfrac{k}{mg}\right)}^2} = \sqrt{\dfrac{kg}{m}}\,dt \Rightarrow \int \dfrac{\sqrt{\dfrac{k}{mg}}\,dv}{1 - \sqrt{v\left(\dfrac{k}{mg}\right)}^2}\,dv$

$= \int \sqrt{\dfrac{kg}{m}}\,dt \Rightarrow \tanh^{-1}\left(\sqrt{\dfrac{k}{mg}}\,v\right) = \sqrt{\dfrac{kg}{m}}\,t + C \Rightarrow v = \sqrt{\dfrac{mg}{k}}\tanh\left(\sqrt{\dfrac{kg}{m}}\,t + C\right); v(0) = 0 \Rightarrow C = 0$

$\Rightarrow v = \sqrt{\dfrac{mg}{k}}\tanh\left(\sqrt{\dfrac{kg}{m}}\,t\right)$

(b) $\lim\limits_{t\to\infty} v = \lim\limits_{t\to\infty}\sqrt{\dfrac{mg}{k}}\tanh\left(\sqrt{\dfrac{kg}{m}}\,t\right) = \sqrt{\dfrac{mg}{k}}\lim\limits_{t\to\infty}\tanh\left(\sqrt{\dfrac{kg}{m}}\,t\right) = \sqrt{\dfrac{mg}{k}}(1) = \sqrt{\dfrac{mg}{k}}$

(c) $\sqrt{\dfrac{160}{0.005}} = \sqrt{\dfrac{1{,}600{,}000}{5}} = \dfrac{400}{\sqrt{5}} = 80\sqrt{5} \approx 178.89$ ft/sec

79. $\dfrac{dy}{dx} = \dfrac{-1}{x\sqrt{1 - x^2}} + \dfrac{x}{\sqrt{1 - x^2}} \Rightarrow y = \int \dfrac{-1}{x\sqrt{1 - x^2}}\,dx + \int \dfrac{x}{\sqrt{1 - x^2}}\,dx \Rightarrow y = \operatorname{sech}^{-1}(x) - \sqrt{1 - x^2} + C$; $x = 1$ and

$y = 0 \Rightarrow C = 0 \Rightarrow y = \operatorname{sech}^{-1}(x) - \sqrt{1 - x^2}$

81. $V = \pi\displaystyle\int_0^2 \left(\cosh^2 x - \sinh^2 x\right)dx = \pi\displaystyle\int_0^2 1\,dx = 2\pi$

83. (a) $y = \dfrac{1}{2}\cosh 2x \Rightarrow y' = \sinh 2x \Rightarrow L = \displaystyle\int_0^{\ln\sqrt{5}} \sqrt{1 + (\sinh 2x)^2}\,dx = \displaystyle\int_0^{\ln\sqrt{5}} \cosh 2x\,dx = \left[\dfrac{1}{2}\sinh 2x\right]_0^{\ln\sqrt{5}}$

$= \left[\dfrac{1}{2}\left(\dfrac{e^{2x} - e^{-2x}}{2}\right)\right]_0^{\ln\sqrt{5}} = \dfrac{1}{4}\left(5 - \dfrac{1}{5}\right) = \dfrac{6}{5}$

(b) $y = \dfrac{1}{a}\cosh ax \Rightarrow 1 + (y')^2 = 1 + \sinh^2 ax \Rightarrow \cosh^2 ax \Rightarrow L = \displaystyle\int_0^b \sqrt{\cosh^2 ax}\,dx = \displaystyle\int_0^b \cosh ax\,dx = \left[\dfrac{\sin ax}{a}\right]_0^b$

$= \dfrac{\sinh ab}{a}$

85. (a) $y = \cosh x \Rightarrow ds = \sqrt{(dx)^2 + (dy)^2} = \sqrt{(dx)^2 + (\sinh^2 x)(dx)^2} = \cosh x\,dx$; $M_x = \displaystyle\int_{-\ln 2}^{\ln 2} y\,ds$

$= \displaystyle\int_{-\ln 2}^{\ln 2} \cosh x\,ds = \displaystyle\int_{-\ln 2}^{\ln 2} \cosh^2 x\,dx = \displaystyle\int_0^{\ln 2} (\cosh 2x + 1)\,dx = \left[\dfrac{\sinh 2x}{2} + x\right]_0^{\ln 2} = \dfrac{1}{4}\left(e^{\ln 4} - e^{-\ln 4}\right) + \ln 2$

$= \dfrac{15}{16} + \ln 2$; $M = 2\displaystyle\int_0^{\ln 2} \sqrt{1 + \sinh^2 x}\,dx = 2\displaystyle\int_0^{\ln 2} \cosh x\,dx = 2[\sinh x]_0^{\ln 2} = e^{\ln 2} - e^{-\ln 2} = 2 - \dfrac{1}{2} = \dfrac{3}{2}$.

Therefore, $\bar{y} = \dfrac{M_x}{M} = \dfrac{\left(\dfrac{15}{16} + \ln 2\right)}{\left(\dfrac{3}{2}\right)} = \dfrac{5}{8} + \dfrac{\ln 4}{3}$, and by symmetry $\bar{x} = 0$.

(b) $\bar{x} = 0$, $\bar{y} \approx 1.09$

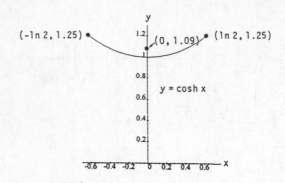

87. (a) $y = \dfrac{H}{W}\cosh\left(\dfrac{W}{H}x\right) \Rightarrow \tan \phi = \dfrac{dy}{dx} = \left(\dfrac{H}{W}\right)\left[\dfrac{W}{H}\sinh\left(\dfrac{W}{H}x\right)\right] = \sinh\left(\dfrac{W}{H}x\right)$

(b) The tension at P is given by $T \cos \phi = H \Rightarrow T = H \sec \phi = H\sqrt{1 + \tan^2 \phi} = H\sqrt{1 + \left(\sinh\dfrac{W}{H}x\right)^2}$

$= H \cosh\left(\dfrac{W}{H}x\right) = w\left(\dfrac{H}{W}\right)\cosh\left(\dfrac{W}{H}x\right) = wy$

89. (a) Since the cable is 32 ft long, $s = 16$ and $x = 15$. From Exercise 88, $x = \dfrac{1}{a}\sinh^{-1} as \Rightarrow 15a = \sinh^{-1} 16a$

$\Rightarrow \sinh 15a = 16a$.

(b) The intersection is near $(0.042, 0.672)$.

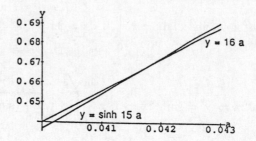

(c) Newton's method indicates that at $a \approx 0.0417525$ the curves $y = 16a$ and $y = \sinh 15a$ intersect.

(d) $T = wy \approx (2\ \text{lb})\left(\dfrac{1}{0.0417525}\right) \approx 47.90\ \text{lb}$

(e) The sag is about 4.8 ft.

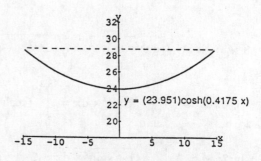

6.11 FIRST ORDER DIFFERENTIAL EQUATIONS

1. (a) $y = e^{-x} \Rightarrow y' = -e^{-x} \Rightarrow 2y' + 3y = 2(-e^{-x}) + 3e^{-x} = e^{-x}$

   (b) $y = e^{-x} + e^{-3x/2} \Rightarrow y' = -e^{-x} - \frac{3}{2}e^{-3x/2} \Rightarrow 2y' + 3y = 2\left(-e^{-x} - \frac{3}{2}e^{-3x/2}\right) + 3\left(e^{-x} + e^{-3x/2}\right) = e^{-x}$

   (c) $y = e^{-x} + Ce^{-3x/2} \Rightarrow y' = -e^{-x} - \frac{3}{2}Ce^{-3x/2} \Rightarrow 2y' + 3y = 2\left(-e^{-x} - \frac{3}{2}Ce^{-3x/2}\right) + 3\left(e^{-x} + Ce^{-3x/2}\right) = e^{-x}$

3. $y = \frac{1}{x}\int_1^x \frac{e^t}{t}\,dt \Rightarrow y' = -\frac{1}{x^2}\int_1^x \frac{e^t}{t}\,dt + \left(\frac{1}{x}\right)\left(\frac{e^x}{x}\right) \Rightarrow x^2 y' = -\int_1^x \frac{e^t}{t}\,dt + e^x = -x\left(\frac{1}{x}\int_1^x \frac{e^t}{t}\,dt\right) + e^x = -xy + e^x$

   $\Rightarrow x^2 y' + xy = e^x$

5. $y = e^{-x}\tan^{-1}(2e^x) \Rightarrow y' = -e^{-x}\tan^{-1}(2e^x) + e^{-x}\left[\frac{1}{1 + (2e^x)^2}\right](2e^x) = -e^{-x}\tan^{-1}(2e^x) + \frac{2}{1 + 4e^{2x}}$

   $\Rightarrow y' = -y + \frac{2}{1 + 4e^{2x}} \Rightarrow y' + y = \frac{2}{1 + 4e^{2x}}$; $y(-\ln 2) = e^{-(-\ln 2)}\tan^{-1}(2e^{-\ln 2}) = 2\tan^{-1} 1 = 2\left(\frac{\pi}{4}\right) = \frac{\pi}{2}$

7. $y = \frac{\cos x}{x} \Rightarrow y' = \frac{-x\sin x - \cos x}{x^2} \Rightarrow y' = -\frac{\sin x}{x} - \frac{1}{x}\left(\frac{\cos x}{x}\right) \Rightarrow y' = -\frac{\sin x}{x} - \frac{y}{x} \Rightarrow xy' = -\sin x - y$

   $\Rightarrow xy' + y = -\sin x$; $y\left(\frac{\pi}{2}\right) = \frac{\cos(\pi/2)}{(\pi/2)} = 0$

9. $\frac{dy}{dx} = 2(x + y^2 x) \Rightarrow \frac{dy}{dx} = 2x(1 + y^2) \Rightarrow \frac{dy}{1 + y^2} = 2x\,dx \Rightarrow \int \frac{dy}{1 + y^2} = \int 2x\,dx \Rightarrow \tan^{-1} y = x^2 + C$

   $\Rightarrow y = \tan(x^2 + C)$

11. $2\sqrt{xy}\,\frac{dy}{dx} = 1 \Rightarrow 2x^{1/2}y^{1/2}\,dy = dx \Rightarrow 2y^{1/2}\,dy = x^{-1/2}\,dx \Rightarrow \int 2y^{1/2}\,dy = \int x^{-1/2}\,dx \Rightarrow 2\left(\frac{2}{3}y^{3/2}\right)$

    $= 2x^{1/2} + C_1 \Rightarrow \frac{2}{3}y^{3/2} - x^{1/2} = C$, where $C = \frac{1}{2}C_1$

13. $\frac{dy}{dx} = e^{x-y} \Rightarrow dy = e^x e^{-y}\,dx \Rightarrow e^y\,dy = e^x\,dx \Rightarrow \int e^y\,dy = \int e^x\,dx \Rightarrow e^y = e^x + C \Rightarrow e^y - e^x = C$

15. $x\frac{dy}{dx} + y = e^x$

    Step 1: $\frac{dy}{dx} + \left(\frac{1}{x}\right)y = \frac{e^x}{x}$, $P(x) = \frac{1}{x}$, $Q(x) = \frac{e^x}{x}$

    Step 2: $\int P(x)\,dx = \int \frac{1}{x}\,dx = \ln|x| = \ln x$, $x > 0$

    Step 3: $v(x) = e^{\int P(x)\,dx} = e^{\ln x} = x$

    Step 4: $y = \frac{1}{v(x)}\int v(x)\,Q(x)\,dx = \frac{1}{x}\int x\left(\frac{e^x}{x}\right)dx = \frac{1}{x}(e^x + C) = \frac{e^x + C}{x}$

17. $xy' + 3y = \dfrac{\sin x}{x^2}$, $x > 0$

    Step 1: $\dfrac{dy}{dx} + \left(\dfrac{3}{x}\right)y = \dfrac{\sin x}{x^3}$, $P(x) = \dfrac{3}{x}$, $Q(x) = \dfrac{\sin x}{x^3}$

    Step 2: $\displaystyle\int \dfrac{3}{x}\,dx = 3\ln|x| = \ln x^3$, $x > 0$

    Step 3: $v(x) = e^{\ln x^3} = x^3$

    Step 4: $y = \dfrac{1}{x^3}\displaystyle\int x^3\left(\dfrac{\sin x}{x^3}\right)dx = \dfrac{1}{x^3}\int \sin x\,dx = \dfrac{1}{x^3}(-\cos x + C) = \dfrac{C - \cos x}{x^3}$, $x > 0$

19. $x\dfrac{dy}{dx} + 2y = 1 - \dfrac{1}{x}$, $x > 0$

    Step 1: $\dfrac{dy}{dx} + \left(\dfrac{2}{x}\right)y = \dfrac{1}{x} - \dfrac{1}{x^2}$, $P(x) = \dfrac{2}{x}$, $Q(x) = \dfrac{1}{x} - \dfrac{1}{x^2}$

    Step 2: $\displaystyle\int \dfrac{2}{x}\,dx = 2\ln|x| = \ln x^2$, $x > 0$

    Step 3: $v(x) = e^{\ln x^2} = x^2$

    Step 4: $y = \dfrac{1}{x^2}\displaystyle\int x^2\left(\dfrac{1}{x} - \dfrac{1}{x^2}\right)dx = \dfrac{1}{x^2}\int (x - 1)\,dx = \dfrac{1}{x^2}\left(\dfrac{x^2}{2} - x + C\right) = \dfrac{1}{2} - \dfrac{1}{x} + \dfrac{C}{x^2}$, $x > 0$

21. $\dfrac{dy}{dx} - \dfrac{1}{2}y = \dfrac{1}{2}e^{x/2} \Rightarrow P(x) = -\dfrac{1}{2}$, $Q(x) = \dfrac{1}{2}e^{x/2} \Rightarrow \displaystyle\int P(x)\,dx = -\dfrac{1}{2}x \Rightarrow v(x) = e^{-x/2}$

    $\Rightarrow y = \dfrac{1}{e^{-x/2}}\displaystyle\int e^{-x/2}\left(\dfrac{1}{2}e^{x/2}\right)dx = e^{x/2}\int \dfrac{1}{2}\,dx = e^{x/2}\left(\dfrac{1}{2}x + C\right) = \dfrac{1}{2}xe^{x/2} + Ce^{x/2}$

23. $\dfrac{dy}{dx} + 2y = 2xe^{-2x} \Rightarrow P(x) = 2$, $Q(x) = 2xe^{-2x} \Rightarrow \displaystyle\int P(x)\,dx = \int 2\,dx = 2x \Rightarrow v(x) = e^{2x}$

    $\Rightarrow y = \dfrac{1}{e^{2x}}\displaystyle\int e^{2x}(2xe^{-2x})\,dx = \dfrac{1}{e^{2x}}\int 2x\,dx = e^{-2x}(x^2 + C) = x^2e^{-2x} + Ce^{-2x}$

25. $\sec x\dfrac{dy}{dx} = e^{y + \sin x} \Rightarrow \sec x\,dy = e^y e^{\sin x}\,dx \Rightarrow e^{-y}\,dy = \dfrac{e^{\sin x}}{\sec x}\,dx \Rightarrow e^{-y}\,dy = e^{\sin x}\cos x\,dx$

    $\Rightarrow \displaystyle\int e^{-y}\,dy = \int e^{\sin x}\cos x\,dx = \int e^u\,du \Rightarrow -e^{-y} = e^{\sin x} + C$, where $u = \sin x$ and $du = \cos x\,dx$

27. $\dfrac{ds}{dt} + \left(\dfrac{4}{t-1}\right)s = \dfrac{t+1}{(t-1)^3} \Rightarrow P(t) = \dfrac{4}{t-1}$, $Q(t) = \dfrac{t+1}{(t-1)^3} \Rightarrow \displaystyle\int P(t)\,dt = \int \dfrac{4}{t-1}\,dt = 4\ln|t-1| = \ln(t-1)^4$

    $\Rightarrow v(t) = e^{\ln(t-1)^4} = (t-1)^4 \Rightarrow s = \dfrac{1}{(t-1)^4}\displaystyle\int (t-1)^4\left[\dfrac{t+1}{(t-1)^3}\right]dt = \dfrac{1}{(t-1)^4}\int (t^2 - 1)\,dt$

    $= \dfrac{1}{(t-1)^4}\left(\dfrac{t^3}{3} - t + C\right) = \dfrac{t^3}{3(t-1)^4} - \dfrac{t}{(t-1)^4} + \dfrac{C}{(t-1)^4}$

29. $\dfrac{\sec^2 \sqrt{x}}{\sqrt{x}}\, dx = dt \Rightarrow \displaystyle\int \dfrac{\sec^2 \sqrt{x}}{\sqrt{x}}\, dx = \int dt \Rightarrow 2 \int \sec^2 u\, du = \int dt \Rightarrow 2\tan u = t + C \Rightarrow 2\tan \sqrt{x} = t + C,$

where $u = \sqrt{x}$ and $du = \frac{1}{2} x^{-1/2}\, dx$

31. $\dfrac{dr}{d\theta} + (\cot \theta)\, r = \sec \theta \Rightarrow P(\theta) = \cot \theta,\ Q(\theta) = \sec \theta \Rightarrow \displaystyle\int P(\theta)\, d\theta = \int \cot \theta\, d\theta = \ln|\sin \theta| \Rightarrow v(\theta) = e^{\ln|\sin \theta|}$

$= \sin \theta$ because $0 < \theta < \dfrac{\pi}{2} \Rightarrow r = \dfrac{1}{\sin \theta} \displaystyle\int (\sin \theta)(\sec \theta)\, d\theta = \dfrac{1}{\sin \theta} \int \tan \theta\, d\theta = \dfrac{1}{\sin \theta}\left(\ln|\sec \theta| + C \right)$

$= (\csc \theta)\left(\ln|\sec \theta| + C \right)$

33. $\dfrac{dy}{dx} + (\tanh x)\, y = e^{-x}\left(\dfrac{1}{\cosh x} \right) \Rightarrow P(x) = \tanh x,\ Q(x) = e^{-x}\left(\dfrac{1}{\cosh x} \right) \Rightarrow \displaystyle\int P(x)\, dx = \int \tanh x\, dx = \ln(\cosh x)$

$\Rightarrow v(x) = e^{\ln(\cosh x)} = \cosh x \Rightarrow y = \dfrac{1}{\cosh x} \displaystyle\int (\cosh x)(e^{-x})\left(\dfrac{1}{\cosh x} \right) dx = \dfrac{1}{\cosh x} \int e^{-x}\, dx$

$= (\operatorname{sech} x)\left(-e^{-x} + C \right) \Rightarrow y = -e^{-x} \operatorname{sech} x + C \operatorname{sech} x$

35. $\dfrac{dy}{dt} + 2y = 3 \Rightarrow P(t) = 2,\ Q(t) = 3 \Rightarrow \displaystyle\int P(t)\, dt = \int 2\, dt = 2t \Rightarrow v(t) = e^{2t} \Rightarrow y = \dfrac{1}{e^{2t}} \int 3e^{2t}\, dt$

$= \dfrac{1}{e^{2t}}\left(\dfrac{3}{2} e^{2t} + C \right);\ y(0) = 1 \Rightarrow \dfrac{3}{2} + C = 1 \Rightarrow C = -\dfrac{1}{2} \Rightarrow y = \dfrac{3}{2} - \dfrac{1}{2} e^{-2t}$

37. $\dfrac{dy}{d\theta} + \left(\dfrac{1}{\theta} \right) y = \dfrac{\sin \theta}{\theta} \Rightarrow P(\theta) = \dfrac{1}{\theta},\ Q(\theta) = \dfrac{\sin \theta}{\theta} \Rightarrow \displaystyle\int P(\theta)\, d\theta = \ln|\theta| \Rightarrow v(\theta) = e^{\ln|\theta|} = |\theta|$

$\Rightarrow y = \dfrac{1}{|\theta|} \displaystyle\int |\theta|\left(\dfrac{\sin \theta}{\theta} \right) d\theta = \dfrac{1}{\theta} \int \theta\left(\dfrac{\sin \theta}{\theta} \right) d\theta$ for $\theta \neq 0 \Rightarrow y = \dfrac{1}{\theta} \displaystyle\int \sin \theta\, d\theta = \dfrac{1}{\theta}(-\cos \theta + C)$

$= -\dfrac{1}{\theta} \cos \theta + \dfrac{C}{\theta};\ y\left(\dfrac{\pi}{2} \right) = 1 \Rightarrow C = \dfrac{\pi}{2} \Rightarrow y = -\dfrac{1}{\theta} \cos \theta + \dfrac{\pi}{2\theta}$

39. $(x+1)\dfrac{dy}{dx} - 2(x^2 + x)y = \dfrac{e^{x^2}}{x+1} \Rightarrow \dfrac{dy}{dx} - 2\left[\dfrac{x(x+1)}{x+1} \right]y = \dfrac{e^{x^2}}{(x+1)^2} \Rightarrow \dfrac{dy}{dx} - 2xy = \dfrac{e^{x^2}}{(x+1)^2} \Rightarrow P(x) = -2x,$

$Q(x) = \dfrac{e^{x^2}}{(x+1)^2} \Rightarrow \displaystyle\int P(x)\, dx = \int -2x\, dx = -x^2 \Rightarrow v(x) = e^{-x^2} \Rightarrow y = \dfrac{1}{e^{-x^2}} \int e^{-x^2}\left[\dfrac{e^{x^2}}{(x+1)^2} \right] dx$

$= e^{x^2} \displaystyle\int \dfrac{1}{(x+1)^2}\, dx = e^{x^2}\left[\dfrac{(x+1)^{-1}}{-1} + C \right] = -\dfrac{e^{x^2}}{x+1} + Ce^{x^2};\ y(0) = 5 \Rightarrow -\dfrac{1}{0+1} + C = 5 \Rightarrow -1 + C = 5$

$\Rightarrow C = 6 \Rightarrow y = 6e^{x^2} - \dfrac{e^{x^2}}{x+1}$

41. $\dfrac{dy}{dt} - ky = 0 \Rightarrow P(t) = -k,\ Q(t) = 0 \Rightarrow \displaystyle\int P(t)\, dt = \int -k\, dt = -kt \Rightarrow v(t) = e^{-kt}$

$\Rightarrow y = \dfrac{1}{e^{-kt}} \displaystyle\int (e^{-kt})(0)\, dt = e^{kt}(0 + C) = Ce^{kt};\ y(0) = y_0 \Rightarrow C = y_0 \Rightarrow y = y_0 e^{kt}$

43. $x \displaystyle\int \dfrac{1}{x}\, dx = x\left(\ln|x| + C \right) = x \ln|x| + Cx \Rightarrow$ (b) is correct

45. (a) $\dfrac{dc}{dt} = \dfrac{G}{100V} - kc \Rightarrow \dfrac{dc}{dt} + kc = \dfrac{G}{100V} \Rightarrow P(t) = k,\ Q(t) = \dfrac{G}{100V} \Rightarrow \displaystyle\int P(t)\,dt = kt \Rightarrow v(t) = e^{kt}$

$\Rightarrow c = \dfrac{1}{e^{kt}} \displaystyle\int e^{kt}\left(\dfrac{G}{100V}\right)dt = \dfrac{1}{e^{kt}}\left[\dfrac{1}{k}\left(\dfrac{G}{100V}\right)e^{kt} + C\right] = \dfrac{1}{k}\left(\dfrac{G}{100V}\right) + Ce^{-kt};\ c(0) = c_0 \Rightarrow C = c_0 - \dfrac{G}{100Vk}$

$\Rightarrow c = \dfrac{G}{100Vk} + \left(c_0 - \dfrac{G}{100Vk}\right)e^{-kt}$

(b) $\displaystyle\lim_{t\to\infty} c = \lim_{t\to\infty}\left[\dfrac{G}{100Vk} + \left(c_0 - \dfrac{G}{100Vk}\right)e^{-kt}\right] = \dfrac{G}{100Vk}$

47. $\dfrac{dy}{dt} = -k\sqrt{y} \Rightarrow y^{-1/2}\,dy = -k\,dt \Rightarrow 2y^{1/2} = -kt + C;\ y(0) = 9 \Rightarrow 2\cdot 9^{1/2} = -k\cdot 0 + C \Rightarrow C = 6$

$\Rightarrow 2y^{1/2} = -\dfrac{1}{10}t + 6 \Rightarrow t = 60 - 20y^{1/2}.$ Therefore, when $y = 0$ the tank is drained and $t = 60$ minutes.

49. (a) distance coasted $= \dfrac{v_0 m}{k} = \dfrac{(22)(5)}{\frac{1}{5}} = 550$ ft

(b) $v = v_0 e^{-(k/m)t} \Rightarrow 1 = 22e^{-(1/25)t} \Rightarrow \ln 1 = \ln 22 - \dfrac{t}{25} \Rightarrow t = 25\ln 22 \approx 77.28$ sec

51. Steady State $= \dfrac{V}{R}$ and we want $i = \dfrac{1}{2}\left(\dfrac{V}{R}\right) \Rightarrow \dfrac{1}{2}\left(\dfrac{V}{R}\right) = \dfrac{V}{R}\left(1 - e^{-Rt/L}\right) \Rightarrow \dfrac{1}{2} = 1 - e^{-Rt/L} \Rightarrow -\dfrac{1}{2} = -e^{-Rt/L}$

$\Rightarrow \ln\dfrac{1}{2} = -\dfrac{Rt}{L} \Rightarrow -\dfrac{L}{R}\ln\dfrac{1}{2} = t \Rightarrow t = \dfrac{L}{R}\ln 2$ sec

53. (a) $t = \dfrac{3L}{R} \Rightarrow i = \dfrac{V}{R}\left(1 - e^{(-R/L)(3L/R)}\right) = \dfrac{V}{R}\left(1 - e^{-3}\right) \approx 0.9502\,\dfrac{V}{R}$ amp, or about 95% of the steady state value

(b) $t = \dfrac{2L}{R} \Rightarrow i = \dfrac{V}{R}\left(1 - e^{(-R/L)(2L/R)}\right) = \dfrac{V}{R}\left(1 - e^{-2}\right) \approx 0.8647\,\dfrac{V}{R}$ amp, or about 86% of the steady state value

55. Let $y(t) =$ the amount of salt in the container and $V(t) =$ the total volume of liquid in the tank at time t.
Then, the departure rate is $\dfrac{y(t)}{V(t)}$ (the outflow rate).

(a) Rate entering $= \dfrac{2\text{ lb}}{\text{gal}}\cdot\dfrac{5\text{ gal}}{\text{min}} = 10$ lb/min

(b) Volume $= V(t) = 100\text{ gal} + (5t\text{ gal} - 4t\text{ gal}) = (100 + t)$ gal

(c) The volume at time t is $(100 + t)$ gal. The amount of salt in the tank at time t is y lbs. So the
concentration at any time t is $\dfrac{y}{100 + t}$ lbs/gal. Then, Rate leaving $= \dfrac{y}{100 + t}$ (lbs/gal) $\cdot 4$ (gal/min)

$= \dfrac{4y}{100 + t}$ lbs/min

(d) $\dfrac{dy}{dt} = 10 - \dfrac{4y}{100 + t} \Rightarrow \dfrac{dy}{dt} + \left(\dfrac{4}{100 + t}\right)y = 10 \Rightarrow P(t) = \dfrac{4}{100 + t},\ Q(t) = 10 \Rightarrow \displaystyle\int P(t)\,dt = \int\dfrac{4}{100 + t}\,dt$

$= 4\ln(100 + t) \Rightarrow v(t) = e^{4\ln(100+t)} = (100 + t)^4 \Rightarrow y = \dfrac{1}{(100 + t)^4}\displaystyle\int (100 + t)^4(10\,dt)$

$= \dfrac{10}{(100 + t)^4}\left(\dfrac{(100 + t)^5}{5} + C\right) = 2(100 + t) + \dfrac{C}{(100 + t)^4};\ y(0) = 50 \Rightarrow 2(100 + 0) + \dfrac{C}{(100 + 0)^4} = 50$

$$\Rightarrow C = -(150)(100)^4 \Rightarrow y = 2(100 + t) - \frac{(150)(100)^4}{(100 + t)^4} \Rightarrow y = 2(100 + t) - \frac{150}{\left(1 + \frac{t}{100}\right)^4}$$

(e) $y(25) = 2(100 + 25) - \frac{(150)(100)^4}{(100 + 25)^4} \approx 188.56$ lbs $\Rightarrow$ concentration $= \frac{y(25)}{\text{volume}} \approx \frac{188.6}{125} \approx 1.5$ lb/gal

57. Let y be the amount of fertilizer in the tank at time t. Then rate entering $= 1 \frac{\text{lb}}{\text{gal}} \cdot 1 \frac{\text{gal}}{\text{min}} = 1 \frac{\text{lb}}{\text{min}}$ and the

volume in the tank at time t is $V(t) = 100 \text{ (gal)} + [1 \text{ (gal/min)} - 3 \text{ (gal/min)}]t \text{ min} = (100 - 2t) \text{ gal}$. Hence

rate out $= \left(\frac{y}{100 - 2t}\right)3 = \frac{3y}{100 - 2t}$ lbs/min $\Rightarrow \frac{dy}{dt} = \left(1 - \frac{3y}{100 - 2t}\right)$ lbs/min $\Rightarrow \frac{dy}{dt} + \left(\frac{3}{100 - 2t}\right)y = 1$

$\Rightarrow P(t) = \frac{3}{100 - 2t}, Q(t) = 1 \Rightarrow \int P(t) \, dt = \int \frac{3}{100 - 2t} \, dt = \frac{3 \ln (100 - 2t)}{-2} \Rightarrow v(t) = e^{(-3 \ln (100 - 2t))2}$

$= (100 - 2t)^{-3/2} \Rightarrow y = \frac{1}{(100 - 2t)^{-3/2}} \int (100 - 2t)^{-3/2} \, dt = (100 - 2t)^{-3/2} \left[\frac{-2(100 - 2t)^{-1/2}}{-2} + C\right]$

$= (100 - 2t) + C(100 - 2t)^{3/2}; y(0) = 0 \Rightarrow [100 - 2(0)] + C[100 - 2(0)]^{3/2} \Rightarrow C(100)^{3/2} = -100$

$\Rightarrow C = -(100)^{-1/2} = -\frac{1}{10} \Rightarrow y = (100 - 2t) - \frac{(100 - 2t)^{3/2}}{10}$. Let $\frac{dy}{dt} = 0 \Rightarrow \frac{dy}{dt} = -2 - \frac{\left(\frac{3}{2}\right)(100 - 2t)^{1/2}(-2)}{10}$

$= -2 + \frac{3\sqrt{100 - 2t}}{10} = 0 \Rightarrow 20 = 3\sqrt{100 - 2t} \Rightarrow 400 = 9(100 - 2t) \Rightarrow 400 = 900 - 18t \Rightarrow -500 = -18t$

$\Rightarrow t \approx 27.8$ min, the time to reach the maximum. The maximum amount is then

$y(27.8) = [100 - 2(27.8)] - \frac{[100 - 2(27.8)]^{3/2}}{10} \approx 14.8$ lb

6.12 EULER'S NUMERICAL METHOD; SLOPE FIELDS

1. $y_1 = y_0 + \left(1 - \frac{y_0}{x_0}\right) dx = -1 + \left(1 - \frac{-1}{2}\right)(.5) = -0.25,$

   $y_2 = y_1 + \left(1 - \frac{y_1}{x_1}\right) dx = -0.25 + \left(1 - \frac{-0.25}{2.5}\right)(.5) = 0.3,$

   $y_3 = y_2 + \left(1 - \frac{y_2}{x_2}\right) dx = 0.3 + \left(1 - \frac{0.3}{3}\right)(.5) = 0.75;$

   $\frac{dy}{dx} + \left(\frac{1}{x}\right)y = 1 \Rightarrow P(x) = \frac{1}{x}, Q(x) = 1 \Rightarrow \int P(x) \, dx = \int \frac{1}{x} \, dx = \ln |x| = \ln x, x > 0 \Rightarrow v(x) = e^{\ln x} = x$

   $\Rightarrow y = \frac{1}{x} \int x \cdot 1 \, dx = \frac{1}{x}\left(\frac{x^2}{2} + C\right); x = 2, y = -1 \Rightarrow -1 = 1 + \frac{C}{2} \Rightarrow C = -4 \Rightarrow y = \frac{x}{2} - \frac{4}{x} \Rightarrow y(3.5) = \frac{3.5}{2} - \frac{4}{3.5}$

   $= \frac{4.25}{7} \approx 0.6071$

3. $y_1 = y_0 + (2x_0y_0 + 2y_0) \, dx = 3 + [2(0)(3) + 2(3)](.2) = 4.2,$

   $y_2 = y_1 + (2x_1y_1 + 2y_1) \, dx = 4.2 + [2(.2)(4.2) + 2(4.2)](.2) = 6.216,$

   $y_3 = y_2 + (2x_2y_2 + 2y_2) \, dx = 6.216 + [2(.4)(6.216) + 2(6.216)](.2) = 9.6969;$

   $\frac{dy}{dx} = 2y(x + 1) \Rightarrow \frac{dy}{y} = 2(x + 1) \, dx \Rightarrow \ln |y| = (x + 1)^2 + C; x = 0, y = 3 \Rightarrow \ln 3 = 1 + C \Rightarrow C = \ln 3 - 1$

$\Rightarrow \ln y = (x+1)^2 + \ln 3 - 1 \Rightarrow y = e^{(x+1)^2 + \ln 3 - 1} = e^{\ln 3} e^{x^2 + 2x} = 3e^{x(x+2)} \Rightarrow y(.6) \approx 14.2765$

5. $y_1 = y_0 + 2x_0 e^{x_0^2} \, dx = 2 + 2(0)(.1) = 2,$

$y_2 = y_1 + 2x_1 e^{x_1^2} \, dx = 2 + 2(.1) e^{.1^2}(.1) = 2.0202,$

$y_3 = y_2 + 2x_2 e^{x_2^2} \, dx = 2.0202 + 2(.2) e^{.2^2}(.1) = 2.0618,$

$dy = 2xe^{x^2} \, dx \Rightarrow y = e^{x^2} + C; \ y(0) = 2 \Rightarrow 2 = 1 + C \Rightarrow C = 1 \Rightarrow y = e^{x^2} + 1 \Rightarrow y(.3) = e^{.3^2} + 1 \approx 2.0942$

7. $y_1 = 1 + 1(.2) = 1.2,$

$y_2 = 1.2 + (1.2)(.2) = 1.44,$

$y_3 = 1.44 + (1.44)(.2) = 1.728,$

$y_4 = 1.728 + (1.728)(.2) = 2.0736,$

$y_5 = 2.0736 + (2.0736)(.2) = 2.48832;$

$\dfrac{dy}{y} = dx \Rightarrow \ln y = x + C_1 \Rightarrow y = Ce^x; \ y(0) = 1 \Rightarrow 1 = Ce^0 \Rightarrow C = 1 \Rightarrow y = e^x \Rightarrow y(1) = e \approx 2.7183$

9. $y_1 = -1 + \left[\dfrac{(-1)^2}{\sqrt{1}} \right](.5) = -.5,$

$y_2 = -.5 + \left[\dfrac{(-.5)^2}{\sqrt{1.5}} \right](.5) = -.39794,$

$y_3 = -.39794 + \left[\dfrac{(-.39794)^2}{\sqrt{2}} \right](.5) = -.34195,$

$y_4 = -.34195 + \left[\dfrac{(-.34195)^2}{\sqrt{2.5}} \right](.5) = -.30497,$

$y_5 = -.27812, \ y_6 = -.25745, \ y_7 = -.24088, \ y_8 = -.2272;$

$\dfrac{dy}{y^2} = \dfrac{dx}{\sqrt{x}} \Rightarrow -\dfrac{1}{y} = 2\sqrt{x} + C; \ y(1) = -1 \Rightarrow 1 = 2 + C \Rightarrow C = -1 \Rightarrow y = \dfrac{1}{1 - 2\sqrt{x}} \Rightarrow y(5) = \dfrac{1}{1 - 2\sqrt{5}} \approx -.2880$

11. b

13. a

15.

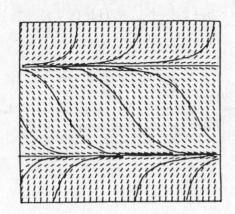

17.

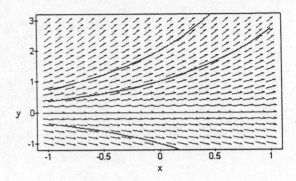

19.

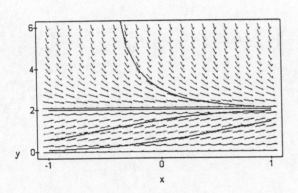

CHAPTER 6 PRACTICE EXERCISES

1. $y = 10e^{-x/5} \Rightarrow \frac{dy}{dx} = (10)\left(-\frac{1}{5}\right)e^{-x/5} = -2e^{-x/5}$

3. $y = \frac{1}{4}xe^{4x} - \frac{1}{16}e^{4x} \Rightarrow \frac{dy}{dx} = \frac{1}{4}\left[x(4e^{4x}) + e^{4x}(1)\right] - \frac{1}{16}(4e^4) = xe^{4x} + \frac{1}{4}e^{4x} - \frac{1}{4}e^{4x} = xe^{4x}$

5. $y = \ln\left(\sin^2\theta\right) \Rightarrow \frac{dy}{d\theta} = \frac{2(\sin\theta)(\cos\theta)}{\sin^2\theta} = \frac{2\cos\theta}{\sin\theta} = 2\cot\theta$

7. $y = \log_2\left(\frac{x^2}{2}\right) = \frac{\ln\left(\frac{x^2}{2}\right)}{\ln 2} \Rightarrow \frac{dy}{dx} = \frac{1}{\ln 2}\left(\frac{x}{\left(\frac{x^2}{2}\right)}\right) = \frac{2}{(\ln 2)x}$

9. $y = 8^{-t} \Rightarrow \frac{dy}{dt} = 8^{-t}(\ln 8)(-1) = -8^{-t}(\ln 8)$ 11. $y = 5x^{3.6} \Rightarrow \frac{dy}{dx} = 5(3.6)x^{2.6} = 18x^{2.6}$

13. $y = (x+2)^{x+2} \Rightarrow \ln y = \ln (x+2)^{x+2} = (x+2) \ln (x+2) \Rightarrow \frac{y'}{y} = (x+2)\left(\frac{1}{x+2}\right) + (1) \ln (x+2)$

$\Rightarrow \frac{dy}{dx} = (x+2)^{x+2}\left[\ln (x+2) + 1\right]$

15. $y = \sin^{-1} \sqrt{1-u^2} = \sin^{-1}(1-u^2)^{1/2} \Rightarrow \frac{dy}{du} = \dfrac{\frac{1}{2}(1-u^2)^{-1/2}(-2u)}{\sqrt{1-\left[(1-u^2)^{1/2}\right]^2}} = \dfrac{-u}{\sqrt{1-u^2}\sqrt{1-(1-u^2)}} = \dfrac{-u}{|u|\sqrt{1-u^2}}$

$= \dfrac{-u}{u\sqrt{1-u^2}} = \dfrac{-1}{\sqrt{1-u^2}}, \ 0 < u < 1$

17. $y = \ln\left(\cos^{-1} x\right) \Rightarrow y' = \dfrac{\left(\dfrac{-1}{\sqrt{1-x^2}}\right)}{\cos^{-1} x} = \dfrac{-1}{\sqrt{1-x^2}\,\cos^{-1} x}$

19. $y = t \tan^{-1} t - \left(\frac{1}{2}\right) \ln t \Rightarrow \frac{dy}{dt} = \tan^{-1} t + t\left(\frac{1}{1+t^2}\right) - \left(\frac{1}{2}\right)\left(\frac{1}{t}\right) = \tan^{-1} t + \frac{t}{1+t^2} - \frac{1}{2t}$

21. $y = z \sec^{-1} z - \sqrt{z^2-1} = z \sec^{-1} z - (z^2-1)^{1/2} \Rightarrow \frac{dy}{dz} = z\left(\dfrac{1}{|z|\sqrt{z^2-1}}\right) + (\sec^{-1} z)(1) - \frac{1}{2}(z^2-1)^{-1/2}(2z)$

$= \dfrac{z}{|z|\sqrt{z^2-1}} - \dfrac{z}{\sqrt{z^2-1}} + \sec^{-1} z = \dfrac{1-z}{\sqrt{z^2-1}} + \sec^{-1} z, \ z > 1$

23. $y = \csc^{-1}(\sec \theta) \Rightarrow \frac{dy}{d\theta} = \dfrac{-\sec \theta \tan \theta}{|\sec \theta|\sqrt{\sec^2 \theta - 1}} = -\dfrac{\tan \theta}{|\tan \theta|} = -1, \ 0 < \theta < \frac{\pi}{2}$

25. $y = \dfrac{2(x^2+1)}{\sqrt{\cos 2x}} \Rightarrow \ln y = \ln\left(\dfrac{2(x^2+1)}{\sqrt{\cos 2x}}\right) = \ln (2) + \ln(x^2+1) - \frac{1}{2} \ln (\cos 2x) \Rightarrow \frac{y'}{y} = 0 + \dfrac{2x}{x^2+1} - \left(\frac{1}{2}\right)\dfrac{(-2\sin 2x)}{\cos 2x}$

$\Rightarrow y' = \left(\dfrac{2x}{x^2+1} + \tan 2x\right) y = \dfrac{2(x^2+1)}{\sqrt{\cos 2x}}\left(\dfrac{2x}{x^2+1} + \tan 2x\right)$

27. $y = \left[\dfrac{(t+1)(t-1)}{(t-2)(t+3)}\right]^5 \Rightarrow \ln y = 5\left[\ln (t+1) + \ln (t-1) - \ln (t-2) - \ln (t+3)\right] \Rightarrow \left(\frac{1}{y}\right)\left(\frac{dy}{dt}\right)$

$= 5\left(\dfrac{1}{t+1} + \dfrac{1}{t-1} - \dfrac{1}{t-2} - \dfrac{1}{t+3}\right) \Rightarrow \frac{dy}{dt} = 5\left[\dfrac{(t+1)(t-1)}{(t-2)(t+3)}\right]^5\left(\dfrac{1}{t+1} + \dfrac{1}{t-1} - \dfrac{1}{t-2} - \dfrac{1}{t+3}\right)$

29. $y = (\sin \theta)^{\sqrt{\theta}} \Rightarrow \ln y = \sqrt{\theta} \ln (\sin \theta) \Rightarrow \left(\frac{1}{y}\right)\left(\frac{dy}{d\theta}\right) = \sqrt{\theta}\left(\dfrac{\cos \theta}{\sin \theta}\right) + \frac{1}{2}\theta^{-1/2} \ln (\sin \theta)$

$\Rightarrow \frac{dy}{d\theta} = (\sin \theta)^{\sqrt{\theta}}\left(\sqrt{\theta} \cot \theta + \dfrac{\ln (\sin \theta)}{2\sqrt{\theta}}\right) = \dfrac{1}{\sqrt{\theta}}(\sin \theta)^{\sqrt{\theta}}(\theta \cot \theta + \ln \sqrt{\sin \theta})$

31. $\displaystyle\int e^x \sin(e^x)\, dx = \int \sin u\, du$, where $u = e^x$ and $du = e^x\, dx$

$= -\cos u + C = -\cos(e^x) + C$

33. $\int e^x \sec^2(e^x - 7) \, dx = \int \sec^2 u \, du$, where $u = e^x - 7$ and $du = e^x \, dx$

$$= \tan u + C = \tan(e^x - 7) + C$$

35. $\int (\sec^2 x) e^{\tan x} \, dx = \int e^u \, du$, where $u = \tan x$ and $du = \sec^2 x \, dx$

$$= e^u + C = e^{\tan x} + C$$

37. $\displaystyle\int_{-1}^{1} \frac{1}{3x - 4} \, dx = \frac{1}{3} \int_{-7}^{-1} \frac{1}{u} \, du$, where $u = 3x - 4$, $du = 3 \, dx$; $x = -1 \Rightarrow u = -7$, $x = 1 \Rightarrow u = -1$

$$= \frac{1}{3} \left[\ln |u| \right]_{-7}^{-1} = \frac{1}{3} \left[\ln |-1| - \ln |-7| \right] = \frac{1}{3} [0 - \ln 7] = -\frac{\ln 7}{3}$$

39. $\displaystyle\int_{0}^{\pi} \tan\left(\frac{x}{3}\right) dx = \int_{0}^{\pi} \frac{\sin\left(\frac{x}{3}\right)}{\cos\left(\frac{x}{3}\right)} \, dx = -3 \int_{1}^{1/2} \frac{1}{u} \, du$, where $u = \cos\left(\frac{x}{3}\right)$, $du = -\frac{1}{3} \sin\left(\frac{x}{3}\right) dx$; $x = 0 \Rightarrow u = 1$, $x = \pi$

$$\Rightarrow u = \frac{1}{2}$$

$$= -3 \left[\ln |u| \right]_{1}^{1/2} = -3 \left[\ln \left|\frac{1}{2}\right| - \ln |1| \right] = -3 \ln \frac{1}{2} = \ln 2^3 = \ln 8$$

41. $\displaystyle\int_{0}^{4} \frac{2t}{t^2 - 25} \, dt = \int_{-25}^{-9} \frac{1}{u} \, du$, where $u = t^2 - 25$, $du = 2t \, dt$; $t = 0 \Rightarrow u = -25$, $t = 4 \Rightarrow u = -9$

$$= \left[\ln |u| \right]_{-25}^{-9} = \ln |-9| - \ln |-25| = \ln 9 - \ln 25 = \ln \frac{9}{25}$$

43. $\int \frac{\tan (\ln v)}{v} \, dv = \int \tan u \, du = \int \frac{\sin u}{\cos u} \, du$, where $u = \ln v$ and $du = \frac{1}{v} \, dv$

$$= -\ln |\cos u| + C = -\ln \left|\cos (\ln v)\right| + C$$

45. $\int \frac{(\ln x)^{-3}}{x} \, dx = \int u^{-3} \, du$, where $u = \ln x$ and $du = \frac{1}{x} \, dx$

$$= \frac{u^{-2}}{-2} + C = -\frac{1}{2} (\ln x)^{-2} + C$$

47. $\int \frac{1}{r} \csc^2 (1 + \ln r) \, dr = \int \csc^2 u \, du$, where $u = 1 + \ln r$ and $du = \frac{1}{r} \, dr$

$$= -\cot u + C = -\cot (1 + \ln r) + C$$

49. $\int x 3^{x^2} \, dx = \frac{1}{2} \int 3^u \, du$, where $u = x^2$ and $du = 2x \, dx$

$$= \frac{1}{2 \ln 3} (3^u) + C = \frac{1}{2 \ln 3} \left(3^{x^2}\right) + C$$

51. $\displaystyle\int_1^7 \frac{3}{x}\, dx = 3\int_1^7 \frac{1}{x}\, dx = 3\big[\ln |x|\big]_1^7 = 3(\ln 7 - \ln 1) = 3\ln 7$

53. $\displaystyle\int_1^4 \left(\frac{x}{8} + \frac{1}{2x}\right) dx = \frac{1}{2}\int_1^4 \left(\frac{1}{4}x + \frac{1}{x}\right) dx = \frac{1}{2}\Big[\frac{1}{8}x^2 + \ln |x|\Big]_1^4 = \frac{1}{2}\Big[\Big(\frac{16}{8} + \ln 4\Big) - \Big(\frac{1}{8} + \ln 1\Big)\Big] = \frac{15}{16} + \frac{1}{2}\ln 4$

$\displaystyle = \frac{15}{16} + \ln\sqrt{4} = \frac{15}{16} + \ln 2$

55. $\displaystyle\int_{-2}^{-1} e^{-(x+1)}\, dx = -\int_1^0 e^u\, du,\ \text{where } u = -(x+1),\ du = -dx;\ x = -2 \Rightarrow u = 1,\ x = -1 \Rightarrow u = 0$

$\displaystyle = -\big[e^u\big]_1^0 = -(e^0 - e^1) = e - 1$

57. $\displaystyle\int_0^{\ln 5} e^r\big(3e^r + 1\big)^{-3/2}\, dr = \frac{1}{3}\int_4^{16} u^{-3/2}\, du,\ \text{where } u = 3e^r + 1,\ du = 3e^r;\ r = 0 \Rightarrow u = 4,\ r = \ln 5 \Rightarrow u = 16$

$\displaystyle = -\frac{2}{3}\big[u^{-1/2}\big]_4^{16} = -\frac{2}{3}(16^{-1/2} - 4^{-1/2}) = \left(-\frac{2}{3}\right)\left(\frac{1}{4} - \frac{1}{2}\right) = \left(-\frac{2}{3}\right)\left(-\frac{1}{4}\right) = \frac{1}{6}$

59. $\displaystyle\int_1^e \frac{1}{x}(1 + 7\ln x)^{-1/3}\, dx = \frac{1}{7}\int_1^8 u^{-1/3}\, du,\ \text{where } u = 1 + 7\ln x,\ du = \frac{7}{x}\, dx,\ x = 1 \Rightarrow u = 1,\ x = e \Rightarrow u = 8$

$\displaystyle = \frac{3}{14}\big[u^{2/3}\big]_1^8 = \frac{3}{14}(8^{2/3} - 1^{2/3}) = \left(\frac{3}{14}\right)(4 - 1) = \frac{9}{14}$

61. $\displaystyle\int_1^3 \frac{[\ln (v+1)]^2}{v+1}\, dv = \int_1^3 [\ln (v+1)]^2 \frac{1}{v+1}\, dv = \int_{\ln 2}^{\ln 4} u^2\, du,\ \text{where } u = \ln (v+1),\ du = \frac{1}{v+1}dv;$

$v = 1 \Rightarrow u = \ln 2,\ v = 3 \Rightarrow u = \ln 4;$

$\displaystyle = \frac{1}{3}\big[u^3\big]_{\ln 2}^{\ln 4} = \frac{1}{3}\big[(\ln 4)^3 - (\ln 2)^3\big] = \frac{1}{3}\big[(2\ln 2)^3 - (\ln 2)^3\big] = \frac{(\ln 2)^3}{3}(8 - 1) = \frac{7}{3}(\ln 2)^3$

63. $\displaystyle\int_1^8 \frac{\log_4 \theta}{\theta}\, d\theta = \frac{1}{\ln 4}\int_1^8 (\ln \theta)\left(\frac{1}{\theta}\right) d\theta = \frac{1}{\ln 4}\int_0^{\ln 8} u\, du,\ \text{where } u = \ln \theta,\ du = \frac{1}{\theta}\, d\theta,\ \theta = 1 \Rightarrow u = 0,\ \theta = 8 \Rightarrow u = \ln 8$

$\displaystyle = \frac{1}{2\ln 4}\big[u^2\big]_0^{\ln 8} = \frac{1}{\ln 16}\big[(\ln 8)^2 - 0^2\big] = \frac{(3\ln 2)^2}{4\ln 2} = \frac{9\ln 2}{4}$

65. $\displaystyle\int_{-3/4}^{3/4} \frac{6}{\sqrt{9 - 4x^2}}\, dx = 3\int_{-3/4}^{3/4} \frac{2}{\sqrt{3^2 - (2x)^2}}\, dx = 3\int_{-3/2}^{3/2} \frac{1}{\sqrt{3^2 - u^2}}\, du,\ \text{where } u = 2x,\ du = 2\, dx;$

$x = -\frac{3}{4} \Rightarrow u = -\frac{3}{2},\ x = \frac{3}{4} \Rightarrow u = \frac{3}{2}$

$\displaystyle = 3\Big[\sin^{-1}\left(\frac{u}{3}\right)\Big]_{-3/2}^{3/2} = 3\Big[\sin^{-1}\left(\frac{1}{2}\right) - \sin^{-1}\left(-\frac{1}{2}\right)\Big] = 3\Big[\frac{\pi}{6} - \left(-\frac{\pi}{6}\right)\Big] = 3\left(\frac{\pi}{3}\right) = \pi$

67. $\displaystyle\int_{-2}^{2} \frac{3}{4+3t^2}\,dt = \sqrt{3}\int_{-2}^{2} \frac{\sqrt{3}}{2^2+\left(\sqrt{3}t\right)^2}\,dt = \sqrt{3}\int_{-2\sqrt{3}}^{2\sqrt{3}} \frac{1}{2^2+u^2}\,du$, where $u = \sqrt{3}t$, $du = \sqrt{3}\,dt$;

$$t = -2 \Rightarrow u = -2\sqrt{3},\ t = 2 \Rightarrow u = 2\sqrt{3}$$

$$= \sqrt{3}\left[\tfrac{1}{2}\tan^{-1}\left(\tfrac{u}{2}\right)\right]_{-2\sqrt{3}}^{2\sqrt{3}} = \frac{\sqrt{3}}{2}\left[\tan^{-1}\left(\sqrt{3}\right) - \tan^{-1}\left(-\sqrt{3}\right)\right] = \frac{\sqrt{3}}{2}\left[\tfrac{\pi}{3} - \left(-\tfrac{\pi}{3}\right)\right] = \frac{\pi}{\sqrt{3}}$$

69. $\displaystyle\int \frac{1}{y\sqrt{4y^2-1}}\,dy = \int \frac{2}{(2y)\sqrt{(2y)^2-1}}\,dy = \int \frac{1}{u\sqrt{u^2-1}}\,du$, where $u = 2y$ and $du = 2\,dy$

$$= \sec^{-1}|u| + C = \sec^{-1}|2y| + C$$

71. $\displaystyle\int_{\sqrt{2}/3}^{2/3} \frac{1}{|y|\sqrt{9y^2-1}}\,dy = \int_{\sqrt{2}/3}^{2/3} \frac{3}{|3y|\sqrt{(3y)^2-1}}\,dy = \int_{\sqrt{2}}^{2} \frac{1}{|u|\sqrt{u^2-1}}\,du$, where $u = 3y$, $du = 3\,dy$;

$$y = \frac{\sqrt{2}}{3} \Rightarrow u = \sqrt{2},\ y = \tfrac{2}{3} \Rightarrow u = 2$$

$$= \left[\sec^{-1}u\right]_{\sqrt{2}}^{2} = \left[\sec^{-1}2 - \sec^{-1}\sqrt{2}\right] = \tfrac{\pi}{3} - \tfrac{\pi}{4} = \tfrac{\pi}{12}$$

73. $\displaystyle\int \frac{1}{\sqrt{-2x-x^2}}\,dx = \int \frac{1}{\sqrt{1-(x^2+2x+1)}}\,dx = \int \frac{1}{\sqrt{1-(x+1)^2}}\,dx = \int \frac{1}{\sqrt{1-u^2}}\,du$, where $u = x+1$ and $du = dx$

$$= \sin^{-1}u + C = \sin^{-1}(x+1) + C$$

75. $\displaystyle\int_{-2}^{-1} \frac{2}{v^2+4v+5}\,dv = 2\int_{-2}^{-1} \frac{1}{1+(v^2+4v+4)}\,dv = 2\int_{-2}^{-1} \frac{1}{1+(v+2)^2}\,dv = 2\int_{0}^{1} \frac{1}{1+u^2}\,du$,

where $u = v+2$, $du = dv$; $v = -2 \Rightarrow u = 0$, $v = -1 \Rightarrow u = 1$

$$= 2\left[\tan^{-1}u\right]_{0}^{1} = 2\left(\tan^{-1}1 - \tan^{-1}0\right) = 2\left(\tfrac{\pi}{4} - 0\right) = \tfrac{\pi}{2}$$

77. $\displaystyle\int \frac{1}{(t+1)\sqrt{t^2+2t-8}}\,dt = \int \frac{1}{(t+1)\sqrt{(t^2+2t+1)-9}}\,dt = \int \frac{1}{(t+1)\sqrt{(t+1)^2-3^2}}\,dt = \int \frac{1}{u\sqrt{u^2-3^2}}\,du$

where $u = t+1$ and $du = dt$

$$= \tfrac{1}{3}\sec^{-1}\left|\tfrac{u}{3}\right| + C = \tfrac{1}{3}\sec^{-1}\left|\tfrac{t+1}{3}\right| + C$$

79. $3^y = 2^{y+1} \Rightarrow \ln 3^y = \ln 2^{y+1} \Rightarrow y(\ln 3) = (y+1)\ln 2 \Rightarrow (\ln 3 - \ln 2)y = \ln 2 \Rightarrow \left(\ln\tfrac{3}{2}\right)y = \ln 2 \Rightarrow y = \dfrac{\ln 2}{\ln\left(\tfrac{3}{2}\right)}$

81. $9e^{2y} = x^2 \Rightarrow e^{2y} = \dfrac{x^2}{9} \Rightarrow \ln e^{2y} = \ln\left(\dfrac{x^2}{9}\right) \Rightarrow 2y(\ln e) = \ln\left(\dfrac{x^2}{9}\right) \Rightarrow y = \tfrac{1}{2}\ln\left(\dfrac{x^2}{9}\right) = \ln\sqrt{\dfrac{x^2}{9}} = \ln\left|\dfrac{x}{3}\right| = \ln|x| - \ln 3$

83. $\ln(y-1) = x + \ln y \Rightarrow e^{\ln(y-1)} = e^{(x+\ln y)} = e^x e^{\ln y} \Rightarrow y - 1 = ye^x \Rightarrow y - ye^x = 1 \Rightarrow y(1-e^x) = 1$

$\Rightarrow y = \dfrac{1}{1-e^x}$

85. The limit leads to the indeterminate form $\frac{0}{0}$: $\displaystyle\lim_{x\to 0} \frac{10^x - 1}{x} = \lim_{x\to 0} \frac{(\ln 10)10^x}{1} = \ln 10$

87. The limit leads to the indeterminate form $\frac{0}{0}$: $\displaystyle\lim_{x\to 0} \frac{2^{\sin x} - 1}{e^x - 1} = \lim_{x\to 0} \frac{2^{\sin x}(\ln 2)(\cos x)}{e^x} = \ln 2$

89. The limit leads to the indeterminate form $\frac{0}{0}$: $\displaystyle\lim_{x\to 0} \frac{5 - 5\cos x}{e^x - x - 1} = \lim_{x\to 0} \frac{5\sin x}{e^x - 1} = \lim_{x\to 0} \frac{5\cos x}{e^x} = 5$

91. The limit leads to the indeterminate form $\frac{0}{0}$: $\displaystyle\lim_{t\to 0^+} \frac{t - \ln(1 + 2t)}{t^2} = \lim_{t\to 0^+} \frac{\left(1 - \dfrac{2}{1 + 2t}\right)}{2t} = -\infty$

93. The limit leads to the indeterminate form $\frac{0}{0}$: $\displaystyle\lim_{t\to 0^+} \left(\frac{e^t}{t} - \frac{1}{t}\right) = \lim_{t\to 0^+} \left(\frac{e^t - 1}{t}\right) = \lim_{t\to 0^+} \frac{e^t}{1} = 1$

95. Let $f(x) = \left(1 + \frac{3}{x}\right)^x \Rightarrow \ln f(x) = \dfrac{\ln(1 + 3x^{-1})}{x^{-1}} \Rightarrow \displaystyle\lim_{x\to\infty} \ln f(x) = \lim_{x\to\infty} \frac{\ln(1 + 3x^{-1})}{x^{-1}}$; the limit leads to the

   indeterminate form $\frac{0}{0}$: $\displaystyle\lim_{x\to\infty} \frac{\left(\dfrac{-3x^{-2}}{1 + 3x^{-1}}\right)}{-x^{-2}} = \lim_{x\to\infty} \frac{3}{1 + \frac{3}{x}} = 3 \Rightarrow \lim_{x\to\infty} \left(1 + \frac{3}{x}\right)^x = \lim_{x\to\infty} e^{\ln f(x)} = e^3$

97. (a) $\displaystyle\lim_{x\to\infty} \frac{\log_2 x}{\log_3 x} = \lim_{x\to\infty} \frac{\left(\dfrac{\ln x}{\ln 2}\right)}{\left(\dfrac{\ln x}{\ln 3}\right)} = \lim_{x\to\infty} \frac{\ln 3}{\ln 2} = \frac{\ln 3}{\ln 2} \Rightarrow$ same rate

   (b) $\displaystyle\lim_{x\to\infty} \frac{x}{x + \left(\frac{1}{x}\right)} = \lim_{x\to\infty} \frac{x^2}{x^2 + 1} = \lim_{x\to\infty} \frac{2x}{2x} = \lim_{x\to\infty} 1 = 1 \Rightarrow$ same rate

   (c) $\displaystyle\lim_{x\to\infty} \frac{\left(\frac{x}{100}\right)}{xe^{-x}} = \lim_{x\to\infty} \frac{xe^x}{100x} = \lim_{x\to\infty} \frac{e^x}{100} = \infty \Rightarrow$ faster

   (d) $\displaystyle\lim_{x\to\infty} \frac{x}{\tan^{-1} x} = \infty \Rightarrow$ faster

   (e) $\displaystyle\lim_{x\to\infty} \frac{\csc^{-1} x}{\left(\frac{1}{x}\right)} = \lim_{x\to\infty} \frac{\sin^{-1}(x^{-1})}{x^{-1}} = \lim_{x\to\infty} \frac{\dfrac{(-x^{-2})}{\sqrt{1 - (x^{-1})^2}}}{-x^{-2}} = \lim_{x\to\infty} \frac{1}{\sqrt{1 - \left(\frac{1}{x^2}\right)}} = 1 \Rightarrow$ same rate

   (f) $\displaystyle\lim_{x\to\infty} \frac{\sinh x}{e^x} = \lim_{x\to\infty} \frac{(e^x - e^{-x})}{2e^x} = \lim_{x\to\infty} \frac{1 - e^{-2x}}{2} = \frac{1}{2} \Rightarrow$ same rate

99. (a) $\dfrac{\left(\dfrac{1}{x^2} + \dfrac{1}{x^4}\right)}{\left(\dfrac{1}{x^2}\right)} = 1 + \dfrac{1}{x^2} \le 2$ for x sufficiently large $\Rightarrow$ true

(b) $\dfrac{\left(\dfrac{1}{x^2} + \dfrac{1}{x^4}\right)}{\left(\dfrac{1}{x^4}\right)} = x^2 + 1 > M$ for any positive integer M whenever $x > \sqrt{M} \Rightarrow$ false

(c) $\lim\limits_{x \to \infty} \dfrac{x}{x + \ln x} = \lim\limits_{x \to \infty} \dfrac{1}{1 + \frac{1}{x}} = 1 \Rightarrow$ the same growth rate $\Rightarrow$ false

(d) $\lim\limits_{x \to \infty} \dfrac{\ln(\ln x)}{\ln x} = \lim\limits_{x \to \infty} \dfrac{\left[\dfrac{\left(\frac{1}{x}\right)}{\ln x}\right]}{\left(\dfrac{1}{x}\right)} = \lim\limits_{x \to \infty} \dfrac{1}{\ln x} = 0 \Rightarrow$ grows slower $\Rightarrow$ true

(e) $\dfrac{\tan^{-1}x}{1} \le \dfrac{\pi}{2}$ for all $x \Rightarrow$ true

(f) $\dfrac{\cosh x}{e^x} = \frac{1}{2}\left(1 + e^{-2x}\right) \le \frac{1}{2}(1 + 1) = 1$ if $x > 0 \Rightarrow$ true

101. $\dfrac{df}{dx} = e^x + 1 \Rightarrow \left(\dfrac{df^{-1}}{dx}\right)_{x=f(\ln 2)} = \dfrac{1}{\left(\dfrac{df}{dx}\right)_{x=\ln 2}} \Rightarrow \left(\dfrac{df^{-1}}{dx}\right)_{x=f(\ln 2)} = \dfrac{1}{\left(e^x + 1\right)_{x=\ln 2}} = \dfrac{1}{2+1} = \dfrac{1}{3}$

103. $y = x \ln 2x - x \Rightarrow y' = x\left(\dfrac{2}{2x}\right) + \ln(2x) - 1 = \ln 2x;$

solving $y' = 0 \Rightarrow x = \frac{1}{2}$; $y' > 0$ for $x > \frac{1}{2}$ and $y' < 0$ for

$x < \frac{1}{2} \Rightarrow$ relative minimum of $-\frac{1}{2}$ at $x = \frac{1}{2}$; $f\left(\dfrac{1}{2e}\right) = -\dfrac{1}{e}$

and $f\left(\dfrac{e}{2}\right) = 0 \Rightarrow$ absolute minimum is $-\frac{1}{2}$ at $x = \frac{1}{2}$ and the

absolute maximum is 0 at $x = \dfrac{e}{2}$

$y = x \ln 2x - x$

(1/e, 0)

(0.5, -0.5)

105. $A = \displaystyle\int_{1}^{e} \dfrac{2\ln x}{x}\, dx = \displaystyle\int_{0}^{1} 2u\, du = \left[u^2\right]_0^1 = 1,$ where

$u = \ln x$ and $du = \frac{1}{x}\, dx$; $x = 1 \Rightarrow u = 0$, $x = e \Rightarrow u = 1$

107. $y = \ln x \Rightarrow \dfrac{dy}{dx} = \dfrac{1}{x}$; $\dfrac{dy}{dt} = \dfrac{dy}{dx}\dfrac{dx}{dt} \Rightarrow \dfrac{dy}{dt} = \left(\dfrac{1}{x}\right)\sqrt{x} = \dfrac{1}{\sqrt{x}} \Rightarrow \left.\dfrac{dy}{dt}\right|_{e^2} = \dfrac{1}{e}$ m/sec

109. $A = xy = xe^{-x^2} \Rightarrow \dfrac{dA}{dx} = e^{-x^2} + (x)(-2x)\,e^{-x^2} = e^{-x^2}\left(1 - 2x^2\right).$ Solving $\dfrac{dA}{dx} = 0 \Rightarrow 1 - 2x^2 = 0$

$\Rightarrow x = \dfrac{1}{\sqrt{2}}$; $\dfrac{dA}{dx} < 0$ for $x > \dfrac{1}{\sqrt{2}}$ and $\dfrac{dA}{dx} > 0$ for $0 < x < \dfrac{1}{\sqrt{2}} \Rightarrow$ absolute maximum of $\dfrac{1}{\sqrt{2}}e^{-1/2} = \dfrac{1}{\sqrt{2e}}$ at

$x = \dfrac{1}{\sqrt{2}}$ units long by $y = e^{-1/2} = \dfrac{1}{\sqrt{e}}$ units high.

111. $K = \ln(5x) - \ln(3x) = \ln 5 + \ln x - \ln 3 - \ln x = \ln 5 - \ln 3 = \ln \dfrac{5}{3}$

113. $\dfrac{\log_4 x}{\log_2 x} = \dfrac{\left(\dfrac{\ln x}{\ln 4}\right)}{\left(\dfrac{\ln x}{\ln 2}\right)} = \dfrac{\ln x}{\ln 4} \cdot \dfrac{\ln 2}{\ln x} = \dfrac{\ln 2}{\ln 4} = \dfrac{\ln 2}{2 \ln 2} = \dfrac{1}{2}$

115. (a) $y = \dfrac{\ln x}{\sqrt{x}} \Rightarrow y' = \dfrac{1}{x\sqrt{x}} - \dfrac{\ln x}{2x^{3/2}} = \dfrac{2 - \ln x}{2x\sqrt{x}}$

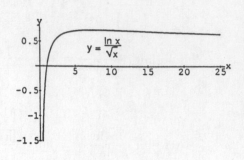

$\Rightarrow y'' = -\dfrac{3}{4} x^{-5/2}(2 - \ln x) - \dfrac{1}{2} x^{-5/2} = x^{-5/2}\left(\dfrac{3}{4} \ln x - 2\right);$

solving $y' = 0 \Rightarrow \ln x = 2 \Rightarrow x = e^2$; $y' < 0$ for $x > e^2$ and

and $y' > 0$ for $x < e^2 \Rightarrow$ a maximum of $\dfrac{2}{e}$; $y'' = 0$

$\Rightarrow \ln x = \dfrac{8}{3} \Rightarrow x = e^{8/3}$; the curve is concave down on

$\left(0, e^{8/3}\right)$ and concave up on $\left(e^{8/3}, \infty\right)$

(b) $y = e^{-x^2} \Rightarrow y' = -2xe^{-x^2} \Rightarrow y'' = -2e^{-x^2} + 4x^2 e^{-x^2}$;

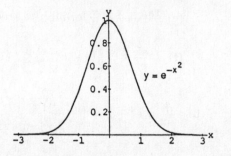

solving $y' = 0 \Rightarrow x = 0$; $y' < 0$ for $x > 0$ and $y' > 0$ for

$x < 0 \Rightarrow$ a maximum at $x = 0$ of $e^0 = 1$; there are points

of inflection at $x = \pm\dfrac{1}{\sqrt{2}}$; the curve is concave down for

$-\dfrac{1}{\sqrt{2}} < x < \dfrac{1}{\sqrt{2}}$ and concave up otherwise

(c) $y = (1 + x)e^{-x} \Rightarrow y' = e^{-x} - (1 + x)e^{-x} = -xe^{-x}$

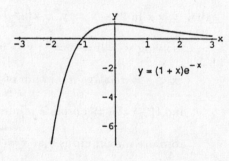

$\Rightarrow y'' = -e^{-x} + xe^{-x} = (x - 1)e^{-x}$; solving $y' = 0$

$\Rightarrow -xe^{-x} = 0 \Rightarrow x = 0$; $y' < 0$ for $x > 0$ and $y' > 0$

for $x < 0 \Rightarrow$ a maximum at $x = 0$ of $(1 + 0)e^0 = 1$;

there is a point of inflection at $x = 1$ and the curve is

concave up for $x > 1$ and concave down for $x < 1$

117. Since the half life is 5700 years and $A(t) = A_0 e^{kt}$ we have $\dfrac{A_0}{2} = A_0 e^{5700k} \Rightarrow \dfrac{1}{2} = e^{5700k} \Rightarrow \ln(0.5) = 5700k$

$\Rightarrow k = \dfrac{\ln(0.5)}{5700}$. With 10% of the original carbon-14 remaining we have $0.1A_0 = A_0 e^{\frac{\ln(0.5)}{5700}t} \Rightarrow 0.1 = e^{\frac{\ln(0.5)}{5700}t}$

$\Rightarrow \ln(0.1) = \dfrac{\ln(0.5)}{5700}t \Rightarrow t = \dfrac{(5700)\ln(0.1)}{\ln(0.5)} \approx 18{,}935$ years (rounded to the nearest year).

119. $\theta = \pi - \cot^{-1}\left(\dfrac{x}{60}\right) - \cot^{-1}\left(\dfrac{5}{3} - \dfrac{x}{30}\right)$, $0 < x < 50 \Rightarrow \dfrac{d\theta}{dx} = \dfrac{\left(\dfrac{1}{60}\right)}{1 + \left(\dfrac{x}{60}\right)^2} + \dfrac{\left(-\dfrac{1}{30}\right)}{1 + \left(\dfrac{50 - x}{30}\right)^2}$

$= 30\left[\dfrac{2}{60^2 + x^2} - \dfrac{1}{30^2 + (50 - x)^2}\right]$; solving $\dfrac{d\theta}{dx} = 0 \Rightarrow x^2 - 200x + 3200 = 0 \Rightarrow x = 100 \pm 20\sqrt{17}$, but

$100 + 20\sqrt{17}$ is not in the domain; $\dfrac{d\theta}{dx} > 0$ for $x < 20\left(5 - \sqrt{17}\right)$ and $\dfrac{d\theta}{dx} < 0$ for $20\left(5 - \sqrt{17}\right) < x < 50$

$\Rightarrow x = 20\left(5 - \sqrt{17}\right) \approx 17.54$ m maximizes θ

121. $\dfrac{dy}{dx} = e^{-x-y-2} \Rightarrow \dfrac{dy}{dx} = e^{-x-2} \cdot e^{-y} \Rightarrow e^y \, dy = e^{-x-2} \, dx = \displaystyle\int e^y \, dy = \int e^{-x-2} \, dx \Rightarrow e^y = -e^{-x-2} + C$; $x = 0$

and $y = -2 \Rightarrow e^{-2} = -e^{-2} + C \Rightarrow C = 2e^{-2} \Rightarrow e^y = -e^{-x-2} + 2e^{-2} \Rightarrow \ln(e^y) = \ln\left(-e^{-x-2} + 2e^{-2}\right)$

$\Rightarrow y = \ln\left(-e^{-x-2} + 2e^{-2}\right)$

123. $\dfrac{dy}{dx} + \left(\dfrac{2}{x+1}\right)y = \dfrac{x}{x+1} \Rightarrow P(x) = \dfrac{2}{x+1}, \ Q(x) = \dfrac{x}{x+1} \Rightarrow \displaystyle\int P(x)\,dx = \int\left(\dfrac{2}{x+1}\right)dx = 2\ln|x+1| = \ln(x+1)^2$

$\Rightarrow v(x) = e^{\ln(x+1)^2} = (x+1)^2 \Rightarrow y = \dfrac{1}{(x+1)^2}\displaystyle\int (x+1)^2\left(\dfrac{x}{x+1}\right)dx = \dfrac{1}{(x+1)^2}\int (x^2+x)\,dx$

$= \left(\dfrac{1}{x+1}\right)^2\left(\dfrac{x^3}{3} + \dfrac{x^2}{2} + C\right)$; $x = 0$ and $y = 1 \Rightarrow 1 = 0 + 0 + C \Rightarrow C = 1 \Rightarrow y = \dfrac{1}{(x+1)^2}\cdot\left(\dfrac{x^3}{3} + \dfrac{x^2}{2} + 1\right)$

125.

x	1	1.2	1.4	1.6	1.8	2.0
y	−1	−0.8	−0.56	−0.28	0.04	0.4

$\dfrac{dy}{dx} = x \Rightarrow dy = x\,dx \Rightarrow y = \dfrac{x^2}{2} + C$; $x = 1$ and $y = -1$

$\Rightarrow -1 = \dfrac{1}{2} + C \Rightarrow C = -\dfrac{3}{2} \Rightarrow y\ (\text{exact}) = \dfrac{x^2}{2} - \dfrac{3}{2}$

$\Rightarrow y(2) = \dfrac{2^2}{2} - \dfrac{3}{2} = \dfrac{1}{2}$ is the exact value

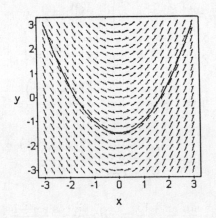

127.

x	1	1.2	1.4	1.6	1.8	2.0
y	−1	−1.2	−1.488	−1.9046	−2.5141	−3.4192

$\dfrac{dy}{dx} = xy \Rightarrow \dfrac{dy}{y} = x\,dx \Rightarrow \ln|y| = \dfrac{x^2}{2} + C \Rightarrow y = e^{\frac{x^2}{2} + C} = e^{x^2/2}\cdot e^C$

$= C_1 e^{x^2/2}$; $x = 1$ and $y = -1 \Rightarrow -1 = C_1 e^{1/2} \Rightarrow C_1 = -e^{-1/2}$

$\Rightarrow y\ (\text{exact}) = -e^{-1/2}\cdot e^{x^2/2} = -e^{(x^2-1)/2} \Rightarrow y(2) = -e^{3/2}$

≈ -4.4817 is the exact value

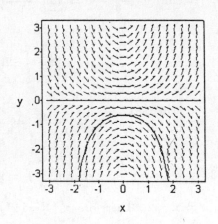

CHAPTER 6   ADDITIONAL EXERCISES–THEORY, EXAMPLES, APPLICATIONS

1. $\displaystyle\lim_{b\to 1^-} \int_0^b \dfrac{1}{\sqrt{1-x^2}}\,dx = \lim_{b\to 1^-}\left[\sin^{-1}x\right]_0^b = \lim_{b\to 1^-}\left(\sin^{-1}b - \sin^{-1}0\right) = \lim_{b\to 1^-}\left(\sin^{-1}b - 0\right) = \lim_{b\to 1^-}\sin^{-1}b = \dfrac{\pi}{2}$

3. $y = (\cos\sqrt{x})^{1/x} \Rightarrow \ln y = \dfrac{1}{x}\ln(\cos\sqrt{x})$ and $\displaystyle\lim_{x\to 0^+}\dfrac{\ln(\cos\sqrt{x})}{x} = \lim_{x\to 0^+}\dfrac{-\sin\sqrt{x}}{2\sqrt{x}\cos\sqrt{x}} = \dfrac{-1}{2}\lim_{x\to 0^+}\dfrac{\tan\sqrt{x}}{\sqrt{x}}$

$= -\dfrac{1}{2}\displaystyle\lim_{x\to 0^+}\dfrac{\frac{1}{2}x^{-1/2}\sec^2\sqrt{x}}{\frac{1}{2}x^{-1/2}} = -\dfrac{1}{2} \Rightarrow \lim_{x\to 0^+}(\cos\sqrt{x})^{1/x} = e^{-1/2} = \dfrac{1}{\sqrt{e}}$

5. $\lim\limits_{x\to\infty}\left(\frac{1}{n+1}+\frac{1}{n+2}+\cdots+\frac{1}{2n}\right)=\lim\limits_{x\to\infty}\left(\left(\frac{1}{n}\right)\left[\frac{1}{1+\left(\frac{1}{n}\right)}\right]+\left(\frac{1}{n}\right)\left[\frac{1}{1+2\left(\frac{1}{n}\right)}\right]+\cdots+\left(\frac{1}{n}\right)\left[\frac{1}{1+n\left(\frac{1}{n}\right)}\right]\right)$

which can be interpreted as a Riemann sum with partitioning $\Delta x=\frac{1}{n}\Rightarrow\lim\limits_{x\to\infty}\left(\frac{1}{n+1}+\frac{1}{n+2}+\cdots+\frac{1}{2n}\right)$

$=\int\limits_0^1\frac{1}{1+x}\,dx=\left[\ln\left(1+x\right)\right]_0^1=\ln 2$

7. $A(t)=\int\limits_0^t e^{-x}\,dx=\left[-e^{-x}\right]_0^t=1-e^{-t}$, $V(t)=\pi\int\limits_0^t e^{-2x}\,dx=\left[-\frac{\pi}{2}e^{-2x}\right]_0^t=\frac{\pi}{2}\left(1-e^{-2t}\right)$

(a) $\lim\limits_{t\to\infty}A(t)=\lim\limits_{t\to\infty}\left(1-e^{-t}\right)=1$

(b) $\lim\limits_{t\to\infty}\dfrac{V(t)}{A(t)}=\lim\limits_{t\to\infty}\dfrac{\frac{\pi}{2}\left(1-e^{-2t}\right)}{1-e^{-t}}=\dfrac{\pi}{2}$

(c) $\lim\limits_{t\to0^+}\dfrac{V(t)}{A(t)}=\lim\limits_{t\to0^+}\dfrac{\frac{\pi}{2}\left(1-e^{-2t}\right)}{1-e^{-t}}=\lim\limits_{t\to0^+}\dfrac{\frac{\pi}{2}\left(1-e^{-t}\right)\left(1+e^{-t}\right)}{\left(1-e^{-t}\right)}=\lim\limits_{t\to0^+}\frac{\pi}{2}\left(1+e^{-t}\right)=\pi$

9. $\lim\limits_{x\to0}\dfrac{\sin ax+bx}{x^3}=\lim\limits_{x\to0}\dfrac{a\cos ax+b}{3x^2}\Rightarrow b=-a$ (the numerator must go to zero) $\Rightarrow\lim\limits_{x\to0}\dfrac{a\cos ax-a}{3x^2}$

$=\lim\limits_{x\to0}\dfrac{-a^2\sin ax}{6x}=\lim\limits_{x\to0}\dfrac{-a^3\cos ax}{6}=-\dfrac{a^3}{6}\Rightarrow-\dfrac{a^3}{6}=-\dfrac{4}{3}\Rightarrow a^3=8\Rightarrow a=2$ and $b=-2$

11. $A_1=\int\limits_1^e\dfrac{2\log_2 x}{x}\,dx=\dfrac{2}{\ln 2}\int\limits_1^e\dfrac{\ln x}{x}\,dx=\left[\dfrac{\left(\ln x\right)^2}{\ln 2}\right]_1^e=\dfrac{1}{\ln 2}$; $A_2=\int\limits_1^e\dfrac{2\log_4 x}{4}\,dx=\dfrac{2}{\ln 4}\int\limits_1^e\dfrac{\ln x}{x}\,dx$

$=\left[\dfrac{\left(\ln x\right)^2}{2\ln 2}\right]_1^e=\dfrac{1}{2\ln 2}\Rightarrow A_1:A_2=2:1$

13. $\ln x^{\left(x^x\right)}=x^x\ln x$ and $\ln\left(x^x\right)^x=x\ln x^x=x^2\ln x$; then, $x^x\ln x=x^2\ln x\Rightarrow x^x=x^2\Rightarrow x\ln x=2\ln x$

$\Rightarrow x=2$. Therefore, $x^{\left(x^x\right)}=\left(x^x\right)^x$ when $x=2$.

15. $f(x)=e^{g(x)}\Rightarrow f'(x)=e^{g(x)}\,g'(x)$, where $g'(x)=\dfrac{x}{1+x^4}\Rightarrow f'(2)=e^0\left(\dfrac{2}{1+16}\right)=\dfrac{2}{17}$

17. Triangle ABD is an isosceles right triangle with its right angle at B and an angle of measure $\frac{\pi}{4}$ at A. We therefore have $\frac{\pi}{4}=\angle DAB=\angle DAE+\angle CAB=\tan^{-1}\frac{1}{3}+\tan^{-1}\frac{1}{2}$.

19. The area of the shaded region is $\int\limits_0^1\sin^{-1}x\,dx=\int\limits_0^1\sin^{-1}y\,dy$, which is the same as the area of the region to

the left of the curve $y=\sin x$ (and part of the rectangle formed by the coordinate axes and dashed lines $y=1$,

$x=\frac{\pi}{2}$). The area of the rectangle is $\frac{\pi}{2}=\int\limits_0^1\sin^{-1}y\,dy+\int\limits_0^{\pi/2}\sin x\,dx$, so we have

$$\frac{\pi}{2} = \int_0^1 \sin^{-1} x \, dx + \int_0^{\pi/2} \sin x \, dx \Rightarrow \int_0^{\pi/2} \sin x \, dx = \frac{\pi}{2} - \int_0^1 \sin^{-1} x \, dx.$$

21. (a) $g(x) + h(x) = 0 \Rightarrow g(x) = -h(x)$; also $g(x) + h(x) = 0 \Rightarrow g(-x) + h(-x) = 0 \Rightarrow g(x) - h(x) = 0$

$\Rightarrow g(x) = h(x)$; therefore $-h(x) = h(x) \Rightarrow h(x) = 0 \Rightarrow g(x) = 0$

(b) $\dfrac{f(x) + f(-x)}{2} = \dfrac{\left[f_E(x) + f_O(x)\right] + \left[f_E(-x) + f_O(-x)\right]}{2} = \dfrac{f_E(x) + f_O(x) + f_E(x) - f_O(x)}{2} = f_E(x);$

$\dfrac{f(x) - f(-x)}{2} = \dfrac{\left[f_E(x) + f_O(x)\right] - \left[f_E(-x) - f_O(-x)\right]}{2} = \dfrac{f_E(x) + f_O(x) - f_E(x) - f_O(x)}{2} = f_O(x)$

23. $M = \displaystyle\int_0^1 \frac{2}{1 + x^2} \, dx = 2\left[\tan^{-1} x\right]_0^1 = \frac{\pi}{2}$ and $M_y = \displaystyle\int_0^1 \frac{2x}{1 + x^2} \, dx = \left[\ln\left(1 + x^2\right)\right]_0^1 = \ln 2 \Rightarrow \overline{x} = \frac{M_y}{M}$

$= \dfrac{\ln 2}{\left(\dfrac{\pi}{2}\right)} = \dfrac{\ln 4}{\pi}$; $\overline{y} = 0$ by symmetry

25. $A(t) = A_0 e^{rt}$; $A(t) = 2A_0 \Rightarrow 2A_0 = A_0 e^{rt} \Rightarrow e^{rt} = 2 \Rightarrow rt = \ln 2 \Rightarrow t = \dfrac{\ln 2}{r} \Rightarrow t \approx \dfrac{.7}{r} = \dfrac{70}{100r} = \dfrac{70}{(r\%)}$

27. (a) $L = k\left(\dfrac{a - b \cot \theta}{R^4} + \dfrac{b \csc \theta}{r^4}\right) \Rightarrow \dfrac{dL}{d\theta} = k\left(\dfrac{b \csc^2 \theta}{R^4} - \dfrac{b \csc \theta \cot \theta}{r^4}\right)$; solving $\dfrac{dL}{d\theta} = 0$

$\Rightarrow r^4 b \csc^2 \theta - b R^4 \csc \theta \cot \theta = 0 \Rightarrow (b \csc \theta)\left(r^4 \csc \theta - R^4 \cot \theta\right) = 0$; but $b \csc \theta \neq 0$ since

$\theta \neq \dfrac{\pi}{2} \Rightarrow r^4 \csc \theta - R^4 \cot \theta = 0 \Rightarrow \cos \theta = \dfrac{r^4}{R^4} \Rightarrow \theta = \cos^{-1}\left(\dfrac{r^4}{R^4}\right)$, the critical value of θ

(b) $\theta = \cos^{-1}\left(\dfrac{5}{6}\right)^4 \approx \cos^{-1}(0.48225) \approx 61°$

NOTES:

CHAPTER 7 TECHNIQUES OF INTEGRATION

7.1 BASIC INTEGRATION FORMULAS

1. $\int \dfrac{16x\,dx}{\sqrt{8x^2+1}}; \begin{bmatrix} u = 8x^2+1 \\ du = 16x\,dx \end{bmatrix} \to \int \dfrac{du}{\sqrt{u}} = 2\sqrt{u}+C = 2\sqrt{8x^2+1}+C$

3. $\int 3\sqrt{\sin v}\,\cos v\,dv; \begin{bmatrix} u = \sin v \\ du = \cos v\,dv \end{bmatrix} \to \int 3\sqrt{u}\,du = 3 \cdot \dfrac{2}{3}u^{3/2}+C = 2(\sin v)^{3/2}+C$

5. $\displaystyle\int_0^1 \dfrac{16x\,dx}{8x^2+2}; \begin{bmatrix} u = 8x^2+2 \\ du = 16x\,dx \\ x=0 \Rightarrow u=2, \quad x=1 \Rightarrow u=10 \end{bmatrix} \to \displaystyle\int_2^{10} \dfrac{du}{u} = \big[\ln|u|\big]_2^{10} = \ln 10 - \ln 2 = \ln 5$

7. $\int \dfrac{dx}{\sqrt{x}\,(\sqrt{x}+1)}; \begin{bmatrix} u = \sqrt{x} \\ du = \dfrac{1}{2\sqrt{x}}\,dx \\ 2\,du = \dfrac{dx}{\sqrt{x}} \end{bmatrix} \to \int \dfrac{2\,du}{u} = 2\ln|u|+C = 2\ln(\sqrt{x}+1)+C$

9. $\int \cot(3-7x)\,dx; \begin{bmatrix} u = 3-7x \\ du = -7\,dx \end{bmatrix} \to -\dfrac{1}{7}\int \cot u\,du = -\dfrac{1}{7}\ln|\sin u|+C = -\dfrac{1}{7}\ln|\sin(3-7x)|+C$

11. $\int e^\theta \csc(e^\theta+1)\,d\theta; \begin{bmatrix} u = e^\theta+1 \\ du = e^\theta\,d\theta \end{bmatrix} \to \int \csc u\,du = -\ln|\csc u + \cot u|+C = -\ln\left|\csc(e^\theta+1)+\cot(e^\theta+1)\right|+C$

13. $\int \sec \dfrac{t}{3}\,dt; \begin{bmatrix} u = \dfrac{t}{3} \\ du = \dfrac{dt}{3} \end{bmatrix} \to \int 3\sec u\,du = 3\ln|\sec u + \tan u|+C = 3\ln\left|\sec\dfrac{t}{3}+\tan\dfrac{t}{3}\right|+C$

15. $\int \csc(s-\pi)\,ds; \begin{bmatrix} u = s-\pi \\ du = ds \end{bmatrix} \to \int \csc u\,du = -\ln|\csc u + \cot u|+C = -\ln\left|\csc(s-\pi)+\cot(s-\pi)\right|+C$

17. $\displaystyle\int_0^{\sqrt{\ln 2}} 2xe^{x^2}\,dx; \begin{bmatrix} u = x^2 \\ du = 2x\,dx \\ x=0 \Rightarrow u=0, \; x=\sqrt{\ln 2} \Rightarrow u=\ln 2 \end{bmatrix} \to \displaystyle\int_0^{\ln 2} e^u\,du = \big[e^u\big]_0^{\ln 2} = e^{\ln 2}-e^0 = 2-1 = 1$

19. $\displaystyle\int e^{\tan v}\sec^2 v\, dv; \begin{bmatrix} u = \tan v \\ du = \sec^2 v\, dv \end{bmatrix} \rightarrow \int e^u\, du = e^u + C = e^{\tan v} + C$

21. $\displaystyle\int 3^{x+1}\, dx; \begin{bmatrix} u = x+1 \\ du = dx \end{bmatrix} \rightarrow \int 3^u\, du = \left(\frac{1}{\ln 3}\right)3^u + C = \frac{3^{(x+1)}}{\ln 3} + C$

23. $\displaystyle\int \frac{2^{\sqrt{w}}\, dw}{2\sqrt{w}}; \begin{bmatrix} u = \sqrt{w} \\ du = \dfrac{dw}{2\sqrt{w}} \end{bmatrix} \rightarrow \int 2^u\, du = \frac{2^u}{\ln 2} + C = \frac{2^{\sqrt{w}}}{\ln 2} + C$

25. $\displaystyle\int \frac{9\, du}{1 + 9u^2}; \begin{bmatrix} x = 3u \\ dx = 3\, du \end{bmatrix} \rightarrow \int \frac{3\, dx}{1 + x^2} = 3\tan^{-1} x + C = 3\tan^{-1} 3u + C$

27. $\displaystyle\int_0^{1/6} \frac{dx}{\sqrt{1 - 9x^2}}; \begin{bmatrix} u = 3x \\ du = 3\, dx \\ x = 0 \Rightarrow u = 0,\ x = \frac{1}{6} \Rightarrow u = \frac{1}{2} \end{bmatrix} \rightarrow \int_0^{1/2} \frac{1}{3}\frac{du}{\sqrt{1 - u^2}} = \left[\frac{1}{3}\sin^{-1} u\right]_0^{1/2} = \frac{1}{3}\left(\frac{\pi}{6} - 0\right) = \frac{\pi}{18}$

29. $\displaystyle\int \frac{2s\, ds}{\sqrt{1 - s^4}}; \begin{bmatrix} u = s^2 \\ du = 2s\, ds \end{bmatrix} \rightarrow \int \frac{du}{\sqrt{1 - u^2}} = \sin^{-1} u + C = \sin^{-1} s^2 + C$

31. $\displaystyle\int \frac{6\, dx}{x\sqrt{25x^2 - 1}} = \int \frac{6\, dx}{5x\sqrt{x^2 - \frac{1}{25}}} = \frac{6}{5}\cdot 5\sec^{-1}|5x| + C = 6\sec^{-1}|5x| + C$

33. $\displaystyle\int \frac{dx}{e^x + e^{-x}} = \int \frac{e^x\, dx}{e^{2x} + 1}; \begin{bmatrix} u = e^x \\ du = e^x\, dx \end{bmatrix} \rightarrow \int \frac{du}{u^2 + 1} = \tan^{-1} u + C = \tan^{-1} e^x + C$

35. $\displaystyle\int_1^{e^{\pi/3}} \frac{dx}{x\cos(\ln x)}; \begin{bmatrix} u = \ln x \\ du = \dfrac{dx}{x} \\ x = 1 \Rightarrow u = 0,\ x = e^{\pi/3} \Rightarrow u = \frac{\pi}{3} \end{bmatrix} \rightarrow \int_0^{\pi/3} \frac{du}{\cos u} = \int_0^{\pi/3} \sec u\, du = \big[\ln|\sec u + \tan u|\big]_0^{\pi/3}$

$= \ln\left|\sec\frac{\pi}{3} + \tan\frac{\pi}{3}\right| - \ln|\sec 0 + \tan 0| = \ln(2 + \sqrt{3}) - \ln(1) = \ln(2 + \sqrt{3})$

37. $\displaystyle\int_1^2 \frac{8\, dx}{x^2 - 2x + 2} = 8\int_1^2 \frac{dx}{1 + (x-1)^2}; \begin{bmatrix} u = x - 1 \\ du = dx \\ x = 1 \Rightarrow u = 0,\ x = 2 \Rightarrow u = 1 \end{bmatrix} \rightarrow 8\int_0^1 \frac{du}{1 + u^2} = 8\big[\tan^{-1} u\big]_0^1$

$= 8(\tan^{-1} 1 - \tan^{-1} 0) = 8\left(\frac{\pi}{4} - 0\right) = 2\pi$

39. $\displaystyle\int \frac{dt}{\sqrt{-t^2 + 4t - 3}} = \int \frac{dt}{\sqrt{1 - (t-2)^2}}; \begin{bmatrix} u = t - 2 \\ du = dt \end{bmatrix} \rightarrow \int \frac{du}{\sqrt{1 - u^2}} = \sin^{-1} u + C = \sin^{-1}(t - 2) + C$

41. $\displaystyle\int \frac{dx}{(x+1)\sqrt{x^2+2x}} = \int \frac{dx}{(x+1)\sqrt{(x+1)^2-1}}; \begin{bmatrix} u = x+1 \\ du = dx \end{bmatrix} \rightarrow \int \frac{du}{u\sqrt{u^2-1}} = \sec^{-1}|u| + C = \sec^{-1}|x+1| + C,$

$|u| = |x+1| > 1$

43. $\displaystyle\int (\sec x + \cot x)^2 \, dx = \int \left(\sec^2 x + 2 \sec x \cot x + \cot^2 x\right) dx = \int \sec^2 x \, dx + \int 2 \csc x \, dx + \int \left(\csc^2 x - 1\right) dx$

$= \tan x - 2 \ln|\csc x + \cot x| - \cot x - x + C$

45. $\displaystyle\int \csc x \sin 3x \, dx = \int (\csc x)(\sin 2x \cos x + \sin x \cos 2x) \, dx = \int (\csc x)\left(2 \sin x \cos^2 x + \sin x \cos 2x\right) dx$

$= \displaystyle\int \left(2 \cos^2 x + \cos 2x\right) dx = \int [(1 + \cos 2x) + \cos 2x] \, dx = \int (1 + 2 \cos 2x) \, dx = x + \sin 2x + C$

47. $\displaystyle\int \frac{x}{x+1} \, dx = \int \left(1 - \frac{1}{x+1}\right) dx = x - \ln|x+1| + C$

49. $\displaystyle\int_{\sqrt{2}}^{3} \frac{2x^3}{x^2-1} \, dx = \int_{\sqrt{2}}^{3} \left(2x + \frac{2x}{x^2-1}\right) dx = \left[x^2 + \ln|x^2-1|\right]_{\sqrt{2}}^{3} = (9 + \ln 8) - (2 + \ln 1) = 7 + \ln 8$

51. $\displaystyle\int \frac{4t^3 - t^2 + 16t}{t^2+4} \, dt = \int \left[(4t-1) + \frac{4}{t^2+4}\right] dt = 2t^2 - t + 2 \tan^{-1}\left(\frac{t}{2}\right) + C$

53. $\displaystyle\int \frac{1-x}{\sqrt{1-x^2}} \, dx = \int \frac{dx}{\sqrt{1-x^2}} - \int \frac{x \, dx}{\sqrt{1-x^2}} = \sin^{-1} x + \sqrt{1-x^2} + C$

55. $\displaystyle\int_{0}^{\pi/4} \frac{1 + \sin x}{\cos^2 x} \, dx = \int_{0}^{\pi/4} \left(\sec^2 x + \sec x \tan x\right) dx = [\tan x + \sec x]_{0}^{\pi/4} = \left(1 + \sqrt{2}\right) - (0 + 1) = \sqrt{2}$

57. $\displaystyle\int \frac{dx}{1 + \sin x} = \int \frac{(1 - \sin x)}{(1 - \sin^2 x)} \, dx = \int \frac{(1 - \sin x)}{\cos^2 x} \, dx = \int \left(\sec^2 x - \sec x \tan x\right) dx = \tan x - \sec x + C$

59. $\displaystyle\int \frac{1}{\sec \theta + \tan \theta} \, d\theta = \int \frac{\cos \theta}{1 + \sin \theta} \, d\theta; \begin{bmatrix} u = 1 + \sin \theta \\ du = \cos \theta \, d\theta \end{bmatrix} \rightarrow \int \frac{du}{u} = \ln|u| + C = \ln|1 + \sin \theta| + C$

61. $\displaystyle\int \frac{1}{1 - \sec x} \, dx = \int \frac{\cos x}{\cos x - 1} \, dx = \int \left(1 + \frac{1}{\cos x - 1}\right) dx = \int \left(1 - \frac{1 + \cos x}{\sin^2 x}\right) dx = \int \left(1 - \csc^2 x - \frac{\cos x}{\sin^2 x}\right) dx$

$= \displaystyle\int \left(1 - \csc^2 x - \csc x \cot x\right) dx = x + \cot x + \csc x + C$

63. $\displaystyle\int_{0}^{2\pi} \sqrt{\frac{1 - \cos x}{2}} \, dx = \int_{0}^{2\pi} \left|\sin \frac{x}{2}\right| dx; \begin{bmatrix} \sin \frac{x}{2} \geq 0 \\ \text{for } 0 \leq \frac{x}{2} \leq \pi \end{bmatrix} \rightarrow \int_{0}^{2\pi} \sin\left(\frac{x}{2}\right) dx = \left[-2 \cos \frac{x}{2}\right]_{0}^{2\pi} = -2(\cos \pi - \cos 0)$

$= (-2)(-2) = 4$

65. $\displaystyle\int_{\pi/2}^{\pi} \sqrt{1 + \cos 2t}\ dt = \int_{\pi/2}^{\pi} \sqrt{2}\ |\cos t|\ dt; \left[\begin{array}{c} \cos t \leq 0 \\ \text{for } \frac{\pi}{2} \leq t \leq \pi \end{array} \right] \to \int_{\pi/2}^{\pi} -\sqrt{2}\ \cos t\ dt = \left[-\sqrt{2}\ \sin t \right]_{\pi/2}^{\pi}$

$\quad = -\sqrt{2}\left(\sin \pi - \sin \frac{\pi}{2} \right) = \sqrt{2}$

67. $\displaystyle\int_{-\pi}^{0} \sqrt{1 - \cos^2 \theta}\ d\theta = \int_{-\pi}^{0} |\sin \theta|\ d\theta; \left[\begin{array}{c} \sin \theta \leq 0 \\ \text{for } -\pi \leq \theta \leq 0 \end{array} \right] \to \int_{-\pi}^{0} -\sin \theta\ d\theta = [\cos \theta]_{-\pi}^{0} = \cos 0 - \cos(-\pi)$

$\quad = 1 - (-1) = 2$

69. $\displaystyle\int_{-\pi/4}^{\pi/4} \sqrt{\tan^2 y + 1}\ dy = \int_{-\pi/4}^{\pi/4} |\sec y|\ dy; \left[\begin{array}{c} \sec y \geq 0 \\ \text{for } -\frac{\pi}{4} \leq y \leq \frac{\pi}{4} \end{array} \right] \to \int_{-\pi/4}^{\pi/4} \sec y\ dy = \left[\ln|\sec y + \tan y| \right]_{-\pi/4}^{\pi/4}$

$\quad = \ln\left| \sqrt{2} + 1 \right| - \ln\left| \sqrt{2} - 1 \right|$

71. $\displaystyle\int_{\pi/4}^{3\pi/4} (\csc x - \cot x)^2\ dx = \int_{\pi/4}^{3\pi/4} \left(\csc^2 x - 2\csc x \cot x + \cot^2 x \right) dx = \int_{\pi/4}^{3\pi/4} \left(2\csc^2 x - 1 - 2\csc x \cot x \right) dx$

$\quad = \left[-2\cot x - x + 2\csc x \right]_{\pi/4}^{3\pi/4} = \left(-2\cot \frac{3\pi}{4} - \frac{3\pi}{4} + 2\csc \frac{3\pi}{4} \right) - \left(-2\cot \frac{\pi}{4} - \frac{\pi}{4} + 2\csc \frac{\pi}{4} \right)$

$\quad = \left[-2(-1) - \frac{3\pi}{4} + 2\left(\sqrt{2} \right) \right] - \left[-2(1) - \frac{\pi}{4} + 2\left(\sqrt{2} \right) \right] = 4 - \frac{\pi}{2}$

73. $\displaystyle\int \cos \theta \csc (\sin \theta)\ d\theta; \left[\begin{array}{c} u = \sin \theta \\ du = \cos \theta\ d\theta \end{array} \right] \to \int \csc u\ du = -\ln|\csc u + \cot u| + C$

$\quad = -\ln|\csc (\sin \theta) + \cot (\sin \theta)| + C$

75. $\displaystyle\int (\csc x - \sec x)(\sin x + \cos x)\ dx = \int (1 + \cot x - \tan x - 1)\ dx = \int \cot x\ dx - \int \tan x\ dx$

$\quad = \ln|\sin x| + \ln|\cos x| + C$

77. $\displaystyle\int \frac{6\ dy}{\sqrt{y}\ (1 + y)}; \left[\begin{array}{c} u = \sqrt{y} \\ du = \frac{1}{2\sqrt{y}}\ dy \end{array} \right] \to \int \frac{12\ du}{1 + u^2} = 12\tan^{-1} u + C = 12\tan^{-1} \sqrt{y} + C$

79. $\displaystyle\int \frac{7\ dx}{(x - 1)\sqrt{x^2 - 2x - 48}} = \int \frac{7\ dx}{(x - 1)\sqrt{(x - 1)^2 - 49}}; \left[\begin{array}{c} u = x - 1 \\ du = dx \end{array} \right] \to \int \frac{7\ du}{u\sqrt{u^2 - 49}} = 7 \cdot \frac{1}{7}\sec^{-1}\left| \frac{u}{7} \right| + C$

$\quad = \sec^{-1}\left| \frac{x - 1}{7} \right| + C$

81. $\displaystyle\int \sec^2 t \tan (\tan t)\ dt; \left[\begin{array}{c} u = \tan t \\ du = \sec^2 t\ dt \end{array} \right] \to \int \tan u\ du = -\ln|\cos u| + C = \ln|\sec u| + C = \ln|\sec (\tan t)| + C$

83. (a) $\int \cos^3 \theta \; d\theta = \int (\cos \theta)(1 - \sin^2 \theta) \, d\theta; \begin{bmatrix} u = \sin \theta \\ du = \cos \theta \; d\theta \end{bmatrix} \rightarrow \int (1 - u^2) \, du = u - \frac{u^3}{3} + C = \sin \theta - \frac{1}{3} \sin^3 \theta + C$

(b) $\int \cos^5 \theta \; d\theta = \int (\cos \theta)(1 - \sin^2 \theta)^2 \, d\theta = \int (1 - u^2)^2 \, du = \int (1 - 2u^2 + u^4) \, du = u - \frac{2}{3} u^3 + \frac{u^5}{5} + C$

$= \sin \theta - \frac{2}{3} \sin^3 \theta + \frac{1}{5} \sin^5 \theta + C$

(c) $\int \cos^9 \theta \; d\theta = \int (\cos^8 \theta)(\cos \theta) \, d\theta = \int (1 - \sin^2 \theta)^4 (\cos \theta) \, d\theta$

85. (a) $\int \tan^3 \theta \; d\theta = \int (\sec^2 \theta - 1)(\tan \theta) \, d\theta = \int \sec^2 \theta \tan \theta \; d\theta - \int \tan \theta \; d\theta = \frac{1}{2} \tan^2 \theta - \int \tan \theta \; d\theta$

$= \frac{1}{2} \tan^2 \theta + \ln |\cos \theta| + C$

(b) $\int \tan^5 \theta \; d\theta = \int (\sec^2 \theta - 1)(\tan^3 \theta) \, d\theta = \int \tan^3 \theta \sec^2 \theta \; d\theta - \int \tan^3 \theta \; d\theta = \frac{1}{4} \tan^4 \theta - \int \tan^3 \theta \; d\theta$

(c) $\int \tan^7 \theta \; d\theta = \int (\sec^2 \theta - 1)(\tan^5 \theta) \, d\theta = \int \tan^5 \theta \sec^2 \theta \; d\theta - \int \tan^5 \theta \; d\theta = \frac{1}{6} \tan^6 \theta - \int \tan^5 \theta \; d\theta$

(d) $\int \tan^{2k+1} \theta \; d\theta = \int (\sec^2 \theta - 1)(\tan^{2k-1} \theta) \, d\theta = \int \tan^{2k-1} \theta \sec^2 \theta \; d\theta - \int \tan^{2k-1} \theta \; d\theta;$

$\begin{bmatrix} u = \tan \theta \\ du = \sec^2 \theta \; d\theta \end{bmatrix} \rightarrow \int u^{2k-1} \, du - \int \tan^{2k-1} \theta \; d\theta = \frac{1}{2k} u^{2k} - \int \tan^{2k-1} \theta \; d\theta = \frac{1}{2k} \tan^{2k} \theta - \int \tan^{2k-1} \theta \; d\theta$

87. $A = \int\limits_{-\pi/4}^{\pi/4} (2 \cos x - \sec x) \, dx = \left[2 \sin x - \ln |\sec x + \tan x| \right]_{-\pi/4}^{\pi/4}$

$= \left[\sqrt{2} - \ln(\sqrt{2} + 1) \right] - \left[-\sqrt{2} - \ln(\sqrt{2} - 1) \right]$

$= 2\sqrt{2} - \ln \left(\frac{\sqrt{2} + 1}{\sqrt{2} - 1} \right) = 2\sqrt{2} - \ln \left(\frac{(\sqrt{2} + 1)^2}{2 - 1} \right)$

$= 2\sqrt{2} - \ln(3 + 2\sqrt{2})$

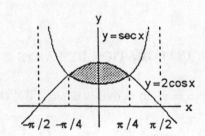

89. $V = \int\limits_{-\pi/4}^{\pi/4} \pi (2 \cos x)^2 \, dx - \int\limits_{-\pi/4}^{\pi/4} \pi \sec^2 x \, dx = 4\pi \int\limits_{-\pi/4}^{\pi/4} \cos^2 x \, dx - \pi \int\limits_{-\pi/4}^{\pi/4} \sec^2 x \, dx$

$= 2\pi \int\limits_{-\pi/4}^{\pi/4} (1 + \cos 2x) \, dx - \pi [\tan x]_{-\pi/4}^{\pi/4} = 2\pi \left[x + \frac{1}{2} \sin 2x \right]_{-\pi/4}^{\pi/4} - \pi [1 - (-1)]$

$= 2\pi \left[\left(\frac{\pi}{4} + \frac{1}{2} \right) - \left(-\frac{\pi}{4} - \frac{1}{2} \right) \right] - 2\pi = 2\pi \left(\frac{\pi}{2} + 1 \right) - 2\pi = \pi^2$

91. $y = \ln(\cos x) \Rightarrow \dfrac{dy}{dx} = -\dfrac{\sin x}{\cos x} \Rightarrow \left(\dfrac{dy}{dx}\right)^2 = \tan^2 x = \sec^2 x - 1; \; L = \displaystyle\int_a^b \sqrt{1 + \left(\dfrac{dy}{dx}\right)^2}\, dx$

$= \displaystyle\int_0^{\pi/3} \sqrt{1 + (\sec^2 x - 1)}\, dx = \displaystyle\int_0^{\pi/3} \sec x\, dx = \big[\ln|\sec x + \tan x|\big]_0^{\pi/3} = \ln\left|2 + \sqrt{3}\right| - \ln|1 + 0| = \ln\left(2 + \sqrt{3}\right)$

93. $M_x = \displaystyle\int_{-\pi/4}^{\pi/4} \left(\tfrac{1}{2}\sec x\right)(\sec x)\, dx = \tfrac{1}{2}\displaystyle\int_{-\pi/4}^{\pi/4} \sec^2 x\, dx$

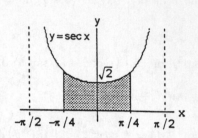

$= \tfrac{1}{2}[\tan x]_{-\pi/4}^{\pi/4} = \tfrac{1}{2}[1 - (-1)] = 1;$

$M = \displaystyle\int_{-\pi/4}^{\pi/4} \sec x\, dx = \big[\ln|\sec x + \tan x|\big]_{-\pi/4}^{\pi/4}$

$= \ln\left|\sqrt{2} + 1\right| - \ln\left|\sqrt{2} - 1\right| = \ln\left(\dfrac{\sqrt{2}+1}{\sqrt{2}-1}\right)$

$= \ln\left(\dfrac{(\sqrt{2}+1)^2}{2-1}\right) = \ln\left(3 + 2\sqrt{2}\right); \; \bar{x} = 0$ by symmetry of the region, and $\bar{y} = \dfrac{M_x}{M} = \dfrac{1}{\ln\left(3 + 2\sqrt{2}\right)}$

95. $\displaystyle\int \csc x\, dx = \displaystyle\int (\csc x)(1)\, dx = \displaystyle\int (\csc x)\left(\dfrac{\csc x + \cot x}{\csc x + \cot x}\right) dx = \displaystyle\int \dfrac{\csc^2 x + \csc x \cot x}{\csc x + \cot x}\, dx;$

$\left[\begin{array}{l} u = \csc x + \cot x \\ du = (-\csc x \cot x - \csc^2 x)\, dx \end{array}\right] \to \displaystyle\int \dfrac{-du}{u} = -\ln|u| + C = -\ln|\csc x + \cot x| + C$

7.2 INTEGRATION BY PARTS

1. $u = x, \; du = dx; \; dv = \sin\tfrac{x}{2}\, dx, \; v = -2\cos\tfrac{x}{2};$

$\displaystyle\int x \sin\tfrac{x}{2}\, dx = -2x \cos\tfrac{x}{2} - \displaystyle\int \left(-2\cos\tfrac{x}{2}\right) dx = -2x \cos\left(\tfrac{x}{2}\right) + 4\sin\left(\tfrac{x}{2}\right) + C$

3.

$$\begin{array}{ll}
 & \cos t \\
t^2 \xrightarrow{\;(+)\;} & \sin t \\
2t \xrightarrow{\;(-)\;} & -\cos t \\
2 \xrightarrow{\;(+)\;} & -\sin t \\
0 &
\end{array}$$

$\displaystyle\int t^2 \cos t\, dt = t^2 \sin t + 2t \cos t - 2\sin t + C$

5. $u = \ln x$, $du = \frac{dx}{x}$; $dv = x\, dx$, $v = \frac{x^2}{2}$;

$$\int_1^2 x \ln x \, dx = \left[\frac{x^2}{2} \ln x\right]_1^2 - \int_1^2 \frac{x^2}{2} \frac{dx}{x} = 2 \ln 2 - \left[\frac{x^2}{4}\right]_1^2 = 2 \ln 2 - \frac{3}{4} = \ln 4 - \frac{3}{4}$$

7. $u = \tan^{-1} y$, $du = \dfrac{dy}{1+y^2}$; $dv = dy$, $v = y$;

$$\int \tan^{-1} y \, dy = y \tan^{-1} y - \int \frac{y\, dy}{\left(1+y^2\right)} = y \tan^{-1} y - \frac{1}{2} \ln\left(1+y^2\right) + C = y \tan^{-1} y - \ln \sqrt{1+y^2} + C$$

9. $u = x$, $du = dx$; $dv = \sec^2 x \, dx$, $v = \tan x$;

$$\int x \sec^2 x \, dx = x \tan x - \int \tan x \, dx = x \tan x + \ln |\cos x| + C$$

11.

$$\int x^3 e^x \, dx = x^3 e^x - 3x^2 e^x + 6x e^x - 6 e^x + C = \left(x^3 - 3x^2 + 6x - 6\right) e^x + C$$

13.

$$\int \left(x^2 - 5x\right) e^x \, dx = \left(x^2 - 5x\right) e^x - (2x - 5)e^x + 2e^x + C = x^2 e^x - 7x e^x + 7 e^x + C$$
$$= \left(x^2 - 7x + 7\right) e^x + C$$

15.

$$\int x^5 e^x \, dx = x^5 e^x - 5x^4 e^x + 20x^3 e^x - 60x^2 e^x + 120x e^x - 120 e^x + C$$
$$= \left(x^5 - 5x^4 + 20x^3 - 60x^2 + 120x - 120\right) e^x + C$$

17.

$$\theta^2 \xrightarrow{(+)} \sin 2\theta$$

$$\theta^2 \xrightarrow{(+)} -\frac{1}{2}\cos 2\theta$$

$$2\theta \xrightarrow{(-)} -\frac{1}{4}\sin 2\theta$$

$$2 \xrightarrow{(+)} \frac{1}{8}\cos 2\theta$$

$$0$$

$$\int_0^{\pi/2} \theta^2 \sin 2\theta \, d\theta = \left[-\frac{\theta^2}{2}\cos 2\theta + \frac{\theta}{2}\sin 2\theta + \frac{1}{4}\cos 2\theta \right]_0^{\pi/2}$$

$$= \left[-\frac{\pi^2}{8}\cdot(-1) + \frac{\pi}{4}\cdot 0 + \frac{1}{4}\cdot(-1) \right] - \left[0 + 0 + \frac{1}{4}\cdot 1 \right] = \frac{\pi^2}{8} - \frac{1}{2} = \frac{\pi^2 - 4}{8}$$

19. $u = \sec^{-1} t, \ du = \dfrac{dt}{t\sqrt{t^2-1}}; \ dv = t \, dt, \ v = \dfrac{t^2}{2};$

$$\int_{2/\sqrt{3}}^{2} t \sec^{-1} t \, dt = \left[\frac{t^2}{2}\sec^{-1} t \right]_{2/\sqrt{3}}^{2} - \int_{2/\sqrt{3}}^{2} \left(\frac{t^2}{2} \right)\frac{dt}{t\sqrt{t^2-1}} = \left(2\cdot\frac{\pi}{3} - \frac{2}{3}\cdot\frac{\pi}{6} \right) - \int_{2/\sqrt{3}}^{2} \frac{t \, dt}{2\sqrt{t^2-1}}$$

$$= \frac{5\pi}{9} - \left[\frac{1}{2}\sqrt{t^2-1} \right]_{2/\sqrt{3}}^{2} = \frac{5\pi}{9} - \frac{1}{2}\left(\sqrt{3} - \sqrt{\frac{4}{3}-1} \right) = \frac{5\pi}{9} - \frac{1}{2}\left(\sqrt{3} - \frac{\sqrt{3}}{3} \right) = \frac{5\pi}{9} - \frac{\sqrt{3}}{3} = \frac{5\pi - 3\sqrt{3}}{9}$$

21. $I = \displaystyle\int e^\theta \sin\theta \, d\theta; \ \left[u = \sin\theta, \ du = \cos\theta \, d\theta; \ dv = e^\theta \, d\theta, \ v = e^\theta \right] \Rightarrow I = e^\theta \sin\theta - \displaystyle\int e^\theta \cos\theta \, d\theta;$

$\left[u = \cos\theta, \ du = -\sin\theta \, d\theta; \ dv = e^\theta \, d\theta, \ v = e^\theta \right] \Rightarrow I = e^\theta \sin\theta - \left(e^\theta \cos\theta + \displaystyle\int e^\theta \sin\theta \, d\theta \right)$

$= e^\theta \sin\theta - e^\theta \cos\theta - I + C' \Rightarrow 2I = \left(e^\theta \sin\theta - e^\theta \cos\theta \right) + C' \Rightarrow I = \frac{1}{2}\left(e^\theta \sin\theta - e^\theta \cos\theta \right) + C,$ where $C = \frac{C'}{2}$ is another arbitrary constant

23. $I = \displaystyle\int e^{2x} \cos 3x \, dx; \ \left[u = \cos 3x; \ du = -3\sin 3x \, dx, \ dv = e^{2x} \, dx; \ v = \frac{1}{2}e^{2x} \right]$

$\Rightarrow I = \frac{1}{2}e^{2x}\cos 3x + \frac{3}{2}\displaystyle\int e^{2x} \sin 3x \, dx; \ \left[u = \sin 3x, \ du = 3\cos 3x, \ dv = e^{2x} \, dx; \ v = \frac{1}{2}e^{2x} \right]$

$\Rightarrow I = \frac{1}{2}e^{2x}\cos 3x + \frac{3}{2}\left(\frac{1}{2}e^{2x}\sin 3x - \frac{3}{2}\displaystyle\int e^{2x}\cos 3x \, dx \right) = \frac{1}{2}e^{2x}\cos 3x + \frac{3}{4}e^{2x}\sin 3x - \frac{9}{4}I + C'$

$\Rightarrow \frac{13}{4}I = \frac{1}{2}e^{2x}\cos 3x + \frac{3}{4}e^{2x}\sin 3x + C' \Rightarrow \frac{e^{2x}}{13}(3\sin 3x + 2\cos 3x) + C,$ where $C = \frac{4}{13}C'$

25. $\displaystyle\int e^{\sqrt{3s+9}} \, ds; \ \begin{bmatrix} 3s + 9 = x^2 \\ ds = \frac{2}{3}x \, dx \end{bmatrix} \rightarrow \displaystyle\int e^x \cdot \frac{2}{3}x \, dx = \frac{2}{3}\displaystyle\int xe^x \, dx; \ \left[u = x, \ du = dx; \ dv = e^x \, dx, \ v = e^x \right];$

$\frac{2}{3}\displaystyle\int xe^x \, dx = \frac{2}{3}\left(xe^x - \displaystyle\int e^x \, dx \right) = \frac{2}{3}(xe^x - e^x) + C = \frac{2}{3}\left(\sqrt{3s+9}\,e^{\sqrt{3s+9}} - e^{\sqrt{3s+9}} \right) + C$

27. $u = x, \; du = dx; \; dv = \tan^2 x \, dx, \; v = \int \tan^2 x \, dx = \int \frac{\sin^2 x}{\cos^2 x} \, dx = \int \frac{1 - \cos^2 x}{\cos^2 x} \, dx = \int \frac{dx}{\cos^2 x} - \int dx$

$= \tan x - x; \quad \int\limits_0^{\pi/3} x \tan^2 x \, dx = \left[x(\tan x - x) \right]_0^{\pi/3} - \int\limits_0^{\pi/3} (\tan x - x) \, dx = \frac{\pi}{3}\left(\sqrt{3} - \frac{\pi}{3} \right) + \left[\ln |\cos x| + \frac{x^2}{2} \right]_0^{\pi/3}$

$= \frac{\pi}{3}\left(\sqrt{3} - \frac{\pi}{3} \right) + \ln \frac{1}{2} + \frac{\pi^2}{18} = \frac{\pi\sqrt{3}}{3} - \ln 2 - \frac{\pi^2}{18}$

29. $\int \sin(\ln x) \, dx; \; \begin{bmatrix} u = \ln x \\ du = \frac{1}{x} dx \\ dx = e^u \, du \end{bmatrix} \rightarrow \int (\sin u) e^u \, du. \;$ From Exercise 21, $\int (\sin u) e^u \, du = e^u\left(\frac{\sin u - \cos u}{2} \right) + C$

$= \frac{1}{2}\left[-x \cos(\ln x) + x \sin(\ln x) \right] + C$

31. (a) $u = x, \; du = dx; \; dv = \sin x \, dx, \; v = -\cos x;$

$S_1 = \int\limits_0^{\pi} x \sin x \, dx = \left[-x \cos x \right]_0^{\pi} + \int\limits_0^{\pi} \cos x \, dx = \pi + \left[\sin x \right]_0^{\pi} = \pi$

(b) $S_2 = -\int\limits_{\pi}^{2\pi} x \sin x \, dx = \left[\left[-x \cos x \right]_{\pi}^{2\pi} + \int\limits_{\pi}^{2\pi} \cos x \, dx \right] = -\left[-3\pi + \left[\sin x \right]_{\pi}^{2\pi} \right] = 3\pi$

(c) $S_3 = \int\limits_{2\pi}^{3\pi} x \sin x \, dx = \left[-x \cos x \right]_{2\pi}^{3\pi} + \int\limits_{2\pi}^{3\pi} \cos x \, dx = 5\pi + \left[\sin x \right]_{2\pi}^{3\pi} = 5\pi$

(d) $S_{n+1} = (-1)^{n+1} \int\limits_{n\pi}^{(n+1)\pi} x \sin x \, dx = (-1)^{n+1}\left[\left[-x \cos x \right]_{n\pi}^{(n+1)\pi} + \left[\sin x \right]_{n\pi}^{(n+1)\pi} \right]$

$= (-1)^{n+1}\left[-(n+1)\pi(-1)^n + n\pi(-1)^{n+1} \right] + 0 = (2n+1)\pi$

33. $V = \int\limits_0^{\ln 2} 2\pi(\ln 2 - x) e^x \, dx = 2\pi \ln 2 \int\limits_0^{\ln 2} e^x \, dx - 2\pi \int\limits_0^{\ln 2} x e^x \, dx$

$= (2\pi \ln 2)\left[e^x \right]_0^{\ln 2} - 2\pi\left(\left[x e^x \right]_0^{\ln 2} - \int\limits_0^{\ln 2} e^x \, dx \right)$

$= 2\pi \ln 2 - 2\pi\left(2 \ln 2 + \left[e^x \right]_0^{\ln 2} \right) = -2\pi \ln 2 + 2 = 2\pi(1 - \ln 2)$

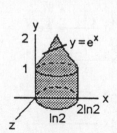

35. (a) $V = \int\limits_0^{\pi/2} 2\pi x \cos x \, dx = 2\pi\left(\left[x \sin x \right]_0^{\pi/2} - \int\limits_0^{\pi/2} \sin x \, dx \right)$

$= 2\pi\left(\frac{\pi}{2} + \left[\cos x \right]_0^{\pi/2} \right) = 2\pi\left(\frac{\pi}{2} + 0 - 1 \right) = \pi(\pi - 2)$

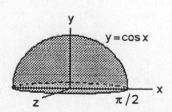

(b) $V = \displaystyle\int_0^{\pi/2} 2\pi\left(\dfrac{\pi}{2} - x\right)\cos x\, dx;\ u = \dfrac{\pi}{2} - x,\ du = -dx;\ dv = \cos x\, dx,\ v = \sin x;$

$V = 2\pi\left[\left(\dfrac{\pi}{2} - x\right)\sin x\right]_0^{\pi/2} + 2\pi\displaystyle\int_0^{\pi/2}\sin x\, dx = 0 + 2\pi[-\cos x]_0^{\pi/2} = 2\pi(0+1) = 2\pi$

37. (a) $A = \displaystyle\int_0^1 x^2 e^x\, dx;$

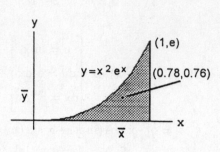

$$
\begin{array}{lcl}
 & & e^x \\
x^2 & \xrightarrow{(+)} & e^x \\
2x & \xrightarrow{(-)} & e^x \\
2 & \xrightarrow{(+)} & e^x \\
0 & &
\end{array}
$$

$\Rightarrow A = \left[x^2 e^x - 2x e^x + 2e^x\right]_0^1 = e - 2;$

$M_y = \displaystyle\int_0^1 xy\, dx = \int_0^1 x^3 e^x\, dx;$

$$
\begin{array}{lcl}
 & & e^x \\
x^3 & \xrightarrow{(+)} & e^x \\
3x^2 & \xrightarrow{(-)} & e^x \\
6x & \xrightarrow{(+)} & e^x \\
6 & \xrightarrow{(-)} & e^x \\
0 & &
\end{array}
$$

$\Rightarrow M_y = \left[x^3 e^x - 3x^2 e^x + 6x e^x - 6e^x\right]_0^1 = (1 - 3 + 6 - 6)\,e - (-6) = 6 - 2e;$

$M_x = \dfrac{1}{2}\displaystyle\int_0^1 y^2\, dx = \int_0^1 x^4 e^{2x}\, dx;$

$$
\begin{array}{lcl}
 & & e^{2x} \\
x^4 & \xrightarrow{(+)} & \tfrac{1}{2}e^{2x} \\
4x^3 & \xrightarrow{(-)} & \tfrac{1}{4}e^{2x} \\
12x^2 & \xrightarrow{(+)} & \tfrac{1}{8}e^{2x} \\
24x & \xrightarrow{(-)} & \tfrac{1}{16}e^{2x} \\
24 & \xrightarrow{(+)} & \tfrac{1}{32}e^{2x} \\
0 & &
\end{array}
$$

$\Rightarrow M_x = \dfrac{1}{2}\left[\dfrac{1}{2}x^4 e^{2x} - x^3 e^{2x} + \dfrac{3}{2}x^2 e^{2x} - \dfrac{3}{2}x e^{2x} + \dfrac{3}{4}e^{2x}\right]_0^1$

$\qquad = \dfrac{1}{2}\left[\left(\dfrac{1}{2} - 1 + \dfrac{3}{2} - \dfrac{3}{2} + \dfrac{3}{4}\right)e^2 - \dfrac{3}{4}\right] = \dfrac{1}{2}\left(\dfrac{1}{4}e^2 - \dfrac{3}{4}\right) = \dfrac{e^2 - 3}{8};$

$\Rightarrow \bar{x} = \dfrac{M_y}{A} = \dfrac{6 - 2e}{e - 2}$ and $\bar{y} = \dfrac{M_x}{A} = \dfrac{e^2 - 3}{8(e - 2)}$

(b) $\bar{x} \approx 0.78$ and $\bar{y} \approx 0.76$

39. $M_y = \int\limits_0^\pi \delta(x)xy \, dx = \int\limits_0^\pi (1+x)x \sin x \, dx = \int\limits_0^\pi x \sin x \, dx + \int\limits_0^\pi x^2 \sin x \, dx$

$= [-x \cos x + \sin x]_0^\pi + [-x^2 \cos x + 2x \sin x + 2 \cos x]_0^\pi = \pi^2 + \pi - 4$

41. (a) $av(y) = \frac{1}{2\pi} \int\limits_0^{2\pi} 2e^{-t} \cos t \, dt$ (b)

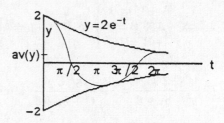

$= \frac{1}{\pi}\left[e^{-t}\left(\frac{\sin t - \cos t}{2}\right)\right]_0^{2\pi}$

(see Exercise 22) $\Rightarrow av(y) = \frac{1}{2\pi}(1 - e^{-2\pi})$

43. $\int \sin^{-1} x \, dx = x \sin^{-1} x - \int \sin y \, dy = x \sin^{-1} x + \cos y + C = x \sin^{-1} x + \cos(\sin^{-1} x) + C$

45. $\int \sec^{-1} x \, dx = x \sec^{-1} x - \int \sec y \, dy = x \sec^{-1} x - \ln|\sec y + \tan y| + C$

$= x \sec^{-1} x - \ln\left|\sec(\sec^{-1} x) + \tan(\sec^{-1} x)\right| + C = x \sec^{-1} x - \ln\left|x + \sqrt{x^2 - 1}\right| + C$

47. Yes, $\cos^{-1} x$ is the angle whose cosine is x which implies $\sin(\cos^{-1} x) = \sqrt{1 - x^2}$.

49. (a) $\int \sinh^{-1} x \, dx = x \sinh^{-1} x - \int \sinh y \, dy = x \sinh^{-1} x - \cosh y + C = x \sinh^{-1} x - \cosh(\sinh^{-1} x) + C;$

check: $d[x \sinh^{-1} x - \cosh(\sinh^{-1} x) + C] = \left[\sinh^{-1} x + \frac{x}{\sqrt{1+x^2}} - \sinh(\sinh^{-1} x)\frac{1}{\sqrt{1+x^2}}\right] dx$

$= \sinh^{-1} x \, dx$

(b) $\int \sinh^{-1} x \, dx = x \sinh^{-1} x - \int x\left(\frac{1}{\sqrt{1+x^2}}\right) dx = x \sinh^{-1} x - \frac{1}{2}\int (1+x^2)^{-1/2} 2x \, dx$

$= x \sinh^{-1} x + (1+x^2)^{1/2} + C$

check: $d\left[x \sinh^{-1} x + (1+x^2)^{1/2} + C\right] = \left[\sinh^{-1} x + \frac{x}{\sqrt{1+x^2}} - \frac{x}{\sqrt{1+x^2}}\right] dx = \sinh^{-1} x \, dx$

7.3 PARTIAL FRACTIONS

1. $\frac{5x - 13}{(x-3)(x-2)} = \frac{A}{x-3} + \frac{B}{x-2} \Rightarrow 5x - 13 = A(x-2) + B(x-3) = (A+B)x - (2A+3B)$

$\Rightarrow \left.\begin{array}{l} A + B = 5 \\ 2A + 3B = 13 \end{array}\right\} \Rightarrow -B = (10-13) \Rightarrow B = 3 \Rightarrow A = 2;$ thus, $\frac{5x-13}{(x-3)(x-2)} = \frac{2}{x-3} + \frac{3}{x-2}$

3. $\dfrac{x+4}{(x+1)^2} = \dfrac{A}{x+1} + \dfrac{B}{(x+1)^2} \Rightarrow x+4 = A(x+1) + B = Ax + (A+B) \Rightarrow \left.\begin{array}{l} A = 1 \\ A+B = 4 \end{array}\right\} \Rightarrow A = 1$ and $B = 3$;

thus, $\dfrac{x+4}{(x+1)^2} = \dfrac{1}{x+1} + \dfrac{3}{(x+1)^2}$

5. $\dfrac{z+1}{z^2(z-1)} = \dfrac{A}{z} + \dfrac{B}{z^2} + \dfrac{C}{z-1} \Rightarrow z+1 = Az(z-1) + B(z-1) + Cz^2 \Rightarrow z+1 = (A+C)z^2 + (-A+B)z - B$

$\Rightarrow \left.\begin{array}{l} A+C = 0 \\ -A+B = 1 \\ -B = 1 \end{array}\right\} \Rightarrow B = -1 \Rightarrow A = -2 \Rightarrow C = 2$; thus, $\dfrac{z+1}{z^2(z-1)} = \dfrac{-2}{z} + \dfrac{-1}{z^2} + \dfrac{2}{z-1}$

7. $\dfrac{t^2+8}{t^2-5t+6} = 1 + \dfrac{5t+2}{t^2-5t+6}$ (after long division); $\dfrac{5t+2}{t^2-5t+6} = \dfrac{5t+2}{(t-3)(t-2)} = \dfrac{A}{t-3} + \dfrac{B}{t-2}$

$\Rightarrow 5t+2 = A(t-2) + B(t-3) = (A+B)t + (-2A-3B) \Rightarrow \left.\begin{array}{l} A+B = 5 \\ -2A-3B = 2 \end{array}\right\} \Rightarrow -B = (10+2) = 12$

$\Rightarrow B = -12 \Rightarrow A = 17$; thus, $\dfrac{t^2+8}{t^2-5t+6} = 1 + \dfrac{17}{t-3} + \dfrac{-12}{t-2}$

9. $\dfrac{1}{1-x^2} = \dfrac{A}{1-x} + \dfrac{B}{1+x} \Rightarrow 1 = A(1+x) + B(1-x);\ x = 1 \Rightarrow A = \dfrac{1}{2};\ x = -1 \Rightarrow B = \dfrac{1}{2}$;

$\displaystyle\int \dfrac{dx}{1-x^2} = \dfrac{1}{2}\int \dfrac{dx}{1-x} + \dfrac{1}{2}\int \dfrac{dx}{1+x} = \dfrac{1}{2}\big[\ln|1+x| - \ln|1-x|\big] + C$

11. $\dfrac{x+4}{x^2+5x-6} = \dfrac{A}{x+6} + \dfrac{B}{x-1} \Rightarrow x+4 = A(x-1) + B(x+6);\ x = 1 \Rightarrow B = \dfrac{5}{7};\ x = -6 \Rightarrow A = \dfrac{-2}{-7} = \dfrac{2}{7}$;

$\displaystyle\int \dfrac{x+4}{x^2+5x-6}\,dx = \dfrac{2}{7}\int \dfrac{dx}{x+6} + \dfrac{5}{7}\int \dfrac{dx}{x-1} = \dfrac{2}{7}\ln|x+6| + \dfrac{5}{7}\ln|x-1| + C = \dfrac{1}{7}\ln\big|(x+6)^2(x-1)^5\big| + C$

13. $\dfrac{y}{y^2-2y-3} = \dfrac{A}{y-3} + \dfrac{B}{y+1} \Rightarrow y = A(y+1) + B(y-3);\ y = -1 \Rightarrow B = \dfrac{-1}{-4} = \dfrac{1}{4};\ y = 3 \Rightarrow A = \dfrac{3}{4}$;

$\displaystyle\int_4^8 \dfrac{y\,dy}{y^2-2y-3} = \dfrac{3}{4}\int_4^8 \dfrac{dy}{y-3} + \dfrac{1}{4}\int_4^8 \dfrac{dy}{y+1} = \left[\dfrac{3}{4}\ln|y-3| + \dfrac{1}{4}\ln|y+1|\right]_4^8 = \left(\dfrac{3}{4}\ln 5 + \dfrac{1}{4}\ln 9\right) - \left(\dfrac{3}{4}\ln 1 + \dfrac{1}{4}\ln 5\right)$

$= \dfrac{1}{2}\ln 5 + \dfrac{1}{2}\ln 3 = \dfrac{\ln 15}{2}$

15. $\dfrac{1}{t^3+t^2-2t} = \dfrac{A}{t} + \dfrac{B}{t+2} + \dfrac{C}{t-1} \Rightarrow 1 = A(t+2)(t-1) + Bt(t-1) + Ct(t+2);\ t = 0 \Rightarrow A = -\dfrac{1}{2};\ t = -2$

$\Rightarrow B = \dfrac{1}{6};\ t = 1 \Rightarrow C = \dfrac{1}{3};\ \displaystyle\int \dfrac{dt}{t^3+t^2-2t} = -\dfrac{1}{2}\int \dfrac{dt}{t} + \dfrac{1}{6}\int \dfrac{dt}{t+2} + \dfrac{1}{3}\int \dfrac{dt}{t-1}$

$= -\dfrac{1}{2}\ln|t| + \dfrac{1}{6}\ln|t+2| + \dfrac{1}{3}\ln|t-1| + C$

17. $\dfrac{x^3}{x^2+2x+1} = (x-2) + \dfrac{3x+2}{(x+1)^2}$ (after long division); $\dfrac{3x+2}{(x+1)^2} = \dfrac{A}{x+1} + \dfrac{B}{(x+1)^2} \Rightarrow 3x+3 = A(x+1) + B$

$= Ax + (A+B) \Rightarrow A = 3, \; A+B = 2 \Rightarrow A = 3, \; B = -1; \; \displaystyle\int_0^1 \dfrac{x^3\,dx}{x^2+2x+1}$

$= \displaystyle\int_0^1 (x-2)\,dx + 3\int_0^1 \dfrac{dx}{x+1} - \int_0^1 \dfrac{dx}{(x+1)^2} = \left[\dfrac{x^2}{2} - 2x + 3\ln|x+1| + \dfrac{1}{x+1}\right]_0^1$

$= \left(\dfrac{1}{2} - 2 + 3\ln 2 + \dfrac{1}{2}\right) - (1) = 3\ln 2 - 2$

19. $\dfrac{1}{(x^2-1)^2} = \dfrac{A}{x+1} + \dfrac{B}{x-1} + \dfrac{C}{(x+1)^2} + \dfrac{D}{(x-1)^2} \Rightarrow 1 = A(x+1)(x-1)^2 + B(x-1)(x+1)^2 + C(x-1)^2 + D(x+1)^2;$

$x = -1 \Rightarrow C = \dfrac{1}{4}; \; x = 1 \Rightarrow D = \dfrac{1}{4};$ coefficient of $x^3 = A + B \Rightarrow A + B = 0;$ constant $= A - B + C + D$

$\Rightarrow A - B + C + D = 1 \Rightarrow A - B = \dfrac{1}{2};$ thus, $A = \dfrac{1}{4} \Rightarrow B = -\dfrac{1}{4}; \; \displaystyle\int \dfrac{dx}{(x^2-1)^2}$

$= \dfrac{1}{4}\displaystyle\int \dfrac{dx}{x+1} - \dfrac{1}{4}\int \dfrac{dx}{x-1} + \dfrac{1}{4}\int \dfrac{dx}{(x+1)^2} + \dfrac{1}{4}\int \dfrac{dx}{(x-1)^2} = \dfrac{1}{4}\ln\left|\dfrac{x+1}{x-1}\right| - \dfrac{x}{2(x^2-1)} + C$

21. $\dfrac{1}{(x+1)(x^2+1)} = \dfrac{A}{x+1} + \dfrac{Bx+C}{x^2+1} \Rightarrow 1 = A(x^2+1) + (Bx+C)(x+1); \; x = -1 \Rightarrow A = \dfrac{1}{2};$ coefficient of x^2

$= A + B \Rightarrow A + B = 0 \Rightarrow B = -\dfrac{1}{2};$ constant $= A + C \Rightarrow A + C = 1 \Rightarrow C = \dfrac{1}{2}; \; \displaystyle\int_0^1 \dfrac{dx}{(x+1)(x^2+1)}$

$= \dfrac{1}{2}\displaystyle\int_0^1 \dfrac{dx}{x+1} + \dfrac{1}{2}\int_0^1 \dfrac{(-x+1)}{x^2+1}\,dx = \left[\dfrac{1}{2}\ln|x+1| - \dfrac{1}{4}\ln(x^2+1) + \dfrac{1}{2}\tan^{-1}x\right]_0^1$

$= \left(\dfrac{1}{2}\ln 2 - \dfrac{1}{4}\ln 2 + \dfrac{1}{2}\tan^{-1}1\right) - \left(\dfrac{1}{2}\ln 1 - \dfrac{1}{4}\ln 1 + \dfrac{1}{2}\tan^{-1}0\right) = \dfrac{1}{4}\ln 2 + \dfrac{1}{2}\left(\dfrac{\pi}{4}\right) = \dfrac{(\pi + 2\ln 2)}{8}$

23. $\dfrac{y^2+2y+1}{(y^2+1)^2} = \dfrac{Ay+B}{y^2+1} + \dfrac{Cy+D}{(y^2+1)^2} \Rightarrow y^2+2y+1 = (Ay+B)(y^2+1) + Cy + D$

$= Ay^3 + By^2 + (A+C)y + (B+D) \Rightarrow A = 0, \; B = 1; \; A+C = 2 \Rightarrow C = 2; \; B+D = 1 \Rightarrow D = 0;$

$\displaystyle\int \dfrac{y^2+2y+1}{(y^2+1)^2}\,dy = \int \dfrac{1}{y^2+1}\,dy + 2\int \dfrac{y}{(y^2+1)^2}\,dy = \tan^{-1}y - \dfrac{1}{y^2+1} + C$

25. $\dfrac{2s+2}{(s^2+1)(s-1)^3} = \dfrac{As+B}{s^2+1} + \dfrac{C}{s-1} + \dfrac{D}{(s-1)^2} + \dfrac{E}{(s-1)^3} \Rightarrow 2s+2$

$= (As+B)(s-1)^3 + C(s^2+1)(s-1)^2 + D(s^2+1)(s-1) + E(s^2+1)$

$= \left[As^4 + (-3A+B)s^3 + (3A-3B)s^2 + (-A+3B)s - B\right] + C(s^4 - 2s^3 + 2s^2 - 2s + 1) + D(s^3 - s^2 + s - 1)$

$\quad + E(s^2+1)$

$= (A+C)s^4 + (-3A+B-2C+D)s^3 + (3A-3B+2C-D+E)s^2 + (-A+3B-2C+D)s + (-B+C-D+E)$

$$\left.\begin{array}{r} A \quad\;\; + C \qquad\qquad = 0 \\ -3A + \;\; B - 2C + D \qquad = 0 \\ \Rightarrow \quad 3A - 3B + 2C - D + E = 0 \\ -A + 3B - 2C + D \qquad = 2 \\ - B + \;\; C - D + E = 2 \end{array}\right\} \text{summing all equations} \Rightarrow 2E = 4 \Rightarrow E = 2;$$

summing eqs (2) and (3) $\Rightarrow -2B + 2 = 0 \Rightarrow B = 1$; summing eqs (3) and (4) $\Rightarrow 2A + 2 = 2 \Rightarrow A = 0; \; C = 0$

from eq (1); then $-1 + 0 - D + 2 = 2$ from eq (5) $\Rightarrow D = -1$;

$$\int \frac{2s+2}{\left(s^2+1\right)(s-1)^3}\,ds = \int \frac{ds}{s^2+1} - \int \frac{ds}{(s-1)^2} + 2\int \frac{ds}{(s-1)^3} = -(s-1)^{-2} + (s-1)^{-1} + \tan^{-1}s + C$$

27. $\dfrac{2\theta^3 + 5\theta^2 + 8\theta + 4}{\left(\theta^2 + 2\theta + 2\right)^2} = \dfrac{A\theta + B}{\theta^2 + 2\theta + 2} + \dfrac{C\theta + D}{\left(\theta^2 + 2\theta + 2\right)^2} \Rightarrow 2\theta^3 + 5\theta^2 + 8\theta + 4 = (A\theta + B)\left(\theta^2 + 2\theta + 2\right) + C\theta + D$

$= A\theta^3 + (2A+B)\theta^2 + (2A + 2B + C)\theta + (2B + D) \Rightarrow A = 2; \; 2A + B = 5 \Rightarrow B = 1; \; 2A + 2B + C = 8 \Rightarrow C = 2;$

$2B + D = 4 \Rightarrow D = 2; \; \displaystyle\int \frac{2\theta^3 + 5\theta^2 + 8\theta + 4}{\left(\theta^2 + 2\theta + 2\right)^2}\,d\theta = \int \frac{2\theta + 1}{\left(\theta^2 + 2\theta + 2\right)}\,d\theta + \int \frac{2\theta + 2}{\left(\theta^2 + 2\theta + 2\right)^2}\,d\theta$

$= \displaystyle\int \frac{2\theta + 2}{\theta^2 + 2\theta + 2}\,d\theta - \int \frac{d\theta}{\theta^2 + 2\theta + 2} + \int \frac{d\left(\theta^2 + 2\theta + 2\right)}{\left(\theta^2 + 2\theta + 2\right)^2} = \int \frac{d\left(\theta^2 + 2\theta + 2\right)}{\theta^2 + 2\theta + 2} - \int \frac{d\theta}{(\theta+1)^2 + 1} - \frac{1}{\theta^2 + 2\theta + 2}$

$= \dfrac{-1}{\theta^2 + 2\theta + 2} + \ln\left(\theta^2 + 2\theta + 2\right) - \tan^{-1}(\theta + 1) + C$

29. $\dfrac{2x^3 - 2x^2 + 1}{x^2 - x} = 2x + \dfrac{1}{x^2 - x} = 2x + \dfrac{1}{x(x-1)}; \; \dfrac{1}{x(x-1)} = \dfrac{A}{x} + \dfrac{B}{x-1} \Rightarrow 1 = A(x-1) + Bx; \; x = 0 \Rightarrow A = -1;$

$x = 1 \Rightarrow B = 1; \; \displaystyle\int \frac{2x^3 - 2x^2 + 1}{x^2 - x} = \int 2x\,dx - \int \frac{dx}{x} + \int \frac{dx}{x-1} = x^2 - \ln|x| + \ln|x-1| + C = x^2 + \ln\left|\frac{x-1}{x}\right| + C$

31. $\dfrac{9x^3 - 3x + 1}{x^3 - x^2} = 9 + \dfrac{9x^2 - 3x + 1}{x^2(x-1)}$ (after long division); $\dfrac{9x^2 - 3x + 1}{x^2(x-1)} = \dfrac{A}{x} + \dfrac{B}{x^2} + \dfrac{C}{x-1}$

$\Rightarrow 9x^2 - 3x + 1 = Ax(x-1) + B(x-1) + Cx^2; \; x = 1 \Rightarrow C = 7; \; x = 0 \Rightarrow B = -1; \; A + C = 9 \Rightarrow A = 2;$

$\displaystyle\int \frac{9x^3 - 3x + 1}{x^3 - x^2}\,dx = \int 9\,dx + 2\int \frac{dx}{x} - \int \frac{dx}{x^2} + 7\int \frac{dx}{x-1} = 9x + 2\ln|x| + \frac{1}{x} + 7\ln|x-1| + C$

33. $\dfrac{y^4 + y^2 - 1}{y^3 + y} = y - \dfrac{1}{y(y^2 + 1)}; \; \dfrac{1}{y(y^2 + 1)} = \dfrac{A}{y} + \dfrac{By + C}{y^2 + 1} \Rightarrow 1 = A(y^2 + 1) + (By + C)y = (A + B)y^2 + Cy + A$

$\Rightarrow A = 1; \; A + B = 0 \Rightarrow B = -1; \; C = 0; \; \displaystyle\int \frac{y^4 + y^2 - 1}{y^3 + y}\,dy = \int y\,dy - \int \frac{dy}{y} + \int \frac{y\,dy}{y^2 + 1}$

$= \dfrac{y^2}{2} - \ln|y| + \dfrac{1}{2}\ln\left(1 + y^2\right) + C$

35. $\displaystyle\int \frac{e^t\,dt}{e^{2t} + 3e^t + 2} = \left[e^t = y\right]\int \frac{dy}{y^2 + 3y + 2} = \int \frac{dy}{y+1} - \int \frac{dy}{y+2} = \ln\left|\frac{y+1}{y+2}\right| + C = \ln\left|\frac{e^t + 1}{e^t + 2}\right| + C$

37. $\int \dfrac{\cos y\, dy}{\sin^2 y + \sin y - 6}$; $[\sin y = t,\ \cos y\, dy = dt] \to \int \dfrac{dy}{t^2 + t - 6} = \dfrac{1}{5}\int\left(\dfrac{1}{t-2} - \dfrac{1}{t+3}\right)dt = \dfrac{1}{5}\ln\left|\dfrac{t-2}{t+3}\right| + C$

$= \dfrac{1}{5}\ln\left|\dfrac{\sin y - 2}{\sin y + 3}\right| + C$

39. $\int \dfrac{(x-2)^2 \tan^{-1}(2x) - 12x^3 - 3x}{(4x^2+1)(x-2)^2}\, dx = \int \dfrac{\tan^{-1}(2x)}{4x^2+1}\, dx - 3\int \dfrac{x}{(x-2)^2}\, dx$

$= \dfrac{1}{2}\int \tan^{-1}(2x)\, d\left(\tan^{-1}(2x)\right) - 3\int \dfrac{dx}{x-2} - 6\int \dfrac{dx}{(x-2)^2} = \dfrac{\left(\tan^{-1}2x\right)^2}{4} - 3\ln|x-2| + \dfrac{6}{x-2} + C$

41. $(t^2 - 3t + 2)\dfrac{dx}{dt} = 1$; $x = \int \dfrac{dt}{t^2 - 3t + 2} = \int \dfrac{dt}{t-2} - \int \dfrac{dt}{t-1} = \ln\left|\dfrac{t-2}{t-1}\right| + C$; $\dfrac{t-2}{t-1} = Ce^x$; $t = 3$ and $x = 0$

$\Rightarrow \dfrac{1}{2} = C \Rightarrow \dfrac{t-2}{t-1} = \dfrac{1}{2}e^x \Rightarrow x = \ln\left|2\left(\dfrac{t-2}{t-1}\right)\right| = \ln|t-2| - \ln|t-1| + \ln 2$

43. $(t^2 + 2t)\dfrac{dx}{dt} = 2x + 2$; $\dfrac{1}{2}\int \dfrac{dx}{x+1} = \int \dfrac{dt}{t^2 + 2t} \Rightarrow \dfrac{1}{2}\ln|x+1| = \dfrac{1}{2}\int \dfrac{dt}{t} - \dfrac{1}{2}\int \dfrac{dt}{t+2} \Rightarrow \ln|x+1| = \ln\left|\dfrac{t}{t+2}\right| + C$;

$t = 1$ and $x = 1 \Rightarrow \ln 2 = \ln\dfrac{1}{3} + C \Rightarrow C = \ln 2 + \ln 3 = \ln 6 \Rightarrow \ln|x+1| = \ln 6\left|\dfrac{t}{t+2}\right| \Rightarrow x + 1 = \dfrac{6t}{t+2}$

$\Rightarrow x = \dfrac{6t}{t+2} - 1$, $t > 0$

45. $V = \pi\displaystyle\int_{0.5}^{2.5} y^2\, dx = \pi\int_{0.5}^{2.5} \dfrac{9}{3x - x^2}\, dx = 3\pi\left(\int_{0.5}^{2.5}\left(-\dfrac{1}{x-3} + \dfrac{1}{x}\right)\right)dx = \left[3\pi\ln\left|\dfrac{x}{x-3}\right|\right]_{0.5}^{2.5} = 3\pi\ln 25$

47. $A = \displaystyle\int_0^{\sqrt{3}} \tan^{-1}x\, dx = \left[x\tan^{-1}x\right]_0^{\sqrt{3}} - \int_0^{\sqrt{3}} \dfrac{x}{1+x^2}\, dx$

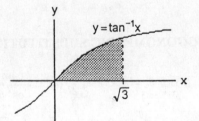

$= \dfrac{\pi\sqrt{3}}{3} - \left[\dfrac{1}{2}\ln(x^2+1)\right]_0^{\sqrt{3}} = \dfrac{\pi\sqrt{3}}{3} - \ln 2$;

$\bar{x} = \dfrac{1}{A}\displaystyle\int_0^{\sqrt{3}} x\tan^{-1}x\, dx = \dfrac{1}{A}\left(\left[\dfrac{1}{2}x^2\tan^{-1}x\right]_0^{\sqrt{3}} - \dfrac{1}{2}\int_0^{\sqrt{3}} \dfrac{x^2}{1+x^2}\, dx\right) = \dfrac{1}{A}\left[\dfrac{\pi}{2} - \left[\dfrac{1}{2}(x - \tan^{-1}x)\right]_0^{\sqrt{3}}\right]$

$= \dfrac{1}{A}\left(\dfrac{\pi}{2} - \dfrac{\sqrt{3}}{2} + \dfrac{\pi}{6}\right) = \dfrac{1}{A}\left(\dfrac{\pi}{2} - \dfrac{\sqrt{3}}{2}\right) \cong 1.10$

49. (a) $\dfrac{dx}{dt} = kx(N - x) \Rightarrow \displaystyle\int \dfrac{dx}{x(N-x)} = \int k\, dt \Rightarrow \dfrac{1}{N}\int \dfrac{dx}{x} + \dfrac{1}{N}\int \dfrac{dx}{N-x} = \int k\, dt \Rightarrow \dfrac{1}{N}\ln\left|\dfrac{x}{N-x}\right| = kt + C$;

$k = \dfrac{1}{250}$, $N = 1000$, $t = 0$ and $x = 2 \Rightarrow \dfrac{1}{1000}\ln\left|\dfrac{2}{998}\right| = C \Rightarrow \dfrac{1}{1000}\ln\left|\dfrac{x}{1000 - x}\right| = \dfrac{t}{250} + \dfrac{1}{1000}\ln\left(\dfrac{1}{499}\right)$

$$\Rightarrow \ln\left|\frac{499x}{1000-x}\right| = 4t \Rightarrow \frac{499x}{1000-x} = e^{4t} \Rightarrow 499x = e^{4t}(1000-x) \Rightarrow (499+e^{4t})x = 1000e^{4t} \Rightarrow x = \frac{1000e^{4t}}{499+e^{4t}}$$

(b) $x = \frac{1}{2}N = 500 \Rightarrow 500 = \frac{1000e^{4t}}{499+e^{4t}} \Rightarrow 500\cdot499 + 500e^{4t} = 1000e^{4t} \Rightarrow e^{4t} = 499 \Rightarrow t = \frac{1}{4}\ln 499 \approx 1.55$ days

51. (a) $\displaystyle\int_0^1 \frac{x^4(x-1)^4}{x^2+1}\,dx = \int_0^1 \left(x^6 - 4x^5 + 5x^4 - 4x^2 + 4 - \frac{4}{x^2+1}\right)dx = \frac{22}{7} - \pi$

(b) $\dfrac{\frac{22}{7}-\pi}{\pi}\cdot 100\% \cong 0.04\%$

(c) The area is less than 0.003

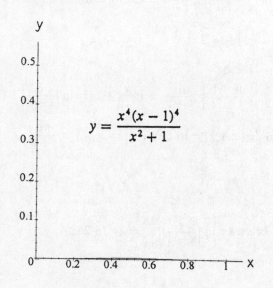
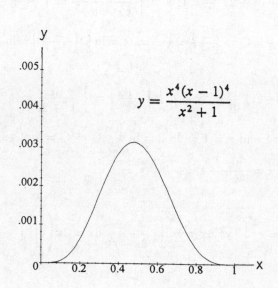

7.4 TRIGONOMETRIC SUBSTITUTIONS

1. $y = 3\tan\theta$, $-\frac{\pi}{2} < \theta < \frac{\pi}{2}$, $dy = \frac{3\,d\theta}{\cos^2\theta}$, $9 + y^2 = 9(1 + \tan^2\theta) = \frac{9}{\cos^2\theta} \Rightarrow \frac{1}{\sqrt{9+y^2}} = \frac{|\cos\theta|}{3} = \frac{\cos\theta}{3}$

$\left(\text{because } \cos\theta > 0 \text{ when } -\frac{\pi}{2} < \theta < \frac{\pi}{2}\right);$

$$\int \frac{dy}{\sqrt{9+y^2}} = 3\int \frac{\cos\theta\,d\theta}{3\cos^2\theta} = \int \frac{d\theta}{\cos\theta} = \ln|\sec\theta + \tan\theta| + C' = \ln\left|\frac{\sqrt{9+y^2}}{3} + \frac{y}{3}\right| + C' = \ln\left|\sqrt{9+y^2} + y\right| + C$$

3. $\displaystyle\int_{-2}^{2} \frac{dx}{4+x^2} = \left[\frac{1}{2}\tan^{-1}\frac{x}{2}\right]_{-2}^{2} = \frac{1}{2}\tan^{-1}1 - \frac{1}{2}\tan^{-1}(-1) = \left(\frac{1}{2}\right)\left(\frac{\pi}{4}\right) - \left(\frac{1}{2}\right)\left(-\frac{\pi}{4}\right) = \frac{\pi}{4}$

5. $\displaystyle\int_{0}^{3/2} \frac{dx}{\sqrt{9-x^2}} = \left[\sin^{-1}\frac{x}{3}\right]_{0}^{3/2} = \sin^{-1}\frac{1}{2} - \sin^{-1}0 = \frac{\pi}{6} - 0 = \frac{\pi}{6}$

7. $t = 5 \sin\theta$, $-\frac{\pi}{2} < \theta < \frac{\pi}{2}$, $dt = 5\cos\theta\, d\theta$, $\sqrt{25 - t^2} = 5\cos\theta$;

$$\int \sqrt{25 - t^2}\, dt = \int (5\cos\theta)(5\cos\theta)\, d\theta = 25 \int \cos^2\theta\, d\theta = 25 \int \frac{1 + \cos 2\theta}{2}\, d\theta = 25\left(\frac{\theta}{2} + \frac{\sin 2\theta}{4}\right) + C$$

$$= \frac{25}{2}(\theta + \sin\theta\cos\theta) + C = \frac{25}{2}\left[\sin^{-1}\left(\frac{t}{5}\right) + \left(\frac{t}{5}\right)\left(\frac{\sqrt{25 - t^2}}{5}\right)\right] + C = \frac{25}{2}\sin^{-1}\left(\frac{t}{5}\right) + \frac{t\sqrt{25 - t^2}}{2} + C$$

9. $x = \frac{7}{2}\sec\theta$, $0 < \theta < \frac{\pi}{2}$, $dx = \frac{7}{2}\sec\theta\tan\theta\, d\theta$, $\sqrt{4x^2 - 49} = \sqrt{49\sec^2\theta - 49} = 7\tan\theta$;

$$\int \frac{dx}{\sqrt{4x^2 - 49}} = \int \frac{\left(\frac{7}{2}\sec\theta\tan\theta\right) d\theta}{7\tan\theta} = \frac{1}{2}\int \sec\theta\, d\theta = \frac{1}{2}\ln|\sec\theta + \tan\theta| + C = \frac{1}{2}\ln\left|\frac{2x}{7} + \frac{\sqrt{4x^2 - 49}}{7}\right| + C$$

11. $y = 7\sec\theta$, $0 < \theta < \frac{\pi}{2}$, $dy = 7\sec\theta\tan\theta\, d\theta$, $\sqrt{y^2 - 49} = 7\tan\theta$;

$$\int \frac{\sqrt{y^2 - 49}}{y}\, dy = \int \frac{(7\tan\theta)(7\sec\theta\tan\theta)\, d\theta}{7\sec\theta} = 7\int \tan^2\theta\, d\theta = 7\int (\sec^2\theta - 1)\, d\theta = 7(\tan\theta - \theta) + C$$

$$= 7\left[\frac{\sqrt{y^2 - 49}}{7} - \sec^{-1}\left(\frac{y}{7}\right)\right] + C$$

13. $x = \sec\theta$, $0 < \theta < \frac{\pi}{2}$, $dx = \sec\theta\tan\theta\, d\theta$, $\sqrt{x^2 - 1} = \tan\theta$;

$$\int \frac{dx}{x^2\sqrt{x^2 - 1}} = \int \frac{\sec\theta\tan\theta\, d\theta}{\sec^2\theta\tan\theta} = \int \frac{d\theta}{\sec\theta} = \sin\theta + C = \tan\theta\cos\theta + C = \frac{\sqrt{x^2 - 1}}{x} + C$$

15. $x = 2\tan\theta$, $-\frac{\pi}{2} < \theta < \frac{\pi}{2}$, $dx = \frac{2\, d\theta}{\cos^2\theta}$, $\sqrt{x^2 + 4} = \frac{2}{\cos\theta}$;

$$\int \frac{x^3\, dx}{\sqrt{x^2 + 4}} = \int \frac{(8\tan^3\theta)(\cos\theta)\, d\theta}{\cos^2\theta} = 8\int \frac{\sin^3\theta\, d\theta}{\cos^4\theta} = 8\int \frac{(\cos^2\theta - 1)(-\sin\theta)\, d\theta}{\cos^4\theta};$$

$$[t = \cos\theta] \to 8\int \frac{t^2 - 1}{t^4}\, dt = 8\int \left(\frac{1}{t^2} - \frac{1}{t^4}\right) dt = 8\left(-\frac{1}{t} + \frac{1}{3t^3}\right) + C = 8\left(-\sec\theta + \frac{\sec^3\theta}{3}\right) + C$$

$$= 8\left(-\frac{\sqrt{x^2 + 4}}{2} + \frac{(x^2 + 4)^{3/2}}{8 \cdot 3}\right) + C = \frac{1}{3}(x^2 + 4)^{3/2} - 4\sqrt{x^2 + 4} + C$$

17. $w = 2\sin\theta$, $-\frac{\pi}{2} < \theta < \frac{\pi}{2}$, $dw = 2\cos\theta\, d\theta$, $\sqrt{4 - w^2} = 2\cos\theta$;

$$\int \frac{8\, dw}{w^2\sqrt{4 - w^2}} = \int \frac{8 \cdot 2\cos\theta\, d\theta}{4\sin^2\theta \cdot 2\cos\theta} = 2\int \frac{d\theta}{\sin^2\theta} = -2\cot\theta + C = \frac{-2\sqrt{4 - w^2}}{w} + C$$

19. $x = \sin\theta$, $0 \le \theta \le \frac{\pi}{3}$, $dx = \cos\theta\, d\theta$, $(1 - x^2)^{3/2} = \cos^3\theta$;

$$\int_0^{\sqrt{3}/2} \frac{4x^2\, dx}{(1 - x^2)^{3/2}} = \int_0^{\pi/3} \frac{4\sin^2\theta\cos\theta\, d\theta}{\cos^3\theta} = 4\int_0^{\pi/3} \left(\frac{1 - \cos^2\theta}{\cos^2\theta}\right) d\theta = 4\int_0^{\pi/3} (\sec^2\theta - 1)\, d\theta$$

$$= 4[\tan \theta - \theta]_0^{\pi/3} = 4\sqrt{3} - \frac{4\pi}{3}$$

21. $x = \sec \theta$, $0 < \theta < \frac{\pi}{2}$, $dx = \sec \theta \tan \theta \, d\theta$, $(x^2 - 1)^{3/2} = \tan^3 \theta$;

$$\int \frac{dx}{(x^2 - 1)^{3/2}} = \int \frac{\sec \theta \tan \theta \, d\theta}{\tan^3 \theta} = \int \frac{\cos \theta \, d\theta}{\sin^2 \theta} = -\frac{1}{\sin \theta} + C = -\left(\frac{1}{\tan \theta}\right)\left(\frac{1}{\cos \theta}\right) + C$$

$$= -\left(\frac{1}{\sqrt{x^2 - 1}}\right)(x) + C = -\frac{x}{\sqrt{x^2 - 1}} + C$$

23. $x = \sin \theta$, $-\frac{\pi}{2} < \theta < \frac{\pi}{2}$, $dx = \cos \theta \, d\theta$, $(1 - x^2)^{3/2} = \cos^3 \theta$;

$$\int \frac{(1 - x^2)^{3/2}}{x^6} \, dx = \int \frac{\cos^3 \theta \cdot \cos \theta \, d\theta}{\sin^6 \theta} = \int \cot^4 \theta \, \csc^2 \theta \, d\theta = -\frac{\cot^5 \theta}{5} + C = -\frac{1}{5}\left(\frac{\sqrt{1 - x^2}}{x}\right)^5 + C$$

25. $x = \frac{1}{2} \tan \theta$, $-\frac{\pi}{2} < \theta < \frac{\pi}{2}$, $dx = \frac{1}{2} \sec^2 \theta \, d\theta$, $(4x^2 + 1)^2 = \sec^4 \theta$;

$$\int \frac{8 \, dx}{(4x^2 + 1)^2} = \int \frac{8\left(\frac{1}{2} \sec^2 \theta\right) d\theta}{\sec^4 \theta} = 4 \int \cos^2 \theta \, d\theta = 2(\theta + \sin \theta \cos \theta) + C = 2(\theta + \tan \theta + \cos^2 \theta) + C$$

$$= 2 \tan^{-1} 2x + \frac{4x}{(4x^2 + 1)} + C$$

27. $v = \sin \theta$, $-\frac{\pi}{2} < \theta < \frac{\pi}{2}$, $dv = \cos \theta \, d\theta$, $(1 - v^2)^{5/2} = \cos^5 \theta$;

$$\int \frac{v^2 \, dv}{(1 - v^2)^{5/2}} = \int \frac{\sin^2 \theta \cos \theta \, d\theta}{\cos^5 \theta} = \int \tan^2 \theta \, \sec^2 \theta \, d\theta = \frac{\tan^3 \theta}{3} + C = \frac{1}{3}\left(\frac{v}{\sqrt{1 - v^2}}\right)^3 + C$$

29. Let $e^t = 3 \tan \theta$, $t = \ln(3 \tan \theta)$, $dt = \frac{\sec^2 \theta}{\tan \theta} \, d\theta$, $\sqrt{e^{2t} + 9} = \sqrt{9 \tan^2 \theta + 9} = 3 \sec \theta$;

$$\int_0^{\ln 4} \frac{e^t \, dt}{\sqrt{e^{2t} + 9}} = \int_{\tan^{-1}(1/3)}^{\tan^{-1}(4/3)} \frac{3 \tan \theta \cdot \sec^2 \theta \, d\theta}{\tan \theta \cdot 3 \sec \theta} = \int_{\tan^{-1}(1/3)}^{\tan^{-1}(4/3)} \sec \theta \, d\theta = \left[\ln |\sec \theta + \tan \theta|\right]_{\tan^{-1}(1/3)}^{\tan^{-1}(4/3)}$$

$$= \ln\left(\frac{5}{3} + \frac{4}{3}\right) - \ln\left(\frac{\sqrt{10}}{3} + \frac{1}{3}\right) = \ln 9 - \ln\left(1 + \sqrt{10}\right)$$

31. $\displaystyle\int_{1/12}^{1/4} \frac{2 \, dt}{\sqrt{t} + 4t\sqrt{t}}$; $\left[u = 2\sqrt{t}, \, du = \frac{1}{\sqrt{t}} \, dt\right] \to \displaystyle\int_{1/\sqrt{3}}^{1} \frac{2 \, du}{1 + u^2}$; $u = \tan \theta$, $\frac{\pi}{6} < \theta < \frac{\pi}{4}$, $du = \sec^2 \theta \, d\theta$, $1 + u^2 = \sec^2 \theta$;

$$\int_{1/\sqrt{3}}^{1} \frac{4 \, du}{u(1 + u^2)} = \int_{\pi/6}^{\pi/4} \frac{2 \sec^2 \theta \, d\theta}{\sec^2 \theta} = [2\theta]_{\pi/6}^{\pi/4} = 2\left(\frac{\pi}{4} - \frac{\pi}{6}\right) = \frac{\pi}{6}$$

33. $x = \sec\theta$, $dx = \sec\theta\tan\theta\,d\theta$, $\sqrt{x^2 - 1} = \sqrt{\sec^2\theta - 1} = \tan\theta$;

$$\int \frac{dx}{x\sqrt{x^2 - 1}} = \int \frac{\sec\theta\tan\theta\,d\theta}{\sec\theta\tan\theta} = \theta + C = \sec^{-1}|x| + C$$

35. $x = \sec\theta$, $dx = \sec\theta\tan\theta\,d\theta$, $\sqrt{x^2 - 1} = \sqrt{\sec^2\theta - 1} = \tan\theta$;

$$\int \frac{x\,dx}{\sqrt{x^2 - 1}} = \int \frac{\sec\theta\cdot\sec\theta\tan\theta\,d\theta}{\tan\theta} = \int \sec^2\theta\,d\theta = \tan\theta + C = \sqrt{x^2 - 1} + C$$

37. $x\dfrac{dy}{dx} = \sqrt{x^2 - 4}$; $dy = \sqrt{x^2 - 4}\dfrac{dx}{x}$; $y = \displaystyle\int \frac{\sqrt{x^2 - 4}}{x}\,dx$; $\begin{bmatrix} x = 2\sec\theta,\ 0 < \theta < \frac{\pi}{2} \\ dx = 2\sec\theta\tan\theta\,d\theta \\ \sqrt{x^2 - 4} = 2\tan\theta \end{bmatrix}$

$$\to y = \int \frac{(2\tan\theta)(2\sec\theta\tan\theta)\,d\theta}{2\sec\theta} = 2\int \tan^2\theta\,d\theta = 2\int \left(\sec^2\theta - 1\right)d\theta = 2(\tan\theta - \theta) + C$$

$$= 2\left[\frac{\sqrt{x^2 - 4}}{2} - \sec^{-1}\left(\frac{x}{2}\right)\right] + C;\ x = 2 \text{ and } y = 0 \Rightarrow 0 = 0 + C \Rightarrow C = 0 \Rightarrow y = 2\left[\frac{\sqrt{x^2 - 4}}{2} - \sec^{-1}\frac{x}{2}\right]$$

39. $(x^2 + 4)\dfrac{dy}{dx} = 3$, $dy = \dfrac{3\,dx}{x^2 + 4}$; $y = 3\displaystyle\int \frac{dx}{x^2 + 4} = \frac{3}{2}\tan^{-1}\frac{x}{2} + C$; $x = 2$ and $y = 0 \Rightarrow 0 = \frac{3}{2}\tan^{-1}1 + C$

$$\Rightarrow C = -\frac{3\pi}{8} \Rightarrow y = \frac{3}{2}\tan^{-1}\left(\frac{x}{2}\right) - \frac{3\pi}{8}$$

41. $A = \displaystyle\int_0^3 \frac{\sqrt{9 - x^2}}{3}\,dx$; $x = 3\sin\theta$, $0 \le \theta \le \frac{\pi}{2}$, $dx = 3\cos\theta\,d\theta$, $\sqrt{9 - x^2} = \sqrt{9 - 9\sin^2\theta} = 3\cos\theta$;

$$A = \int_0^{\pi/2} \frac{3\cos\theta\cdot 3\cos\theta\,d\theta}{3} = 3\int_0^{\pi/2} \cos^2\theta\,d\theta = \frac{3}{2}\left[\theta + \sin\theta\cos\theta\right]_0^{\pi/2} = \frac{3\pi}{4}$$

43. $\displaystyle\int \frac{dx}{1 - \sin x} = \int \frac{\left(\dfrac{2\,dz}{1 + z^2}\right)}{1 - \left(\dfrac{2z}{1 + z^2}\right)} = \int \frac{2\,dz}{(1 - z)^2} = \frac{2}{1 - z} + C = \frac{2}{1 - \tan\left(\frac{x}{2}\right)} + C$

45. $\displaystyle\int_0^{\pi/2} \frac{dx}{1 + \sin x} = \int_0^1 \frac{\left(\dfrac{2\,dz}{1 + z^2}\right)}{1 + \left(\dfrac{2z}{1 + z^2}\right)} = \int_0^1 \frac{2\,dz}{(1 + z)^2} = -\left[\frac{2}{1 + z}\right]_0^1 = -(1 - 2) = 1$

47. $\displaystyle\int_0^{\pi/2} \frac{d\theta}{2 + \cos\theta} = \int_0^1 \frac{\left(\dfrac{2\,dz}{1 + z^2}\right)}{2 + \left(\dfrac{1 - z^2}{1 + z^2}\right)} = \int_0^1 \frac{2\,dz}{2 + 2z^2 + 1 - z^2} = \int_0^1 \frac{2\,dz}{z^2 + 3} = \frac{2}{\sqrt{3}}\left[\tan^{-1}\frac{z}{\sqrt{3}}\right]_0^1 = \frac{2}{\sqrt{3}}\tan^{-1}\frac{1}{\sqrt{3}}$

$$= \frac{\pi}{3\sqrt{3}} = \frac{\sqrt{3}\pi}{9}$$

49. $\displaystyle\int \frac{dt}{\sin t - \cos t} = \int \frac{\left(\dfrac{2\,dz}{1+z^2}\right)}{\left(\dfrac{2z}{1+z^2} - \dfrac{1-z^2}{1+z^2}\right)} = \int \frac{2\,dz}{2z - 1 + z^2} = \int \frac{2\,dz}{(z+1)^2 - 2} = \frac{1}{\sqrt{2}} \ln \left| \frac{z+1-\sqrt{2}}{z+1+\sqrt{2}} \right| + C$

$\displaystyle = \frac{1}{\sqrt{2}} \ln \left| \frac{\tan\left(\frac{t}{2}\right) + 1 - \sqrt{2}}{\tan\left(\frac{t}{2}\right) + 1 + \sqrt{2}} \right| + C$

51. $\displaystyle\int \sec\theta\, d\theta = \int \frac{d\theta}{\cos\theta} = \int \frac{\left(\dfrac{2\,dz}{1+z^2}\right)}{\left(\dfrac{1-z^2}{1+z^2}\right)} = \int \frac{2\,dz}{1-z^2} = \int \frac{2\,dz}{(1+z)(1-z)} = \int \frac{dz}{1+z} + \int \frac{dz}{1-z}$

$\displaystyle = \ln|1+z| - \ln|1-z| + C = \ln \left| \frac{1+\tan\left(\frac{\theta}{2}\right)}{1-\tan\left(\frac{\theta}{2}\right)} \right| + C$

7.5 INTEGRAL TABLES AND *CAS*

1. $\displaystyle\int \frac{dx}{x\sqrt{x-3}} = \frac{2}{\sqrt{3}} \tan^{-1} \sqrt{\frac{x-3}{3}} + C$

   (We used FORMULA 13(a) with $a = 1$, $b = -3$)

3. $\displaystyle\int \frac{x\,dx}{\sqrt{x-2}} = \int \frac{(x-2)\,dx}{\sqrt{x-2}} + 2\int \frac{dx}{\sqrt{x-2}} = \int \left(\sqrt{x-2}\right)^1 dx + 2\int \left(\sqrt{x-2}\right)^{-1} dx$

   $\displaystyle = \left(\frac{2}{1}\right)\frac{\left(\sqrt{x-2}\right)^3}{3} + 2\left(\frac{2}{1}\right)\frac{\left(\sqrt{x-2}\right)^1}{1} = \sqrt{x-2}\left[\frac{2(x-2)}{3} + 4\right] + C$

   (We used FORMULA 11 with $a = 1$, $b = -2$, $n = 1$ and $a = 1$, $b = -2$, $n = -1$)

5. $\displaystyle\int x\sqrt{2x-3}\, dx = \frac{1}{2}\int (2x-3)\sqrt{2x-3}\, dx + \frac{3}{2}\int \sqrt{2x-3}\, dx = \frac{1}{2}\int \left(\sqrt{2x-3}\right)^3 dx + \frac{3}{2}\int \left(\sqrt{2x-3}\right)^1 dx$

   $\displaystyle = \left(\frac{1}{2}\right)\left(\frac{2}{2}\right)\frac{\left(\sqrt{2x-3}\right)^5}{5} + \left(\frac{3}{2}\right)\left(\frac{2}{2}\right)\frac{\left(\sqrt{2x-3}\right)^3}{3} + C = \frac{(2x-3)^{3/2}}{2}\left[\frac{2x-3}{5} + 1\right] + C = \frac{(2x-3)^{3/2}(x+1)}{5} + C$

   (We used FORMULA 11 with $a = 2$, $b = -3$, $n = 3$ and $a = 2$, $b = -3$, $n = 1$)

7. $\displaystyle\int \frac{\sqrt{9-4x}}{x^2}\, dx = -\frac{\sqrt{9-4x}}{x} + \frac{(-4)}{2}\int \frac{dx}{x\sqrt{9-4x}} + C$

   (We used FORMULA 14 with $a = -4$, $b = 9$)

   $\displaystyle = -\frac{\sqrt{9-4x}}{x} - 2\left(\frac{1}{\sqrt{9}}\right) \ln \left| \frac{\sqrt{9-4x} - \sqrt{9}}{\sqrt{9-4x} + \sqrt{9}} \right| + C$

   (We used FORMULA 13(b) with $a = -4$, $b = 9$)

$$= \frac{-\sqrt{9-4x}}{x} - \frac{2}{3} \ln \left| \frac{\sqrt{9-4x}-3}{\sqrt{9-4x}+3} \right| + C$$

9. $\int x\sqrt{4x-x^2}\, dx = \int x\sqrt{2\cdot 2x - x^2}\, dx = \dfrac{(x+2)(2x-3\cdot 2)\sqrt{2\cdot 2\cdot x - x^2}}{6} + \dfrac{2^3}{2}\sin^{-1}\left(\dfrac{x-2}{2}\right)+C$

$$= \frac{(x+2)(2x-6)\sqrt{4x-x^2}}{6} + 4\,\sin^{-1}\left(\frac{x-2}{2}\right)+C$$

(We used FORMULA 51 with a = 2)

11. $\int \dfrac{dx}{x\sqrt{7+x^2}} = \int \dfrac{dx}{x\sqrt{\left(\sqrt{7}\right)^2 + x^2}} = -\dfrac{1}{\sqrt{7}}\ln\left| \dfrac{\sqrt{7}+\sqrt{\left(\sqrt{7}\right)^2 + x^2}}{x} \right| + C = -\dfrac{1}{\sqrt{7}}\ln\left| \dfrac{\sqrt{7}+\sqrt{7+x^2}}{x} \right| + C$

(We used FORMULA 26 with $a = \sqrt{7}$)

13. $\int \dfrac{\sqrt{4-x^2}}{x}\, dx = \int \dfrac{\sqrt{2^2-x^2}}{x}\, dx = \sqrt{2^2-x^2} - 2\ln\left| \dfrac{2+\sqrt{2^2-x^2}}{x} \right| + C = \sqrt{4-x^2} - 2\ln\left| \dfrac{2+\sqrt{4-x^2}}{x} \right| + C$

(We used FORMULA 31 with a = 2)

15. $\int \sqrt{25-p^2}\, dp = \int \sqrt{5^2-p^2}\, dp = \dfrac{p}{2}\sqrt{5^2-p^2} + \dfrac{5^2}{2}\sin^{-1}\dfrac{p}{5} + C = \dfrac{p}{2}\sqrt{25-p^2} + \dfrac{25}{2}\sin^{-1}\dfrac{p}{5} + C$

(We used FORMULA 29 with a = 5)

17. $\int \dfrac{r^2}{\sqrt{4-r^2}}\, dr = \int \dfrac{r^2}{\sqrt{2^2-r^2}}\, dr = \dfrac{2^2}{2}\sin^{-1}\left(\dfrac{r}{2}\right) - \dfrac{1}{2}r\sqrt{2^2-r^2} + C = 2\sin^{-1}\left(\dfrac{r}{2}\right) - \dfrac{1}{2}r\sqrt{4-r^2} + C$

(We used FORMULA 33 with a = 2)

19. $\int \dfrac{d\theta}{5+4\sin 2\theta} = \dfrac{-2}{2\sqrt{25-16}}\tan^{-1}\left[\sqrt{\dfrac{5-4}{5+4}}\tan\left(\dfrac{\pi}{4} - \dfrac{2\theta}{2}\right) \right] + C = -\dfrac{1}{3}\tan^{-1}\left[\dfrac{1}{3}\tan\left(\dfrac{\pi}{4} - \theta\right) \right] + C$

(We used FORMULA 70 with b = 5, c = 4, a = 2)

21. $\int e^{2t}\cos 3t\, dt = \dfrac{e^{2t}}{2^2+3^2}(2\cos 3t + 3\sin 3t) + C = \dfrac{e^{2t}}{13}(2\cos 3t + 3\sin 3t) + C$

(We used FORMULA 108 with a = 2, b = 3)

23. $\int x\cos^{-1}x\, dx = \int x^1\cos^{-1}x\, dx = \dfrac{x^{1+1}}{1+1}\cos^{-1}x + \dfrac{1}{1+1}\int \dfrac{x^{1+1}\, dx}{\sqrt{1-x^2}} = \dfrac{x^2}{2}\cos^{-1}x + \dfrac{1}{2}\int \dfrac{x^2\, dx}{\sqrt{1-x^2}}$

(We used FORMULA 100 with a = 1, n = 1)

$$= \frac{x^2}{2}\cos^{-1}x + \frac{1}{2}\left(\frac{1}{2}\sin^{-1}x\right) - \frac{1}{2}\left(\frac{1}{2}x\sqrt{1-x^2}\right) + C = \frac{x^2}{2}\cos^{-1}x + \frac{1}{4}\sin^{-1}x - \frac{1}{4}x\sqrt{1-x^2} + C$$

(We used FORMULA 33 with a = 1)

25. $\int \dfrac{ds}{\left(9-s^2\right)^2} = \int \dfrac{ds}{\left(3^3-s^2\right)^2} = \dfrac{s}{2\cdot 3^2 \cdot \left(3^2-s^2\right)} + \dfrac{1}{2\cdot 3^2} \int \dfrac{ds}{3^2-s^2}$

(We used FORMULA 19 with a = 3)

$= \dfrac{s}{18\left(9-s^2\right)} + \dfrac{1}{18}\left(\dfrac{1}{2\cdot 3}\ln\left|\dfrac{s+3}{s-3}\right|\right) + C = \dfrac{s}{18\left(9-s^2\right)} + \dfrac{1}{108}\ln\left|\dfrac{s+3}{s-3}\right| + C$

(We used FORMULA 18 with a = 3)

27. $\int \dfrac{\sqrt{4x+9}}{x^2}\,dx = -\dfrac{\sqrt{4x+9}}{x} + \dfrac{4}{2}\int \dfrac{dx}{x\sqrt{4x+9}}$

(We used FORMULA 14 with a = 4, b = 9)

$= -\dfrac{\sqrt{4x+9}}{x} + 2\left(\dfrac{1}{\sqrt{9}}\ln\left|\dfrac{\sqrt{4x+9}-\sqrt{9}}{\sqrt{4x+9}+\sqrt{9}}\right|\right) + C = -\dfrac{\sqrt{4x+9}}{x} + \dfrac{2}{3}\ln\left|\dfrac{\sqrt{4x+9}-3}{\sqrt{4x+9}+3}\right| + C$

(We used FORMULA 13(b) with a = 4, b = 9)

29. $\int \dfrac{\sqrt{3t-4}}{t}\,dt = 2\sqrt{3t-4} + (-4)\int \dfrac{dt}{t\sqrt{3t-4}}$

(We used FORMULA 12 with a = 3, b = −4)

$= 2\sqrt{3t-4} - 4\left(\dfrac{2}{\sqrt{4}}\tan^{-1}\sqrt{\dfrac{3t-4}{4}}\right) + C = 2\sqrt{3t-4} - 4\tan^{-1}\sqrt{\dfrac{3t-4}{4}} + C$

(We used FORMULA 13(a) with a = 3, b = −4)

31. $\int x^2 \tan^{-1}x\,dx = \dfrac{x^{2+1}}{2+1}\tan^{-1}x - \dfrac{1}{2+1}\int \dfrac{x^{2+1}}{1+x^2}\,dx = \dfrac{x^3}{3}\tan^{-1}x - \dfrac{1}{3}\int \dfrac{x^3}{1+x^2}\,dx$

(We used FORMULA 101 with a = 1, n = 2);

$\int \dfrac{x^3}{1+x^2}\,dx = \int x\,dx - \int \dfrac{x\,dx}{1+x^2} = \dfrac{x^2}{2} - \dfrac{1}{2}\ln\left(1+x^2\right) + C \Rightarrow \int x^2 \tan^{-1}x\,dx$

$= \dfrac{x^3}{3}\tan^{-1}x - \dfrac{x^2}{6} + \dfrac{1}{6}\ln\left(1+x^2\right) + C$

33. $\int \sin 3x \cos 2x\,dx = -\dfrac{\cos 5x}{10} - \dfrac{\cos x}{2} + C$

(We used FORMULA 62(a) with a = 3, b = 2)

35. $\int 8\sin 4t \sin \dfrac{t}{2}\,dx = \dfrac{8}{7}\sin\left(\dfrac{7t}{2}\right) - \dfrac{8}{9}\sin\left(\dfrac{9t}{2}\right) + C = 8\left[\dfrac{\sin\left(\dfrac{7t}{2}\right)}{7} - \dfrac{\sin\left(\dfrac{9t}{2}\right)}{9}\right] + C$

(We used FORMULA 62(b) with a = 4, b = $\frac{1}{2}$)

37. $\int \cos \dfrac{\theta}{3}\cos \dfrac{\theta}{4}\,d\theta = 6\sin\left(\dfrac{\theta}{12}\right) + \dfrac{6}{7}\sin\left(\dfrac{7\theta}{12}\right) + C$

(We used FORMULA 62(c) with a = $\frac{1}{3}$, b = $\frac{1}{4}$)

39. $\int \dfrac{x^3 + x + 1}{(x^2+1)^2}\,dx = \int \dfrac{x\,dx}{x^2+1} + \int \dfrac{dx}{(x^2+1)^2} = \dfrac{1}{2}\int \dfrac{d(x^2+1)}{x^2+1} + \int \dfrac{dx}{(x^2+1)^2}$

$= \dfrac{1}{2}\ln|x^2+1| + \dfrac{x}{2(1+x^2)} + \dfrac{1}{2}\tan^{-1}x + C$

(For the second integral we used FORMULA 17 with $a=1$)

41. $\displaystyle\int \sin^{-1}\sqrt{x}\,dx;\ \begin{bmatrix} u=\sqrt{x} \\ x=u^2 \\ dx=2u\,du \end{bmatrix} \rightarrow 2\int u^1 \sin^{-1}u\,du = 2\left(\dfrac{u^{1+1}}{1+1}\sin^{-1}u - \dfrac{1}{1+1}\int \dfrac{u^{1+1}}{\sqrt{1-u^2}}\,du\right)$

$= u^2\sin^{-1}u - \displaystyle\int \dfrac{u^2\,du}{\sqrt{1-u^2}}$

(We used FORMULA 99 with $a=1$, $n=1$)

$= u^2\sin^{-1}u - \left(\dfrac{1}{2}\sin^{-1}u - \dfrac{1}{2}u\sqrt{1-u^2}\right) + C = \left(u^2 - \dfrac{1}{2}\right)\sin^{-1}u + \dfrac{1}{2}u\sqrt{1-u^2} + C$

(We used FORMULA 33 with $a=1$)

$= \left(x - \dfrac{1}{2}\right)\sin^{-1}\sqrt{x} + \dfrac{1}{2}\sqrt{x-x^2} + C$

43. $\displaystyle\int \dfrac{\sqrt{x}}{\sqrt{1-x}}\,dx;\ \begin{bmatrix} u=\sqrt{x} \\ x=u^2 \\ dx=2u\,du \end{bmatrix} \rightarrow \int \dfrac{u\cdot 2u}{\sqrt{1-u^2}}\,du = 2\int \dfrac{u^2}{\sqrt{1-u^2}}\,du = 2\left(\dfrac{1}{2}\sin^{-1}u - \dfrac{1}{2}u\sqrt{1-u^2}\right) + C$

$= \sin^{-1}u - u\sqrt{1-u^2} + C$

(We used FORMULA 33 with $a=1$)

$= \sin^{-1}\sqrt{x} - \sqrt{x}\sqrt{1-x} + C = \sin^{-1}\sqrt{x} - \sqrt{x-x^2} + C$

45. $\displaystyle\int (\cot t)\sqrt{1-\sin^2 t}\,dt = \int \dfrac{\sqrt{1-\sin^2 t}\,(\cos t)\,dt}{\sin t};\ \begin{bmatrix} u=\sin t \\ du=\cos t\,dt \end{bmatrix} \rightarrow \int \dfrac{\sqrt{1-u^2}\,du}{u}$

$= \sqrt{1-u^2} - \ln\left|\dfrac{1+\sqrt{1-u^2}}{u}\right| + C$

(We used FORMULA 31 with $a=1$)

$= \sqrt{1-\sin^2 t} - \ln\left|\dfrac{1+\sqrt{1-\sin^2 t}}{\sin t}\right| + C$

47. $\displaystyle\int \dfrac{dy}{y\sqrt{3+(\ln y)^2}};\ \begin{bmatrix} u=\ln y \\ y=e^u \\ dy=e^u\,du \end{bmatrix} \rightarrow \int \dfrac{e^u\,du}{e^u\sqrt{3+u^2}} = \int \dfrac{du}{\sqrt{3+u^2}} = \ln\left|u+\sqrt{3+u^2}\right| + C$

$= \ln\left|\ln y + \sqrt{3+(\ln y)^2}\right| + C$

(We used FORMULA 20 with $a=\sqrt{3}$)

49. $\int \dfrac{3\,dr}{\sqrt{9r^2-1}}; \begin{bmatrix} u = 3r \\ du = 3\,dr \end{bmatrix} \to \int \dfrac{du}{\sqrt{u^2-1}} = \ln\left|u + \sqrt{u^2-1}\right| + C = \ln\left|3r + \sqrt{9r^2-1}\right| + C$

(We used FORMULA 36 with a = 1)

51. $\int \cos^{-1}\sqrt{x}\,dx; \begin{bmatrix} t = \sqrt{x} \\ x = t^2 \\ dx = 2t\,dt \end{bmatrix} \to 2\int t\cos^{-1}t\,dt = 2\left(\dfrac{t^2}{2}\cos^{-1}t + \dfrac{1}{2}\int \dfrac{t^2}{\sqrt{1-t^2}}\,dt\right) = t^2\cos^{-1}t + \int \dfrac{t^2}{\sqrt{1-t^2}}\,dt$

(We used FORMULA 100 with a = 1, n = 1)

$= t^2\cos^{-1}t + \dfrac{1}{2}\sin^{-1}t - \dfrac{1}{2}t\sqrt{1-t^2} + C$

(We used FORMULA 33 with a = 1)

$= x\cos^{-1}\sqrt{x} + \dfrac{1}{2}\sin^{-1}\sqrt{x} - \dfrac{1}{2}\sqrt{x}\sqrt{1-x} + C = x\cos^{-1}\sqrt{x} + \dfrac{1}{2}\sin^{-1}\sqrt{x} - \dfrac{1}{2}\sqrt{x-x^2} + C$

53. $\int \sin^5 2x\,dx = -\dfrac{\sin^4 2x\cos 2x}{5\cdot 2} + \dfrac{5-1}{5}\int \sin^3 2x\,dx = -\dfrac{\sin^4 2x\cos 2x}{10} + \dfrac{4}{5}\left[-\dfrac{\sin^2 2x\cos 2x}{3\cdot 2} + \dfrac{3-1}{3}\int \sin 2x\,dx\right]$

(We used FORMULA 60 with a = 2, n = 5 and a = 2, n = 3)

$= -\dfrac{\sin^4 2x\cos 2x}{10} - \dfrac{2}{15}\sin^2 2x\cos 2x + \dfrac{8}{15}\left(-\dfrac{1}{2}\right)\cos 2x + C = -\dfrac{\sin^4 2x\cos 2x}{10} - \dfrac{2\sin^2 2x\cos 2x}{15} - \dfrac{4\cos 2x}{15} + C$

55. $\int 8\cos^4 2\pi t\,dt = 8\left(\dfrac{\cos^3 2\pi t\sin 2\pi t}{4\cdot 2\pi} + \dfrac{4-1}{4}\int \cos^2 2\pi t\,dt\right)$

(We used FORMULA 61 with a = 2π, n = 4)

$= \dfrac{\cos^3 2\pi t\sin 2\pi t}{\pi} + 6\left[\dfrac{t}{2} + \dfrac{\sin(2\cdot 2\pi\cdot t)}{4\cdot 2\pi}\right] + C$

(We used FORMULA 59 with a = 2π)

$= \dfrac{\cos^3 2\pi t\sin 2\pi t}{\pi} + 3t + \dfrac{3\sin 4\pi t}{4\pi} + C = \dfrac{\cos^3 2\pi t\sin 2\pi t}{\pi} + \dfrac{3\cos 2\pi t\sin 2\pi t}{2\pi} + 3t + C$

57. $\int \sin^2 2\theta\cos^3 2\theta\,d\theta = \dfrac{\sin^3 2\theta\cos^2 2\theta}{2(2+3)} + \dfrac{3-1}{3+2}\int \sin^2 2\theta\cos 2\theta\,d\theta$

(We used FORMULA 69 with a = 2, m = 3, n = 2)

$= \dfrac{\sin^3 2\theta\cos^2 2\theta}{10} + \dfrac{2}{5}\int \sin^2 2\theta\cos 2\theta\,d\theta = \dfrac{\sin^3 2\theta\cos^2 2\theta}{10} + \dfrac{2}{5}\left[\dfrac{1}{2}\int \sin^2 2\theta\,d(\sin 2\theta)\right] = \dfrac{\sin^3 2\theta\cos^2 2\theta}{10} + \dfrac{\sin^3 2\theta}{15} + C$

59. $\int 2\sin^2 t\sec^4 t\,dt = \int 2\sin^2 t\cos^{-4}t\,dt = 2\left(-\dfrac{\sin t\cos^{-3}t}{2-4} + \dfrac{2-1}{2-4}\int \cos^{-4}t\,dt\right)$

(We used FORMULA 68 with a = 1, n = 2, m = −4)

$= \sin t\cos^{-3}t - \int \cos^{-4}t\,dt = \sin t\cos^{-3}t - \int \sec^4 t\,dt = \sin t\cos^{-3}t - \left(\dfrac{\sec^2 t\tan t}{4-1} + \dfrac{4-2}{4-1}\int \sec^2 t\,dt\right)$

(We used FORMULA 92 with a = 1, n = 4)

$= \sin t\cos^{-3}t - \left(\dfrac{\sec^2 t\tan t}{3}\right) - \dfrac{2}{3}\tan t + C = \dfrac{2}{3}\sec^2 t\tan t - \dfrac{2}{3}\tan t + C = \dfrac{2}{3}\tan t\left(\sec^2 t - 1\right) + C$

$$= \tfrac{2}{3} \tan^3 t + C$$

An easy way to find the integral using substitution:

$$\int 2 \sin^2 t \cos^{-4} t \; dt = \int 2 \tan^2 t \sec^2 t \; dt = \int 2 \tan^2 t \; d(\tan t) = \tfrac{2}{3} \tan^3 t + C$$

61. $\displaystyle \int 4 \tan^3 2x \; dx = 4 \left(\frac{\tan^2 2x}{2 \cdot 2} - \int \tan 2x \; dx \right) = \tan^2 2x - 4 \int \tan 2x \; dx$

    (We used FORMULA 86 with n = 3, a = 2)

    $= \tan^2 2x - \tfrac{4}{2} \ln |\sec 2x| + C = \tan^2 2x - 2 \ln |\sec 2x| + C$

63. $\displaystyle \int 8 \cot^4 t \; dt = 8 \left(-\frac{\cot^3 t}{3} - \int \cot^2 t \; dt \right)$

    (We used FORMULA 87 with a = 1, n = 4)

    $= 8 \left(-\tfrac{1}{3} \cot^3 t + \cot t + t \right) + C$

    (We used FORMULA 85 with a = 1)

65. $\displaystyle \int 2 \sec^3 \pi x \; dx = 2 \left[\frac{\sec \pi x \tan \pi x}{\pi (3-1)} + \frac{3-2}{3-1} \int \sec \pi x \; dx \right]$

    (We used FORMULA 92 with n = 3, a = π)

    $= \tfrac{1}{\pi} \sec \pi x \tan \pi x + \tfrac{1}{\pi} \ln |\sec \pi x + \tan \pi x| + C$

    (We used FORMULA 88 with a = π)

67. $\displaystyle \int 3 \sec^4 3x \; dx = 3 \left[\frac{\sec^2 3x \tan 3x}{3(4-1)} + \frac{4-2}{4-1} \int \sec^2 3x \; dx \right]$

    (We used FORMULA 92 with n = 4, a = 3)

    $= \dfrac{\sec^2 3x \tan 3x}{3} + \dfrac{2}{3} \tan 3x + C$

    (We used FORMULA 90 with a = 3)

69. $\displaystyle \int \csc^5 x \; dx = -\frac{\csc^3 x \cot x}{5-1} + \frac{5-2}{5-1} \int \csc^3 x \; dx = -\frac{\csc^3 x \cot x}{4} + \frac{3}{4} \left(-\frac{\csc x \cot x}{3-1} + \frac{3-2}{3-1} \int \csc x \; dx \right)$

    (We used FORMULA 93 with n = 5, a = 1 and n = 3, a = 1)

    $= -\tfrac{1}{4} \csc^3 x \cot x - \tfrac{3}{8} \csc x \cot x - \tfrac{3}{8} \ln |\csc x + \cot x| + C$

    (We used FORMULA 89 with a = 1)

71. $\displaystyle \int 16 x^3 (\ln x)^2 \; dx = 16 \left[\frac{x^4 (\ln x)^2}{4} - \frac{2}{4} \int x^3 \ln x \; dx \right] = 16 \left[\frac{x^4 (\ln x)^2}{4} - \frac{1}{2} \left[\frac{x^4 (\ln x)}{4} - \frac{1}{4} \int x^3 \; dx \right] \right]$

    (We used FORMULA 110 with a = 1, n = 3, m = 2 and a = 1, n = 3, m = 1)

    $= 16 \left(\dfrac{x^4 (\ln x)^2}{4} - \dfrac{x^4 (\ln x)}{8} + \dfrac{x^4}{32} \right) + C = 4 x^4 (\ln x)^2 - 2 x^4 \ln x + \dfrac{x^4}{2} + C$

73. $\displaystyle\int xe^{3x}\,dx = \frac{e^{3x}}{3^2}(3x-1)+C = \frac{e^{3x}}{9}(3x-1)+C$

(We used FORMULA 104 with $a=3$)

75. $\displaystyle\int x^3e^{x/2}\,dx = 2x^3e^{x/2}-3\cdot 2\int x^2e^{x/2}\,dx = 2x^3e^{x/2}-6\left(2x^2e^{x/2}-2\cdot 2\int xe^{x/2}\,dx\right)$

$\displaystyle = 2x^3e^{x/2}-12x^2e^{x/2}+24\cdot 4e^{x/2}\left(\frac{x}{2}-1\right)+C = 2x^3e^{x/2}-12x^2e^{x/2}+96e^{x/2}\left(\frac{x}{2}-1\right)+C$

$\left(\text{We used FORMULA 105 with } a=\frac{1}{2} \text{ twice and FORMULA 104 with } a=\frac{1}{2}\right)$

77. $\displaystyle\int x^2 2^x\,dx = \frac{x^2 2^x}{\ln 2}-\frac{2}{\ln 2}\int x2^x\,dx = \frac{x^2 2^x}{\ln 2}-\frac{2}{\ln 2}\left(\frac{x2^x}{\ln 2}-\frac{1}{\ln 2}\int 2^x\,dx\right) = \frac{x^2 2^x}{\ln 2}-\frac{2}{\ln 2}\left[\frac{x2^x}{\ln 2}-\frac{2^x}{(\ln 2)^2}\right]+C$

(We used FORMULA 106 with $a=1$, $b=2$)

79. $\displaystyle\int x\pi^x\,dx = \frac{x\pi^x}{\ln \pi}-\frac{1}{\ln \pi}\int \pi^x\,dx = \frac{x\pi^x}{\ln \pi}-\frac{1}{\ln \pi}\left(\frac{\pi^x}{\ln \pi}\right)+C = \frac{x\pi^x}{\ln \pi}-\frac{\pi^x}{\ln \pi}-\frac{\pi^x}{(\ln \pi)^2}+C$

(We used FORMULA 106 with $n=1$, $b=\pi$, $a=1$)

81. $\displaystyle\int e^t \sec^3(e^t-t)\,dt;\ \begin{bmatrix}x=e^t-1\\dx=e^t\,dt\end{bmatrix}\rightarrow \int \sec^3 x\,dx = \frac{\sec x \tan x}{3-1}+\frac{3-2}{3-1}\int \sec x\,dx$

(We used FORMULA 92 with $a=1$, $n=3$)

$\displaystyle = \frac{\sec x \tan x}{2}+\frac{1}{2}\ln|\sec x+\tan x|+C = \frac{1}{2}\left[\sec(e^t-1)\tan(e^t-1)+\ln\left|\sec(e^t-1)+\tan(e^t-1)\right|\right]+C$

83. $\displaystyle\int_0^1 2\sqrt{x^2+1}\,dx;\ [x=\tan t]\rightarrow 2\int_0^{\pi/4}\sec t\cdot\sec^2 t\,dt = 2\int_0^{\pi/4}\sec^3 t\,dt = 2\left[\left[\frac{\sec t\cdot\tan t}{3-1}\right]_0^{\pi/4}+\frac{3-2}{3-1}\int_0^{\pi/4}\sec t\,dt\right]$

(We used FORMULA 92 with $n=3$, $a=1$)

$\displaystyle = \left[\sec t\cdot\tan t+\ln|\sec t+\tan t|\right]_0^{\pi/4} = \sqrt{2}+\ln\left(\sqrt{2}+1\right)$

85. $\displaystyle\int_1^2 \frac{(r^2-1)^{3/2}}{r}\,dr;\ [r=\sec\theta]\rightarrow \int_0^{\pi/3}\frac{\tan^3\theta}{\sec\theta}(\sec\theta\tan\theta)\,d\theta = \int_0^{\pi/3}\tan^4\theta\,d\theta = \left[\frac{\tan^3\theta}{4-1}\right]_0^{\pi/3}-\int_0^{\pi/3}\tan^2\theta\,d\theta$

$\displaystyle = \left[\frac{\tan^3\theta}{3}-\tan\theta+\theta\right]_0^{\pi/3} = \frac{3\sqrt{3}}{3}-\sqrt{3}+\frac{\pi}{3} = \frac{\pi}{3}$

(We used FORMULA 86 with $a=1$, $n=4$ and FORMULA 84 with $a=1$)

87. $\displaystyle\int \frac{1}{8}\sinh^5 3x\,dx = \frac{1}{8}\left(\frac{\sinh^4 3x\cosh 3x}{5\cdot 3}-\frac{5-1}{5}\int \sinh^3 3x\,dx\right)$

$\displaystyle = \frac{\sinh^4 3x\cosh 3x}{120}-\frac{1}{10}\left(\frac{\sinh 3x\cosh 3x}{3\cdot 3}-\frac{3-1}{3}\int \sinh 3x\,dx\right)$

(We used FORMULA 117 with $a=3$, $n=5$ and $a=1$, $n=3$)

$$= \frac{\sinh^4 3x \cosh 3x}{120} - \frac{\sinh 3x \cosh 3x}{90} + \frac{2}{30}\left(\frac{1}{3}\cosh 3x\right) + C$$

$$= \frac{1}{120}\sinh^4 3x \cosh 3x - \frac{1}{90}\sinh 3x \cosh 3x + \frac{2}{90}\cosh 3x + C$$

89. $\displaystyle\int x^2 \cosh 3x\,dx = \frac{x^2}{3}\sinh 3x - \frac{2}{3}\int x \sinh 3x\,dx = \frac{x^2}{3}\sinh 3x - \frac{2}{3}\left(\frac{x}{3}\cosh 3x - \frac{1}{3}\int \cosh 3x\,dx\right)$

  (We used FORMULA 122 with $a = 3$, $n = 2$ and FORMULA 121 with $a = 3$, $n = 1$)

  $= \displaystyle\frac{x^2}{3}\sinh 3x - \frac{2x}{9}\cosh 3x + \frac{2}{27}\sinh 3x + C$

91. $\displaystyle\int \text{sech}^7 x \tanh x\,dx = -\frac{\text{sech}^7 x}{7} + C$

  (We used FORMULA 135 with $a = 1$, $n = 7$)

93. $u = ax + b \Rightarrow x = \dfrac{u - b}{a} \Rightarrow dx = \dfrac{du}{a}$;

  $\displaystyle\int \frac{x\,dx}{(ax + b)^2} = \int \frac{(u - b)}{au^2}\frac{du}{a} = \frac{1}{a^2}\int\left(\frac{1}{u} - \frac{b}{u^2}\right)du = \frac{1}{a^2}\left[\ln|u| + \frac{b}{u}\right] + C = \frac{1}{a^2}\left[\ln|ax + b| + \frac{b}{ax + b}\right] + C$

95. $x = a\sin\theta \Rightarrow a^2 - x^2 = a^2\cos^2\theta \Rightarrow -2x\,dx = -2a^2\cos\theta\sin\theta\,d\theta \Rightarrow dx = a\cos\theta\,d\theta$;

  $\displaystyle\int\sqrt{a^2 - x^2}\,dx = \int a\cos\theta(a\cos\theta)\,d\theta = a^2\int\cos^2\theta\,d\theta = \frac{a^2}{2}\int(1 + \cos 2\theta)\,d\theta = \frac{a^2}{2}\left(\theta + \frac{\sin 2\theta}{2}\right) + C$

  $= \displaystyle\frac{a^2}{2}(\theta + \cos\theta\sin\theta) + C = \frac{a^2}{2}\left(\theta + \sqrt{1 - \sin^2\theta}\cdot\sin\theta\right) + C = \frac{a^2}{2}\left(\sin^{-1}\frac{x}{a} + \frac{\sqrt{a^2 - x^2}}{a}\cdot\frac{x}{a}\right) + C$

  $= \displaystyle\frac{a^2}{2}\sin^{-1}\frac{x}{a} + \frac{x}{2}\sqrt{a^2 - x^2} + C$

97. $\displaystyle\int x^n \sin ax\,dx = -\int x^n\left(\frac{1}{a}\right)d(\cos ax) = (\cos ax)x^n\left(-\frac{1}{a}\right) + \frac{1}{a}\int \cos ax\cdot nx^{n-1}\,dx$

  $= \displaystyle -\frac{x^n}{a}\cos ax + \frac{n}{a}\int x^{n-1}\cos ax\,dx$

  $\left(\text{We used integration by parts }\displaystyle\int u\,dv = uv - \int v\,du \text{ with } u = x^n,\ v = -\frac{1}{a}\cos ax\right)$

99. $\displaystyle\int x^n \sin^{-1} ax\,dx = \int \sin^{-1} ax\,d\left(\frac{x^{n+1}}{n + 1}\right) = \frac{x^{n+1}}{n + 1}\sin^{-1} ax - \int\left(\frac{x^{n+1}}{n + 1}\right)\frac{a}{\sqrt{1 - (ax)^2}}\,dx$

  $= \displaystyle\frac{x^{n+1}}{n + 1}\sin^{-1} ax - \frac{a}{n + 1}\int \frac{x^{n+1}\,dx}{\sqrt{1 - a^2 x^2}},\ n \ne -1$

  $\left(\text{We used integration by parts }\displaystyle\int u\,dv = uv - \int v\,du \text{ with } u = \sin^{-1} ax,\ v = \frac{x^{n+1}}{n + 1}\right)$

101. $S = \displaystyle\int_0^{\sqrt{2}} 2\pi y \sqrt{1 + (y')^2}\, dx = 2\pi \int_0^{\sqrt{2}} \sqrt{x^2 + 2}\,\sqrt{1 + \dfrac{x^2}{x^2 + 2}}\, dx$

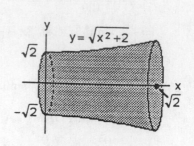

$= 2\sqrt{2}\pi \displaystyle\int_0^{\sqrt{2}} \sqrt{x^2 + 1}\, dx = 2\sqrt{2}\pi \left[\dfrac{x\sqrt{x^2+1}}{2} + \dfrac{1}{2}\ln\left|x + \sqrt{x^2+1}\right|\right]_0^{\sqrt{2}}$

(We used FORMULA 21 with $a = 1$)

$= \sqrt{2}\pi\left[\sqrt{6} + \ln\left(\sqrt{2} + \sqrt{3}\right)\right] = 2\pi\sqrt{3} + \pi\sqrt{2}\ln\left(\sqrt{2} + \sqrt{3}\right)$

103. $A = \displaystyle\int_0^3 \dfrac{dx}{\sqrt{x+1}} = \left[2\sqrt{x+1}\right]_0^3 = 2;\ \ \bar{x} = \dfrac{1}{A}\int_0^3 \dfrac{x\, dx}{\sqrt{x+1}}$

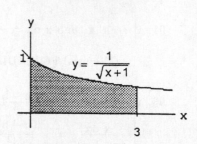

$= \dfrac{1}{A}\displaystyle\int_0^3 \sqrt{x+1}\, dx - \dfrac{1}{A}\int_0^3 \dfrac{dx}{\sqrt{x+1}} = \dfrac{1}{2}\cdot\dfrac{2}{3}\left[(x+1)^{3/2}\right]_0^3 - 1 = \dfrac{4}{3};$

(We used FORMULA 11 with $a = 1$, $b = 1$, $n = 1$ and $a = 1$, $b = 1$, $n = -1$)

$\bar{y} = \dfrac{1}{2A}\displaystyle\int_0^3 \dfrac{dx}{x+1} = \dfrac{1}{4}\left[\ln(x+1)\right]_0^3 = \dfrac{1}{4}\ln 4 = \dfrac{1}{2}\ln 2 = \ln\sqrt{2}$

105. $S = 2\pi \displaystyle\int_{-1}^1 x^2\sqrt{1 + 4x^2}\, dx;\ \begin{bmatrix} u = 2x \\ du = 2\, dx \end{bmatrix} \to \dfrac{\pi}{4}\int_{-2}^2 u^2\sqrt{1 + u^2}\, du$

$= \dfrac{\pi}{4}\left[\dfrac{u}{8}(1 + 2u^2)\sqrt{1 + u^2} - \dfrac{1}{8}\ln\left(u + \sqrt{1 + u^2}\right)\right]_{-2}^2$

(We used FORMULA 22 with $a = 1$)

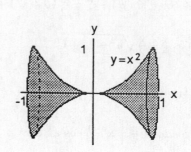

$= \dfrac{\pi}{4}\left[\dfrac{2}{8}(1 + 2\cdot 4)\sqrt{1 + 4} - \dfrac{1}{8}\ln\left(2 + \sqrt{1 + 4}\right)\right.$

$\left. + \dfrac{2}{8}(1 + 2\cdot 4)\sqrt{1 + 4} + \dfrac{1}{8}\ln\left(-2 + \sqrt{1 + 4}\right)\right]$

$= \dfrac{\pi}{4}\left[\dfrac{9}{2}\sqrt{5} - \dfrac{1}{8}\ln\left(\dfrac{2 + \sqrt{5}}{-2 + \sqrt{5}}\right)\right] \approx 7.62$

107. The integrand $f(x) = \sqrt{x - x^2}$ is nonnegative, so the integral is maximized by integrating over the function's entire domain, which runs from $x = 0$ to $x = 1$

$\Rightarrow \displaystyle\int_0^1 \sqrt{x - x^2}\, dx = \int_0^1 \sqrt{2\cdot\tfrac{1}{2}x - x^2}\, dx = \left[\dfrac{\left(x - \tfrac{1}{2}\right)}{2}\sqrt{2\cdot\tfrac{1}{2}x - x^2} + \dfrac{\left(\tfrac{1}{2}\right)^2}{2}\sin^{-1}\left(\dfrac{x - \tfrac{1}{2}}{\tfrac{1}{2}}\right)\right]_0^1$

$\left(\text{We used FORMULA 48 with } a = \tfrac{1}{2}\right)$

$= \left[\dfrac{\left(x - \tfrac{1}{2}\right)}{2}\sqrt{x - x^2} + \dfrac{1}{8}\sin^{-1}(2x - 1)\right]_0^1 = \dfrac{1}{8}\cdot\dfrac{\pi}{2} - \dfrac{1}{8}\left(-\dfrac{\pi}{2}\right) = \dfrac{\pi}{8}$

CAS EXPLORATIONS

For MAPLE use the int(f(x),x) command, and for MATHEMATICA use the command Integrate[f(x),x], as discussed in the text.

109. (e) $\int x^n \ln x \, dx = \dfrac{x^{n+1} \ln x}{n+1} - \dfrac{1}{n+1} \int x^n \, dx, \; n \neq -1$

(We used FORMULA 110 with a = 1, m = 1)

$= \dfrac{x^{n+1} \ln x}{n+1} - \dfrac{x^{n+1}}{(n+1)^2} + C = \dfrac{x^{n+1}}{n+1}\left(\ln x - \dfrac{1}{n+1}\right) + C$

111. (a) Neither MAPLE nor MATHEMATICA can find this integral for arbitrary n.

(b) MAPLE and MATHEMATICA get stuck at about n = 5.

(c) Let $x = \dfrac{\pi}{2} - u \Rightarrow dx = -du; \; x = 0 \Rightarrow u = \dfrac{\pi}{2}, \; x = \dfrac{\pi}{2} \Rightarrow u = 0;$

$$I = \int_0^{\pi/2} \frac{\sin^n x \, dx}{\sin^n x + \cos^n x} = \int_{\pi/2}^0 \frac{-\sin^n\left(\frac{\pi}{2} - u\right) du}{\sin^n\left(\frac{\pi}{2} - u\right) + \cos^n\left(\frac{\pi}{2} - u\right)} = \int_0^{\pi/2} \frac{\cos^n u \, du}{\cos^n u + \sin^n u} = \int_0^{\pi/2} \frac{\cos^n x \, dx}{\cos^n x + \sin^n x}$$

$$\Rightarrow I + I = \int_0^{\pi/2} \left(\frac{\sin^n x + \cos^n x}{\sin^n x + \cos^n x}\right) dx = \int_0^{\pi/2} dx = \frac{\pi}{2} \Rightarrow I = \frac{\pi}{4}$$

7.6 IMPROPER INTEGRALS

1. $\displaystyle\int_0^\infty \frac{dx}{x^2+1} = \lim_{b\to\infty} \int_0^b \frac{dx}{x^2+1} = \lim_{b\to\infty} \left[\tan^{-1} x\right]_0^b = \lim_{b\to\infty} \left(\tan^{-1} b - \tan^{-1} 0\right) = \frac{\pi}{2} - 0 = \frac{\pi}{2}$

3. $\displaystyle\int_0^1 \frac{dx}{\sqrt{x}} = \lim_{b\to 0^+} \int_b^1 x^{-1/2} \, dx = \lim_{b\to 0^+} \left[2x^{1/2}\right]_b^1 = \lim_{b\to 0^+} \left(2 - 2\sqrt{b}\right) = 2 - 0 = 2$

5. $\displaystyle\int_{-1}^1 \frac{dx}{x^{2/3}} = \int_{-1}^0 \frac{dx}{x^{2/3}} + \int_0^1 \frac{dx}{x^{2/3}} = \lim_{b\to 0^-} \left[3x^{1/3}\right]_{-1}^b + \lim_{c\to 0^+} \left[3x^{1/3}\right]_c^1$

$= \lim_{b\to 0^-} \left[3b^{1/3} - 3(-1)^{1/3}\right] + \lim_{c\to 0^+} \left[3(1)^{1/3} - 3c^{1/3}\right] = (0+3) + (3-0) = 6$

7. $\displaystyle\int_0^1 \frac{dx}{\sqrt{1-x^2}} = \lim_{b\to 1^-} \left[\sin^{-1} x\right]_0^b = \lim_{b\to 1^-} \left(\sin^{-1} b - \sin^{-1} 0\right) = \frac{\pi}{2} - 0 = \frac{\pi}{2}$

9. $\displaystyle\int_{-\infty}^{-2} \frac{2\,dx}{x^2+1} = \int_{-\infty}^{-2} \frac{dx}{x-1} - \int_{-\infty}^{-2} \frac{dx}{x+1} = \lim_{b\to-\infty}\Big[\ln|x-1|\Big]_b^{-2} - \lim_{b\to-\infty}\Big[\ln|x+1|\Big]_b^{-2} = \lim_{b\to-\infty}\left[\ln\left|\frac{x-1}{x+1}\right|\right]_b^{-2}$

$\displaystyle = \lim_{b\to-\infty}\left(\ln\left|\frac{-3}{-1}\right| - \ln\left|\frac{b-1}{b+1}\right|\right) = \ln 3 - \ln\left(\lim_{b\to-\infty}\frac{b-1}{b+1}\right) = \ln 3 - \ln 1 = \ln 3$

11. $\displaystyle\int_{2}^{\infty} \frac{2\,dv}{v^2-v} = \lim_{b\to\infty}\left[2\ln\left|\frac{v-1}{v}\right|\right]_2^b = \lim_{b\to\infty}\left(2\ln\left|\frac{b-1}{b}\right| - 2\ln\left|\frac{2-1}{2}\right|\right) = 2\ln(1) - 2\ln\left(\tfrac{1}{2}\right) = 0 + 2\ln 2 = \ln 4$

13. $\displaystyle\int_{-\infty}^{\infty} \frac{2x\,dx}{(x^2+1)^2} = \int_{-\infty}^{0} \frac{2x\,dx}{(x^2+1)^2} + \int_{0}^{\infty} \frac{2x\,dx}{(x^2+1)^2}; \begin{bmatrix} u = x^2+1 \\ du = 2x\,dx \end{bmatrix} \to \int_{\infty}^{1} \frac{du}{u^2} + \int_{1}^{\infty} \frac{du}{u^2} = \lim_{b\to\infty}\left[-\frac{1}{u}\right]_b^1 + \lim_{c\to\infty}\left[-\frac{1}{u}\right]_1^c$

$\displaystyle = \lim_{b\to\infty}\left(-1+\frac{1}{b}\right) + \lim_{c\to\infty}\left[-\frac{1}{c}-(-1)\right] = (-1+0) + (0+1) = 0$

15. $\displaystyle\int_{0}^{1} \frac{\theta+1}{\sqrt{\theta^2+2\theta}}\,d\theta; \begin{bmatrix} u = \theta^2+2\theta \\ du = 2(\theta+1)\,d\theta \end{bmatrix} \to \int_{0}^{3} \frac{du}{2\sqrt{u}} = \lim_{b\to 0^+}\int_{b}^{3} \frac{du}{2\sqrt{u}} = \lim_{b\to 0^+}\Big[\sqrt{u}\Big]_b^3 = \lim_{b\to 0^+}\left(\sqrt{3}-\sqrt{b}\right)$

$\displaystyle = \sqrt{3} - 0 = \sqrt{3}$

17. $\displaystyle\int_{0}^{\infty} \frac{dx}{(1+x)\sqrt{x}}; \begin{bmatrix} u = \sqrt{x} \\ du = \dfrac{dx}{2\sqrt{x}} \end{bmatrix} \to \int_{0}^{\infty} \frac{2\,du}{u^2+1} = \lim_{b\to\infty}\int_{0}^{b} \frac{2\,du}{u^2+1} = \lim_{b\to\infty}\Big[2\tan^{-1}u\Big]_0^b$

$\displaystyle = \lim_{b\to\infty}\left(2\tan^{-1}b - 1\tan^{-1}0\right) = 2\left(\frac{\pi}{2}\right) - 2(0) = \pi$

19. $\displaystyle\int_{0}^{\infty} \frac{dv}{(1+v^2)(1+\tan^{-1}v)} = \lim_{b\to\infty}\Big[\ln|1+\tan^{-1}v|\Big]_0^b = \lim_{b\to\infty}\Big[\ln|1+\tan^{-1}b|\Big] - \ln|1+\tan^{-1}0|$

$\displaystyle = \ln\left(1+\frac{\pi}{2}\right) - \ln(1+0) = \ln\left(1+\frac{\pi}{2}\right)$

21. $\displaystyle\int_{-\infty}^{0} \theta e^{\theta}\,d\theta = \lim_{b\to-\infty}\Big[\theta e^{\theta} - e^{\theta}\Big]_b^0 = (0\cdot e^0 - e^0) - \lim_{b\to-\infty}\Big[be^b - e^b\Big] = -1 - \lim_{b\to-\infty}\left(\frac{b-1}{e^{-b}}\right)$

$\displaystyle = -1 - \lim_{b\to-\infty}\left(\frac{1}{-e^{-b}}\right)$ (l'Hôpital's rule for $\frac{\infty}{\infty}$ form)

$\displaystyle = -1 - 0 = -1$

23. $\displaystyle\int_{-\infty}^{0} e^{-|x|}\,dx = \int_{-\infty}^{0} e^x\,dx = \lim_{b\to-\infty}\Big[e^x\Big]_b^0 = \lim_{b\to-\infty}\left(1-e^b\right) = (1-0) = 1$

25. $\displaystyle\int_0^1 x \ln x \, dx = \lim_{b \to 0^+} \left[\frac{x^2}{2} \ln x - \frac{x^2}{4} \right]_b^1 = \left(\frac{1}{2} \ln 1 - \frac{1}{4} \right) - \lim_{b \to 0^+} \left(\frac{b^2}{2} \ln b - \frac{b^2}{4} \right) = -\frac{1}{4} - \lim_{b \to 0^+} \frac{\ln b}{\left(\frac{2}{b^2} \right)} + 0$

$= -\frac{1}{4} - \lim_{b \to 0^+} \frac{\left(\frac{1}{b} \right)}{\left(-\frac{4}{b^3} \right)} = -\frac{1}{4} + \lim_{b \to 0^+} \left(\frac{b^2}{4} \right) = -\frac{1}{4} + 0 = -\frac{1}{4}$

27. $\displaystyle\int_0^2 \frac{ds}{\sqrt{4 - s^2}} = \lim_{b \to 2^-} \left[\sin^{-1} \frac{s}{2} \right]_0^b = \lim_{b \to 2^-} \left(\sin^{-1} \frac{b}{2} \right) - \sin^{-1} 0 = \frac{\pi}{2} - 0 = \frac{\pi}{2}$

29. $\displaystyle\int_1^2 \frac{ds}{s\sqrt{s^2 - 1}} = \lim_{b \to 1^+} \left[\sec^{-1} s \right]_b^2 = \sec^{-1} 2 - \lim_{b \to 1^+} \sec^{-1} b = \frac{\pi}{3} - 0 = \frac{\pi}{3}$

31. $\displaystyle\int_{-1}^4 \frac{dx}{\sqrt{|x|}} = \lim_{b \to 0^-} \int_{-1}^b \frac{dx}{\sqrt{-x}} + \lim_{c \to 0^+} \int_c^4 \frac{dx}{\sqrt{x}} = \lim_{b \to 0^-} \left[-2\sqrt{-x} \right]_{-1}^b + \lim_{c \to 0^+} \left[2\sqrt{x} \right]_c^4$

$= \lim_{b \to 0^-} \left(-2\sqrt{-b} \right) - \left(-2\sqrt{-(-1)} \right) + 2\sqrt{4} - 6 \lim_{c \to 0^+} 2\sqrt{c} = 0 + 2 + 2 \cdot 2 - 0 = 6$

33. $\displaystyle\int_{-1}^\infty \frac{d\theta}{\theta^2 + 5\theta + 6} = \lim_{b \to \infty} \left[\ln \left| \frac{\theta + 2}{\theta + 3} \right| \right]_{-1}^b = \lim_{b \to \infty} \left[\ln \left| \frac{b + 2}{b + 3} \right| \right] - \ln \left| \frac{-1 + 2}{-1 + 3} \right| = 0 - \ln\left(\frac{1}{2} \right) = \ln 2$

35. $\displaystyle\int_0^{\pi/2} \tan \theta \, d\theta = \lim_{b \to \frac{\pi}{2}^-} \left[-\ln |\cos \theta| \right]_0^b = \lim_{b \to \frac{\pi}{2}^-} \left[-\ln |\cos b| \right] + \ln 1 = \lim_{b \to \frac{\pi}{2}^-} \left[-\ln |\cos b| \right] = -\infty,$

the integral diverges

37. $\displaystyle\int_0^\pi \frac{\sin \theta \, d\theta}{\sqrt{\pi - \theta}}; \; [\pi - \theta = x] \to -\int_\pi^0 \frac{\sin x \, dx}{\sqrt{x}} = \int_0^\pi \frac{\sin x \, dx}{\sqrt{x}}.$ Since $0 \le \frac{\sin x}{\sqrt{x}} \le \frac{1}{\sqrt{x}}$ for all $0 \le x \le \pi$ and $\displaystyle\int_0^\pi \frac{dx}{\sqrt{x}}$

converges, then $\displaystyle\int_0^\pi \frac{\sin x}{\sqrt{x}} dx$ converges by the Direct Comparison Test.

39. $\displaystyle\int_0^{\ln 2} x^{-2} e^{-1/x} \, dx; \; \left[\frac{1}{x} = y \right] \to \int_\infty^{1/\ln 2} \frac{y^2 e^{-y} \, dy}{-y^2} = \int_{1/\ln 2}^\infty e^{-y} \, dy = \lim_{b \to \infty} \left[-e^{-y} \right]_{1/\ln 2}^b = \lim_{b \to \infty} \left[-e^{-b} \right] - \left[-e^{-1/\ln 2} \right]$

$= 0 + e^{-1/\ln 2} = e^{-1/\ln 2}$, so the integral converges.

41. $\displaystyle\int_0^\pi \frac{dt}{\sqrt{t} + \sin t}.$ Since for $0 \le t \le \pi$, $0 \le \frac{1}{\sqrt{t} + \sin t} \le \frac{1}{\sqrt{t}}$ and $\displaystyle\int_0^\pi \frac{dt}{\sqrt{t}}$ converges, then the original integral

converges as well by the Direct Comparison Test.

43. $\int_{0}^{2} \frac{dx}{1-x^2} = \int_{0}^{1} \frac{dx}{1-x^2} + \int_{1}^{2} \frac{dx}{1-x^2}$ and $\int_{0}^{1} \frac{dx}{1-x^2} = \lim_{b\to1^-} \left[\frac{1}{2}\ln\left|\frac{1+x}{1-x}\right|\right]_{0}^{b} = \lim_{b\to1^-}\left[\frac{1}{2}\ln\left|\frac{1+b}{1-b}\right|\right] - 0 = \infty$, which

diverges $\Rightarrow \int_{0}^{2} \frac{dx}{1-x^2}$ diverges as well.

45. $\int_{-1}^{1} \ln|x|\,dx = \int_{-1}^{0} \ln(-x)\,dx + \int_{0}^{1} \ln x\,dx;$ $\int_{0}^{1} \ln x\,dx = \lim_{b\to0^+} [x\ln x - x]_{b}^{1} = [1\cdot0 - 1] - \lim_{b\to0^+} [b\ln b - b]$

$= -1 - 0 = -1;$ $\int_{-1}^{0} \ln(-x)\,dx = -1 \Rightarrow \int_{-1}^{1} \ln|x|\,dx = -2$ converges.

47. $\int_{1}^{\infty} \frac{dx}{1+x^3}; 0 \le \frac{1}{x^3+1} \le \frac{1}{x^3}$ for $1 \le x < \infty$ and $\int_{1}^{\infty} \frac{dx}{x^3}$ converges $\Rightarrow \int_{1}^{\infty} \frac{dx}{1+x^3}$ converges by the Direct

Comparison Test.

49. $\int_{2}^{\infty} \frac{dv}{\sqrt{v-1}}; \lim_{v\to\infty} \frac{\left(\frac{1}{\sqrt{v-1}}\right)}{\left(\frac{1}{\sqrt{v}}\right)} = \lim_{v\to\infty} \frac{\sqrt{v}}{\sqrt{v-1}} = \lim_{v\to\infty} \frac{1}{\sqrt{1-\frac{1}{v}}} = \frac{1}{\sqrt{1-0}} = 1$ and $\int_{2}^{\infty} \frac{dv}{\sqrt{v}} = \lim_{b\to\infty} [2\sqrt{v}]_{2}^{b} = \infty,$

which diverges $\Rightarrow \int_{2}^{\infty} \frac{dv}{\sqrt{v-1}}$ diverges by the Limit Comparison Test.

51. $\int_{0}^{\infty} \frac{dx}{\sqrt{x^6+1}} = \int_{0}^{1} \frac{dx}{\sqrt{x^6+1}} + \int_{1}^{\infty} \frac{dx}{\sqrt{x^6+1}} < \int_{0}^{1} \frac{dx}{\sqrt{x^6+1}} + \int_{1}^{\infty} \frac{dx}{x^3}$ and $\int_{1}^{\infty} \frac{dx}{x^3} = \lim_{b\to\infty} \left[-\frac{1}{2x^2}\right]_{1}^{b}$

$= \lim_{b\to\infty} \left(-\frac{1}{2b^2} + \frac{1}{2}\right) = \frac{1}{2} \Rightarrow \int_{0}^{\infty} \frac{dx}{\sqrt{x^6+1}}$ converges by the Direct Comparison Test.

53. $\int_{1}^{\infty} \frac{\sqrt{x+1}}{x^2}\,dx; \lim_{x\to\infty} \frac{\left(\frac{\sqrt{x}}{x^2}\right)}{\left(\frac{\sqrt{x+1}}{x^2}\right)} = \lim_{x\to\infty} \frac{\sqrt{x}}{\sqrt{x+1}} = \lim_{x\to\infty} \frac{1}{\sqrt{1+\frac{1}{x}}} = 1;$ $\int_{1}^{\infty} \frac{\sqrt{x}}{x^2}\,dx = \int_{1}^{\infty} \frac{dx}{x^{3/2}}$

$= \lim_{b\to\infty} \left[-2x^{-1/2}\right]_{1}^{b} = \lim_{b\to\infty} \left(\frac{-2}{\sqrt{x}} + 2\right) = 2 \Rightarrow \int_{1}^{\infty} \frac{\sqrt{x+1}}{x^2}\,dx$ converges by the Limit Comparison Test.

55. $\int_{\pi}^{\infty} \frac{2+\cos x}{x}\,dx; 0 < \frac{1}{x} \le \frac{2+\cos x}{x}$ for $x \ge \pi$ and $\int_{\pi}^{\infty} \frac{dx}{x} = \lim_{b\to\infty} [\ln x]_{\pi}^{b} = \infty$, which diverges

$\Rightarrow \int_{\pi}^{\infty} \frac{2+\cos x}{x}\,dx$ diverges by the Direct Comparison Test.

57. $\int_4^\infty \frac{2\,dt}{t^{3/2}+1}$; $\lim\limits_{t\to\infty} \frac{t^{3/2}}{t^{3/2}+1} = 1$ and $\int_4^\infty \frac{2\,dt}{t^{3/2}} = \lim\limits_{b\to\infty}\left[-4t^{-1/2}\right]_4^b = \lim\limits_{b\to\infty}\left(\frac{-4}{\sqrt{b}}+2\right) = 2 \Rightarrow \int_4^\infty \frac{2\,dt}{t^{3/2}}$ converges

$\Rightarrow \int_4^\infty \frac{2\,dt}{t^{3/2}+1}$ converges by the Direct Comparison Test.

59. $\int_1^\infty \frac{e^x}{x}\,dx$; $0 < \frac{1}{x} < \frac{e^x}{x}$ for $x > 1$ and $\int_1^\infty \frac{dx}{x}$ diverges $\Rightarrow \int_1^\infty \frac{e^x\,dx}{x}$ diverges by the Direct Comparison Test.

61. $\int_1^\infty \frac{dx}{\sqrt{e^x - x}}$; $\lim\limits_{x\to\infty} \frac{\left(\frac{1}{\sqrt{e^x-x}}\right)}{\left(\frac{1}{\sqrt{e^x}}\right)} = \lim\limits_{x\to\infty} \frac{\sqrt{e^x}}{\sqrt{e^x-x}} = \lim\limits_{x\to\infty} \frac{1}{\sqrt{1-\frac{x}{e^x}}} = \frac{1}{\sqrt{1-0}} = 1$; $\int_1^\infty \frac{dx}{\sqrt{e^x}} = \int_1^\infty e^{-x/2}\,dx$

$= \lim\limits_{b\to\infty}\left[-2e^{-x/2}\right]_1^b = \lim\limits_{b\to\infty}\left(-2e^{-b/2}+2e^{-1/2}\right) = \frac{2}{\sqrt{e}} \Rightarrow \int_1^\infty e^{-x/2}\,dx$ converges $\Rightarrow \int_1^\infty \frac{dx}{\sqrt{e^x - x}}$ converges

by the Limit Comparison Test.

63. $\int_{-\infty}^\infty \frac{dx}{\sqrt{x^4+1}} = 2\int_0^\infty \frac{dx}{\sqrt{x^4+1}}$; $\lim\limits_{x\to\infty} \frac{x^2}{\sqrt{x^4+1}} = 1$; $\int_0^\infty \frac{dx}{\sqrt{x^4+1}} = \int_0^1 \frac{dx}{\sqrt{x^4+1}} + \int_1^\infty \frac{dx}{\sqrt{x^4+1}}$

$< \int_0^1 \frac{dx}{\sqrt{x^4+1}} + \int_1^\infty \frac{dx}{x^2}$ and $\int_1^\infty \frac{dx}{x^2} = \lim\limits_{b\to\infty}\left[-\frac{1}{x}\right]_1^b = \lim\limits_{b\to\infty}\left(-\frac{1}{b}+1\right) = 1 \Rightarrow \int_{-\infty}^\infty \frac{dx}{\sqrt{x^4+1}}$ converges by the

Direct Comparison Test.

65. (a) $\int_1^\infty e^{-3x}\,dx = \lim\limits_{b\to\infty}\left[-\frac{1}{3}e^{-3x}\right]_3^b = \lim\limits_{b\to\infty}\left(-\frac{1}{3}e^{-3b}\right)-\left(-\frac{1}{3}e^{-3\cdot3}\right) = 0 + \frac{1}{3}\cdot e^{-9} = \frac{1}{3}e^{-9}$

$\approx 0.0000411 < 0.000042$. Since $e^{-x^2} \le e^{-3x}$ for $x > 3$, then $\int_3^\infty e^{-x^2}\,dx < 0.000042$ and therefore

$\int_0^\infty e^{-x^2}\,dx$ can be replaced by $\int_0^3 e^{-x^2}\,dx$ without introducing an error greater than 0.000042.

(b) $\int_0^3 e^{-x^2}\,dx \cong 0.88621$

67. (a) $\int_1^\infty \frac{dx}{x^p} = \lim\limits_{b\to\infty}\left[\frac{1}{1-p}x^{1-p}\right]_1^b = \lim\limits_{b\to\infty}\left(\frac{b^{1-p}}{1-p}-\frac{1}{1-p}\right) = \begin{cases} \frac{1}{p-1}, & p > 1 \\ \infty, & p < 1 \end{cases}$

(b) $\int_0^1 \frac{dx}{x^p} = \lim\limits_{b\to0^+}\left[\frac{1}{1-p}x^{1-p}\right]_b^1 = \lim\limits_{b\to0^+}\left(\frac{1}{1-p}-\frac{b^{1-p}}{1-p}\right)$, $p \ne 1$; if $p < 1$, then $\lim\limits_{b\to0^+} b^{1-p} = 0$ since $1 - p > 0$;

if $p > 1$, then $\lim\limits_{b\to0^+} b^{1-p} = \lim\limits_{b\to0^+} \frac{1}{b^{p-1}} = \infty$; for $p = 1$: $\int_0^1 \frac{dx}{x} = \lim\limits_{b\to0^+}[\ln x]_b^1 = \lim\limits_{b\to0^+}(\ln 1 - \ln b) = \infty$;

in conclusion, $\displaystyle\int_0^1 \frac{dx}{x^p} = \begin{cases} \dfrac{1}{1-p}, & p < 1 \\ \text{diverges}, & p \geq 1 \end{cases}$

69. $A = \displaystyle\int_0^\infty e^{-x}\, dx = \lim_{b\to\infty}\left[-e^{-x}\right]_0^b = \lim_{b\to\infty}\left(-e^{-b}\right)-\left(-e^{-0}\right)$

$= 0 + 1 = 1$

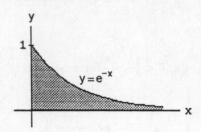

71. $V = \displaystyle\int_0^\infty 2\pi x e^{-x}\, dx = 2\pi\int_0^\infty x e^{-x}\, dx = 2\pi \lim_{b\to\infty}\left[-xe^{-x}-e^{-x}\right]_0^b = 2\pi\left[\lim_{b\to\infty}\left(-be^{-b}-e^{-b}\right)-1\right] = 2\pi$

73. $A = \displaystyle\int_0^{\pi/2} (\sec x - \tan x)\, dx = \lim_{b\to\frac{\pi}{2}^-}\left[\ln|\sec x + \tan x| - \ln|\sec x|\right]_0^b = \lim_{b\to\frac{\pi}{2}^-}\left(\ln\left|1 + \frac{\tan b}{\sec b}\right| - \ln|1+0|\right)$

$= \displaystyle\lim_{b\to\frac{\pi}{2}^-} \ln|1 + \sin b| = \ln 2$

75. $\displaystyle\int_0^\infty \frac{2x\, dx}{x^2+1} = \lim_{b\to\infty}\left[\ln(x^2+1)\right]_0^b = \lim_{b\to\infty}\left[\ln(b^2+1)\right]-0 = \lim_{b\to\infty}\ln(b^2+1) = \infty \Rightarrow$ the integral $\displaystyle\int_{-\infty}^\infty \frac{2x}{x^2+1}\, dx$

diverges. But $\displaystyle\lim_{b\to\infty}\int_{-b}^b \frac{2x\, dx}{x^2+1} = \lim_{b\to\infty}\left[\ln(x^2+1)\right]_{-b}^b = \lim_{b\to\infty}\left[\ln(b^2+1) - \ln(b^2+1)\right] = \lim_{b\to\infty}\ln\left(\frac{b^2+1}{b^2+1}\right)$

$= \displaystyle\lim_{b\to\infty}(\ln 1) = 0$

77. (a) The statement is true since $\displaystyle\int_{-\infty}^b f(x)\, dx = \int_{-\infty}^a f(x)\, dx + \int_a^b f(x)\, dx, \int_b^\infty f(x)\, dx = \int_a^\infty f(x)\, dx - \int_a^b f(x)\, dx$

and $\displaystyle\int_a^b f(x)\, dx$ exists since $f(x)$ is integrable on every interval $[a, b]$.

(b) $\displaystyle\int_{-\infty}^a f(x)\, dx + \int_a^\infty f(x)\, dx = \int_{-\infty}^a f(x)\, dx + \int_a^b f(x)\, dx - \int_a^b f(x)\, dx + \int_a^\infty f(x)\, dx$

$= \displaystyle\int_{-\infty}^b f(x)\, dx + \int_b^a f(x)\, dx + \int_a^\infty f(x)\, dx = \int_{-\infty}^b f(x)\, dx + \int_b^\infty f(x)\, dx$

79. $\displaystyle\int_{-\infty}^{\infty} \frac{dx}{\sqrt{x^2+1}} = \int_{-\infty}^{1} \frac{dx}{\sqrt{x^2+1}} + \int_{1}^{\infty} \frac{dx}{\sqrt{x^2+1}}$; $\displaystyle\int_{1}^{\infty} \frac{dx}{\sqrt{x^2+1}}$ diverges because $\displaystyle\lim_{x\to\infty} \frac{\left(\frac{1}{x}\right)}{\left(\frac{1}{\sqrt{x^2+1}}\right)}$

$\displaystyle = \lim_{x\to\infty} \frac{\sqrt{x^2+1}}{x} = \lim_{x\to\infty} \sqrt{1+\frac{1}{x^2}} = 1$ and $\displaystyle\int_{1}^{\infty} \frac{dx}{x}$ diverges; therefore, $\displaystyle\int_{-\infty}^{\infty} \frac{dx}{\sqrt{x^2+1}}$ diverges

81. $\displaystyle\int_{-\infty}^{\infty} \frac{e^x \, dx}{e^{2x}+1}$; $\displaystyle\frac{e^x}{e^{2x}+1} = \frac{1}{e^x+e^{-x}} < \frac{1}{e^x}$ and $\displaystyle\int_{0}^{\infty} \frac{dx}{e^x} = \lim_{c\to\infty} \left[-e^{-x}\right]_0^c = \lim_{c\to\infty} \left(-e^{-c}+1\right) = 1$

$\displaystyle \Rightarrow \int_{-\infty}^{\infty} \frac{e^x \, dx}{e^{2x}+1} = 2\int_{0}^{\infty} \frac{dx}{e^x+e^{-x}}$ converges

83. $\displaystyle\int_{-\infty}^{\infty} e^{-|x|} \, dx = 2\int_{0}^{\infty} e^{-x} \, dx = 2\lim_{b\to\infty} \int_{0}^{b} e^{-x} \, dx = -2\lim_{b\to\infty} \left[e^{-x}\right]_0^b = 2$, so the integral converges.

85. $\displaystyle\int_{-\infty}^{\infty} \frac{|\sin x|+|\cos x|}{|x|+1} \, dx = 2\int_{0}^{\infty} \frac{|\sin x|+|\cos x|}{x+1} \, dx \geq 2\int_{0}^{\infty} \frac{\sin^2 x + \cos^2 x}{x+1} \, dx = 2\lim_{b\to\infty} \int_{0}^{b} \frac{dx}{x+1} \, dx$

$\displaystyle = 2\lim_{b\to\infty} \left[\ln|x+1|\right]_0^b = \infty$, which diverges $\Rightarrow \int_{-\infty}^{\infty} \frac{|\sin x|+|\cos x|}{|x|+1} \, dx$ diverges

87-89.   Use the MAPLE or MATHEMATICA integration commands, as discussed in the text.

91. (a)

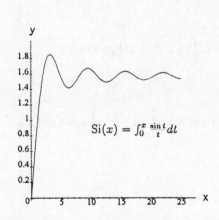

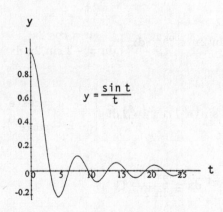

(b) > int((sin(t))/t, t=0..infinity); $\left(\text{answer is } \frac{\pi}{2}\right)$

CHAPTER 7 PRACTICE EXERCISES

1. $\displaystyle\int x\sqrt{4x^2-9} \, dx$; $\begin{bmatrix} u = 4x^2-9 \\ du = 8x \, dx \end{bmatrix} \rightarrow \frac{1}{8}\int \sqrt{u} \, du = \frac{1}{8}\cdot\frac{2}{3}u^{3/2} + C = \frac{1}{12}\left(4x^2-9\right)^{3/2} + C$

3. $\int x(2x+1)^{1/2}\,dx; \begin{bmatrix} u = 2x+1 \\ du = 2\,dx \end{bmatrix} \to \frac{1}{2}\int \left(\frac{u-1}{2}\right)\sqrt{u}\,du = \frac{1}{4}\left(\int u^{3/2}\,du - \int u^{1/2}\,du\right) = \frac{1}{4}\left(\frac{2}{5}u^{5/2} - \frac{2}{3}u^{3/2}\right) + C$

$= \dfrac{(2x+1)^{5/2}}{10} - \dfrac{(2x+1)^{3/2}}{6} + C$

5. $\int \dfrac{x\,dx}{\sqrt{8x^2+1}}; \begin{bmatrix} u = 8x^2+1 \\ du = 16x\,dx \end{bmatrix} \to \frac{1}{16}\int \dfrac{du}{\sqrt{u}} = \frac{1}{16}\cdot 2u^{1/2} + C = \dfrac{\sqrt{8x^2+1}}{8} + C$

7. $\int \dfrac{y\,dy}{25+y^2}; \begin{bmatrix} u = 25+y^2 \\ du = 2y\,dy \end{bmatrix} \to \frac{1}{2}\int \dfrac{du}{u} = \frac{1}{2}\ln|u| + C = \frac{1}{2}\ln(25+y^2) + C$

9. $\int \dfrac{t^3\,dt}{\sqrt{9-4t^4}}; \begin{bmatrix} u = 9-4t^4 \\ du = -16t^3\,dt \end{bmatrix} \to -\frac{1}{16}\int \dfrac{du}{\sqrt{u}} = -\frac{1}{16}\cdot 2u^{1/2} + C = -\dfrac{\sqrt{9-4t^4}}{8} + C$

11. $\int z^{2/3}\left(z^{5/3}+1\right)^{2/3}\,dz; \begin{bmatrix} u = z^{5/3}+1 \\ du = \frac{5}{3}z^{2/3}\,dz \end{bmatrix} \to \frac{3}{5}\int u^{2/3}\,du = \frac{3}{5}\cdot\frac{3}{5}u^{5/3} + C = \frac{9}{25}\left(z^{5/3}+1\right)^{5/3} + C$

13. $\int \dfrac{\sin 2\theta\,d\theta}{(1-\cos 2\theta)^2}; \begin{bmatrix} u = 1-\cos 2\theta \\ du = 2\sin 2\theta\,d\theta \end{bmatrix} \to \frac{1}{2}\int \dfrac{du}{u^2} = -\frac{1}{2u} + C = -\dfrac{1}{2(1-\cos 2\theta)} + C$

15. $\int \dfrac{\sin t\,dt}{3+4\cos t}; \begin{bmatrix} u = 3+4\cos t \\ du = -4\sin t\,dt \end{bmatrix} \to -\frac{1}{4}\int \dfrac{du}{u} = -\frac{1}{4}\ln|u| + C = -\frac{1}{4}\ln|3+4\cos t| + C$

17. $\int (\sin 2x)\,e^{\cos 2x}\,dx; \begin{bmatrix} u = \cos 2x \\ du = -2\sin 2x\,dx \end{bmatrix} \to -\frac{1}{2}\int e^u\,du = -\frac{1}{2}e^u + C = -\frac{1}{2}e^{\cos 2x} + C$

19. $\int e^\theta \sin(e^\theta)\cos^2(e^\theta)\,d\theta; \begin{bmatrix} u = \cos(e^\theta) \\ du = -\sin(e^\theta)\cdot e^\theta\,d\theta \end{bmatrix} \to \int -u^2\,du = -\frac{1}{3}u^3 + C = -\frac{1}{3}\cos^3(e^\theta) + C$

21. $\int 2^{x-1}\,dx = \dfrac{2^{x-1}}{\ln 2} + C$

23. $\int \dfrac{dv}{v\ln v}; \begin{bmatrix} u = \ln v \\ du = \frac{1}{v}\,dv \end{bmatrix} \to \int \dfrac{du}{u} = \ln|u| + C = \ln|\ln v| + C$

25. $\int \dfrac{dx}{(x^2+1)(2+\tan^{-1}x)}; \begin{bmatrix} u = 2+\tan^{-1}x \\ du = \dfrac{dx}{x^2+1} \end{bmatrix} \to \int \dfrac{du}{u} = \ln|u| + C = \ln\left|2+\tan^{-1}x\right| + C$

27. $\int \dfrac{2\,dx}{\sqrt{1-4x^2}}; \begin{bmatrix} u = 2x \\ du = 2\,dx \end{bmatrix} \to \int \dfrac{du}{\sqrt{1-u^2}} = \sin^{-1}u + C = \sin^{-1}(2x) + C$

29. $\int \frac{dt}{\sqrt{16-9t^2}} = \frac{1}{4}\int \frac{dt}{\sqrt{1-\left(\frac{3t}{4}\right)^2}}; \left[\begin{array}{c} u = \frac{3}{4}t \\ du = \frac{3}{4}\,dt \end{array}\right] \to \frac{1}{3}\int \frac{du}{\sqrt{1-u^2}} = \frac{1}{3}\sin^{-1}u + C = \frac{1}{3}\sin^{-1}\left(\frac{3t}{4}\right) + C$

31. $\int \frac{dt}{9+t^2} = \frac{1}{9}\int \frac{dt}{1+\left(\frac{t}{3}\right)^2}; \left[\begin{array}{c} u = \frac{1}{3}t \\ du = \frac{1}{3}\,dt \end{array}\right] \to \frac{1}{3}\int \frac{du}{1+u^2} = \frac{1}{3}\tan^{-1}u + C = \frac{1}{3}\tan^{-1}\left(\frac{t}{3}\right) + C$

33. $\int \frac{4\,dx}{5x\sqrt{25x^2-16}} = \frac{4}{25}\int \frac{dx}{x\sqrt{x^2-\frac{16}{25}}} = \frac{1}{5}\sec^{-1}\left|\frac{5x}{4}\right| + C$

35. $\int \frac{dx}{\sqrt{4x-x^2}} = \int \frac{d(x-2)}{\sqrt{4-(x-2)^2}} = \sin^{-1}\left(\frac{x-2}{2}\right) + C$

37. $\int \frac{dy}{y^2-4y+8} = \int \frac{d(y-2)}{(y-2)^2+4} = \frac{1}{2}\tan^{-1}\left(\frac{y-2}{2}\right) + C$

39. $\int \frac{dx}{(x-1)\sqrt{x^2-2x}} = \int \frac{d(x-1)}{(x-1)\sqrt{(x-1)^2-1}} = \sec^{-1}|x-1| + C$

41. $\int \sin^2 x\,dx = \int \frac{1-\cos 2x}{2}\,dx = \frac{x}{2} - \frac{\sin 2x}{4} + C$

43. $\int \sin^3\frac{\theta}{2}\,d\theta = \int \left(1-\cos^2\frac{\theta}{2}\right)\left(\sin\frac{\theta}{2}\right)d\theta; \left[\begin{array}{c} u = \cos\frac{\theta}{2} \\ du = -\frac{1}{2}\sin\frac{\theta}{2}\,d\theta \end{array}\right] \to -2\int \left(1-u^2\right)du = \frac{2u^3}{3} - 2u + C$

$= \frac{2}{3}\cos^3\frac{\theta}{2} - 2\cos\frac{\theta}{2} + C$

45. $\int \tan^3 2t\,dt = \int (\tan 2t)\left(\sec^2 2t - 1\right)dt = \int \tan 2t\,\sec^2 2t\,dt - \int \tan 2t\,dt; \left[\begin{array}{c} u = 2t \\ du = 2\,dt \end{array}\right]$

$\to \frac{1}{2}\int \tan u\,\sec^2 u\,du - \frac{1}{2}\int \tan u\,du = \frac{1}{4}\tan^2 u + \frac{1}{2}\ln|\cos u| + C = \frac{1}{4}\tan^2 2t + \frac{1}{2}\ln|\cos 2t| + C$

$= \frac{1}{4}\tan^2 2t - \frac{1}{2}\ln|\sec 2t| + C$

47. $\int \frac{dx}{2\sin x \cos x} = \int \frac{dx}{\sin 2x} = \int \csc 2x\,dx = -\frac{1}{2}\ln|\csc 2x + \cot 2x| + C$

49. $\int_{\pi/4}^{\pi/2} \sqrt{\csc^2 y - 1}\,dy = \int_{\pi/4}^{\pi/2} \cot y\,dy = \left[\ln|\sin y|\right]_{\pi/4}^{\pi/2} = \ln 1 - \ln\frac{1}{\sqrt{2}} = \ln\sqrt{2}$

51. $\int\limits_{0}^{\pi} \sqrt{1-\cos^2 2x}\,dx = \int\limits_{0}^{\pi} |\sin 2x|\,dx = \int\limits_{0}^{\pi/2} \sin 2x\,dx - \int\limits_{\pi/2}^{\pi} \sin 2x\,dx = -\left[\frac{\cos 2x}{2}\right]_{0}^{\pi/2} + \left[\frac{\cos 2x}{2}\right]_{\pi/2}^{\pi}$

$= -\left(-\frac{1}{2} - \frac{1}{2}\right) + \left[\frac{1}{2} - \left(-\frac{1}{2}\right)\right] = 2$

53. $\int\limits_{-\pi/2}^{\pi/2} \sqrt{1-\cos 2t}\,dt = \sqrt{2}\int\limits_{-\pi/2}^{\pi/2} |\sin t|\,dt = 2\sqrt{2}\int\limits_{0}^{\pi/2} \sin t\,dt = \left[-2\sqrt{2}\,\cos t\right]_{0}^{\pi/2} = 2\sqrt{2}\,[0-(-1)] = 2\sqrt{2}$

55. $\int \frac{x^2\,dx}{x^2+4} = x - \int \frac{4\,dx}{x^2+4} = x - 2\,\tan^{-1}\left(\frac{x}{2}\right) + C$

57. $\int \frac{4x^2+3}{2x-1}\,dx = \int\left[(2x+1) + \frac{4}{2x-1}\right]dx = x + x^2 + 2\,\ln|2x-1| + C$

59. $\int \frac{2y-1}{y^2+4}\,dy = \int \frac{2y\,dy}{y^2+4} - \int \frac{dy}{y^2+4} = \ln\left(y^2+4\right) - \frac{1}{2}\,\tan^{-1}\left(\frac{y}{2}\right) + C$

61. $\int \frac{t+2}{\sqrt{4-t^2}}\,dt = \int \frac{t\,dt}{\sqrt{4-t^2}} + 2\int \frac{dt}{\sqrt{4-t^2}} = -\sqrt{4-t^2} + 2\,\sin^{-1}\left(\frac{t}{2}\right) + C$

63. $\int \frac{\tan x\,dx}{\tan x + \sec x} = \int \frac{\sin x\,dx}{\sin x + 1} = \int \frac{(\sin x)(1-\sin x)}{1-\sin^2 x}\,dx = \int \frac{\sin x - 1 + \cos^2 x}{\cos^2 x}\,dx$

$= -\int \frac{d(\cos x)}{\cos^2 x} - \int \frac{dx}{\cos^2 x} + \int dx = \frac{1}{\cos x} - \tan x + x + C = x - \tan x + \sec x + C$

65. $\int \sec(5-3x)\,dx; \begin{bmatrix} y = 5-3x \\ dy = -3\,dx \end{bmatrix} \rightarrow \int \sec y \cdot \left(-\frac{dy}{3}\right) = -\frac{1}{3}\int \sec y\,dy = -\frac{1}{3}\,\ln|\sec y + \tan y| + C$

$= -\frac{1}{3}\,\ln|\sec(5-3x) + \tan(5-3x)| + C$

67. $\int \cot\left(\frac{x}{4}\right)dx = 4\int \cot\left(\frac{x}{4}\right)d\left(\frac{x}{4}\right) = 4\,\ln\left|\sin\left(\frac{x}{4}\right)\right| + C$

69. $\int x\sqrt{1-x}\,dx; \begin{bmatrix} u = 1-x \\ du = -dx \end{bmatrix} \rightarrow -\int (1-u)\sqrt{u}\,du = \int\left(u^{3/2} - u^{1/2}\right)du = \frac{2}{5}u^{5/2} - \frac{2}{3}u^{3/2} + C$

$= \frac{2}{5}(1-x)^{5/2} - \frac{2}{3}(1-x)^{3/2} + C = -2\left[\frac{\left(\sqrt{1-x}\right)^3}{3} - \frac{\left(\sqrt{1-x}\right)^5}{5}\right] + C$

71. $\int \sqrt{z^2 + 1}\, dz;\ \begin{bmatrix} z = \tan\theta \\ dz = \sec^2\theta\, d\theta \end{bmatrix} \rightarrow \int \sqrt{\tan^2\theta + 1}\cdot\sec^2\theta\, d\theta = \int \sec^3\theta\, d\theta$

$\qquad = \dfrac{\sec t \tan\theta}{3-1} + \dfrac{3-2}{3-1}\int \sec\theta\, d\theta \qquad \text{(FORMULA 92)}$

$\qquad = \dfrac{\sin\theta}{2\cos^2\theta} + \dfrac{1}{2}\ln|\sec\theta + \tan\theta| + C \ = \dfrac{z\sqrt{z^2+1}}{2} + \dfrac{1}{2}\ln\left|z + \sqrt{1+z^2}\right| + C$

73. $\int \dfrac{dy}{\sqrt{25+y^2}} = \dfrac{1}{5}\int \dfrac{dy}{\sqrt{1+\left(\frac{y}{5}\right)^2}} = \int \dfrac{du}{\sqrt{1+u^2}},\left[u = \dfrac{y}{5}\right];\begin{bmatrix} u = \tan\theta \\ du = \sec^2\theta\, d\theta \end{bmatrix} \rightarrow \int \dfrac{\sec^2\theta\, d\theta}{\sqrt{1+\tan^2\theta}} = \int \sec\theta\, d\theta$

$\qquad = \ln|\sec\theta + \tan\theta| + C_1 = \ln\left|\sqrt{1+u^2} + u\right| + C_1 = \ln\left|\sqrt{1+\left(\frac{y}{5}\right)^2} + \dfrac{y}{5}\right| + C_1 = \ln\left|\dfrac{\sqrt{25+y^2}+y}{5}\right| + C_1$

$\qquad = \ln\left|y + \sqrt{25+y^2}\right| + C$

75. $\int \dfrac{dx}{x^2\sqrt{1-x^2}};\ \begin{bmatrix} x = \sin\theta \\ dx = \cos\theta\, d\theta \end{bmatrix} \rightarrow \int \dfrac{\cos\theta\, d\theta}{\sin^2\theta\cos\theta} = \int \csc^2\theta\, d\theta = -\cot\theta + C = -\dfrac{\cos\theta}{\sin\theta} + C = \dfrac{-\sqrt{1-x^2}}{x} + C$

77. $\int \dfrac{x^2\, dx}{\sqrt{1-x^2}};\ \begin{bmatrix} x = \sin\theta \\ dx = \cos\theta\, d\theta \end{bmatrix} \rightarrow \int \dfrac{\sin^2\theta\cos\theta\, d\theta}{\cos\theta} = \int \sin^2\theta\, d\theta = \int \dfrac{1-\cos 2\theta}{2}\, d\theta = \dfrac{1}{2}\theta - \dfrac{1}{4}\sin 2\theta$

$\qquad = \dfrac{1}{2}\theta - \dfrac{1}{2}\sin\theta\cos\theta = \dfrac{\sin^{-1}x}{2} - \dfrac{x\sqrt{1-x^2}}{2} + C$

79. $\int \dfrac{dx}{\sqrt{x^2-9}};\ \begin{bmatrix} x = 3\sec\theta \\ dx = 3\sec\theta\tan\theta\, d\theta \end{bmatrix} \rightarrow \int \dfrac{3\sec\theta\tan\theta\, d\theta}{\sqrt{9\sec^2\theta - 9}} = \int \dfrac{3\sec\theta\tan\theta\, d\theta}{3\tan\theta} = \int \sec\theta\, d\theta$

$\qquad = \ln|\sec\theta + \tan\theta| + C_1 = \ln\left|\dfrac{x}{3} + \sqrt{\left(\frac{x}{3}\right)^2 - 1}\right| + C_1 - \ln\left|\dfrac{x + \sqrt{x^2-9}}{3}\right| + C_1 = \ln\left|x + \sqrt{x^2-9}\right| + C$

81. $\int \dfrac{\sqrt{w^2-1}}{w}\, dw;\ \begin{bmatrix} w = \sec\theta \\ dw = \sec\theta\tan\theta\, d\theta \end{bmatrix} \rightarrow \int \left(\dfrac{\tan\theta}{\sec\theta}\right)\cdot\sec\theta\tan\theta\, d\theta = \int \tan^2\theta\, d\theta = \int \left(\sec^2\theta - 1\right) d\theta$

$\qquad = \tan\theta - \theta + C = \sqrt{w^2-1} - \sec^{-1}w + C$

83. $u = \ln(x+1),\ du = \dfrac{dx}{x+1};\ dv = dx,\ v = x;$

$\qquad \int \ln(x+1)\, dx = x\ln(x+1) - \int \dfrac{x}{x+1}\, dx = x\ln(x+1) - \int dx + \int \dfrac{dx}{x+1} = x\ln(x+1) - x + \ln(x+1) + C_1$

$\qquad = (x+1)\ln(x+1) - x + C_1 = (x+1)\ln(x+1) - (x+1) + C,\ \text{where } C = C_1 + 1$

85. $u = \tan^{-1} 3x$, $du = \dfrac{3\,dx}{1 + 9x^2}$; $dv = dx$, $v = x$;

$$\int \tan^{-1} 3x \, dx = x \tan^{-1} 3x - \int \frac{3x\,dx}{1 + 9x^2}; \begin{bmatrix} y = 1 + 9x^2 \\ dy = 18x\,dx \end{bmatrix} \to x \tan^{-1} 3x - \frac{1}{6} \int \frac{dy}{y}$$

$$= x \tan^{-1}(3x) - \frac{1}{6} \ln(1 + 9x^2) + C$$

87.

$$
\begin{array}{lll}
 & & e^x \\
(x+1)^2 & \xrightarrow{(+)} & e^x \\
2(x+1) & \xrightarrow{(-)} & e^x \\
2 & \xrightarrow{(+)} & e^x \\
0 & &
\end{array}
\quad \Rightarrow \int (x+1)^2 e^x \, dx = \left[(x+1)^2 - 2(x+1) + 2 \right] e^x + C
$$

89. $u = \cos 2x$, $du = -2 \sin 2x \, dx$; $dv = e^x \, dx$, $v = e^x$;

$$I = \int e^x \cos 2x \, dx = e^x \cos 2x + 2 \int e^x \sin 2x \, dx;$$

$u = \sin 2x$, $du = 2 \cos 2x \, dx$; $dv = e^x \, dx$, $v = e^x$;

$$I = e^x \cos 2x + 2 \left[e^x \sin 2x - 2 \int e^x \cos 2x \, dx \right] = e^x \cos 2x + 2 e^x \sin 2x - 4I \Rightarrow I = \frac{e^x \cos 2x}{5} + \frac{2 e^x \sin 2x}{5} + C$$

91. $\displaystyle \int \frac{x\,dx}{x^2 - 3x + 2} = \int \frac{2\,dx}{x - 2} - \int \frac{dx}{x - 1} = 2 \ln |x - 2| - \ln |x - 1| + C$

93. $\displaystyle \int \frac{dx}{x(x+1)^2} = \int \left(\frac{1}{x} - \frac{2}{x+1} + \frac{x}{(x+1)^2} \right) dx = \ln|x| - 2 \ln|x+1| + \left(\ln|x+1| + \frac{1}{x+1} \right) + C$

$$= \ln|x| - \ln|x+1| + \frac{1}{x+1} + C$$

95. $\displaystyle \int \frac{\sin \theta \, d\theta}{\cos^2 \theta + \cos \theta - 2}; \; [\cos \theta = y] \to -\int \frac{dy}{y^2 + y - 2} = -\frac{1}{3} \int \frac{dy}{y - 1} + \frac{1}{3} \int \frac{dy}{y + 2} = \frac{1}{3} \ln \left| \frac{y+2}{y-1} \right| + C$

$$= \frac{1}{3} \ln \left| \frac{\cos \theta + 2}{\cos \theta - 1} \right| + C = -\frac{1}{3} \ln \left| \frac{\cos \theta - 1}{\cos \theta + 2} \right| + C$$

97. $\displaystyle \int \frac{3x^2 + 4x + 4}{x^3 + x} \, dx = \int \frac{4}{x} \, dx - \int \frac{x - 4}{x^2 + 1} \, dx = 4 \ln|x| - \frac{1}{2} \ln(x^2 + 1) + 4 \tan^{-1} x + C$

99. $\displaystyle \int \frac{(v+3)\,dv}{2v^3 - 8v} = \frac{1}{2} \int \left(-\frac{3}{4v} + \frac{5}{8(v-2)} + \frac{1}{8(v+2)} \right) dv = -\frac{3}{8} \ln|v| + \frac{5}{16} \ln|v - 2| + \frac{1}{16} \ln|v + 2| + C$

$$= \frac{1}{16} \ln \left| \frac{(v-2)^5 (v+2)}{v^6} \right| + C$$

101. $\displaystyle \int \frac{dt}{t^4 + 4t^2 + 3} = \frac{1}{2} \int \frac{dt}{t^2 + 1} - \frac{1}{2} \int \frac{dt}{t^2 + 3} = \frac{1}{2} \tan^{-1} t - \frac{1}{2\sqrt{3}} \tan^{-1} \left(\frac{t}{\sqrt{3}} \right) + C = \frac{1}{2} \tan^{-1} t - \frac{\sqrt{3}}{6} \tan^{-1} \frac{t}{\sqrt{3}} + C$

103. $\int \dfrac{x^3 + x^2}{x^2 + x - 2}\, dx = \int \left(x + \dfrac{2x}{x^2 + x - 2}\right) dx = \int x\, dx + \dfrac{2}{3}\int \dfrac{dx}{x - 1} + \dfrac{4}{3}\int \dfrac{dx}{x + 2}$

$= \dfrac{x^2}{2} + \dfrac{4}{3}\ln|x + 2| + \dfrac{2}{3}\ln|x - 1| + C$

105. $\int \dfrac{x^3 + 4x^2}{x^2 + 4x + 3}\, dx = \int \left(x - \dfrac{3x}{x^2 + 4x + 3}\right) dx = \int x\, dx + \dfrac{3}{2}\int \dfrac{dx}{x + 1} - \dfrac{9}{2}\int \dfrac{dx}{x + 3}$

$= \dfrac{x^2}{2} - \dfrac{9}{2}\ln|x + 3| + \dfrac{3}{2}\ln|x + 1| + C$

107. $\int \dfrac{dx}{x(3\sqrt{x + 1})};\ \begin{bmatrix} u = \sqrt{x + 1} \\ du = \dfrac{dx}{2\sqrt{x + 1}} \\ dx = 2u\, du \end{bmatrix} \to \dfrac{2}{3}\int \dfrac{u\, du}{(u^2 - 1)u} = \dfrac{1}{3}\int \dfrac{du}{u - 1} - \dfrac{1}{3}\int \dfrac{du}{u + 1} = \dfrac{1}{3}\ln|u - 1| - \dfrac{1}{3}\ln|u + 1| + C$

$= \dfrac{1}{3}\ln\left|\dfrac{\sqrt{x + 1} - 1}{\sqrt{x + 1} + 1}\right| + C$

109. $\int \dfrac{ds}{e^s - 1};\ \begin{bmatrix} u = e^s - 1 \\ du = e^s\, ds \\ ds = \dfrac{du}{u + 1} \end{bmatrix} \to \int \dfrac{du}{u(u - 1)} = \int \dfrac{du}{u - 1} - \int \dfrac{du}{u} = \ln\left|\dfrac{u - 1}{u}\right| + C = \ln\left|\dfrac{e^s - 1}{e^s}\right| + C = \ln|1 - e^{-s}| + C$

111. $\displaystyle\int_0^3 \dfrac{dx}{\sqrt{9 - x^2}} = \lim_{b \to 3^-} \int_0^b \dfrac{dx}{\sqrt{9 - x^2}} = \lim_{b \to 3^-} \left[\sin^{-1}\left(\dfrac{x}{3}\right)\right]_0^b = \lim_{b \to 3^-} \sin^{-1}\left(\dfrac{b}{3}\right) - \sin^{-1}\left(\dfrac{0}{3}\right) = \dfrac{\pi}{2} - 0 = \dfrac{\pi}{2}$

113. $\displaystyle\int_{-1}^1 \dfrac{dy}{y^{2/3}} = \int_{-1}^0 \dfrac{dy}{y^{2/3}} + \int_0^1 \dfrac{dy}{y^{2/3}} = 2\int_0^1 \dfrac{dy}{y^{2/3}} = 2 \cdot 3 \lim_{b \to 0^+} \left[y^{1/3}\right]_b^1 = 6\left(1 - \lim_{b \to 0^+} b^{1/3}\right) = 6$

115. $\displaystyle\int_3^\infty \dfrac{2\, du}{u^2 - 2u} = \int_3^\infty \dfrac{du}{u - 2} - \int_3^\infty \dfrac{du}{u} = \lim_{b \to \infty} \left[\ln\left|\dfrac{u - 2}{u}\right|\right]_3^b = \lim_{b \to \infty} \left[\ln\left|\dfrac{b - 2}{b}\right|\right] - \ln\left|\dfrac{3 - 2}{3}\right| = 0 - \ln\left(\dfrac{1}{3}\right) = \ln 3$

117. $\displaystyle\int_0^\infty x^2 e^{-x}\, dx = \lim_{b \to \infty} \left[-x^2 e^{-x} - 2x e^{-x} - 2e^{-x}\right]_0^b = \lim_{b \to \infty} \left(-b^2 e^{-b} - 2b e^{-b} - 2e^{-b}\right) - (-2) = 0 + 2 = 2$

119. $\displaystyle\int_{-\infty}^\infty \dfrac{dx}{4x^2 + 9} = 2\int_0^\infty \dfrac{dx}{4x^2 + 9} = \dfrac{1}{2}\int_0^\infty \dfrac{dx}{x^2 + \dfrac{9}{4}} = \dfrac{1}{2}\lim_{b \to \infty} \left[\dfrac{2}{3}\tan^{-1}\left(\dfrac{2x}{3}\right)\right]_0^b = \dfrac{1}{2}\lim_{b \to \infty} \left[\dfrac{2}{3}\tan^{-1}\left(\dfrac{2b}{3}\right)\right] - \dfrac{1}{3}\tan^{-1}(0)$

$= \dfrac{1}{2}\left(\dfrac{2}{3} \cdot \dfrac{\pi}{2}\right) - 0 = \dfrac{\pi}{6}$

121. $\displaystyle\lim_{\theta\to\infty} \frac{\theta}{\sqrt{\theta^2+1}} = 1$ and $\displaystyle\int_6^\infty \frac{d\theta}{\theta}$ diverges $\Rightarrow \displaystyle\int_6^\infty \frac{d\theta}{\sqrt{\theta^2+1}}$ diverges

123. $\displaystyle\int_1^\infty \frac{\ln z}{z}\, dz = \int_1^e \frac{\ln z}{z}\, dz + \int_e^\infty \frac{\ln z}{z}\, dz = \left[(\ln z)^2\right]_1^e + \lim_{b\to\infty}\left[(\ln z)^2\right]_e^b = \left(1^2 - 0\right) + \lim_{b\to\infty}\left[(\ln b)^2 - 1\right]$

$= \infty \Rightarrow$ diverges

125. $0 < \dfrac{e^{-x}}{3+e^{-2x}} = \dfrac{1}{3e^x + e^{-x}} < \dfrac{1}{e^x + e^{-x}}$ and $\displaystyle\int_{-\infty}^\infty \frac{dx}{e^x+e^{-x}} = 2\int_0^\infty \frac{dx}{e^x+e^{-x}} < \int_0^\infty \frac{2\,dx}{e^x}$ converges

$\Rightarrow \displaystyle\int_{-\infty}^\infty \frac{e^{-x}}{3+e^{-2x}}\, dx$ converges

127. (a) $\displaystyle\int \frac{y\,dy}{\sqrt{16-y^2}} = -\frac{1}{2}\int \frac{d\left(16-y^2\right)}{\sqrt{16-y^2}} = -\sqrt{16-y^2} + C$

(b) $\displaystyle\int \frac{y\,dy}{\sqrt{16-y^2}};\ [y = 4\sin x] \to 4\int \frac{\sin x \cos x\, dx}{\cos x} = -4\cos x + C = -\frac{4\sqrt{16-y^2}}{4} + C = -\sqrt{16-y^2} + C$

129. (a) $\displaystyle\int \frac{x\,dx}{4-x^2} = -\frac{1}{2}\int \frac{d\left(4-x^2\right)}{4-x^2} = -\frac{1}{2}\ln\left|4-x^2\right| + C$

(b) $\displaystyle\int \frac{x\,dx}{4-x^2};\ [x = 2\sin\theta] \to \int \frac{2\sin\theta \cdot 2\cos\theta\, d\theta}{4\cos^2\theta} = \int \tan\theta\, d\theta = -\ln|\cos\theta| + C = -\ln\sqrt{4-x^2} + C$

$= -\frac{1}{2}\ln\left|4-x^2\right| + C$

131. $\displaystyle\int \frac{x\,dx}{9-x^2};\ \begin{bmatrix} u = 9-x^2 \\ du = -2x\,dx \end{bmatrix} \to -\frac{1}{2}\int \frac{du}{u} = -\frac{1}{2}\ln|u| + C = \ln\frac{1}{\sqrt{u}} + C = \ln\frac{1}{\sqrt{9-x^2}} + C$

133. $\displaystyle\int \frac{dx}{9-x^2} = \frac{1}{6}\int \frac{dx}{3-x} + \frac{1}{6}\int \frac{dx}{3+x} = -\frac{1}{6}\ln|3-x| + \frac{1}{6}\ln|3+x| + C = \frac{1}{6}\ln\left|\frac{x+3}{x-3}\right| + C$

135. $\displaystyle\int \frac{x\,dx}{1+\sqrt{x}};\ \begin{bmatrix} u = \sqrt{x} \\ du = \dfrac{dx}{2\sqrt{x}} \end{bmatrix} \to \int \frac{u^2 \cdot 2u\, du}{1+u} = \int \left(2u^2 - 2u + 2 - \frac{2}{1+u}\right) du = \frac{2}{3}u^3 - u^2 + 2u - 2\ln|1+u| + C$

$= \frac{2x^{3/2}}{3} - x + 2\sqrt{x} - 2\ln\left(1+\sqrt{x}\right) + C$

137. $\displaystyle\int \frac{dx}{x\left(x^2+1\right)^2};\ \begin{bmatrix} x = \tan\theta \\ dx = \sec^2\theta\, d\theta \end{bmatrix} \to \int \frac{\sec^2\theta\, d\theta}{\tan\theta\sec^4\theta} = \int \frac{\cos^3\theta}{\sin\theta}\, d\theta = \int \left(\frac{1-\sin^2\theta}{\sin\theta}\right) d(\sin\theta)$

$= \ln|\sin\theta| - \frac{1}{2}\sin^2\theta + C = \ln\left|\frac{x}{\sqrt{x^2+1}}\right| - \frac{1}{2}\left(\frac{x}{\sqrt{x^2+1}}\right)^2 + C$

139. $\displaystyle \int \frac{dx}{\sqrt{-2x-x^2}} = \int \frac{d(x+1)}{\sqrt{1-(x+1)^2}} = \sin^{-1}(x+1) + C$

141. $\displaystyle \int \frac{du}{\sqrt{1+u^2}}; \; [u = \tan\theta] \to \int \frac{\sec^2\theta \; d\theta}{\sec\theta} = \ln|\sec\theta + \tan\theta| + C = \ln\left|\sqrt{1+u^2} + u\right| + C$

143. $\displaystyle \int \frac{2 - \cos x + \sin x}{\sin^2 x}\, dx = \int 2\csc^2 x\, dx - \int \frac{\cos x\, dx}{\sin^2 x} + \int \csc x\, dx = -2\cot x + \frac{1}{\sin x} - \ln|\csc x + \cot x| + C$

$\displaystyle = -2\cot x + \csc x - \ln|\csc x + \cot x| + C$

145. $\displaystyle \int \frac{9\; dv}{81 - v^4} = \frac{1}{2}\int \frac{dv}{v^2+9} + \frac{1}{12}\int \frac{dv}{3-v} + \frac{1}{12}\int \frac{dv}{3+v} = \frac{1}{12}\ln\left|\frac{3+v}{3-v}\right| + \frac{1}{6}\tan^{-1}\frac{v}{3} + C$

147.

$$\cos(2\theta+1)$$

$$\theta \xrightarrow{\;(+)\;} \tfrac{1}{2}\sin(2\theta+1)$$

$$1 \xrightarrow{\;(-)\;} -\tfrac{1}{4}\cos(2\theta+1)$$

$$0 \qquad \int \theta \cos(2\theta+1)\, d\theta = \frac{\theta}{2}\sin(2\theta+1) + \frac{1}{4}\cos(2\theta+1) + C$$

149. $\displaystyle \int \frac{x^3\, dx}{x^2 - 2x + 1} = \int \left(x + 2 + \frac{3x+2}{x^2 - 2x + 1}\right) dx = \int (x+2)\, dx + 3\int \frac{dx}{x-1} + \int \frac{dx}{(x-1)^2}$

$\displaystyle = \frac{x^2}{2} + 2x + 3\ln|x-1| - \frac{1}{x-1} + C$

151. $\displaystyle \int \frac{2\sin\sqrt{x}\; dx}{\sqrt{x}\,\sec\sqrt{x}}; \; \begin{bmatrix} y = \sqrt{x} \\ dy = \dfrac{dx}{2\sqrt{x}} \end{bmatrix} \to \int \frac{2\sin y \cdot 2y\, dy}{y\,\sec y} = \int 2\sin 2y\, dy = -\cos(2y) + C = -\cos(2\sqrt{x}) + C$

153. $\displaystyle \int \frac{dy}{\sin y \cos y} = \int \frac{2\, dy}{\sin 2y} = \int 2\csc(2y)\, dy = -\ln|\csc(2y) + \cot(2y)| + C$

155. $\displaystyle \int \frac{\tan x}{\cos^2 x}\, dx = \int \tan x \sec^2 x\, dx = \int \tan x \cdot d(\tan x) = \frac{1}{2}\tan^2 x + C$

157. $\displaystyle \int \frac{(r+2)\, dr}{\sqrt{-r^2 - 4r}} = \int \frac{(r+2)\, dr}{\sqrt{4 - (r+2)^2}}; \; \begin{bmatrix} u = 4 - (r+2)^2 \\ du = -2(r+2)\, dr \end{bmatrix} \to -\int \frac{du}{2\sqrt{u}} = -\sqrt{u} + C = -\sqrt{4 - (r+2)^2} + C$

159. $\displaystyle \int \frac{\sin 2\theta\; d\theta}{(1 + \cos 2\theta)^2} = -\frac{1}{2}\int \frac{d(1 + \cos 2\theta)}{(1 + \cos 2\theta)^2} = \frac{1}{2(1 + \cos 2\theta)} + C = \frac{1}{4}\sec^2\theta + C$

161. $\displaystyle\int_{\pi/4}^{\pi/2} \sqrt{1 + \cos 4x}\, dx = -\sqrt{2} \int_{\pi/4}^{\pi/2} \cos 2x\, dx = \left[-\frac{\sqrt{2}}{2} \sin 2x\right]_{\pi/4}^{\pi/2} = \frac{\sqrt{2}}{2}$

163. $\displaystyle\int \frac{x\, dx}{\sqrt{2-x}}; \begin{bmatrix} y = 2 - x \\ dy = -dx \end{bmatrix} \to -\int \frac{(2-y)\, dy}{\sqrt{y}} = \frac{2}{3} y^{3/2} - 4y^{1/2} + C = \frac{2}{3}(2-x)^{3/2} - 4(2-x)^{1/2} + C$

$\displaystyle = 2\left[\frac{\left(\sqrt{2-x}\right)^3}{3} - 2\sqrt{2-x}\right] + C$

165. $\displaystyle\int \frac{dy}{y^2 - 2y + 2} = \int \frac{d(y-1)}{(y-1)^2 + 1} = \tan^{-1}(y-1) + C$

167. $\displaystyle\int \theta^2 \tan(\theta^3)\, d\theta = \frac{1}{3} \int \tan(\theta^3)\, d(\theta^3) = \frac{1}{3} \ln\left|\sec \theta^3\right| + C$

169. $\displaystyle\int \frac{z+1}{z^2(z^2+4)}\, dz = \frac{1}{4} \int \left(\frac{1}{z} + \frac{1}{z^2} - \frac{z+1}{z^2+4}\right) dz = \frac{1}{4} \ln|z| - \frac{1}{4z} - \frac{1}{8} \ln(z^2+4) - \frac{1}{8} \tan^{-1} \frac{z}{2} + C$

171. $\displaystyle\int \frac{t\, dt}{\sqrt{9 - 4t^2}} = -\frac{1}{8} \int \frac{d(9 - 4t^2)}{\sqrt{9 - 4t^2}} = -\frac{1}{4} \sqrt{9 - 4t^2} + C$

173. $\displaystyle\int \frac{\cot \theta\, d\theta}{1 + \sin^2 \theta} = \int \frac{\cos \theta\, d\theta}{(\sin \theta)(1 + \sin^2 \theta)}; \begin{bmatrix} x = \sin \theta \\ dx = \cos \theta\, d\theta \end{bmatrix} \to \int \frac{dx}{x(1 + x^2)} = \int \frac{dx}{x} - \int \frac{x\, dx}{x^2 + 1}$

$\displaystyle = \ln|\sin \theta| - \frac{1}{2} \ln(1 + \sin^2 \theta) + C$

175. $\displaystyle\int \frac{\tan \sqrt{y}\, dy}{2\sqrt{y}}; \left[\sqrt{y} = x\right] \to \int \frac{\tan x \cdot 2x\, dx}{2x} = \ln|\sec x| + C = \ln\left|\sec \sqrt{y}\right| + C$

177. $\displaystyle\int \frac{\theta^2\, d\theta}{4 - \theta^2} = \int \left(-1 + \frac{4}{4 - \theta^2}\right) d\theta = -\int d\theta - \int \frac{d\theta}{\theta - 2} + \int \frac{d\theta}{\theta + 2} = -\theta - \ln|\theta - 2| + \ln|\theta + 2| + C$

$\displaystyle = -\theta + \ln\left|\frac{\theta + 2}{\theta - 2}\right| + C$

179. $\displaystyle\int \frac{\cos(\sin^{-1} x)\, dx}{\sqrt{1 - x^2}}; \begin{bmatrix} u = \sin^{-1} x \\ du = \dfrac{dx}{\sqrt{1 - x^2}} \end{bmatrix} \to \int \cos u\, du = \sin u + C = \sin(\sin^{-1} x) + C = x + C$

181. $\displaystyle\int \sin \frac{x}{2} \cos \frac{x}{2}\, dx = \int \frac{1}{2} \sin\left(\frac{x}{2} + \frac{x}{2}\right) dx = \frac{1}{2} \int \sin x\, dx = -\frac{1}{2} \cos x + C$

183. $\displaystyle\int \frac{e^t\, dt}{1 + e^t} = \ln(1 + e^t) + C$

185. $\displaystyle\int_1^\infty \frac{\ln y \, dy}{y^3}; \begin{bmatrix} x = \ln y \\ dx = \dfrac{dy}{y} \\ dy = e^x \, dx \end{bmatrix} \to \int_0^\infty \frac{x \cdot e^x}{e^{3x}} \, dx = \int_0^\infty xe^{-2x} \, dx = \lim_{b\to\infty} \left[-\frac{x}{2}e^{-2x} - \frac{1}{4}e^{-2x} \right]_0^b$

$= \lim_{b\to\infty} \left(\frac{-b}{2e^{2b}} - \frac{1}{4e^{2b}} \right) - \left(0 - \frac{1}{4} \right) = \frac{1}{4}$

187. $\displaystyle\int \frac{\cot v \, dv}{\ln(\sin v)} = \int \frac{\cos v \, dv}{(\sin v)\ln(\sin v)}; \begin{bmatrix} u = \ln(\sin v) \\ du = \dfrac{\cos v \, dv}{\sin v} \end{bmatrix} \to \int \frac{du}{u} = \ln|u| + C = \ln\left|\ln(\sin v)\right| + C$

189. $\displaystyle\int e^{\ln \sqrt{x}} \, dx = \int \sqrt{x} \, dx = \frac{2}{3}x^{3/2} + C$

191. $\displaystyle\int \frac{\sin 5t \, dt}{1 + (\cos 5t)^2}; \begin{bmatrix} u = \cos 5t \\ du = -5\sin 5t \, dt \end{bmatrix} \to -\frac{1}{5}\int \frac{du}{1+u^2} = -\frac{1}{5}\tan^{-1}u + C = -\frac{1}{5}\tan^{-1}(\cos 5t) + C$

193. $\displaystyle\int (27)^{3\theta+1} \, d\theta = \frac{1}{3}\int (27)^{3\theta+1} \, d(3\theta+1) = \frac{1}{3\ln 27}(27)^{3\theta+1} + C = \frac{1}{3}\left(\frac{27^{3\theta+1}}{\ln 27} \right) + C$

195. $\displaystyle\int \frac{dr}{1 + \sqrt{r}}; \begin{bmatrix} u = \sqrt{r} \\ du = \dfrac{dr}{2\sqrt{r}} \end{bmatrix} \to \int \frac{2u \, du}{1+u} = \int \left(2 - \frac{2}{1+u} \right) du = 2u - 2\ln|1+u| + C = 2\sqrt{r} - 2\ln(1 + \sqrt{r}) + C$

197. $\displaystyle\int \frac{8 \, dy}{y^3(y+2)} = \int \frac{dy}{y} - \int \frac{2 \, dy}{y^2} + \int \frac{4 \, dy}{y^3} - \int \frac{dy}{(y+2)} = \ln\left| \frac{y}{y+2} \right| + \frac{2}{y} - \frac{2}{y^2} + C$

199. $\displaystyle\int \frac{8 \, dm}{m\sqrt{49m^2 - 4}} = \frac{8}{7}\int \frac{dm}{m\sqrt{m^2 - \left(\frac{2}{7}\right)^2}} = 4\sec^{-1}\left(\frac{7m}{2} \right) + C$

201. If $u = \displaystyle\int_0^x \sqrt{1 + (t-1)^4} \, dt$ and $dv = 3(x-1)^2 \, dx$, then $du = \sqrt{1 + (x-1)^4} \, dx$, and $v = (x-1)^3$ so integration

by parts $\Rightarrow \displaystyle\int_0^1 3(x-1)^2 \left[\int_0^x \sqrt{1 + (t-1)^4} (x-1)^3 \right] dx = \left[(x-1)^3 \int_0^x \sqrt{1 + (t-1)^4} \, dt \right]_0^1$

$- \displaystyle\int_0^1 (x-1)^3 \sqrt{1 + (x-1)^4} \, dx = \left[-\frac{1}{6}\left(1 + (x-1)^4 \right)^{3/2} \right]_0^1 = \frac{\sqrt{8} - 1}{6}$

203. $u = f(x)$, $du = f'(x)\, dx$; $dv = dx$, $v = x$;

$$\int_{\pi/2}^{3\pi/2} f(x)\, dx = \left[x\, f(x)\right]_{\pi/2}^{3\pi/2} - \int_{\pi/2}^{3\pi/2} xf'(x)\, dx = \left[\frac{3\pi}{2} f\!\left(\frac{3\pi}{2}\right) - \frac{\pi}{2} f\!\left(\frac{\pi}{2}\right)\right] - \int_{\pi/2}^{3\pi/2} \cos x\, dx$$

$$= \left(\frac{3\pi}{2} - \frac{\pi}{2}\right) - [\sin x]_{\pi/2}^{3\pi/2} = \pi - [(-1) - 1] = \pi + 2$$

CHAPTER 7 ADDITIONAL EXERCISES–THEORY, EXAMPLES, APPLICATIONS

1. $u = \left(\sin^{-1} x\right)^2$, $du = \dfrac{2\sin^{-1} x\, dx}{\sqrt{1 - x^2}}$; $dv = dx$, $v = x$;

$$\int \left(\sin^{-1} x\right)^2 dx = x\left(\sin^{-1} x\right)^2 - \int \frac{2x \sin^{-1} x\, dx}{\sqrt{1 - x^2}};$$

$u = \sin^{-1} x$, $du = \dfrac{dx}{\sqrt{1 - x^2}}$; $dv = -\dfrac{2x\, dx}{\sqrt{1 - x^2}}$, $v = 2\sqrt{1 - x^2}$;

$$\int \frac{2x \sin^{-1} x\, dx}{\sqrt{1 - x^2}} = 2\left(\sin^{-1} x\right)\sqrt{1 - x^2} - \int 2\, dx = 2\left(\sin^{-1} x\right)\sqrt{1 - x^2} - 2x + C; \text{ therefore}$$

$$\int \left(\sin^{-1} x\right)^2 dx = x\left(\sin^{-1} x\right)^2 + 2\left(\sin^{-1} x\right)\sqrt{1 - x^2} - 2x + C$$

3. $u = \sin^{-1} x$, $du = \dfrac{dx}{\sqrt{1 - x^2}}$; $dv = x\, dx$, $v = \dfrac{x^2}{2}$;

$$\int x \sin^{-1} x\, dx = \frac{x^2}{2} \sin^{-1} x - \int \frac{x^2\, dx}{2\sqrt{1 - x^2}}; \begin{bmatrix} x = \sin\theta \\ dx = \cos\theta\, d\theta \end{bmatrix} \rightarrow \int x \sin^{-1} x\, dx = \frac{x^2}{2} \sin^{-1} x - \int \frac{\sin^2\theta \cos\theta\, d\theta}{2\cos\theta}$$

$$= \frac{x^2}{2} \sin^{-1} x - \frac{1}{2}\int \sin^2\theta\, d\theta = \frac{x^2}{2} \sin^{-1} x - \frac{1}{2}\left(\frac{\theta}{2} - \frac{\sin 2\theta}{4}\right) + C = \frac{x^2}{2} \sin^{-1} x + \frac{\sin\theta \cos\theta - \theta}{4} + C$$

$$= \frac{x^2}{2} \sin^{-1} x + \frac{x\sqrt{1 - x^2} - \sin^{-1} x}{4} + C$$

5. $\displaystyle\int \frac{d\theta}{1 - \tan^2\theta} = \int \frac{\cos^2\theta}{\cos^2\theta - \sin^2\theta}\, d\theta = \int \frac{1 + \cos 2\theta}{2\cos 2\theta}\, d\theta = \frac{1}{2}\int (\sec 2\theta + 1)\, d\theta = \frac{\ln|\sec 2\theta + \tan 2\theta| + 2\theta}{4} + C$

7. $\displaystyle\int \frac{dt}{t - \sqrt{1 - t^2}}; \begin{bmatrix} t = \sin\theta \\ dt = \cos\theta\, d\theta \end{bmatrix} \rightarrow \int \frac{\cos\theta\, d\theta}{\sin\theta - \cos\theta} = \int \frac{d\theta}{\tan\theta - 1}; \begin{bmatrix} u = \tan\theta \\ du = \sec^2\theta\, d\theta \\ d\theta = \dfrac{du}{u^2 + 1} \end{bmatrix} \rightarrow \int \frac{du}{(u - 1)(u^2 + 1)}$

$$= \frac{1}{2}\int \frac{du}{u - 1} - \frac{1}{2}\int \frac{du}{u^2 + 1} - \frac{1}{2}\int \frac{u\, du}{u^2 + 1} = \frac{1}{2}\ln\left|\frac{u - 1}{\sqrt{u^2 + 1}}\right| - \frac{1}{2}\tan^{-1} u + C = \frac{1}{2}\ln\left|\frac{\tan\theta - 1}{\sec\theta}\right| - \frac{1}{2}\theta + C$$

$$= \frac{1}{2}\ln\left(t - \sqrt{1 - t^2}\right) - \frac{1}{2}\sin^{-1} t + C$$

9. $\int \dfrac{1}{x^4+4}\,dx = \int \dfrac{1}{\left(x^2+2\right)^2 - 4x^2}\,dx = \int \dfrac{1}{\left(x^2+2x+2\right)\left(x^2-2x+2\right)}\,dx$

$= \dfrac{1}{16}\int \left[\dfrac{2x+2}{x^2+2x+2} + \dfrac{2}{(x+1)^2+1} - \dfrac{2x-2}{x^2-2x+2} + \dfrac{2}{(x-1)^2+1}\right]dx$

$= \dfrac{1}{16}\ln\left|\dfrac{x^2+2x+2}{x^2-2x+2}\right| + \dfrac{1}{8}\left[\tan^{-1}(x+1) + \tan^{-1}(x-1)\right] + C$

11. $\displaystyle\lim_{x\to\infty}\int_{-x}^{x}\sin t\,dt = \lim_{x\to\infty}\left[-\cos t\right]_{-x}^{x} = \lim_{x\to\infty}\left[-\cos x + \cos(-x)\right] = \lim_{x\to\infty}(-\cos x + \cos x) = \lim_{x\to\infty} 0 = 0$

13. $\displaystyle\lim_{n\to\infty}\sum_{k=1}^{n}\ln\sqrt[n]{1+\frac{k}{n}} = \lim_{n\to\infty}\sum_{k=1}^{n}\ln\left(1+k\left(\tfrac{1}{n}\right)\right)\left(\tfrac{1}{n}\right) = \int_{0}^{1}\ln(1+x)\,dx;\ \left[\begin{array}{l} u = 1+x,\ du = dx \\ x = 0 \Rightarrow u = 1,\ x = 1 \Rightarrow u = 2 \end{array}\right]$

$\to \displaystyle\int_{1}^{2}\ln u\,du = \left[u\ln u - u\right]_{1}^{2} = (2\ln 2 - 2) - (\ln 1 - 1) = 2\ln 2 - 1 = \ln 4 - 1$

15. $\dfrac{dy}{dx} = \sqrt{\cos 2x} \Rightarrow 1 + \left(\dfrac{dy}{dx}\right)^2 = 1 + \cos 2x = 2\cos^2 x;\ L = \displaystyle\int_{0}^{\pi/4}\sqrt{1 + \left(\sqrt{\cos 2t}\right)^2}\,dt = \sqrt{2}\int_{0}^{\pi/4}\sqrt{\cos^2 t}\,dt$

$= \sqrt{2}\,[\sin t]_{0}^{\pi/4} = 1$

17. $V = \displaystyle\int_{a}^{b} 2\pi\binom{\text{shell}}{\text{radius}}\binom{\text{shell}}{\text{height}}dx = \int_{0}^{1} 2\pi xy\,dx$

$= 6\pi \displaystyle\int_{0}^{1} x^2\sqrt{1-x}\,dx;\ \left[\begin{array}{l} u = 1-x \\ du = -dx \\ x^2 = (1-u)^2 \end{array}\right]$

$\to -6\pi \displaystyle\int_{1}^{0}(1-u)^2\sqrt{u}\,du = -6\pi\int_{1}^{0}\left(u^{1/2} - 2u^{3/2} + u^{5/2}\right)du$

$= -6\pi\left[\dfrac{2}{3}u^{3/2} - \dfrac{4}{5}u^{5/2} + \dfrac{2}{7}u^{7/2}\right]_{1}^{0} = 6\pi\left(\dfrac{2}{3} - \dfrac{4}{5} + \dfrac{2}{7}\right) = 6\pi\left(\dfrac{70-84+30}{105}\right) = 6\pi\left(\dfrac{16}{105}\right) = \dfrac{32\pi}{35}$

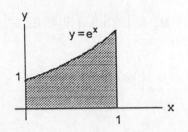

19. $V = \displaystyle\int_{a}^{b} 2\pi\binom{\text{shell}}{\text{radius}}\binom{\text{shell}}{\text{height}}dx = \int_{0}^{1} 2\pi xe^x\,dx$

$= 2\pi\left[xe^x - e^x\right]_{0}^{1} = 2\pi$

21. (a) $V = \int\limits_1^e \pi\left[1 - (\ln x)^2\right] dx$

$$= \pi\left[x - x(\ln x)^2\right]_1^e - 2\pi \int\limits_1^e \ln x \, dx \qquad \text{(FORMULA 110)}$$

$$= \pi\left[x - x(\ln x)^2 + 2(x \ln x - x)\right]_1^e$$

$$= \pi\left[-x - x(\ln x)^2 + 2x \ln x\right]_1^e = \pi\left[-e - e + 2e - (-1)\right] = \pi$$

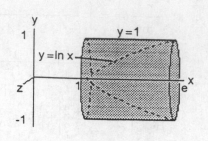

(b) $V = \int\limits_1^e \pi(1 - \ln x)^2 \, dx = \pi \int\limits_1^e \left[1 - 2 \ln x + (\ln x)^2\right] dx$

$$= \pi\left[x - 2(x \ln x - x) + x(\ln x)^2\right]_1^e - 2\pi \int\limits_1^e \ln x \, dx$$

$$= \pi\left[x - 2(x \ln x - x) + x(\ln x)^2 - 2(x \ln x - x)\right]_1^e$$

$$= \pi\left[5x - 4x \ln x + x(\ln x)^2\right]_1^e = \pi\left[(5e - 4e + e) - (5)\right]$$

$$= \pi(2e - 5)$$

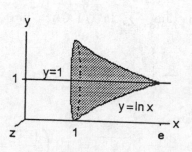

23. (a) $\lim\limits_{x \to 0^+} x \ln x = 0 \Rightarrow \lim\limits_{x \to 0^+} f(x) = 0 = f(0) \Rightarrow f$ is continuous

(b) $V = \int\limits_0^2 \pi x^2 (\ln x)^2 \, dx; \begin{bmatrix} u = (\ln x)^2 \\ du = (2 \ln x)\dfrac{dx}{x} \\ dv = x^2 \\ v = \dfrac{x^3}{3} \end{bmatrix} \rightarrow \pi\left(\lim\limits_{b \to 0^+} \left[\dfrac{x^3}{3}(\ln x)^2\right]_b^2 - \int\limits_0^2 \left(\dfrac{x^3}{3}\right)(2 \ln x)\dfrac{dx}{x}\right)$

$$= \pi\left[\left(\dfrac{8}{3}\right)(\ln 2)^2 - \left(\dfrac{2}{3}\right)\lim\limits_{b \to 0^+}\left[\dfrac{x^3}{3}\ln x - \dfrac{x^3}{9}\right]_b^2\right] = \pi\left[\dfrac{8(\ln 2)^2}{3} - \dfrac{16(\ln 2)}{9} + \dfrac{16}{27}\right]$$

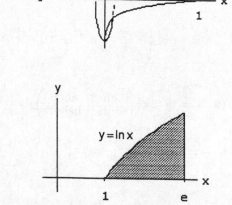

25. $M = \int\limits_1^e \ln x \, dx = [x \ln x - x]_1^e = (e - e) - (0 - 1) = 1;$

$$M_x = \int\limits_1^e (\ln x)\left(\dfrac{\ln x}{2}\right) dx = \dfrac{1}{2}\int\limits_1^e (\ln x)^2 \, dx$$

$$= \dfrac{1}{2}\left(\left[x(\ln x)^2\right]_1^e - 2\int\limits_1^e \ln x \, dx\right) = \dfrac{1}{2}(e - 2);$$

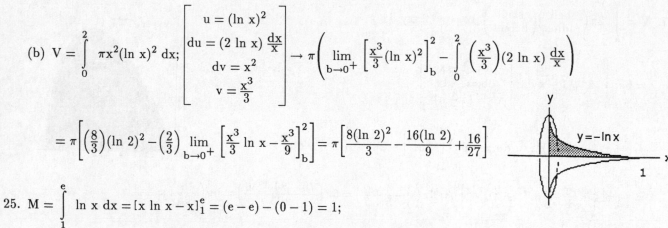

$$M_y = \int_1^e x \ln x \, dx = \left[\frac{x^2 \ln x}{2}\right]_1^e - \frac{1}{2}\int_1^e x \, dx = \frac{1}{2}\left[x^2 \ln x - \frac{x^2}{2}\right]_1^e$$

$$= \frac{1}{2}\left[\left(e^2 - \frac{e^2}{2}\right) + \frac{1}{2}\right] = \frac{1}{4}(e^2 + 1); \text{ therefore, } \bar{x} = \frac{M_y}{M} = \frac{e^2 + 1}{4} \text{ and } \bar{y} = \frac{M_x}{M} = \frac{e-2}{2}$$

27. $$L = \int_1^e \sqrt{1 + \frac{1}{x^2}} \, dx = \int_1^e \frac{\sqrt{x^2 + 1}}{x} \, dx; \begin{bmatrix} x = \tan \theta \\ dx = \sec^2 \theta \, d\theta \end{bmatrix} \rightarrow L = \int_{\pi/4}^{\tan^{-1} e} \frac{\sec \theta \cdot \sec^2 \theta \, d\theta}{\tan \theta}$$

$$= \int_{\pi/4}^{\tan^{-1} e} \frac{(\sec \theta)(\tan^2 \theta + 1)}{\tan \theta} \, d\theta = \int_{\pi/4}^{\tan^{-1} e} (\tan \theta \sec \theta + \csc \theta) \, d\theta = \left[\sec \theta - \ln|\csc \theta + \cot \theta|\right]_{\pi/4}^{\tan^{-1} e}$$

$$= \left(\sqrt{1 + e^2} - \ln\left|\frac{\sqrt{1 + e^2}}{e} + \frac{1}{e}\right|\right) = \left[\sqrt{2} - \ln(1 + \sqrt{2})\right] = \sqrt{1 + e^2} - \ln\left(\frac{\sqrt{1 + e^2}}{e} + \frac{1}{e}\right) - \sqrt{2} + \ln(1 + \sqrt{2})$$

29. $$L = 4\int_0^1 \sqrt{1 + \left(\frac{dy}{dx}\right)^2} \, dy; \ x^{2/3} + y^{2/3} = 1 \Rightarrow y = \left(1 - x^{2/3}\right)^{3/2} \Rightarrow \frac{dy}{dx} = -\frac{3}{2}\left(1 - x^{2/3}\right)^{1/2}\left(x^{-1/3}\right)\left(\frac{2}{3}\right)$$

$$\Rightarrow \left(\frac{dy}{dx}\right)^2 = \frac{1 - x^{2/3}}{x^{2/3}} \Rightarrow L = 4\int_0^1 \sqrt{1 + \left(\frac{1 - x^{2/3}}{x^{2/3}}\right)} \, dx = 4\int_0^1 \frac{dx}{x^{1/3}} = 6\left[x^{2/3}\right]_0^1 = 6$$

31. $$\left(\frac{dy}{dx}\right)^2 = \frac{1}{4x} \Rightarrow \frac{dy}{dx} = \frac{\pm 1}{2\sqrt{x}} \Rightarrow y = \pm\sqrt{x}, \ 0 \le x \le 4$$

33. (b) $$\int_{-\infty}^{\infty} e^{(x - e^x)} \, dx = \int_{-\infty}^{\infty} e^{(-e^x)} e^x \, dx$$

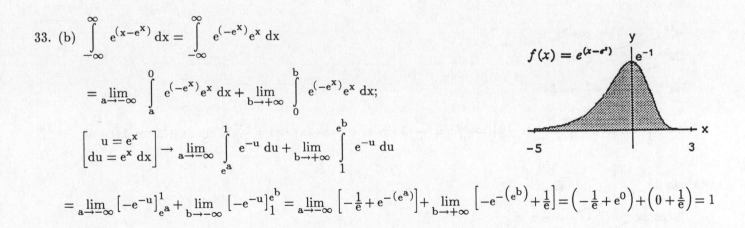

$$= \lim_{a \to -\infty} \int_a^0 e^{(-e^x)} e^x \, dx + \lim_{b \to +\infty} \int_0^b e^{(-e^x)} e^x \, dx;$$

$$\begin{bmatrix} u = e^x \\ du = e^x \, dx \end{bmatrix} \rightarrow \lim_{a \to -\infty} \int_{e^a}^1 e^{-u} \, du + \lim_{b \to +\infty} \int_1^{e^b} e^{-u} \, du$$

$$= \lim_{a \to -\infty} \left[-e^{-u}\right]_{e^a}^1 + \lim_{b \to -\infty} \left[-e^{-u}\right]_1^{e^b} = \lim_{a \to -\infty} \left[-\frac{1}{e} + e^{-(e^a)}\right] + \lim_{b \to +\infty} \left[-e^{-(e^b)} + \frac{1}{e}\right] = \left(-\frac{1}{e} + e^0\right) + \left(0 + \frac{1}{e}\right) = 1$$

35. $u = x^2 - a^2 \Rightarrow du = 2x \, dx;$

$$\int x\left(\sqrt{x^2 - a^2}\right)^n \, dx = \frac{1}{2}\int \left(\sqrt{u}\right)^n \, du = \frac{1}{2}\int u^{n/2} \, du = \frac{1}{2}\left(\frac{u^{n/2 + 1}}{\frac{n}{2} + 1}\right) + C, \ n \neq -2$$

$$= \frac{u^{(n+2)/2}}{n + 2} + C = \frac{\left(\sqrt{u}\right)^{n+2}}{n + 2} + C = \frac{\left(\sqrt{x^2 - a^2}\right)^{n+2}}{n + 2} + C$$

37. $\int\limits_{1}^{\infty}\left(\dfrac{ax}{x^2+1}-\dfrac{1}{2x}\right)dx = \lim\limits_{b\to\infty}\int\limits_{1}^{b}\left(\dfrac{ax}{x^2+1}-\dfrac{1}{2x}\right)dx = \lim\limits_{b\to\infty}\left[\dfrac{a}{2}\ln(x^2+1)-\dfrac{1}{2}\ln x\right]_1^b = \lim\limits_{b\to\infty}\left[\dfrac{1}{2}\ln\dfrac{(x^2+1)^a}{x}\right]_1^b$

$= \lim\limits_{b\to\infty}\dfrac{1}{2}\left[\ln\dfrac{(b^2+1)^a}{b}-\ln 2^a\right]$; $\lim\limits_{b\to\infty}\dfrac{(b^2+1)^a}{b} > \lim\limits_{b\to\infty}\dfrac{b^{2a}}{b} = \lim\limits_{b\to\infty}b^{2\left(a-\frac{1}{2}\right)} = \infty$ if $a > \dfrac{1}{2} \Rightarrow$ the improper

integral diverges if $a > \dfrac{1}{2}$; for $a = \dfrac{1}{2}$: $\lim\limits_{b\to\infty}\dfrac{\sqrt{b^2+1}}{b} = \lim\limits_{b\to\infty}\sqrt{1+\dfrac{1}{b^2}} = 1 \Rightarrow \lim\limits_{b\to\infty}\dfrac{1}{2}\left[\ln\dfrac{(b^2+1)^{1/2}}{b}-\ln 2^{1/2}\right]$

$= \dfrac{1}{2}\left(\ln 1 - \dfrac{1}{2}\ln 2\right) = -\dfrac{\ln 2}{4}$; if $a < \dfrac{1}{2}$: $0 \le \lim\limits_{b\to\infty}\dfrac{(b^2+1)^a}{b} < \lim\limits_{b\to\infty}\dfrac{(b+1)^{2a}}{b+1} = \lim\limits_{b\to\infty}(b+1)^{2a-1} = 0$

$\Rightarrow \lim\limits_{b\to\infty}\ln\dfrac{(b^2+1)^a}{b} = -\infty \Rightarrow$ the improper integral diverges if $a < \dfrac{1}{2}$; in summary, the improper integral

$\int\limits_{1}^{\infty}\left(\dfrac{ax}{x^2+1}-\dfrac{1}{2x}\right)dx$ converges only when $a = \dfrac{1}{2}$ and has the value $-\dfrac{\ln 2}{4}$

39. $A = \int\limits_{1}^{\infty}\dfrac{dx}{x^p}$ converges if $p > 1$ and diverges if $p \le 1$ (Exercise 67 in Section 7.6). Thus, $p \le 1$ for infinite area.

The volume of the solid of revolution about the x-axis is $V = \int\limits_{1}^{\infty}\pi\left(\dfrac{1}{x^p}\right)^2 dx = \pi\int\limits_{1}^{\infty}\dfrac{dx}{x^{2p}}$ which converges if

$2p > 1$ and diverges if $2p \le 1$. Thus we want $p > \dfrac{1}{2}$ for finite volume. In conclusion, the curve $y = x^{-p}$ gives

infinite area and finite volume for values of p satisfying $\dfrac{1}{2} < p \le 1$.

41. $e^{2x} \qquad (+) \qquad \cos 3x$

$2e^{2x} \qquad (-) \qquad \dfrac{1}{3}\sin 3x$

$4e^{2x} \qquad (+) \qquad -\dfrac{1}{9}\cos 3x$

$I = \dfrac{e^{2x}}{3}\sin 3x + \dfrac{2e^{2x}}{9}\cos 3x - \dfrac{4}{9}I \Rightarrow \dfrac{13}{9}I = \dfrac{e^{2x}}{9}(3\sin 3x + 2\cos 3x) \Rightarrow I = \dfrac{e^{2x}}{13}(3\sin 3x + 2\cos 3x) + C$

43. $\sin 3x \qquad (+) \qquad \sin x$

$3\cos 3x \qquad (-) \qquad -\cos x$

$-9\sin 3x \qquad (+) \qquad -\sin x$

$I = -\sin 3x\cos x + 3\cos 3x\sin x + 9I \Rightarrow -8I = -\sin 3x\cos x + 3\cos 3x\sin x$

$\Rightarrow I = \dfrac{\sin 3x\cos x - 3\cos 3x\sin x}{8} + C$

45. e^{ax} (+)   sin bx

ae^{ax} (−) $-\dfrac{1}{b}\cos bx$

$a^2 e^{ax}$ (+) $-\dfrac{1}{b^2}\sin bx$

$$I = -\frac{e^{ax}}{b}\cos bx + \frac{ae^{ax}}{b^2}\sin bx - \frac{a^2}{b^2}I \Rightarrow \left(\frac{a^2+b^2}{b^2}\right)I = \frac{e^{ax}}{b^2}(a\sin bx - b\cos bx)$$

$$\Rightarrow I = \frac{e^{ax}}{a^2+b^2}(a\sin bx - b\cos bx) + C$$

47. $\ln(ax)$ (+)   1

$\dfrac{1}{x}$ (−)   x

$$I = x\ln(ax) - \int\left(\frac{1}{x}\right)x\,dx = x\ln(ax) - x + C$$

49. (a) $\Gamma(1) = \displaystyle\int_0^\infty e^{-t}\,dt = \lim_{b\to\infty}\int_0^b e^{-t}\,dt = \lim_{b\to\infty}\left[-e^{-t}\right]_0^b = \lim_{b\to\infty}\left[-\frac{1}{e^b} - (-1)\right] = 0 + 1 = 1$

   (b) $u = t^x$, $du = xt^{x-1}\,dt$; $dv = e^{-t}\,dt$, $v = -e^{-t}$; x = fixed positive real

$$\Rightarrow \Gamma(x+1) = \int_0^\infty t^x e^{-t}\,dt = \lim_{b\to\infty}\left[-t^x e^{-t}\right]_0^b + x\int_0^\infty t^{x-1}e^{-t}\,dt = \lim_{b\to\infty}\left(-\frac{b^x}{e^b} + 0^x e^0\right) + x\Gamma(x) = x\Gamma(x)$$

   (c) $\Gamma(n+1) = n\Gamma(n) = n!$:

       $n = 0$: $\Gamma(0+1) = \Gamma(1) = 0!$;

       $n = k$: Assume $\Gamma(k+1) = k!$ for some $k > 0$;

       $n = k+1$: $\Gamma(k+1+1) = (k+1)\,\Gamma(k+1)$ from part (b)

                       $= (k+1)k!$ induction hypothesis

                       $= (k+1)!$ definition of factorial

    Thus, $\Gamma(n+1) = n\Gamma(n) = n!$ for every positive integer n.

NOTES:

CHAPTER 8 INFINITE SERIES

8.1 LIMITS OF SEQUENCES OF NUMBERS

1. $a_1 = \dfrac{1-1}{1^2} = 0$, $a_2 = \dfrac{1-2}{2^2} = -\dfrac{1}{4}$, $a_3 = \dfrac{1-3}{3^2} = -\dfrac{2}{9}$, $a_4 = \dfrac{1-4}{4^2} = -\dfrac{3}{16}$

3. $a_1 = \dfrac{(-1)^2}{2-1} = 1$, $a_2 = \dfrac{(-1)^3}{4-1} = -\dfrac{1}{3}$, $a_3 = \dfrac{(-1)^4}{6-1} = \dfrac{1}{5}$, $a_4 = \dfrac{(-1)^5}{8-1} = -\dfrac{1}{7}$

5. $a_1 = \dfrac{2}{2^2} = \dfrac{1}{2}$, $a_2 = \dfrac{2^2}{2^3} = \dfrac{1}{2}$, $a_3 = \dfrac{2^3}{2^4} = \dfrac{1}{2}$, $a_4 = \dfrac{2^4}{2^5} = \dfrac{1}{2}$

7. $a_1 = 1$, $a_2 = 1 + \dfrac{1}{2} = \dfrac{3}{2}$, $a_3 = \dfrac{3}{2} + \dfrac{1}{2^2} = \dfrac{7}{4}$, $a_4 = \dfrac{7}{4} + \dfrac{1}{2^3} = \dfrac{15}{8}$, $a_5 = \dfrac{15}{8} + \dfrac{1}{2^4} = \dfrac{31}{16}$, $a_6 = \dfrac{63}{32}$,

$a_7 = \dfrac{127}{64}$, $a_8 = \dfrac{255}{128}$, $a_9 = \dfrac{511}{256}$, $a_{10} = \dfrac{1023}{512}$

9. $a_1 = 2$, $a_2 = \dfrac{(-1)^2(2)}{2} = 1$, $a_3 = \dfrac{(-1)^3(1)}{2} = -\dfrac{1}{2}$, $a_4 = \dfrac{(-1)^4\left(-\frac{1}{2}\right)}{2} = -\dfrac{1}{4}$, $a_5 = \dfrac{(-1)^5\left(-\frac{1}{4}\right)}{2} = \dfrac{1}{8}$,

$a_6 = \dfrac{1}{16}$, $a_7 = -\dfrac{1}{32}$, $a_8 = -\dfrac{1}{64}$, $a_9 = \dfrac{1}{128}$, $a_{10} = \dfrac{1}{256}$

11. $a_1 = 1$, $a_2 = 1$, $a_3 = 1 + 1 = 2$, $a_4 = 2 + 1 = 3$, $a_5 = 3 + 2 = 5$, $a_6 = 8$, $a_7 = 13$, $a_8 = 21$, $a_9 = 34$, $a_{10} = 55$

13. $a_n = (-1)^{n+1}$, $n = 1, 2, \ldots$

15. $a_n = (-1)^{n+1}n^2$, $n = 1, 2, \ldots$

17. $a_n = n^2 - 1$, $n = 1, 2, \ldots$

19. $a_n = 4n - 3$, $n = 1, 2, \ldots$

21. $a_n = \dfrac{1 + (-1)^{n+1}}{2}$, $n = 1, 2, \ldots$

23. $\left| \sqrt[n]{0.5} - 1 \right| < 10^{-3} \Rightarrow -\dfrac{1}{1000} < \left(\dfrac{1}{2}\right)^{1/n} - 1 < \dfrac{1}{1000} \Rightarrow \left(\dfrac{999}{1000}\right)^n < \dfrac{1}{2} < \left(\dfrac{1001}{1000}\right)^n \Rightarrow n > \dfrac{\ln\left(\frac{1}{2}\right)}{\ln\left(\frac{999}{1000}\right)} \Rightarrow n > 692.8$

$\Rightarrow N = 692$; $a_n = \left(\dfrac{1}{2}\right)^{1/n}$ and $\lim\limits_{n \to \infty} a_n = 1$

25. $(0.9)^n < 10^{-3} \Rightarrow n \ln(0.9) < -3 \ln 10 \Rightarrow n > \dfrac{-3 \ln 10}{\ln(0.9)} \approx 65.54 \Rightarrow N = 65$; $a_n = \left(\dfrac{9}{10}\right)^n$ and $\lim\limits_{n \to \infty} a_n = 0$

27. (a) $f(x) = x^2 - a \Rightarrow f'(x) = 2x \Rightarrow x_{n+1} = x_n - \dfrac{x_n^2 - a}{2x_n} \Rightarrow x_{n+1} = \dfrac{2x_n^2 - \left(x_n^2 - a\right)}{2x_n} = \dfrac{x_n^2 + a}{2x_n} = \dfrac{\left(x_n + \frac{a}{x_n}\right)}{2}$

   (b) $x_1 = 2$, $x_2 = 1.75$, $x_3 = 1.732142857$, $x_4 = 1.73205081$, $x_5 = 1.732050808$; we are finding the positive number where $x^2 - 3 = 0$; that is, where $x^2 = 3$, $x > 0$, or where $x = \sqrt{3}$.

29. $x_1 = 1$, $x_2 = 1 + \cos(1) = 1.540302306$, $x_3 = 1.540302306 + \cos(1 + \cos(1)) = 1.570791601$,

$x_4 = 1.570791601 + \cos(1.570791601) = 1.570796327 = \frac{\pi}{2}$ to 9 decimal places. After a few steps, the

$\text{arc}(x_{n-1})$ and line segment $\cos(x_{n-1})$ are nearly the same as the quarter circle.

31. $a_{n+1} \geq a_n \Rightarrow \dfrac{3(n+1)+1}{(n+1)+1} > \dfrac{3n+1}{n+1} \Rightarrow \dfrac{3n+4}{n+2} > \dfrac{3n+1}{n+1} \Rightarrow 3n^2 + 3n + 4n + 4 > 3n^2 + 6n + n + 2$

$\Rightarrow 4 > 2$; the steps are reversible so the sequence is nondecreasing; $\dfrac{3n+1}{n+1} < 3 \Rightarrow 3n+1 < 3n+3$

$\Rightarrow 1 < 3$; the steps are reversible so the sequence is bounded above by 3

33. $a_{n+1} \leq a_n \Rightarrow \dfrac{2^{n+1}3^{n+1}}{(n+1)!} \leq \dfrac{2^n 3^n}{n!} \Rightarrow \dfrac{2^{n+1}3^{n+1}}{2^n 3^n} \leq \dfrac{(n+1)!}{n!} \Rightarrow 2 \cdot 3 \leq n+1$ which is true for $n \geq 5$; the steps are

reversible so the sequence is decreasing after a_5, but it is not nondecreasing for all its terms; $a_1 = 6$, $a_2 = 18$,

$a_3 = 36$, $a_4 = 54$, $a_5 = \dfrac{324}{5} = 64.8 \Rightarrow$ the sequence is bounded from above by 64.8

35. $a_n = 1 - \dfrac{1}{n}$ converges because $\dfrac{1}{n} \to 0$ by Example 2; also it is a nondecreasing sequence bounded above by 1

37. $a_n = \dfrac{2^n - 1}{2^n} = 1 - \dfrac{1}{2^n}$ and $0 < \dfrac{1}{2^n} < \dfrac{1}{n}$; since $\dfrac{1}{n} \to 0$ (by Example 2) $\Rightarrow \dfrac{1}{2^n} \to 0$, the sequence converges; also it is

a nondecreasing sequence bounded above by 1

39. $a_n = \left((-1)^n + 1\right)\left(\dfrac{n+1}{n}\right)$ diverges because $a_n = 0$ for n odd, while for n even $a_n = 2\left(1 + \dfrac{1}{n}\right)$ converges to 2; it

diverges by definition of divergence

41. If $\{a_n\}$ is nonincreasing with lower bound M, then $\{-a_n\}$ is a nondecreasing sequence with upper bound $-M$.

By Theorem 1, $\{-a_n\}$ converges and hence $\{a_n\}$ converges. If $\{a_n\}$ has no lower bound, then $\{-a_n\}$ has no

upper bound and therefore diverges. Hence, $\{a_n\}$ also diverges.

43. $a_n \geq a_{n+1} \Leftrightarrow \dfrac{1 + \sqrt{2n}}{\sqrt{n}} \geq \dfrac{1 + \sqrt{2(n+1)}}{\sqrt{n+1}} \Leftrightarrow \sqrt{n+1} + \sqrt{2n^2 + 2n} \geq \sqrt{n} + \sqrt{2n^2 + 2n} \Leftrightarrow \sqrt{n+1} \geq \sqrt{n}$

and $\dfrac{1 + \sqrt{2n}}{\sqrt{n}} \geq \sqrt{2}$; thus the sequence is nonincreasing and bounded below by $\sqrt{2} \Rightarrow$ it converges

45. $\dfrac{4^{n+1} + 3^n}{4^n} = 4 + \left(\dfrac{3}{4}\right)^n$ so $a_n \geq a_{n+1} \Leftrightarrow 4 + \left(\dfrac{3}{4}\right)^n \geq 4 + \left(\dfrac{3}{4}\right)^{n+1} \Leftrightarrow \left(\dfrac{3}{4}\right)^n \geq \left(\dfrac{3}{4}\right)^{n+1} \Leftrightarrow 1 \geq \dfrac{3}{4}$ and

$4 + \left(\dfrac{3}{4}\right)^n \geq 4$; thus the sequence is nonincreasing and bounded below by $4 \Rightarrow$ it converges

47. Let $0 < M < 1$ and let N be an integer greater than $\dfrac{M}{1-M}$. Then $n > N \Rightarrow n > \dfrac{M}{1-M} \Rightarrow n - nM > M$

$\Rightarrow n > M + nM \Rightarrow n > M(n+1) \Rightarrow \dfrac{n}{n+1} > M$.

49. The sequence $a_n = 1 + \dfrac{(-1)^n}{2}$ is the sequence $\dfrac{1}{2}, \dfrac{3}{2}, \dfrac{1}{2}, \dfrac{3}{2}, \ldots$. This sequence is bounded above by $\dfrac{3}{2}$,

but it clearly does not converge, by definition of convergence.

51. Given an $\epsilon > 0$, by definition of convergence there corresponds an N such that for all $n > N$,

$|L_1 - a_n| < \epsilon$ and $|L_2 - a_n| < \epsilon$. Now $|L_2 - L_1| = |L_2 - a_n + a_n - L_1| \leq |L_2 - a_n| + |a_n - L_1| < \epsilon + \epsilon = 2\epsilon$.

$|L_2 - L_1| < 2\epsilon$ says that the difference between two fixed values is smaller than any positive number 2ϵ.

The only nonnegative number smaller than every positive number is 0, so $|L_1 - L_2| = 0$ or $L_1 = L_2$.

53. $a_{2k} \to L \Leftrightarrow$ given an $\epsilon > 0$ there corresponds an N_1 such that $\left[2k > N_1 \Rightarrow |a_{2k} - L| < \epsilon\right]$. Similarly,

$a_{2k+1} \to L \Leftrightarrow \left[2k + 1 > N_2 \Rightarrow |a_{2k+1} - L| < \epsilon\right]$. Let $N = \max\{N_1, N_2\}$. Then $n > N \Rightarrow |a_n - L| < \epsilon$ whether

n is even or odd, and hence $a_n \to L$.

8.2   THEOREMS FOR CALCULATING LIMITS OF SEQUENCES

1. $\lim\limits_{n \to \infty} 2 + (0.1)^n = 2 \Rightarrow$ converges   (Table 8.1, #4)

3. $\lim\limits_{n \to \infty} \dfrac{1 - 2n}{1 + 2n} = \lim\limits_{n \to \infty} \dfrac{\left(\frac{1}{n}\right) - 2}{\left(\frac{1}{n}\right) + 2} = \lim\limits_{n \to \infty} \dfrac{-2}{2} = -1 \Rightarrow$ converges

5. $\lim\limits_{n \to \infty} \dfrac{1 - 5n^4}{n^4 + 8n^3} = \lim\limits_{n \to \infty} \dfrac{\left(\frac{1}{n^4}\right) - 5}{1 + \left(\frac{8}{n}\right)} = -5 \Rightarrow$ converges

7. $\lim\limits_{n \to \infty} \dfrac{n^2 - 2n + 1}{n - 1} = \lim\limits_{n \to \infty} \dfrac{(n - 1)(n - 1)}{n - 1} = \lim\limits_{n \to \infty} (n - 1) = \infty \Rightarrow$ diverges

9. $\lim\limits_{n \to \infty} \left(1 + (-1)^n\right)$ does not exist $\Rightarrow$ diverges

11. $\lim\limits_{n \to \infty} \left(\dfrac{n + 1}{2n}\right)\left(1 - \dfrac{1}{n}\right) = \lim\limits_{n \to \infty} \left(\dfrac{1}{2} + \dfrac{1}{2n}\right)\left(1 - \dfrac{1}{n}\right) = \dfrac{1}{2} \Rightarrow$ converges

13. $\lim\limits_{n \to \infty} \dfrac{(-1)^{n+1}}{2n - 1} = 0 \Rightarrow$ converges

15. $\lim\limits_{n \to \infty} \sqrt{\dfrac{2n}{n + 1}} = \sqrt{\lim\limits_{n \to \infty} \dfrac{2n}{n + 1}} = \sqrt{\lim\limits_{n \to \infty} \left(\dfrac{2}{1 + \frac{1}{n}}\right)} = \sqrt{2} \Rightarrow$ converges

17. $\lim\limits_{n \to \infty} \sin\left(\dfrac{\pi}{2} + \dfrac{1}{n}\right) = \sin\left(\lim\limits_{n \to \infty} \left(\dfrac{\pi}{2} + \dfrac{1}{n}\right)\right) = \sin\dfrac{\pi}{2} = 1 \Rightarrow$ converges

19. $\lim\limits_{n \to \infty} \dfrac{\sin n}{n} = 0$ because $-\dfrac{1}{n} \leq \dfrac{\sin n}{n} \leq \dfrac{1}{n} \Rightarrow$ converges by the Sandwich Theorem for sequences

21. $\lim\limits_{n \to \infty} \dfrac{n}{2^n} = \lim\limits_{n \to \infty} \dfrac{1}{2^n \ln 2} = 0 \Rightarrow$ converges (using l'Hôpital's rule)

23. $\lim\limits_{n\to\infty} \dfrac{\ln(n+1)}{\sqrt{n}} = \lim\limits_{n\to\infty} \dfrac{\left(\frac{1}{n+1}\right)}{\left(\frac{1}{2\sqrt{n}}\right)} = \lim\limits_{n\to\infty} \dfrac{2\sqrt{n}}{n+1} = \lim\limits_{n\to\infty} \dfrac{\left(\frac{2}{\sqrt{n}}\right)}{1+\left(\frac{1}{n}\right)} = 0 \Rightarrow$ converges

25. $\lim\limits_{n\to\infty} 8^{1/n} = 1 \Rightarrow$ converges   (Table 8.1, #3)

27. $\lim\limits_{n\to\infty} \left(1+\dfrac{7}{n}\right)^n = e^7 \Rightarrow$ converges   (Table 8.1, #5)

29. $\lim\limits_{n\to\infty} \sqrt[n]{10n} = \lim\limits_{n\to\infty} 10^{1/n} \cdot n^{1/n} = 1\cdot 1 = 1 \Rightarrow$ converges   (Table 8.1, #3 and #2)

31. $\lim\limits_{n\to\infty} \left(\dfrac{3}{n}\right)^{1/n} = \dfrac{\lim\limits_{n\to\infty} 3^{1/n}}{\lim\limits_{n\to\infty} n^{1/n}} = \dfrac{1}{1} = 1 \Rightarrow$ converges   (Table 8.1, #3 and #2)

33. $\lim\limits_{n\to\infty} \dfrac{\ln n}{n^{1/n}} = \dfrac{\lim\limits_{n\to\infty} \ln n}{\lim\limits_{n\to\infty} n^{1/n}} = \dfrac{\infty}{1} = \infty \Rightarrow$ diverges   (Table 8.1, #2)

35. $\lim\limits_{n\to\infty} \sqrt[n]{4^n n} = \lim\limits_{n\to\infty} 4 \sqrt[n]{n} = 4\cdot 1 = 4 \Rightarrow$ converges   (Table 8.1, #2)

37. $\lim\limits_{n\to\infty} \dfrac{n!}{n^n} = \lim\limits_{n\to\infty} \dfrac{1\cdot 2\cdot 3\cdots(n-1)(n)}{n\cdot n\cdot n\cdots n\cdot n} \le \lim\limits_{n\to\infty} \left(\dfrac{1}{n}\right) = 0$ and $\dfrac{n!}{n^n} \ge 0 \Rightarrow \lim\limits_{n\to\infty} \dfrac{n!}{n^n} = 0 \Rightarrow$ converges

39. $\lim\limits_{n\to\infty} \dfrac{n!}{10^{6n}} = \lim\limits_{n\to\infty} \dfrac{1}{\left(\frac{(10^6)^n}{n!}\right)} = \infty \Rightarrow$ diverges   (Table 8.1, #6)

41. $\lim\limits_{n\to\infty} \left(\dfrac{1}{n}\right)^{1/(\ln n)} = \lim\limits_{n\to\infty} \exp\left(\dfrac{1}{\ln n}\ln\left(\dfrac{1}{n}\right)\right) = \lim\limits_{n\to\infty} \exp\left(\dfrac{\ln 1 - \ln n}{\ln n}\right) = e^{-1} \Rightarrow$ converges

43. $\lim\limits_{n\to\infty} \left(\dfrac{3n+1}{3n-1}\right)^n = \lim\limits_{n\to\infty} \exp\left(n\ln\left(\dfrac{3n+1}{3n-1}\right)\right) = \lim\limits_{n\to\infty} \exp\left(\dfrac{\ln(3n+1)-\ln(3n-1)}{\frac{1}{n}}\right)$

$= \lim\limits_{n\to\infty} \exp\left(\dfrac{\frac{3}{3n+1}-\frac{3}{3n-1}}{\left(-\frac{1}{n^2}\right)}\right) = \lim\limits_{n\to\infty} \exp\left(\dfrac{6n^2}{(3n+1)(3n-1)}\right) = \exp\left(\dfrac{6}{9}\right) = e^{2/3} \Rightarrow$ converges

45. $\lim\limits_{n\to\infty} \left(\dfrac{x^n}{2n+1}\right)^{1/n} = \lim\limits_{n\to\infty} x\left(\dfrac{1}{2n+1}\right)^{1/n} = x\lim\limits_{n\to\infty} \exp\left(\dfrac{1}{n}\ln\left(\dfrac{1}{2n+1}\right)\right) = x\lim\limits_{n\to\infty} \exp\left(\dfrac{-\ln(2n+1)}{n}\right)$

$= x\lim\limits_{n\to\infty} \exp\left(\dfrac{-2}{2n+1}\right) = xe^0 = x,\ x > 0 \Rightarrow$ converges

47. $\lim\limits_{n\to\infty} \dfrac{3^n\cdot 6^n}{2^{-n}\cdot n!} = \lim\limits_{n\to\infty} \dfrac{36^n}{n!} = 0 \Rightarrow$ converges   (Table 8.1, #6)

49. $\lim\limits_{n\to\infty} \tanh n = \lim\limits_{n\to\infty} \dfrac{e^n - e^{-n}}{e^n + e^{-n}} = \lim\limits_{n\to\infty} \dfrac{e^{2n} - 1}{e^{2n} + 1} = \lim\limits_{n\to\infty} \dfrac{2e^{2n}}{2e^{2n}} = \lim\limits_{n\to\infty} 1 = 1 \Rightarrow$ converges

51. $\lim\limits_{n\to\infty} \dfrac{n^2 \sin\left(\frac{1}{n}\right)}{2n - 1} = \lim\limits_{n\to\infty} \dfrac{\sin\left(\frac{1}{n}\right)}{\left(\frac{2}{n} - \frac{1}{n^2}\right)} = \lim\limits_{n\to\infty} \dfrac{-\left(\cos\left(\frac{1}{n}\right)\right)\left(\frac{1}{n^2}\right)}{\left(-\frac{2}{n^2} + \frac{2}{n^3}\right)} = \lim\limits_{n\to\infty} \dfrac{-\cos\left(\frac{1}{n}\right)}{-2 + \left(\frac{2}{n}\right)} = \frac{1}{2} \Rightarrow$ converges

53. $\lim\limits_{n\to\infty} \tan^{-1} n = \frac{\pi}{2} \Rightarrow$ converges

55. $\lim\limits_{n\to\infty} \left(\frac{1}{3}\right)^n + \dfrac{1}{\sqrt{2^n}} = \lim\limits_{n\to\infty} \left(\left(\frac{1}{3}\right)^n + \left(\frac{1}{\sqrt{2}}\right)^n\right) = 0 \Rightarrow$ converges   (Table 8.1, #4)

57. $\lim\limits_{n\to\infty} \dfrac{(\ln n)^{200}}{n} = \lim\limits_{n\to\infty} \dfrac{200 (\ln n)^{199}}{n} = \lim\limits_{n\to\infty} \dfrac{200 \cdot 199 (\ln n)^{198}}{n} = \ldots = \lim\limits_{n\to\infty} \dfrac{200!}{n} = 0 \Rightarrow$ converges

59. $\lim\limits_{n\to\infty} (n - \sqrt{n^2 - n}) = \lim\limits_{n\to\infty} (n - \sqrt{n^2 - n})\left(\dfrac{n + \sqrt{n^2 - n}}{n + \sqrt{n^2 - n}}\right) = \lim\limits_{n\to\infty} \dfrac{n}{n + \sqrt{n^2 - n}} = \lim\limits_{n\to\infty} \dfrac{1}{1 + \sqrt{1 - \frac{1}{n}}}$

$= \frac{1}{2} \Rightarrow$ converges

61. $\lim\limits_{n\to\infty} \dfrac{1}{n} \displaystyle\int_1^n \dfrac{1}{x}\, dx = \lim\limits_{n\to\infty} \dfrac{\ln n}{n} = \lim\limits_{n\to\infty} \dfrac{1}{n} = 0 \Rightarrow$ converges   (Table 8.1, #1)

63. $1, 1, 2, 4, 8, 16, 32, \ldots = 1, 2^0, 2^1, 2^2, 2^3, 2^4, 2^5, \ldots \Rightarrow x_1 = 1$ and $x_n = 2^{n-2}$ for $n \ge 2$

65. (a) $f(x) = x^2 - 2$; the sequence converges to $1.414213562 \approx \sqrt{2}$

    (b) $f(x) = \tan(x) - 1$; the sequence converges to $0.7853981635 \approx \frac{\pi}{4}$

    (c) $f(x) = e^x$; the sequence $1, 0, -1, -2, -3, -4, -5, \ldots$ diverges

67. (a) If $a = 2n + 1$, then $b = \left\lfloor \dfrac{a^2}{2} \right\rfloor = \left\lfloor \dfrac{4n^2 + 4n + 1}{2} \right\rfloor = \left\lfloor 2n^2 + 2n + \frac{1}{2} \right\rfloor = 2n^2 + 2n$, $c = \left\lceil \dfrac{a^2}{2} \right\rceil = \left\lceil 2n^2 + 2n + \frac{1}{2} \right\rceil$

    $= 2n^2 + 2n + 1$ and $a^2 + b^2 = (2n + 1)^2 + (2n^2 + 2n)^2 = 4n^2 + 4n + 1 + 4n^4 + 8n^3 + 4n^2$

    $= 4n^4 + 8n^3 + 8n^2 + 4n + 1 = (2n^2 + 2n + 1)^2 = c^2$.

    (b) $\lim\limits_{a\to\infty} \dfrac{\left\lfloor \frac{a^2}{2} \right\rfloor}{\left\lceil \frac{a^2}{2} \right\rceil} = \lim\limits_{a\to\infty} \dfrac{2n^2 + 2n}{2n^2 + 2n + 1} = 1$ or $\lim\limits_{a\to\infty} \dfrac{\left\lfloor \frac{a^2}{2} \right\rfloor}{\left\lceil \frac{a^2}{2} \right\rceil} = \lim\limits_{a\to\infty} \sin\theta = \lim\limits_{\theta\to\pi/2} \sin\theta = 1$

69. (a) $\lim\limits_{n\to\infty} \dfrac{\ln n}{n^c} = \lim\limits_{n\to\infty} \dfrac{\left(\frac{1}{n}\right)}{cn^{c-1}} = \lim\limits_{n\to\infty} \dfrac{1}{cn^c} = 0$

    (b) For all $\epsilon > 0$, there exists an $N = e^{-(\ln \epsilon)/c}$ such that $n > e^{-(\ln \epsilon)/c} \Rightarrow \ln n > -\dfrac{\ln \epsilon}{c} \Rightarrow \ln n^c > \ln\left(\frac{1}{\epsilon}\right)$

    $\Rightarrow n^c > \frac{1}{\epsilon} \Rightarrow \frac{1}{n^c} < \epsilon \Rightarrow \left|\frac{1}{n^c} - 0\right| < \epsilon \Rightarrow \lim\limits_{n\to\infty} \frac{1}{n^c} = 0$

71. $\displaystyle\lim_{n\to\infty} n^{1/n} = \lim_{n\to\infty} \exp\left(\frac{1}{n}\ln n\right) = \lim_{n\to\infty} \exp\left(\frac{1}{n}\right) = e^0 = 1$

73. Assume the hypotheses of the theorem and let ϵ be a positive number. For all ϵ there exists a N_1 such that when $n > N_1$ then $|a_n - L| < \epsilon \Rightarrow -\epsilon < a_n - L < \epsilon \Rightarrow L - \epsilon < a_n$, and there exists a N_2 such that when $n > N_2$ then $|c_n - L| < \epsilon \Rightarrow -\epsilon < c_n - L < \epsilon \Rightarrow c_n < L + \epsilon$. If $n > \max\{N_1, N_2\}$, then $L - \epsilon < a_n \le b_n \le c_n < L + \epsilon \Rightarrow |b_n - L| < \epsilon \Rightarrow \displaystyle\lim_{n\to\infty} b_n = L.$

75. $g(x) = \sqrt{x}$; $2 \to 1.00000132$ in 20 iterations; $.1 \to 0.9999956$ in 20 iterations; a root is 1

77. $g(x) = -\cos x$; $x_0 = .1 \to 0.73908456$ in 35 iterations

79. $g(x) = 0.1 + \sin x$; $x_0 = -2 \to 0.853748068$ in 43 iterations

81. $x_0 = $ initial guess $> 0 \Rightarrow x_1 = \sqrt{x_0} = (x_0)^{1/2} \Rightarrow x_2 = \sqrt{x_0^{1/2}} = x_0^{1/4}, \dots \Rightarrow x_n = x_0^{1/(2n)} \Rightarrow x_n \to 1$ as $n \to \infty$

83. $g(x) = 2x + 3 \Rightarrow g^{-1}(x) = \dfrac{x-3}{2}$ and when the iterative method is applied to $g^{-1}(x)$ we have $x_0 = 2$ $\to -2.99999881$ in 23 iterations $\Rightarrow -3$ is the fixed point

8.3 INFINITE SERIES

1. $s_n = \dfrac{a(1-r^n)}{(1-r)} = \dfrac{2\left(1-\left(\frac{1}{3}\right)^n\right)}{1-\left(\frac{1}{3}\right)} \Rightarrow \displaystyle\lim_{n\to\infty} s_n = \dfrac{2}{1-\left(\frac{1}{3}\right)} = 3$

3. $s_n = \dfrac{a(1-r^n)}{(1-r)} = \dfrac{1-\left(-\frac{1}{2}\right)^n}{1-\left(-\frac{1}{2}\right)} \Rightarrow \displaystyle\lim_{n\to\infty} s_n = \dfrac{1}{\left(\frac{3}{2}\right)} = \dfrac{2}{3}$

5. $\dfrac{1}{(n+1)(n+2)} = \dfrac{1}{n+1} - \dfrac{1}{n+2} \Rightarrow s_n = \left(\frac{1}{2}-\frac{1}{3}\right) + \left(\frac{1}{3}-\frac{1}{4}\right) + \dots + \left(\frac{1}{n+1}-\frac{1}{n+2}\right) = \frac{1}{2} - \frac{1}{n+2} \Rightarrow \displaystyle\lim_{n\to\infty} s_n = \frac{1}{2}$

7. $1 - \frac{1}{4} + \frac{1}{16} - \frac{1}{64} + \dots$, the sum of this geometric series is $\dfrac{1}{1-\left(-\frac{1}{4}\right)} = \dfrac{1}{1+\left(\frac{1}{4}\right)} = \dfrac{4}{5}$

9. $\frac{7}{4} + \frac{7}{16} + \frac{7}{64} + \dots$, the sum of this geometric series is $\dfrac{\left(\frac{7}{4}\right)}{1-\left(\frac{1}{4}\right)} = \dfrac{7}{3}$

11. $(5+1) + \left(\frac{5}{2}+\frac{1}{3}\right) + \left(\frac{5}{4}+\frac{1}{9}\right) + \left(\frac{5}{8}+\frac{1}{27}\right) + \dots$, is the sum of two geometric series; the sum is
$\dfrac{5}{1-\left(\frac{1}{2}\right)} + \dfrac{1}{1-\left(\frac{1}{3}\right)} = 10 + \frac{3}{2} = \frac{23}{2}$

13. $(1+1) + \left(\frac{1}{2} - \frac{1}{5}\right) + \left(\frac{1}{4} + \frac{1}{25}\right) + \left(\frac{1}{8} - \frac{1}{125}\right) + \ldots,$ is the sum of two geometric series; the sum is

$$\frac{1}{1 - \left(\frac{1}{2}\right)} + \frac{1}{1 + \left(\frac{1}{5}\right)} = 2 + \frac{5}{6} = \frac{17}{6}$$

15. $\frac{4}{(4n-3)(4n+1)} = \frac{1}{4n-3} - \frac{1}{4n+1} \Rightarrow s_n = \left(1 - \frac{1}{5}\right) + \left(\frac{1}{5} - \frac{1}{9}\right) + \left(\frac{1}{9} - \frac{1}{13}\right) + \ldots + \left(\frac{1}{4n-7} - \frac{1}{4n-3}\right)$

$+ \left(\frac{1}{4n-3} - \frac{1}{4n+1}\right) = 1 - \frac{1}{4n+1} \Rightarrow \lim_{n\to\infty} s_n = \lim_{n\to\infty} \left(1 - \frac{1}{4n+1}\right) = 1$

17. $\frac{40n}{(2n-1)^2(2n+1)^2} = \frac{A}{(2n-1)} + \frac{B}{(2n-1)^2} + \frac{C}{(2n+1)} + \frac{D}{(2n+1)^2}$

$= \frac{A(2n-1)(2n+1)^2 + B(2n+1)^2 + C(2n+1)(2n-1)^2 + D(2n-1)^2}{(2n-1)^2(2n+1)^2}$

$\Rightarrow A(2n-1)(2n+1)^2 + B(2n+1)^2 + C(2n+1)(2n-1)^2 + D(2n-1)^2 = 40n$

$\Rightarrow A(8n^3 + 4n^2 - 2n - 1) + B(4n^2 + 4n + 1) + C(8n^3 - 4n^2 - 2n + 1) = D(4n^2 - 4n + 1) = 40n$

$\Rightarrow (8A + 8C)n^3 + (4A + 4B - 4C + 4D)n^2 + (-2A + 4B - 2C - 4D)n + (-A + B + C + D) = 40n$

$\Rightarrow \begin{cases} 8A + 8C = 0 \\ 4A + 4B - 4C + 4D = 0 \\ -2A + 4B - 2C - 4D = 40 \\ -A + B + C + D = 0 \end{cases} \Rightarrow \begin{cases} 8A + 8C = 0 \\ A + B - C + D = 0 \\ -A + 2B - C - 2D = 20 \\ -A + B + C + D = 0 \end{cases} \Rightarrow \begin{cases} B + D = 0 \\ 2B - 2D = 20 \end{cases} \Rightarrow 4B = 20 \Rightarrow B = 5$ and

$D = -5 \Rightarrow \begin{cases} A + C = 0 \\ -A + 5 + C - 5 = 0 \end{cases} \Rightarrow C = 0$ and $A = 0$. Hence, $\sum_{n=1}^{k} \left[\frac{40n}{(2n-1)^2(2n+1)^2}\right]$

$= 5 \sum_{n=1}^{k} \left[\frac{1}{(2n-1)^2} - \frac{1}{(2n+1)^2}\right] = 5\left(\frac{1}{1} - \frac{1}{9} + \frac{1}{9} - \frac{1}{25} + \frac{1}{25} - \ldots - \frac{1}{(2(k-1)+1)^2} + \frac{1}{(2k-1)^2} - \frac{1}{(2k+1)^2}\right)$

$= 5\left(1 - \frac{1}{(2k+1)^2}\right) \Rightarrow$ the sum is $\lim_{n\to\infty} 5\left(1 - \frac{1}{(2k+1)^2}\right) = 5$

19. $s_n = \left(1 - \frac{1}{\sqrt{2}}\right) + \left(\frac{1}{\sqrt{2}} - \frac{1}{\sqrt{3}}\right) + \left(\frac{1}{\sqrt{3}} - \frac{1}{\sqrt{4}}\right) + \ldots + \left(\frac{1}{\sqrt{n-1}} + \frac{1}{\sqrt{n}}\right) + \left(\frac{1}{\sqrt{n}} - \frac{1}{\sqrt{n+1}}\right) = 1 - \frac{1}{\sqrt{n+1}}$

$\Rightarrow \lim_{n\to\infty} s_n = \lim_{n\to\infty} \left(1 - \frac{1}{\sqrt{n+1}}\right) = 1$

21. $s_n = \left(\frac{1}{\ln 3} - \frac{1}{\ln 2}\right) + \left(\frac{1}{\ln 4} - \frac{1}{\ln 3}\right) + \left(\frac{1}{\ln 5} - \frac{1}{\ln 4}\right) + \ldots + \left(\frac{1}{\ln(n+1)} - \frac{1}{\ln n}\right) + \left(\frac{1}{\ln(n+2)} - \frac{1}{\ln(n+1)}\right)$

$= -\frac{1}{\ln 2} + \frac{1}{\ln(n+2)} \Rightarrow \lim_{n\to\infty} s_n = -\frac{1}{\ln 2}$

23. convergent geometric series with sum $\frac{1}{1 - \left(\frac{1}{\sqrt{2}}\right)} = \frac{\sqrt{2}}{\sqrt{2} - 1} = 2 + \sqrt{2}$

25. convergent geometric series with sum $\dfrac{\left(\frac{3}{2}\right)}{1-\left(-\frac{1}{2}\right)} = 1$

27. $\lim\limits_{n\to\infty}\ \cos(n\pi) = \lim\limits_{n\to\infty}\ (-1)^n \neq 0 \Rightarrow$ diverges

29. convergent geometric series with sum $\dfrac{1}{1-\left(\frac{1}{e^2}\right)} = \dfrac{e^2}{e^2-1}$

31. convergent geometric series with sum $\dfrac{2}{1-\left(\frac{1}{10}\right)} - 2 = \dfrac{20}{9} - \dfrac{18}{9} = \dfrac{2}{9}$

33. difference of two geometric series with sum $\dfrac{1}{1-\left(\frac{2}{3}\right)} - \dfrac{1}{1-\left(\frac{1}{3}\right)} = 3 - \dfrac{3}{2} = \dfrac{3}{2}$

35. $\lim\limits_{n\to\infty}\ \dfrac{n!}{1000^n} = \infty \neq 0 \Rightarrow$ diverges

37. $\sum\limits_{n=1}^{\infty} \ln\left(\dfrac{n}{n+1}\right) = \sum\limits_{n=1}^{\infty} \left[\ln(n) - \ln(n+1)\right] \Rightarrow s_n = \left[\ln(1) - \ln(2)\right] + \left[\ln(2) - \ln(3)\right] + \left[\ln(3) - \ln(4)\right] + \dots$

$+ \left[\ln(n-1) - \ln(n)\right] + \left[\ln(n) - \ln(n+1)\right] = \ln(1) - \ln(n+1) = -\ln(n+1) \Rightarrow \lim\limits_{n\to\infty} s_n = -\infty, \Rightarrow$ diverges

39. convergent geometric series with sum $\dfrac{1}{1-\left(\frac{e}{\pi}\right)} = \dfrac{\pi}{\pi - e}$

41. $\sum\limits_{n=0}^{\infty} (-1)^n x^n = \sum\limits_{n=0}^{\infty} (-x)^n;\ a = 1,\ r = -x;$ converges to $\dfrac{1}{1-(-x)} = \dfrac{1}{1+x}$ for $|x| < 1$

43. $a = 3,\ r = \dfrac{x-1}{2};$ converges to $\dfrac{3}{1-\left(\frac{x-1}{2}\right)} = \dfrac{6}{3-x}$ for $-1 < \dfrac{x-1}{2} < 1$ or $-1 < x < 3$

45. $a = 1,\ r = 2x;$ converges to $\dfrac{1}{1-2x}$ for $|2x| < 1$ or $|x| < \dfrac{1}{2}$

47. $a = 1,\ r = -(x+1)^n;$ converges to $\dfrac{1}{1+(x+1)} = \dfrac{1}{2+x}$ for $|x+1| < 1$ or $-2 < x < 0$

49. $a = 1,\ r = \sin x;$ converges to $\dfrac{1}{1-\sin x}$ for $x \neq (2k+1)\dfrac{\pi}{2}$, k an integer

51. $0.\overline{23} = \sum\limits_{n=0}^{\infty} \dfrac{23}{100}\left(\dfrac{1}{10^2}\right)^n = \dfrac{\left(\frac{23}{100}\right)}{1-\left(\frac{1}{100}\right)} = \dfrac{23}{99}$ 53. $0.\overline{7} = \sum\limits_{n=0}^{\infty} \dfrac{7}{10}\left(\dfrac{1}{10}\right)^n = \dfrac{\left(\frac{7}{10}\right)}{1-\left(\frac{1}{10}\right)} = \dfrac{7}{9}$

55. $0.0\overline{6} = \sum\limits_{n=0}^{\infty} \left(\dfrac{1}{10}\right)\left(\dfrac{6}{10}\right)\left(\dfrac{1}{10}\right)^n = \dfrac{\left(\frac{6}{100}\right)}{1-\left(\frac{1}{10}\right)} = \dfrac{6}{90} = \dfrac{1}{15}$

57. $1.24\overline{123} = \dfrac{124}{100} + \sum\limits_{n=0}^{\infty} \dfrac{123}{10^5}\left(\dfrac{1}{10^3}\right)^n = \dfrac{124}{100} + \dfrac{\left(\dfrac{123}{10^5}\right)}{1 - \left(\dfrac{1}{10^3}\right)} = \dfrac{124}{100} + \dfrac{123}{10^5 - 10^2} = \dfrac{124}{100} + \dfrac{123}{99,900} = \dfrac{123,753}{99,900} = \dfrac{41,251}{33,300}$

59. (a) $\sum\limits_{n=-2}^{\infty} \dfrac{1}{(n+4)(n+5)}$ (b) $\sum\limits_{n=0}^{\infty} \dfrac{1}{(n+2)(n+3)}$ (c) $\sum\limits_{n=5}^{\infty} \dfrac{1}{(n-3)(n-2)}$

61. (a) one example is $\dfrac{1}{2} + \dfrac{1}{4} + \dfrac{1}{8} + \dfrac{1}{16} + \ldots = \dfrac{\left(\dfrac{1}{2}\right)}{1 - \left(\dfrac{1}{2}\right)} = 1$

(b) one example is $-\dfrac{3}{2} - \dfrac{3}{4} - \dfrac{3}{8} - \dfrac{3}{16} - \ldots = \dfrac{\left(-\dfrac{3}{2}\right)}{1 - \left(\dfrac{1}{2}\right)} = -3$

(c) one example is $1 - \dfrac{1}{2} - \dfrac{1}{4} - \dfrac{1}{8} - \dfrac{1}{16} - \ldots$; the series $\dfrac{k}{2} + \dfrac{k}{4} + \dfrac{k}{8} + \ldots = \dfrac{\left(\dfrac{k}{2}\right)}{1 - \left(\dfrac{1}{2}\right)} = k$ where k is any positive or

negative number.

63. Let $a_n = b_n = \left(\dfrac{1}{2}\right)^n$. Then $\sum\limits_{n=1}^{\infty} a_n = \sum\limits_{n=1}^{\infty} b_n = \sum\limits_{n=1}^{\infty} \left(\dfrac{1}{2}\right)^n = 1$, while $\sum\limits_{n=1}^{\infty} \left(\dfrac{a_n}{b_n}\right) = \sum\limits_{n=1}^{\infty} (1)$ diverges.

65. Let $a_n = \left(\dfrac{1}{4}\right)^n$ and $b_n = \left(\dfrac{1}{2}\right)^n$. Then $A = \sum\limits_{n=1}^{\infty} a_n = \dfrac{1}{3}$, $B = \sum\limits_{n=1}^{\infty} b_n = 1$ and $\sum\limits_{n=1}^{\infty} \left(\dfrac{a_n}{b_n}\right) = \sum\limits_{n=1}^{\infty} \left(\dfrac{1}{2}\right)^n = 1 \neq \dfrac{A}{B}$.

67. Since the sum of a finite number of terms is finite, adding or subtracting a finite number of terms from a series that diverges does not change the divergence of the series.

69. (a) $\dfrac{2}{1-r} = 5 \Rightarrow \dfrac{2}{5} = 1 - r \Rightarrow r = \dfrac{3}{5}$; $2 + 2\left(\dfrac{3}{5}\right) + 2\left(\dfrac{3}{5}\right)^2 + \ldots$

(b) $\dfrac{\left(\dfrac{13}{2}\right)}{1-r} = 5 \Rightarrow \dfrac{13}{10} = 1 - r \Rightarrow r = -\dfrac{3}{10}$; $\dfrac{13}{2} - \dfrac{13}{2}\left(\dfrac{3}{10}\right) + \dfrac{13}{2}\left(\dfrac{3}{10}\right)^2 - \dfrac{13}{2}\left(\dfrac{3}{10}\right)^3 + \ldots$

71. $s_n = 1 + 2r + r^2 + 2r^3 + r^4 + 2r^5 + \ldots + r^{2n} + 2r^{2n+1}$, $n = 0, 1, \ldots$

$\Rightarrow s_n = \left(1 + r^2 + r^4 + \ldots + r^{2n}\right) + \left(2r + 2r^3 + 2r^5 + \ldots + 2r^{2n+1}\right) \Rightarrow \lim\limits_{n\to\infty} s_n = \dfrac{1}{1-r^2} + \dfrac{2r}{1-r^2}$

$= \dfrac{1+2r}{1-r^2}$, if $|r^2| < 1$ or $|r| < 1$

73. distance $= 4 + 2\left[(4)\left(\dfrac{3}{4}\right) + (4)\left(\dfrac{3}{4}\right)^2 + \ldots\right] = 4 + 2\left(\dfrac{3}{1 - \left(\dfrac{3}{4}\right)}\right) = 28$ m

75. area $= 2^2 + \left(\sqrt{2}\right)^2 + (1)^2 + \left(\dfrac{1}{\sqrt{2}}\right)^2 + \ldots = 4 + 2 + 1 + \dfrac{1}{2} + \ldots = \dfrac{4}{1 - \dfrac{1}{2}} = 8$ m^2

77. (a) $L_1 = 3$, $L_2 = 3\left(\frac{4}{3}\right)$, $L_3 = 3\left(\frac{4}{3}\right)^2$, ..., $L_n = 3\left(\frac{4}{3}\right)^{n-1} \Rightarrow \lim_{n\to\infty} L_n = \lim_{n\to\infty} 3\left(\frac{4}{3}\right)^{n-1} = \infty$

(b) $A_1 = \frac{1}{2}(1)\left(\frac{\sqrt{3}}{2}\right) = \frac{\sqrt{3}}{4}$, $A_2 = A_1 + 3\left(\frac{1}{2}\right)\left(\frac{1}{3}\right)\left(\frac{\sqrt{3}}{6}\right) = \frac{\sqrt{3}}{4} + \frac{\sqrt{3}}{12}$, $A_3 = A_2 + 12\left(\frac{1}{2}\right)\left(\frac{1}{9}\right)\left(\frac{\sqrt{3}}{18}\right)$

$= \frac{\sqrt{3}}{4} + \frac{\sqrt{3}}{12} + \frac{\sqrt{3}}{27}$, $A_4 = A_3 + 48\left(\frac{1}{2}\right)\left(\frac{1}{27}\right)\left(\frac{\sqrt{3}}{54}\right)$, ..., $A_n = \frac{\sqrt{3}}{4} + \frac{27\sqrt{3}}{64}\left(\frac{4}{9}\right)^2 + \frac{27\sqrt{3}}{64}\left(\frac{4}{9}\right)^3 + \cdots$

$= \frac{\sqrt{3}}{4} + \sum_{n=2}^{\infty} \frac{27\sqrt{3}}{64}\left(\frac{4}{9}\right)^n = \frac{\sqrt{3}}{4} + \frac{\left(\frac{27\sqrt{3}}{64}\right)\left(\frac{4}{9}\right)^2}{1 - \left(\frac{4}{9}\right)} = \frac{\sqrt{3}}{4} + \frac{\left(\frac{27\sqrt{3}}{64}\right)\left(\frac{16}{9}\right)}{9 - 4} = \frac{\sqrt{3}}{4} + \frac{3\sqrt{3}}{4 \cdot 5} = \frac{5\sqrt{3} + 3\sqrt{3}}{20} = \frac{2\sqrt{3}}{5}$

8.4 THE INTEGRAL TEST FOR SERIES OF NONNEGATIVE TERMS

1. converges; a geometric series with $r = \frac{1}{10} < 1$

3. diverges; by the nth-Term Test for Divergence, $\lim_{n\to\infty} \frac{n}{n+1} = 1 \neq 0$

5. diverges; $\sum_{n=1}^{\infty} \frac{3}{\sqrt{n}} = 3 \sum_{n=1}^{\infty} \frac{1}{\sqrt{n}}$, which is a divergent p-series

7. converges; a geometric series with $r = \frac{1}{8} < 1$

9. diverges by the Integral Test: $\int_2^n \frac{\ln x}{x}\, dx = \frac{1}{2}\left(\ln^2 n - \ln 2\right) \Rightarrow \int_2^\infty \frac{\ln x}{x}\, dx \to \infty$

11. converges; a geometric series with $r = \frac{2}{3} < 1$

13. diverges; $\sum_{n=0}^{\infty} \frac{-2}{n+1} = -2 \sum_{n=0}^{\infty} \frac{1}{n+1}$, which diverges by the Integral Test

15. diverges; $\lim_{n\to\infty} a_n = \lim_{n\to\infty} \frac{2^n}{n+1} = \lim_{n\to\infty} \frac{2^n \ln 2}{1} = \infty \neq 0$

17. diverges; $\lim_{n\to\infty} \frac{\sqrt{n}}{\ln n} = \lim_{n\to\infty} \frac{\left(\frac{1}{2\sqrt{n}}\right)}{\left(\frac{1}{n}\right)} = \lim_{n\to\infty} \frac{\sqrt{n}}{2} = \infty \neq 0$

19. diverges; a geometric series with $r = \frac{1}{\ln 2} \approx 1.44 > 1$

21. converges by the Integral Test: $\int_3^\infty \frac{\left(\frac{1}{x}\right)}{(\ln x)\sqrt{(\ln x)^2 - 1}}\, dx; \begin{bmatrix} u = \ln x \\ du = \frac{1}{x}\, dx \end{bmatrix} \to \int_{\ln 3}^\infty \frac{1}{u\sqrt{u^2 - 1}}\, du$

$= \lim_{b\to\infty} \left[\sec^{-1}|u|\right]_{\ln 3}^b = \lim_{b\to\infty} \left[\sec^{-1} b - \sec^{-1}(\ln 3)\right] = \lim_{b\to\infty} \left[\cos^{-1}\left(\frac{1}{b}\right) - \sec^{-1}(\ln 3)\right]$

$= \cos^{-1}(0) - \sec^{-1}(\ln 3) = \frac{\pi}{2} - \sec^{-1}(\ln 3) \approx 1.1439$

23. diverges by the nth-Term Test for divergence; $\displaystyle\lim_{n \to \infty} n \sin\left(\frac{1}{n}\right) = \lim_{n \to \infty} \frac{\sin\left(\frac{1}{n}\right)}{\left(\frac{1}{n}\right)} = \lim_{x \to 0} \frac{\sin x}{x} = 1 \neq 0$

25. converges by the Integral Test: $\displaystyle\int_1^\infty \frac{e^x}{1 + e^{2x}}\, dx; \begin{bmatrix} u = e^x \\ du = e^x\, dx \end{bmatrix} \to \int_e^\infty \frac{1}{1 + u^2}\, du = \lim_{n \to \infty} \left[\tan^{-1} u\right]_e^b$

$= \displaystyle\lim_{b \to \infty} \left(\tan^{-1} b - \tan^{-1} e\right) = \frac{\pi}{2} - \tan^{-1} e \approx 0.35$

27. converges by the Integral Test: $\displaystyle\int_1^\infty \frac{8 \tan^{-1} x}{1 + x^2}\, dx; \begin{bmatrix} u = \tan^{-1} x \\ du = \frac{dx}{1 + x^2} \end{bmatrix} \to \int_{\pi/4}^{\pi/2} 8u\, du = \left[4u^2\right]_{\pi/4}^{\pi/2} = 4\left(\frac{\pi^2}{4} - \frac{\pi^2}{16}\right) = \frac{3\pi^2}{4}$

29. converges by the Integral Test: $\displaystyle\int_1^\infty \operatorname{sech} x\, dx = 2 \lim_{b \to \infty} \int_1^b \frac{e^x}{1 + \left(e^x\right)^2}\, dx = 2 \lim_{b \to \infty} \left[\tan^{-1} e^x\right]_1^b$

$= 2 \displaystyle\lim_{b \to \infty} \left(\tan^{-1} e^b - \tan^{-1} e\right) = \pi - 2 \tan^{-1} e$

31. $\displaystyle\int_1^\infty \left(\frac{a}{x + 2} - \frac{1}{x + 4}\right) dx = \lim_{b \to \infty} \left[a \ln|x + 2| - \ln|x + 4|\right]_1^b = \lim_{b \to \infty} \ln \frac{(b + 2)^a}{b + 4} - \ln\left(\frac{3^a}{5}\right);$

$\displaystyle\lim_{b \to \infty} \frac{(b + 2)^a}{b + 4} = a \lim_{b \to \infty} (b + 2)^{a-1} = \begin{cases} \infty, & a > 1 \\ 1, & a = 1 \end{cases} \Rightarrow$ the series converges to $\ln\left(\frac{5}{3}\right)$ if $a = 1$ and diverges to ∞ if

$a > 1$. If $a < 1$, the terms of the series eventually become negative and the Integral Test does not apply. From that point on, however, the series behaves like a negative multiple of the harmonic series, and so it diverges.

33. (a)

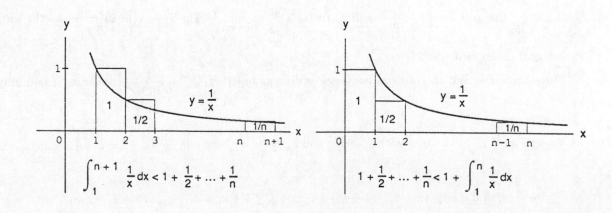

(b) There are $(13)(365)(24)(60)(60)(10^9)$ seconds in 13 billion years; by part (a) $s_n \le 1 + \ln n$ where

$n = (13)(365)(24)(60)(60)(10^9) \Rightarrow s_n \le 1 + \ln\left((13)(365)(24)(60)(60)(10^9)\right)$

$= 1 + \ln(13) + \ln(365) + \ln(24) + 2\ln(60) + 9\ln(10) \approx 41.55$

35. Yes. If $\sum\limits_{n=1}^{\infty} a_n$ is a divergent series of positive numbers, then $\left(\frac{1}{2}\right)\sum\limits_{n=1}^{\infty} a_n = \sum\limits_{n=1}^{\infty}\left(\frac{a_n}{2}\right)$ also diverges and $\frac{a_n}{2} < a_n$.

There is no "smallest" divergent series of positive numbers: for any divergent series $\sum\limits_{n=1}^{\infty} a_n$ of positive

numbers $\sum\limits_{n=1}^{\infty}\left(\frac{a_n}{2}\right)$ has smaller terms and still diverges.

37. Let $A_n = \sum\limits_{k=1}^{n} a_k$ and $B_n = \sum\limits_{k=1}^{n} 2^k a_{\left(2^k\right)}$, where $\{a_k\}$ is a nonincreasing sequence of positive terms converging to

0. Note that $\{A_n\}$ and $\{B_n\}$ are nondecreasing sequences of positive terms. Now,

$B_n = 2a_2 + 4a_4 + 8a_8 + \ldots + 2^n a_{\left(2^n\right)} = 2a_2 + (2a_4 + 2a_4) + (2a_8 + 2a_8 + 2a_8 + 2a_8) + \ldots$

$+\underbrace{\left(2a_{\left(2^n\right)} + 2a_{\left(2^n\right)} + \ldots + 2a_{\left(2^n\right)}\right)}_{2^{n-1} \text{ terms}} \le 2a_1 + 2a_2 + (2a_3 + 2a_4) + (2a_5 + 2a_6 + 2a_7 + 2a_8) + \ldots$

$+\left(2a_{\left(2^{n-1}\right)} + 2a_{\left(2^{n-1}+1\right)} + \ldots + 2a_{\left(2^n\right)}\right) = 2A_{\left(2^n\right)} \le 2\sum\limits_{k=1}^{\infty} a_k$. Therefore if $\sum a_k$ converges,

then $\{B_n\}$ is bounded above $\Rightarrow \sum 2^k a_{\left(2^k\right)}$ converges. Conversely,

$A_n = a_1 + (a_2 + a_3) + (a_4 + a_5 + a_6 + a_7) + \ldots + a_n < a_1 + 2a_2 + 4a_4 + \ldots + 2^n a_{\left(2^n\right)} = a_1 + B_n < a_1 + \sum\limits_{k=1}^{\infty} 2^k a_{\left(2^k\right)}$.

Therefore, if $\sum\limits_{k=1}^{\infty} 2^k a_{\left(2^k\right)}$ converges, then $\{A_n\}$ is bounded above and hence converges.

39. (a) $\int\limits_{2}^{\infty} \frac{dx}{x(\ln x)^p}; \left[\begin{array}{l} u = \ln x \\ du = \frac{dx}{x} \end{array}\right] \rightarrow \int\limits_{\ln 2}^{\infty} u^{-p}\, du = \lim\limits_{b \to \infty}\left[\frac{u^{-p+1}}{-p+1}\right]_{\ln 2}^{b} = \lim\limits_{b \to \infty}\left(\frac{1}{1-p}\right)\left[b^{-p+1} - (\ln 2)^{-p+1}\right]$

$= \begin{cases} \frac{1}{p-1}(\ln 2)^{-p+1}, & p > 1 \\ \infty, & p < 1 \end{cases} \Rightarrow$ the improper integral converges if $p > 1$ and diverges

if $p < 1$. For $p = 1$: $\int\limits_{2}^{\infty} \frac{dx}{x\ln x} = \lim\limits_{b \to \infty}\left[\ln(\ln x)\right]_{2}^{b} = \lim\limits_{b \to \infty}\left[\ln(\ln b) - \ln(\ln 2)\right] = \infty$, so the improper

integral diverges if $p = 1$.

(b) Since the series and the integral converge or diverge together, $\sum\limits_{n=2}^{\infty} \frac{1}{n(\ln n)^p}$ converges if and only if $p > 1$.

41. (a) From Fig. 8.13 in the text with $f(x) = \frac{1}{x}$ and $a_k = \frac{1}{k}$, we have $\int\limits_{1}^{n+1} \frac{1}{x}\, dx \le 1 + \frac{1}{2} + \frac{1}{3} + \ldots + \frac{1}{n}$

$\le 1 + \int\limits_{1}^{n} f(x)\, dx \Rightarrow \ln(n+1) \le 1 + \frac{1}{2} + \frac{1}{3} + \ldots + \frac{1}{n} \le 1 + \ln n \Rightarrow 0 \le \ln(n+1) - \ln n$

$\leq \left(1 + \frac{1}{2} + \frac{1}{3} + \ldots + \frac{1}{n}\right) - \ln n \leq 1$. Therefore the sequence $\left\{\left(1 + \frac{1}{2} + \frac{1}{3} + \ldots + \frac{1}{n}\right) - \ln n\right\}$ is bounded above by 1 and below by 0.

(b) From the graph in Fig. 8.13(a) with $f(x) = \frac{1}{x}$, $\dfrac{1}{n+1} < \displaystyle\int_{n}^{n+1} \frac{1}{x}\, dx = \ln(n+1) - \ln n$

$\Rightarrow 0 > \dfrac{1}{n+1} - [\ln(n+1) - \ln n] = \left(1 + \frac{1}{2} + \frac{1}{3} + \ldots + \frac{1}{n+1} - \ln(n+1)\right) - \left(1 + \frac{1}{2} + \frac{1}{3} + \ldots + \frac{1}{n} - \ln n\right).$

If we define $a_n = 1 + \frac{1}{2} = \frac{1}{3} + \frac{1}{n} - \ln n$, then $0 > a_{n+1} - a_n \Rightarrow a_{n+1} < a_n \Rightarrow \{a_n\}$ is a decreasing sequence of nonnegative terms.

8.5 COMPARISON TESTS FOR SERIES OF NONNEGATIVE TERMS

1. diverges by the Limit Comparison Test (part 1) when compared with $\displaystyle\sum_{n=1}^{\infty} \frac{1}{\sqrt{n}}$, a divergent p-series:

$$\lim_{n\to\infty} \frac{\left(\dfrac{1}{2\sqrt{n} + \sqrt[3]{n}}\right)}{\left(\dfrac{1}{\sqrt{n}}\right)} = \lim_{n\to\infty} \frac{\sqrt{n}}{2\sqrt{n} + \sqrt[3]{n}} = \lim_{n\to\infty} \left(\frac{1}{2 + n^{-1/6}}\right) = \frac{1}{2}$$

3. converges by the Direct Comparison Test; $\dfrac{\sin^2 n}{2^n} \leq \dfrac{1}{2^n}$, which is the nth term of a convergent geometric series

5. diverges since $\displaystyle\lim_{n\to\infty} \frac{2n}{3n-1} = \frac{2}{3} \neq 0$

7. converges by the Direct Comparison Test; $\left(\dfrac{n}{3n+1}\right)^n < \left(\dfrac{n}{3n}\right)^n < \left(\dfrac{1}{3}\right)^n$, the nth term of a convergent geometric series

9. diverges by the Direct Comparison Test; $n > \ln n \Rightarrow \ln n > \ln \ln n \Rightarrow \dfrac{1}{\ln n} < \dfrac{1}{\ln(\ln n)}$ and the series $\displaystyle\sum_{n=3}^{\infty} \frac{1}{n}$ diverges

11. converges by the Limit Comparison Test (part 2) when compared with $\displaystyle\sum_{n=1}^{\infty} \frac{1}{n^2}$, a convergent p-series:

$$\lim_{n\to\infty} \frac{\left[\dfrac{(\ln n)^2}{n^3}\right]}{\left(\dfrac{1}{n^2}\right)} = \lim_{n\to\infty} \frac{(\ln n)^2}{n} = \lim_{n\to\infty} \frac{2(\ln n)\left(\frac{1}{n}\right)}{1} = 2 \lim_{n\to\infty} \frac{\ln n}{n} = 0 \qquad \text{(Table 8.1)}$$

13. diverges by the Limit Comparison Test (part 3) with $\frac{1}{n}$, the nth term of the divergent harmonic series:

$$\lim_{n\to\infty} \frac{\left[\dfrac{1}{\sqrt{n}\,\ln n}\right]}{\left(\dfrac{1}{n}\right)} = \lim_{n\to\infty} \frac{\sqrt{n}}{\ln n} = \lim_{n\to\infty} \frac{\left(\dfrac{1}{2\sqrt{n}}\right)}{\left(\dfrac{1}{n}\right)} = \lim_{n\to\infty} \frac{\sqrt{n}}{2} = \infty$$

15. diverges by the Limit Comparison Test (part 3) with $\frac{1}{n}$, the nth term of the divergent harmonic series:

$$\lim_{n \to \infty} \frac{\left(\frac{1}{1+\ln n}\right)}{\left(\frac{1}{n}\right)} = \lim_{n \to \infty} \frac{n}{1+\ln n} = \lim_{n \to \infty} \frac{1}{\left(\frac{1}{n}\right)} = \lim_{n \to \infty} n = \infty$$

17. diverges by the Integral Test: $\int_{2}^{\infty} \frac{\ln(x+1)}{x+1}\,dx = \int_{\ln 3}^{\infty} u\,du = \lim_{b \to \infty}\left[\frac{1}{2}u^2\right]_{\ln 3}^{b} = \lim_{b \to \infty} \frac{1}{2}\left(b^2 - \ln^2 3\right) = \infty$

19. converges by the Direct Comparison Test with $\frac{1}{n^{3/2}}$, the nth term of a convergent p-series: $n^2 - 1 > n$ for

$n \geq 2 \Rightarrow n^2(n^2-1) > n^3 \Rightarrow n\sqrt{n^2-1} > n^{3/2} \Rightarrow \frac{1}{n^{3/2}} > \frac{1}{n\sqrt{n^2-1}}$

21. converges because $\sum_{n=1}^{\infty} \frac{1-n}{n2^n} = \sum_{n=1}^{\infty} \frac{1}{n2^n} + \sum_{n=1}^{\infty} \frac{-1}{2^n}$ which is the sum of two convergent series:

$\sum_{n=1}^{\infty} \frac{1}{n2^n}$ converges by the Direct Comparison Test since $\frac{1}{n2^n} < \frac{1}{2^n}$, and $\sum_{n=1}^{\infty} \frac{-1}{2^n}$ is a convergent geometric series

23. converges by the Direct Comparison Test: $\frac{1}{3^{n-1}+1} < \frac{1}{3^{n-1}}$, which is the nth term of a convergent geometric series

25. diverges by the Limit Comparison Test (part 1) with $\frac{1}{n}$, the nth term of the divergent harmonic series:

$$\lim_{n \to \infty} \frac{\left(\sin\frac{1}{n}\right)}{\left(\frac{1}{n}\right)} = \lim_{x \to 0} \frac{\sin x}{x} = 1$$

27. converges by the Limit Comparison Test (part 1) with $\frac{1}{n^2}$, the nth term of a convergent p-series:

$$\lim_{n \to \infty} \frac{\left(\frac{10n+1}{n(n+1)(n+2)}\right)}{\left(\frac{1}{n^2}\right)} = \lim_{n \to \infty} \frac{10n^2+n}{n^2+3n+2} = \lim_{n \to \infty} \frac{20n+1}{2n+3} = \lim_{n \to \infty} \frac{20}{2} = 10$$

29. converges by the Direct Comparison Test: $\frac{\tan^{-1} n}{n^{1.1}} < \frac{\frac{\pi}{2}}{n^{1.1}}$ and $\sum_{n=1}^{\infty} \frac{\frac{\pi}{2}}{n^{1.1}} = \frac{\pi}{2} \sum_{n=1}^{\infty} \frac{1}{n^{1.1}}$ is the product of a convergent p-series and a nonzero constant

31. converges by the Limit Comparison Test (part 1) with $\frac{1}{n^2}$: $\lim_{n \to \infty} \frac{\left(\frac{\coth n}{n^2}\right)}{\left(\frac{1}{n^2}\right)} = \lim_{n \to \infty} \coth n = \lim_{n \to \infty} \frac{e^n + e^{-n}}{e^n - e^{-n}}$

$= \lim_{n \to \infty} \frac{1+e^{-2n}}{1-e^{-2n}} = 1$

33. diverges: $f(x) = \sqrt[x]{x} \Rightarrow f'(x) > 0$ when $1 < x < e$ and $f'(x) < 0$ when $x > e \Rightarrow \sqrt[e]{e} > \sqrt[n]{n}$ for all $n \geq 3$; also

$3^e > e \Rightarrow 3 > \sqrt[e]{e}$. Consequently, $3n > n\sqrt[n]{n} \Rightarrow \frac{1}{3n} < \frac{1}{n\sqrt[n]{n}} \Rightarrow \sum_{n=1}^{\infty} \frac{1}{n\sqrt[n]{n}}$ diverges by the Direct Comparison Test

35. $\frac{1}{1+2+3+\ldots+n} = \frac{1}{\left(\frac{n(n+1)}{2}\right)} = \frac{2}{n(n+1)} \leq \frac{1}{n^2} \Rightarrow$ the series converges by the Direct Comparison Test

37. (a) If $\lim\limits_{n\to\infty} \frac{a_n}{b_n} = 0$, then there exists an integer N such that for all $n > N$, $\left|\frac{a_n}{b_n} - 0\right| < 1 \Rightarrow -1 < \frac{a_n}{b_n} < 1$

$\Rightarrow a_n < b_n$. Thus, if $\sum b_n$ converges, then $\sum a_n$ converges by the Direct Comparison Test.

(b) If $\lim\limits_{n\to\infty} \frac{a_n}{b_n} = \infty$, then there exists an integer N such that for all $n > N$, $\frac{a_n}{b_n} > 1 \Rightarrow a_n > b_n$. Thus, if

$\sum b_n$ diverges, then $\sum a_n$ diverges by the Direct Comparison Test.

39. $\lim\limits_{n\to\infty} \frac{a_n}{b_n} = \infty \Rightarrow$ there exists an integer N such that for all $n > N$, $\frac{a_n}{b_n} > 1 \Rightarrow a_n > b_n$. If $\sum a_n$ converges,

then $\sum b_n$ converges by the Direct Comparison Test

8.6 THE RATIO AND ROOT TESTS FOR SERIES WITH NONNEGATIVE TERMS

1. converges by the Ratio Test: $\lim\limits_{n\to\infty} \frac{a_{n+1}}{a_n} = \lim\limits_{n\to\infty} \frac{\left[\frac{(n+1)^{\sqrt{2}}}{2^{n+1}}\right]}{\left[\frac{n^{\sqrt{2}}}{2^n}\right]} = \lim\limits_{n\to\infty} \frac{(n+1)^{\sqrt{2}}}{2^{n+1}} \cdot \frac{2^n}{n^{\sqrt{2}}}$

$= \lim\limits_{n\to\infty} \left(1+\frac{1}{n}\right)^{\sqrt{2}} \left(\frac{1}{2}\right) = \frac{1}{2} < 1$

3. diverges by the Ratio Test: $\lim\limits_{n\to\infty} \frac{a_{n+1}}{a_n} = \lim\limits_{n\to\infty} \frac{\left(\frac{(n+1)!}{e^{n+1}}\right)}{\left(\frac{n!}{e^n}\right)} = \lim\limits_{n\to\infty} \frac{(n+1)!}{e^{n+1}} \cdot \frac{e^n}{n!} = \lim\limits_{n\to\infty} \frac{n+1}{e} = \infty$

5. converges by the Ratio Test: $\lim\limits_{n\to\infty} \frac{a_{n+1}}{a_n} = \lim\limits_{n\to\infty} \frac{\left(\frac{(n+1)^{10}}{10^{n+1}}\right)}{\left(\frac{n^{10}}{10^n}\right)} = \lim\limits_{n\to\infty} \frac{(n+1)^{10}}{10^{n+1}} \cdot \frac{10^n}{n^{10}} = \lim\limits_{n\to\infty} \left(1+\frac{1}{n}\right)^{10} \left(\frac{1}{10}\right)$

$= \frac{1}{10} < 1$

7. converges by the Direct Comparison Test: $\frac{2+(-1)^n}{(1.25)^n} = \left(\frac{4}{5}\right)^n [2+(-1)^n] \leq \left(\frac{4}{5}\right)^n (3)$

9. diverges; $\lim\limits_{n\to\infty} a_n = \lim\limits_{n\to\infty} \left(1-\frac{3}{n}\right)^n = \lim\limits_{n\to\infty} \left(1+\frac{-3}{n}\right)^n = e^{-3} \approx 0.05 \neq 0$

11. converges by the Direct Comparison Test: $\dfrac{\ln n}{n^3} < \dfrac{n}{n^3} = \dfrac{1}{n^2}$ for $n \ge 2$

13. diverges by the Direct Comparison Test: $\dfrac{1}{n} - \dfrac{1}{n^2} = \dfrac{n-1}{n^2} > \dfrac{1}{2}\left(\dfrac{1}{n}\right)$ for $n > 2$

15. diverges by the Direct Comparison Test: $\dfrac{\ln n}{n} > \dfrac{1}{n}$ for $n \ge 3$

17. converges by the Ratio Test: $\displaystyle\lim_{n\to\infty} \dfrac{a_{n+1}}{a_n} = \lim_{n\to\infty} \dfrac{(n+2)(n+3)}{(n+1)!} \cdot \dfrac{n!}{(n+1)(n+2)} = 0 < 1$

19. converges by the Ratio Test: $\displaystyle\lim_{n\to\infty} \dfrac{a_{n+1}}{a_n} = \lim_{n\to\infty} \dfrac{(n+4)!}{3!\,(n+1)!\,3^{n+1}} \cdot \dfrac{3!\,n!\,3^n}{(n+3)!} = \lim_{n\to\infty} \dfrac{n+4}{3(n+1)} = \dfrac{1}{3} < 1$

21. converges by the Ratio Test: $\displaystyle\lim_{n\to\infty} \dfrac{a_{n+1}}{a_n} = \lim_{n\to\infty} \dfrac{(n+1)!}{(2n+3)!} \cdot \dfrac{(2n+1)!}{n!} = \lim_{n\to\infty} \dfrac{n+1}{(2n+3)(2n+2)} = 0 < 1$

23. converges by the Root Test: $\displaystyle\lim_{n\to\infty} \sqrt[n]{a_n} = \lim_{n\to\infty} \sqrt[n]{\dfrac{n}{(\ln n)^n}} = \lim_{n\to\infty} \dfrac{\sqrt[n]{n}}{\ln n} = \lim_{n\to\infty} \dfrac{1}{\ln n} = 0 < 1$

25. converges by the Direct Comparison Test: $\dfrac{n!\,\ln n}{n(n+2)!} = \dfrac{\ln n}{n(n+1)(n+2)} < \dfrac{n}{n(n+1)(n+2)} = \dfrac{1}{(n+1)(n+2)} < \dfrac{1}{n^2}$

which is the nth-term of a convergent p-series

27. converges by the Ratio Test: $\displaystyle\lim_{n\to\infty} \dfrac{a_{n+1}}{a_n} = \lim_{n\to\infty} \dfrac{\left(\dfrac{1+\sin n}{n}\right)a_n}{a_n} = 0 < 1$

29. diverges by the Ratio Test: $\displaystyle\lim_{n\to\infty} \dfrac{a_{n+1}}{a_n} = \lim_{n\to\infty} \dfrac{\left(\dfrac{3n-1}{2n+1}\right)a_n}{a_n} = \lim_{n\to\infty} \dfrac{3n-1}{2n+1} = \dfrac{3}{2} > 1$

31. converges by the Ratio Test: $\displaystyle\lim_{n\to\infty} \dfrac{a_{n+1}}{a_n} = \lim_{n\to\infty} \dfrac{\left(\dfrac{2}{n}\right)a_n}{a_n} = \lim_{n\to\infty} \dfrac{2}{n} = 0 < 1$

33. converges by the Ratio Test: $\displaystyle\lim_{n\to\infty} \dfrac{a_{n+1}}{a_n} = \lim_{n\to\infty} \dfrac{\left(\dfrac{1+\ln n}{n}\right)a_n}{a_n} = \lim_{n\to\infty} \dfrac{1+\ln n}{n} = \lim_{n\to\infty} \dfrac{1}{n} = 0 < 1$

35. diverges by the nth-Term Test: $a_1 = \dfrac{1}{3}$, $a_2 = \sqrt[2]{\dfrac{1}{3}}$, $a_3 = \sqrt[3]{\sqrt[2]{\dfrac{1}{3}}} = \sqrt[6]{\dfrac{1}{3}}$, $a_4 = \sqrt[4]{\sqrt[3]{\sqrt[2]{\dfrac{1}{3}}}} = \sqrt[4!]{\dfrac{1}{3}}, \ldots$,

$a_n = \sqrt[n!]{\dfrac{1}{3}} \Rightarrow \displaystyle\lim_{n\to\infty} a_n = 1$ because $\left\{\sqrt[n!]{\dfrac{1}{3}}\right\}$ is a subsequence of $\left\{\sqrt[n]{\dfrac{1}{3}}\right\}$ whose limit is 1 by Table 8.1

37. converges by the Ratio Test: $\lim\limits_{n\to\infty} \dfrac{a_{n+1}}{a_n} = \lim\limits_{n\to\infty} \dfrac{2^{n+1}(n+1)!(n+1)!}{(2n+2)!} \cdot \dfrac{(2n)!}{2^n n! \, n!} = \lim\limits_{n\to\infty} \dfrac{2(n+1)(n+1)}{(2n+2)(2n+1)}$

$= \lim\limits_{n\to\infty} \dfrac{n+1}{2n+1} = \dfrac{1}{2} < 1$

39. diverges by the Root Test: $\lim\limits_{n\to\infty} \sqrt[n]{a_n} \equiv \lim\limits_{n\to\infty} \sqrt[n]{\dfrac{(n!)^n}{(n^n)^2}} = \lim\limits_{n\to\infty} \dfrac{n!}{n^2} = \infty > 1$

41. converges by the Root Test: $\lim\limits_{n\to\infty} \sqrt[n]{a_n} = \lim\limits_{n\to\infty} \sqrt[n]{\dfrac{n^n}{2^{n^2}}} = \lim\limits_{n\to\infty} \dfrac{n}{2^n} = \lim\limits_{n\to\infty} \dfrac{1}{2^n \ln 2} = 0 < 1$

43. converges by the Ratio Test: $\lim\limits_{n\to\infty} \dfrac{a_{n+1}}{a_n} = \lim\limits_{n\to\infty} \dfrac{1 \cdot 3 \cdot \cdots \cdot (2n-1)(2n+1)}{4^{n+1}2^{n+1}(n+1)!} \cdot \dfrac{4^n 2^n n!}{1 \cdot 3 \cdot \cdots \cdot (2n-1)}$

$= \lim\limits_{n\to\infty} \dfrac{2n+1}{(4 \cdot 2)(n+1)} = \dfrac{1}{4} < 1$

45. Ratio: $\lim\limits_{n\to\infty} \dfrac{a_{n+1}}{a_n} = \lim\limits_{n\to\infty} \dfrac{1}{(n+1)^P} \cdot \dfrac{n^P}{1} = \lim\limits_{n\to\infty} \left(\dfrac{n}{n+1}\right)^P = 1^P = 1 \Rightarrow$ no conclusion

Root: $\lim\limits_{n\to\infty} \sqrt[n]{a_n} = \lim\limits_{n\to\infty} \sqrt[n]{\dfrac{1}{n^P}} = \lim\limits_{n\to\infty} \dfrac{1}{\left(\sqrt[n]{n}\right)^P} = \dfrac{1}{(1)^P} = 1 \Rightarrow$ no conclusion

47. $a_n \le \dfrac{n}{2^n}$ for every n and the series $\sum\limits_{n=1}^{\infty} \dfrac{n}{2^n}$ converges by the Ratio Test since $\lim\limits_{n\to\infty} \dfrac{(n+1)}{2^{n+1}} \cdot \dfrac{2^n}{n} = \dfrac{1}{2} < 1$

$\Rightarrow \sum\limits_{n=1}^{\infty} a_n$ converges by the Direct Comparison Test

8.7 ALTERNATING SERIES, ABSOLUTE AND CONDITIONAL CONVERGENCE

1. converges absolutely $\Rightarrow$ converges by the Absolute Convergence Test since $\sum\limits_{n=1}^{\infty} |a_n| = \sum\limits_{n=1}^{\infty} \dfrac{1}{n^2}$ which is a convergent p-series

3. diverges by the nth-Term Test since for $n > 10 \Rightarrow \dfrac{n}{10} > 1 \Rightarrow \lim\limits_{n\to\infty} \left(\dfrac{n}{10}\right)^n \ne 0 \Rightarrow \sum\limits_{n=1}^{\infty} (-1)^{n+1}\left(\dfrac{n}{10}\right)^n$ diverges

5. converges by the Alternating Series Test because $f(x) = \ln x$ is an increasing function of $x \Rightarrow \dfrac{1}{\ln x}$ is decreasing

$\Rightarrow u_n \ge u_{n+1}$ for $n \ge 1$; also $u_n \ge 0$ for $n \ge 1$ and $\lim\limits_{n\to\infty} \dfrac{1}{\ln n} = 0$

7. diverges by the nth-Term Test since $\lim\limits_{n\to\infty} \dfrac{\ln n}{\ln n^2} = \lim\limits_{n\to\infty} \dfrac{\ln n}{2 \ln n} = \lim\limits_{n\to\infty} \dfrac{1}{2} = \dfrac{1}{2} \ne 0$

9. converges by the Alternating Series Test since $f(x) = \dfrac{\sqrt{x}+1}{x+1} \Rightarrow f'(x) = \dfrac{1 - x - 2\sqrt{x}}{2\sqrt{x}(x+1)^2} < 0 \Rightarrow f(x)$ is decreasing

$\Rightarrow u_n \ge u_{n+1}$; also $u_n \ge 0$ for $n \ge 1$ and $\lim\limits_{n\to\infty} u_n = \lim\limits_{n\to\infty} \dfrac{\sqrt{n}+1}{n+1} = 0$

11. converges absolutely since $\sum\limits_{n=1}^{\infty} |a_n| = \sum\limits_{n=1}^{\infty} \left(\frac{1}{10}\right)^n$ a convergent geometric series

13. converges conditionally since $\frac{1}{\sqrt{n}} > \frac{1}{\sqrt{n+1}} > 0$ and $\lim\limits_{n\to\infty} \frac{1}{\sqrt{n}} = 0 \Rightarrow$ convergence; but $\sum\limits_{n=1}^{\infty} |a_n| = \sum\limits_{n=1}^{\infty} \frac{1}{n^{1/2}}$

   is a divergent p-series

15. converges absolutely since $\sum\limits_{n=1}^{\infty} |a_n| = \sum\limits_{n=1}^{\infty} \frac{n}{n^3+1}$ and $\frac{n}{n^3+1} < \frac{1}{n^2}$ which is the nth-term of a converging p-series

17. converges conditionally since $\frac{1}{n+3} > \frac{1}{(n+1)+3} > 0$ and $\lim\limits_{n\to\infty} \frac{1}{n+3} = 0 \Rightarrow$ convergence; but $\sum\limits_{n=1}^{\infty} |a_n|$

   $= \sum\limits_{n=1}^{\infty} \frac{1}{n+3}$ diverges because $\frac{1}{n+3} \geq \frac{1}{4n}$ and $\sum\limits_{n=1}^{\infty} \frac{1}{n}$ is a divergent series

19. diverges by the nth-Term Test since $\lim\limits_{n\to\infty} \frac{3+n}{5+n} = 1 \neq 0$

21. converges conditionally since $f(x) = \frac{1}{x^2} + \frac{1}{x} \Rightarrow f'(x) = -\left(\frac{2}{x^3} + \frac{1}{x^2}\right) < 0 \Rightarrow f(x)$ is decreasing and hence

   $u_n > u_{n+1} > 0$ for $n \geq 1$ and $\lim\limits_{n\to\infty} \left(\frac{1}{n^2} + \frac{1}{n}\right) = 0 \Rightarrow$ convergence; but $\sum\limits_{n=1}^{\infty} |a_n| = \sum\limits_{n=1}^{\infty} \frac{1+n}{n^2}$

   $= \sum\limits_{n=1}^{\infty} \frac{1}{n^2} + \sum\limits_{n=1}^{\infty} \frac{1}{n}$ is the sum of a convergent and divergent series, and hence diverges

23. converges absolutely by the Ratio Test: $\lim\limits_{n\to\infty} \left(\frac{u_{n+1}}{u_n}\right) = \lim\limits_{n\to\infty} \left[\frac{(n+1)^2 \left(\frac{2}{3}\right)^{n+1}}{n^2 \left(\frac{2}{3}\right)^n}\right] = \frac{2}{3} < 1$

25. converges absolutely by the Integral Test since $\int\limits_{1}^{\infty} \left(\tan^{-1} x\right)\left(\frac{1}{1+x^2}\right) dx = \lim\limits_{b\to\infty} \left[\frac{\left(\tan^{-1} x\right)^2}{2}\right]_1^b$

   $= \lim\limits_{b\to\infty} \left[\left(\tan^{-1} b\right)^2 - \left(\tan^{-1} 1\right)^2\right] = \frac{1}{2}\left[\left(\frac{\pi}{2}\right)^2 - \left(\frac{\pi}{4}\right)^2\right] = \frac{3\pi^2}{32}$

27. diverges by the nth-Term Test since $\lim\limits_{n\to\infty} \frac{n}{n+1} = 1 \neq 0$

29. converges absolutely by the Ratio Test: $\lim\limits_{n\to\infty} \left(\frac{u_{n+1}}{u_n}\right) = \lim\limits_{n\to\infty} \frac{(100)^{n+1}}{(n+1)!} \cdot \frac{n!}{(100)^n} = \lim\limits_{n\to\infty} \frac{100}{n+1} = 0 < 1$

31. converges absolutely by the Direct Comparison Test since $\sum\limits_{n=1}^{\infty} |a_n| = \sum\limits_{n=1}^{\infty} \frac{1}{n^2+2n+1}$ and

   $\frac{1}{n^2+2n+1} < \frac{1}{n^2}$ which is the nth-term of a convergent p-series

33. converges absolutely since $\sum\limits_{n=1}^{\infty} |a_n| = \sum\limits_{n=1}^{\infty} \left| \dfrac{(-1)^n}{n\sqrt{n}} \right| = \sum\limits_{n=1}^{\infty} \dfrac{1}{n^{3/2}}$ is a convergent p-series

35. converges absolutely by the Root Test: $\lim\limits_{n\to\infty} \sqrt[n]{|a_n|} = \lim\limits_{n\to\infty} \left(\dfrac{(n+1)^n}{(2n)^n} \right)^{1/n} = \lim\limits_{n\to\infty} \dfrac{n+1}{2n} = \dfrac{1}{2}$

37. diverges by the nth-Term Test since $\lim\limits_{n\to\infty} |a_n| = \lim\limits_{n\to\infty} \dfrac{(2n)!}{2^n n! \, n} = \lim\limits_{n\to\infty} \dfrac{(n+1)(n+2)\cdots(2n)}{2^n n}$

$= \lim\limits_{n\to\infty} \dfrac{(n+1)(n+2)\cdots(n+(n-1))}{2^{n-1}} > \lim\limits_{n\to\infty} \left(\dfrac{n+1}{2} \right)^{n-1} = \infty \neq 0$

39. converges conditionally since $\dfrac{\sqrt{n+1} - \sqrt{n}}{1} \cdot \dfrac{\sqrt{n+1} + \sqrt{n}}{\sqrt{n+1} + \sqrt{n}} = \dfrac{1}{\sqrt{n+1} + \sqrt{n}}$ and $\left\{ \dfrac{1}{\sqrt{n+1} + \sqrt{n}} \right\}$ is a

decreasing sequence of positive terms which converges to $0 \Rightarrow \sum\limits_{n=1}^{\infty} \dfrac{(-1)^n}{\sqrt{n+1} + \sqrt{n}}$ converges; but $n > \dfrac{1}{3} \Rightarrow 3n > 1$

$\Rightarrow 4n > n+1 \Rightarrow 2\sqrt{n} > \sqrt{n+1} \Rightarrow 3\sqrt{n} > \sqrt{n+1} + \sqrt{n} \Rightarrow \dfrac{1}{3\sqrt{n}} < \dfrac{1}{\sqrt{n+1} + \sqrt{n}} \Rightarrow \sum\limits_{n=1}^{\infty} \dfrac{1}{\sqrt{n+1} + \sqrt{n}}$

diverges by the Direct Comparison Test

41. diverges by the nth-Term Test since $\lim\limits_{n\to\infty} \left(\sqrt{n + \sqrt{n}} - \sqrt{n} \right) = \lim\limits_{n\to\infty} \left[\left(\sqrt{n + \sqrt{n}} - \sqrt{n} \right) \left(\dfrac{\sqrt{n + \sqrt{n}} + \sqrt{n}}{\sqrt{n + \sqrt{n}} + \sqrt{n}} \right) \right]$

$= \lim\limits_{n\to\infty} \dfrac{\sqrt{n}}{\sqrt{n + \sqrt{n}} + \sqrt{n}} = \lim\limits_{n\to\infty} \dfrac{1}{\sqrt{1 + \dfrac{1}{\sqrt{n}}} + 1} = \dfrac{1}{2} \neq 0$

43. converges absolutely by the Direct Comparison Test since $\text{sech}\,(n) = \dfrac{2}{e^n + e^{-n}} = \dfrac{2e^n}{e^{2n} + 1} < \dfrac{2e^n}{e^{2n}} = \dfrac{2}{e^n}$ which is the nth term of a convergent geometric series

45. $|\,\text{error}\,| < \left| (-1)^6 \left(\dfrac{1}{5} \right) \right| = 0.2$ 47. $|\,\text{error}\,| < \left| (-1)^6 \dfrac{(0.01)^5}{5} \right| = 2 \times 10^{-11}$

49. $\dfrac{1}{(2n)!} < \dfrac{5}{10^6} \Rightarrow (2n)! > \dfrac{10^6}{5} = 200{,}000 \Rightarrow n \geq 5 \Rightarrow 1 - \dfrac{1}{2!} + \dfrac{1}{4!} - \dfrac{1}{6!} + \dfrac{1}{8!} \approx 0.54030$

51. (a) $a_n \geq a_{n+1}$ fails since $\dfrac{1}{3} < \dfrac{1}{2}$

(b) Since $\sum\limits_{n=1}^{\infty} |a_n| = \sum\limits_{n=1}^{\infty} \left[\left(\dfrac{1}{3} \right)^n + \left(\dfrac{1}{2} \right)^n \right] = \sum\limits_{n=1}^{\infty} \left(\dfrac{1}{3} \right)^n + \sum\limits_{n=1}^{\infty} \left(\dfrac{1}{2} \right)^n$ is the sum of two absolutely convergent

series, we can rearrange the terms of the original series to find its sum:

$\left(\dfrac{1}{3} + \dfrac{1}{9} + \dfrac{1}{27} + \ldots \right) - \left(\dfrac{1}{2} + \dfrac{1}{4} + \dfrac{1}{8} + \ldots \right) = \dfrac{\left(\dfrac{1}{3} \right)}{1 - \left(\dfrac{1}{3} \right)} - \dfrac{\left(\dfrac{1}{2} \right)}{1 - \left(\dfrac{1}{2} \right)} = \dfrac{1}{2} - 1 = -\dfrac{1}{2}$

53. The unused terms are $\displaystyle\sum_{j=n+1}^{\infty} (-1)^{j+1} a_j = (-1)^{n+1}(a_{n+1}-a_{n+2}) + (-1)^{n+3}(a_{n+3}-a_{n+4}) + \cdots$

$= (-1)^{n+1}\left[(a_{n+1}-a_{n+2}) + (a_{n+3}-a_{n+4}) + \cdots\right].$ Each grouped term is positive, so the remainder

has the same sign as $(-1)^{n+1}$, which is the sign of the first unused term.

55. Using the Direct Comparison Test, since $|a_n| \geq a_n$ and $\displaystyle\sum_{n=1}^{\infty} a_n$ diverges we must have that $\displaystyle\sum_{n=1}^{\infty} |a_n|$ diverges.

57. (a) $\displaystyle\sum_{n=1}^{\infty} |a_n + b_n|$ converges by the Direct Comparison Test since $|a_n + b_n| \leq |a_n| + |b_n|$ and hence

$\displaystyle\sum_{n=1}^{\infty} (a_n + b_n)$ converges absolutely

(b) $\displaystyle\sum_{n=1}^{\infty} |b_n|$ converges $\Rightarrow \displaystyle\sum_{n=1}^{\infty} -b_n$ converges absolutely; since $\displaystyle\sum_{n=1}^{\infty} a_n$ converges absolutely and

$\displaystyle\sum_{n=1}^{\infty} -b_n$ converges absolutely, we have $\displaystyle\sum_{n=1}^{\infty} \left[a_n + (-b_n)\right] = \displaystyle\sum_{n=1}^{\infty} (a_n - b_n)$ converges absolutely by part (a)

(c) $\displaystyle\sum_{n=1}^{\infty} |a_n|$ converges $\Rightarrow |k| \displaystyle\sum_{n=1}^{\infty} |a_n| = \displaystyle\sum_{n=1}^{\infty} |ka_n|$ converges $\Rightarrow \displaystyle\sum_{n=1}^{\infty} ka_n$ converges absolutely

59. $s_1 = -\frac{1}{2}$, $s_2 = -\frac{1}{2} + 1 = \frac{1}{2}$,

$s_3 = -\frac{1}{2} + 1 - \frac{1}{4} - \frac{1}{6} - \frac{1}{8} - \frac{1}{10} - \frac{1}{12} - \frac{1}{14} - \frac{1}{16} - \frac{1}{18} - \frac{1}{20} - \frac{1}{22} \approx -0.5099,$

$s_4 = s_3 + \frac{1}{3} \approx -0.1766,$

$s_5 = s_4 - \frac{1}{24} - \frac{1}{26} - \frac{1}{28} - \frac{1}{30} - \frac{1}{32} - \frac{1}{34} - \frac{1}{36} - \frac{1}{38} - \frac{1}{40} - \frac{1}{42} - \frac{1}{44} \approx -0.512,$

$s_6 = s_5 + \frac{1}{5} \approx -0.312,$

$s_7 = s_6 - \frac{1}{46} - \frac{1}{48} - \frac{1}{50} - \frac{1}{52} - \frac{1}{54} - \frac{1}{56} - \frac{1}{58} - \frac{1}{60} - \frac{1}{62} - \frac{1}{64} - \frac{1}{66} \approx -0.51106$

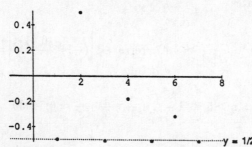

61. (a) If $\displaystyle\sum_{n=1}^{\infty} |a_n|$ converges, then $\displaystyle\sum_{n=1}^{\infty} a_n$ converges and $\frac{1}{2}\displaystyle\sum_{n=1}^{\infty} a_n + \frac{1}{2}\displaystyle\sum_{n=1}^{\infty} |a_n| = \displaystyle\sum_{n=1}^{\infty} \frac{a_n + |a_n|}{2}$

converges where $b_n = \dfrac{a_n + |a_n|}{2} = \begin{cases} a_n, & \text{if } a_n \geq 0 \\ 0, & \text{if } a_n < 0 \end{cases}.$

(b) If $\sum\limits_{n=1}^{\infty} |a_n|$ converges, then $\sum\limits_{n=1}^{\infty} a_n$ converges and $\frac{1}{2}\sum\limits_{n=1}^{\infty} a_n - \frac{1}{2}\sum\limits_{n=1}^{\infty} |a_n| = \sum\limits_{n=1}^{\infty} \frac{a_n - |a_n|}{2}$

converges where $c_n = \dfrac{a_n - |a_n|}{2} = \begin{cases} 0, & \text{if } a_n \geq 0 \\ a_n, & \text{if } a_n < 0 \end{cases}$.

63. Here is an example figure when N = 5. Notice that
$u_3 > u_2 > u_1$ and $u_3 > u_5 > u_4$, but $u_n \geq u_{n+1}$ for $n \geq 5$.

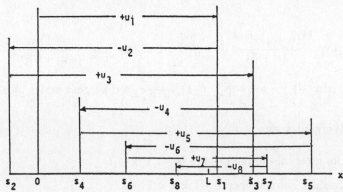

8.8 POWER SERIES

1. $\lim\limits_{n\to\infty} \left|\dfrac{u_{n+1}}{u_n}\right| < 1 \Rightarrow \lim\limits_{n\to\infty} \left|\dfrac{x^{n+1}}{x^n}\right| < 1 \Rightarrow |x| < 1 \Rightarrow -1 < x < 1$; when x = −1 we have $\sum\limits_{n=1}^{\infty} (-1)^n$, a divergent

series; when x = 1 we have $\sum\limits_{n=1}^{\infty} 1$, a divergent series
   (a) the radius is 1; the interval of convergence is −1 < x < 1
   (b) the interval of absolute convergence is −1 < x < 1
   (c) there are no values for which the series converges conditionally

3. $\lim\limits_{n\to\infty} \left|\dfrac{u_{n+1}}{u_n}\right| < 1 \Rightarrow \lim\limits_{n\to\infty} \left|\dfrac{(4x+1)^{n+1}}{(4x+1)^n}\right| < 1 \Rightarrow |4x+1| < 1 \Rightarrow -1 < 4x+1 < 1 \Rightarrow -\frac{1}{2} < x < 0$; when $x = -\frac{1}{2}$ we

have $\sum\limits_{n=1}^{\infty} (-1)^n(-1)^n = \sum\limits_{n=1}^{\infty} (-1)^{2n} = \sum\limits_{n=1}^{\infty} 1^n$, a divergent series; when x = 0 we have $\sum\limits_{n=1}^{\infty} (-1)^n(1)^n$

$= \sum\limits_{n=1}^{\infty} (-1)^n$, a divergent series
   (a) the radius is $\frac{1}{4}$; the interval of convergence is $-\frac{1}{2} < x < 0$
   (b) the interval of absolute convergence is $-\frac{1}{2} < x < 0$
   (c) there are no values for which the series converges conditionally

5. $\lim\limits_{n\to\infty}\left|\dfrac{u_{n+1}}{u_n}\right|<1\Rightarrow\lim\limits_{n\to\infty}\left|\dfrac{(x-2)^{n+1}}{10^{n+1}}\cdot\dfrac{10^n}{(x-2)^n}\right|<1\Rightarrow\dfrac{|x-2|}{10}<1\Rightarrow|x-2|<10\Rightarrow-10<x-2<10$

$\Rightarrow-8<x<12$; when $x=-8$ we have $\sum\limits_{n=1}^{\infty}(-1)^n$, a divergent series; when $x=12$ we have $\sum\limits_{n=1}^{\infty}1$, a divergent series

   (a) the radius is 10; the interval of convergence is $-8<x<12$

   (b) the interval of absolute convergence is $-8<x<12$

   (c) there are no values for which the series converges conditionally

7. $\lim\limits_{n\to\infty}\left|\dfrac{u_{n+1}}{u_n}\right|<1\Rightarrow\lim\limits_{n\to\infty}\left|\dfrac{(n+1)x^{n+1}}{(n+3)}\cdot\dfrac{(n+2)}{nx^n}\right|<1\Rightarrow|x|\lim\limits_{n\to\infty}\dfrac{(n+1)(n+2)}{(n+3)(n)}<1\Rightarrow|x|<1$

$\Rightarrow-1<x<1$; when $x=-1$ we have $\sum\limits_{n=1}^{\infty}(-1)^n\dfrac{n}{n+2}$, a divergent series by the nth-term Test; when $x=1$ we

have $\sum\limits_{n=1}^{\infty}\dfrac{n}{n+2}$, a divergent series

   (a) the radius is 1; the interval of convergence is $-1<x<1$

   (b) the interval of absolute convergence is $-1<x<1$

   (c) there are no values for which the series converges conditionally

9. $\lim\limits_{n\to\infty}\left|\dfrac{u_{n+1}}{u_n}\right|<1\Rightarrow\lim\limits_{n\to\infty}\left|\dfrac{x^{n+1}}{(n+1)\sqrt{n+1}\,3^{n+1}}\cdot\dfrac{n\sqrt{n}\,3^n}{x^n}\right|<1\Rightarrow\dfrac{|x|}{3}\left(\lim\limits_{n\to\infty}\dfrac{n}{n+1}\right)\left(\sqrt{\lim\limits_{n\to\infty}\dfrac{n}{n+1}}\right)<1$

$\Rightarrow\dfrac{|x|}{3}(1)(1)<1\Rightarrow|x|<3\Rightarrow-3<x<3$; when $x=-3$ we have $\sum\limits_{n=1}^{\infty}\dfrac{(-1)^n}{n^{3/2}}$, an absolutely convergent series;

when $x=3$ we have $\sum\limits_{n=1}^{\infty}\dfrac{1}{n^{3/2}}$, a convergent p-series

   (a) the radius is 3; the interval of convergence is $-3\le x\le 3$

   (b) the interval of absolute convergence is $-3\le x\le 3$

   (c) there are no values for which the series converges conditionally

11. $\lim\limits_{n\to\infty}\left|\dfrac{u_{n+1}}{u_n}\right|<1\Rightarrow\lim\limits_{n\to\infty}\left|\dfrac{x^{n+1}}{(n+1)!}\cdot\dfrac{n!}{x^n}\right|<1\Rightarrow|x|\lim\limits_{n\to\infty}\left(\dfrac{1}{n+1}\right)<1$ for all x

   (a) the radius is ∞; the series converges for all x

   (b) the series converges absolutely for all x

   (c) there are no values for which the series converges conditionally

13. $\lim\limits_{n\to\infty}\left|\dfrac{u_{n+1}}{u_n}\right|<1\Rightarrow\lim\limits_{n\to\infty}\left|\dfrac{x^{2n+3}}{(n+1)!}\cdot\dfrac{n!}{x^{2n+1}}\right|<1\Rightarrow x^2\lim\limits_{n\to\infty}\left(\dfrac{1}{n+1}\right)<1$ for all x

   (a) the radius is ∞; the series converges for all x

   (b) the series converges absolutely for all x

   (c) there are no values for which the series converges conditionally

15. $\lim\limits_{n\to\infty}\left|\dfrac{u_{n+1}}{u_n}\right|<1\Rightarrow\lim\limits_{n\to\infty}\left|\dfrac{x^{n+1}}{\sqrt{(n+1)^2+3}}\cdot\dfrac{\sqrt{n^2+3}}{x^n}\right|<1\Rightarrow|x|\sqrt{\lim\limits_{n\to\infty}\dfrac{n^2+3}{n^2+2n+4}}<1\Rightarrow|x|<1$

$\Rightarrow -1<x<1$; when $x=-1$ we have $\sum\limits_{n=1}^{\infty}\dfrac{(-1)^n}{\sqrt{n^2+3}}$, a conditionally convergent series; when $x=1$ we have

$\sum\limits_{n=1}^{\infty}\dfrac{1}{n^2+3}$, a divergent series

(a) the radius is 1; the interval of convergence is $-1\le x<1$

(b) the interval of absolute convergence is $-1<x<1$

(c) the series converges conditionally at $x=-1$

17. $\lim\limits_{n\to\infty}\left|\dfrac{u_{n+1}}{u_n}\right|<1\Rightarrow\lim\limits_{n\to\infty}\left|\dfrac{(n+1)(x+3)^{n+1}}{5^{n+1}}\cdot\dfrac{5^n}{n(x+3)^n}\right|<1\Rightarrow\dfrac{|x+3|}{5}\lim\limits_{n\to\infty}\left(\dfrac{n+1}{n}\right)<1\Rightarrow\dfrac{|x+3|}{5}<1$

$\Rightarrow|x+3|<5\Rightarrow-5<x+3<5\Rightarrow-8<x<2$; when $x=-8$ we have $\sum\limits_{n=1}^{\infty}\dfrac{n(-5)^n}{5^n}=\sum\limits_{n=1}^{\infty}(-1)^n n$, a divergent

series; when $x=2$ we have $\sum\limits_{n=1}^{\infty}\dfrac{n5^n}{5^n}=\sum\limits_{n=1}^{\infty}n$, a divergent series

(a) the radius is 5; the interval of convergence is $-8<x<2$

(b) the interval of absolute convergence is $-8<x<2$

(c) there are no values for which the series converges conditionally

19. $\lim\limits_{n\to\infty}\left|\dfrac{u_{n+1}}{u_n}\right|<1\Rightarrow\lim\limits_{n\to\infty}\left|\dfrac{\sqrt{n+1}\,x^{n+1}}{3^{n+1}}\cdot\dfrac{3^n}{\sqrt{n}\,x^n}\right|<1\Rightarrow\dfrac{|x|}{3}\sqrt{\lim\limits_{n\to\infty}\left(\dfrac{n+1}{n}\right)}<1\Rightarrow\dfrac{|x|}{3}<1\Rightarrow|x|<3$

$\Rightarrow -3<x<3$; when $x=-3$ we have $\sum\limits_{n=1}^{\infty}(-1)^n\sqrt{n}$, a divergent series; when $x=3$ we have

$\sum\limits_{n=1}^{\infty}\sqrt{n}$, a divergent series

(a) the radius is 3; the interval of convergence is $-3<x<3$

(b) the interval of absolute convergence is $-3<x<3$

(c) there are no values for which the series converges conditionally

21. $\lim\limits_{n\to\infty}\left|\dfrac{u_{n+1}}{u_n}\right|<1\Rightarrow\lim\limits_{n\to\infty}\left|\dfrac{\left(1+\frac{1}{n+1}\right)^{n+1}x^{n+1}}{\left(1+\frac{1}{n}\right)^n x^n}\right|<1\Rightarrow|x|\left(\dfrac{\lim\limits_{t\to\infty}\left(1+\frac{1}{t}\right)^t}{\lim\limits_{n\to\infty}\left(1+\frac{1}{n}\right)^n}\right)<1\Rightarrow|x|\left(\dfrac{e}{e}\right)<1\Rightarrow|x|<1$

$\Rightarrow -1<x<1$; when $x=-1$ we have $\sum\limits_{n=1}^{\infty}(-1)^n\left(1+\dfrac{1}{n}\right)^n$, a divergent series by the nth-Term Test since

$\lim\limits_{n\to\infty}\left(1+\dfrac{1}{n}\right)^n=e\ne 0$; when $x=1$ we have $\sum\limits_{n=1}^{\infty}\left(1+\dfrac{1}{n}\right)^n$, a divergent series

(a) the radius is 1; the interval of convergence is $-1<x<1$

(b) the interval of absolute convergence is $-1<x<1$

(c) there are no values for which the series converges conditionally

23. $\lim\limits_{n\to\infty}\left|\dfrac{u_{n+1}}{u_n}\right|<1\Rightarrow\lim\limits_{n\to\infty}\left|\dfrac{(n+1)^{n+1}x^{n+1}}{n^n x^n}\right|<1\Rightarrow|x|\left(\lim\limits_{n\to\infty}\left(1+\dfrac{1}{n}\right)^n\right)\left(\lim\limits_{n\to\infty}(n+1)\right)<1$

$\Rightarrow e|x|\lim\limits_{n\to\infty}(n+1)<1\Rightarrow$ only $x=0$ satisfies this inequality

(a) the radius is 0; the series converges only for $x=0$

(b) the series converges absolutely only for $x=0$

(c) there are no values for which the series converges conditionally

25. $\lim\limits_{n\to\infty}\left|\dfrac{u_{n+1}}{u_n}\right|<1\Rightarrow\lim\limits_{n\to\infty}\left|\dfrac{(x+2)^{n+1}}{(n+1)\,2^{n+1}}\cdot\dfrac{n2^n}{(x+2)^n}\right|<1\Rightarrow\dfrac{|x+2|}{2}\lim\limits_{n\to\infty}\left(\dfrac{n}{n+1}\right)<1\Rightarrow\dfrac{|x+2|}{2}<1\Rightarrow|x+2|<2$

$\Rightarrow-2<x+2<2\Rightarrow-4<x<0$; when $x=-4$ we have $\sum\limits_{n=1}^{\infty}\dfrac{-1}{n}$, a divergent series; when $x=0$ we have

$\sum\limits_{n=1}^{\infty}\dfrac{(-1)^{n+1}}{n}$, the alternating harmonic series which converges conditionally

(a) the radius is 2; the interval of convergence is $-4<x\le0$

(b) the interval of absolute convergence is $-4<x<0$

(c) the series converges conditionally at $x=0$

27. $\lim\limits_{n\to\infty}\left|\dfrac{u_{n+1}}{u_n}\right|<1\Rightarrow\lim\limits_{n\to\infty}\left|\dfrac{x^{n+1}}{(n+1)\left(\ln(n+1)\right)^2}\cdot\dfrac{n(\ln n)^2}{x^n}\right|<1\Rightarrow|x|\left(\lim\limits_{n\to\infty}\dfrac{n}{n+1}\right)\left(\lim\limits_{n\to\infty}\dfrac{\ln n}{\ln(n+1)}\right)^2<1$

$\Rightarrow|x|(1)\left(\lim\limits_{n\to\infty}\dfrac{\left(\frac{1}{n}\right)}{\left(\frac{1}{n+1}\right)}\right)^2<1\Rightarrow|x|\left(\lim\limits_{n\to\infty}\dfrac{n+1}{n}\right)^2<1\Rightarrow|x|<1\Rightarrow-1<x<1$; when $x=-1$ we have

$\sum\limits_{n=1}^{\infty}\dfrac{(-1)^n}{n(\ln n)^2}$ which converges absolutely; when $x=1$ we have $\sum\limits_{n=1}^{\infty}\dfrac{1}{n(\ln n)^2}$ which converges

(a) the radius is 1; the interval of convergence is $-1\le1\le1$

(b) the interval of absolute convergence is $-1\le x\le1$

(c) there are no values for which the series converges conditionally

29. $\lim\limits_{n\to\infty}\left|\dfrac{u_{n+1}}{u_n}\right|<1\Rightarrow\lim\limits_{n\to\infty}\left|\dfrac{(4x-5)^{2n+3}}{(n+1)^{3/2}}\cdot\dfrac{n^{3/2}}{(4x-5)^{2n+1}}\right|<1\Rightarrow(4x-5)^2\left(\lim\limits_{n\to\infty}\dfrac{n}{n+1}\right)^{3/2}<1\Rightarrow(4x-5)^2<1$

$\Rightarrow|4x-5|<1\Rightarrow-1<4x-5<1\Rightarrow1<x<\dfrac{3}{2}$; when $x=1$ we have $\sum\limits_{n=1}^{\infty}\dfrac{(-1)^{2n+1}}{n^{3/2}}=\sum\limits_{n=1}^{\infty}\dfrac{-1}{n^{3/2}}$ which is

absolutely convergent; when $x=\dfrac{3}{2}$ we have $\sum\limits_{n=1}^{\infty}\dfrac{(1)^{2n+1}}{n^{3/2}}$, a convergent p-series

(a) the radius is $\dfrac{1}{4}$; the interval of convergence is $1\le x\le\dfrac{3}{2}$

(b) the interval of absolute convergence is $1\le x\le\dfrac{3}{2}$

(c) there are no values for which the series converges conditionally

31. $\lim_{n \to \infty} \left| \frac{u_{n+1}}{u_n} \right| < 1 \Rightarrow \lim_{n \to \infty} \left| \frac{(x+\pi)^{n+1}}{\sqrt{n+1}} \cdot \frac{\sqrt{n}}{(x+\pi)^n} \right| < 1 \Rightarrow |x+\pi| \lim_{n \to \infty} \left| \sqrt{\frac{n}{n+1}} \right| < 1$

$\Rightarrow |x+\pi| \sqrt{\lim_{n \to \infty} \left(\frac{n}{n+1} \right)} < 1 \Rightarrow |x+\pi| < 1 \Rightarrow -1 < x+\pi < 1 \Rightarrow -1-\pi < x < 1-\pi;$

when $x = -1-\pi$ we have $\sum_{n=1}^{\infty} \frac{(-1)^n}{\sqrt{n}} = \sum_{n=1}^{\infty} \frac{(-1)^n}{n^{1/2}}$, a conditionally convergent series; when $x = 1-\pi$ we have

$\sum_{n=1}^{\infty} \frac{1^n}{\sqrt{n}} = \sum_{n=1}^{\infty} \frac{1}{n^{1/2}}$, a divergent p-series

(a) the radius is 1; the interval of convergence is $(-1-\pi) \le x < (1-\pi)$

(b) the interval of absolute convergence is $-1-\pi < x < 1-\pi$

(c) the series converges conditionally at $x = -1-\pi$

33. $\lim_{n \to \infty} \left| \frac{u_{n+1}}{u_n} \right| < 1 \Rightarrow \lim_{n \to \infty} \left| \frac{(x-1)^{2n+2}}{4^{n+1}} \cdot \frac{4^n}{(x-1)^{2n}} \right| < 1 \Rightarrow \frac{(x-1)^2}{4} \lim_{n \to \infty} |1| < 1 \Rightarrow (x-1)^2 < 4 \Rightarrow |x-1| < 2$

$\Rightarrow -2 < x-1 < 2 \Rightarrow -1 < x < 3$; at $x = -1$ we have $\sum_{n=0}^{\infty} \frac{(-2)^{2n}}{4^n} = \sum_{n=0}^{\infty} \frac{4^n}{4^n} = \sum_{n=0}^{\infty} 1$, which diverges; at $x = 3$

we have $\sum_{n=0}^{\infty} \frac{2^{2n}}{4^n} = \sum_{n=0}^{\infty} \frac{4^n}{4^n} = \sum_{n=0}^{\infty} 1$, a divergent series; the interval of convergence is $-1 < x < 3$; the series

$\sum_{n=0}^{\infty} \frac{(x-1)^{2n}}{4^n} = \sum_{n=0}^{\infty} \left(\left(\frac{x-1}{2} \right)^2 \right)^n$ is a convergent geometric series when $-1 < x < 3$ and the sum is

$\frac{1}{1 - \left(\frac{x-1}{2} \right)^2} = \frac{1}{\left[\frac{4-(x-1)^2}{4} \right]} = \frac{4}{4-x^2+2x-1} = \frac{4}{3+2x-x^2}$

35. $\lim_{n \to \infty} \left| \frac{u_{n+1}}{u_n} \right| < 1 \Rightarrow \lim_{n \to \infty} \left| \frac{(\sqrt{x}-2)^{n+1}}{2^{n+1}} \cdot \frac{2^n}{(\sqrt{x}-2)^n} \right| < 1 \Rightarrow \left| \sqrt{x}-2 \right| < 2 \Rightarrow -2 < \sqrt{x}-2 < 2 \Rightarrow 0 < \sqrt{x} < 4$

$\Rightarrow 0 < x < 16$; when $x = 0$ we have $\sum_{n=0}^{\infty} (-1)^n$, a divergent series; when $x = 16$ we have $\sum_{n=0}^{\infty} (1)^n$, a divergent

series; the interval of convergence is $0 < x < 16$; the series $\sum_{n=0}^{\infty} \left(\frac{\sqrt{x}-2}{2} \right)^n$ is a convergent geometric series when

$0 < x < 16$ and its sum is $\dfrac{1}{1 - \left(\frac{\sqrt{x}-2}{2} \right)} = \dfrac{1}{\left(\frac{2-\sqrt{x}+2}{2} \right)} = \dfrac{2}{4-\sqrt{x}}$

37. $\lim_{n \to \infty} \left| \frac{u_{n+1}}{u_n} \right| < 1 \Rightarrow \lim_{n \to \infty} \left| \left(\frac{x^2+1}{3} \right)^{n+1} \cdot \left(\frac{3}{x^2+1} \right)^n \right| < 1 \Rightarrow \frac{(x^2+1)}{3} \lim_{n \to \infty} |1| < 1 \Rightarrow \frac{x^2+1}{3} < 1 \Rightarrow x^2 < 2$

$\Rightarrow |x| < \sqrt{2} \Rightarrow -\sqrt{2} < x < \sqrt{2}$; at $x = \pm\sqrt{2}$ we have $\sum_{n=0}^{\infty} (1)^n$ which diverges; the interval of convergence is

$-\sqrt{2} < x < \sqrt{2}$; the series $\sum_{n=0}^{\infty} \left(\frac{x^2+1}{3} \right)^n$ is a convergent geometric series when $-\sqrt{2} < x < \sqrt{2}$ and its sum is

$\dfrac{1}{1 - \left(\frac{x^2+1}{3} \right)} = \dfrac{1}{\left(\frac{3-x^2-1}{3} \right)} = \dfrac{3}{2-x^2}$

39. $\lim\limits_{n\to\infty}\left|\dfrac{(x-3)^{n+1}}{2^{n+1}}\cdot\dfrac{2^n}{(x-3)^n}\right|<1\Rightarrow|x-3|<2\Rightarrow1<x<5$; when $x=1$ we have $\sum\limits_{n=1}^{\infty}(1)^n$ which diverges;

when $x=5$ we have $\sum\limits_{n=1}^{\infty}(-1)^n$ which also diverges; the interval of convergence is $1<x<5$; the sum of this

convergent geometric series is $\dfrac{1}{1+\left(\frac{x-3}{2}\right)}=\dfrac{2}{x-1}$. If $f(x)=1-\frac{1}{2}(x-3)+\frac{1}{4}(x-3)^2+\ldots+\left(-\frac{1}{2}\right)^n(x-3)^n+\ldots$

$=\dfrac{2}{x-1}$ then $f'(x)=-\frac{1}{2}+\frac{1}{2}(x-3)+\ldots+\left(-\frac{1}{2}\right)^n n(x-3)^{n-1}+\ldots$ is convergent when $1<x<5$, and diverges

when $x=1$ or 5. The sum for $f'(x)$ is $\dfrac{-2}{(x-1)^2}$, the derivative of $\dfrac{2}{x-1}$.

41. (a) Differentiate the series for $\sin x$ to get $\cos x=1-\dfrac{3x^2}{3!}+\dfrac{5x^4}{5!}-\dfrac{7x^6}{7!}+\dfrac{9x^8}{9!}-\dfrac{11x^{10}}{11!}+\ldots$

$=1-\dfrac{x^2}{2!}+\dfrac{x^4}{4!}-\dfrac{x^6}{6!}+\dfrac{x^8}{8!}-\dfrac{x^{10}}{10!}+\ldots$. The series converges for all values of x since

$\lim\limits_{n\to\infty}\left|\dfrac{x^{n+1}}{(n+1)!}\cdot\dfrac{n!}{x^n}\right|=|x|\lim\limits_{n\to\infty}\left(\dfrac{1}{n+1}\right)=0<1$ for all x

(b) $\sin 2x=2x-\dfrac{2^3x^3}{3!}+\dfrac{2^5x^5}{5!}-\dfrac{2^7x^7}{7!}+\dfrac{2^9x^9}{9!}-\dfrac{2^{11}x^{11}}{11!}+\ldots=2x-\dfrac{8x^3}{3!}+\dfrac{32x^5}{5!}-\dfrac{128x^7}{7!}+\dfrac{512x^9}{9!}-\dfrac{2048x^{11}}{11!}+\ldots$

(c) $2\sin x\cos x=2\Big[(0\cdot1)+(0\cdot0+1\cdot1)x+\left(0\cdot\dfrac{-1}{2}+1\cdot0+0\cdot1\right)x^2+\left(0\cdot0-1\cdot\dfrac{1}{2}+0\cdot0-1\dfrac{1}{3!}\right)x^3$

$+\left(0\cdot\dfrac{1}{4!}+1\cdot0-0\cdot\dfrac{1}{2}-0\cdot\dfrac{1}{3!}+0\cdot1\right)x^4+\left(0\cdot0+1\cdot\dfrac{1}{4!}+0\cdot0+\dfrac{1}{2}\cdot\dfrac{1}{3!}+0\cdot0+1\dfrac{1}{5!}\right)x^5$

$+\left(0\cdot\dfrac{1}{6!}+1\cdot0+0\cdot\dfrac{1}{4!}+0\cdot\dfrac{1}{3!}+0\cdot\dfrac{1}{2}+0\cdot\dfrac{1}{5!}+0\cdot1\right)x^6+\ldots\Big]=2\left[x-\dfrac{4x^3}{3!}+\dfrac{16x^5}{5!}-\ldots\right]$

$=2x-\dfrac{2^3x^3}{3!}+\dfrac{2^5x^5}{5!}-\dfrac{2^7x^7}{7!}+\dfrac{2^9x^9}{9!}-\dfrac{2^{11}x^{11}}{11!}+\ldots$

43. (a) $\ln|\sec x|+C=\displaystyle\int\tan x\,dx=\int\left(x+\dfrac{x^3}{3}+\dfrac{2x^5}{15}+\dfrac{17x^7}{315}+\dfrac{62x^9}{2835}+\ldots\right)dx$

$=\dfrac{x^2}{2}+\dfrac{x^4}{12}+\dfrac{x^6}{45}+\dfrac{17x^8}{2520}+\dfrac{31x^{10}}{14,175}+\ldots+C$; $x=0\Rightarrow C=0\Rightarrow\ln|\sec x|=\dfrac{x^2}{2}+\dfrac{x^4}{12}+\dfrac{x^6}{45}+\dfrac{17x^8}{2520}+\dfrac{31x^{10}}{14,175}+\ldots$,

converges when $-\dfrac{\pi}{2}<x<\dfrac{\pi}{2}$

(b) $\sec^2 x=\dfrac{d(\tan x)}{dx}=\dfrac{d}{dx}\left(x+\dfrac{x^3}{3}+\dfrac{2x^5}{15}+\dfrac{17x^7}{315}+\dfrac{62x^9}{2835}+\ldots\right)=1+x^2+\dfrac{2x^4}{3}+\dfrac{17x^6}{45}+\dfrac{62x^8}{315}+\ldots$, converges

when $-\dfrac{\pi}{2}<x<\dfrac{\pi}{2}$

(c) $\sec^2 x=(\sec x)(\sec x)=\left(1+\dfrac{x^2}{2}+\dfrac{5x^4}{24}+\dfrac{61x^6}{720}+\ldots\right)\left(1+\dfrac{x^2}{2}+\dfrac{5x^4}{24}+\dfrac{61x^6}{720}+\ldots\right)$

$=1+\left(\dfrac{1}{2}+\dfrac{1}{2}\right)x^2+\left(\dfrac{5}{24}+\dfrac{1}{4}+\dfrac{5}{24}\right)x^4+\left(\dfrac{61}{720}+\dfrac{5}{48}+\dfrac{5}{48}+\dfrac{61}{720}\right)x^6+\ldots$

$=1+x^2+\dfrac{2x^4}{3}+\dfrac{17x^6}{45}+\dfrac{62x^8}{315}+\ldots$, $-\dfrac{\pi}{2}<x<\dfrac{\pi}{2}$

45. (a) If $f(x) = \sum\limits_{n=0}^{\infty} a_n x^n$, then $f^{(k)}(x) = \sum\limits_{n=k}^{\infty} n(n-1)(n-2)\cdots(n-(k-1))\, a_n x^{n-k}$ and $f^{(k)}(0) = k!a_k$

$\Rightarrow a_k = \dfrac{f^{(k)}(0)}{k!}$; likewise if $f(x) = \sum\limits_{n=0}^{\infty} b_n x^n$, then $b_k = \dfrac{f^{(k)}(0)}{k!} \Rightarrow a_k = b_k$ for every nonnegative integer k

(b) If $f(x) = \sum\limits_{n=0}^{\infty} a_n x^n = 0$ for all x, then $f^{(k)}(x) = 0$ for all x $\Rightarrow$ from part (a) that $a_k = 0$ for every

nonnegative integer k

47. The series $\sum\limits_{n=1}^{\infty} \dfrac{x^n}{n}$ converges conditionally at the left-hand endpoint of its interval of convergence $[-1, 1]$; the

series $\sum\limits_{n=1}^{\infty} \dfrac{x^n}{(n^2)}$ converges absolutely at the left-hand endpoint of its interval of convergence $[-1, 1]$

8.9 TAYLOR AND MACLAURIN SERIES

1. $f(x) = \ln x,\ f'(x) = \frac{1}{x},\ f''(x) = -\frac{1}{x^2},\ f'''(x) = \frac{2}{x^3};\ f(1) = \ln 1 = 0,\ f'(1) = 1,\ f''(1) = -1,\ f'''(1) = 2 \Rightarrow P_0(x) = 0,$

$P_1(x) = (x-1),\ P_2(x) = (x-1) - \frac{1}{2}(x-1)^2,\ P_3(x) = (x-1) - \frac{1}{2}(x-1)^2 + \frac{1}{3}(x-1)^3$

3. $f(x) = \frac{1}{x} = x^{-1},\ f'(x) = -x^{-2},\ f''(x) = 2x^{-3},\ f'''(x) = -6x^{-4};\ f(2) = \frac{1}{2},\ f'(2) = -\frac{1}{4},\ f''(2) = \frac{1}{4},\ f'''(x) = -\frac{3}{8}$

$\Rightarrow P_0(x) = \frac{1}{2},\ P_1(x) = \frac{1}{2} - \frac{1}{4}(x-2),\ P_2(x) = \frac{1}{2} - \frac{1}{4}(x-2) + \frac{1}{8}(x-2)^2,$

$P_3(x) = \frac{1}{2} - \frac{1}{4}(x-2) + \frac{1}{8}(x-2)^2 - \frac{1}{16}(x-2)^3$

5. $f(x) = \sin x,\ f'(x) = \cos x,\ f''(x) = -\sin x,\ f'''(x) = -\cos x;\ f\left(\frac{\pi}{4}\right) = \sin \frac{\pi}{4} = \frac{\sqrt{2}}{2},\ f'\left(\frac{\pi}{4}\right) = \cos \frac{\pi}{4} = \frac{\sqrt{2}}{2},$

$f''\left(\frac{\pi}{4}\right) = -\sin \frac{\pi}{4} = -\frac{\sqrt{2}}{2},\ f'''\left(\frac{\pi}{4}\right) = -\cos \frac{\pi}{4} = -\frac{\sqrt{2}}{2} \Rightarrow P_0 = \frac{\sqrt{2}}{2},\ P_1(x) = \frac{\sqrt{2}}{2} + \frac{\sqrt{2}}{2}\left(x - \frac{\pi}{4}\right),$

$P_2(x) = \frac{\sqrt{2}}{2} + \frac{\sqrt{2}}{2}\left(x - \frac{\pi}{4}\right) - \frac{\sqrt{2}}{4}\left(x - \frac{\pi}{4}\right)^2,\ P_3(x) = \frac{\sqrt{2}}{2} + \frac{\sqrt{2}}{2}\left(x - \frac{\pi}{4}\right) - \frac{\sqrt{2}}{4}\left(x - \frac{\pi}{4}\right)^2 - \frac{\sqrt{2}}{12}\left(x - \frac{\pi}{4}\right)^3$

7. $f(x) = \sqrt{x} = x^{1/2},\ f'(x) = \left(\frac{1}{2}\right)x^{-1/2},\ f''(x) = \left(-\frac{1}{4}\right)x^{-3/2},\ f'''(x) = \left(\frac{3}{8}\right)x^{-5/2};\ f(4) = \sqrt{4} = 2,$

$f'(4) = \left(\frac{1}{2}\right)4^{-1/2} = \frac{1}{4},\ f''(4) = \left(-\frac{1}{4}\right)4^{-3/2} = -\frac{1}{32},\ f'''(4) = \left(\frac{3}{8}\right)4^{-5/2} = \frac{3}{256} \Rightarrow P_0(x) = 2,\ P_1(x) = 2 + \frac{1}{4}(x-4),$

$P_2(x) = 2 + \frac{1}{4}(x-4) - \frac{1}{64}(x-4)^2,\ P_3(x) = 2 + \frac{1}{4}(x-4) - \frac{1}{64}(x-4)^2 + \frac{1}{512}(x-4)^3$

9. $e^x = \sum\limits_{n=0}^{\infty} \frac{x^n}{n!} \Rightarrow e^{-x} = \sum\limits_{n=0}^{\infty} \frac{(-x)^n}{n!} = 1 - x + \frac{x^2}{2!} - \frac{x^3}{3!} + \frac{x^4}{4!} - \cdots$

11. $f(x) = (1+x)^{-1} \Rightarrow f'(x) = -(1+x)^{-2}, f''(x) = 2(1+x)^{-3}, f'''(x) = -3!(1+x)^{-4} \Rightarrow \dots f^{(k)}(x)$

$= (-1)^k k! (1+x)^{-k-1}; f(0) = 1, f'(0) = -1, f''(0) = 2, f'''(0) = -3!, \dots, f^{(k)}(0) = (-1)^k k!$

$\Rightarrow \frac{1}{1+x} = 1 - x + x^2 - x^3 + \dots = \sum_{n=0}^{\infty} (-x)^n = \sum_{n=0}^{\infty} (-1)^n x^n$

13. $\sin x = \sum_{n=0}^{\infty} \frac{(-1)^n x^{2n+1}}{(2n+1)!} \Rightarrow \sin 3x = \sum_{n=0}^{\infty} \frac{(-1)^n (3x)^{2n+1}}{(2n+1)!} = \sum_{n=0}^{\infty} \frac{(-1)^n 3^{2n+1} x^{2n+1}}{(2n+1)!} = 3x - \frac{3^3 x^3}{3!} + \frac{3^5 x^5}{5!} - \dots$

15. $7\cos(-x) = 7\cos x = 7 \sum_{n=0}^{\infty} \frac{(-1)^n x^{2n}}{(2n)!} = 7 - \frac{7x^2}{2!} + \frac{7x^4}{4!} - \frac{7x^6}{6!} + \dots$, since the cosine is an even function

17. $\cosh x = \frac{e^x + e^{-x}}{2} = \frac{1}{2}\left[\left(1 + x^2 + \frac{x^2}{2!} + \frac{x^3}{3!} + \frac{x^4}{4!} + \dots\right) + \left(1 - x + \frac{x^2}{2!} - \frac{x^3}{3!} + \frac{x^4}{4!} - \dots\right)\right] = 1 + \frac{x^2}{2!} + \frac{x^4}{4!} + \frac{x^6}{6!} + \dots$

$= \sum_{n=0}^{\infty} \frac{x^{2n}}{(2n)!}$

19. $f(x) = x^4 - 2x^3 - 5x + 4 \Rightarrow f'(x) = 4x^3 - 6x^2 - 5, f''(x) = 12x^2 - 12x, f'''(x) = 24x - 12, f^{(4)}(x) = 24$

$\Rightarrow f^{(n)}(x) = 0$ if $n \geq 5$; $f(0) = 4, f'(0) = -5, f''(0) = 0, f'''(0) = -12, f^{(4)}(0) = 24, f^{(n)}(0) = 0$ if $n \geq 5$

$\Rightarrow x^4 - 2x^3 - 5x + 4 = 4 - 5x - \frac{12}{3!}x^3 + \frac{24}{4!}x^4 = x^4 - 2x^3 - 5x + 4$ itself

21. $f(x) = x^3 - 2x + 4 \Rightarrow f'(x) = 3x^2 - 2, f''(x) = 6x, f'''(x) = 6 \Rightarrow f^{(n)}(x) = 0$ if $n \geq 4$; $f(2) = 8, f'(2) = 10,$

$f''(2) = 12, f'''(2) = 6, f^{(n)}(2) = 0$ if $n \geq 4 \Rightarrow x^3 - 2x + 4 = 8 + 10(x - 2) + \frac{12}{2!}(x - 2)^2 + \frac{6}{3!}(x - 2)^3$

$= 8 + 10(x - 2) + 6(x - 2)^2 + (x - 2)^3$

23. $f(x) = x^4 + x^2 + 1 \Rightarrow f'(x) = 4x^3 + 2x, f''(x) = 12x^2 + 2, f'''(x) = 24x, f^{(4)}(x) = 24, f^{(n)}(x) = 0$ if $n \geq 5$;

$f(-2) = 21, f'(-2) = -36, f''(-2) = 50, f'''(-2) = -48, f^{(4)}(-2) = 24, f^{(n)}(-2) = 0$ if $n \geq 5 \Rightarrow x^4 + x^2 + 1$

$= 21 - 36(x + 2) + \frac{50}{2!}(x + 2)^2 - \frac{48}{3!}(x + 2)^3 + \frac{24}{4!}(x + 2)^4 = 21 - 36(x + 2) + 25(x + 2)^2 - 8(x + 2)^3 + (x + 2)^4$

25. $f(x) = x^{-2} \Rightarrow f'(x) = -2x^{-3}, f''(x) = 3!x^{-4}, f'''(x) = -4!x^{-5} \Rightarrow f^{(n)}(x) = (-1)^n(n+1)!x^{-n-2}$;

$f(1) = 1, f'(1) = -2, f''(1) = 3!, f'''(1) = -4!, f^{(n)}(1) = (-1)^n(n+1)! \Rightarrow \frac{1}{x^2}$

$= 1 - 2(x - 1) + 3(x - 1)^2 - 4(x - 1)^3 + \dots = \sum_{n=0}^{\infty} (-1)^n(n+1)(x - 1)^n$

27. $f(x) = e^x \Rightarrow f'(x) = e^x, f''(x) = e^x \Rightarrow f^{(n)}(x) = e^x; f(2) = e^2, f'(2) = e^2, \dots f^{(n)}(2) = e^2$

$\Rightarrow e^x = e^2 + e^2(x - 2) + \frac{e^2}{2}(x - 2)^2 + \frac{e^2}{3!}(x - 2)^3 + \dots = \sum_{n=0}^{\infty} \frac{e^2}{n!}(x - 2)^n$

29. If $e^x = \sum_{n=0}^{\infty} \frac{f^{(n)}(a)}{n!}(x - a)^n$ and $f(x) = e^x$, we have $f^{(n)}(a) = e^a$ f or all $n = 0, 1, 2, 3, \dots$

$\Rightarrow e^x = e^a\left[\frac{(x-a)^0}{0!} + \frac{(x-a)^1}{1!} + \frac{(x-a)^2}{2!} + \dots\right] = e^a\left[1 + (x - a) + \frac{(x-a)^2}{2!} + \dots\right]$ at $x = a$

31. $f(x) = f(a) + f'(a)(x-a) + \dfrac{f''(a)}{2}(x-a)^2 + \dfrac{f'''(a)}{3!}(x-a)^3 + \ldots \Rightarrow f'(x)$

$= f'(a) + f''(a)(x-a) + \dfrac{f'''(a)}{3!}3(x-a)^2 + \ldots \Rightarrow f''(x) = f''(a) + f'''(a)(x-a) + \dfrac{f^{(4)}(a)}{4!}4\cdot3(x-a)^2 + \ldots$

$\Rightarrow f^{(n)}(x) = f^{(n)}(a) + f^{(n+1)}(a)(x-a) + \dfrac{f^{(n+2)}(a)}{2}(x-a)^2 + \ldots$

$\Rightarrow f(a) = f(a) + 0,\ f'(a) = f'(a) + 0,\ \ldots,\ f^{(n)}(a) = f^{(n)}(a) + 0$

33. $f(x) = \ln(\cos x) \Rightarrow f'(x) = -\tan x$ and $f''(x) = -\sec^2 x$; $f(0) = 0$, $f'(0) = 0$, $f''(0) = -1$

$\Rightarrow L(x) = 0$ and $Q(x) = -\dfrac{x^2}{2}$

35. $f(x) = \left(1-x^2\right)^{-1/2} \Rightarrow f'(x) = x\left(1-x^2\right)^{-3/2}$ and $f''(x) = \left(1-x^2\right)^{-3/2} + 3x^2\left(1-x^2\right)^{-5/2}$; $f(0) = 1$,

$f'(0) = 0$, $f''(0) = 1 \Rightarrow L(x) = 1$ and $Q(x) = 1 = \dfrac{x^2}{2}$

37. $f(x) = \sin x \Rightarrow f'(x) = \cos x$ and $f''(x) = -\sin x$; $f(0) = 0$, $f'(0) = 1$, $f''(0) = 0 \Rightarrow L(x) = x$ and $Q(x) = x$

8.10 CONVERGENCE OF TAYLOR SERIES; ERROR ESTIMATES

1. $e^x = 1 + x + \dfrac{x^2}{2!} + \ldots = \sum\limits_{n=0}^{\infty} \dfrac{x^n}{n!} \Rightarrow e^{-5x} = 1 + (-5x) + \dfrac{(-5x)^2}{2!} + \ldots = 1 - 5x + \dfrac{5^2 x^2}{2!} - \dfrac{5^3 x^3}{3!} + \ldots = \sum\limits_{n=0}^{\infty} \dfrac{(-1)^n 5^n x^n}{n!}$

3. $\sin x = x - \dfrac{x^3}{3!} + \dfrac{x^5}{5!} - \ldots = \sum\limits_{n=0}^{\infty} \dfrac{(-1)^n x^{2n+1}}{(2n+1)!} \Rightarrow 5\sin(-x) = 5\left[(-x) - \dfrac{(-x)^3}{3!} + \dfrac{(-x)^5}{5!} - \ldots\right]$

$= \sum\limits_{n=0}^{\infty} \dfrac{5(-1)^{n+1} x^{2n+1}}{(2n+1)!}$

5. $\cos x = \sum\limits_{n=0}^{\infty} \dfrac{(-1)^n x^{2n}}{(2n)!} \Rightarrow \cos\sqrt{x} = \sum\limits_{n=0}^{\infty} \dfrac{(-1)^n \left(x^{1/2}\right)^{2n}}{(2n)!} = \sum\limits_{n=0}^{\infty} \dfrac{(-1)^n x^n}{(2n)!} = 1 - \dfrac{x}{2!} + \dfrac{x^2}{4!} - \dfrac{x^3}{6!} + \ldots$

7. $e^x = \sum\limits_{n=0}^{\infty} \dfrac{x^n}{n!} \Rightarrow xe^x = x\left(\sum\limits_{n=0}^{\infty} \dfrac{x^n}{n!}\right) = \sum\limits_{n=0}^{\infty} \dfrac{x^{n+1}}{n!} = x + x^2 + \dfrac{x^3}{2!} + \dfrac{x^4}{3!} + \dfrac{x^5}{4!} + \ldots$

9. $\cos x = \sum\limits_{n=0}^{\infty} \dfrac{(-1)^n x^{2n}}{(2n)!} \Rightarrow \dfrac{x^2}{2} - 1 + \cos x = \dfrac{x^2}{2} - 1 + \sum\limits_{n=0}^{\infty} \dfrac{(-1)^n x^{2n}}{(2n)!} = \dfrac{x^2}{2} - 1 + 1 - \dfrac{x^2}{2} + \dfrac{x^4}{4!} - \dfrac{x^6}{6!} + \dfrac{x^8}{8!} - \dfrac{x^{10}}{10!} + \ldots$

$= \dfrac{x^4}{4!} - \dfrac{x^6}{6!} + \dfrac{x^8}{8!} - \dfrac{x^{10}}{10!} + \ldots = \sum\limits_{n=2}^{\infty} \dfrac{(-1)^n x^{2n}}{(2n)!}$

11. $\cos x = \sum\limits_{n=0}^{\infty} \dfrac{(-1)^n x^{2n}}{(2n)!} \Rightarrow x\cos\pi x = x\sum\limits_{n=0}^{\infty} \dfrac{(-1)^n (\pi x)^{2n}}{(2n)!} = \sum\limits_{n=0}^{\infty} \dfrac{(-1)^n \pi^{2n} x^{2n+1}}{(2n)!} = x - \dfrac{\pi^2 x^3}{2!} + \dfrac{\pi^4 x^5}{4!} - \dfrac{\pi^6 x^7}{6!} + \ldots$

13. $\cos^2 x = \frac{1}{2} + \frac{\cos 2x}{2} = \frac{1}{2} + \frac{1}{2} \sum_{n=0}^{\infty} \frac{(-1)^n (2x)^{2n}}{(2n)!} = \frac{1}{2} + \frac{1}{2}\left[1 - \frac{(2x)^2}{2!} + \frac{(2x)^4}{4!} - \frac{(2x)^6}{6!} + \frac{(2x)^8}{8!} - \ldots\right]$

$= 1 - \frac{(2x)^2}{2 \cdot 2!} + \frac{(2x)^4}{2 \cdot 4!} - \frac{(2x)^6}{2 \cdot 6!} + \frac{(2x)^8}{2 \cdot 8!} - \ldots = 1 + \sum_{n=1}^{\infty} \frac{(-1)^n (2x)^{2n}}{2 \cdot (2n)!}$

15. $\frac{x^2}{1 - 2x} = x^2\left(\frac{1}{1 - 2x}\right) = x^2 \sum_{n=0}^{\infty} (2x)^n = \sum_{n=0}^{\infty} 2^n x^{n+2} = x^2 + 2x^3 + 2^2 x^4 + 2^3 x^5 + \ldots$

17. $\frac{1}{1 - x} = \sum_{n=0}^{\infty} x^n = 1 + x + x^2 + x^3 + \ldots \Rightarrow \frac{d}{dx}\left(\frac{1}{1 - x}\right) = \frac{1}{(1 - x)^2} = 1 + 2x + 3x^2 + \ldots = \sum_{n=1}^{\infty} nx^{n-1}$

$= \sum_{n=0}^{\infty} (n + 1)x^n$

19. By the Alternating Series Estimation Theorem, the error is less than $\frac{|x|^5}{5!} \Rightarrow |x|^5 < (5!)(5 \times 10^{-4})$

$\Rightarrow |x|^5 < 600 \times 10^{-4} \Rightarrow |x| < \sqrt[5]{6 \times 10^{-2}} \approx 0.56968$

21. If $\sin x = x$ and $|x| < 10^{-3}$, then the $|\text{error}| = |R_2(x)| = \left|\frac{-\cos c}{3!} x^3\right| < \frac{(10^{-3})^3}{3!} \approx 1.67 \times 10^{-10}$, where c is

between 0 and x. The Alternating Series Estimation Theorem says $R_2(x)$ has the same sign as $-\frac{x^3}{3!}$. Moreover

$x < \sin x \Rightarrow 0 < \sin x - x = R_2(x) \Rightarrow x < 0 \Rightarrow -10^{-3} < x < 0$.

23. $|R_2(x)| = \left|\frac{e^c x^3}{3!}\right| < \frac{3^{(0.1)}(0.1)^3}{3!} < 1.87 \times 10^{-5}$, where c is between 0 and x

25. $|R_4(x)| < \left|\frac{\cosh c}{5!} x^5\right| = \left|\frac{e^c + e^{-c}}{2} \frac{x^5}{5!}\right| < \frac{1.65 + \frac{1}{1.65}}{2} \cdot \frac{(0.5)^5}{5!} = (1.3)\frac{(0.5)^5}{5!} \approx 0.000293653$

27. $|R_1| = \left|\frac{1}{(1 + c)^2} \frac{x^2}{2!}\right| < \frac{x^2}{2} = \left|\frac{x}{2}\right||x| < .01\,|x| = (1\%)|x| \Rightarrow \left|\frac{x}{2}\right| < .01 \Rightarrow 0 < |x| < .02$

29. (a) $\sin x = x - \frac{x^3}{3!} + \frac{x^5}{5!} - \frac{x^7}{7!} + \ldots \Rightarrow \frac{\sin x}{x} = 1 - \frac{x^2}{3!} + \frac{x^4}{5!} - \frac{x^6}{7!} + \ldots$, $s_1 = 1$ and $s_2 = 1 - \frac{x^2}{6}$; if L is the sum of the

series representing $\frac{\sin x}{x}$, then by the Alternating Series Estimation Theorem, $L - s_1 = \frac{\sin x}{x} - 1 < 0$ and

$L - s_2 = \frac{\sin x}{x} - \left(1 - \frac{x^2}{6}\right) > 0$. Therefore $1 - \frac{x^2}{6} < \frac{\sin x}{x} < 1$

(b) The graph of $y = \frac{\sin x}{x}$, $x \neq 0$, is bounded below by the

graph of $y = 1 - \frac{x^2}{6}$ and above by the graph of $y = 1$ as

derived in part (a).

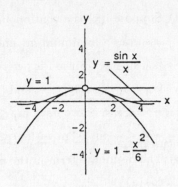

31. $\sin x$ when $x = 0.1$; the sum is $\sin(0.1) \approx 0.099833416$

33. $\tan^{-1} x$ when $x = \frac{\pi}{3}$; the sum is $\tan^{-1}\left(\frac{\pi}{3}\right) = \sqrt{3} \approx 0.808448$

35. $e^x \sin x = 0 + x + x^2 + x^3\left(-\frac{1}{3!} + \frac{1}{2!}\right) + x^4\left(-\frac{1}{3!} + \frac{1}{3!}\right) + x^5\left(\frac{1}{5!} - \frac{1}{2!}\frac{1}{3!} + \frac{1}{4!}\right) + x^6\left(\frac{1}{5!} - \frac{1}{3!}\frac{1}{3!} + \frac{1}{5!}\right) + \cdots$

$= x + x^2 + \frac{1}{3}x^3 = \frac{1}{30}x^5 - \frac{1}{90}x^6 + \cdots$

37. $\sin^2 x = \left(\frac{1 - \cos 2x}{2}\right) = \frac{1}{2} - \frac{1}{2}\cos 2x = \frac{1}{2} - \frac{1}{2}\left(1 - \frac{(2x)^2}{2!} + \frac{(2x)^4}{4!} - \frac{(2x)^6}{6!} + \cdots\right) = \frac{2x^2}{2!} - \frac{2^3 x^4}{4!} + \frac{2^5 x^6}{6!} - \cdots$

$\Rightarrow \frac{d}{dx}(\sin^2 x) = \frac{d}{dx}\left(\frac{2x^2}{2!} - \frac{2^3 x^4}{4!} + \frac{2^5 x^6}{6!} - \cdots\right) = 2x - \frac{(2x)^3}{3!} + \frac{(2x)^5}{5!} - \frac{(2x)^7}{7!} + \cdots \Rightarrow 2\sin x \cos x$

$= 2x - \frac{(2x)^3}{3!} + \frac{(2x)^5}{5!} - \frac{(2x)^7}{7!} + \cdots = \sin 2x$, which checks

39. A special case of Taylor's Formula is $f(x) = f(a) + f'(c)(x - a)$. Let $x = b$ and this becomes

$f(b) - f(a) = f'(c)(b - a)$, the Mean Value Theorem

41. (a) $f'' \leq 0$, $f'(a) = 0$ and $x = a$ interior to the interval $I \Rightarrow f(x) - f(a) = \frac{f''(c_2)}{2}(x - a)^2 \leq 0$ throughout I

$\Rightarrow f(x) \leq f(a)$ throughout $I \Rightarrow f$ has a local maximum at $x = a$

(b) similar reasoning gives $f(x) - f(a) = \frac{f''(c_2)}{2}(x - a)^2 \geq 0$ throughout $I \Rightarrow f(x) \geq f(a)$ throughout $I \Rightarrow f$ has a

local minimum at $x = a$

43. (a) $f(x) = (1 + x)^k \Rightarrow f'(x) = k(1 + x)^{k-1} \Rightarrow f''(x) = k(k - 1)(1 + x)^{k-2}$; $f(0) = 1$, $f'(0) = k$, and $f''(0) = k(k - 1)$

$\Rightarrow Q(x) = 1 + kx + \frac{k(k - 1)}{2}x^2$

(b) $\left| R_2(x) \right| = \left| \frac{3 \cdot 2 \cdot 1}{3!}x^3 \right| < \frac{1}{100} \Rightarrow \left| x^3 \right| < \frac{1}{100} \Rightarrow 0 < x < \frac{1}{100^{1/3}}$ or $0 < x < .21544$

45. If $f(x) = \sum\limits_{n=0}^{\infty} a_n x^n$, then $f^{(k)}(x) = \sum\limits_{n=k}^{\infty} n(n - 1)(n - 2)\cdots(n - k + 1)a_n x^{n-k}$ and $f^{(k)}(0) = k! \, a_k$

$\Rightarrow a_k = \frac{f^{(k)}(0)}{k!}$ for k a nonnegative integer. Therefore, the coefficients of $f(x)$ are identical with the

corresponding coefficients in the Maclaurin series of $f(x)$ and the statement follows.

47. (a) Suppose $f(x)$ is a continuous periodic function with period p. Let x_0 be an arbitrary real number. Then f assumes a minimum m_1 and a maximum m_2 in the interval $[x_0, x_0 + p]$; i.e., $m_1 \leq f(x) \leq m_2$ for all x in $[x_0, x_0 + p]$. Since f is periodic it has exactly the same values on all other intervals $[x_0 + p, x_0 + 2p]$, $[x_0 + 2p, x_0 + 3p]$, ..., and $[x_0 - p, x_0]$, $[x_0 - 2p, x_0 - p]$, ..., and so forth. That is, for all real numbers $-\infty < x < \infty$ we have $m_1 \leq f(x) \leq m_2$. Now choose $M = \max\{|m_1|, |m_2|\}$. Then

$$-M \leq -|m_1| \leq m_1 \leq f(x) \leq m_2 \leq |m_2| \leq M \Rightarrow |f(x)| \leq M \text{ for all x.}$$

(b) The dominate term in the nth order Taylor polynomial generated by cos x about $x = a$ is $\dfrac{\sin(a)}{n!}(x-a)^n$ or $\dfrac{\cos(a)}{n!}(x-a)^n$. In both cases, as $|x|$ increases the absolute value of these dominate terms tends to ∞, causing the graph of $P_n(x)$ to move away from cos x.

49. (a) $e^{-i\pi} = \cos(-\pi) + i\sin(-\pi) = -1 + i(0) = -1$

   (b) $e^{i\pi/4} = \cos\left(\dfrac{\pi}{4}\right) + i\sin\left(\dfrac{\pi}{4}\right) = \dfrac{1}{\sqrt{2}} + \dfrac{i}{\sqrt{2}} = \left(\dfrac{1}{\sqrt{2}}\right)(1+i)$

   (c) $e^{-i\pi/2} = \cos\left(-\dfrac{\pi}{2}\right) + i\sin\left(-\dfrac{\pi}{2}\right) = 0 + i(-1) = -i$

51. $e^x = 1 + x + \dfrac{x^2}{2!} + \dfrac{x^3}{3!} + \dfrac{x^4}{4!} + \ldots \Rightarrow e^{i\theta} = 1 + i\theta + \dfrac{(i\theta)^2}{2!} + \dfrac{(i\theta)^3}{3!} + \dfrac{(i\theta)^4}{4!} + \ldots$ and

$$e^{-i\theta} = 1 - i\theta + \dfrac{(-i\theta)^2}{2!} + \dfrac{(-i\theta)^3}{3!} + \dfrac{(-i\theta)^4}{4!} + \ldots = 1 - i\theta + \dfrac{(i\theta)^2}{2!} - \dfrac{(i\theta)^3}{3!} + \dfrac{(i\theta)^4}{4!} - \ldots$$

$$\Rightarrow \dfrac{e^{i\theta} + e^{-i\theta}}{2} = \dfrac{\left(1 + i\theta + \dfrac{(i\theta)^2}{2!} + \dfrac{(i\theta)^3}{3!} + \dfrac{(i\theta)^4}{4!} + \ldots\right) + \left(1 - i\theta + \dfrac{(i\theta)^2}{2!} - \dfrac{(i\theta)^3}{3!} + \dfrac{(i\theta)^4}{4!} - \ldots\right)}{2}$$

$$= 1 - \dfrac{\theta^2}{2!} + \dfrac{\theta^4}{4!} - \dfrac{\theta^6}{6!} + \ldots = \cos\theta;$$

$$\dfrac{e^{i\theta} - e^{-i\theta}}{2} = \dfrac{\left(1 + i\theta + \dfrac{(i\theta)^2}{2!} + \dfrac{(i\theta)^3}{3!} + \dfrac{(i\theta)^4}{4!} + \ldots\right) - \left(1 - i\theta + \dfrac{(i\theta)^2}{2!} - \dfrac{(i\theta)^3}{3!} + \dfrac{(i\theta)^4}{4!} - \ldots\right)}{2}$$

$$= \theta - \dfrac{\theta^3}{3!} + \dfrac{\theta^5}{5!} - \dfrac{\theta^7}{7!} + \ldots = \sin\theta$$

53. $e^x \sin x = \left(1 + x + \dfrac{x^2}{2!} + \dfrac{x^3}{3!} + \dfrac{x^4}{4!} + \ldots\right)\left(x - \dfrac{x^3}{3!} + \dfrac{x^5}{5!} - \dfrac{x^7}{7!} + \ldots\right)$

$= (1)x + (1)x^2 + \left(-\dfrac{1}{6} + \dfrac{1}{2}\right)x^3 + \left(-\dfrac{1}{6} + \dfrac{1}{6}\right)x^4 + \left(\dfrac{1}{120} - \dfrac{1}{12} + \dfrac{1}{24}\right)x^5 + \ldots = x + x^2 + \dfrac{1}{3}x^3 - \dfrac{1}{30}x^5 + \ldots;$

$e^x \cdot e^{ix} = e^{(1+i)x} = e^x(\cos x + i\sin x) = e^x\cos x + i(e^x\sin x) \Rightarrow e^x\sin x$ is the series of the imaginary part

of $e^{(1+i)x}$ which we calculate next; $e^{(1+i)x} = \displaystyle\sum_{n=0}^{\infty} \dfrac{(x+ix)^n}{n!} = 1 + (x+ix) + \dfrac{(x+ix)^2}{2!} + \dfrac{(x+ix)^3}{3!} + \dfrac{(x+ix)^4}{4!} + \ldots$

$= 1 + x + ix + \dfrac{1}{2!}(2ix^2) + \dfrac{1}{3!}(2ix^3 - 2x^3) + \dfrac{1}{4!}(-4x^4) + \dfrac{1}{5!}(-4x^5 - 4ix^5) + \dfrac{1}{6!}(-8ix^6) + \ldots \Rightarrow$ the imaginary part

of $e^{(1+i)x}$ is $x + \dfrac{2}{2!}x^2 + \dfrac{2}{3!}x^3 - \dfrac{4}{5!}x^5 - \dfrac{8}{6!}x^6 + \ldots = x + x^2 + \dfrac{1}{3}x^3 - \dfrac{1}{30}x^5 - \dfrac{1}{90}x^6 + \ldots$ in agreement with our

product calculation

55. (a) $e^{i\theta_1}e^{i\theta_2} = (\cos\theta_1 + i\sin\theta_1)(\cos\theta_2 + i\sin\theta_2) = (\cos\theta_1\cos\theta_2 - \sin\theta_1\sin\theta_2) + i(\sin\theta_1\cos\theta_2 + \sin\theta_2\cos\theta_1)$

$$= \cos(\theta_1 + \theta_2) + i\sin(\theta_1 + \theta_2) = e^{i(\theta_1+\theta_2)}$$

(b) $e^{-i\theta} = \cos(-\theta) + i\sin(-\theta) = \cos\theta - i\sin\theta = (\cos\theta - i\sin\theta)\left(\dfrac{\cos\theta + i\sin\theta}{\cos\theta + i\sin\theta}\right) = \dfrac{1}{\cos\theta + i\sin\theta} = \dfrac{1}{e^{i\theta}}$

8.11 APPLICATIONS OF POWER SERIES

1. $(1+x)^{1/2} = 1 + \dfrac{1}{2}x + \dfrac{\left(\frac{1}{2}\right)\left(-\frac{1}{2}\right)x^2}{2!} + \dfrac{\left(\frac{1}{2}\right)\left(-\frac{1}{2}\right)\left(-\frac{3}{2}\right)x^3}{3!} + \ldots = 1 + \dfrac{1}{2}x - \dfrac{1}{8}x^2 + \dfrac{1}{16}x^3 - \ldots$

3. $(1-x)^{-1/2} = 1 - \dfrac{1}{2}(-x) + \dfrac{\left(-\frac{1}{2}\right)\left(-\frac{3}{2}\right)(-x)^2}{2!} + \dfrac{\left(-\frac{1}{2}\right)\left(-\frac{3}{2}\right)\left(-\frac{5}{2}\right)(-x)^3}{3!} + \ldots = 1 + \dfrac{1}{2}x + \dfrac{3}{8}x^2 + \dfrac{5}{16}x^3 + \ldots$

5. $\left(1+\dfrac{x}{2}\right)^{-2} = 1 - 2\left(\dfrac{x}{2}\right) + \dfrac{(-2)(-3)\left(\frac{x}{2}\right)^2}{2!} + \dfrac{(-2)(-3)(-4)\left(\frac{x}{2}\right)^3}{3!} + \ldots = 1 - x + \dfrac{3}{4}x^2 - \dfrac{1}{2}x^3$

7. $(1+x^3)^{-1/2} = 1 - \dfrac{1}{2}x^3 + \dfrac{\left(-\frac{1}{2}\right)\left(-\frac{3}{2}\right)(x^3)^2}{2!} + \dfrac{\left(-\frac{1}{2}\right)\left(-\frac{3}{2}\right)\left(-\frac{5}{2}\right)(x^3)^3}{3!} + \ldots = 1 - \dfrac{1}{2}x^3 + \dfrac{3}{8}x^6 - \dfrac{5}{16}x^9 + \ldots$

9. $\left(1+\dfrac{1}{x}\right)^{1/2} = 1 + \dfrac{1}{2}\left(\dfrac{1}{x}\right) + \dfrac{\left(\frac{1}{2}\right)\left(-\frac{1}{2}\right)\left(\frac{1}{x}\right)^2}{2!} + \dfrac{\left(\frac{1}{2}\right)\left(-\frac{1}{2}\right)\left(-\frac{3}{2}\right)\left(\frac{1}{x}\right)^3}{3!} + \ldots = 1 + \dfrac{1}{2x} - \dfrac{1}{8x^2} + \dfrac{1}{16x^3}$

11. $(1+x)^4 = 1 + 4x + \dfrac{(4)(3)x^2}{2!} + \dfrac{(4)(3)(2)x^3}{3!} + \dfrac{(4)(3)(2)x^4}{4!} = 1 + 4x + 6x^2 + 4x^3 + x^4$

13. $(1-2x)^3 = 1 + 3(-2x) + \dfrac{(3)(2)(-2x)^2}{2!} + \dfrac{(3)(2)(1)(-2x)^3}{3!} = 1 - 6x + 12x^2 - 8x^3$

15. Assume the solution has the form $y = a_0 + a_1 x + a_2 x^2 + \ldots + a_{n-1}x^{n-1} + a_n x^n + \ldots$

$\Rightarrow \dfrac{dy}{dx} = a_1 + 2a_2 x + \ldots + na_n x^{n-1} + \ldots$

$\Rightarrow \dfrac{dy}{dx} + y = (a_1 + a_0) + (2a_2 + a_1)x + (3a_3 + a_2)x^2 + \ldots + (na_n + a_{n-1})x^{n-1} + \ldots = 0$

$\Rightarrow a_1 + a_0 = 0,\ 2a_2 + a_1 = 0,\ 3a_3 + a_2 = 0$ and in general $na_n + a_{n-1} = 0$. Since $y = 1$ when $x = 0$ we have

$a_0 = 1$. Therefore $a_1 = -1$, $a_2 = \dfrac{-a_1}{2\cdot 1} = \dfrac{1}{2}$, $a_3 = \dfrac{-a_2}{3} = -\dfrac{1}{3\cdot 2}, \ldots, a_n = \dfrac{-a_{n-1}}{n} = \dfrac{(-1)^n}{n!}$

$\Rightarrow y = 1 - x + \dfrac{1}{2}x^2 - \dfrac{1}{3!}x^3 + \ldots + \dfrac{(-1)^n}{n!}x^n + \ldots = \sum_{n=0}^{\infty} \dfrac{(-1)^n x^n}{n!} = e^{-x}$

17. Assume the solution has the form $y = a_0 + a_1 x + a_2 x^2 + \ldots + a_{n-1}x^{n-1} + a_n x^n + \ldots$

$\Rightarrow \dfrac{dy}{dx} = a_1 + 2a_2 x + \ldots + na_n x^{n-1} + \ldots$

$\Rightarrow \dfrac{dy}{dx} - y = (a_1 - a_0) + (2a_2 - a_1)x + (3a_3 - a_2)x^2 + \ldots + (na_n - a_{n-1})x^{n-1} + \ldots = 1$

$\Rightarrow a_1 - a_0 = 1,\ 2a_2 - a_1 = 0,\ 3a_3 - a_2 = 0$ and in general $na_n - a_{n-1} = 0$. Since $y = 0$ when $x = 0$ we have

$a_0 = 0$. Therefore $a_1 = 1,\ a_2 = \dfrac{a_1}{2} = \dfrac{1}{2},\ a_3 = \dfrac{a_2}{3} = \dfrac{1}{3 \cdot 2},\ a_4 = \dfrac{a_3}{4} = \dfrac{1}{4 \cdot 3 \cdot 2},\ \ldots,\ a_n = \dfrac{a_{n-1}}{n} = \dfrac{1}{n!}$

$\Rightarrow y = 0 + 1x + \dfrac{1}{2}x^2 + \dfrac{1}{3 \cdot 2}x^3 + \dfrac{1}{4 \cdot 3 \cdot 2}x^4 + \ldots + \dfrac{1}{n!}x^n + \ldots$

$= \left(1 + 1x + \dfrac{1}{2}x^2 + \dfrac{1}{3 \cdot 2}x^3 + \dfrac{1}{4 \cdot 3 \cdot 2}x^4 + \ldots + \dfrac{1}{n!}x^n + \ldots\right) - 1 = \displaystyle\sum_{n=0}^{\infty} \dfrac{x^n}{n!} - 1 = e^x - 1$

19. Assume the solution has the form $y = a_0 + a_1x + a_2x^2 + \ldots + a_{n-1}x^{n-1} + a_nx^n + \ldots$

$\Rightarrow \dfrac{dy}{dx} = a_1 + 2a_2x + \ldots + na_nx^{n-1} + \ldots$

$\Rightarrow \dfrac{dy}{dx} - y = (a_1 - a_0) + (2a_2 - a_1)x + (3a_3 - a_2)x^2 + \ldots + (na_n - a_{n-1})x^{n-1} + \ldots = x$

$\Rightarrow a_1 - a_0 = 0,\ 2a_2 - a_1 = 1,\ 3a_3 - a_2 = 0$ and in general $na_n - a_{n-1} = 0$. Since $y = 0$ when $x = 0$ we have

$a_0 = 0$. Therefore $a_1 = 0,\ a_2 = \dfrac{1 + a_1}{2} = \dfrac{1}{2},\ a_3 = \dfrac{a_2}{3} = \dfrac{1}{3 \cdot 2},\ a_4 = \dfrac{a_3}{4} = \dfrac{1}{4 \cdot 3 \cdot 2},\ \ldots,\ a_n = \dfrac{a_{n-1}}{n} = \dfrac{1}{n!}$

$\Rightarrow y = 0 + 0x + \dfrac{1}{2}x^2 + \dfrac{1}{3 \cdot 2}x^3 + \dfrac{1}{4 \cdot 3 \cdot 2}x^4 + + \ldots + \dfrac{1}{n!}x^n + \ldots$

$= \left(1 + 1x + \dfrac{1}{2}x^2 + \dfrac{1}{3 \cdot 2}x^3 + \dfrac{1}{4 \cdot 3 \cdot 2}x^4 + \ldots + \dfrac{1}{n!}x^n + \ldots\right) - 1 - x = \displaystyle\sum_{n=0}^{\infty} \dfrac{x^n}{n!} - 1 - x = e^x - x - 1$.

21. $y' - xy = a_1 + (2a_2 - a_0)x + (3a_3 - a_1)x + \ldots + (na_n - a_{n-2})x^{n-1} + \ldots = 0 \Rightarrow a_1 = 0,\ 2a_2 - a_0 = 0,\ 3a_3 - a_1 = 0,$

$4a_4 - a_2 = 0$ and in general $na_n - a_{n-2} = 0$. Since $y = 1$ when $x = 0$, we have $a_0 = 1$. Therefore $a_2 = \dfrac{a_0}{2} = \dfrac{1}{2},$

$a_3 = \dfrac{a_1}{3} = 0,\ a_4 = \dfrac{a_2}{4} = \dfrac{1}{2 \cdot 4},\ a_5 = \dfrac{a_3}{5} = 0,\ \ldots,\ a_{2n} = \dfrac{1}{2 \cdot 4 \cdot 6 \cdots 2n}$ and $a_{2n+1} = 0$

$\Rightarrow y = 1 + \dfrac{1}{2}x^2 + \dfrac{1}{2 \cdot 4}x^4 + \dfrac{1}{2 \cdot 4 \cdot 6}x^6 + \ldots + \dfrac{1}{2 \cdot 4 \cdot 6 \cdots 2n}x^{2n} + \ldots = \displaystyle\sum_{n=0}^{\infty} \dfrac{x^{2n}}{2^n n!} = \displaystyle\sum_{n=0}^{\infty} \dfrac{\left(\dfrac{x^2}{2}\right)^n}{n!} = e^{x^2/2}$

23. $(1 - x)y' - y = (a_1 - a_0) + (2a_2 - a_1 - a_1)x + (3a_3 - 2a_2 - a_2)x^2 + (4a_4 - 3a_3 - a_3)x^3 + \ldots$

$+ (na_n - (n-1)a_{n-1} - a_{n-1})x^{n-1} + \ldots = 0 \Rightarrow a_1 - a_0 = 0,\ 2a_2 - 2a_1 = 0,\ 3a_3 - 3a_2 = 0$ and in

general $(na_n - na_{n-1}) = 0$. Since $y = 2$ when $x = 0$, we have $a_0 = 2$. Therefore

$a_1 = 2,\ a_2 = 2,\ \ldots,\ a_n = 2 \Rightarrow y = 2 + 2x + 2x^2 + \ldots = \displaystyle\sum_{n=0}^{\infty} 2x^n = \dfrac{2}{1 - x}$

25. $y = a_0 + a_1x + a_2x^2 + \ldots + a_nx^n + \ldots \Rightarrow y'' = 2a_2 + 3 \cdot 2a_3x + \ldots + n(n-1)a_nx^{n-2} + \ldots \Rightarrow y'' - y$

$= (2a_2 - a_0) + (3 \cdot 2a_3 - a_1)x + (4 \cdot 3a_4 - a_2)x^2 + \ldots + (n(n-1)a_n - a_{n-2})x^{n-2} + \ldots = 0 \Rightarrow 2a_2 - a_0 = 0,$

$3 \cdot 2a_3 - a_1 = 0,\ 4 \cdot 3a_4 - a_2 = 0$ and in general $n(n-1)a_n - a_{n-2} = 0$. Since $y' = 1$ and $y = 0$ when $x = 0$,

we have $a_0 = 0$ and $a_1 = 1$. Therefore $a_2 = 0,\ a_3 = \dfrac{1}{3 \cdot 2},\ a_4 = 0,\ a_5 = \dfrac{1}{5 \cdot 4 \cdot 3 \cdot 2},\ \ldots,\ a_{2n+1} = \dfrac{1}{(2n+1)!}$ and

$a_{2n} = 0 \Rightarrow y = x + \dfrac{1}{3!}x^3 + \dfrac{1}{5!}x^5 + \ldots = \displaystyle\sum_{n=0}^{\infty} \dfrac{x^{2n+1}}{(2n+1)!} = \sinh x$

27. $y = a_0 + a_1x + a_2x^2 + \ldots + a_nx^n + \ldots \Rightarrow y'' = 2a_2 + 3 \cdot 2a_3x + \ldots + n(n-1)a_nx^{n-2} + \ldots \Rightarrow y'' + y$

$= (2a_2 + a_0) + (3 \cdot 2a_3 + a_1)x + (4 \cdot 3a_4 + a_2)x^2 + \ldots + (n(n-1)a_n + a_{n-2})x^{n-2} + \ldots = x \Rightarrow 2a_2 + a_0 = 0,$

$3 \cdot 2a_3 + a_1 = 1$, $4 \cdot 3a_4 + a_2 = 0$ and in general $n(n-1)a_n + a_{n-2} = 0$. Since $y' = 1$ and $y = 2$ when $x = 0$, we have $a_0 = 2$ and $a_1 = 1$. Therefore $a_2 = -1$, $a_3 = 0$, $a_4 = \dfrac{1}{4 \cdot 3}$, $a_5 = 0$, $\ldots$, $a_{2n} = -2 \cdot \dfrac{(-1)^{n+1}}{(2n)!}$ and

$a_{2n+1} = 0 \Rightarrow y = 2 + x - x^2 + 2 \cdot \dfrac{x^4}{4!} + \ldots = 2 + x - 2 \displaystyle\sum_{n=1}^{\infty} \dfrac{(-1)^{n+1}x^{2n}}{(2n)!}$

29. $y = a_0 + a_1 x + a_2 x^2 + \ldots + a_n x^n + \ldots \Rightarrow y'' = 2a_2 + 3 \cdot 2a_3 x + \ldots + n(n-1)a_n x^{n-2} + \ldots \Rightarrow y'' - y$

$= (2a_2 - a_0) + (3 \cdot 2a_3 - a_1)(x - 2) + (4 \cdot 3a_4 - a_2)(x - 2)^2 + \ldots + (n(n-1)a_n - a_{n-2})(x-2)^{n-2} + \ldots = 0$

$\Rightarrow 2a_2 - a_0 = 0$, $3 \cdot 2a_3 - a_1 = 0$, $4 \cdot 3a_4 - a_2 = 0$ and in general $n(n-1)a_n - a_{n-2} = 0$. Since $y' = -2$ and

$y = 0$ when $x = 2$, we have $a_0 = 0$ and $a_1 = -2$. Therefore $a_2 = 0$, $a_3 = \dfrac{-2}{3 \cdot 2}$, $a_4 = 0$, $a_5 = \dfrac{-2}{5!}$, $\ldots$, $a_{2n} = 0$, and

$a_{2n+1} = \dfrac{-2}{(2n+1)!} \Rightarrow y = -2(x-2) - \dfrac{2}{3!}(x-2)^3 - \ldots = \displaystyle\sum_{n=0}^{\infty} \dfrac{-2(x-2)^{2n+1}}{(2n+1)!}$

31. $y'' + x^2 y = 2a_2 + 6a_3 x + (4 \cdot 3a_4 + a_0)x^2 + \ldots + (n(n-1)a_n + a_{n-4})x^{n-2} + \ldots = x \Rightarrow 2a_2 = 0$, $6a_3 = 1$,

$4 \cdot 3a_4 + a_0 = 0$, $5 \cdot 4a_5 + a_1 = 0$, and in general $n(n-1)a_n + a_{n-4} = 0$. Since $y' = b$ and $y = a$ when $x = 0$,

we have $a_0 = a$ and $a_1 = b$. Therefore $a_2 = 0$, $a_3 = \dfrac{1}{2 \cdot 3}$, $a_4 = -\dfrac{a}{3 \cdot 4}$, $a_5 = -\dfrac{b}{4 \cdot 5}$, $a_6 = 0$, $a_7 = \dfrac{1}{2 \cdot 3 \cdot 6 \cdot 7}$

$\Rightarrow y = a + bx + \dfrac{1}{2 \cdot 3}x^3 - \dfrac{a}{3 \cdot 4}x^4 - \dfrac{b}{4 \cdot 5}x^5 - \dfrac{1}{2 \cdot 3 \cdot 6 \cdot 7}x^7 + \dfrac{ax^8}{3 \cdot 4 \cdot 7 \cdot 8} + \dfrac{bx^9}{4 \cdot 5 \cdot 8 \cdot 9} + \ldots$

33. $\displaystyle\int_0^{0.2} \sin x^2 \, dx = \int_0^{0.2} \left(x^2 - \dfrac{x^6}{3!} + \dfrac{x^{10}}{5!} - \ldots \right) dx = \left[\dfrac{x^3}{3} - \dfrac{x^7}{7 \cdot 3!} + \ldots \right]_0^{0.2} \approx \left[\dfrac{x^3}{3} \right]_0^{0.2} \approx 0.00267$ with error

$|E| \leq \dfrac{(.2)^7}{7 \cdot 3!} \approx 0.0000003$

35. $\displaystyle\int_0^{0.1} \dfrac{1}{\sqrt{1 + x^4}} \, dx = \int_0^{0.1} \left(1 - \dfrac{x^4}{2} + \dfrac{3x^8}{8} - \ldots \right) dx = \left[x - \dfrac{x^5}{10} + \ldots \right]_0^{0.1} \approx [x]_0^{0.1} \approx 0.1$ with error

$|E| \leq \dfrac{(0.1)^5}{10} = 0.000001$

37. $\displaystyle\int_0^{0.1} \dfrac{\sin x}{x} \, dx = \int_0^{0.1} \left(1 - \dfrac{x^2}{3!} + \dfrac{x^4}{5!} - \dfrac{x^6}{7!} + \ldots \right) dx = \left[x - \dfrac{x^3}{3 \cdot 3!} + \dfrac{x^5}{5 \cdot 5!} - \dfrac{x^7}{7 \cdot 7!} + \ldots \right]_0^{0.1} \approx \left[x - \dfrac{x^3}{3 \cdot 3!} + \dfrac{x^5}{5 \cdot 5!} \right]_0^{0.1}$

≈ 0.0999444611

39. $(1 + x^4)^{1/2} = (1)^{1/2} + \dfrac{\left(\frac{1}{2}\right)}{1}(1)^{-1/2}(x^4) + \dfrac{\left(\frac{1}{2}\right)\left(-\frac{1}{2}\right)}{2!}(1)^{-3/2}(x^4)^2 + \dfrac{\left(\frac{1}{2}\right)\left(-\frac{1}{2}\right)\left(-\frac{3}{2}\right)}{3!}(1)^{-5/2}(x^4)^3$

$+ \dfrac{\left(\frac{1}{2}\right)\left(-\frac{1}{2}\right)\left(-\frac{3}{2}\right)\left(-\frac{5}{2}\right)}{4!}(1)^{-7/2}(x^4)^4 + \ldots = 1 + \dfrac{x^4}{2} - \dfrac{x^8}{8} + \dfrac{x^{12}}{16} - \dfrac{5x^{16}}{128} + \ldots$

$\Rightarrow \displaystyle\int_0^{0.1} \left(1 + \dfrac{x^4}{2} - \dfrac{x^8}{8} + \dfrac{x^{12}}{16} - \dfrac{5x^{16}}{128} + \ldots \right) dx = \left[x + \dfrac{x^5}{10} - \dfrac{x^9}{72} + \dfrac{x^{13}}{208} - \dfrac{5x^{17}}{2176} + \ldots \right]_0^{0.1} \approx 0.100001$

41. $\displaystyle\int_0^1 \cos t^2 \, dt = \int_0^1 \left(1 - \frac{t^4}{2} + \frac{t^8}{4!} - \frac{t^{12}}{6!} + \dots\right) dt = \left[t - \frac{t^5}{10} + \frac{t^9}{9 \cdot 4!} - \frac{t^{13}}{13 \cdot 6!} + \dots\right]_0^1 \Rightarrow |\text{error}| < \frac{1}{13 \cdot 6!} \approx .00011$

43. $\displaystyle F(x) = \int_0^x \left(t^2 - \frac{t^6}{3!} + \frac{t^{10}}{5!} - \frac{t^{14}}{7!} + \dots\right) dt = \left[\frac{t^3}{3} - \frac{t^7}{7 \cdot 3!} + \frac{t^{11}}{11 \cdot 5!} - \frac{t^{15}}{15 \cdot 7!} + \dots\right]_0^x \approx \frac{x^3}{3} - \frac{x^7}{7 \cdot 3!} + \frac{x^{11}}{11 \cdot 5!}$

   $\Rightarrow |\text{error}| < \frac{1}{15 \cdot 7!} \approx 0.00002$

45. (a) $\displaystyle F(x) = \int_0^x \left(t - \frac{t^3}{3} + \frac{t^5}{5} - \frac{t^7}{7} + \dots\right) dt = \left[\frac{t^2}{2} - \frac{t^4}{12} + \frac{t^6}{30} - \dots\right]_0^x \approx \frac{x^2}{2} - \frac{x^4}{12} \Rightarrow |\text{error}| < \frac{(0.5)^6}{30} \approx .00052$

   (b) $|\text{error}| < \frac{1}{33 \cdot 34} \approx .00089$ so $F(x) \approx \frac{x^2}{2} - \frac{x^4}{3 \cdot 4} + \frac{x^6}{5 \cdot 6} - \frac{x^8}{7 \cdot 8} + \dots + (-1)^{15} \frac{x^{32}}{31 \cdot 32}$

47. $\frac{1}{x^2}\left(e^x - (1+x)\right) = \frac{1}{x^2}\left(\left(1 + x + \frac{x^2}{2} + \frac{x^3}{3!} + \dots\right) - 1 - x\right) = \frac{1}{2} + \frac{x}{3!} + \frac{x^2}{4!} + \dots \Rightarrow \lim_{x \to 0} \frac{e^x - (1+x)}{x^2}$

   $= \lim_{x \to 0} \left(\frac{1}{2} + \frac{x}{3!} + \frac{x^2}{4!} + \dots\right) = \frac{1}{2}$

49. $\frac{1}{t^4}\left(1 - \cos t - \frac{t^2}{2}\right) = \frac{1}{t^4}\left[1 - \frac{t^2}{2} - \left(1 - \frac{t^2}{2} + \frac{t^4}{4!} - \frac{t^6}{6!} + \dots\right)\right] = -\frac{1}{4!} + \frac{t^2}{6!} - \frac{t^4}{8!} + \dots \Rightarrow \lim_{t \to 0} \frac{1 - \cos t - \left(\frac{t^2}{2}\right)}{t^4}$

   $= \lim_{t \to 0} \left(-\frac{1}{4!} + \frac{t^2}{6!} - \frac{t^4}{8!} + \dots\right) = -\frac{1}{24}$

51. $\frac{1}{y^3}(y - \tan^{-1} y) = \frac{1}{y^3}\left[y - \left(y - \frac{y^3}{3} + \frac{y^5}{5} - \dots\right)\right] = \frac{1}{3} - \frac{y^2}{5} + \frac{y^4}{7} - \dots \Rightarrow \lim_{y \to 0} \frac{y - \tan^{-1} y}{y^3} = \lim_{y \to 0} \left(\frac{1}{3} - \frac{y^2}{5} + \frac{y^4}{7} - \dots\right)$

   $= \frac{1}{3}$

53. $x^2\left(-1 + e^{-1/x^2}\right) = x^2\left(-1 + 1 - \frac{1}{x^2} + \frac{1}{2x^4} - \frac{1}{6x^6} + \dots\right) = -1 + \frac{1}{2x^2} - \frac{1}{6x^4} + \dots \Rightarrow \lim_{x \to \infty} x^2\left(e^{-1/x^2} - 1\right)$

   $= \lim_{x \to \infty} \left(-1 + \frac{1}{2x^2} - \frac{1}{6x^4} + \dots\right) = -1$

55. $\frac{\ln(1 + x^2)}{1 - \cos x} = \frac{\left(x^2 - \frac{x^4}{2} + \frac{x^6}{3} - \dots\right)}{1 - \left(1 - \frac{x^2}{2!} + \frac{x^4}{4!} - \dots\right)} = \frac{\left(1 - \frac{x^2}{2} + \frac{x^4}{3} - \dots\right)}{\left(\frac{1}{2!} - \frac{x^2}{4!} + \dots\right)} \Rightarrow \lim_{x \to 0} \frac{\ln(1 + x^2)}{1 - \cos x} = \lim_{x \to 0} \frac{\left(1 - \frac{x^2}{2} + \frac{x^4}{3} - \dots\right)}{\left(\frac{1}{2!} - \frac{x^2}{4!} + \dots\right)} = 2!$

   $= 2$

57. $\ln\left(\frac{1+x}{1-x}\right) = \ln(1+x) - \ln(1-x) = \left(x - \frac{x^2}{2} + \frac{x^3}{3} - \frac{x^4}{4} + \dots\right) - \left(-x - \frac{x^2}{2} - \frac{x^3}{3} - \frac{x^4}{4} - \dots\right) = 2\left(x + \frac{x^3}{3} + \frac{x^5}{5} + \dots\right)$

59. $\tan^{-1} x = x - \dfrac{x^3}{3} + \dfrac{x^5}{5} - \dfrac{x^7}{7} + \dfrac{x^9}{9} - \cdots + \dfrac{(-1)^{n-1}x^{2n-1}}{2n-1} + \cdots \Rightarrow |\text{error}| = \left| \dfrac{(-1)^{n-1}x^{2n-1}}{2n-1} \right| = \dfrac{1}{2n-1}$ when $x = 1$;

$\dfrac{1}{2n-1} < \dfrac{1}{10^3} \Rightarrow n > \dfrac{1001}{2} = 500.5 \Rightarrow$ the first term not used is the $501^{\text{st}} \Rightarrow$ we must use 500 terms

61. $\tan^{-1} x = x - \dfrac{x^3}{3} + \dfrac{x^5}{5} - \dfrac{x^7}{7} + \dfrac{x^9}{9} - \cdots + \dfrac{(-1)^{n-1}x^{2n-1}}{2n-1} + \cdots$ and when the series representing $48 \tan^{-1}\left(\dfrac{1}{18}\right)$ has an

error of magnitude less than 10^{-6}, then the series representing the sum

$48 \tan^{-1}\left(\dfrac{1}{18}\right) + 32 \tan^{-1}\left(\dfrac{1}{57}\right) - 20 \tan^{-1}\left(\dfrac{1}{239}\right)$ also has an error of magnitude less than 10^{-6}; thus

$|\text{error}| = \dfrac{\left(\dfrac{1}{18}\right)^{2n-1}}{2n-1} < \dfrac{1}{10^6} \Rightarrow n \geq 3$ using a calculator $\Rightarrow$ 3 terms

63. (a) $\left(1-x^2\right)^{-1/2} \approx 1 + \dfrac{x^2}{2} + \dfrac{3x^4}{8} + \dfrac{5x^6}{16} \Rightarrow \sin^{-1} x \approx x + \dfrac{x^3}{6} + \dfrac{3x^5}{40} + \dfrac{5x^7}{112}$;

$\displaystyle\lim_{n\to\infty} \left| \dfrac{1\cdot 3\cdot 5\cdots(2n-1)(2n+1)x^{2n+3}}{2\cdot 4\cdot 6\cdots(2n)(2n+2)(2n+3)} \cdot \dfrac{2\cdot 4\cdot 6\cdots(2n)(2n+1)}{1\cdot 3\cdot 5\cdots(2n-1)x^{2n+1}} \right| < 1 \Rightarrow x^2 \lim_{n\to\infty} \left| \dfrac{(2n+1)(2n+1)}{(2n+2)(2n+3)} \right| < 1$

$\Rightarrow |x| < 1 \Rightarrow$ the radius of convergence is 1

(b) $\dfrac{d}{dx}\left(\cos^{-1} x\right) = -\left(1-x^2\right)^{-1/2} \Rightarrow \cos^{-1} x = \dfrac{\pi}{2} - \sin^{-1} x \approx \dfrac{\pi}{2} - \left(x + \dfrac{x^3}{6} + \dfrac{3x^5}{40} + \dfrac{5x^7}{112}\right) \approx \dfrac{\pi}{2} - x - \dfrac{x^3}{6} - \dfrac{3x^5}{40} - \dfrac{5x^7}{112}$

65. $\dfrac{-1}{1+x} = -\dfrac{1}{1-(-x)} = -1 + x - x^2 + x^3 - \cdots \Rightarrow \dfrac{d}{dx}\left(\dfrac{-1}{1+x}\right) = \dfrac{1}{1+x^2} = \dfrac{d}{dx}\left(-1 + x - x^2 + x^3 - \cdots\right)$

$= 1 - 2x + 3x^2 - 4x^3 + \cdots$

67. Wallis' formula gives the approximation $\pi \approx 4\left[\dfrac{2\cdot 4\cdot 4\cdot 6\cdot 6\cdot 8\cdots(2n-2)\cdot(2n)}{3\cdot 3\cdot 5\cdot 5\cdot 7\cdot 7\cdots(2n-1)\cdot(2n-1)}\right]$ to produce the table

n	$\sim \pi$
10	3.221088998
20	3.181104886
30	3.167880758
80	3.151425420
90	3.150331383
93	3.150049112
94	3.149959030
95	3.149870848
100	3.149456425

At $n = 1929$ we obtain the first approximation accurate to 3 decimals: 3.141999845. At $n = 30,000$ we still do not obtain accuracy to 4 decimals: 3.141617732, so the convergence to π is very slow. Here is a <u>Maple</u> CAS procedure to produce these approximations:

```
pie :=
   proc(n)
   local i,j;
       a(2) := evalf(8/9);
       for i from 3 to n do a(i) := evalf(2*(2*i-2)*i/(2*i-1)^2*a(i-1)) od;
       [[j,4*a(j)] $ (j = n-5 .. n)]
   end
```

69. $(1-x^2)^{-1/2} = (1+(-x^2))^{-1/2} = (1)^{-1/2} + \left(-\frac{1}{2}\right)(1)^{-3/2}(-x^2) + \dfrac{\left(-\frac{1}{2}\right)\left(-\frac{3}{2}\right)(1)^{-5/2}(-x^2)^2}{2!}$

$+ \dfrac{\left(-\frac{1}{2}\right)\left(-\frac{3}{2}\right)\left(-\frac{5}{2}\right)(1)^{-7/2}(-x^2)^3}{3!} + \ldots = 1 + \dfrac{x^2}{2} + \dfrac{1 \cdot 3 x^4}{2^2 \cdot 2!} + \dfrac{1 \cdot 3 \cdot 5 x^6}{2^3 \cdot 3!} + \ldots = 1 + \sum_{n=1}^{\infty} \dfrac{1 \cdot 3 \cdot 5 \cdots (2n-1)x^{2n}}{2^n \cdot n!}$

$\Rightarrow \sin^{-1} x = \int_0^x (1-t^2)^{-1/2} \, dt = \int_0^x \left(1 + \sum_{n=1}^{\infty} \dfrac{1 \cdot 3 \cdot 5 \cdots (2n-1)x^{2n}}{2^n \cdot n!}\right) dt = x + \sum_{n=1}^{\infty} \dfrac{1 \cdot 3 \cdot 5 \cdots (2n-1)x^{2n+1}}{2 \cdot 4 \cdots (2n)(2n+1)},$

where $|x| < 1$

71. (a) $\tan\left(\tan^{-1}(n+1) - \tan^{-1}(n-1)\right) = \dfrac{\tan\left(\tan^{-1}(n+1)\right) - \tan\left(\tan^{-1}(n-1)\right)}{1 + \tan\left(\tan^{-1}(n+1)\right)\tan\left(\tan^{-1}(n-1)\right)} = \dfrac{(n+1)-(n-1)}{1+(n+1)(n-1)} = \dfrac{2}{n^2}$

   (b) $\sum_{n=1}^{N} \tan^{-1}\left(\dfrac{2}{n^2}\right) = \sum_{n=1}^{N} \left[\tan^{-1}(n+1) - \tan^{-1}(n-1)\right] = \left(\tan^{-1}2 - \tan^{-1}0\right) + \left(\tan^{-1}3 - \tan^{-1}1\right)$

   $+ \left(\tan^{-1}4 - \tan^{-1}2\right) + \ldots + \left(\tan^{-1}(N+1) - \tan^{-1}(N-1)\right) = \tan^{-1}(N+1) + \tan^{-1}N - \dfrac{\pi}{4}$

   (c) $\sum_{n=1}^{\infty} \tan^{-1}\left(\dfrac{2}{n^2}\right) = \lim_{n \to \infty} \left[\tan^{-1}(N+1) + \tan^{-1}N - \dfrac{\pi}{4}\right] = \dfrac{\pi}{2} + \dfrac{\pi}{2} - \dfrac{\pi}{4} = \dfrac{3\pi}{4}$

CHAPTER 8 PRACTICE EXERCISES

1. converges to 1, since $\lim_{n \to \infty} a_n = \lim_{n \to \infty} \left(1 + \dfrac{(-1)^n}{n}\right) = 1$

3. converges to -1, since $\lim_{n \to \infty} a_n = \lim_{n \to \infty} \left(\dfrac{1-2^n}{2^n}\right) = \lim_{n \to \infty} \left(\dfrac{1}{2^n} - 1\right) = -1$

5. diverges, since $\left\{\sin \dfrac{n\pi}{2}\right\} = \{0, 1, 0, -1, 0, 1, \ldots\}$

7. converges to 0, since $\lim_{n \to \infty} a_n = \lim_{n \to \infty} \dfrac{\ln n^2}{n} = 2 \lim_{n \to \infty} \dfrac{\left(\frac{1}{n}\right)}{1} = 0$

9. converges to 1, since $\lim_{n \to \infty} a_n = \lim_{n \to \infty} \left(\dfrac{n + \ln n}{n}\right) = \lim_{n \to \infty} \dfrac{1 + \left(\frac{1}{n}\right)}{1} = 1$

11. converges to e^{-5}, since $\lim_{n \to \infty} a_n = \lim_{n \to \infty} \left(\dfrac{n-5}{n}\right)^n = \lim_{n \to \infty} \left(1 + \dfrac{(-5)}{n}\right)^n = e^{-5}$ by Table 8.1

13. converges to 3, since $\lim_{n \to \infty} a_n = \lim_{n \to \infty} \left(\dfrac{3^n}{n}\right)^{1/n} = \lim_{n \to \infty} \dfrac{3}{n^{1/n}} = \dfrac{3}{1} = 3$ by Table 8.1

15. converges to $\ln 2$, since $\lim_{n \to \infty} a_n = \lim_{n \to \infty} n\left(2^{1/n} - 1\right) = \lim_{n \to \infty} \dfrac{2^{1/n} - 1}{\left(\frac{1}{n}\right)} = \lim_{n \to \infty} \dfrac{\left[\frac{(-2^{1/n}\ln 2)}{n^2}\right]}{\left(\frac{-1}{n^2}\right)} = \lim_{n \to \infty} 2^{1/n} \ln 2$
    $= 2^0 \cdot \ln 2 = \ln 2$

17. diverges, since $\lim\limits_{n\to\infty} a_n = \lim\limits_{n\to\infty} \dfrac{(n+1)!}{n!} = \lim\limits_{n\to\infty} (n+1) = \infty$

19. $\dfrac{1}{(2n-3)(2n-1)} = \dfrac{\left(\frac{1}{2}\right)}{2n-3} - \dfrac{\left(\frac{1}{2}\right)}{2n-1} \Rightarrow s_n = \left[\dfrac{\left(\frac{1}{2}\right)}{3} - \dfrac{\left(\frac{1}{2}\right)}{5}\right] + \left[\dfrac{\left(\frac{1}{2}\right)}{5} - \dfrac{\left(\frac{1}{2}\right)}{7}\right] + \cdots + \left[\dfrac{\left(\frac{1}{2}\right)}{2n-3} - \dfrac{\left(\frac{1}{2}\right)}{2n-1}\right] = \dfrac{\left(\frac{1}{2}\right)}{3} - \dfrac{\left(\frac{1}{2}\right)}{2n-1}$

$\Rightarrow \lim\limits_{n\to\infty} s_n = \lim\limits_{n\to\infty} \left[\dfrac{1}{6} - \dfrac{\left(\frac{1}{2}\right)}{2n-1}\right] = \dfrac{1}{6}$

21. $\dfrac{9}{(3n-1)(3n+2)} = \dfrac{3}{3n-1} - \dfrac{3}{3n+2} \Rightarrow s_n = \left(\dfrac{3}{2} - \dfrac{3}{5}\right) + \left(\dfrac{3}{5} - \dfrac{3}{8}\right) + \left(\dfrac{3}{8} - \dfrac{3}{11}\right) + \cdots + \left(\dfrac{3}{3n-1} - \dfrac{3}{3n+2}\right)$

$= \dfrac{3}{2} - \dfrac{3}{3n+2} \Rightarrow \lim\limits_{n\to\infty} s_n = \lim\limits_{n\to\infty} \left(\dfrac{3}{2} - \dfrac{3}{3n+2}\right) = \dfrac{3}{2}$

23. $\sum\limits_{n=0}^{\infty} e^{-n} = \sum\limits_{n=0}^{\infty} \dfrac{1}{e^n}$, a convergent geometric series with $r = \dfrac{1}{e}$ and $a = 1 \Rightarrow$ the sum is $\dfrac{1}{1 - \left(\frac{1}{e}\right)} = \dfrac{e}{e-1}$

25. diverges, a p-series with $p = \dfrac{1}{2}$

27. Since $f(x) = \dfrac{1}{x^{1/2}} \Rightarrow f'(x) = -\dfrac{1}{2x^{3/2}} < 0 \Rightarrow f(x)$ is decreasing $\Rightarrow a_{n+1} < a_n$, and $\lim\limits_{n\to\infty} a_n = \lim\limits_{n\to\infty} \dfrac{(-1)}{\sqrt{n}} = 0$, the

series $\sum\limits_{n=1}^{\infty} \dfrac{(-1)^n}{\sqrt{n}}$ converges by the Alternating Series Test. Since $\sum\limits_{n=1}^{\infty} \dfrac{1}{\sqrt{n}}$ diverges, the given series converges

conditionally.

29. The given series does not converge absolutely by the Direct Comparison Test since $\dfrac{1}{\ln(n+1)} > \dfrac{1}{n+1}$, which is

the nth term of a divergent series. Since $f(x) = \dfrac{1}{\ln(x+1)} \Rightarrow f'(x) = -\dfrac{1}{(\ln(x+1))^2(x+1)} < 0 \Rightarrow f(x)$ is

decreasing $\Rightarrow a_{n+1} < a_n$, and $\lim\limits_{n\to\infty} a_n = \lim\limits_{n\to\infty} \dfrac{1}{\ln(n+1)} = 0$, the given series converges conditionally by the

Alternating Series Test.

31. converges absolutely by the Direct Comparison Test since $\dfrac{\ln n}{n^3} < \dfrac{n}{n^3} = \dfrac{1}{n^2}$, the nth term of a convergent p-series

33. $\lim\limits_{n\to\infty} \dfrac{\left(\dfrac{1}{n\sqrt{n^2+1}}\right)}{\left(\dfrac{1}{n^2}\right)} = \sqrt{\lim\limits_{n\to\infty} \dfrac{n^2}{n^2+1}} = \sqrt{1} = 1 \Rightarrow$ converges absolutely by the Limit Comparison Test

35. converges absolutely by the Ratio Test since $\lim\limits_{n\to\infty} \left[\dfrac{n+2}{(n+1)!} \cdot \dfrac{n!}{n+1}\right] = \lim\limits_{n\to\infty} \dfrac{n+2}{(n+1)^2} = 0 < 1$

37. converges absolutely by the Ratio Test since $\lim\limits_{n\to\infty} \left[\dfrac{3^{n+1}}{(n+1)!} \cdot \dfrac{n!}{3^n}\right] = \lim\limits_{n\to\infty} \dfrac{3}{n+1} = 0 < 1$

39. converges absolutely by the Limit Comparison Test since $\lim\limits_{n\to\infty} \dfrac{\left(\frac{1}{n^{3/2}}\right)}{\left(\frac{1}{\sqrt{n(n+1)(n+2)}}\right)} = \sqrt{\lim\limits_{n\to\infty}\dfrac{n(n+1)(n+2)}{n^3}}$

$= 1$

41. $\lim\limits_{n\to\infty}\left|\dfrac{u_{n+1}}{u_n}\right| < 1 \Rightarrow \lim\limits_{n\to\infty}\left|\dfrac{(x+4)^{n+1}}{(n+1)3^{n+1}}\cdot\dfrac{n3^n}{(x+4)^n}\right| < 1 \Rightarrow \dfrac{|x+4|}{3}\lim\limits_{n\to\infty}\left(\dfrac{n}{n+1}\right) < 1 \Rightarrow \dfrac{|x+4|}{3} < 1$

$\Rightarrow |x+4| < 3 \Rightarrow -3 < x+4 < 3 \Rightarrow -7 < x < -1$; at $x = -7$ we have $\sum\limits_{n=1}^{\infty}\dfrac{(-1)^n 3^n}{n3^n} = \sum\limits_{n=1}^{\infty}\dfrac{(-1)^n}{n}$, the

alternating harmonic series, which converges conditionally; at $x = -1$ we have $\sum\limits_{n=1}^{\infty}\dfrac{3^n}{n3^n} = \sum\limits_{n=1}^{\infty}\dfrac{1}{n}$, the divergent

harmonic series

(a) the radius is 3; the interval of convergence is $-7 \leq x < -1$

(b) the interval of absolute convergence is $-7 < x < -1$

(c) the series converges conditionally at $x = -7$

43. $\lim\limits_{n\to\infty}\left|\dfrac{u_{n+1}}{u_n}\right| < 1 \Rightarrow \lim\limits_{n\to\infty}\left|\dfrac{(3x-1)^{n+1}}{(n+1)^2}\cdot\dfrac{n^2}{(3x-1)^n}\right| < 1 \Rightarrow |3x-1|\lim\limits_{n\to\infty}\dfrac{n^2}{(n+1)^2} < 1 \Rightarrow |3x-1| < 1$

$\Rightarrow -1 < 3x-1 < 1 \Rightarrow 0 < 3x < 2 \Rightarrow 0 < x < \dfrac{2}{3}$; at $x = 0$ we have $\sum\limits_{n=1}^{\infty}\dfrac{(-1)^{n-1}(-1)^n}{n^2} = \sum\limits_{n=1}^{\infty}\dfrac{(-1)^{2n-1}}{n^2}$

$= -\sum\limits_{n=1}^{\infty}\dfrac{1}{n^2}$, a nonzero constant multiple of a convergent p-series, which is absolutely convergent; at $x = \dfrac{2}{3}$ we

have $\sum\limits_{n=1}^{\infty}\dfrac{(-1)^{n-1}(1)^n}{n^2} = \sum\limits_{n=1}^{\infty}\dfrac{(-1)^{n-1}}{n^2}$, which converges absolutely

(a) the radius is $\dfrac{1}{3}$; the interval of convergence is $0 \leq x \leq \dfrac{2}{3}$

(b) the interval of absolute convergence is $0 \leq x \leq \dfrac{2}{3}$

(c) there are no values for which the series converges conditionally

45. $\lim\limits_{n\to\infty}\left|\dfrac{u_{n+1}}{u_n}\right| < 1 \Rightarrow \lim\limits_{n\to\infty}\left|\dfrac{x^{n+1}}{(n+1)^{n+1}}\cdot\dfrac{n^n}{x^n}\right| < 1 \Rightarrow |x|\lim\limits_{n\to\infty}\left|\left(\dfrac{n}{n+1}\right)^n\left(\dfrac{1}{n+1}\right)\right| < 1 \Rightarrow \dfrac{|x|}{e}\lim\limits_{n\to\infty}\left(\dfrac{1}{n+1}\right) < 1$

$\Rightarrow \dfrac{|x|}{e}\cdot 0 < 1$, which holds for all x

(a) the radius is ∞; the series converges for all x

(b) the series converges absolutely for all x

(c) there are no values for which the series converges conditionally

47. $\lim\limits_{n\to\infty}\left|\dfrac{u_{n+1}}{u_n}\right| < 1 \Rightarrow \lim\limits_{n\to\infty}\left|\dfrac{(n+2)x^{2n+1}}{3^{n+1}}\cdot\dfrac{3^n}{(n+1)x^{2n-1}}\right| < 1 \Rightarrow \dfrac{x^2}{3}\lim\limits_{n\to\infty}\left(\dfrac{n+2}{n+1}\right) < 1 \Rightarrow -\sqrt{3} < x < \sqrt{3}$;

the series $\sum\limits_{n=1}^{\infty} -\dfrac{n+1}{\sqrt{3}}$ and $\sum\limits_{n=1}^{\infty}\dfrac{n+1}{\sqrt{3}}$, obtained with $x = \pm\sqrt{3}$, both diverge

(a) the radius is $\sqrt{3}$; the interval of convergence is $-\sqrt{3} < x < \sqrt{3}$

(b) the interval of absolute convergence is $-\sqrt{3} < x < \sqrt{3}$

(c) there are no values for which the series converges conditionally

49. $\lim\limits_{n\to\infty}\left|\dfrac{u_{n+1}}{u_n}\right| < 1 \Rightarrow \lim\limits_{n\to\infty}\left|\dfrac{\operatorname{csch}(n+1)x^{n+1}}{\operatorname{csch}(n)x^n}\right| < 1 \Rightarrow |x|\lim\limits_{n\to\infty}\left|\dfrac{\left(\dfrac{2}{e^{n+1}-e^{-n-1}}\right)}{\left(\dfrac{2}{e^n-e^{-n}}\right)}\right| < 1$

$\Rightarrow |x|\lim\limits_{n\to\infty}\left|\dfrac{e^{-1}-e^{-2n-1}}{1-e^{-2n-2}}\right| < 1 \Rightarrow \dfrac{|x|}{e} < 1 \Rightarrow -e < x < e$; the series $\sum\limits_{n=1}^{\infty}(\pm e)^n\operatorname{csch} n$, obtained with $x = \pm e$,

both diverge since $\lim\limits_{n\to\infty}(\pm e)^n\operatorname{csch} n \neq 0$

(a) the radius is e; the interval of convergence is $-e < x < e$

(b) the interval of absolute convergence is $-e < x < e$

(c) there are no values for which the series converges conditionally

51. The given series has the form $1 - x + x^2 - x^3 + \ldots + (-x)^n + \ldots = \dfrac{1}{1+x}$, where $x = \dfrac{1}{4}$; the sum is $\dfrac{1}{1+\left(\frac{1}{4}\right)} = \dfrac{4}{5}$

53. The given series has the form $x - \dfrac{x^3}{3!} + \dfrac{x^5}{5!} - \ldots + (-1)^n\dfrac{x^{2n+1}}{(2n+1)!} + \ldots = \sin x$, where $x = \pi$; the sum is
$\sin \pi = 0$

55. The given series has the form $1 + x + \dfrac{x^2}{2!} + \dfrac{x^2}{3!} + \ldots + \dfrac{x^n}{n!} + \ldots = e^x$, where $x = \ln 2$; the sum is $e^{\ln(2)} = 2$

57. Consider $\dfrac{1}{1-2x}$ as the sum of a convergent geometric series with $a = 1$ and $r = 2x \Rightarrow \dfrac{1}{1-2x}$

$= 1 + (2x) + (2x)^2 + (2x)^3 + \ldots = \sum\limits_{n=0}^{\infty}(2x)^n = \sum\limits_{n=0}^{\infty}2^n x^n$ where $|2x| < 1 \Rightarrow |x| < \dfrac{1}{2}$

59. $\sin x = \sum\limits_{n=0}^{\infty}\dfrac{(-1)^n x^{2n+1}}{(2n+1)!} \Rightarrow \sin \pi x = \sum\limits_{n=0}^{\infty}\dfrac{(-1)^n(\pi x)^{2n+1}}{(2n+1)!} = \sum\limits_{n=0}^{\infty}\dfrac{(-1)^n\pi^{2n+1}x^{2n+1}}{(2n+1)!}$

61. $\cos x = \sum\limits_{n=0}^{\infty}\dfrac{(-1)^n x^{2n}}{(2n)!} \Rightarrow \cos\left(x^{5/2}\right) = \sum\limits_{n=0}^{\infty}\dfrac{(-1)^n\left(x^{5/2}\right)^{2n}}{(2n)!} = \sum\limits_{n=0}^{\infty}\dfrac{(-1)^n x^{5n}}{(2n)!}$

63. $e^x = \sum\limits_{n=0}^{\infty}\dfrac{x^n}{n!} \Rightarrow e^{(\pi x/2)} = \sum\limits_{n=0}^{\infty}\dfrac{\left(\frac{\pi x}{2}\right)^n}{n!} = \sum\limits_{n=0}^{\infty}\dfrac{\pi^n x^n}{2^n n!}$

65. $f(x) = \sqrt{3+x^2} = \left(3+x^2\right)^{1/2} \Rightarrow f'(x) = x\left(3+x^2\right)^{-1/2} \Rightarrow f''(x) = -x^2\left(3+x^2\right)^{-3/2} + \left(3+x^2\right)^{-1/2}$

$\Rightarrow f'''(x) = 3x^3\left(3+x^2\right)^{-5/2} - 3x\left(3+x^2\right)^{-3/2}$; $f(-1) = 2$, $f'(-1) = -\dfrac{1}{2}$, $f''(-1) = -\dfrac{1}{8} + \dfrac{1}{2} = \dfrac{3}{8}$,

$f'''(-1) = -\dfrac{3}{32} + \dfrac{3}{8} = \dfrac{9}{32} \Rightarrow \sqrt{3+x^2} = 2 - \dfrac{(x+1)}{2\cdot 1!} + \dfrac{3(x+1)^2}{2^3\cdot 2!} + \dfrac{9(x+1)^3}{2^5\cdot 3!} + \ldots$

67. $f(x) = \frac{1}{x+1} = (x+1)^{-1} \Rightarrow f'(x) = -(x+1)^{-2} \Rightarrow f''(x) = 2(x+1)^{-3} \Rightarrow f'''(x) = -6(x+1)^{-4}$; $f(3) = \frac{1}{4}$,

$f'(3) = -\frac{1}{4^2}$, $f''(3) = \frac{2}{4^3}$, $f'''(2) = \frac{-6}{4^4} \Rightarrow \frac{1}{x+1} = \frac{1}{4} - \frac{1}{4^2}(x-3) + \frac{1}{4^3}(x-3)^2 - \frac{1}{4^4}(x-3)^3 + \dots$

69. Assume the solution has the form $y = a_0 + a_1x + a_2x^2 + \dots + a_{n-1}x^{n-1} + a_nx^n + \dots$

$\Rightarrow \frac{dy}{dx} = a_1 + 2a_2x + \dots + na_nx^{n-1} + \dots \Rightarrow \frac{dy}{dx} + y$

$= (a_1 + a_0) + (2a_2 + a_1)x + (3a_3 + a_2)x^2 + \dots + (na_n + a_{n-1})x^{n-1} + \dots = 0 \Rightarrow a_1 + a_0 = 0$, $2a_2 + a_1 = 0$,

$3a_3 + a_2 = 0$ and in general $na_n + a_{n-1} = 0$. Since $y = -1$ when $x = 0$ we have $a_0 = -1$. Therefore $a_1 = 1$,

$a_2 = \frac{-a_1}{2 \cdot 1} = -\frac{1}{2}$, $a_3 = \frac{-a_2}{3} = \frac{1}{3 \cdot 2}$, $a_4 = \frac{-a_3}{4} = -\frac{1}{4 \cdot 3 \cdot 2}$, $\dots$, $a_n = \frac{-a_{n-1}}{n} = \frac{-1}{n}\frac{(-1)^n}{(n-1)!} = \frac{(-1)^{n+1}}{n!}$

$\Rightarrow y = -1 + x - \frac{1}{2}x^2 + \frac{1}{3 \cdot 2}x^3 - \dots + \frac{(-1)^{n+1}}{n!}x^n + \dots = -\sum_{n=0}^{\infty} \frac{(-1)^n x^n}{n!} = -e^{-x}$

71. Assume the solution has the form $y = a_0 + a_1x + a_2x^2 + \dots + a_{n-1}x^{n-1} + a_nx^n + \dots$

$\Rightarrow \frac{dy}{dx} = a_1 + 2a_2x + \dots + na_nx^{n-1} + \dots \Rightarrow \frac{dy}{dx} + 2y$

$= (a_1 + 2a_0) + (2a_2 + 2a_1)x + (3a_3 + 2a_2)x^2 + \dots + (na_n + 2a_{n-1})x^{n-1} + \dots = 0$. Since $y = 3$ when $x = 0$ we

have $a_0 = 3$. Therefore $a_1 = -2a_0 = -2(3) = -3(2)$, $a_2 = -\frac{2}{2}a_1 = -\frac{2}{2}(-2 \cdot 3) = 3\left(\frac{2^2}{2}\right)$, $a_3 = -\frac{2}{3}a_2$

$= -\frac{2}{3}\left[3\left(\frac{2^2}{2}\right)\right] = -3\left(\frac{2^3}{3 \cdot 2}\right)$, $\dots$, $a_n = \left(-\frac{2}{n}\right)a_{n-1} = \left(-\frac{2}{n}\right)\left(3\left(\frac{(-1)^{n-1}2^{n-1}}{(n-1)!}\right)\right) = 3\left(\frac{(-1)^n 2^n}{n!}\right)$

$\Rightarrow y = 3 - 3(2x) + 3\frac{(2)^2}{2}x^2 - 3\frac{(2)^3}{3 \cdot 2}x^3 + \dots + 3\frac{(-1)^n 2^n}{n!}x^n + \dots$

$= 3\left[1 - (2x) + \frac{(2x)^2}{2!} - \frac{(2x)^3}{3!} + \dots + \frac{(-1)^n(2x)^n}{n!} + \dots\right] = 3\sum_{n=0}^{\infty} \frac{(-1)^n(2x)^n}{n!} = 3e^{-2x}$

73. Assume the solution has the form $y = a_0 + a_1x + a_2x^2 + \dots + a_{n-1}x^{n-1} + a_nx^n + \dots$

$\Rightarrow \frac{dy}{dx} = a_1 + 2a_2x + \dots + na_nx^{n-1} + \dots \Rightarrow \frac{dy}{dx} - y$

$= (a_1 - a_0) + (2a_2 - a_1)x + (3a_3 - a_2)x^2 + \dots + (na_n - a_{n-1})x^{n-1} + \dots = 3x \Rightarrow a_1 - a_0 = 0$, $2a_2 - a_1 = 3$,

$3a_3 - a_2 = 0$ and in general $na_n - a_{n-1} = 0$ for $n > 2$. Since $y = -1$ when $x = 0$ we have $a_0 = -1$. Therefore

$a_1 = -1$, $a_2 = \frac{3 + a_1}{2} = \frac{2}{2}$, $a_3 = \frac{a_2}{3} = \frac{2}{3 \cdot 2}$, $a_4 = \frac{a_3}{4} = \frac{2}{4 \cdot 3 \cdot 2}$, $\dots$, $a_n = \frac{a_{n-1}}{n} = \frac{2}{n!}$

$\Rightarrow y = -1 - x + \left(\frac{2}{2}\right)x^2 + \frac{3}{3 \cdot 2}x^3 + \frac{2}{4 \cdot 3 \cdot 2}x^4 + \dots + \frac{2}{n!}x^n + \dots$

$= 2\left(1 + x + \frac{1}{2}x^2 + \frac{1}{3 \cdot 2}x^3 + \frac{1}{4 \cdot 3 \cdot 2}x^4 + \dots + \frac{1}{n!}x^n + \dots\right) - 3 - 3x = 2\sum_{n=0}^{\infty} \frac{x^n}{n!} - 3 - 3x = 2e^x - 3x - 3$

75. Assume the solution has the form $y = a_0 + a_1x + a_2x^2 + \dots + a_{n-1}x^{n-1} + a_nx^n + \dots$

$\Rightarrow \frac{dy}{dx} = a_1 + 2a_2x + \dots + na_nx^{n-1} + \dots \Rightarrow \frac{dy}{dx} - y$

$= (a_1 - a_0) + (2a_2 - a_1)x + (3a_3 - a_2)x^2 + \dots + (na_n - a_{n-1})x^{n-1} + \dots = x \Rightarrow a_1 - a_0 = 0$, $2a_2 - a_1 = 1$,

$3a_3 - a_2 = 0$ and in general $na_n - a_{n-1} = 0$ for $n > 2$. Since $y = 1$ when $x = 0$ we have $a_0 = 1$. Therefore

$$a_1 = 1, \ a_2 = \frac{1+a_1}{2} = \frac{2}{2}, \ a_3 = \frac{a_2}{3} = \frac{2}{3 \cdot 2}, \ a_4 = \frac{a_3}{4} = \frac{2}{4 \cdot 3 \cdot 2}, \ \ldots, \ a_n = \frac{a_{n-1}}{n} = \frac{2}{n!}$$

$$\Rightarrow y = 1 + x + \left(\frac{2}{2}\right)x^2 + \frac{2}{3 \cdot 2}x^3 + \frac{2}{4 \cdot 2 \cdot 2}x^4 + \ldots + \frac{2}{n!}x^n + \ldots$$

$$= 2\left(1 + x + \frac{1}{2}x^2 + \frac{1}{3 \cdot 2}x^3 + \frac{1}{4 \cdot 3 \cdot 2}x^4 + \ldots + \frac{1}{n!}x^n + \ldots\right) - 1 - x = 2\sum_{n=0}^{\infty} \frac{x^n}{n!} - 1 - x = 2e^x - x - 1$$

77. $$\int_0^{1/2} \exp(-x^3)\,dx = \int_0^{1/2}\left(1 - x^3 + \frac{x^6}{2!} - \frac{x^9}{3!} + \frac{x^{12}}{4!} + \ldots\right)dx = \left[x - \frac{x^4}{4} + \frac{x^7}{7 \cdot 2!} - \frac{x^{10}}{10 \cdot 3!} + \frac{x^{13}}{13 \cdot 4!} - \ldots\right]_0^{1/2}$$

$$\approx \frac{1}{2} - \frac{1}{2^4 \cdot 4} + \frac{1}{2^7 \cdot 7 \cdot 2!} - \frac{1}{2^{10} \cdot 10 \cdot 3!} + \frac{1}{2^{13} \cdot 13 \cdot 4!} - \frac{1}{2^{16} \cdot 16 \cdot 5!} \approx 0.484917143$$

79. $$\int_1^{1/2} \frac{\tan^{-1}x}{x}\,dx = \int_1^{1/2}\left(1 - \frac{x^2}{3} + \frac{x^4}{5} - \frac{x^6}{7} + \frac{x^8}{9} - \frac{x^{10}}{11} + \ldots\right)dx = \left[x - \frac{x^3}{9} + \frac{x^5}{25} - \frac{x^7}{49} + \frac{x^9}{81} - \frac{x^{11}}{121} + \ldots\right]_0^{1/2}$$

$$\approx \frac{1}{2} - \frac{1}{9 \cdot 2^3} + \frac{1}{5^2 \cdot 2^5} - \frac{1}{7^2 \cdot 2^7} + \frac{1}{9^2 \cdot 2^9} - \frac{1}{11^2 \cdot 2^{11}} + \frac{1}{13^2 \cdot 2^{13}} - \frac{1}{15^2 \cdot 2^{15}} + \frac{1}{17^2 \cdot 2^{17}} - \frac{1}{19^2 \cdot 2^{19}} + \frac{1}{21^2 \cdot 2^{21}}$$

$$\approx 0.4872223583$$

81. $$\lim_{x \to 0} \frac{7\sin x}{e^{2x} - 1} = \lim_{x \to 0} \frac{7\left(x - \frac{x^3}{3!} + \frac{x^5}{5!} - \ldots\right)}{\left(2x + \frac{2^2 x^2}{2!} + \frac{2^3 x^3}{3!} + \ldots\right)} = \lim_{x \to 0} \frac{7\left(1 - \frac{x^2}{3!} + \frac{x^4}{5!} - \ldots\right)}{\left(2 + \frac{2^2 x}{2!} + \frac{2^3 x^2}{3!} + \ldots\right)} = \frac{7}{2}$$

83. $$\lim_{t \to 0}\left(\frac{1}{2 - 2\cos t} - \frac{1}{t^2}\right) = \lim_{t \to 0} \frac{t^2 - 2 + 2\cos t}{2t^2(1 - \cos t)} = \lim_{t \to 0} \frac{t^2 - 2 + 2\left(1 - \frac{t^2}{2} + \frac{t^4}{4!} - \ldots\right)}{2t^2\left(1 - 1 + \frac{t^2}{2} - \frac{t^4}{4!} + \ldots\right)} = \lim_{t \to 0} \frac{2\left(\frac{t^4}{4!} - \frac{t^6}{6!} + \ldots\right)}{\left(t^4 - \frac{2t^6}{4!} + \ldots\right)}$$

$$= \lim_{t \to 0} \frac{2\left(\frac{1}{4!} - \frac{t^2}{6!} + \ldots\right)}{\left(1 - \frac{2t^2}{4!} + \ldots\right)} = \frac{1}{12}$$

85. $$\lim_{z \to 0} \frac{1 - \cos^2 z}{\ln(1 - z) + \sin z} = \lim_{z \to 0} \frac{1 - \left(1 - z^2 + \frac{z^4}{3} - \ldots\right)}{\left(-z - \frac{z^2}{2} - \frac{z^3}{3} - \ldots\right) + \left(z - \frac{z^3}{3!} + \frac{z^5}{5!} - \ldots\right)} = \lim_{z \to 0} \frac{\left(z^2 - \frac{z^4}{3} + \ldots\right)}{\left(-\frac{z^2}{2} - \frac{2z^3}{3} - \frac{z^4}{4} - \ldots\right)}$$

$$= \lim_{z \to 0} \frac{\left(1 - \frac{z^2}{3} + \ldots\right)}{\left(-\frac{1}{2} - \frac{2z}{3} - \frac{z^2}{4} - \ldots\right)} = -2$$

87. $\lim\limits_{x\to 0}\left(\dfrac{\sin 3x}{x^3}+\dfrac{r}{x^2}+s\right)=\lim\limits_{x\to 0}\left[\dfrac{\left(3x-\dfrac{(3x)^3}{6}+\dfrac{(3x)^5}{120}-\cdots\right)}{x^3}+\dfrac{r}{x^2}+s\right]=\lim\limits_{x\to 0}\left(\dfrac{3}{x^2}-\dfrac{9}{2}+\dfrac{81x^2}{40}+\cdots+\dfrac{r}{x^2}+s\right)=0$

$\Rightarrow \dfrac{r}{x^2}+\dfrac{3}{x^2}=0$ and $s-\dfrac{9}{2}=0 \Rightarrow r=-3$ and $s=\dfrac{9}{2}$

89. (a) $\displaystyle\sum_{n=1}^{\infty}\left(\sin\dfrac{1}{2n}-\sin\dfrac{1}{2n+1}\right)=\left(\sin\dfrac{1}{2}-\sin\dfrac{1}{3}\right)+\left(\sin\dfrac{1}{4}-\sin\dfrac{1}{5}\right)+\left(\sin\dfrac{1}{6}-\sin\dfrac{1}{7}\right)+\ldots+\left(\sin\dfrac{1}{2n}-\sin\dfrac{1}{2n+1}\right)$

$+\ldots=\displaystyle\sum_{n=2}^{\infty}(-1)^n\sin\dfrac{1}{n};\ f(x)=\sin\dfrac{1}{x}\Rightarrow f'(x)=\dfrac{-\cos\left(\frac{1}{x}\right)}{x^2}<0$ if $x\geq 2 \Rightarrow \sin\dfrac{1}{n+1}<\sin\dfrac{1}{n}$, and

$\lim\limits_{n\to\infty}\sin\dfrac{1}{n}=0 \Rightarrow \displaystyle\sum_{n=2}^{\infty}(-1)^n\sin\dfrac{1}{n}$ converges by the Alternating Series Test

(b) $|\text{error}|<\left|\sin\dfrac{1}{42}\right|\approx 0.02381$ and the sum is an underestimate because the remainder is positive

91. $\lim\limits_{n\to\infty}\left|\dfrac{2\cdot 5\cdot 8\cdots(3n-1)(3n+2)x^{n+1}}{2\cdot 4\cdot 6\cdots(2n)(2n+2)}\cdot\dfrac{2\cdot 4\cdot 6\cdots(2n)}{2\cdot 5\cdot 8\cdots(3n-1)x^n}\right|<1 \Rightarrow |x|\lim\limits_{n\to\infty}\left|\dfrac{3n+2}{2n+2}\right|<1 \Rightarrow |x|<\dfrac{2}{3}$

$\Rightarrow$ the radius of convergence is $\dfrac{2}{3}$

93. $\displaystyle\sum_{k=2}^{n}\ln\left(1-\dfrac{1}{k^2}\right)=\sum_{k=2}^{n}\left[\ln\left(1+\dfrac{1}{k}\right)+\ln\left(1-\dfrac{1}{k}\right)\right]=\sum_{k=2}^{n}\left[\ln(k+1)-\ln k+\ln(k-1)-\ln k\right]$

$=\left[\ln 3-\ln 2+\ln 1-\ln 2\right]+\left[\ln 4-\ln 3+\ln 2-\ln 3\right]+\left[\ln 5-\ln 4+\ln 3-\ln 4\right]+\left[\ln 6-\ln 5+\ln 4-\ln 5\right]$

$+\ldots+\left[\ln(n+1)-\ln n+\ln(n-1)-\ln n\right]=\left[\ln 1-\ln 2\right]+\left[\ln(n+1)-\ln n\right]$ after cancellation

$\Rightarrow\displaystyle\sum_{k=2}^{n}\ln\left(1-\dfrac{1}{k^2}\right)=\ln\left(\dfrac{n+1}{2n}\right)\Rightarrow\sum_{k=2}^{\infty}\ln\left(1-\dfrac{1}{k^2}\right)=\lim\limits_{n\to\infty}\ln\left(\dfrac{n+1}{2n}\right)=\ln\dfrac{1}{2}$ is the sum

95. (a) $\lim\limits_{n\to\infty}\left|\dfrac{1\cdot 4\cdot 7\cdots(3n-2)(3n+1)x^{3n+3}}{(3n+3)!}\cdot\dfrac{(3n)!}{1\cdot 4\cdot 7\cdots(3n-2)x^{3n}}\right|<1\Rightarrow|x^3|\lim\limits_{n\to\infty}\dfrac{(3n+1)}{(3n+1)(3n+2)(3n+3)}$

$=|x^3|\cdot 0<1\Rightarrow$ the radius of convergence is ∞

(b) $y=1+\displaystyle\sum_{n=1}^{\infty}\dfrac{1\cdot 4\cdot 7\cdots(3n-2)}{(3n)!}x^{3n}\Rightarrow\dfrac{dy}{dx}=\sum_{n=1}^{\infty}\dfrac{1\cdot 4\cdot 7\cdots(3n-2)}{(3n-1)!}x^{3n-1}$

$\Rightarrow\dfrac{d^2y}{dx^2}=\displaystyle\sum_{n=1}^{\infty}\dfrac{1\cdot 4\cdot 7\cdots(3n-2)}{(3n-2)!}x^{3n-2}=x+\sum_{n=2}^{\infty}\dfrac{1\cdot 4\cdot 7\cdots(3n-5)}{(3n-3)!}x^{3n-2}$

$=x\left(1+\displaystyle\sum_{n=1}^{\infty}\dfrac{1\cdot 4\cdot 7\cdots(3n-2)}{(3n)!}x^{3n}\right)=xy+0\Rightarrow a=1$ and $b=0$

97. Yes, the series $\displaystyle\sum_{n=1}^{\infty}a_n b_n$ converges as we now show. Since $\displaystyle\sum_{n=1}^{\infty}a_n$ converges it follows that $a_n\to 0\Rightarrow a_n<1$

for $n>$ some index $N\Rightarrow a_n b_n<b_n$ for $n>N\Rightarrow\displaystyle\sum_{n=1}^{\infty}a_n b_n$ converges by the Direct Comparison Test with

$\displaystyle\sum_{n=1}^{\infty}b_n$

99. $\sum_{n=1}^{\infty} (x_{n+1} - x_n) = \lim_{n\to\infty} \sum_{k=1}^{\infty} (x_{k+1} - x_k) = \lim_{n\to\infty} (x_{n+1} - x_1) = \lim_{n\to\infty} (x_{n+1}) - x_1 \Rightarrow$ both the series and

sequence must either converge or diverge.

101. Newton's method gives $x_{n+1} = x_n - \dfrac{(x_n - 1)^{40}}{40(x_n - 1)^{39}} = \dfrac{39}{40} x_n + \dfrac{1}{40}$, and if the sequence $\{x_n\}$ has the limit L, then

$L = \dfrac{39}{40} L + \dfrac{1}{40} \Rightarrow L = 1$

103. (a) $T = \dfrac{\left(\frac{1}{2}\right)}{2}\left(0 + 2\left(\frac{1}{2}\right)^2 e^{1/2} + e\right) = \frac{1}{8} e^{1/2} + \frac{1}{4} e \approx 0.885660616$

(b) $x^2 e^x = x^2\left(1 + x + \dfrac{x^2}{2} + \dots\right) = x^2 + x^3 + \dfrac{x^4}{2} + \dots \Rightarrow \displaystyle\int_0^1 \left(x^2 + x^3 + \dfrac{x^4}{2}\right) dx = \left[\dfrac{x^3}{3} + \dfrac{x^4}{4} + \dfrac{x^5}{10}\right]_0^1 = \dfrac{41}{60} = 0.68333$

(c) If the second derivative is positive, the curve is concave upward and the polygonal line segments used in the trapezoidal rule lie above the curve. The trapezoidal approximation is therefore greater than the actual area under the graph.

(d) All terms in the Maclaurin series are positive. If we truncate the series, we are omitting positive terms and hence the estimate is too small.

(e) $\displaystyle\int_0^1 x^2 e^x \, dx = \left[x^2 e^x - 2x e^x + 2 e^x\right]_0^1 = e - 2e + 2e - 2 = e - 2 \approx 0.7182818285$

CHAPTER 8 ADDITIONAL EXERCISES–THEORY, EXAMPLES, APPLICATIONS

1. converges since $\dfrac{1}{(3n-2)^{(2n+1)/2}} < \dfrac{1}{(3n-2)^{3/2}}$ and $\displaystyle\sum_{n=1}^{\infty} \dfrac{1}{(3n-2)^{3/2}}$ converges by the Limit Comparison Test:

$\lim_{n\to\infty} \dfrac{\left(\dfrac{1}{n^{3/2}}\right)}{\left(\dfrac{1}{(3n-2)^{3/2}}\right)} = \lim_{n\to\infty} \left(\dfrac{3n-2}{n}\right)^{3/2} = 3^{3/2}$

3. diverges by the nth-Term Test since $\lim_{n\to\infty} a_n = \lim_{n\to\infty} (-1)^n \tanh n = \lim_{b\to\infty} (-1)^n\left(\dfrac{1 - e^{-2n}}{1 + e^{-2n}}\right) = \lim_{n\to\infty} (-1)^n$

does not exist

5. converges by the Direct Comparison Test: $a_1 = 1 = \dfrac{12}{(1)(3)(2)^2}$, $a_2 = \dfrac{1 \cdot 2}{3 \cdot 4} = \dfrac{12}{(2)(4)(3)^2}$, $a_3 = \left(\dfrac{2 \cdot 3}{4 \cdot 5}\right)\left(\dfrac{1 \cdot 2}{3 \cdot 4}\right)$

$= \dfrac{12}{(3)(5)(4)^2}$, $a_4 = \left(\dfrac{3 \cdot 4}{5 \cdot 6}\right)\left(\dfrac{2 \cdot 3}{4 \cdot 5}\right)\left(\dfrac{1 \cdot 2}{3 \cdot 4}\right) = \dfrac{12}{(4)(6)(5)^2}$, $\dots \Rightarrow 1 + \displaystyle\sum_{n=1}^{\infty} \dfrac{12}{(n+1)(n+3)(n+2)^2}$ represents the

given series and $\dfrac{12}{(n+1)(n+3)(n+2)^2} < \dfrac{12}{n^4}$, which is the nth-term of a convergent p-series

7. diverges by the nth-Term Test since if $a_n \to L$ as $n \to \infty$, then $L = \dfrac{1}{1+L} \Rightarrow L^2 + L - 1 = 0 \Rightarrow L = \dfrac{-1 \pm \sqrt{5}}{2}$

$\neq 0$

9. $f(x) = \cos x$ with $a = \frac{\pi}{3} \Rightarrow f\left(\frac{\pi}{3}\right) = 0.5$, $f'\left(\frac{\pi}{3}\right) = -\frac{\sqrt{3}}{2}$, $f''\left(\frac{\pi}{3}\right) = -0.5$, $f'''\left(\frac{\pi}{3}\right) = \frac{\sqrt{3}}{2}$, $f^{(4)}\left(\frac{\pi}{3}\right) = 0.5$;

$$\cos x = \frac{1}{2} - \frac{\sqrt{3}}{2}\left(x - \frac{\pi}{3}\right) - \frac{1}{4}\left(x - \frac{\pi}{3}\right)^2 + \frac{\sqrt{3}}{12}\left(x - \frac{\pi}{3}\right)^3 + \ldots$$

11. $e^x = 1 + x + \frac{x^2}{2!} + \frac{x^3}{2!} + \ldots$ with $a = 0$

13. $f(x) = \cos x$ with $a = 22\pi \Rightarrow f(22\pi) = 1$, $f'(22\pi) = 0$, $f''(22\pi) = -1$, $f'''(22\pi) = 0$, $f^{(4)}(22\pi) = 1$,

$f^{(5)}(22\pi) = 0$, $f^{(6)}(22\pi) = -1$; $\cos x = 1 - \frac{1}{2}(x - 22\pi)^2 + \frac{1}{4!}(x - 22\pi)^4 - \frac{1}{6!}(x - 22\pi)^6 + \ldots$

15. Yes, the sequence converges: $c_n = (a^n + b^n)^{1/n} \Rightarrow c_n = b\left(\left(\frac{a}{b}\right)^n + 1\right)^{1/n} \Rightarrow \lim_{n \to \infty} c_n = \lim_{n \to \infty} b\left(\left(\frac{a}{b}\right)^n + 1\right)^{1/n} = b$

since $0 < a < b$

17. $s_n = \sum_{k=0}^{n-1} \int_{k}^{k+1} \frac{dx}{1 + x^2} \Rightarrow s_n = \int_0^1 \frac{dx}{1 + x^2} + \int_1^2 \frac{dx}{1 + x^2} + \ldots + \int_{n-1}^{n} \frac{dx}{1 + x^2} \Rightarrow s_n = \int_0^n \frac{dx}{1 + x^2}$

$\Rightarrow \lim_{n \to \infty} s_n = \lim_{n \to \infty} \left(\tan^{-1} n - \tan^{-1} 0\right) = \frac{\pi}{2}$

19. (a) Each A_{n+1} fits into the corresponding upper triangular region, whose vertices are:

$(n, f(n) - f(n+1))$, $(n+1, f(n+1))$ and $(n, f(n))$ along the line whose slope is $f(n+2) - f(n+1)$.

All the A_n's fit into the first upper triangular region whose area is $\frac{f(1) - f(2)}{2} \Rightarrow \sum_{n=1}^{\infty} A_n < \frac{f(1) - f(2)}{2}$

(b) If $A_k = \frac{f(k+1) + f(k)}{2} - \int_k^{k+1} f(x) \, dx$, then

$\sum_{k=1}^{n-1} A_k = \frac{f(1) + f(2) + f(2) + f(3) + f(3) + \ldots + f(n-1) + f(n)}{2} - \int_1^2 f(x) \, dx - \int_2^3 f(x) \, dx - \ldots - \int_{n-1}^n f(x) \, dx$

$= \frac{f(1) + f(n)}{2} + \sum_{k=2}^{n-1} f(k) - \int_1^n f(x) \, dx \Rightarrow \sum_{k=1}^{n-1} A_k = \sum_{k=1}^{n} f(k) - \frac{f(1) + f(n)}{2} - \int_1^n f(x) \, dx < \frac{f(1) - f(2)}{2}$, from

part (a). The sequence $\left\{\sum_{k=1}^{n-1} A_k\right\}$ is bounded above and increasing, so it converges and the limit in

question must exist.

(c) From part (b) we have $\sum_{k=1}^{\infty} f(k) - \int_1^n f(x) \, dx < f(1) - \frac{f(2)}{2} + \frac{f(n)}{2}$

$\Rightarrow \lim_{n \to \infty} \left[\sum_{k=1}^{n} f(k) - \int_1^n f(x) \, dx\right] < \lim_{n \to \infty} \left[f(1) - \frac{f(2)}{2} + \frac{f(n)}{2}\right] = f(1) - \frac{f(2)}{2}$. The sequence

$\left\{\sum_{k=1}^{n} f(k) - \int_1^n f(x) \, dx\right\}$ is bounded and increasing, so it converges and the limit in question

must exist.

21. (a) No, the limit does not appear to depend on the value of the constant a

(b) Yes, the limit depends on the value of b

(c) $s = \left(1 - \frac{\cos\left(\frac{a}{n}\right)}{n}\right)^n \Rightarrow \log s = \frac{\log\left(1 - \frac{\cos\left(\frac{a}{n}\right)}{n}\right)}{\left(\frac{1}{n}\right)} \Rightarrow \lim_{n\to\infty} \log s = \frac{\left(\frac{1}{1 - \frac{\cos\left(\frac{a}{n}\right)}{n}}\right)\left(\frac{-\frac{a}{n}\sin\left(\frac{a}{n}\right) + \cos\left(\frac{a}{n}\right)}{n^2}\right)}{\left(-\frac{1}{n^2}\right)}$

$= \lim_{n\to\infty} \frac{\frac{a}{n}\sin\left(\frac{a}{n}\right) - \cos\left(\frac{a}{n}\right)}{1 - \frac{\cos\left(\frac{a}{n}\right)}{n}} = \frac{0-1}{1-0} = -1 \Rightarrow \lim_{n\to\infty} s = e^{-1} \approx 0.3678794412$; similarly,

$\lim_{n\to\infty} \left(1 - \frac{\cos\left(\frac{a}{n}\right)}{bn}\right)^n = e^{-1/b}$

23. $\lim_{n\to\infty} \left|\frac{u_{n+1}}{u_n}\right| < 1 \Rightarrow \lim_{n\to\infty} \left|\frac{b^{n+1}x^{n+1}}{\ln(n+1)} \cdot \frac{\ln n}{b^n x^n}\right| < 1 \Rightarrow |bx| < 1 \Rightarrow -\frac{1}{b} < x < \frac{1}{b} = 5 \Rightarrow b = \pm\frac{1}{5}$

25. $\lim_{x\to 0} \frac{\sin(ax) - \sin x - x}{x^3} = \lim_{x\to 0} \frac{\left(ax - \frac{a^3 x^3}{3!} + \ldots\right) - \left(x - \frac{x^3}{3!} + \ldots\right) - x}{x^3}$

$= \lim_{x\to 0} \left[\frac{a-2}{x^2} - \frac{a^3}{3!} + \frac{1}{3!} - \left(\frac{a^5}{5!} - \frac{1}{5!}\right)x^2 + \ldots\right]$ is finite if $a - 2 = 0 \Rightarrow a = 2$;

$\lim_{x\to 0} \frac{\sin 2x - \sin x - x}{x^3} = -\frac{2^3}{3!} + \frac{1}{3!} = -\frac{7}{6}$

27. (a) $\frac{u_n}{u_{n+1}} = \frac{(n+1)^2}{n^2} = 1 + \frac{2}{n} + \frac{1}{n^2} \Rightarrow C = 2 > 1$ and $\sum_{n=1}^{\infty} \frac{1}{n^2}$ converges

(b) $\frac{u_n}{u_{n+1}} = \frac{n+1}{n} = 1 + \frac{1}{n} + \frac{0}{n^2} \Rightarrow C = 1 \leq 1$ and $\sum_{n=1}^{\infty} \frac{1}{n}$ diverges

29. (a) $\sum_{n=1}^{\infty} a_n = L \Rightarrow a_n^2 \leq a_n \sum_{n=1}^{\infty} a_n = a_n L \Rightarrow \sum_{n=1}^{\infty} a_n^2$ converges by the Direct Comparison Test

(b) converges by the Limit Comparison Test: $\lim_{n\to\infty} \frac{\left(\frac{a_n}{1 - a_n}\right)}{a_n} = \lim_{n\to\infty} \frac{1}{1 - a_n} = 1$ since $\sum_{n=1}^{\infty} a_n$ converges and

therefore $\lim_{x\to\infty} a_n = 0$

31. $(1 - x)^{-1} = 1 + \sum_{n=1}^{\infty} x^n$ where $|x| < 1 \Rightarrow \frac{1}{(1-x)^2} = \frac{d}{dx}(1-x)^{-1} = \sum_{n=1}^{\infty} n x^{n-1}$ and when $x = \frac{1}{2}$ we have

$4 = 1 + 2\left(\frac{1}{2}\right) + 3\left(\frac{1}{2}\right)^2 + 4\left(\frac{1}{2}\right)^3 + \ldots + n\left(\frac{1}{2}\right)^{n-1} + \ldots$

33. The sequence $\{x_n\}$ converges to $\frac{\pi}{2}$ from below so $\epsilon_n = \frac{\pi}{2} - x_n > 0$ for each n. By the Alternating Series

Estimation Theorem $\epsilon_{n+1} \approx \frac{1}{3!}(\epsilon_n)^3$ with $|\text{error}| < \frac{1}{5!}(\epsilon_n)^5$, and since the remainder is negative this is an overestimate $\Rightarrow 0 < \epsilon_{n+1} < \frac{1}{6}(\epsilon_n)^3$.

35. (a) $\dfrac{1}{(1-x)^2} = \dfrac{d}{dx}\left(\dfrac{1}{1-x}\right) = \dfrac{d}{dx}\left(1 + x + x^2 + x^3 + \ldots\right) = 1 + 2x + 3x^2 + 4x^3 + \ldots = \displaystyle\sum_{n=1}^{\infty} nx^{n-1}$

(b) from part (a) we have $\displaystyle\sum_{n=1}^{\infty} n\left(\dfrac{5}{6}\right)^{n-1}\left(\dfrac{1}{6}\right) = \left(\dfrac{1}{6}\right)\left[\dfrac{1}{1 - \left(\frac{5}{6}\right)}\right] = 6$

(c) from part (a) we have $\displaystyle\sum_{n=1}^{\infty} np^{n-1}q = \dfrac{q}{(1-p)^2} = \dfrac{q}{q^2} = \dfrac{1}{q}$

37. (a) $R_n = C_0 e^{-kt_0} + C_0 e^{-2kt_0} + \ldots + C_0 e^{-nkt_0} = \dfrac{C_0 e^{-kt_0}\left(1 - e^{-nkt_0}\right)}{1 - e^{-kt_0}} \Rightarrow R = \displaystyle\lim_{n\to\infty} R_n = \dfrac{C_0 e^{-kt_0}}{1 - e^{-kt_0}} = \dfrac{C_0}{e^{kt_0} - 1}$

(b) $R_n = \dfrac{e^{-1}\left(1 - e^{-n}\right)}{1 - e^{-1}} \Rightarrow R_1 = e^{-1} \approx 0.36787944$ and $R_{10} = \dfrac{e^{-1}\left(1 - e^{-10}\right)}{1 - e^{-1}} \approx 0.58195028;$

$R = \dfrac{1}{e-1} \approx 0.58197671;\ R - R_{10} \approx 0.00002643 \Rightarrow \dfrac{R - R_{10}}{R} < 0.0001$

(c) $R_n = \dfrac{e^{-.1}\left(1 - e^{-.1n}\right)}{1 - e^{-.1}},\ \dfrac{R}{2} = \dfrac{1}{2}\left(\dfrac{1}{e^{.1} - 1}\right) \approx 4.7541659;\ R_n > \dfrac{R}{2} \Rightarrow \dfrac{1 - e^{-.1n}}{e^{.1} - 1} > \left(\dfrac{1}{2}\right)\left(\dfrac{1}{e^{.1} - 1}\right)$

$\Rightarrow 1 - e^{-n/10} > \dfrac{1}{2} \Rightarrow e^{-n/10} < \dfrac{1}{2} \Rightarrow -\dfrac{n}{10} < \ln\left(\dfrac{1}{2}\right) \Rightarrow \dfrac{n}{10} > -\ln\left(\dfrac{1}{2}\right) \Rightarrow n > 6.93 \Rightarrow n = 7$

39. The convergence of $\displaystyle\sum_{n=1}^{\infty} |a_n|$ implies that $\displaystyle\lim_{n\to\infty} |a_n| = 0$. Let $N > 0$ be such that $|a_n| < \frac{1}{2} \Rightarrow 1 - |a_n| > \frac{1}{2}$

$\Rightarrow \dfrac{|a_n|}{1 - |a_n|} < 2|a_n|$ for all $n > N$. Now $\left|\ln(1 + a_n)\right| = \left|a_n - \dfrac{a_n^2}{2} + \dfrac{a_n^3}{3} - \dfrac{a_n^4}{4} + \ldots\right| \le |a_n| + \left|\dfrac{a_n^2}{2}\right| + \left|\dfrac{a_n^3}{3}\right| + \left|\dfrac{a_n^4}{4}\right| + \ldots$

$< |a_n| + |a_n|^2 + |a_n|^3 + |a_n|^4 + \ldots = \dfrac{|a_n|}{1 - |a_n|} < 2|a_n|$. Therefore $\displaystyle\sum_{n=1}^{\infty} \ln(1 + a_n)$ converges by the Direct

Comparison Test since $\displaystyle\sum_{n=1}^{\infty} |a_n|$ converges.

41. (a) $s_{2n+1} = \dfrac{c_1}{1} + \dfrac{c_2}{2} + \dfrac{c_3}{3} + \ldots + \dfrac{c_{2n+1}}{2n+1} = \dfrac{t_1}{1} + \dfrac{t_2 - t_1}{2} + \dfrac{t_3 - t_2}{3} + \ldots + \dfrac{t_{2n+1} - t_{2n}}{2n+1}$

$= t_1\left(1 - \dfrac{1}{2}\right) + t_2\left(\dfrac{1}{2} - \dfrac{1}{3}\right) + \ldots + t_{2n}\left(\dfrac{1}{2n} - \dfrac{1}{2n+1}\right) + \dfrac{t_{2n+1}}{2n+1} = \displaystyle\sum_{k=1}^{2n} \dfrac{t_k}{k(k+1)} + \dfrac{t_{2n+1}}{2n+1}.$

(b) $\{c_n\} = \{(-1)^n\} \Rightarrow \displaystyle\sum_{n=1}^{\infty} \dfrac{(-1)^n}{n}$ converges

(c) $\{c_n\} = \{1, -1, -1, 1, 1, -1, -1, 1, 1, \ldots\} \Rightarrow$ the series $1 - \dfrac{1}{2} - \dfrac{1}{3} + \dfrac{1}{4} + \dfrac{1}{5} - \dfrac{1}{6} - \dfrac{1}{7} + \ldots$ converges

NOTES:

CHAPTER 9 CONIC SECTIONS, PARAMETRIZED CURVES, AND POLAR COORDINATES

9.1 CONIC SECTIONS AND QUADRATIC EQUATIONS

1. $x = \dfrac{y^2}{8} \Rightarrow 4p = 8 \Rightarrow p = 2$; focus is $(2, 0)$, directrix is $x = -2$

3. $y = -\dfrac{x^2}{6} \Rightarrow 4p = 6 \Rightarrow p = \dfrac{3}{2}$; focus is $\left(0, -\dfrac{3}{2}\right)$, directrix is $y = \dfrac{3}{2}$

5. $\dfrac{x^2}{4} - \dfrac{y^2}{9} = 1 \Rightarrow c = \sqrt{4 + 9} = \sqrt{13} \Rightarrow$ foci are $\left(\pm\sqrt{13}, 0\right)$; vertices are $\left(\pm 2, 0\right)$; asymptotes are $y = \pm\dfrac{3}{2}x$

7. $\dfrac{x^2}{2} + y^2 = 1 \Rightarrow c = \sqrt{2 - 1} = 1 \Rightarrow$ foci are $\left(\pm 1, 0\right)$; vertices are $\left(\pm\sqrt{2}, 0\right)$

9. $y^2 = 12x \Rightarrow x = \dfrac{y^2}{12} \Rightarrow 4p = 12 \Rightarrow p = 3$;

 focus is $(3, 0)$, directrix is $x = -3$

11. $x^2 = -8y \Rightarrow y = \dfrac{x^2}{-8} \Rightarrow 4p = 8 \Rightarrow p = 2$;

 focus is $(0, -2)$, directrix is $y = 2$

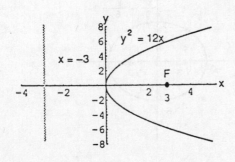

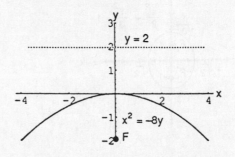

13. $y = 4x^2 \Rightarrow y = \dfrac{x^2}{\left(\frac{1}{4}\right)} \Rightarrow 4p = \dfrac{1}{4} \Rightarrow p = \dfrac{1}{16}$;

 focus is $\left(0, \dfrac{1}{16}\right)$, directrix is $y = -\dfrac{1}{16}$

15. $x = -3y^2 \Rightarrow x = -\dfrac{y^2}{\left(\frac{1}{3}\right)} \Rightarrow 4p = \dfrac{1}{3} \Rightarrow p = \dfrac{1}{12}$;

 focus is $\left(-\dfrac{1}{12}, 0\right)$, directrix is $x = \dfrac{1}{12}$

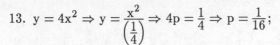

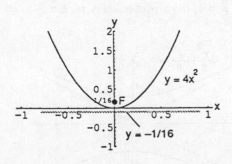

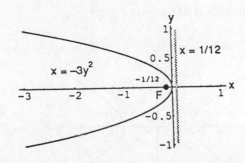

17. $16x^2 + 25y^2 = 400 \Rightarrow \frac{x^2}{25} + \frac{y^2}{16} = 1$

$\Rightarrow c = \sqrt{a^2 - b^2} = \sqrt{25 - 16} = 3$

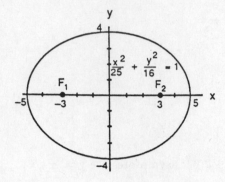

19. $2x^2 + y^2 = 2 \Rightarrow x^2 + \frac{y^2}{2} = 1$

$\Rightarrow c = \sqrt{a^2 - b^2} = \sqrt{2 - 1} = 1$

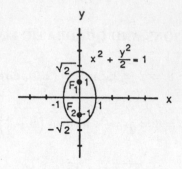

21. $3x^2 + 2y^2 = 6 \Rightarrow \frac{x^2}{2} + \frac{y^2}{3} = 1$

$\Rightarrow c = \sqrt{a^2 - b^2} = \sqrt{3 - 2} = 1$

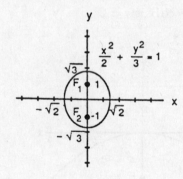

23. $6x^2 + 9y^2 = 54 \Rightarrow \frac{x^2}{9} + \frac{y^2}{6} = 1$

$\Rightarrow c = \sqrt{a^2 - b^2} = \sqrt{9 - 6} = \sqrt{3}$

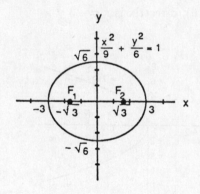

25. Foci: $\left(\pm\sqrt{2}, 0\right)$, Vertices: $\left(\pm 2, 0\right) \Rightarrow a = 2, c = \sqrt{2} \Rightarrow b^2 = a^2 - c^2 = 4 - \left(\sqrt{2}\right)^2 = 2 \Rightarrow \frac{x^2}{4} + \frac{y^2}{2} = 1$

27. $x^2 - y^2 = 1 \Rightarrow c = \sqrt{a^2 + b^2} = \sqrt{1 + 1} = \sqrt{2}$;

asymptotes are $y = \pm x$

29. $y^2 - x^2 = 8 \Rightarrow \frac{y^2}{8} - \frac{x^2}{8} = 1 \Rightarrow c = \sqrt{a^2 + b^2}$

$= \sqrt{8 + 8} = 4$; asymptotes are $y = \pm x$

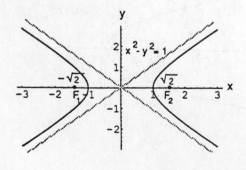

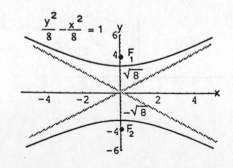

31. $8x^2 - 2y^2 = 16 \Rightarrow \dfrac{x^2}{2} - \dfrac{y^2}{8} = 1 \Rightarrow c = \sqrt{a^2 + b^2}$

    $= \sqrt{2 + 8} = \sqrt{10}$; asymptotes are $y = \pm 2x$

33. $8y^2 - 2x^2 = 16 \Rightarrow \dfrac{y^2}{2} - \dfrac{x^2}{8} = 1 \Rightarrow c = \sqrt{a^2 + b^2}$

    $= \sqrt{2 + 8} = \sqrt{10}$; asymptotes are $y = \pm \dfrac{x}{2}$

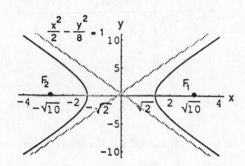

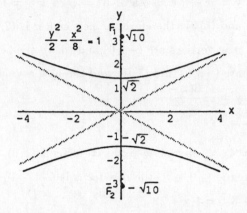

35. Foci: $\left(0, \pm\sqrt{2}\right)$, Asymptotes: $y = \pm x \Rightarrow c = \sqrt{2}$ and $\dfrac{b}{a} = 1 \Rightarrow a = b \Rightarrow c^2 = a^2 + b^2 = 2a^2 \Rightarrow 2 = 2a^2$

    $\Rightarrow a = 1 \Rightarrow b = 1 \Rightarrow y^2 - x^2 = 1$

37. Vertices: $(\pm 3, 0)$, Asymptotes: $y = \pm\dfrac{4}{3}x \Rightarrow a = 3$ and $\dfrac{b}{a} = \dfrac{4}{3} \Rightarrow b = \dfrac{4}{3}(3) = 4 \Rightarrow \dfrac{x^2}{9} - \dfrac{y^2}{16} = 1$

39. (a) $y^2 = 8x \Rightarrow 4p = 8 \Rightarrow p = 2 \Rightarrow$ directrix is $x = -2$,

    focus is $(2, 0)$, and vertex is $(0, 0)$; therefore the new

    directrix is $x = -1$, the new focus is $(3, -2)$, and the

    new vertex is $(1, -2)$

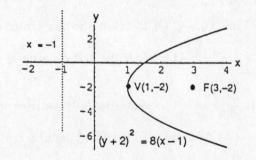

41. (a) $\dfrac{x^2}{16} + \dfrac{y^2}{9} = 1 \Rightarrow$ center is $(0, 0)$, vertices are $(-4, 0)$

    and $(4, 0)$; $c = \sqrt{a^2 - b^2} = \sqrt{7} \Rightarrow$ foci are $\left(\sqrt{7}, 0\right)$

    and $\left(-\sqrt{7}, 0\right)$; therefore the new center is $(4, 3)$, the

    new vertices are $(0, 3)$ and $(8, 3)$, and the new foci are

    $\left(4 \pm \sqrt{7}, 3\right)$

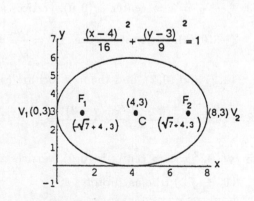

43. (a) $\dfrac{x^2}{16} - \dfrac{y^2}{9} = 1 \Rightarrow$ center is $(0,0)$, vertices are $(-4,0)$

and $(4,0)$, and the asymptotes are $\dfrac{x}{4} = \pm\dfrac{y}{3}$ or

$y = \pm\dfrac{3x}{4}$; $c = \sqrt{a^2 + b^2} = \sqrt{25} = 5 \Rightarrow$ foci are $(-5,0)$

and $(5,0)$; therefore the new center is $(2,0)$, the

new vertices are $(-2,0)$ and $(6,0)$, the new foci

are $(-3,0)$ and $(7,0)$, and the new asymptotes are

$y = \pm\dfrac{3(x-2)}{4}$

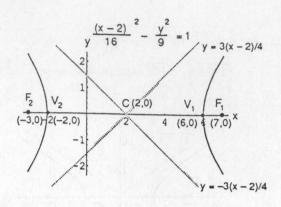

45. $y^2 = 4x \Rightarrow 4p = 4 \Rightarrow p = 1 \Rightarrow$ focus is $(1,0)$, directrix is $x = -1$, and vertex is $(0,0)$; therefore the new

vertex is $(-2,-3)$, the new focus is $(-1,-3)$, and the new directrix is $x = -3$; the new equation is

$(y+3)^2 = 4(x+2)$

47. $x^2 = 8y \Rightarrow 4p = 8 \Rightarrow p = 2 \Rightarrow$ focus is $(0,2)$, directrix is $y = -2$, and vertex is $(0,0)$; therefore the new

vertex is $(1,-7)$, the new focus is $(1,-5)$, and the new directrix is $y = -9$; the new equation is

$(x-1)^2 = 8(y+7)$

49. $\dfrac{x^2}{6} + \dfrac{y^2}{9} = 1 \Rightarrow$ center is $(0,0)$, vertices are $(0,3)$ and $(0,-3)$; $c = \sqrt{a^2 - b^2} = \sqrt{9-6} = \sqrt{3} \Rightarrow$ foci are $\left(0, \sqrt{3}\right)$

and $\left(0, -\sqrt{3}\right)$; therefore the new center is $(-2,-1)$, the new vertices are $(-2,2)$ and $(-2,-4)$, and the new foci

are $\left(-2, -1 \pm \sqrt{3}\right)$; the new equation is $\dfrac{(x+2)^2}{6} + \dfrac{(y+1)^2}{9} = 1$

51. $\dfrac{x^2}{3} + \dfrac{y^2}{2} = 1 \Rightarrow$ center is $(0,0)$, vertices are $\left(\sqrt{3},0\right)$ and $\left(-\sqrt{3},0\right)$; $c = \sqrt{a^2 - b^2} = \sqrt{3-2} = 1 \Rightarrow$ foci are

$(-1,0)$ and $(1,0)$; therefore the new center is $(2,3)$, the new vertices are $\left(2 \pm \sqrt{3}, 3\right)$, and the new foci

are $(1,3)$ and $(3,3)$; the new equation is $\dfrac{(x-2)^2}{3} + \dfrac{(y-3)^2}{2} = 1$

53. $\dfrac{x^2}{4} - \dfrac{y^2}{5} = 1 \Rightarrow$ center is $(0,0)$, vertices are $(2,0)$ and $(-2,0)$; $c = \sqrt{a^2 + b^2} = \sqrt{4+5} = 3 \Rightarrow$ foci are $(3,0)$ and

$(-3,0)$; the asymptotes are $\pm\dfrac{x}{2} = \dfrac{y}{\sqrt{5}} \Rightarrow y = \pm\dfrac{\sqrt{5}x}{2}$; therefore the new center is $(2,2)$, the new vertices are

$(4,2)$ and $(0,2)$, and the new foci are $(5,2)$ and $(-1,2)$; the new asymptotes are $y - 2 = \pm\dfrac{\sqrt{5}(x-2)}{2}$; the new

equation is $\dfrac{(x-2)^2}{4} - \dfrac{(y-2)^2}{5} = 1$

55. $y^2 - x^2 = 1 \Rightarrow$ center is $(0,0)$, vertices are $(0,1)$ and $(0,-1)$; $c = \sqrt{a^2 + b^2} = \sqrt{1+1} = \sqrt{2} \Rightarrow$ foci are

$\left(0, \pm\sqrt{2}\right)$; the asymptotes are $y = \pm x$; therefore the new center is $(-1,-1)$, the new vertices are $(-1,0)$ and

$(-1,-2)$, and the new foci are $\left(-1, -1 \pm \sqrt{2}\right)$; the new asymptotes are $y + 1 = \pm(x+1)$; the new equation is

$(y+1)^2 - (x+1)^2 = 1$

57. $x^2 + 4x + y^2 = 12 \Rightarrow x^2 + 4x + 4 + y^2 = 12 + 4 \Rightarrow (x+2)^2 + y^2 = 16$; this is a circle: center at $C(-2, 0)$, $a = 4$

59. $x^2 + 2x + 4y - 3 = 0 \Rightarrow x^2 + 2x + 1 = -4y + 3 + 1 \Rightarrow (x+1)^2 = -4(y-1)$; this is a parabola: $V(-1, 1)$, $F(-1, 0)$

61. $x^2 + 5y^2 + 4x = 1 \Rightarrow x^2 + 4x + 4 + 5y^2 = 5 \Rightarrow (x+2)^2 + 5y^2 = 5 \Rightarrow \dfrac{(x+2)^2}{5} + y^2 = 1$; this is an ellipse: the

center is $(-2, 0)$, the vertices are $\left(-2 \pm \sqrt{5}, 0\right)$; $c = \sqrt{a^2 - b^2} = \sqrt{5 - 1} = 2 \Rightarrow$ the foci are $(-4, 0)$ and $(0, 0)$

63. $x^2 + 2y^2 - 2x - 4y = -1 \Rightarrow x^2 - 2x + 1 + 2\left(y^2 - 2y + 1\right) = 2 \Rightarrow (x-1)^2 + 2(y-1)^2 = 2$

$\Rightarrow \dfrac{(x-1)^2}{2} + (y-1)^2 = 1$; this is an ellipse: the center is $(1, 1)$, the vertices are $\left(1 \pm \sqrt{2}, 1\right)$;

$c = \sqrt{a^2 - b^2} = \sqrt{2 - 1} = 1 \Rightarrow$ the foci are $(2, 1)$ and $(0, 1)$

65. $x^2 - y^2 - 2x + 4y = 4 \Rightarrow x^2 - 2x + 1 - \left(y^2 - 4y + 4\right) = 1 \Rightarrow (x-1)^2 - (y-2)^2 = 1$; this is a hyperbola:

the center is $(1, 2)$, the vertices are $(2, 2)$ and $(0, 2)$; $c = \sqrt{a^2 + b^2} = \sqrt{1 + 1} = \sqrt{2} \Rightarrow$ the foci are $\left(1 \pm \sqrt{2}, 2\right)$;

the asymptotes are $y - 2 = \pm (x - 1)$

67. $2x^2 - y^2 + 6y = 3 \Rightarrow 2x^2 - \left(y^2 - 6y + 9\right) = -6 \Rightarrow \dfrac{(y-3)^2}{6} - \dfrac{x^2}{3} = 1$; this is a hyperbola: the center is $(0, 3)$,

the vertices are $\left(0, 3 \pm \sqrt{6}\right)$; $c = \sqrt{a^2 + b^2} = \sqrt{6 + 3} = 3 \Rightarrow$ the foci are $(0, 6)$ and $(0, 0)$; the asymptotes are

$\dfrac{y-3}{\sqrt{6}} = \pm \dfrac{x}{\sqrt{3}} \Rightarrow y = \pm \sqrt{2}x + 3$

69.

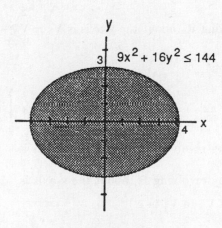

71.

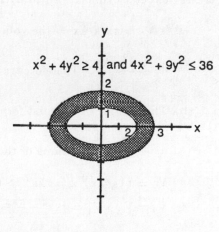

73.

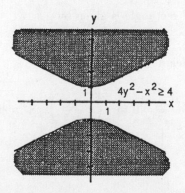

75. Volume of the Parabolic Solid: $V_1 = \displaystyle\int_0^{b/2} 2\pi x\left(h - \frac{4h}{b^2}x^2\right)dx = 2\pi h \int_0^{b/2}\left(x - \frac{4x^3}{b^2}\right)dx = 2\pi h\left[\frac{x^2}{2} - \frac{x^4}{b^2}\right]_0^{b/2}$

$= \frac{\pi h b^2}{8}$; Volume of the Cone: $V_2 = \frac{1}{3}\pi\left(\frac{b}{2}\right)^2 h = \frac{1}{3}\pi\left(\frac{b^2}{4}\right)h = \frac{\pi h b^2}{12}$; therefore $V_1 = \frac{3}{2}V_2$

77. A general equation of the circle is $x^2 + y^2 + ax + by + c = 0$, so we will substitute the three given points into

this equation and solve the resulting system: $\left.\begin{array}{r} a \quad\;\;\; + c = -1 \\ b + c = -1 \\ 2a + 2b + c = -8 \end{array}\right\} \Rightarrow c = \frac{4}{3}$ and $a = b = -\frac{7}{3}$; therefore

$3x^2 + 3y^2 - 7x - 7y + 4 = 0$ represents the circle

79. $r^2 = (-2-1)^2 + (1-3)^2 = 13 \Rightarrow (x+2)^2 + (y-1)^2 = 13$ is an equation of the circle; the distance from the

center to $(1.1, 2.8)$ is $\sqrt{(-2-1.1)^2 + (1-2.8)^2} = \sqrt{12.85} < \sqrt{13}$, the radius $\Rightarrow$ the point is inside the circle

81. (a) $y^2 = kx \Rightarrow x = \frac{y^2}{k}$; the volume of the solid formed by

revolving R_1 about the y-axis is $V_1 = \displaystyle\int_0^{\sqrt{kx}} \pi\left(\frac{y^2}{k}\right)^2 dy$

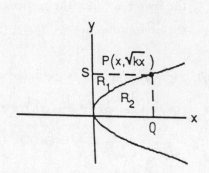

$= \frac{\pi}{k^2}\displaystyle\int_0^{\sqrt{kx}} y^4\, dy = \frac{\pi x^2\sqrt{kx}}{5}$; the volume of the right

circular cylinder formed by revolving PQ about the y-axis

is $V_2 = \pi x^2\sqrt{kx} \Rightarrow$ the volume of the solid formed by revolving R_2 about the y-axis is $V_3 = V_2 - V_1$

$= \frac{4\pi x^2\sqrt{kx}}{5}$. Therefore we can see the ratio of V_3 to V_1 is 4:1.

(b) The volume of the solid formed by revolving R_2 about the x-axis is $V_1 = \displaystyle\int_0^x \pi\left(\sqrt{kt}\right)^2 dt = \pi k\int_0^x t\,dt$

$= \frac{\pi k x^2}{2}$. The volume of the right circular cylinder formed by revolving PS about the x-axis is

$V_2 = \pi\left(\sqrt{kx}\right)^2 x = \pi k x^2 \Rightarrow$ the volume of the solid formed by revolving R_1 about the x-axis is
$V_3 = V_2 - V_1 = \pi k x^2 - \frac{\pi k x^2}{2} = \frac{\pi k x^2}{2}$. Therefore the ratio of V_3 to V_1 is 1:1.

83. Let $y = \sqrt{1 - \frac{x^2}{4}}$ on the interval $0 \le x \le 2$. The area of the inscribed rectangle is given by

$A(x) = 2x\left(2\sqrt{1 - \frac{x^2}{4}}\right) = 4x\sqrt{1 - \frac{x^2}{4}}$ (since the length is 2x and the height is 2y)

$\Rightarrow A'(x) = 4\sqrt{1 - \frac{x^2}{4}} - \frac{x^2}{\sqrt{1 - \frac{x^2}{4}}}$. Thus $A'(x) = 0 \Rightarrow 4\sqrt{1 - \frac{x^2}{4}} - \frac{x^2}{\sqrt{1 - \frac{x^2}{4}}} = 0 \Rightarrow 4\left(1 - \frac{x^2}{4}\right) - x^2 = 0 \Rightarrow x^2 = 2$

$\Rightarrow x = \sqrt{2}$ (only the positive square root lies in the interval). Since $A(0) = A(2) = 0$ we have that $A\left(\sqrt{2}\right) = 4$ is the maximum area when the length is $2\sqrt{2}$ and the height is $\sqrt{2}$.

85. $9x^2 - 4y^2 = 36 \Rightarrow y^2 = \dfrac{9x^2 - 36}{4} \Rightarrow y = \pm\dfrac{3}{2}\sqrt{x^2 - 4}$ on the interval $2 \le x \le 4 \Rightarrow V = \displaystyle\int_2^4 \pi\left(\dfrac{3}{2}\sqrt{x^2 - 4}\right)^2 dx$

$= \dfrac{9\pi}{4}\displaystyle\int_2^4 \left(x^2 - 4\right) dx = \dfrac{9\pi}{4}\left[\dfrac{x^3}{3} - 4x\right]_2^4 = \dfrac{9\pi}{4}\left[\left(\dfrac{64}{3} - 16\right) - \left(\dfrac{8}{3} - 8\right)\right] = \dfrac{9\pi}{4}\left(\dfrac{56}{3} - 8\right) = \dfrac{3\pi}{4}(56 - 24) = 24\pi$

87. Let $y = \sqrt{16 - \dfrac{16}{9}x^2}$ on the interval $-3 \le x \le 3$. Since the plate is symmetric about the y-axis, $\bar{x} = 0$. For a

vertical strip: $\left(\widetilde{x}, \widetilde{y}\right) = \left(x, \dfrac{\sqrt{16 - \dfrac{16}{9}x^2}}{2}\right)$, length $= \sqrt{16 - \dfrac{16}{9}x^2}$, width $= dx \Rightarrow$ area $= dA = \sqrt{16 - \dfrac{16}{9}x^2}\, dx$

$\Rightarrow$ mass $= dm = \delta\, dA = \delta\sqrt{16 - \dfrac{16}{9}x^2}\, dx$. Moment of the strip about the x-axis:

$\widetilde{y}\, dm = \dfrac{\sqrt{16 - \dfrac{16}{9}x^2}}{2}\left(\delta\sqrt{16 - \dfrac{16}{9}x^2}\right) dx = \delta\left(8 - \dfrac{8}{9}x^2\right) dx$ so the moment of the plate about the x-axis is

$M_x = \displaystyle\int \widetilde{y}\, dm = \int_{-3}^3 \delta\left(8 - \dfrac{8}{9}x^2\right) dx = \delta\left[8x - \dfrac{8}{27}x^3\right]_{-3}^3 = 32\delta$; also the mass of the plate is

$M = \displaystyle\int_{-3}^3 \delta\sqrt{16 - \dfrac{16}{9}x^2}\, dx = \int_{-3}^3 4\delta\sqrt{1 - \left(\dfrac{1}{3}x\right)^2}\, dx = 4\delta\int_{-1}^1 3\sqrt{1 - u^2}\, du$ where $u = \dfrac{x}{3} \Rightarrow 3\, du = dx$; $x = -3$

$\Rightarrow u = -1$ and $x = 3 \Rightarrow u = 1$. Hence, $4\delta\displaystyle\int_{-1}^1 3\sqrt{1 - u^2}\, du = 12\delta\int_{-1}^1 \sqrt{1 - u^2}\, du$

$= 12\delta\left[\dfrac{1}{2}\left(u\sqrt{1 - u^2} + \sin^{-1} u\right)\right]_{-1}^1 = 6\pi\delta \Rightarrow \bar{y} = \dfrac{M_x}{M} = \dfrac{32\delta}{6\pi\delta} = \dfrac{16}{3\pi}$. Therefore the center of mass is $\left(0, \dfrac{16}{3\pi}\right)$.

89. $\dfrac{dr_A}{dt} = \dfrac{dr_B}{dt} \Rightarrow \dfrac{d}{dt}\left(r_A - r_B\right) = 0 \Rightarrow r_A - r_B = C$, a constant $\Rightarrow$ the points $P(t)$ lie on a hyperbola with foci at A and B

91. PF will always equal PB because the string has constant length $AB = FP + PA = AP + PB$.

93. $x^2 = 4py$ and $y = p \Rightarrow x^2 = 4p^2 \Rightarrow x = \pm 2p$. Therefore the line $y = p$ cuts the parabola at points $(-2p, p)$ and

$(2p, p)$, and these points are $\sqrt{[2p - (-2p)]^2 + (p - p)^2} = 4p$ units apart.

9.2 CLASSIFYING CONIC SECTIONS BY ECCENTRICITY

1. $16x^2 + 25y^2 = 400 \Rightarrow \frac{x^2}{25} + \frac{y^2}{16} = 1 \Rightarrow c = \sqrt{a^2 - b^2}$

   $= \sqrt{25 - 16} = 3 \Rightarrow e = \frac{c}{a} = \frac{3}{5}; \ F(\pm 3, 0);$

   directrices are $x = 0 \pm \frac{a}{e} = \pm \dfrac{5}{\left(\frac{3}{5}\right)} = \pm \frac{25}{3}$

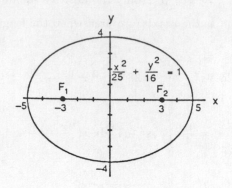

3. $2x^2 + y^2 = 2 \Rightarrow x^2 + \frac{y^2}{2} = 1 \Rightarrow c = \sqrt{a^2 - b^2}$

   $= \sqrt{2 - 1} = 1 \Rightarrow e = \frac{c}{a} = \frac{1}{\sqrt{2}}; \ F(0, \pm 1);$

   directrices are $y = 0 \pm \frac{a}{e} = \pm \dfrac{\sqrt{2}}{\left(\frac{1}{\sqrt{2}}\right)} = \pm 2$

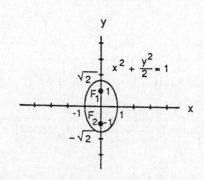

5. $3x^2 + 2y^2 = 6 \Rightarrow \frac{x^2}{2} + \frac{y^2}{3} = 1 \Rightarrow c = \sqrt{a^2 - b^2}$

   $= \sqrt{3 - 2} = 1 \Rightarrow e = \frac{c}{a} = \frac{1}{\sqrt{3}}; \ F(0, \pm 1);$

   directrices are $y = 0 \pm \frac{a}{e} = \pm \dfrac{\sqrt{3}}{\left(\frac{1}{\sqrt{3}}\right)} = \pm 3$

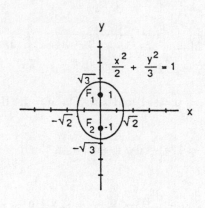

7. $6x^2 + 9y^2 = 54 \Rightarrow \frac{x^2}{9} + \frac{y^2}{6} = 1 \Rightarrow c = \sqrt{a^2 - b^2}$

   $= \sqrt{9 - 6} = \sqrt{3} \Rightarrow e = \frac{c}{a} = \frac{\sqrt{3}}{3}; \ F(\pm \sqrt{3}, 0);$

   directrices are $x = 0 \pm \frac{a}{e} = \pm \dfrac{3}{\left(\frac{\sqrt{3}}{3}\right)} = \pm 3\sqrt{3}$

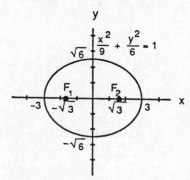

9. Foci: $(0, \pm 3)$, $e = 0.5 \Rightarrow c = 3$ and $a = \frac{c}{e} = \frac{3}{0.5} = 6 \Rightarrow b^2 = 36 - 9 = 27 \Rightarrow \frac{x^2}{27} + \frac{y^2}{36} = 1$

11. Vertices: $(0, \pm 70)$, $e = 0.1 \Rightarrow a = 70$ and $c = ae = 70(0.1) = 7 \Rightarrow b^2 = 4900 - 49 = 4851 \Rightarrow \frac{x^2}{4851} + \frac{y^2}{4900} = 1$

13. Focus: $\left(\sqrt{5},0\right)$, Directrix: $x = \dfrac{9}{\sqrt{5}} \Rightarrow c = ae = \sqrt{5}$ and $\dfrac{a}{e} = \dfrac{9}{\sqrt{5}} \Rightarrow \dfrac{ae}{e^2} = \dfrac{9}{\sqrt{5}} \Rightarrow \dfrac{\sqrt{5}}{e^2} = \dfrac{9}{\sqrt{5}} \Rightarrow e^2 = \dfrac{5}{9}$

$\Rightarrow e = \dfrac{\sqrt{5}}{3}$. Then $PF = \dfrac{\sqrt{5}}{3} PD \Rightarrow \sqrt{\left(x - \sqrt{5}\right)^2 + (y - 0)^2} = \dfrac{\sqrt{5}}{3}\left| x - \dfrac{9}{\sqrt{5}} \right| \Rightarrow \left(x - \sqrt{5}\right)^2 + y^2 = \dfrac{5}{9}\left(x - \dfrac{9}{\sqrt{5}}\right)^2$

$\Rightarrow x^2 - 2\sqrt{5}\,x + 5 + y^2 = \dfrac{5}{9}\left(x^2 - \dfrac{18}{\sqrt{5}}x + \dfrac{81}{5}\right) \Rightarrow \dfrac{4}{9}x^2 + y^2 = 4 \Rightarrow \dfrac{x^2}{9} + \dfrac{y^2}{4} = 1$

15. Focus: $(-4, 0)$, Directrix: $x = -16 \Rightarrow c = ae = 4$ and $\dfrac{a}{e} = 16 \Rightarrow \dfrac{ae}{e^2} = 16 \Rightarrow \dfrac{4}{e^2} = 16 \Rightarrow e^2 = \dfrac{1}{4} \Rightarrow e = \dfrac{1}{2}$. Then

$PF = \dfrac{1}{2}PD \Rightarrow \sqrt{(x+4)^2 + (y-0)^2} = \dfrac{1}{2}|x + 16| \Rightarrow (x+4)^2 + y^2 = \dfrac{1}{4}(x + 16)^2 \Rightarrow x^2 + 8x + 16 + y^2$

$= \dfrac{1}{4}\left(x^2 + 32x + 256\right) \Rightarrow \dfrac{3}{4}x^2 + y^2 = 48 \Rightarrow \dfrac{x^2}{64} + \dfrac{y^2}{48} = 1$

17. $e = \dfrac{4}{5} \Rightarrow$ take $c = 4$ and $a = 5$; $c^2 = a^2 - b^2$

$\Rightarrow 16 = 25 - b^2 \Rightarrow b^2 = 9 \Rightarrow b = 3$; therefore

$\dfrac{x^2}{25} + \dfrac{y^2}{9} = 1$

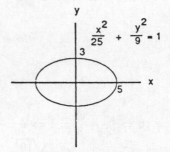

19. One axis is from $A(1,1)$ to $B(1,7)$ and is 6 units long; the
other axis is from $C(3,4)$ to $D(-1,4)$ and is 4 units long.
Therefore $a = 3$, $b = 2$ and the major axis is vertical. The
center is the point $C(1,4)$ and the ellipse is given by

$\dfrac{(x-1)^2}{4} + \dfrac{(y-4)^2}{9} = 1$; $c^2 = a^2 - b^2 = 3^2 - 2^2 = 5$

$\Rightarrow c = \sqrt{5}$; therefore the foci are $F\left(1, 4 \pm \sqrt{5}\right)$, the

eccentricity is $e = \dfrac{c}{a} = \dfrac{\sqrt{5}}{3}$, and the directrices are

$y = 4 \pm \dfrac{a}{e} = 4 \pm \dfrac{3}{\left(\dfrac{\sqrt{5}}{3}\right)} = 4 \pm \dfrac{9\sqrt{5}}{5}$.

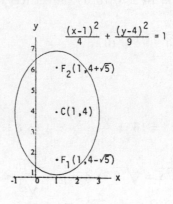

21. The ellipse must pass through $(0,0) \Rightarrow c = 0$; the point $(-1, 2)$ lies on the ellipse $\Rightarrow -a + 2b = -8$. The ellipse
is tangent to the x-axis $\Rightarrow$ its center is on the y-axis, so $a = 0$ and $b = -4 \Rightarrow$ the equation is $4x^2 + y^2 - 4y = 0$.

Next, $4x^2 + y^2 - 4y + 16 = 16 \Rightarrow 4x^2 + (y-4)^2 = 16 \Rightarrow \dfrac{x^2}{4} + \dfrac{(y-4)^2}{16} = 1 \Rightarrow a = 4$ and $b = 2$ (now using the

standard symbols) $\Rightarrow c^2 = a^2 - b^2 = 16 - 4 = 12 \Rightarrow c = \sqrt{12} \Rightarrow e = \dfrac{c}{a} = \dfrac{\sqrt{12}}{4} = \dfrac{\sqrt{3}}{2}$.

23. $x^2 - y^2 = 1 \Rightarrow c = \sqrt{a^2 + b^2} = \sqrt{1+1} = \sqrt{2} \Rightarrow e = \frac{c}{a}$

$= \frac{\sqrt{2}}{1} = \sqrt{2}$; asymptotes are $y = \pm x$; $F\left(\pm \sqrt{2}, 0 \right)$;

directrices are $x = 0 \pm \frac{a}{e} = \pm \frac{1}{\sqrt{2}}$

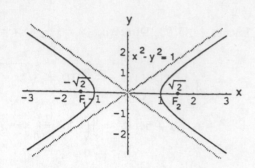

25. $y^2 - x^2 = 8 \Rightarrow \frac{y^2}{8} - \frac{x^2}{8} = 1 \Rightarrow c = \sqrt{a^2 + b^2}$

$= \sqrt{8+8} = 4 \Rightarrow e = \frac{c}{a} = \frac{4}{\sqrt{8}} = \sqrt{2}$; asymptotes are

$y = \pm x$; $F\left(0, \pm 4 \right)$; directrices are $y = 0 \pm \frac{a}{e}$

$= \pm \frac{\sqrt{8}}{\sqrt{2}} = \pm 2$

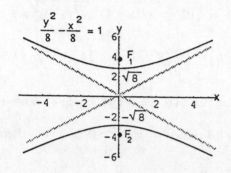

27. $8x^2 - 2y^2 = 16 \Rightarrow \frac{x^2}{2} - \frac{y^2}{8} = 1 \Rightarrow c = \sqrt{a^2 + b^2}$

$= \sqrt{2+8} = \sqrt{10} \Rightarrow e = \frac{c}{a} = \frac{\sqrt{10}}{\sqrt{2}} = \sqrt{5}$; asymptotes are

$y = \pm 2x$; $F\left(\pm \sqrt{10}, 0 \right)$; directrices are $x = 0 \pm \frac{a}{e}$

$= \pm \frac{\sqrt{2}}{\sqrt{5}} = \pm \frac{2}{\sqrt{10}}$

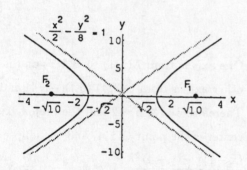

29. $8y^2 - 2x^2 = 16 \Rightarrow \frac{y^2}{2} - \frac{x^2}{8} = 1 \Rightarrow c = \sqrt{a^2 + b^2}$

$= \sqrt{2+8} = \sqrt{10} \Rightarrow e = \frac{c}{a} = \frac{\sqrt{10}}{\sqrt{2}} = \sqrt{5}$; asymptotes are

$y = \pm \frac{x}{2}$; $F\left(0, \pm \sqrt{10} \right)$; directrices are $y = 0 \pm \frac{a}{e}$

$= \pm \frac{\sqrt{2}}{\sqrt{5}} = \pm \frac{2}{\sqrt{10}}$

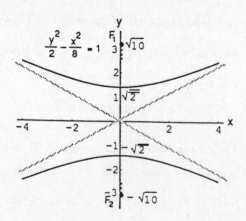

31. Vertices $\left(0, \pm 1 \right)$ and $e = 3 \Rightarrow a = 1$ and $e = \frac{c}{a} = 3 \Rightarrow c = 3a = 3 \Rightarrow b^2 = c^2 - a^2 = 9 - 1 = 8 \Rightarrow y^2 - \frac{x^2}{8} = 1$

33. Foci $\left(\pm 3, 0 \right)$ and $e = 3 \Rightarrow c = 3$ and $e = \frac{c}{a} = 3 \Rightarrow c = 3a \Rightarrow a = 1 \Rightarrow b^2 = c^2 - a^2 = 9 - 1 = 8 \Rightarrow x^2 - \frac{y^2}{8} = 1$

35. Focus $(4,0)$ and Directrix $x = 2 \Rightarrow c = ae = 4$ and $\frac{a}{e} = 2 \Rightarrow \frac{ae}{e^2} = 2 \Rightarrow \frac{4}{e^2} = 2 \Rightarrow e^2 = 2 \Rightarrow e = \sqrt{2}$. Then

$$PF = \sqrt{2}\,PD \Rightarrow \sqrt{(x-4)^2 + (y-0)^2} = \sqrt{2}\,|x-2| \Rightarrow (x-4)^2 + y^2 = 2(x-2)^2 \Rightarrow x^2 - 8x + 16 + y^2$$

$$= 2(x^2 - 4x + 4) \Rightarrow -x^2 + y^2 = -8 \Rightarrow \frac{x^2}{8} - \frac{y^2}{8} = 1$$

37. Focus $(-2,0)$ and Directrix $x = -\frac{1}{2} \Rightarrow c = ae = 2$ and $\frac{a}{e} = \frac{1}{2} \Rightarrow \frac{ae}{e^2} = \frac{1}{2} \Rightarrow \frac{2}{e^2} = \frac{1}{2} \Rightarrow e^2 = 4 \Rightarrow e = 2$. Then

$$PF = 2PD \Rightarrow \sqrt{(x+2)^2 + (y-0)^2} = 2\left|x + \frac{1}{2}\right| \Rightarrow (x+2)^2 + y^2 = 4\left(x + \frac{1}{2}\right)^2 \Rightarrow x^2 + 4x + 4 + y^2 = 4\left(x^2 + x + \frac{1}{4}\right)$$

$$\Rightarrow -3x^2 + y^2 = -3 \Rightarrow x^2 - \frac{y^2}{3} = 1$$

39. $\sqrt{(x-1)^2 + (y+3)^2} = \frac{3}{2}|y-2| \Rightarrow x^2 - 2x + 1 + y^2 + 6y + 9 = \frac{9}{4}(y^2 - 4y + 4) \Rightarrow 4x^2 - 5y^2 - 8x + 60y + 4 = 0$

$$\Rightarrow 4(x^2 - 2x + 1) - 5(y^2 - 12y + 36) = -4 + 4 - 180 \Rightarrow \frac{(y-6)^2}{36} - \frac{(x-1)^2}{45} = 1$$

41. To prove the reflective property for hyperbolas:

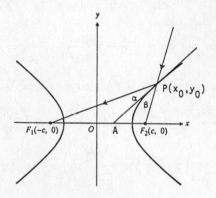

$\frac{x^2}{a^2} - \frac{y^2}{b^2} = 1 \Rightarrow a^2 y^2 = b^2 x^2 - a^2 b^2$ and $\frac{dy}{dx} = \frac{xb^2}{ya^2}$.

Let $P(x_0, y_0)$ be a point of tangency (see the accompanying figure). The slope from P to $F(-c, 0)$ is $\frac{y_0}{x_0 + c}$ and from

P to $F_2(c, 0)$ it is $\frac{y_0}{x_0 - c}$. Let the tangent through P meet the x-axis in point A, and define the angles $\angle F_1 PA = \alpha$ and $\angle F_2 PA = \beta$. We will show that $\tan \alpha = \tan \beta$. From the preliminary result in Exercise 22,

$$\tan \alpha = \frac{\left(\dfrac{x_0 b^2}{y_0 a^2} - \dfrac{y_0}{x_0 + c}\right)}{1 + \left(\dfrac{x_0 b^2}{y_0 a^2}\right)\left(\dfrac{y_0}{x_0 + c}\right)} = \frac{x_0^2 b^2 + x_0 b^2 c - y_0^2 a^2}{x_0 y_0 a^2 + y_0 a^2 c + x_0 y_0 b^2} = \frac{a^2 b^2 + x_0 b^2 c}{x_0 y_0 c^2 + y_0 a^2 c} = \frac{b^2}{y_0 c}.$$ In a similar manner,

$$\tan \beta = \frac{\left(\dfrac{y_0}{x_0 - c} - \dfrac{x_0 b^2}{y_0 a^2}\right)}{1 + \left(\dfrac{y_0}{x_0 - c}\right)\left(\dfrac{x_0 b^2}{y_0 a^2}\right)} = \frac{b^2}{y_0 c}.$$ Since $\tan \alpha = \tan \beta$, and α and β are acute angles, we have $\alpha = \beta$.

9.3 QUADRATIC EQUATIONS AND ROTATIONS

1. $x^2 - 3xy + y^2 - x = 0 \Rightarrow B^2 - 4AC = (-3)^2 - 4(1)(1) = 5 > 0 \Rightarrow$ Hyperbola

3. $3x^2 - 7xy + \sqrt{17}y^2 = 1 \Rightarrow B^2 - 4AC = (-7)^2 - 4(3)\sqrt{17} \approx -0.477 < 0 \Rightarrow$ Ellipse

5. $x^2 + 2xy + y^2 + 2x - y + 2 = 0 \Rightarrow B^2 - 4AC = 2^2 - 4(1)(1) = 0 \Rightarrow$ Parabola

7. $x^2 + 4xy + 4y^2 - 3x = 6 \Rightarrow B^2 - 4AC = 4^2 - 4(1)(4) = 0 \Rightarrow$ Parabola

9. $xy + y^2 - 3x = 5 \Rightarrow B^2 - 4AC = 1^2 - 4(0)(1) = 1 > 0 \Rightarrow$ Hyperbola

11. $3x^2 - 5xy + 2y^2 - 7x - 14y = -1 \Rightarrow B^2 - 4AC = (-5)^2 - 4(3)(2) = 1 > 0 \Rightarrow$ Hyperbola

13. $x^2 - 3xy + 3y^2 + 6y = 7 \Rightarrow B^2 - 4AC = (-3)^2 - 4(1)(3) = -3 < 0 \Rightarrow$ Ellipse

15. $6x^2 + 3xy + 2y^2 + 17y + 2 = 0 \Rightarrow B^2 - 4AC = 3^2 - 4(6)(2) = -39 < 0 \Rightarrow$ Ellipse

17. $\cot 2\alpha = \dfrac{A-C}{B} = \dfrac{0}{1} = 0 \Rightarrow 2\alpha = \dfrac{\pi}{2} \Rightarrow \alpha = \dfrac{\pi}{4}$; therefore $x = x' \cos\alpha - y' \sin\alpha$,

$y = x' \sin\alpha + y' \cos\alpha \Rightarrow x = x'\dfrac{\sqrt{2}}{2} - y'\dfrac{\sqrt{2}}{2}$, $y = x'\dfrac{\sqrt{2}}{2} + y'\dfrac{\sqrt{2}}{2}$

$\Rightarrow \left(\dfrac{\sqrt{2}}{2}x' - \dfrac{\sqrt{2}}{2}y'\right)\left(\dfrac{\sqrt{2}}{2}x' + \dfrac{\sqrt{2}}{2}y'\right) = 2 \Rightarrow \dfrac{1}{2}x'^2 - \dfrac{1}{2}y'^2 = 2 \Rightarrow x'^2 - y'^2 = 4 \Rightarrow$ Hyperbola

19. $\cot 2\alpha = \dfrac{A-C}{B} = \dfrac{3-1}{2\sqrt{3}} = \dfrac{1}{\sqrt{3}} \Rightarrow 2\alpha = \dfrac{\pi}{3} \Rightarrow \alpha = \dfrac{\pi}{6}$; therefore $x = x' \cos\alpha - y' \sin\alpha$,

$y = x' \sin\alpha + y' \cos\alpha \Rightarrow x = \dfrac{\sqrt{3}}{2}x' - \dfrac{1}{2}y'$, $y = \dfrac{1}{2}x' + \dfrac{\sqrt{3}}{2}y'$

$\Rightarrow 3\left(\dfrac{\sqrt{3}}{2}x' - \dfrac{1}{2}y'\right)^2 + 2\sqrt{3}\left(\dfrac{\sqrt{3}}{2}x' + \dfrac{1}{2}y'\right)\left(\dfrac{1}{2}x' + \dfrac{\sqrt{3}}{2}y'\right) + \left(\dfrac{1}{2}x' + \dfrac{\sqrt{3}}{2}y'\right)^2 - 8\left(\dfrac{\sqrt{3}}{2}x' - \dfrac{1}{2}y'\right)$

$+ 8\sqrt{3}\left(\dfrac{1}{2}x' + \dfrac{\sqrt{3}}{2}y'\right) = 0 \Rightarrow 4x'^2 + 16y' = 0 \Rightarrow$ Parabola

21. $\cot 2\alpha = \dfrac{A-C}{B} = \dfrac{1-1}{-2} = 0 \Rightarrow 2\alpha = \dfrac{\pi}{2} \Rightarrow \alpha = \dfrac{\pi}{4}$; therefore $x = x' \cos\alpha - y' \sin\alpha$,

$y = x' \sin\alpha + y' \cos\alpha \Rightarrow x = \dfrac{\sqrt{2}}{2}x' - \dfrac{\sqrt{2}}{2}y'$, $y = \dfrac{\sqrt{2}}{2}x' + \dfrac{\sqrt{2}}{2}y'$

$\Rightarrow \left(\dfrac{\sqrt{2}}{2}x' - \dfrac{\sqrt{2}}{2}y'\right)^2 - 2\left(\dfrac{\sqrt{2}}{2}x' - \dfrac{\sqrt{2}}{2}y'\right)\left(\dfrac{\sqrt{2}}{2}x' + \dfrac{\sqrt{2}}{2}y'\right) + \left(\dfrac{\sqrt{2}}{2}x' + \dfrac{\sqrt{2}}{2}y'\right)^2 = 2 \Rightarrow y'^2 = 1$

$\Rightarrow$ Parallel horizontal lines

23. $\cot 2\alpha = \dfrac{A-C}{B} = \dfrac{\sqrt{2}-\sqrt{2}}{2\sqrt{2}} = 0 \Rightarrow 2\alpha = \dfrac{\pi}{2} \Rightarrow \alpha = \dfrac{\pi}{4}$; therefore $x = x' \cos\alpha - y' \sin\alpha$,

$y = x' \sin\alpha + y' \cos\alpha \Rightarrow x = \dfrac{\sqrt{2}}{2}x' - \dfrac{\sqrt{2}}{2}y'$, $y = \dfrac{\sqrt{2}}{2}x' + \dfrac{\sqrt{2}}{2}y'$

$\Rightarrow \sqrt{2}\left(\dfrac{\sqrt{2}}{2}x' - \dfrac{\sqrt{2}}{2}y'\right)^2 + 2\sqrt{2}\left(\dfrac{\sqrt{2}}{2}x' - \dfrac{\sqrt{2}}{2}y'\right)\left(\dfrac{\sqrt{2}}{2}x' + \dfrac{\sqrt{2}}{2}y'\right) + \sqrt{2}\left(\dfrac{\sqrt{2}}{2}x' + \dfrac{\sqrt{2}}{2}y'\right)^2$

$- 8\left(\dfrac{\sqrt{2}}{2}x' - \dfrac{\sqrt{2}}{2}y'\right) + 8\left(\dfrac{\sqrt{2}}{2}x' + \dfrac{\sqrt{2}}{2}y'\right) = 0 \Rightarrow 2\sqrt{2}x'^2 + 8\sqrt{2}y' = 0 \Rightarrow$ Parabola

25. $\cot 2\alpha = \frac{A-C}{B} = \frac{3-3}{2} = 0 \Rightarrow 2\alpha = \frac{\pi}{2} \Rightarrow \alpha = \frac{\pi}{4}$; therefore $x = x' \cos \alpha - y' \sin \alpha$,

$y = x' \sin \alpha + y' \cos \alpha \Rightarrow x = \frac{\sqrt{2}}{2} x' - \frac{\sqrt{2}}{2} y', \ y = \frac{\sqrt{2}}{2} x' + \frac{\sqrt{2}}{2} y'$

$\Rightarrow 3 \left(\frac{\sqrt{2}}{2} x' - \frac{\sqrt{2}}{2} y' \right)^2 + 2 \left(\frac{\sqrt{2}}{2} x' - \frac{\sqrt{2}}{2} y' \right) \left(\frac{\sqrt{2}}{2} x' + \frac{\sqrt{2}}{2} y' \right) + 3 \left(\frac{\sqrt{2}}{2} x' + \frac{\sqrt{2}}{2} y' \right)^2 = 19 \Rightarrow 4x'^2 + 2y'^2 = 19$

$\Rightarrow$ Ellipse

27. $\cot 2\alpha = \frac{14-2}{16} = \frac{3}{4} \Rightarrow \cos 2\alpha = \frac{3}{5}$ (if we choose 2α in Quadrant I); thus $\sin \alpha = \sqrt{\frac{1 - \cos 2\alpha}{2}} = \sqrt{\frac{1 - \left(\frac{3}{5} \right)}{2}} = \frac{1}{\sqrt{5}}$

and $\cos \alpha = \sqrt{\frac{1 + \cos 2\alpha}{2}} = \sqrt{\frac{1 + \left(\frac{3}{5} \right)}{2}} = \frac{2}{\sqrt{5}}$ (or $\sin \alpha = -\frac{2}{\sqrt{5}}$ and $\cos \alpha = \frac{1}{\sqrt{5}}$)

29. $\tan 2\alpha = \frac{-1}{1-3} = \frac{1}{2} \Rightarrow 2\alpha \approx 26.57° \Rightarrow \alpha \approx 13.28° \Rightarrow \sin \alpha \approx 0.23, \cos \alpha \approx 0.97$; then $A' \approx 0.88, B' \approx 0.00$,

$C' \approx 3.10, D' \approx 0.74, E' \approx -1.20$, and $F' = -3 \Rightarrow 0.88x'^2 + 3.10y'^2 + 0.74x' - 1.20y' - 3 = 0$, an ellipse

31. $\tan 2\alpha = \frac{-4}{1-4} = \frac{4}{3} \Rightarrow 2\alpha \approx 53.13° \Rightarrow \alpha \approx 26.56° \Rightarrow \sin \alpha \approx 0.45, \cos \alpha \approx 0.89$; then $A' \approx 0.00, B' \approx 0.00$,

$C' \approx 5.00, D' \approx 0, E' \approx 0$, and $F' = -5 \Rightarrow 5.00y'^2 - 5 = 0$ or $y' = \pm 1.00$, parallel lines

33. $\tan 2\alpha = \frac{5}{3-2} = 5 \Rightarrow 2\alpha \approx 78.69° \Rightarrow \alpha \approx 39.34° \Rightarrow \sin \alpha \approx 0.63, \cos \alpha \approx 0.77$; then $A' \approx 5.05, B' \approx 0.00$,

$C' \approx -0.05, D' \approx -5.07, E' \approx -6.18$, and $F' = -1 \Rightarrow 5.05x'^2 - 0.05y'^2 - 5.07x' - 6.18y' - 1 = 0$, a hyperbola

35. $\alpha = 90° \Rightarrow x = x' \cos 90° - y' \sin 90° = -y'$ and $y = x' \sin 90° + y' \cos 90° = x'$

(a) $\frac{x'^2}{b^2} + \frac{y'^2}{a^2} = 1$ (b) $\frac{y'^2}{a^2} - \frac{x'^2}{b^2} = 1$ (c) $x'^2 + y'^2 = a^2$

(d) $y = mx \Rightarrow y - mx = 0 \Rightarrow D = -m$ and $E = 1; \ \alpha = 90° \Rightarrow D' = 1$ and $E' = m \Rightarrow my' + x' = 0 \Rightarrow y' = -\frac{1}{m} x'$

(e) $y = mx + b \Rightarrow y - mx - b = 0 \Rightarrow D = -m$ and $E = 1; \ \alpha = 90° \Rightarrow D' = 1, \ E' = m$ and $F' = -b$

$\Rightarrow my' + x' - b = 0 \Rightarrow y' = -\frac{1}{m} x' + \frac{b}{m}$

37. (a) $A' = \cos 45° \sin 45° = \left(\frac{\sqrt{2}}{2} \right) \left(\frac{\sqrt{2}}{2} \right) = \frac{1}{2}, \ B' = 0, \ C' = -\cos 45° \sin 45° = -\frac{1}{2}, \ F' = -1$

$\Rightarrow \frac{1}{2} x'^2 - \frac{1}{2} y'^2 = 1 \Rightarrow x'^2 - y'^2 = 2$

(b) $A' = \frac{1}{2}, \ C' = -\frac{1}{2}$ (see part (a) above), $D' = E' = B' = 0, \ F' = -a \Rightarrow \frac{1}{2} x'^2 - \frac{1}{2} y'^2 = a \Rightarrow x'^2 - y'^2 = 2a$

39. Yes, the graph is a hyperbola: with $AC < 0$ we have $-4AC > 0$ and $B^2 - 4AC > 0$.

41. Let α be any angle. Then $A' = \cos^2 \alpha + \sin^2 \alpha = 1, \ B' = 0, \ C' = \sin^2 \alpha + \cos^2 \alpha = 1, \ D' = E' = 0$ and $F' = -a^2$

$\Rightarrow x'^2 + y'^2 = a^2$.

43. (a) $B^2 - 4AC = 4^2 - 4(1)(4) = 0$, so the discriminant indicates this conic is a parabola

(b) The left-hand side of $x^2 + 4xy + 4y^2 + 6x + 12y + 9 = 0$ factors as a perfect square: $(x + 2y + 3)^2 = 0$

$\Rightarrow x + 2y + 3 = 0 \Rightarrow 2y = -x - 3$; thus the curve is a degenerate parabola (i.e., a straight line).

45. (a) $B^2 - 4AC = 1 - 4(0)(0) = 1 \Rightarrow$ hyperbola

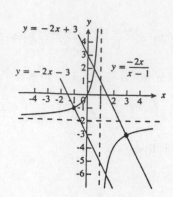

(b) $xy + 2x - y = 0 \Rightarrow y(x-1) = -2x \Rightarrow y = \dfrac{-2x}{x-1}$

(c) $y = \dfrac{-2x}{x-1} \Rightarrow \dfrac{dy}{dx} = \dfrac{2}{(x-1)^2}$ and we want $\dfrac{-1}{\left(\dfrac{dy}{dx}\right)} = -2,$

the slope of $y = -2x \Rightarrow -2 = -\dfrac{(x-1)^2}{2} \Rightarrow (x-1)^2 = 4$

$\Rightarrow x = 3$ or $x = -1;\ x = 3 \Rightarrow y = -3 \Rightarrow (3, -3)$ is a

point on the hyperbola where the line with slope $m = -2$

is normal $\Rightarrow$ the line is $y + 3 = -2(x-3)$ or

$y = -2x + 3;\ x = -1 \Rightarrow y = -1 \Rightarrow (-1, -1)$ is a point on the hyperbola where the line with slope $m = -2$ is

normal $\Rightarrow$ the line is $y + 1 = -2(x+1)$ or $y = -2x - 3$

47. Assume the ellipse has been rotated to eliminate the xy-term $\Rightarrow$ the new equation is $A'x'^2 + C'y'^2 = 1 \Rightarrow$ the

semi-axes are $\sqrt{\dfrac{1}{A'}}$ and $\sqrt{\dfrac{1}{C'}} \Rightarrow$ the area is $\pi\left(\sqrt{\dfrac{1}{A'}}\right)\left(\sqrt{\dfrac{1}{C'}}\right) = \dfrac{\pi}{\sqrt{A'C'}} = \dfrac{2\pi}{\sqrt{4A'C'}}$. Since $B^2 - 4AC$

$= B'^2 - 4A'C' = -4A'C'$ (because $B' = 0$) we find that the area is $\dfrac{2\pi}{\sqrt{4AC - B^2}}$ as claimed.

49. $B'^2 - 4A'C'$

$= \left(B\cos 2\alpha + (C-A)\sin 2\alpha\right)^2 - 4\left(A\cos^2\alpha + B\cos\alpha\sin\alpha + C\sin^2\alpha\right)\left(A\sin^2\alpha - B\cos\alpha\sin\alpha + C\cos^2\alpha\right)$

$= B^2\cos^2 2\alpha + 2B(C-A)\sin 2\alpha\cos 2\alpha + (C-A)^2\sin^2 2\alpha - 4A^2\cos^2\alpha\sin^2\alpha + 4AB\cos^3\alpha\sin\alpha$

$\quad - 4AC\cos^4\alpha - 4AB\cos\alpha\sin^3\alpha + 4B^2\cos^2\alpha\sin^2\alpha - 4BC\cos^3\alpha\sin\alpha - 4AC\sin^4\alpha + 4BC\cos\alpha\sin^3\alpha$

$\quad - 4C^2\cos^2\alpha\sin^2\alpha$

$= B^2\cos^2 2\alpha + 2BC\sin 2\alpha\cos 2\alpha - 2AB\sin 2\alpha\cos 2\alpha + C^2\sin^2 2\alpha - 2AC\sin^2 2\alpha + A^2\sin^2 2\alpha$

$\quad - 4A^2\cos^2\alpha\sin^2\alpha + 4AB\cos^3\alpha\sin\alpha - 4AC\cos^4\alpha - 4AB\cos\alpha\sin^3\alpha + B^2\sin^2 2\alpha - 4BC\cos^3\alpha\sin\alpha$

$\quad - 4AC\sin^4\alpha + 4BC\cos\alpha\sin^3\alpha - 4C^2\cos^2\alpha\sin^2\alpha$

$= B^2 + 2BC(2\sin\alpha\cos\alpha)\left(\cos^2\alpha - \sin^2\alpha\right) - 2AB(2\sin\alpha\cos\alpha)\left(\cos^2\alpha - \sin^2\alpha\right) + C^2\left(4\sin^2\alpha\cos^2\alpha\right)$

$\quad - 2AC\left(4\sin^2\alpha\cos^2\alpha\right) + A^2\left(4\sin^2\alpha\cos^2\alpha\right) - 4A^2\cos^2\alpha\sin^2\alpha + 4AB\cos^3\alpha\sin\alpha - 4AC\cos^4\alpha$

$\quad - 4AB\cos\alpha\sin^3\alpha - 4BC\cos^3\alpha\sin\alpha - 4AC\sin^4\alpha + 4BC\cos\alpha\sin^3\alpha - 4C^2\cos^2\alpha\sin^2\alpha$

$= B^2 - 8AC\sin^2\alpha\cos^2\alpha - 4AC\cos^4\alpha - 4AC\sin^4\alpha$

$= B^2 - 4AC\left(\cos^4\alpha + 2\sin^2\alpha\cos^2\alpha + \sin^4\alpha\right)$

$= B^2 - 4AC\left(\cos^2\alpha + \sin^2\alpha\right)^2$

$= B^2 - 4AC$

9.4 PARAMETRIZATIONS OF CURVES

1. $x = \cos t$, $y = \sin t$, $0 \le t \le \pi$

   $\Rightarrow \cos^2 t + \sin^2 t = 1 \Rightarrow x^2 + y^2 = 1$

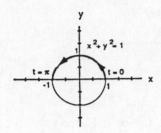

3. $x = \sin(2\pi(1-t))$, $y = \cos(2\pi(1-t))$, $0 \le t \le 1$

   $\Rightarrow \sin^2(2\pi(1-t)) + \cos^2(2\pi(1-t)) = 1$

   $\Rightarrow x^2 + y^2 = 1$

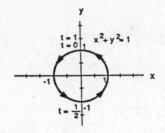

5. $x = 4\cos t$, $y = 2\sin t$, $0 \le t \le 2\pi$

   $\Rightarrow \dfrac{16\cos^2 t}{16} + \dfrac{4\sin^2 t}{4} = 1 \Rightarrow \dfrac{x^2}{16} + \dfrac{y^2}{4} = 1$

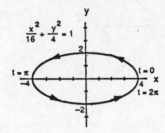

7. $x = 4\cos t$, $y = 5\sin t$, $0 \le t \le \pi$

   $\Rightarrow \dfrac{16\cos^2 t}{16} + \dfrac{25\sin^2 t}{25} = 1 \Rightarrow \dfrac{x^2}{16} + \dfrac{y^2}{25} = 1$

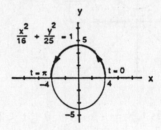

9. $x = 3t$, $y = 9t^2$, $-\infty < t < \infty \Rightarrow y = x^2$

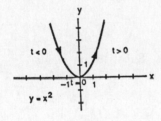

11. $x = t$, $y = \sqrt{t}$, $t \ge 0 \Rightarrow y = \sqrt{x}$

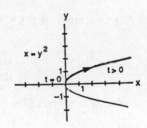

13. $x = -\sec t$, $y = \tan t$, $-\dfrac{\pi}{2} < t < \dfrac{\pi}{2}$

   $\Rightarrow \sec^2 t - \tan^2 t = 1 \Rightarrow x^2 - y^2 = 1$

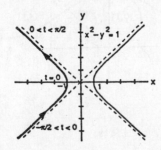

15. $x = 2t - 5$, $y = 4t - 7$, $-\infty < t < \infty$

   $\Rightarrow x + 5 = 2t \Rightarrow 2(x + 5) = 4t$

   $\Rightarrow y = 2(x + 5) - 7 \Rightarrow y = 2x + 3$

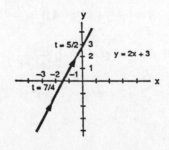

17. $x = t$, $y = 1 - t$, $0 \le t \le 1$

$\Rightarrow y = 1 - x$

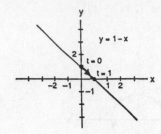

19. $x = t$, $y = \sqrt{1 - t^2}$, $-1 \le t \le 0$

$\Rightarrow y = \sqrt{1 - x^2}$

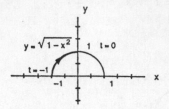

21. $x = t^2$, $y = \sqrt{t^4 + 1}$, $t \ge 0$

$\Rightarrow y = \sqrt{x^2 + 1}$, $x \ge 0$

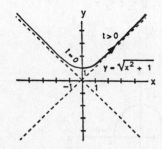

23. $x = -\cosh t$, $y = \sinh t$, $-\infty < 1 < \infty$

$\Rightarrow \cosh^2 t - \sinh^2 t = 1 \Rightarrow x^2 - y^2 = 1$

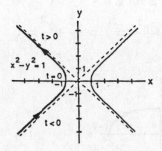

25. (a) $x = a \cos t$, $y = -a \sin t$, $0 \le t \le 2\pi$

   (b) $x = a \cos t$, $y = a \sin t$, $0 \le t \le 2\pi$

   (c) $x = a \cos t$, $y = -a \sin t$, $0 \le t \le 4\pi$

   (d) $x = a \cos t$, $y = a \sin t$, $0 \le t \le 4\pi$

27. $x^2 + y^2 = a^2 \Rightarrow 2x + 2y \dfrac{dy}{dx} = 0 \Rightarrow \dfrac{dy}{dx} = -\dfrac{x}{y}$; let $t = \dfrac{dy}{dx} \Rightarrow -\dfrac{x}{y} = t \Rightarrow x = -yt$. Substitution yields

$y^2 t^2 + y^2 = a^2 \Rightarrow y = \dfrac{a}{\sqrt{1 + t^2}}$ and $x = \dfrac{-at}{\sqrt{1 + t}}$, $-\infty < t < \infty$

29. Extend the vertical line through A to the x-axis and

let C be the point of intersection. Then $OC = AQ = x$

and $\tan t = \dfrac{2}{OC} = \dfrac{2}{x} \Rightarrow x = \dfrac{2}{\tan t} = 2 \cot t$; $\sin t = \dfrac{2}{OA}$

$\Rightarrow OA = \dfrac{2}{\sin t}$; and $(AB)(OA) = (AQ)^2 \Rightarrow AB\left(\dfrac{2}{\sin t}\right) = x^2$

$\Rightarrow AB\left(\dfrac{2}{\sin t}\right) = \left(\dfrac{2}{\tan t}\right)^2 \Rightarrow AB = \dfrac{2 \sin t}{\tan^2 t}$. Next

$y = 2 - AB \sin t \Rightarrow y = 2 - \left(\dfrac{2 \sin t}{\tan^2 t}\right) \sin t =$

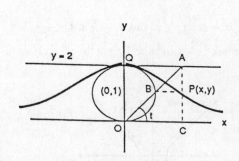

$2 - \dfrac{2 \sin^2 t}{\tan^2 t} = 2 - 2 \cos^2 t = 2 \sin^2 t$. Therefore let $x = 2 \cot t$ and $y = 2 \sin^2 t$, $0 < t < \pi$.

31. (a) $x = x_0 + (x_1 - x_0)t$ and $y = y_0 + (y_1 - y_0)t \Rightarrow t = \frac{x - x_0}{x_1 - x_0} \Rightarrow y = y_0 + (y_1 - y_0)\left(\frac{x - x_0}{x_1 - x_0}\right)$

$\Rightarrow y - y_0 = \left(\frac{y_1 - y_0}{x_1 - x_0}\right)(x - x_0)$ which is an equation of the line through the points (x_0, y_0) and (x_1, y_1)

(b) Let $x_0 = y_0 = 0$ in (a) $\Rightarrow x = x_1 t$, $y = y_1 t$ (the answer is not unique)

(c) Let $(x_0, y_0) = (-1, 0)$ and $(x_1, y_1) = (0, 1)$ or let $(x_0, y_0) = (0, 1)$ and $(x_1, y_1) = (-1, 0)$ in part (a)

$\Rightarrow x = -1 + t$, $y = t$ or $x = -t$, $y = 1 - t$ (the answer is not unique)

33. Arc $PF = $ Arc AF since each is the distance rolled and

$\frac{\text{Arc } PF}{b} = \angle FCP \Rightarrow \text{Arc } PF = b(\angle FCP); \frac{\text{Arc } AF}{a} = \theta$

$\Rightarrow \text{Arc } AF = a\theta \Rightarrow a\theta = b(\angle FCP) \Rightarrow \angle FCP = \frac{a}{b}\theta;$

$\angle OCG = \frac{\pi}{2} - \theta; \angle OCG = \angle OCP + \angle PCE$

$= \angle OCP + \left(\frac{\pi}{2} - \alpha\right).$ Now $\angle OCP = \pi - \angle FCP$

$= \pi - \frac{a}{b}\theta.$ Thus $\angle OCG = \pi - \frac{a}{b}\theta + \frac{\pi}{2} - \alpha \Rightarrow \frac{\pi}{2} - \theta$

$= \pi - \frac{a}{b}\theta + \frac{\pi}{2} - \alpha \Rightarrow \alpha = \pi - \frac{a}{b}\theta + \theta = \pi - \left(\frac{a - b}{b}\theta\right).$

Then $x = OG - BG = OG - PE = (a - b)\cos\theta - b\cos\alpha = (a - b)\cos\theta - b\cos\left(\pi - \frac{a - b}{b}\theta\right)$

$= (a - b)\cos\theta + b\cos\left(\frac{a - b}{b}\theta\right).$ Also $y = EG = CG - CE = (a - b)\sin\theta - b\sin\alpha$

$= (a - b)\sin\theta - b\sin\left(\pi - \frac{a - b}{b}\theta\right) = (a - b)\sin\theta - b\sin\left(\frac{a - b}{b}\theta\right).$ Therefore

$x = (a - b)\cos\theta + b\cos\left(\frac{a - b}{b}\theta\right)$ and $y = (a - b)\sin\theta - b\sin\left(\frac{a - b}{b}\theta\right).$

If $b = \frac{a}{4}$, then $x = \left(a - \frac{a}{4}\right)\cos\theta + \frac{a}{4}\cos\left(\frac{a - \left(\frac{a}{4}\right)}{\left(\frac{a}{4}\right)}\theta\right)$

$= \frac{3a}{4}\cos\theta + \frac{a}{4}\cos 3\theta = \frac{3a}{4}\cos\theta + \frac{a}{4}(\cos\theta\cos 2\theta - \sin\theta\sin 2\theta)$

$= \frac{3a}{4}\cos\theta + \frac{a}{4}\left((\cos\theta)(\cos^2\theta - \sin^2\theta) - (\sin\theta)(2\sin\theta\cos\theta)\right)$

$= \frac{3a}{4}\cos\theta + \frac{a}{4}\cos^3\theta - \frac{a}{4}\cos\theta\sin^2\theta - \frac{2a}{4}\sin^2\theta\cos\theta$

$= \frac{3a}{4}\cos\theta + \frac{a}{4}\cos^3\theta - \frac{3a}{4}(\cos\theta)(1 - \cos^2\theta) = a\cos^3\theta;$

$y = \left(a - \frac{a}{4}\right)\sin\theta - \frac{a}{4}\sin\left(\frac{a - \left(\frac{a}{4}\right)}{\left(\frac{a}{4}\right)}\theta\right) = \frac{3a}{4}\sin\theta - \frac{a}{4}\sin 3\theta = \frac{3a}{4}\sin\theta - \frac{a}{4}(\sin\theta\cos 2\theta + \cos\theta\sin 2\theta)$

$= \frac{3a}{4}\sin\theta - \frac{a}{4}\left((\sin\theta)(\cos^2\theta - \sin^2\theta) + (\cos\theta)(2\sin\theta\cos\theta)\right)$

$= \frac{3a}{4}\sin\theta - \frac{a}{4}\sin\theta\cos^2\theta + \frac{a}{4}\sin^3\theta - \frac{2a}{4}\cos^2\theta\sin\theta$

$= \frac{3a}{4}\sin\theta - \frac{3a}{4}\sin\theta\cos^2\theta + \frac{a}{4}\sin^3\theta$

$= \frac{3a}{4}\sin\theta - \frac{3a}{4}(\sin\theta)(1 - \sin^2\theta) + \frac{a}{4}\sin^3\theta = a\sin^3\theta.$

35. Draw line AM in the figure and note that $\angle AMO$ is a right angle since it is an inscribed angle which spans the diameter of a circle. Then $AN^2 = MN^2 + AM^2$. Now, $OA = a$, $\frac{AN}{a} = \tan t$, and $\frac{AM}{a} = \sin t$. Next $MN = OP \Rightarrow OP^2 = AN^2 - AM^2$

$$= a^2 \tan^2 t - a^2 \sin^2 t \Rightarrow OP = \sqrt{a^2 \tan^2 t - a^2 \sin^2 t}$$

$$= (a \sin t)\sqrt{\sec^2 t - 1} = \frac{a \sin^2 t}{\cos t}. \text{ In triangle BPO,}$$

$$x = OP \sin t = \frac{a \sin^3 t}{\cos t} = a \sin^2 t \tan t \text{ and}$$

$$y = OP \cos t = a \sin^2 t \Rightarrow x = a \sin^2 t \tan t \text{ and } y = a \sin^2 t.$$

37. $D = \sqrt{(x-2)^2 + \left(y - \frac{1}{2}\right)^2} \Rightarrow D^2 = (x-2)^2 + \left(y - \frac{1}{2}\right)^2 = (t-2)^2 + \left(t^2 - \frac{1}{2}\right)^2 \Rightarrow D^2 = t^4 - 4t + \frac{17}{4}$

$\Rightarrow \frac{d(D^2)}{dt} = 4t^3 - 4 = 0 \Rightarrow t = 1$. The second derivative is always positive for $t \neq 0 \Rightarrow t = 1$ gives a local minimum for D^2 (and hence D) which is an absolute minimum since it is the only extremum $\Rightarrow$ the closest point on the parabola is $(1, 1)$.

39. (a)   (b)   (c)

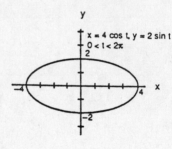   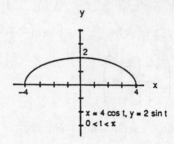   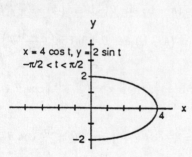

41.

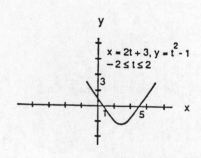

43. (a)

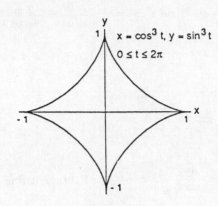

(b)

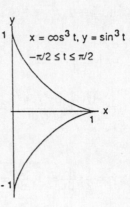

45. (a)

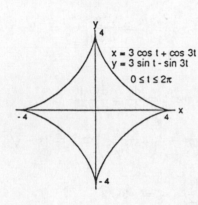

(b)

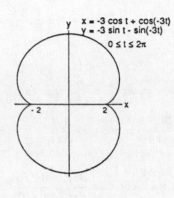

47. (a)

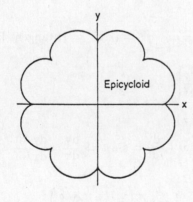

(b)

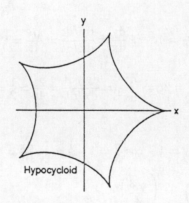

(c)

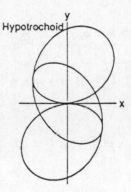

9.5 CALCULUS WITH PARAMETRIZED CURVES

1. $t = \frac{\pi}{4} \Rightarrow x = 2 \cos \frac{\pi}{4} = \sqrt{2}, \; y = 2 \sin \frac{\pi}{4} = \sqrt{2}; \; \frac{dx}{dt} = -2 \sin t, \; \frac{dy}{dt} = 2 \cos t \Rightarrow \frac{dy}{dx} = \frac{dy/dt}{dx/dt} = \frac{2 \cos t}{-2 \sin t} = -\cot t$

$\Rightarrow \left. \frac{dy}{dx} \right|_{t = \frac{\pi}{4}} = -\cot \frac{\pi}{4} = -1$; tangent line is $y - \sqrt{2} = -1 \left(x - \sqrt{2} \right)$ or $y = -x + 2\sqrt{2}; \; \frac{dy'}{dt} = \csc^2 t$

$\Rightarrow \frac{d^2 y}{dx^2} = \frac{dy'/dt}{dx/dt} = \frac{\csc^2 t}{-2 \sin t} = -\frac{1}{2 \sin^3 t} \Rightarrow \left. \frac{d^2 y}{dx^2} \right|_{t = \frac{\pi}{4}} = -\sqrt{2}$

3. $t = \frac{\pi}{4} \Rightarrow x = 4 \sin \frac{\pi}{4} = 2\sqrt{2}$, $y = 2 \cos \frac{\pi}{4} = \sqrt{2}$; $\frac{dx}{dt} = 4 \cos t$, $\frac{dy}{dt} = -2 \sin t \Rightarrow \frac{dy}{dx} = \frac{dy/dt}{dx/dt} = \frac{-2 \sin t}{4 \cos t}$

$= -\frac{1}{2} \tan t \Rightarrow \frac{dy}{dx}\Big|_{t=\frac{\pi}{4}} = -\frac{1}{2} \tan \frac{\pi}{4} = -\frac{1}{2}$; tangent line is $y - \sqrt{2} = -\frac{1}{2}(x - 2\sqrt{2})$ or $y = -\frac{1}{2}x + 2\sqrt{2}$;

$\frac{dy'}{dt} = -\frac{1}{2} \sec^2 t \Rightarrow \frac{d^2y}{dx^2} = \frac{dy'/dt}{dx/dt} = \frac{-\frac{1}{2} \sec^2 t}{4 \cos t} = -\frac{1}{8 \cos^3 t} \Rightarrow \frac{d^2y}{dx^2}\Big|_{t=\frac{\pi}{4}} = -\frac{\sqrt{2}}{4}$

5. $t = \frac{1}{4} \Rightarrow x = \frac{1}{4}$, $y = \frac{1}{2}$; $\frac{dx}{dt} = 1$, $\frac{dy}{dt} = \frac{1}{2\sqrt{t}} \Rightarrow \frac{dy}{dx} = \frac{dy/dt}{dx/dt} = \frac{1}{2\sqrt{t}} \Rightarrow \frac{dy}{dx}\Big|_{t=\frac{1}{4}} = \frac{1}{2\sqrt{\frac{1}{4}}} = 1$; tangent line is

$y - \frac{1}{2} = 1 \cdot \left(x - \frac{1}{4}\right)$ or $y = x + \frac{1}{4}$; $\frac{dy'}{dt} = -\frac{1}{4} t^{-3/2} \Rightarrow \frac{d^2y}{dx^2} = \frac{dy'/dt}{dx/dt} = -\frac{1}{4} t^{-3/2} \Rightarrow \frac{d^2y}{dx^2}\Big|_{t=\frac{1}{4}} = -2$

7. $t = \frac{\pi}{6} \Rightarrow x = \sec \frac{\pi}{6} = \frac{2}{\sqrt{3}}$, $y = \tan \frac{\pi}{6} = \frac{1}{\sqrt{3}}$; $\frac{dx}{dt} = \sec t \tan t$, $\frac{dy}{dt} = \sec^2 t \Rightarrow \frac{dy}{dx} = \frac{dy/dt}{dx/dt}$

$= \frac{\sec^2 t}{\sec t \tan t} = \csc t \Rightarrow \frac{dy}{dx}\Big|_{t=\frac{\pi}{6}} = \csc \frac{\pi}{6} = 2$; tangent line is $y - \frac{1}{\sqrt{3}} = 2\left(x - \frac{2}{\sqrt{3}}\right)$ or $y = 2x - \sqrt{3}$;

$\frac{dy'}{dt} = -\csc t \cot t \Rightarrow \frac{d^2y}{dx^2} = \frac{dy'/dt}{dx/dt} = \frac{-\csc t \cot t}{\sec t \tan t} = -\cot^3 t \Rightarrow \frac{d^2y}{dx^2}\Big|_{t=\frac{\pi}{6}} = -3\sqrt{3}$

9. $t = -1 \Rightarrow x = 5$, $y = 1$; $\frac{dx}{dt} = 4t$, $\frac{dy}{dt} = 4t^3 \Rightarrow \frac{dy}{dx} = \frac{dy/dt}{dx/dt} = \frac{4t^3}{4t} = t^2 \Rightarrow \frac{dy}{dx}\Big|_{t=-1} = (-1)^2 = 1$; tangent line is

$y - 1 = 1 \cdot (x - 5)$ or $y = x - 4$; $\frac{dy'}{dt} = 2t \Rightarrow \frac{d^2y}{dx^2} = \frac{dy'/dt}{dx/dt} = \frac{2t}{4t} = \frac{1}{2} \Rightarrow \frac{d^2y}{dx^2}\Big|_{t=-1} = \frac{1}{2}$

11. $t = \frac{\pi}{3} \Rightarrow x = \frac{\pi}{3} - \sin \frac{\pi}{3} = \frac{\pi}{3} - \frac{\sqrt{3}}{2}$, $y = 1 - \cos \frac{\pi}{3} = 1 - \frac{1}{2} = \frac{1}{2}$; $\frac{dx}{dt} = 1 - \cos t$, $\frac{dy}{dt} = \sin t \Rightarrow \frac{dy}{dx} = \frac{dy/dt}{dx/dt}$

$= \frac{\sin t}{1 - \cos t} \Rightarrow \frac{dy}{dx}\Big|_{t=\frac{\pi}{3}} = \frac{\sin\left(\frac{\pi}{3}\right)}{1 - \cos\left(\frac{\pi}{3}\right)} = \frac{\left(\frac{\sqrt{3}}{2}\right)}{\left(\frac{1}{2}\right)} = \sqrt{3}$; tangent line is $y - \frac{1}{2} = \sqrt{3}\left(x - \frac{\pi}{3} + \frac{\sqrt{3}}{2}\right)$

$\Rightarrow y = \sqrt{3}x - \frac{\pi\sqrt{3}}{3} + 2$; $\frac{dy'}{dt} = \frac{(1 - \cos t)(\cos t) - (\sin t)(\sin t)}{(1 - \cos t)^2} = \frac{-1}{1 - \cos t} \Rightarrow \frac{d^2y}{dx^2} = \frac{dy'/dt}{dx/dt} = \frac{\left(\frac{-1}{1 - \cos t}\right)}{1 - \cos t}$

$= \frac{-1}{(1 - \cos t)^2} \Rightarrow \frac{d^2y}{dx^2}\Big|_{t=\frac{\pi}{3}} = -4$

13. $x^2 - 2tx + 2t^2 = 4 \Rightarrow 2x \frac{dx}{dt} - 2x - 2t \frac{dx}{dt} + 4t = 0 \Rightarrow (2x - 2t) \frac{dx}{dt} = 2x - 4t \Rightarrow \frac{dx}{dt} = \frac{2x - 4t}{2x - 2t} = \frac{x - 2t}{x - t}$;

$2y^3 - 3t^2 = 4 \Rightarrow 6y^2 \frac{dy}{dt} - 6t = 0 \Rightarrow \frac{dy}{dt} = \frac{6t}{6y^2} = \frac{t}{y^2}$; thus $\frac{dy}{dx} = \frac{dy/dt}{dx/dt} = \frac{\left(\frac{t}{y^2}\right)}{\left(\frac{x - 2t}{x - t}\right)} = \frac{t(x - t)}{y^2(x - 2t)}$; $t = 2$

$\Rightarrow x^2 - 2(2)x + 2(2)^2 = 4 \Rightarrow x^2 - 4x + 4 = 0 \Rightarrow (x-2)^2 = 0 \Rightarrow x = 2; \ t = 2 \Rightarrow 2y^3 - 3(2)^2 = 4$

$\Rightarrow 2y^3 = 16 \Rightarrow y^3 = 8 \Rightarrow y = 2;$ therefore $\left.\dfrac{dy}{dx}\right|_{t=2} = \dfrac{2(2-2)}{(2)^2(2-2(2))} = 0$

15. $x + 2x^{3/2} = t^2 + t \Rightarrow \dfrac{dx}{dt} + 3x^{1/2}\dfrac{dx}{dt} = 2t + 1 \Rightarrow \left(1 + 3x^{1/2}\right)\dfrac{dx}{dt} = 2t + 1 \Rightarrow \dfrac{dx}{dt} = \dfrac{2t+1}{1+3x^{1/2}}; \ y\sqrt{t+1} + 2t\sqrt{y} = 4$

$\Rightarrow \dfrac{dy}{dt}\sqrt{t+1} + y\left(\dfrac{1}{2}\right)(t+1)^{-1/2} + 2\sqrt{y} + 2t\left(\dfrac{1}{2}y^{-1/2}\right)\dfrac{dy}{dt} = 0 \Rightarrow \dfrac{dy}{dt}\sqrt{t+1} + \dfrac{y}{2\sqrt{t+1}} + 2\sqrt{y} + \left(\dfrac{t}{\sqrt{y}}\right)\dfrac{dy}{dt} = 0$

$\Rightarrow \left(\sqrt{t+1} + \dfrac{t}{\sqrt{y}}\right)\dfrac{dy}{dt} = \dfrac{-y}{2\sqrt{t+1}} - 2\sqrt{y} \Rightarrow \dfrac{dy}{dt} = \dfrac{\left(\dfrac{-y}{2\sqrt{t+1}} - 2\sqrt{y}\right)}{\left(\sqrt{t+1} + \dfrac{t}{\sqrt{y}}\right)} = \dfrac{-y\sqrt{y} - 4y\sqrt{t+1}}{2\sqrt{y}(t+1) + 2t\sqrt{t+1}};$ thus

$\dfrac{dy}{dx} = \dfrac{dy/dt}{dx/dt} = \dfrac{\left(\dfrac{-y\sqrt{y} - 4y\sqrt{t+1}}{2\sqrt{y}(t+1) + 2t\sqrt{t+1}}\right)}{\left(\dfrac{2t+1}{1+3x^{1/2}}\right)}; \ t = 0 \Rightarrow x + 2x^{3/2} = 0 \Rightarrow x\left(1 + 2x^{1/2}\right) = 0 \Rightarrow x = 0; \ t = 0$

$\Rightarrow y\sqrt{0+1} + 2(0)\sqrt{y} = 4 \Rightarrow y = 4;$ therefore $\left.\dfrac{dy}{dx}\right|_{t=0} = \dfrac{\left(\dfrac{-4\sqrt{4} - 4(4)\sqrt{0+1}}{2\sqrt{4}(0+1) + 2(0)\sqrt{0+1}}\right)}{\left(\dfrac{2(0)+1}{1+3(0)^{1/2}}\right)} = -6$

17. $\dfrac{dx}{dt} = -\sin t$ and $\dfrac{dy}{dt} = 1 + \cos t \Rightarrow \sqrt{\left(\dfrac{dx}{dt}\right)^2 + \left(\dfrac{dy}{dt}\right)^2} = \sqrt{(-\sin t)^2 + (1+\cos t)^2} = \sqrt{2 + 2\cos t}$

$\Rightarrow$ Length $= \displaystyle\int_0^\pi \sqrt{2 + 2\cos t} \ dt = \sqrt{2}\int_0^\pi \sqrt{\left(\dfrac{1-\cos t}{1-\cos t}\right)(1+\cos t)} \ dt = \sqrt{2}\int_0^\pi \sqrt{\dfrac{\sin^2 t}{1-\cos t}} \ dt$

$= \sqrt{2}\displaystyle\int_0^\pi \dfrac{\sin t}{\sqrt{1-\cos t}} \ dt$ (since $\sin t \geq 0$ on $[0, \pi]$); $[u = 1 - \cos t \Rightarrow du = \sin t \ dt; \ t = 0 \Rightarrow u = 0,$

$t = \pi \Rightarrow u = 2] \rightarrow \sqrt{2}\displaystyle\int_0^2 u^{-1/2} \ du = \sqrt{2}\left[2u^{1/2}\right]_0^2 = 4$

19. $\dfrac{dx}{dt} = t$ and $\dfrac{dy}{dt} = (2t+1)^{1/2} \Rightarrow \sqrt{\left(\dfrac{dx}{dt}\right)^2 + \left(\dfrac{dy}{dt}\right)^2} = \sqrt{t^2 + (2t+1)} = \sqrt{(t+1)^2} = |t+1| = t + 1$ since $0 \leq t \leq 4$

$\Rightarrow$ Length $= \displaystyle\int_0^4 (t+1) \ dt = \left[\dfrac{t^2}{2} + t\right]_0^4 = (8 + 4) = 12$

21. $\frac{dx}{dt} = 8t \cos t$ and $\frac{dy}{dt} = 8t \sin t \Rightarrow \sqrt{\left(\frac{dx}{dt}\right)^2 + \left(\frac{dy}{dt}\right)^2} = \sqrt{(8t \cos t)^2 + (8t \sin t)^2} = \sqrt{64t^2 \cos^2 t + 64t^2 \sin^2 t}$

$= |8t| = 8t$ since $0 \le t \le \frac{\pi}{2} \Rightarrow$ Length $= \int_0^{\pi/2} 8t \, dt = \left[4t^2\right]_0^{\pi/2} = \pi^2$

23. $\frac{dx}{dt} = -\sin t$ and $\frac{dy}{dt} = \cos t \Rightarrow \sqrt{\left(\frac{dx}{dt}\right)^2 + \left(\frac{dy}{dt}\right)^2} = \sqrt{(-\sin t)^2 + (\cos t)^2} = 1 \Rightarrow$ Area $= \int 2\pi y \, ds$

$= \int_0^{2\pi} 2\pi(2 + \sin t)(1) \, dt = 2\pi[2t - \cos t]_0^{2\pi} = 2\pi[(4\pi - 1) - (0 - 1)] = 8\pi^2$

25. $\frac{dx}{dt} = 1$ and $\frac{dy}{dt} = t + \sqrt{2} \Rightarrow \sqrt{\left(\frac{dx}{dt}\right)^2 + \left(\frac{dy}{dt}\right)^2} = \sqrt{1^2 + \left(t + \sqrt{2}\right)^2} = \sqrt{t^2 + 2\sqrt{2}t + 3} \Rightarrow$ Area $= \int 2\pi x \, ds$

$= \int_{-\sqrt{2}}^{\sqrt{2}} 2\pi\left(t + \sqrt{2}\right)\sqrt{t^2 + 2\sqrt{2}t + 3} \, dt; \left[u = t^2 + 2\sqrt{2}t + 3 \Rightarrow du = \left(2t + 2\sqrt{2}\right) dt; \, t = -\sqrt{2} \Rightarrow u = 1,\right.$

$\left[t = \sqrt{2} \Rightarrow u = 9\right] \rightarrow \int_1^9 \pi\sqrt{u} \, du = \left[\frac{2}{3}\pi u^{3/2}\right]_1^9 = \frac{2\pi}{3}(27 - 1) = \frac{52\pi}{3}$

27. $\frac{dx}{dt} = 2$ and $\frac{dy}{dt} = 1 \Rightarrow \sqrt{\left(\frac{dx}{dt}\right)^2 + \left(\frac{dy}{dt}\right)^2} = \sqrt{2^2 + 1^2} = \sqrt{5} \Rightarrow$ Area $= \int 2\pi y \, ds = \int_0^1 2\pi(t + 1)\sqrt{5} \, dt$

$= 2\pi\sqrt{5}\left[\frac{t^2}{2} + t\right]_0^1 = 3\pi\sqrt{5}$. Check: slant height is $\sqrt{5} \Rightarrow$ Area is $\pi(1 + 2)\sqrt{5} = 3\pi\sqrt{5}$.

29. (a) Let the density be $\delta = 1$. Then $x = \cos t + t \sin t \Rightarrow \frac{dx}{dt} = t \cos t$, and $y = \sin t - t \cos t \Rightarrow \frac{dy}{dt} = t \sin t$

$\Rightarrow dm = 1 \cdot ds = \sqrt{\left(\frac{dx}{dt}\right)^2 + \left(\frac{dy}{dt}\right)^2} \, dt = \sqrt{(t \cos t)^2 + (t \sin t)^2} = |t| \, dt = t \, dt$ since $0 \le t \le \frac{\pi}{2}$. The curve's

mass is $M = \int dm = \int_0^{\pi/2} t \, dt = \frac{\pi^2}{8}$. Also $M_x = \int \widetilde{y} \, dm = \int_0^{\pi/2} (\sin t - t \cos t) t \, dt$

$= \int_0^{\pi/2} t \sin t \, dt - \int_0^{\pi/2} t^2 \cos t \, dt = [\sin t - t \cos t]_0^{\pi/2} - [t^2 \sin t - 2 \sin t + 2t \cos t]_0^{\pi/2} = 3 - \frac{\pi^2}{4}$, where

we integrated by parts. Therefore, $\bar{y} = \frac{M_x}{M} = \frac{\left(3 - \frac{\pi^2}{4}\right)}{\left(\frac{\pi^2}{8}\right)} = \frac{24}{\pi^2} - 2$. Next, $M_y = \int \widetilde{x} \, dm$

$= \int_0^{\pi/2} (\cos t + t \sin t) t \, dt = \int_0^{\pi/2} t \cos t \, dt + \int_0^{\pi/2} t^2 \sin t \, dt$

$$= \Big[\cos t + t \sin t\Big]_0^{\pi/2} + \Big[-t^2 \cos t + 2 \cos t + 2t \sin t\Big]_0^{\pi/2} = \frac{3\pi}{2} - 3, \text{ again integrating by parts.}$$

Hence $\overline{x} = \dfrac{M_y}{M} = \dfrac{\left(\dfrac{3\pi}{2} - 3\right)}{\left(\dfrac{\pi^2}{8}\right)} = \dfrac{12}{\pi} - \dfrac{24}{\pi^2}$. Therefore $(\overline{x}, \overline{y}) = \left(\dfrac{12}{\pi} - \dfrac{24}{\pi^2}, \dfrac{24}{\pi^2} - 2\right)$

(b) $(\overline{x}, \overline{y}) \approx (1.4, 0.4)$

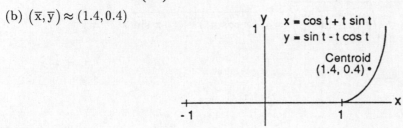

31. (a) Let the density be $\delta = 1$. Then $x = \cos t \Rightarrow \dfrac{dx}{dt} = -\sin t$, and $y = t + \sin t \Rightarrow \dfrac{dy}{dt} = 1 + \cos t$

$$\Rightarrow dm = 1 \cdot ds = \sqrt{\left(\dfrac{dx}{dt}\right)^2 + \left(\dfrac{dy}{dt}\right)^2}\, dt = \sqrt{(-\sin t)^2 + (1 + \cos t)^2}\, dt = \sqrt{2 + 2\cos t}\, dt. \text{ The curve's mass}$$

is $M = \displaystyle\int dm = \int_0^\pi \sqrt{2 + 2\cos t}\, dt = \sqrt{2} \int_0^\pi \sqrt{1 + \cos t}\, dt = \sqrt{2}\int_0^\pi \sqrt{2\cos^2\left(\dfrac{t}{2}\right)}\, dt = 2\int_0^\pi \left|\cos\left(\dfrac{t}{2}\right)\right| dt$

$$= 2\int_0^\pi \cos\left(\dfrac{t}{2}\right) dt \left(\text{since } 0 \le t \le \pi \Rightarrow 0 \le \dfrac{t}{2} \le \dfrac{\pi}{2}\right) = 2\Big[2\sin\left(\dfrac{t}{2}\right)\Big]_0^\pi = 4. \text{ Also } M_x = \int \widetilde{y}\, dm$$

$$= \int_0^\pi (t + \sin t)\left(2\cos\dfrac{t}{2}\right) dt = \int_0^\pi 2t\cos\left(\dfrac{t}{2}\right) dt + \int_0^\pi 2\sin t \cos\left(\dfrac{t}{2}\right) dt$$

$$= 2\Big[4\cos\left(\dfrac{t}{2}\right) + 2t\sin\left(\dfrac{t}{2}\right)\Big]_0^\pi + 2\Big[-\dfrac{1}{3}\cos\left(\dfrac{3}{2}t\right) - \cos\left(\dfrac{1}{2}t\right)\Big]_0^\pi = 4\pi - \dfrac{16}{3} \Rightarrow \overline{y} = \dfrac{M_x}{M} = \dfrac{\left(4\pi - \dfrac{16}{3}\right)}{4} = \pi - \dfrac{4}{3}.$$

Next $M_y = \displaystyle\int \widetilde{x}\, dm = \int_0^\pi (\cos t)\left(2\cos\dfrac{t}{2}\right) dt = 2\int_0^\pi \cos t \cos\left(\dfrac{t}{2}\right) dt = 2\Big[\sin\left(\dfrac{t}{2}\right) + \dfrac{\sin\left(\dfrac{3}{2}t\right)}{3}\Big]_0^\pi = 2 - \dfrac{2}{3}$

$$= \dfrac{4}{3} \Rightarrow \overline{x} = \dfrac{M_y}{M} = \dfrac{\left(\dfrac{4}{3}\right)}{4} = \dfrac{1}{3}. \text{ Therefore } (\overline{x}, \overline{y}) = \left(\dfrac{1}{3}, \pi - \dfrac{4}{3}\right)$$

(b) $(\overline{x}, \overline{y}) \approx (0.33, 1.81)$

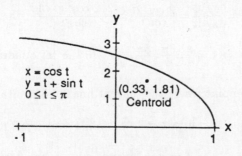

33. (a) $\frac{dx}{dt} = -2\sin 2t$ and $\frac{dy}{dt} = 2\cos 2t \Rightarrow \sqrt{\left(\frac{dx}{dt}\right)^2 + \left(\frac{dy}{dt}\right)^2} = \sqrt{(-2\sin 2t)^2 + (2\cos 2t)^2} = 2$

$\Rightarrow$ Length $= \displaystyle\int_0^{\pi/2} 2 \, dt = [2t]_0^{\pi/2} = \pi$

(b) $\frac{dx}{dt} = \pi\cos\pi t$ and $\frac{dy}{dt} = -\pi\sin\pi t = \sqrt{\left(\frac{dx}{dt}\right)^2 + \left(\frac{dy}{dt}\right)^2} = \sqrt{(\pi\cos\pi t)^2 + (-\pi\sin\pi t)^2} = \pi$

$\Rightarrow$ Length $= \displaystyle\int_{-1/2}^{1/2} \pi \, dt = [\pi t]_{-1/2}^{1/2} = \pi$

35. $x = x \Rightarrow \frac{dy}{dx} = 1$, and $y = f(x) \Rightarrow \frac{dy}{dx} = f'(x)$; then Length $= \displaystyle\int_a^b \sqrt{\left(\frac{dx}{dt}\right)^2 + \left(\frac{dy}{dt}\right)^2} \, dt = \int_a^b \sqrt{\left(\frac{dx}{dx}\right)^2 + \left(\frac{dy}{dx}\right)^2} \, dx$

$= \displaystyle\int_a^b \sqrt{1 + \left(\frac{dy}{dx}\right)^2} \, dx = \int_a^b \sqrt{1 + [f'(x)]^2} \, dx$

37. For one arch of the cycloid we use the interval $0 \le \theta \le 2\pi$. Then, $A = \displaystyle\int_0^{2\pi} y(\theta) \, dx = \int_0^{2\pi} a(1 - \cos\theta)\left(\frac{dx}{d\theta}\right) d\theta$

and $\frac{dx}{d\theta} = a(1 - \cos\theta) \Rightarrow A = \displaystyle\int_0^{2\pi} a^2(1 - \cos\theta)^2 \, d\theta = a^2 \int_0^{2\pi} \left(1 - 2\cos\theta + \cos^2\theta\right) d\theta$

$= a^2\left[\displaystyle\int_0^{2\pi} d\theta - 2\int_0^{2\pi} \cos\theta \, d\theta + \int_0^{2\pi} \tfrac{1}{2}(1 + \cos 2\theta) \, d\theta\right] = a^2\left([\theta]_0^{2\pi} - 2[\sin\theta]_0^{2\pi} + \tfrac{1}{2}\left[\theta + \tfrac{1}{2}\sin 2\theta\right]_0^{2\pi}\right)$

$= a^2\left[(2\pi - 0) - 2(0 - 0) + \tfrac{1}{2}(2\pi - 0)\right] = 3\pi a^2$

39. $x = \theta - \sin\theta$ and $y = 1 - \cos\theta$, $0 \le \theta \le 2\pi \Rightarrow ds = \sqrt{(1 - \cos\theta)^2 + \sin^2\theta} \, d\theta = \sqrt{2 - 2\cos\theta} \, d\theta \Rightarrow S = \displaystyle\int 2\pi y \, ds$

$= 2\sqrt{2}\displaystyle\int_0^{2\pi} \pi(1 - \cos\theta)^{3/2} \, d\theta = 2\sqrt{2}\int_0^{2\pi} \pi\left(\sqrt{2}\sin\tfrac{\theta}{2}\right)^3 d\theta = 8\pi\int_0^{2\pi} \left(1 - \cos^2\tfrac{\theta}{2}\right)\left(\sin\tfrac{\theta}{2}\right) d\theta$

$= 8\pi\left[-2\cos\tfrac{\theta}{2} + \tfrac{2}{3}\cos^3\tfrac{\theta}{2}\right]_0^{2\pi} = \frac{64\pi}{3}$

41. $\frac{dx}{dt} = \cos t$ and $\frac{dy}{dt} = 2\cos 2t \Rightarrow \frac{dy}{dx} = \frac{dy/dt}{dx/dt} = \frac{2\cos 2t}{\cos t} = \frac{2(2\cos^2 t - 1)}{\cos t}$; then $\frac{dy}{dx} = 0 \Rightarrow \frac{2(2\cos^2 t - 1)}{\cos t} = 0$

$\Rightarrow 2\cos^2 t - 1 = 0 \Rightarrow \cos t = \pm\frac{1}{\sqrt{2}} \Rightarrow t = \frac{\pi}{4}, \frac{3\pi}{4}, \frac{5\pi}{4}, \frac{7\pi}{4}$. In the 1st quadrant: $t = \frac{\pi}{4} \Rightarrow x = \sin\frac{\pi}{4} = \frac{\sqrt{2}}{2}$ and

$y = \sin 2\left(\frac{\pi}{4}\right) = 1 \Rightarrow \left(\frac{\sqrt{2}}{2}, 1\right)$ is the point where the tangent line is horizontal. At the origin: $x = 0$ and $y = 0$

$\Rightarrow \sin t = 0 \Rightarrow t = 0$ or $t = \pi$ and $\sin 2t = 0 \Rightarrow t = 0, \frac{\pi}{2}, \pi, \frac{3\pi}{2}$; thus $t = 0$ and $t = \pi$ give the tangent lines at

the origin. Tangents at origin: $\frac{dy}{dx}\Big|_{t=0} = 2 \Rightarrow y = 2x$ and $\frac{dy}{dx}\Big|_{t=\pi} = -2 \Rightarrow y = -2x$

43.
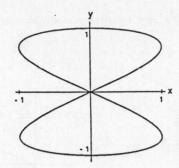

45.
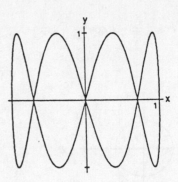

47.
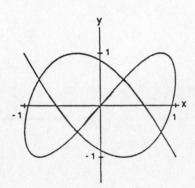

49.
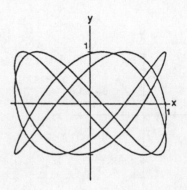

9.6 POLAR COORDINATES

1. a, e; b, g; c, h; d, f

3. (a) $\left(2, \frac{\pi}{2} + 2n\pi\right)$ and $\left(-2, \frac{\pi}{2} + (2n+1)\pi\right)$, n an integer

   (b) $(2, 2n\pi)$ and $(-2, (2n+1)\pi)$, n an integer

   (c) $\left(2, \frac{3\pi}{2} + 2n\pi\right)$ and $\left(-2, \frac{3\pi}{2} + (2n+1)\pi\right)$, n an integer

   (d) $(2, (2n+1)\pi)$ and $(-2, 2n\pi)$, n an integer

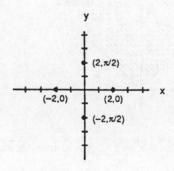

5. (a) $x = r \cos \theta = 3 \cos 0 = 3$, $y = r \sin \theta = 3 \sin 0 = 0 \Rightarrow$ Cartesian coordinates are $(3, 0)$

   (b) $x = r \cos \theta = -3 \cos 0 = -3$, $y = r \sin \theta = -3 \sin 0 = 0 \Rightarrow$ Cartesian coordinates are $(-3, 0)$

   (c) $x = r \cos \theta = 2 \cos \frac{2\pi}{3} = -1$, $y = r \sin \theta = 2 \sin \frac{2\pi}{3} = \sqrt{3} \Rightarrow$ Cartesian coordinates are $\left(-1, \sqrt{3}\right)$

   (d) $x = r \cos \theta = 2 \cos \frac{7\pi}{3} = 1$, $y = r \sin \theta = 2 \sin \frac{7\pi}{3} = \sqrt{3} \Rightarrow$ Cartesian coordinates are $\left(1, \sqrt{3}\right)$

   (e) $x = r \cos \theta = -3 \cos \pi = 3$, $y = r \sin \theta = -3 \sin \pi = 0 \Rightarrow$ Cartesian coordinates are $(3, 0)$

   (f) $x = r \cos \theta = 2 \cos \frac{\pi}{3} = 1$, $y = r \sin \theta = 2 \sin \frac{\pi}{3} = \sqrt{3} \Rightarrow$ Cartesian coordinates are $\left(1, \sqrt{3}\right)$

   (g) $x = r \cos \theta = -3 \cos 2\pi = -3$, $y = r \sin \theta = -3 \sin 2\pi = 0 \Rightarrow$ Cartesian coordinates are $(-3, 0)$

(h) $x = r \cos \theta = -2 \cos\left(-\frac{\pi}{3}\right) = -1$, $y = r \sin \theta = -2 \sin\left(-\frac{\pi}{3}\right) = \sqrt{3}$ $\Rightarrow$ Cartesian coordinates are $\left(-1, \sqrt{3}\right)$

7.
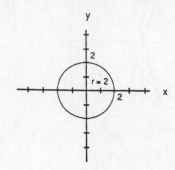

9.
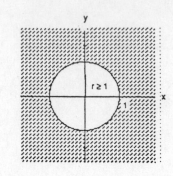

11.
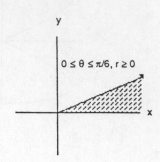

13.
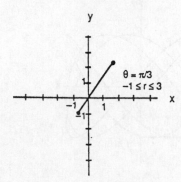

15.
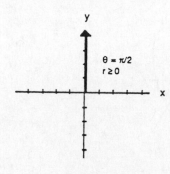

17.
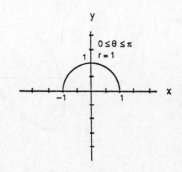

19.
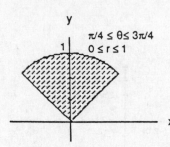

21.
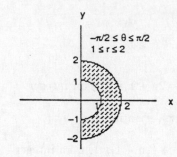

23. $r \cos \theta = 2 \Rightarrow x = 2$, vertical line through $(2, 0)$ 25. $r \sin \theta = 0 \Rightarrow y = 0$, the x-axis

27. $r = 4 \csc \theta \Rightarrow r = \dfrac{4}{\sin \theta} \Rightarrow r \sin \theta = 4 \Rightarrow y = 4$, a horizontal line through $(0, 4)$

29. $r \cos \theta + r \sin \theta = 1 \Rightarrow x + y = 1$, line with slope $m = -1$ and intercept $b = 1$

31. $r^2 = 1 \Rightarrow x^2 + y^2 = 1$, circle with center $C = (0, 0)$ and radius 1

33. $r = \dfrac{5}{\sin \theta - 2 \cos \theta} \Rightarrow r \sin \theta - 2r \cos \theta = 5 \Rightarrow y - 2x = 5$, line with slope $m = 2$ and intercept $b = 5$

35. $r = \cot \theta \csc \theta = \left(\dfrac{\cos \theta}{\sin \theta}\right)\left(\dfrac{1}{\sin \theta}\right) \Rightarrow r \sin^2 \theta = \cos \theta \Rightarrow r^2 \sin^2 \theta = r \cos \theta \Rightarrow y^2 = x$, parabola with vertex $(0, 0)$ which opens to the right

37. $r = (\csc \theta)\, e^{r \cos \theta} \Rightarrow r \sin \theta = e^{r \cos \theta} \Rightarrow y = e^x$, graph of the natural exponential function

39. $r^2 + 2r^2 \cos \theta \sin \theta = 1 \Rightarrow x^2 + y^2 + 2xy = 1 \Rightarrow x^2 + 2xy + y^2 = 1 \Rightarrow (x + y)^2 = 1 \Rightarrow x + y = \pm 1$, two parallel straight lines of slope -1 and y-intercepts $b = \pm 1$

41. $r^2 = -4r \cos \theta \Rightarrow x^2 + y^2 = -4x \Rightarrow x^2 + 4x + y^2 = 0 \Rightarrow x^2 + 4x + 4 + y^2 = 4 \Rightarrow (x + 2)^2 + y^2 = 4$, a circle with center $C(-2, 0)$ and radius 2

43. $r = 8 \sin \theta \Rightarrow r^2 = 8r \sin \theta \Rightarrow x^2 + y^2 = 8y \Rightarrow x^2 + y^2 - 8y = 0 \Rightarrow x^2 + y^2 - 8y + 16 = 16$
    $\Rightarrow x^2 + (y - 4)^2 = 16$, a circle with center $C(0, 4)$ and radius 4

45. $r = 2 \cos \theta + 2 \sin \theta \Rightarrow r^2 = 2r \cos \theta + 2r \sin \theta \Rightarrow x^2 + y^2 = 2x + 2y \Rightarrow x^2 - 2x + y^2 - 2y = 0$
    $\Rightarrow (x - 1)^2 + (y - 1)^2 = 2$, a circle with center $C(1, 1)$ and radius $\sqrt{2}$

47. $r \sin \left(\theta + \frac{\pi}{6} \right) = 2 \Rightarrow r \left(\sin \theta \cos \frac{\pi}{6} + \cos \theta \sin \frac{\pi}{6} \right) = 2 \Rightarrow \frac{\sqrt{3}}{2} r \sin \theta + \frac{1}{2} r \cos \theta = 2 \Rightarrow \frac{\sqrt{3}}{2} y + \frac{1}{2} x = 2$
    $\Rightarrow \sqrt{3} y + x = 4$, line with slope $m = -\frac{1}{\sqrt{3}}$ and intercept $b = \frac{4}{\sqrt{3}}$

49. $x = 7 \Rightarrow r \cos \theta = 7$ 51. $x = y \Rightarrow r \cos \theta = r \sin \theta \Rightarrow \theta = \frac{\pi}{4}$

53. $x^2 + y^2 = 4 \Rightarrow r^2 = 4 \Rightarrow r = 2$ or $r = -2$

55. $\frac{x^2}{9} + \frac{y^2}{4} = 1 \Rightarrow 4x^2 + 9y^2 = 36 \Rightarrow 4r^2 \cos^2 \theta + 9r^2 \sin^2 \theta = 36$

57. $y^2 = 4x \Rightarrow r^2 \sin^2 \theta = 4r \cos \theta \Rightarrow r \sin^2 \theta = 4 \cos \theta$

59. $x^2 + (y - 2)^2 = 4 \Rightarrow x^2 + y^2 - 4y + 4 = 4 \Rightarrow x^2 + y^2 = 4y \Rightarrow r^2 = 4r \sin \theta \Rightarrow r = 4 \sin \theta$

61. $(x - 3)^2 + (y + 1)^2 = 4 \Rightarrow x^2 - 6x + 9 + y^2 + 2y + 1 = 4 \Rightarrow x^2 + y^2 = 6x - 2y - 6 \Rightarrow r^2 = 6r \cos \theta - 2r \sin \theta - 6$

63. $(0, \theta)$ where θ is any angle

9.7 GRAPHING IN POLAR COORDINATES

1. $1 + \cos (-\theta) = 1 + \cos \theta = r \Rightarrow$ symmetric about the x-axis;
   $1 + \cos (-\theta) \neq -r$ and $1 + \cos (\pi - \theta) = 1 - \cos \theta \neq r$
   $\Rightarrow$ not symmetric about the y-axis; therefore not symmetric
   about the origin

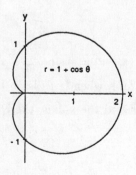

3. $1 - \sin(-\theta) = 1 + \sin\theta \neq r$ and $1 - \sin(\pi - \theta)$

    $= 1 - \sin\theta \neq -r \Rightarrow$ not symmetric about the x-axis;

    $1 - \sin(\pi - \theta) = 1 - \sin\theta = r \Rightarrow$ symmetric about

    the y-axis; therefore not symmetric about the origin

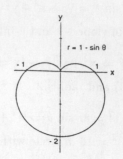

5. $2 + \sin(-\theta) = 2 - \sin\theta \neq r$ and $2 + \sin(\pi - \theta)$

    $= 2 + \sin\theta \neq r \Rightarrow$ not symmetric about the x-axis;

    $2 + \sin(\pi - \theta) = 2 + \sin\theta = r \Rightarrow$ symmetric about the

    y-axis; therefore not symmetric about the origin

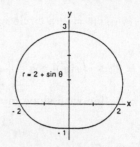

7. $\sin\left(-\frac{\theta}{2}\right) = -\sin\left(\frac{\theta}{2}\right) = -r \Rightarrow$ symmetric about the y-axis;

    $\sin\left(-\frac{\theta}{2}\right) = -\sin\left(\frac{\theta}{2}\right) = -r \neq r$ and $\sin\left(\frac{\pi - \theta}{2}\right) = \sin\left(\frac{\pi}{2} - \frac{\theta}{2}\right)$

    $= \cos\left(\frac{\theta}{2}\right) \neq -r$, but clearly the graph is symmetric about the

    x-axis and the origin. The symmetry tests as stated do not

    necessarily tell when a graph is not symmetric. Note that

    $\sin\left(\frac{2\pi - \theta}{2}\right) = \sin\left(\frac{\theta}{2}\right)$, so the graph is symmetric about the

    x-axis, and hence the origin.

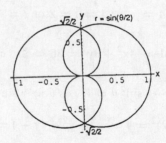

9. $\cos(-\theta) = \cos\theta = r^2 \Rightarrow (r, -\theta)$ and $(-r, -\theta)$ are on the graph

    when (r, θ) is on the graph $\Rightarrow$ symmetric about the x-axis and

    the y-axis; therefore symmetric about the origin

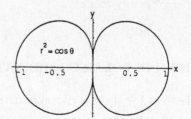

11. $-\sin(\pi - \theta) = -\sin\theta = r^2 \Rightarrow (r, \pi - \theta)$ and $(-r, \pi - \theta)$ are on

    the graph when (r, θ) is on the graph $\Rightarrow$ symmetric about the

    y-axis and the x-axis; therefore symmetric about the origin

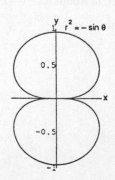

13. Since $(\pm r, -\theta)$ are on the graph when (r, θ) is on the graph
$\left((\pm r)^2 = 4\cos 2(-\theta) \Rightarrow r^2 = 4\cos 2\theta\right)$, the graph is
symmetric about the x-axis and the y-axis $\Rightarrow$ the graph is
symmetric about the origin

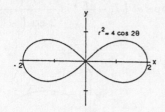

15. Since (r, θ) on the graph $\Rightarrow (-r, \theta)$ is on the graph
$\left((\pm r)^2 = -\sin 2\theta \Rightarrow r^2 = -\sin 2\theta\right)$, the graph is
symmetric about the origin. But $-\sin 2(-\theta) = -(-\sin 2\theta)$
$\sin 2\theta \neq r^2$ and $-\sin 2(\pi - \theta) = -\sin(2\pi - 2\theta)$
$= -\sin(-2\theta) = -(-\sin 2\theta) = \sin 2\theta \neq r^2 \Rightarrow$ the graph
is not symmetric about the x-axis; therefore the graph is
not symmetric about the y-axis

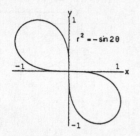

17. $\theta = \frac{\pi}{2} \Rightarrow r = -1 \Rightarrow \left(-1, \frac{\pi}{2}\right)$, and $\theta = -\frac{\pi}{2} \Rightarrow r = -1$
$\Rightarrow \left(-1, -\frac{\pi}{2}\right)$; $r' = \frac{dr}{d\theta} = -\sin\theta$; Slope $= \dfrac{r'\sin\theta + r\cos\theta}{r'\cos\theta - r\sin\theta}$

$= \dfrac{-\sin^2\theta + r\cos\theta}{-\sin\theta\cos\theta - r\sin\theta} \Rightarrow$ Slope at $\left(-1, \frac{\pi}{2}\right)$ is

$\dfrac{-\sin^2\left(\frac{\pi}{2}\right) + (-1)\cos\frac{\pi}{2}}{-\sin\frac{\pi}{2}\cos\frac{\pi}{2} - (-1)\sin\frac{\pi}{2}} = -1$; Slope at $\left(-1, -\frac{\pi}{2}\right)$ is

$\dfrac{-\sin^2\left(-\frac{\pi}{2}\right) + (-1)\cos\left(-\frac{\pi}{2}\right)}{-\sin\left(-\frac{\pi}{2}\right)\cos\left(-\frac{\pi}{2}\right) - (-1)\sin\left(-\frac{\pi}{2}\right)} = 1$

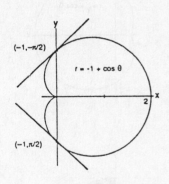

19. $\theta = \frac{\pi}{4} \Rightarrow r = 1 \Rightarrow \left(1, \frac{\pi}{4}\right)$; $\theta = -\frac{\pi}{4} \Rightarrow r = -1 \Rightarrow \left(-1, -\frac{\pi}{4}\right)$;
$\theta = \frac{3\pi}{4} \Rightarrow r = -1 \Rightarrow \left(-1, \frac{3\pi}{4}\right)$; $\theta = -\frac{3\pi}{4} \Rightarrow r = 1 \Rightarrow \left(1, -\frac{3\pi}{4}\right)$;
$r' = \frac{dr}{d\theta} = 2\cos 2\theta$;

Slope $= \dfrac{r'\sin\theta + r\cos\theta}{r'\cos\theta - r\sin\theta} = \dfrac{2\cos 2\theta\sin\theta + r\cos\theta}{2\cos 2\theta\cos\theta - r\sin\theta}$

$\Rightarrow$ Slope at $\left(1, \frac{\pi}{4}\right)$ is $\dfrac{2\cos\left(\frac{\pi}{2}\right)\sin\left(\frac{\pi}{4}\right) + (1)\cos\left(\frac{\pi}{4}\right)}{2\cos\left(\frac{\pi}{2}\right)\cos\left(\frac{\pi}{4}\right) - (1)\sin\left(\frac{\pi}{4}\right)} = 1$;

Slope at $\left(-1, -\frac{\pi}{4}\right)$ is $\dfrac{2\cos\left(-\frac{\pi}{2}\right)\sin\left(-\frac{\pi}{4}\right) + (-1)\cos\left(-\frac{\pi}{4}\right)}{2\cos\left(-\frac{\pi}{2}\right)\cos\left(-\frac{\pi}{4}\right) - (-1)\sin\left(-\frac{\pi}{4}\right)} = 1$;

Slope at $\left(-1, \frac{3\pi}{4}\right)$ is $\dfrac{2\cos\left(\frac{3\pi}{2}\right)\sin\left(\frac{3\pi}{4}\right) + (-1)\cos\left(\frac{3\pi}{4}\right)}{2\cos\left(\frac{3\pi}{2}\right)\cos\left(\frac{3\pi}{4}\right) - (-1)\sin\left(\frac{3\pi}{4}\right)} = 1$;

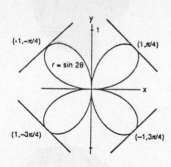

Slope at $\left(1, -\frac{3\pi}{4}\right)$ is $\dfrac{2\cos\left(-\frac{3\pi}{2}\right)\sin\left(-\frac{3\pi}{4}\right) + (1)\cos\left(-\frac{3\pi}{4}\right)}{2\cos\left(-\frac{3\pi}{2}\right)\cos\left(-\frac{3\pi}{4}\right) - (1)\sin\left(-\frac{3\pi}{4}\right)} = -1$

21. (a)

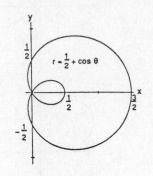

(b)

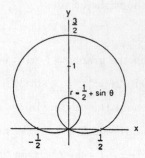

23. (a)

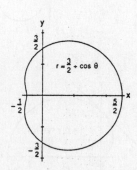

(b)

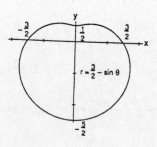

25.

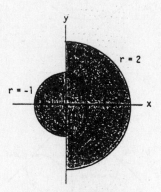

27.

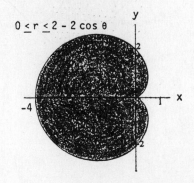

29. $\left(2, \frac{3\pi}{4}\right)$ is the same point as $\left(-2, -\frac{\pi}{4}\right)$; $r = 2 \sin 2\left(-\frac{\pi}{4}\right) = 2 \sin\left(-\frac{\pi}{2}\right) = -2 \Rightarrow \left(-2, -\frac{\pi}{4}\right)$ is on the graph

$\Rightarrow \left(2, \frac{3\pi}{4}\right)$ is on the graph

31. $1 + \cos\theta = 1 - \cos\theta \Rightarrow \cos\theta = 0 \Rightarrow \theta = \frac{\pi}{2}, \frac{3\pi}{2} \Rightarrow r = 1$;

points of intersection are $\left(1, \frac{\pi}{2}\right)$ and $\left(1, \frac{3\pi}{2}\right)$. The point of

intersection $(0, 0)$ is found by graphing.

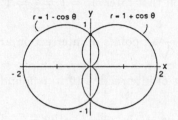

33. $2 \sin\theta = 2 \sin 2\theta \Rightarrow \sin\theta = \sin 2\theta \Rightarrow \sin\theta$

$\quad = 2 \sin\theta\cos\theta \Rightarrow \sin\theta - 2 \sin\theta\cos\theta = 0$

$\Rightarrow (\sin\theta)(1 - 2\cos\theta) = 0 \Rightarrow \sin\theta = 0$ or $\cos\theta = \frac{1}{2}$

$\Rightarrow \theta = 0, \frac{\pi}{3},$ or $-\frac{\pi}{3}$; $\theta = 0 \Rightarrow r = 0$, $\theta = \frac{\pi}{3} \Rightarrow r = \sqrt{3}$,

and $\theta = -\frac{\pi}{3} \Rightarrow r = -\sqrt{3}$; points of intersection are

$(0, 0), \left(\sqrt{3}, \frac{\pi}{3}\right),$ and $\left(-\sqrt{3}, -\frac{\pi}{3}\right)$

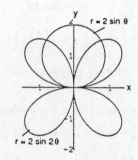

35. $\left(\sqrt{2}\right)^2 = 4 \sin\theta \Rightarrow \frac{1}{2} = \sin\theta \Rightarrow \theta = \frac{\pi}{6}, \frac{5\pi}{6}$; points

of intersection are $\left(\sqrt{2}, \frac{\pi}{6}\right)$ and $\left(\sqrt{2}, \frac{5\pi}{6}\right)$. The

points $\left(\sqrt{2}, -\frac{\pi}{6}\right)$ and $\left(\sqrt{2}, -\frac{5\pi}{6}\right)$ are found by

graphing.

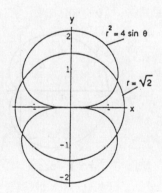

37. $1 = 2 \sin 2\theta \Rightarrow \sin 2\theta = \frac{1}{2} \Rightarrow 2\theta = \frac{\pi}{6}, \frac{5\pi}{6}, \frac{13\pi}{6}, \frac{17\pi}{6}$

$\Rightarrow \theta = \frac{\pi}{12}, \frac{5\pi}{12}, \frac{13\pi}{12}, \frac{17\pi}{12}$; points of intersection are

$\left(1, \frac{\pi}{12}\right), \left(1, \frac{5\pi}{12}\right), \left(1, \frac{13\pi}{12}\right),$ and $\left(1, \frac{17\pi}{12}\right)$. No other

points are found by graphing.

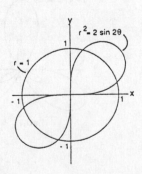

39. $r^2 = \sin 2\theta$ and $r^2 = \cos 2\theta$ are generated completely for

$0 \le \theta \le \frac{\pi}{2}$. Then $\sin 2\theta = \cos 2\theta \Rightarrow 2\theta = \frac{\pi}{4}$ is the only

solution on that interval $\Rightarrow \theta = \frac{\pi}{8} \Rightarrow r^2 = \sin 2\left(\frac{\pi}{8}\right) = \frac{1}{\sqrt{2}}$

$\Rightarrow r = \pm \frac{1}{\sqrt[4]{2}}$; points of intersection are $\left(\pm \frac{1}{\sqrt[4]{2}}, \frac{\pi}{8}\right)$.

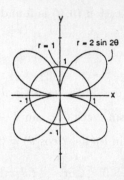

The point of intersection $(0,0)$ is found by graphing.

41. $1 = 2\sin 2\theta \Rightarrow \sin 2\theta = \frac{1}{2} \Rightarrow 2\theta = \frac{\pi}{6}, \frac{5\pi}{6}, \frac{13\pi}{6}, \frac{17\pi}{6}$

$\Rightarrow \theta = \frac{\pi}{12}, \frac{5\pi}{12}, \frac{13\pi}{12}, \frac{17\pi}{12}$; points of intersection are

$\left(1, \frac{\pi}{12}\right), \left(1, \frac{5\pi}{12}\right), \left(1, \frac{13\pi}{12}\right)$, and $\left(1, \frac{17\pi}{12}\right)$. The points

of intersection $\left(1, \frac{7\pi}{12}\right), \left(1, \frac{11\pi}{12}\right), \left(1, \frac{19\pi}{12}\right)$ and

$\left(1, \frac{23\pi}{12}\right)$ are found by graphing and symmetry.

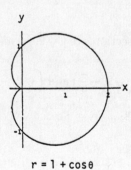

43. Note that (r, θ) and $(-r, \theta + \pi)$ describe the same point in the plane. Then $r = 1 - \cos\theta \Leftrightarrow -1 - \cos(\theta + \pi)$

$= -1 - (\cos\theta \cos\pi - \sin\theta \sin\pi) = -1 + \cos\theta = -(1 - \cos\theta) = -r$; therefore (r, θ) is on the graph of

$r = 1 - \cos\theta \Leftrightarrow (-r, \theta + \pi)$ is on the graph of $r = -1 - \cos\theta \Rightarrow$ the answer is (a).

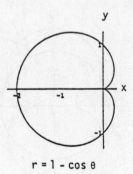

$r = 1 - \cos\theta$

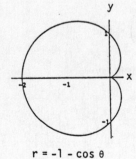

$r = -1 - \cos\theta$

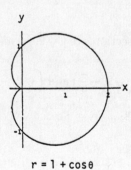

$r = 1 + \cos\theta$

45.

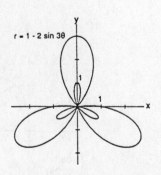

47. (a)   (b)   (c)   (d)

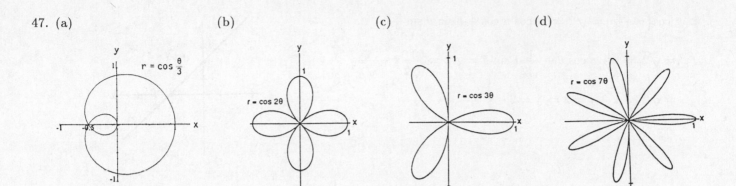

49. (a) $r^2 = -4 \cos \theta \Rightarrow \cos \theta = -\dfrac{r^2}{4}$; $r = 1 - \cos \theta \Rightarrow r = 1 - \left(-\dfrac{r^2}{4}\right) \Rightarrow 0 = r^2 - 4r + 4 \Rightarrow (r-2)^2 = 0$

$\Rightarrow r = 2$; therefore $\cos \theta = -\dfrac{2^2}{4} = -1 \Rightarrow \theta = \pi \Rightarrow (2, \pi)$ is a point of intersection

(b) $r = 0 \Rightarrow 0^2 = 4 \cos \theta \Rightarrow \cos \theta = 0 \Rightarrow \theta = \dfrac{\pi}{2}, \dfrac{3\pi}{2} \Rightarrow \left(0, \dfrac{\pi}{2}\right)$ or $\left(0, \dfrac{3\pi}{2}\right)$ is on the graph; $r = 0 \Rightarrow 0 = 1 - \cos \theta$

$\Rightarrow \cos \theta = 1 \Rightarrow \theta = 0 \Rightarrow (0,0)$ is on the graph. Since $(0,0) = \left(0, \dfrac{\pi}{2}\right)$ for polar coordinates, the graphs

intersect at the origin.

51. The maximum width of the petal of the rose which lies along the x-axis is twice the largest y value of the curve

on the interval $0 \le \theta \le \dfrac{\pi}{4}$. So we wish to maximize $2y = 2r \sin \theta = 2 \cos 2\theta \sin \theta$ on $0 \le \theta \le \dfrac{\pi}{4}$. Let

$f(\theta) = 2 \cos 2\theta \sin \theta = 2\left(1 - 2 \sin^2 \theta\right)(\sin \theta) = 2 \sin \theta - 4 \sin^3 \theta \Rightarrow f'(\theta) = 2 \cos \theta - 12 \sin^2 \theta \cos \theta$. Then

$f'(\theta) = 0 \Rightarrow 2 \cos \theta - 12 \sin^2 \theta \cos \theta = 0 \Rightarrow (\cos \theta)\left(1 - 6 \sin^2 \theta\right) = 0 \Rightarrow \cos \theta = 0$ or $1 - 6 \sin^2 \theta = 0 \Rightarrow \theta = \dfrac{\pi}{2}$ or

$\sin \theta = \dfrac{\pm 1}{\sqrt{6}}$. Since we want $0 \le \theta \le \dfrac{\pi}{4}$, we choose $\theta = \sin^{-1}\left(\dfrac{1}{\sqrt{6}}\right) \Rightarrow f(\theta) = 2 \sin \theta - 2 \sin^3 \theta$

$= 2\left(\dfrac{1}{\sqrt{6}}\right) - 4 \cdot \dfrac{1}{6\sqrt{6}} = \dfrac{2\sqrt{6}}{9}$. We can see from the graph of $r = \cos 2\theta$ that a maximum does occur in the

interval $0 \le \theta \le \dfrac{\pi}{4}$. Therefore the maximum width occurs at $\theta = \sin^{-1}\left(\dfrac{1}{\sqrt{6}}\right)$, and the maximum width

is $\dfrac{2\sqrt{6}}{9}$.

9.8 POLAR EQUATIONS OF CONIC SECTIONS

1. $r \cos\left(\theta - \dfrac{\pi}{6}\right) = 5 \Rightarrow r\left(\cos \theta \cos \dfrac{\pi}{6} + \sin \theta \sin \dfrac{\pi}{6}\right) = 5 \Rightarrow \dfrac{\sqrt{3}}{2} r \cos \theta + \dfrac{1}{2} r \sin \theta = 5 \Rightarrow \dfrac{\sqrt{3}}{2} x + \dfrac{1}{2} y = 5 \Rightarrow \sqrt{3}\, x + y$

$= 10 \Rightarrow y = -\sqrt{3}\, x + 10$

3. $r \cos\left(\theta - \dfrac{4\pi}{3}\right) = 3 \Rightarrow r\left(\cos \theta \cos \dfrac{4\pi}{3} + \sin \theta \sin \dfrac{4\pi}{3}\right) = 3 \Rightarrow -\dfrac{1}{2} r \cos \theta - \dfrac{\sqrt{3}}{2} r \sin \theta = 3$

$\Rightarrow -\dfrac{1}{2} x - \dfrac{\sqrt{3}}{2} y = 3 \Rightarrow x + \sqrt{3}\, y = -6 \Rightarrow y = -\dfrac{\sqrt{3}}{3} x - 2\sqrt{3}$

5. $r \cos\left(\theta - \frac{\pi}{4}\right) = \sqrt{2} \Rightarrow r\left(\cos\theta \cos\frac{\pi}{4} + \sin\theta \sin\frac{\pi}{4}\right)$

$= \sqrt{2} \Rightarrow \frac{1}{\sqrt{2}}r\cos\theta + \frac{1}{\sqrt{2}}r\sin\theta = \sqrt{2} \Rightarrow \frac{1}{\sqrt{2}}x + \frac{1}{\sqrt{2}}y$

$= \sqrt{2} \Rightarrow x + y = 2 \Rightarrow y = 2 - x$

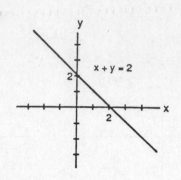

7. $r\cos\left(\theta - \frac{2\pi}{3}\right) = 3 \Rightarrow r\left(\cos\theta\cos\frac{2\pi}{3} + \sin\theta\sin\frac{2\pi}{3}\right) = 3$

$\Rightarrow -\frac{1}{2}r\cos\theta + \frac{\sqrt{3}}{2}r\sin\theta = 3 \Rightarrow -\frac{1}{2}x + \frac{\sqrt{3}}{2}y = 3$

$\Rightarrow -x + \sqrt{3}y = 6 \Rightarrow y = \frac{\sqrt{3}}{3}x + 2\sqrt{3}$

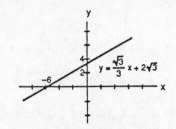

9. $\sqrt{2}x + \sqrt{2}y = 6 \Rightarrow \sqrt{2}r\cos\theta + \sqrt{2}r\sin\theta = 6 \Rightarrow r\left(\frac{\sqrt{2}}{2}\cos\theta + \frac{\sqrt{2}}{2}\sin\theta\right) = 3 \Rightarrow r\left(\cos\frac{\pi}{4}\cos\theta + \sin\frac{\pi}{4}\sin\theta\right)$

$= 3 \Rightarrow r\cos\left(\theta - \frac{\pi}{4}\right) = 3$

11. $y = -5 \Rightarrow r\sin\theta = -5 \Rightarrow -r\sin\theta = 5 \Rightarrow r\sin(-\theta) = 5 \Rightarrow r\cos\left(\frac{\pi}{2} - (-\theta)\right) = 5 \Rightarrow r\cos\left(\theta + \frac{\pi}{2}\right) = 5$

13. $r = 2(4)\cos\theta = 8\cos\theta$

15. $r = 2\sqrt{2}\sin\theta$

17.

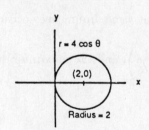

19.

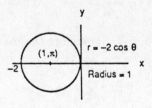

21. $(x - 6)^2 + y^2 = 36 \Rightarrow C = (6, 0),\ a = 6$

$\Rightarrow r = 12\cos\theta$ is the polar equation

23. $x^2 + (y - 5)^2 = 25 \Rightarrow C = (0, 5),\ a = 5$

$\Rightarrow r = 10\sin\theta$ is the polar equation

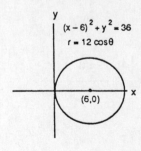

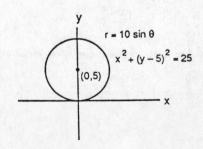

25. $x^2 + 2x + y^2 = 0 \Rightarrow (x+1)^2 + y^2 = 1$

$\Rightarrow C = (-1, 0), a = 1 \Rightarrow r = -2\cos\theta$ is

the polar equation

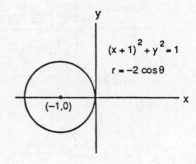

27. $x^2 + y^2 + y = 0 \Rightarrow x^2 + \left(y + \frac{1}{2}\right)^2 = \frac{1}{4}$

$\Rightarrow C = \left(0, -\frac{1}{2}\right), a = \frac{1}{2} \Rightarrow r = -\sin\theta$ is the

polar equation

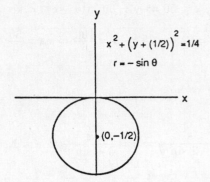

29. $e = 1, x = 2 \Rightarrow k = 2 \Rightarrow r = \dfrac{2(1)}{1 + (1)\cos\theta} = \dfrac{2}{1 + \cos\theta}$

31. $e = 5, y = -6 \Rightarrow k = 6 \Rightarrow r = \dfrac{6(5)}{1 - 5\sin\theta} = \dfrac{30}{1 - 5\sin\theta}$

33. $e = \frac{1}{2}, x = 1 \Rightarrow k = 1 \Rightarrow r = \dfrac{\left(\frac{1}{2}\right)(1)}{1 + \left(\frac{1}{2}\right)\cos\theta} = \dfrac{1}{2 + \cos\theta}$

35. $e = \frac{1}{5}, x = -10 \Rightarrow k = 10 \Rightarrow r = \dfrac{\left(\frac{1}{5}\right)(10)}{1 - \left(\frac{1}{5}\right)\sin\theta} = \dfrac{10}{5 - \sin\theta}$

37. $r = \dfrac{1}{1 + \cos\theta} \Rightarrow e = 1, k = 1 \Rightarrow x = 1$

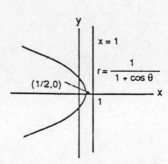

39. $r = \dfrac{25}{10 - 5\cos\theta} \Rightarrow r = \dfrac{\left(\frac{25}{10}\right)}{1 - \left(\frac{5}{10}\right)\cos\theta} = \dfrac{\left(\frac{5}{2}\right)}{1 - \left(\frac{1}{2}\right)\cos\theta}$

$\Rightarrow e = \frac{1}{2}, k = 5 \Rightarrow x = -5; a(1 - e^2) = ke \Rightarrow a\left[1 - \left(\frac{1}{2}\right)^2\right]$

$= \frac{5}{2} \Rightarrow \frac{3}{4}a = \frac{5}{2} \Rightarrow a = \frac{10}{3} \Rightarrow ea = \frac{5}{3}$

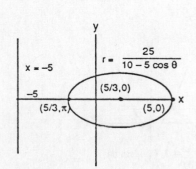

41. $r = \dfrac{400}{16 + 8\sin\theta} \Rightarrow r = \dfrac{\left(\dfrac{400}{16}\right)}{1 + \left(\dfrac{8}{16}\right)\sin\theta} \Rightarrow r = \dfrac{25}{1 + \left(\dfrac{1}{2}\right)\sin\theta}$

$e = \dfrac{1}{2},\ k = 50 \Rightarrow y = 50;\ a(1 - e^2) = ke \Rightarrow a\left[1 - \left(\dfrac{1}{2}\right)^2\right]$

$= 25 \Rightarrow \dfrac{3}{4}a = 25 \Rightarrow a = \dfrac{100}{3} \Rightarrow ea = \dfrac{50}{3}$

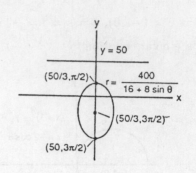

43. $r = \dfrac{8}{2 - 2\sin\theta} \Rightarrow r = \dfrac{4}{1 - \sin\theta} \Rightarrow e = 1,$

$k = 4 \Rightarrow y = -4$

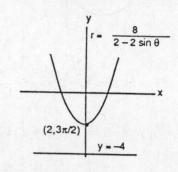

45.

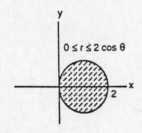

47.

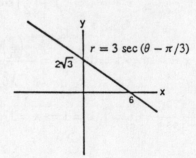

49.

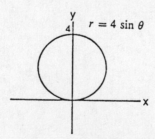

51.

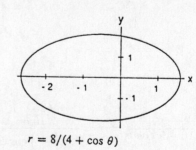

$r = 8/(4 + \cos\theta)$

53.

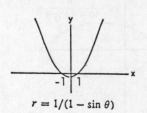

$r = 1/(1 - \sin\theta)$

55.

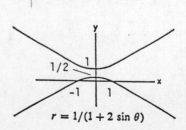

$r = 1/(1 + 2\sin\theta)$

57. (a) Perihelion $= a - ae = a(1 - e)$, Aphelion $= ea + a = a(1 + e)$

(b)

Planet	Perihelion	Aphelion
Mercury	0.3075 AU	0.4667 AU
Venus	0.7184 AU	0.7282 AU
Earth	0.9833 AU	1.0167 AU
Mars	1.3817 AU	1.6663 AU
Jupiter	4.9512 AU	5.4548 AU
Satrun	9.0210 AU	10.0570 AU
Uranus	18.2977 AU	20.0623 AU
Neptune	29.8135 AU	30.3065 AU
Pluto	29.6549 AU	49.2251 AU

59. (a) $r = 4 \sin \theta \Rightarrow r^2 = 4r \sin \theta \Rightarrow x^2 + y^2 = 4y$;

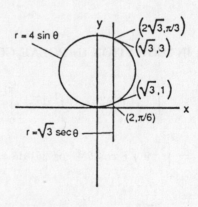

$r = \sqrt{3} \sec \theta \Rightarrow r = \dfrac{\sqrt{3}}{\cos \theta} \Rightarrow r \cos \theta = \sqrt{3}$

$\Rightarrow x = \sqrt{3}; \ x = \sqrt{3} \Rightarrow \left(\sqrt{3}\right)^2 + y^2 = 4y$

$\Rightarrow y^2 - 4y + 3 = 0 \Rightarrow (y - 3)(y - 1) = 0 \Rightarrow y = 3$

or $y = 1$. Therefore in Cartesian coordinates, the points

of intersection are $\left(\sqrt{3}, 3\right)$ and $\left(\sqrt{3}, 1\right)$. In polar

coordinates, $4 \sin \theta = \sqrt{3} \sec \theta \Rightarrow 4 \sin \theta \cos \theta = \sqrt{3}$

$\Rightarrow 2 \sin \theta \cos \theta = \dfrac{\sqrt{3}}{2} \Rightarrow \sin 2\theta = \dfrac{\sqrt{3}}{2} \Rightarrow 2\theta = \dfrac{\pi}{3}$ or

$\dfrac{2\pi}{3} \Rightarrow \theta = \dfrac{\pi}{6}$ or $\dfrac{\pi}{3}$; $\theta = \dfrac{\pi}{6} \Rightarrow r = 2$, and $\theta = \dfrac{\pi}{3} \Rightarrow r = 2\sqrt{3} \Rightarrow \left(2, \dfrac{\pi}{6}\right)$ and $\left(2\sqrt{3}, \dfrac{\pi}{3}\right)$ are the points of

intersection in polar coordinates.

61. $r \cos \theta = 4 \Rightarrow x = 4 \Rightarrow k = 4$: parabola $\Rightarrow e = 1 \Rightarrow r = \dfrac{4}{1 + \cos \theta}$

63. (a) Let the ellipse be the orbit, with the Sun at one focus.

Then $r_{max} = a + c$ and $r_{min} = a - c \Rightarrow \dfrac{r_{max} - r_{min}}{r_{max} + r_{min}}$

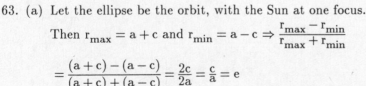

$= \dfrac{(a + c) - (a - c)}{(a + c) + (a - c)} = \dfrac{2c}{2a} = \dfrac{c}{a} = e$

(b) Let F_1, F_2 be the foci. Then $PF_1 + PF_2 = 10$ where

P is any point on the ellipse. If P is a vertex, then

$PF_1 = a + c$ and $PF_2 = a - c \Rightarrow (a + c) + (a - c) = 10$

$\Rightarrow 2a = 10 \Rightarrow a = 5$. Since $e = \dfrac{c}{a}$ we have $0.2 = \dfrac{c}{5} \Rightarrow c = 1.0 \Rightarrow$ the pins should be 2 inches apart.

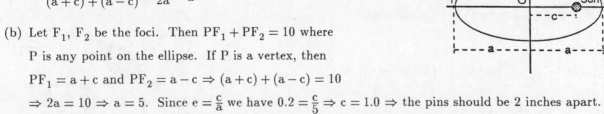

65. $x^2 + y^2 - 2ay = 0 \Rightarrow (r \cos \theta)^2 + (r \sin \theta)^2 - 2ar \sin \theta = 0$

$\Rightarrow r^2 \cos^2 \theta + r^2 \sin^2 \theta - 2ar \sin \theta = 0 \Rightarrow r^2 = 2ar \sin \theta$

$\Rightarrow r = 2a \sin \theta$

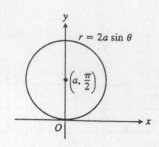

67. $x \cos \alpha + y \sin \alpha = p \Rightarrow r \cos \theta \cos \alpha + r \sin \theta \sin \alpha = p$

$\Rightarrow r(\cos \theta \cos \alpha + \sin \theta \sin \alpha) = p \Rightarrow r \cos(\theta - \alpha) = p$

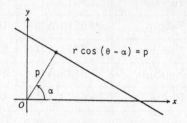

9.9 INTEGRATION IN POLAR COORDINATES

1. $A = \int_0^{2\pi} \frac{1}{2}(4 + 2 \cos \theta)^2 \, d\theta = \int_0^{2\pi} \frac{1}{2}(16 + 16 \cos \theta + 4 \cos^2 \theta) \, d\theta = \int_0^{2\pi} \left[8 + 8 \cos \theta + 2\left(\frac{1 + \cos 2\theta}{2}\right)\right] d\theta$

$= \int_0^{2\pi} (9 + 8 \cos \theta + \cos 2\theta) \, d\theta = \left[9\theta + 8 \sin \theta + \frac{1}{2} \sin 2\theta\right]_0^{2\pi} = 18\pi$

3. $A = 2 \int_0^{\pi/4} \frac{1}{2} \cos^2 2\theta \, d\theta = \int_0^{\pi/4} \frac{1 + \cos 4\theta}{2} \, d\theta = \frac{1}{2}\left[\theta + \frac{\sin 4\theta}{4}\right]_0^{\pi/4} = \frac{\pi}{8}$

5. $A = 2 \int_0^{\pi/2} \frac{1}{2}(4 \sin 2\theta) \, d\theta = \int_0^{\pi/2} 2 \sin 2\theta \, d\theta = [-\cos 2\theta]_0^{\pi/2} = 2$

7. $r = 2 \cos \theta$ and $r = 2 \sin \theta \Rightarrow 2 \cos \theta = 2 \sin \theta$

$\Rightarrow \cos \theta = \sin \theta \Rightarrow \theta = \frac{\pi}{4}$; therefore

$A = 2 \int_0^{\pi/4} \frac{1}{2}(2 \sin \theta)^2 \, d\theta = \int_0^{\pi/4} 4 \sin^2 \theta \, d\theta$

$= \int_0^{\pi/4} 4\left(\frac{1 - \cos 2\theta}{2}\right) d\theta = \int_0^{\pi/4} (2 - 2 \cos 2\theta) \, d\theta$

$= [2\theta - \sin 2\theta]_0^{\pi/4} = \frac{\pi}{2} - 1$

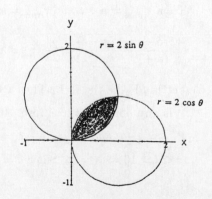

9. r = 2 and r = 2(1 − cos θ) ⇒ 2 = 2(1 − cos θ) ⇒ cos θ = 0

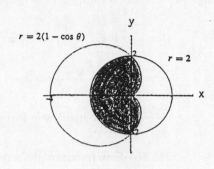

$\Rightarrow \theta = \pm\frac{\pi}{2}$; therefore $A = 2 \int_{0}^{\pi/2} \frac{1}{2}[2(1-\cos\theta)]^2 \, d\theta$

$+\frac{1}{2}$ area of the circle $= \int_{0}^{\pi/2} 4\left(1 - 2\cos\theta + \cos^2\theta\right) d\theta + \left(\frac{1}{2}\pi\right)(2)^2$

$= \int_{0}^{\pi/2} 4\left(1 - 2\cos\theta + \frac{1+\cos 2\theta}{2}\right) d\theta + 2\pi$

$= \int_{0}^{\pi/2} (4 - 8\cos\theta + 2 + 2\cos 2\theta) \, d\theta + 2\pi$

$= \left[6\theta - 8\sin\theta + \sin 2\theta\right]_{0}^{\pi/2} + 2\pi = 5\pi - 8$

11. r = √3 and r² = 6 cos 2θ ⇒ 3 = 6 cos 2θ ⇒ cos 2θ = $\frac{1}{2}$

$\Rightarrow \theta = \frac{\pi}{6}$ (in the 1st quadrant); we use symmetry of the

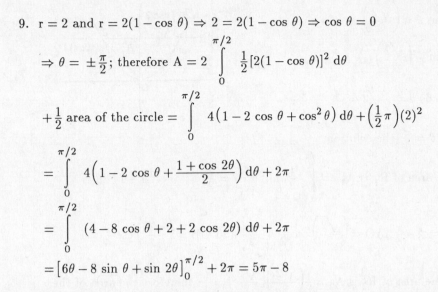

graph to find the area, so $A = 4 \int_{0}^{\pi/6} \left[\frac{1}{2}(6\cos 2\theta) - \frac{1}{2}\left(\sqrt{3}\right)^2\right] d\theta$

$= 2 \int_{0}^{\pi/6} (6\cos 2\theta - 3) \, d\theta = 2[3\sin 2\theta - 3\theta]_{0}^{\pi/6} = 3\sqrt{3} - \pi$

13. r = 1 and r = −2 cos θ ⇒ 1 = −2 cos θ ⇒ cos θ = $-\frac{1}{2}$

$\Rightarrow \theta = \frac{2\pi}{3}$ in quadrant II; therefore

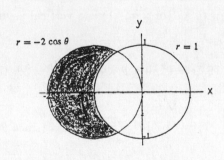

$A = 2 \int_{2\pi/3}^{\pi} \frac{1}{2}[(-2\cos\theta)^2 - 1^2] \, d\theta = \int_{2\pi/3}^{\pi} \left(4\cos^2\theta - 1\right) d\theta$

$= \int_{2\pi/3}^{\pi} [2(1 + \cos 2\theta) - 1] \, d\theta = \int_{2\pi/3}^{\pi} (1 + 2\cos 2\theta) \, d\theta$

$= \left[\theta + \sin 2\theta\right]_{2\pi/3}^{\pi} = \frac{\pi}{3} + \frac{\sqrt{3}}{2}$

15. r = 6 and r = 3 csc θ ⇒ 6 sin θ = 3 ⇒ sin θ = $\frac{1}{2}$ ⇒ θ = $\frac{\pi}{6}$

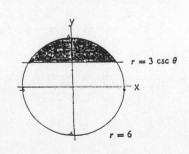

or $\frac{5\pi}{6}$; therefore $A = \int_{\pi/6}^{5\pi/6} \frac{1}{2}\left(6^2 - 9\csc^2\theta\right) d\theta$

$= \int_{\pi/6}^{5\pi/6} \left(18 - \frac{9}{2}\csc^2\theta\right) d\theta = \left[18\theta + \frac{9}{2}\cot\theta\right]_{\pi/6}^{5\pi/6}$

$= \left(15\pi - \frac{9}{2}\sqrt{3}\right) - \left(3\pi + \frac{9}{2}\sqrt{3}\right) = 12\pi - 9\sqrt{3}$

17. (a) $r = \tan \theta$ and $r = \left(\dfrac{\sqrt{2}}{2}\right) \csc \theta \Rightarrow \tan \theta = \left(\dfrac{\sqrt{2}}{2}\right) \csc \theta$

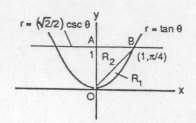

$\Rightarrow \sin^2 \theta = \left(\dfrac{\sqrt{2}}{2}\right) \cos \theta \Rightarrow 1 - \cos^2 \theta = \left(\dfrac{\sqrt{2}}{2}\right) \cos \theta$

$\Rightarrow \cos^2 \theta + \left(\dfrac{\sqrt{2}}{2}\right) \cos \theta - 1 = 0 \Rightarrow \cos \theta = -\sqrt{2}$ or

$\dfrac{\sqrt{2}}{2}$ (use the quadratic formula) $\Rightarrow \theta = \dfrac{\pi}{4}$ (the solution

in the first quadrant); therefore the area of R_1 is $A_1 = \displaystyle\int_0^{\pi/4} \dfrac{1}{2} \tan^2 \theta \; d\theta = \dfrac{1}{2} \int_0^{\pi/4} (\sec^2 \theta - 1) \; d\theta$

$= \dfrac{1}{2}[\tan \theta - \theta]_0^{\pi/4} = \dfrac{1}{2}\left(\tan \dfrac{\pi}{4} - \dfrac{\pi}{4}\right) = \dfrac{1}{2} - \dfrac{\pi}{8}$; $AO = \left(\dfrac{\sqrt{2}}{2}\right) \csc \dfrac{\pi}{2} = \dfrac{\sqrt{2}}{2}$ and $OB = \left(\dfrac{\sqrt{2}}{2}\right) \csc \dfrac{\pi}{4} = 1$

$\Rightarrow AB = \sqrt{1^2 - \left(\dfrac{\sqrt{2}}{2}\right)^2} = \dfrac{\sqrt{2}}{2} \Rightarrow$ the area of R_2 is $A_2 = \dfrac{1}{2}\left(\dfrac{\sqrt{2}}{2}\right)\left(\dfrac{\sqrt{2}}{2}\right) = \dfrac{1}{4}$; therefore the area of the

region shaded in the text is $2\left(\dfrac{1}{2} - \dfrac{\pi}{8} + \dfrac{1}{4}\right) = \dfrac{3}{2} - \dfrac{\pi}{4}$. Note: The area must be found this way since no

common interval generates the region. For example, the interval $0 \leq \theta \leq \dfrac{\pi}{4}$ generates the arc OB of

$r = \tan \theta$ but does not generate the segment AB of the line $r = \dfrac{\sqrt{2}}{2} \csc \theta$. Instead the interval generates

the half-line from B to $+\infty$ on the line $r = \dfrac{\sqrt{2}}{2} \csc \theta$.

(b) $\displaystyle\lim_{\theta \to \pi/2^-} \tan \theta = \infty$ and the line $x = 1$ is $r = \sec \theta$ in polar coordinates; then $\displaystyle\lim_{\theta \to \pi/2^-} (\tan \theta - \sec \theta)$

$= \displaystyle\lim_{\theta \to \pi/2^-} \left(\dfrac{\sin \theta}{\cos \theta} - \dfrac{1}{\cos \theta}\right) = \lim_{\theta \to \pi/2^-} \left(\dfrac{\sin \theta - 1}{\cos \theta}\right) = \lim_{\theta \to \pi/2^-} \left(\dfrac{\cos \theta}{-\sin \theta}\right) = 0 \Rightarrow r = \tan \theta$ approaches

$r = \sec \theta$ as $\theta \to \dfrac{\pi^-}{2} \Rightarrow r = \sec \theta$ (or $x = 1$) is a vertical asymptote of $r = \tan \theta$. Similarly, $r = -\sec \theta$

(or $x = -1$) is a vertical asymptote of $r = \tan \theta$.

19. $r = \theta^2$, $0 \leq \theta \leq \sqrt{5} \Rightarrow \dfrac{dr}{d\theta} = 2\theta$; therefore Length $= \displaystyle\int_0^{\sqrt{5}} \sqrt{\left(\theta^2\right)^2 + (2\theta)^2} \; d\theta = \int_0^{\sqrt{5}} \sqrt{\theta^4 + 4\theta^2} \; d\theta$

$= \displaystyle\int_0^{\sqrt{5}} |\theta| \sqrt{\theta^2 + 4} \; d\theta = \text{(since } \theta \geq 0) \int_0^{\sqrt{5}} \theta\sqrt{\theta^2 + 4} \; d\theta; \left[u = \theta^2 + 4 \Rightarrow \dfrac{1}{2} du = \theta \; d\theta; \theta = 0 \Rightarrow u = 4,\right.$

$\theta = \sqrt{5} \Rightarrow u = 9] \to \displaystyle\int_4^9 \dfrac{1}{2}\sqrt{u} \; du = \dfrac{1}{2}\left[\dfrac{2}{3}u^{3/2}\right]_4^9 = \dfrac{19}{3}$

21. $r = 1 + \cos \theta \Rightarrow \dfrac{dr}{d\theta} = -\sin \theta$; therefore Length $= \displaystyle\int_0^{2\pi} \sqrt{(1 + \cos \theta)^2 + (-\sin \theta)^2} \; d\theta$

$= 2 \displaystyle\int_0^{\pi} \sqrt{2 + 2 \cos \theta} \; d\theta = 2 \int_0^{\pi} \sqrt{\dfrac{4(1 + \cos \theta)}{2}} \; d\theta = 4 \int_0^{\pi} \sqrt{\dfrac{1 + \cos \theta}{2}} \; d\theta = 4 \int_0^{\pi} \cos\left(\dfrac{\theta}{2}\right) d\theta = 4\left[2 \sin \dfrac{\theta}{2}\right]_0^{\pi} = 8$

23. $r = \dfrac{6}{1 + \cos\theta}$, $0 \le \theta \le \dfrac{\pi}{2} \Rightarrow \dfrac{dr}{d\theta} = \dfrac{6\sin\theta}{(1+\cos\theta)^2}$; therefore Length $= \displaystyle\int_0^{\pi/2} \sqrt{\left(\dfrac{6}{1+\cos\theta}\right)^2 + \left(\dfrac{6\sin\theta}{(1+\cos\theta)^2}\right)^2}\, d\theta$

$= \displaystyle\int_0^{\pi/2} \sqrt{\dfrac{36}{(1+\cos\theta)^2} + \dfrac{36\sin^2\theta}{(1+\cos^2\theta)^4}}\, d\theta = 6\displaystyle\int_0^{\pi/2} \left|\dfrac{1}{1+\cos\theta}\right| \sqrt{1 + \dfrac{\sin^2\theta}{(1+\cos\theta)^2}}\, d\theta$

$= \left(\text{since } \dfrac{1}{1+\cos\theta} > 0 \text{ on } 0 \le \theta \le \dfrac{\pi}{2}\right) 6\displaystyle\int_0^{\pi/2} \left(\dfrac{1}{1+\cos\theta}\right) \sqrt{\dfrac{1 + 2\cos\theta + \cos^2\theta + \sin^2\theta}{(1+\cos\theta)^2}}\, d\theta$

$= 6\displaystyle\int_0^{\pi/2} \left(\dfrac{1}{1+\cos\theta}\right) \sqrt{\dfrac{2 + 2\cos\theta}{(1+\cos\theta)^2}}\, d\theta = 6\sqrt{2}\displaystyle\int_0^{\pi/2} \dfrac{d\theta}{(1+\cos\theta)^{3/2}} = 6\sqrt{2}\displaystyle\int_0^{\pi/2} \dfrac{d\theta}{\left(2\cos^2\frac{\theta}{2}\right)^{3/2}} = 6\displaystyle\int_0^{\pi/2} \left|\sec^3\dfrac{\theta}{2}\right|\, d\theta$

$= 6\displaystyle\int_0^{\pi/2} \sec^3\dfrac{\theta}{2}\, d\theta = 12\displaystyle\int_0^{\pi/4} \sec^3 u\, du = (\text{use tables})\ 6\left(\left[\dfrac{\sec u\tan u}{2}\right]_0^{\pi/4} + \dfrac{1}{2}\displaystyle\int_0^{\pi/4} \sec u\, du\right)$

$= 6\left(\dfrac{1}{\sqrt{2}} + \left[\dfrac{1}{2}\ln|\sec u + \tan u|\right]_0^{\pi/4}\right) = 3\left[\sqrt{2} + \ln\left(1 + \sqrt{2}\right)\right]$

25. $r = \cos^3\dfrac{\theta}{3} \Rightarrow \dfrac{dr}{d\theta} = -\sin\dfrac{\theta}{3}\cos^2\dfrac{\theta}{3}$; therefore Length $= \displaystyle\int_0^{\pi/4} \sqrt{\left(\cos^3\dfrac{\theta}{3}\right)^2 + \left(-\sin\dfrac{\theta}{3}\cos^2\dfrac{\theta}{3}\right)^2}\, d\theta$

$= \displaystyle\int_0^{\pi/4} \sqrt{\cos^6\left(\dfrac{\theta}{3}\right) + \sin^2\left(\dfrac{\theta}{3}\right)\cos^4\left(\dfrac{\theta}{3}\right)}\, d\theta = \displaystyle\int_0^{\pi/4} \left(\cos^2\dfrac{\theta}{3}\right)\sqrt{\cos^2\left(\dfrac{\theta}{3}\right) + \sin^2\left(\dfrac{\theta}{3}\right)}\, d\theta = \displaystyle\int_0^{\pi/4} \cos^2\left(\dfrac{\theta}{3}\right)\, d\theta$

$= \displaystyle\int_0^{\pi/4} \dfrac{1 + \cos\left(\frac{2\theta}{3}\right)}{2}\, d\theta = \dfrac{1}{2}\left[\theta + \dfrac{3}{2}\sin\dfrac{2\theta}{3}\right]_0^{\pi/4} = \dfrac{\pi}{8} + \dfrac{3}{8}$

27. $r = \sqrt{1 + \cos 2\theta} \Rightarrow \dfrac{dr}{d\theta} = \dfrac{1}{2}(1 + \cos 2\theta)^{-1/2}(-2\sin 2\theta)$; therefore Length $= \displaystyle\int_0^{\pi\sqrt{2}} \sqrt{(1 + \cos 2\theta) + \dfrac{\sin^2 2\theta}{(1 + \cos 2\theta)}}\, d\theta$

$= \displaystyle\int_0^{\pi\sqrt{2}} \sqrt{\dfrac{1 + 2\cos 2\theta + \cos^2 2\theta + \sin^2 2\theta}{1 + \cos 2\theta}}\, d\theta = \displaystyle\int_0^{\pi\sqrt{2}} \sqrt{\dfrac{2 + 2\cos 2\theta}{1 + \cos 2\theta}}\, d\theta = \displaystyle\int_0^{\pi\sqrt{2}} \sqrt{2}\, d\theta = \left[\sqrt{2}\,\theta\right]_0^{\pi\sqrt{2}} = 2\pi$

29. $r = \sqrt{\cos 2\theta}$, $0 \le \theta \le \dfrac{\pi}{4} \Rightarrow \dfrac{dr}{d\theta} = \dfrac{1}{2}(\cos 2\theta)^{-1/2}(-\sin 2\theta)(2) = \dfrac{-\sin 2\theta}{\sqrt{\cos 2\theta}}$; therefore Surface Area

$= \displaystyle\int_0^{\pi/4} (2\pi r\cos\theta)\sqrt{\left(\sqrt{\cos 2\theta}\right)^2 + \left(\dfrac{-\sin 2\theta}{\sqrt{\cos 2\theta}}\right)^2}\, d\theta = \displaystyle\int_0^{\pi/4} \left(2\pi\sqrt{\cos 2\theta}\right)(\cos\theta)\sqrt{\cos 2\theta + \dfrac{\sin^2 2\theta}{\cos 2\theta}}\, d\theta$

$= \displaystyle\int_0^{\pi/4} \left(2\pi\sqrt{\cos 2\theta}\right)(\cos\theta)\sqrt{\dfrac{1}{\cos 2\theta}}\, d\theta = \displaystyle\int_0^{\pi/4} 2\pi\cos\theta\, d\theta = [2\pi\sin\theta]_0^{\pi/4} = \pi\sqrt{2}$

31. $r^2 = \cos 2\theta \Rightarrow r = \pm \sqrt{\cos 2\theta}$; use $r = \sqrt{\cos 2\theta}$ on $\left[0, \frac{\pi}{4}\right] \Rightarrow \frac{dr}{d\theta} = \frac{1}{2}(\cos 2\theta)^{-1/2}(-\sin 2\theta)(2) = \frac{-\sin 2\theta}{\sqrt{\cos 2\theta}}$;

therefore Surface Area $= 2\displaystyle\int_0^{\pi/4} \left(2\pi\sqrt{\cos 2\theta}\right)(\sin\theta)\sqrt{\cos 2\theta + \frac{\sin^2 2\theta}{\cos 2\theta}}\, d\theta = 4\pi\displaystyle\int_0^{\pi/4}\sqrt{\cos 2\theta}\,(\sin\theta)\sqrt{\frac{1}{\cos 2\theta}}\, d\theta$

$= 4\pi\displaystyle\int_0^{\pi/4}\sin\theta\, d\theta = 4\pi[-\cos\theta]_0^{\pi/4} = 4\pi\left[-\frac{\sqrt{2}}{2} - (-1)\right] = 2\pi\left(2 - \sqrt{2}\right)$

33. Let $r = f(\theta)$. Then $x = f(\theta)\cos\theta \Rightarrow \frac{dx}{d\theta} = f'(\theta)\cos\theta - f(\theta)\sin\theta \Rightarrow \left(\frac{dx}{d\theta}\right)^2 = \left[f'(\theta)\cos\theta - f(\theta)\sin\theta\right]^2$

$= \left[f'(\theta)\right]^2\cos^2\theta - 2f'(\theta)\,f(\theta)\sin\theta\cos\theta + [f(\theta)]^2\sin^2\theta;\ y = f(\theta)\sin\theta \Rightarrow \frac{dy}{d\theta} = f'(\theta)\sin\theta + f(\theta)\cos\theta$

$\Rightarrow \left(\frac{dy}{d\theta}\right)^2 = \left[f'(\theta)\sin\theta + f(\theta)\cos\theta\right]^2 = \left[f'(\theta)\right]^2\sin^2\theta + 2f'(\theta)f(\theta)\sin\theta\cos\theta + [f(\theta)]^2\cos^2\theta$. Therefore

$\left(\frac{dx}{d\theta}\right)^2 + \left(\frac{dy}{d\theta}\right)^2 = \left[f'(\theta)\right]^2\left(\cos^2\theta + \sin^2\theta\right) + [f(\theta)]^2\left(\cos^2\theta + \sin^2\theta\right) = \left[f'(\theta)\right]^2 + [f(\theta)]^2 = r^2 + \left(\frac{dr}{d\theta}\right)^2$.

Thus, $L = \displaystyle\int_\alpha^\beta\sqrt{\left(\frac{dx}{d\theta}\right)^2 + \left(\frac{dy}{d\theta}\right)^2}\, d\theta = \displaystyle\int_\alpha^\beta\sqrt{r^2 + \left(\frac{dr}{d\theta}\right)^2}\, d\theta$.

35. $r = 2f(\theta),\ \alpha \le \theta \le \beta \Rightarrow \frac{dr}{d\theta} = 2f'(\theta) \Rightarrow r^2 + \left(\frac{dr}{d\theta}\right)^2 = [2f(\theta)]^2 + [2f'(\theta)]^2 \Rightarrow$ Length $= \displaystyle\int_\alpha^\beta\sqrt{4[f(\theta)]^2 + 4\left[f'(\theta)\right]^2}\, d\theta$

$= 2\displaystyle\int_\alpha^\beta\sqrt{[f(\theta)]^2 + \left[f'(\theta)\right]^2}\, d\theta$ which is twice the length of the curve $r = f(\theta)$ for $\alpha \le \theta \le \beta$.

37. $\bar{x} = \dfrac{\frac{2}{3}\displaystyle\int_0^{2\pi} r^3\cos\theta\, d\theta}{\displaystyle\int_0^{2\pi} r^2\, d\theta} = \dfrac{\frac{2}{3}\displaystyle\int_0^{2\pi}[a(1 + \cos\theta)]^3(\cos\theta)\, d\theta}{\displaystyle\int_0^{2\pi}[a(1 + \cos\theta)]^2\, d\theta} = \dfrac{\frac{2}{3}a^3\displaystyle\int_0^{2\pi}\left(1 + 3\cos\theta + 3\cos^2\theta + \cos^3\theta\right)(\cos\theta)\, d\theta}{a^2\displaystyle\int_0^{2\pi}\left(1 + 2\cos\theta + \cos^2\theta\right)\, d\theta}$

$= \dfrac{\frac{2}{3}a\displaystyle\int_0^{2\pi}\left[\cos\theta + 3\left(\frac{1 + \cos 2\theta}{2}\right) + 3\left(1 - \sin^2\theta\right)(\cos\theta) + \left(\frac{1 + \cos 2\theta}{2}\right)^2\right]\, d\theta}{\displaystyle\int_0^{2\pi}\left[1 + 2\cos\theta + \left(\frac{1 + \cos 2\theta}{2}\right)\right]\, d\theta} =$ (After considerable algebra using

the identity $\cos^2 A = \dfrac{1 + \cos 2A}{2}$) $\dfrac{a\displaystyle\int_0^{2\pi}\left(\frac{15}{12} + \frac{8}{3}\cos\theta + \frac{4}{3}\cos 2\theta - 2\cos\theta\sin^2\theta + \frac{1}{12}\cos 4\theta\right)\, d\theta}{\displaystyle\int_0^{2\pi}\left(\frac{3}{2} + 2\cos\theta + \frac{1}{2}\cos 2\theta\right)\, d\theta}$

$$= \frac{a\left[\frac{15}{12}\theta + \frac{8}{3}\sin\theta + \frac{2}{3}\sin 2\theta - \frac{2}{3}\sin^3\theta + \frac{1}{48}\sin 4\theta\right]_0^{2\pi}}{\left[\frac{3}{2}\theta + 2\sin\theta + \frac{1}{4}\sin 2\theta\right]_0^{2\pi}} = \frac{a\left(\frac{15}{6}\pi\right)}{3\pi} = \frac{5}{6}a;$$

$$\overline{y} = \frac{\frac{2}{3}\int_0^{2\pi} r^3 \sin\theta\, d\theta}{\int_0^{2\pi} r^2\, d\theta} = \frac{\frac{2}{3}\int_0^{2\pi}[a(1+\cos\theta)]^3(\sin\theta)\, d\theta}{3\pi}; \left[u = a(1+\cos\theta) \Rightarrow -\frac{1}{a}\,du = \sin\theta\, d\theta; \theta = 0 \Rightarrow u = 2a;\right.$$

$$\theta = 2\pi \Rightarrow u = 2a\right] \rightarrow \frac{\frac{2}{3}\int_{2a}^{2a} -\frac{1}{a}u^3\, du}{3\pi} = \frac{0}{3\pi} = 0. \text{ Therefore the centroid is } (\overline{x}, \overline{y}) = \left(\frac{5}{6}a, 0\right)$$

CHAPTER 9 PRACTICE EXERCISES

1. $x^2 = -4y \Rightarrow y = -\frac{x^2}{4} \Rightarrow 4p = 4 \Rightarrow p = 1;$

    therefore Focus is $(0,-1)$, Directrix is $y = 1$

3. $y^2 = 3x \Rightarrow x = \frac{y^2}{3} \Rightarrow 4p = 3 \Rightarrow p = \frac{3}{4};$

    therefore Focus is $\left(\frac{3}{4},0\right)$, Directrix is $x = -\frac{3}{4}$

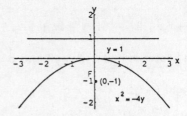

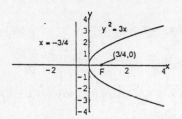

5. $16x^2 + 7y^2 = 112 \Rightarrow \frac{x^2}{7} + \frac{y^2}{16} = 1$

    $\Rightarrow c^2 = 16 - 7 = 9 \Rightarrow c = 3; e = \frac{c}{a} = \frac{3}{4}$

7. $3x^2 - y^2 = 3 \Rightarrow x^2 - \frac{y^2}{3} = 1 \Rightarrow c^2 = 1 + 3 = 4$

    $\rightarrow c - 2; e - \frac{c}{a} - \frac{2}{1} = 2;$ the asymptotes are

    $y = \pm\sqrt{3}\,x$

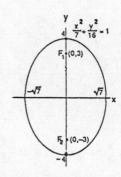

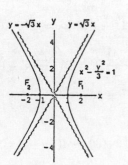

9. $x^2 = -12y \Rightarrow -\frac{x^2}{12} = y \Rightarrow 4p = 12 \Rightarrow p = 3 \Rightarrow$ focus is $(0,-3)$, directrix is $y = 3$, vertex is $(0,0)$; therefore new

    vertex is $(2,3)$, new focus is $(2,0)$, new directrix is $y = 6$, and the new equation is $(x-2)^2 = -12(y-3)$

11. $\frac{x^2}{9} + \frac{y^2}{25} = 1 \Rightarrow a = 5$ and $b = 3 \Rightarrow c = \sqrt{25-9} = 4 \Rightarrow$ foci are $(0, \pm 4)$, vertices are $(0, \pm 5)$, center is

$(0,0)$; therefore the new center is $(-3,-5)$, new foci are $(-3,-1)$ and $(-3,-9)$, new vertices are $(-3,-10)$ and

$(-3,0)$, and the new equation is $\frac{(x+3)^2}{9} + \frac{(y+5)^2}{25} = 1$

13. $\frac{y^2}{8} - \frac{x^2}{2} = 1 \Rightarrow a = 2\sqrt{2}$ and $b = \sqrt{2} \Rightarrow c = \sqrt{8+2} = \sqrt{10} \Rightarrow$ foci are $(0, \pm\sqrt{10})$, vertices are

$(0, \pm 2\sqrt{2})$, center is $(0,0)$, and the asymptotes are $y = \pm 2x$; therefore the new center is $(2, 2\sqrt{2})$, new foci are

$(2, 2\sqrt{2} \pm \sqrt{10})$, new vertices are $(2, 4\sqrt{2})$ and $(2,0)$, the new asymptotes are $y = 2x - 4 + 2\sqrt{2}$ and

$y = -2x + 4 + 2\sqrt{2}$; the new equation is $\frac{(y-2\sqrt{2})^2}{8} - \frac{(x-2)^2}{2} = 1$

15. $x^2 - 4x - 4y^2 = 0 \Rightarrow x^2 - 4x + 4 - 4y^2 = 4 \Rightarrow (x-2)^2 - 4y^2 = 4 \Rightarrow \frac{(x-2)^2}{4} - y^2 = 1$, a hyperbola; $a = 2$ and

$b = 1 \Rightarrow c = \sqrt{1+4} = \sqrt{5}$; the center is $(2,0)$, the vertices are $(0,0)$ and $(4,0)$; the foci are $(2 \pm \sqrt{5}, 0)$ and

the asymptotes are $y = \pm \frac{x-2}{2}$

17. $y^2 - 2y + 16x = -49 \Rightarrow y^2 - 2y + 1 = -16x - 48 \Rightarrow (y-1)^2 = -16(x+3)$, a parabola; the vertex is $(-3,1)$;

$4p = 16 \Rightarrow p = 4 \Rightarrow$ the focus is $(-7,1)$ and the directrix is $x = 1$

19. $9x^2 + 16y^2 + 54x - 64y = -1 \Rightarrow 9(x^2 + 6x) + 16(y^2 - 4y) = -1 \Rightarrow 9(x^2 + 6x + 9) + 16(y^2 - 4y + 4) = 144$

$\Rightarrow 9(x+3)^2 + 16(y-2)^2 = 144 \Rightarrow \frac{(x+3)^2}{16} + \frac{(y-2)^2}{9} = 1$, an ellipse; the center is $(-3,2)$; $a = 4$ and $b = 3$

$\Rightarrow c = \sqrt{16-9} = \sqrt{7}$; the foci are $(-3 \pm \sqrt{7}, 2)$; the vertices are $(1,2)$ and $(-7,2)$

21. $x^2 + y^2 - 2x - 2y = 0 \Rightarrow x^2 - 2x + 1 + y^2 - 2y + 1 = 2 \Rightarrow (x-1)^2 + (y-1)^2 = 2$, a circle with center $(1,1)$ and

radius $= \sqrt{2}$

23. $B^2 - 4AC = 1 - 4(1)(1) = -3 < 0 \Rightarrow$ ellipse   25. $B^2 - 4AC = 3^2 - 4(1)(2) = 1 > 0 \Rightarrow$ hyperbola

27. $x^2 - 2xy + y^2 = 0 \Rightarrow (x-y)^2 = 0 \Rightarrow x - y = 0$ or $y = x$, a straight line

29. $B^2 - 4AC = 1^2 - 4(2)(2) = -15 < 0 \Rightarrow$ ellipse; $\cot 2\alpha = \frac{A-C}{B} = 0 \Rightarrow 2\alpha = \frac{\pi}{2} \Rightarrow \alpha = \frac{\pi}{4}$; $x = \frac{\sqrt{2}}{2}x' - \frac{\sqrt{2}}{2}y'$ and

$y = \frac{\sqrt{2}}{2}x' + \frac{\sqrt{2}}{2}y' \Rightarrow 2\left(\frac{\sqrt{2}}{2}x' - \frac{\sqrt{2}}{2}y'\right)^2 + \left(\frac{\sqrt{2}}{2}x' - \frac{\sqrt{2}}{2}y'\right)\left(\frac{\sqrt{2}}{2}x' + \frac{\sqrt{2}}{2}y'\right) + 2\left(\frac{\sqrt{2}}{2}x' + \frac{\sqrt{2}}{2}y'\right)^2 - 15 = 0$

$\Rightarrow 5x'^2 + 3y'^2 = 30$

31. $B^2 - 4AC = (2\sqrt{3})^2 - 4(1)(-1) = 16 \Rightarrow$ hyperbola; $\cot 2\alpha = \frac{A-C}{B} = \frac{1}{\sqrt{3}} \Rightarrow 2\alpha = \frac{\pi}{3} \Rightarrow \alpha = \frac{\pi}{6}$; $x = \frac{\sqrt{3}}{2}x' - \frac{1}{2}y'$

and $y = \frac{1}{2}x' + \frac{\sqrt{3}}{2}y' \Rightarrow \left(\frac{\sqrt{3}}{2}x' - \frac{1}{2}y'\right)^2 + 2\sqrt{3}\left(\frac{\sqrt{3}}{2}x' - \frac{1}{2}y'\right)\left(\frac{1}{2}x' + \frac{\sqrt{3}}{2}y'\right) - \left(\frac{1}{2}x' + \frac{\sqrt{3}}{2}y'\right)^2 = 4$

$\Rightarrow 2x'^2 - 2y'^2 = 4 \Rightarrow x'^2 - y'^2 = 2$

33. $x = \frac{t}{2}$ and $y = t + 1 \Rightarrow 2x = t \Rightarrow y = 2x + 1$

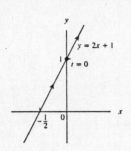

35. $x = \frac{1}{2} \tan t$ and $y = \frac{1}{2} \sec t \Rightarrow x^2 = \frac{1}{4} \tan^2 t$

and $y^2 = \frac{1}{4} \sec^2 t \Rightarrow 4x^2 = \tan^2 t$ and

$4y^2 = \sec^2 t \Rightarrow 4x^2 + 1 = 4y^2 \Rightarrow 4y^2 - 4x^2 = 1$

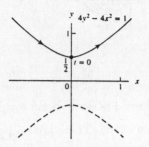

37. $x = -\cos t$ and $y = \cos^2 t \Rightarrow y = (-x)^2 = x^2$

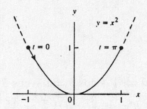

39. $16x^2 + 9y^2 = 144 \Rightarrow \dfrac{x^2}{9} + \dfrac{y^2}{16} = 1 \Rightarrow a = 3$ and $b = 4 \Rightarrow x = 3 \cos t$ and $y = 4 \sin t$, $0 \le t \le 2\pi$

41. $x = \frac{1}{2} \tan t$, $y = \frac{1}{2} \sec t \Rightarrow \dfrac{dy}{dx} = \dfrac{dy/dt}{dx/dt} = \dfrac{\frac{1}{2} \sec t \tan t}{\frac{1}{2} \sec^2 t} = \dfrac{\tan t}{\sec t} = \sin t \Rightarrow \dfrac{dy}{dx}\bigg|_{t = \pi/3} = \sin \dfrac{\pi}{3} = \dfrac{\sqrt{3}}{2}$; $t = \dfrac{\pi}{3}$

$\Rightarrow x = \frac{1}{2} \tan \dfrac{\pi}{3} = \dfrac{\sqrt{3}}{2}$ and $y = \frac{1}{2} \sec \dfrac{\pi}{3} = 1 \Rightarrow y = \dfrac{\sqrt{3}}{2} x + \dfrac{1}{4}$; $\dfrac{d^2 y}{dx^2} = \dfrac{dy'/dt}{dx/dt} = \dfrac{\cos t}{\frac{1}{2} \sec^2 t} = 2 \cos^3 t \Rightarrow \dfrac{d^2 y}{dx^2}\bigg|_{t = \pi/3}$

$= 2 \cos^3 \left(\dfrac{\pi}{3} \right) = \dfrac{1}{4}$

43. $x = e^{2t} - \dfrac{t}{8}$ and $y = e^t$, $0 \le t \le \ln 2 \Rightarrow \dfrac{dx}{dt} = 2e^{2t} - \dfrac{1}{8}$ and $\dfrac{dy}{dt} = e^t \Rightarrow$ Length $= \displaystyle\int_0^{\ln 2} \sqrt{\left(2e^{2t} - \dfrac{1}{8} \right)^2 + \left(e^t \right)^2} \, dt$

$= \displaystyle\int_0^{\ln 2} \sqrt{4e^{4t} + \dfrac{1}{2} e^{2t} + \dfrac{1}{64}} \, dt = \displaystyle\int_0^{\ln 2} \sqrt{\left(2e^{2t} + \dfrac{1}{8} \right)^2} \, dt = \displaystyle\int_0^{\ln 2} \left(2e^{2t} + \dfrac{1}{8} \right) dt = \left[e^{2t} + \dfrac{t}{8} \right]_0^{\ln 2} = 3 + \dfrac{\ln 2}{8}$

45. $x = \dfrac{t^2}{2}$ and $y = 2t$, $0 \le t \le \sqrt{5} \Rightarrow \dfrac{dx}{dt} = t$ and $\dfrac{dy}{dt} = 2 \Rightarrow$ Surface Area $= \displaystyle\int_0^{\sqrt{5}} 2\pi (2t) \sqrt{t^2 + 4} \, dt = \displaystyle\int_4^9 2\pi u^{1/2} \, du$

$= 2\pi \left[\dfrac{2}{3} u^{3/2} \right]_4^9 = \dfrac{76\pi}{3}$, where $u = t^2 + 4 \Rightarrow du = 2t \, dt$; $t = 0 \Rightarrow u = 4$, $t = \sqrt{5} \Rightarrow u = 9$

47.

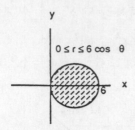

49. d   51. l   53. k   55. i

57. $r = \sin\theta$ and $r = 1 + \sin\theta \Rightarrow \sin\theta = 1 + \sin\theta \Rightarrow 0 = 1$

so no solutions exist. There are no points of intersection
found by solving the system. The point of intersection
$(0,0)$ is found by graphing.

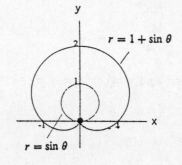

59. $r = 1 + \cos\theta$ and $r = 1 - \cos\theta \Rightarrow 1 + \cos\theta = 1 - \cos\theta$

$\Rightarrow 2\cos\theta = 0 \Rightarrow \cos\theta = 0 \Rightarrow \theta = \frac{\pi}{2}, \frac{3\pi}{2}; \theta = \frac{\pi}{2}$ or $\frac{3\pi}{2}$

$\Rightarrow r = 1$. The points of intersection are $\left(1, \frac{\pi}{2}\right)$ and $\left(1, \frac{3\pi}{2}\right)$.

The point of intersection $(0,0)$ is found by graphing.

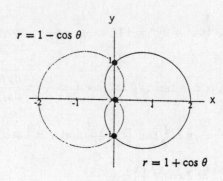

61. $r = 1 + \sin\theta$ and $r = -1 + \sin\theta$ intersect at all points of
$r = 1 + \sin\theta$ because the graphs coincide. This can be
seen by graphing them.

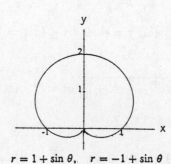

63. $r = \sec\theta$ and $r = 2\sin\theta \Rightarrow \sec\theta = 2\sin\theta$

$\Rightarrow 1 = 2\sin\theta\cos\theta \Rightarrow 1 = \sin 2\theta \Rightarrow 2\theta = \frac{\pi}{2} \Rightarrow \theta = \frac{\pi}{4}$

$\Rightarrow r = 2\sin\frac{\pi}{4} = \sqrt{2} \Rightarrow$ the point of intersection is

$\left(\sqrt{2}, \frac{\pi}{4}\right)$. No other points of intersection exist.

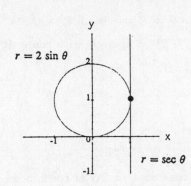

65. $r^2 = \cos 2\theta \Rightarrow r = 0$ when $\cos 2\theta = 0 \Rightarrow 2\theta = \frac{\pi}{2}, \frac{3\pi}{2} \Rightarrow \theta = \frac{\pi}{4}, \frac{3\pi}{4}; \ \theta_1 = \frac{\pi}{4} \Rightarrow m_1 = \tan\frac{\pi}{4} = 1 \Rightarrow y = x$ is one

tangent line; $\theta_2 = \frac{3\pi}{4} \Rightarrow m_2 = \tan\frac{3\pi}{4} = -1 \Rightarrow y = -x$ is the other tangent line

67. The tips of the petals are at $\theta = \frac{\pi}{4}, \frac{3\pi}{4}, \frac{5\pi}{4}, \frac{7\pi}{4}$ and $r = 1$ at those values of θ. Then for $\theta = \frac{\pi}{4}$, the tangent line

is $r\cos\left(\theta - \frac{\pi}{4}\right) = 1$; for $\theta = \frac{3\pi}{4}$, $r\cos\left(\theta - \frac{3\pi}{4}\right) = 1$; for $\theta = \frac{5\pi}{4}$, $r\cos\left(\theta - \frac{5\pi}{4}\right) = 1$; and for $\theta = \frac{7\pi}{4}$,

$r\cos\left(\theta - \frac{7\pi}{4}\right) = 1$.

69. $r\cos\left(\theta + \frac{\pi}{3}\right) = 2\sqrt{3} \Rightarrow r\left(\cos\theta\cos\frac{\pi}{3} - \sin\theta\sin\frac{\pi}{3}\right)$

$= 2\sqrt{3} \Rightarrow \frac{1}{2}r\cos\theta - \frac{\sqrt{3}}{2}r\sin\theta = 2\sqrt{3}$

$\Rightarrow r\cos\theta - \sqrt{3}r\sin\theta = 4\sqrt{3} \Rightarrow x - \sqrt{3}y = 4\sqrt{3}$

$\Rightarrow y = \frac{\sqrt{3}}{3}x - 4$

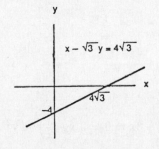

71. $r = 2\sec\theta \Rightarrow r = \frac{2}{\cos\theta} \Rightarrow r\cos\theta = 2 \Rightarrow x = 2$

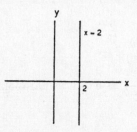

73. $r = -\frac{3}{2}\csc\theta \Rightarrow r\sin\theta = -\frac{3}{2} \Rightarrow y = -\frac{3}{2}$

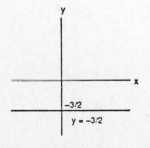

75. $r = -4 \sin \theta \Rightarrow r^2 = -4r \sin \theta \Rightarrow x^2 + y^2 + 4y = 0$

   $\Rightarrow x^2 + (y + 2)^2 = 4$; circle with center $(0, 2)$ and

   radius 2.

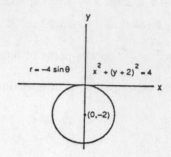

77. $r = 2\sqrt{2} \cos \theta \Rightarrow r^2 = 2\sqrt{2}\, r \cos \theta \Rightarrow x^2 + y^2 - 2\sqrt{2}\, x = 0$

   $\Rightarrow \left(x - \sqrt{2}\right)^2 + y^2 = 2$; circle with center $\left(\sqrt{2}, 0\right)$ and

   radius $\sqrt{2}$

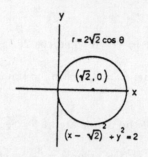

79. $x^2 + y^2 + 5y = 0 \Rightarrow x^2 + \left(y + \frac{5}{2}\right)^2 = \frac{25}{4} \Rightarrow C = \left(0, -\frac{5}{2}\right)$

   and $a = \frac{5}{2}$; $r^2 + 5r \sin \theta = 0 \Rightarrow r = -5 \sin \theta$

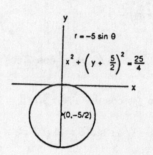

81. $x^2 + y^2 - 3x = 0 \Rightarrow \left(x - \frac{3}{2}\right)^2 + y^2 = \frac{9}{4} \Rightarrow C = \left(\frac{3}{2}, 0\right)$ and

   $a = \frac{3}{2}$; $r^2 - 3r \cos \theta = 0 \Rightarrow r = 3 \cos \theta$

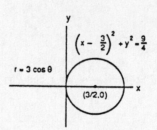

83. $r = \dfrac{2}{1 + \cos \theta} \Rightarrow e = 1 \Rightarrow$ parabola with vertex at $(1, 0)$

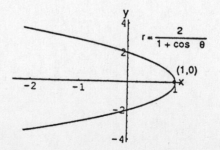

85. $r = \dfrac{6}{1 - 2\cos\theta} \Rightarrow e = 2 \Rightarrow$ hyperbola; $ke = 6 \Rightarrow 2k = 6$

$\Rightarrow k = 3 \Rightarrow$ vertices are $(2, \pi)$ and $(6, \pi)$

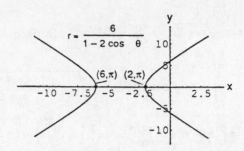

87. $e = 2$ and $r\cos\theta = 2 \Rightarrow x = 2$ is directrix $\Rightarrow k = 2$; the conic is a hyperbola; $r = \dfrac{ke}{1 + e\cos\theta} \Rightarrow r = \dfrac{(2)(2)}{1 + 2\cos\theta}$

$\Rightarrow r = \dfrac{4}{1 + 2\cos\theta}$

89. $e = \dfrac{1}{2}$ and $r\sin\theta = 2 \Rightarrow y = 2$ is directrix $\Rightarrow k = 2$; the conic is an ellipse; $r = \dfrac{ke}{1 + e\sin\theta} \Rightarrow r = \dfrac{(2)\left(\frac{1}{2}\right)}{1 + \left(\frac{1}{2}\right)\sin\theta}$

$\Rightarrow r = \dfrac{2}{2 + \sin\theta}$

91. $A = 2 \displaystyle\int_{0}^{\pi} \dfrac{1}{2}r^2 \, d\theta = \int_{0}^{\pi} (2 - \cos\theta)^2 \, d\theta = \int_{0}^{\pi} \left(4 - 2\cos\theta + \cos^2\theta\right) d\theta = \int_{0}^{\pi} \left(4 - 2\cos\theta + \dfrac{1 + \cos 2\theta}{2}\right) d\theta$

$= \displaystyle\int_{0}^{\pi} \left(\dfrac{9}{2} - 2\cos\theta + \dfrac{\cos 2\theta}{2}\right) d\theta = \left[\dfrac{9}{2}\theta - 2\sin\theta + \dfrac{\sin 2\theta}{4}\right]_{0}^{\pi} = \dfrac{9}{2}\pi$

93. $r = 1 + \cos 2\theta$ and $r = 1 \Rightarrow 1 = 1 + \cos 2\theta \Rightarrow 0 = \cos 2\theta \Rightarrow 2\theta = \dfrac{\pi}{2} \Rightarrow \theta = \dfrac{\pi}{4}$; therefore

$A = 4 \displaystyle\int_{0}^{\pi/4} \dfrac{1}{2}\left[(1 + \cos 2\theta)^2 - 1^2\right] d\theta = 2 \int_{0}^{\pi/4} \left(1 + 2\cos 2\theta + \cos^2 2\theta - 1\right) d\theta$

$= 2 \displaystyle\int_{0}^{\pi/4} \left(2\cos 2\theta + \dfrac{1}{2} + \dfrac{\cos 4\theta}{2}\right) d\theta = 2\left[\sin 2\theta + \dfrac{1}{2}\theta + \dfrac{\sin 4\theta}{8}\right]_{0}^{\pi/4} = 2\left(1 + \dfrac{\pi}{8} + 0\right) = 2 + \dfrac{\pi}{4}$

95. $r = -1 + \cos\theta \Rightarrow \dfrac{dr}{d\theta} = -\sin\theta$; Length $= \displaystyle\int_{0}^{2\pi} \sqrt{(-1 + \cos\theta)^2 + (-\sin\theta)^2} \, d\theta = \int_{0}^{2\pi} \sqrt{2 - 2\cos\theta} \, d\theta$

$= \displaystyle\int_{0}^{2\pi} \sqrt{\dfrac{4(1 - \cos\theta)}{2}} \, d\theta = \int_{0}^{2\pi} 2\sin\dfrac{\theta}{2} \, d\theta = \left[-4\cos\dfrac{\theta}{2}\right]_{0}^{2\pi} = (-4)(-1) - (-4)(1) = 8$

97. $r = 8\sin^3\left(\dfrac{\theta}{3}\right)$, $0 \le \theta \le \dfrac{\pi}{4} \Rightarrow \dfrac{dr}{d\theta} = 8\sin^2\left(\dfrac{\theta}{3}\right)\cos\left(\dfrac{\theta}{3}\right)$; $r^2 + \left(\dfrac{dr}{d\theta}\right)^2 = \left[8\sin^3\left(\dfrac{\theta}{3}\right)\right]^2 + \left[8\sin^2\left(\dfrac{\theta}{3}\right)\cos\left(\dfrac{\theta}{3}\right)\right]^2$

$= 64\sin^4\left(\dfrac{\theta}{3}\right) \Rightarrow L = \displaystyle\int_{0}^{\pi/4} \sqrt{64\sin^4\left(\dfrac{\theta}{3}\right)} \, d\theta = \int_{0}^{\pi/4} 8\sin^2\left(\dfrac{\theta}{3}\right) d\theta = \int_{0}^{\pi/4} 8\left[\dfrac{1 - \cos\left(\frac{2\theta}{3}\right)}{2}\right] d\theta$

$= \displaystyle\int_{0}^{\pi/4} \left[4 - 4\cos\left(\dfrac{2\theta}{3}\right)\right] d\theta = \left[4\theta - 6\sin\left(\dfrac{2\theta}{3}\right)\right]_{0}^{\pi/4} = 4\left(\dfrac{\pi}{4}\right) - 6\sin\left(\dfrac{\pi}{6}\right) - 0 = \pi - 3$

99. $r = \sqrt{\cos 2\theta} \Rightarrow \frac{dr}{d\theta} = \frac{-\sin 2\theta}{\sqrt{\cos 2\theta}}$; Surface Area $= \int_0^{\pi/4} 2\pi(r \sin \theta) \sqrt{r^2 + \left(\frac{dr}{d\theta}\right)^2} \, d\theta$

$= \int_0^{\pi/4} 2\pi \sqrt{\cos 2\theta} (\sin \theta) \sqrt{\cos 2\theta + \frac{\sin^2 2\theta}{\cos 2\theta}} \, d\theta = \int_0^{\pi/4} 2\pi \sqrt{\cos 2\theta} (\sin \theta) \sqrt{\frac{1}{\cos 2\theta}} \, d\theta = \int_0^{\pi/4} 2\pi \sin \theta \, d\theta$

$= \left[2\pi(-\cos \theta) \right]_0^{\pi/4} = 2\pi \left(1 - \frac{\sqrt{2}}{2} \right) = \left(2 - \sqrt{2} \right) \pi$

101. (a) Around the x-axis: $9x^2 + 4y^2 = 36 \Rightarrow y^2 = 9 - \frac{9}{4}x^2 \Rightarrow y = \pm \sqrt{9 - \frac{9}{4}x^2}$ and we use the positive root:

$$V = 2 \int_0^2 \pi \left(\sqrt{9 - \frac{9}{4}x^2} \right)^2 dx = 2 \int_0^2 \pi \left(9 - \frac{9}{4}x^2 \right) dx = 2\pi \left[9x - \frac{3}{4}x^3 \right]_0^2 = 24\pi$$

(b) Around the y-axis: $9x^2 + 4y^2 = 36 \Rightarrow x^2 = 4 - \frac{4}{9}y^2 \Rightarrow x = \pm \sqrt{4 - \frac{4}{9}y^2}$ and we use the positive root:

$$V = 2 \int_0^3 \pi \left(\sqrt{4 - \frac{4}{9}y^2} \right)^2 dy = 2 \int_0^3 \pi \left(4 - \frac{4}{9}y^2 \right) dy = 2\pi \left[4y - \frac{4}{27}y^3 \right]_0^3 = 16\pi$$

103. Each portion of the wave front reflects to the other focus, and since the wave front travels at a constant speed as it expands, the different portions of the wave arrive at the second focus simultaneously, from all directions, causing a spurt at the second focus.

105. The time for the bullet to hit the target remains constant, say $t = t_0$. Let the time it takes for sound to travel from the target to the listener be t_2. Since the listener hears the sounds simultaneously, $t_1 = t_0 + t_2$ where t_1 is the time for the sound to travel from the rifle to the listener. If v is the velocity of sound, then $vt_1 = vt_0 + vt_2$ or $vt_1 - vt_2 = vt_0$. Now vt_1 is the distance from the rifle to the listener and vt_2 is the distance from the target to the listener. Therefore the difference of the distances is constant since vt_0 is constant so the listener is on a branch of a hyperbola with foci at the rifle and the target. The branch is the one with target as focus.

107. (a) $r = \frac{k}{1 + e \cos \theta} \Rightarrow r + er \cos \theta = k \Rightarrow \sqrt{x^2 + y^2} + ex = k \Rightarrow \sqrt{x^2 + y^2} = k - ex \Rightarrow x^2 + y^2$

$= k^2 - 2kex + e^2x^2 \Rightarrow x^2 - e^2x^2 + y^2 + 2kex - k^2 = 0 \Rightarrow \left(1 - e^2 \right)x^2 + y^2 + 2kex - k^2 = 0$

(b) $c = 0 \Rightarrow x^2 + y^2 - k^2 = 0 \Rightarrow x^2 + y^2 = k^2 \Rightarrow$ circle;

$0 < e < 1 \Rightarrow e^2 < 1 \Rightarrow e^2 - 1 < 0 \Rightarrow B^2 - 4AC = 0^2 - 4\left(1 - e^2 \right)(1) = 4\left(e^2 - 1 \right) < 0 \Rightarrow$ ellipse;

$e = 1 \Rightarrow B^2 - 4AC = 0^2 - 4(0)(1) = 0 \Rightarrow$ parabola;

$e > 1 \Rightarrow e^2 > 1 \Rightarrow B^2 - 4AC = 0^2 - 4\left(1 - e^2 \right)(1) = 4e^2 - 4 > 0 \Rightarrow$ hyperbola

109. $\beta = \psi_2 - \psi_1 \Rightarrow \tan \beta = \tan (\psi_2 - \psi_1) = \frac{\tan \psi_2 - \tan \psi_1}{1 + \tan \psi_2 \tan \psi_1}$;

the curves will be orthogonal when $\tan \beta$ is undefined, or

when $\tan \psi_2 = \frac{-1}{\tan \psi_1} \Rightarrow \frac{r}{g'(\theta)} = \frac{-1}{\left[\frac{r}{f'(\theta)} \right]} \Rightarrow r^2 = -f'(\theta) \, g'(\theta)$

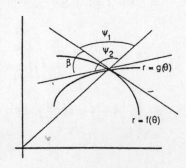

111. $r = 2a \sin 3\theta \Rightarrow \frac{dr}{d\theta} = 6a \cos 3\theta \Rightarrow \tan \psi = \frac{r}{\left(\frac{dr}{d\theta}\right)} = \frac{2a \sin 3\theta}{6a \cos 3\theta} = \frac{1}{3} \tan 3\theta$; when $\theta = \frac{\pi}{6}$, $\tan \psi = \frac{1}{3} \tan \frac{\pi}{2}$

$\Rightarrow \psi = \frac{\pi}{2}$

113. $\tan \psi_1 = \frac{\sqrt{3} \cos \theta}{-\sqrt{3} \sin \theta} = -\cot \theta$ is $-\frac{1}{\sqrt{3}}$ at $\theta = \frac{\pi}{3}$; $\tan \psi_2 = \frac{\sin \theta}{\cos \theta} = \tan \theta$ is $\sqrt{3}$ at $\theta = \frac{\pi}{3}$; since the product of

these slopes is -1, the tangents are perpendicular

115. $r_1 = \frac{1}{1 - \cos \theta} \Rightarrow \frac{dr_1}{d\theta} = -\frac{\sin \theta}{(1 - \cos \theta)^2}$; $r_2 = \frac{3}{1 + \cos \theta} \Rightarrow \frac{dr_2}{d\theta} = \frac{3 \sin \theta}{(1 + \cos \theta)^2}$; $\frac{1}{1 - \cos \theta} = \frac{3}{1 + \cos \theta}$

$\Rightarrow 1 + \cos \theta = 3 - 3 \cos \theta \Rightarrow 4 \cos \theta = 2 \Rightarrow \cos \theta = \frac{1}{2} \Rightarrow \theta = \pm \frac{\pi}{3} \Rightarrow r_1 = r_2 = 2 \Rightarrow$ the curves intersect at the

points $\left(2, \pm \frac{\pi}{3}\right)$; $\tan \psi_1 = \frac{\left(\frac{1}{1 - \cos \theta}\right)}{\left[\frac{-\sin \theta}{(1 - \cos \theta)^2}\right]} = -\frac{1 - \cos \theta}{\sin \theta}$ is $-\frac{1}{\sqrt{3}}$ at $\theta = \frac{\pi}{3}$; $\tan \psi_2 = \frac{\left(\frac{3}{1 + \cos \theta}\right)}{\left[\frac{3 \sin \theta}{(1 + \cos \theta)^2}\right]} = \frac{1 + \cos \theta}{\sin \theta}$ is

$\sqrt{3}$ at $\theta = \frac{\pi}{3}$; therefore $\tan \beta$ is undefined at $\theta = \frac{\pi}{3}$ since $1 + \tan \psi_1 \tan \psi_2 = 1 + \left(-\frac{1}{\sqrt{3}}\right)(\sqrt{3}) = 0 \Rightarrow \beta = \frac{\pi}{2}$;

$\tan \psi_1\big|_{\theta = -\pi/3} = -\frac{1 - \cos\left(-\frac{\pi}{3}\right)}{\sin\left(-\frac{\pi}{3}\right)} = \frac{1}{\sqrt{3}}$ and $\tan \psi_2\big|_{\theta = -\pi/3} = \frac{1 + \cos\left(-\frac{\pi}{3}\right)}{\sin\left(-\frac{\pi}{3}\right)} = -\sqrt{3} \Rightarrow \tan \beta$ is also undefined

at $\theta = -\frac{\pi}{3} \Rightarrow \beta = \frac{\pi}{2}$

117. $r_1 = \frac{a}{1 + \cos \theta} \Rightarrow \frac{dr_1}{d\theta} = \frac{a \sin \theta}{(1 + \cos \theta)^2}$ and $r_2 = \frac{b}{1 - \cos \theta} \Rightarrow \frac{dr_2}{d\theta} = -\frac{b \sin \theta}{(1 - \cos \theta)^2}$; then

$\tan \psi_1 = \frac{\left(\frac{a}{1 + \cos \theta}\right)}{\left[\frac{a \sin \theta}{(1 + \cos \theta)^2}\right]} = \frac{1 + \cos \theta}{\sin \theta}$ and $\tan \psi_2 = \frac{\left(\frac{b}{1 - \cos \theta}\right)}{\left[\frac{-b \sin \theta}{(1 - \cos \theta)^2}\right]} = \frac{1 - \cos \theta}{-\sin \theta} \Rightarrow 1 + \tan \psi_1 \tan \psi_2$

$= 1 + \left(\frac{1 + \cos \theta}{\sin \theta}\right)\left(\frac{1 - \cos \theta}{-\sin \theta}\right) = 1 - \frac{1 - \cos^2 \theta}{\sin^2 \theta} = 0 \Rightarrow \beta$ is undefined $\Rightarrow$ the parabolas are orthogonal at each

point of intersection

119. $r = 3 \sec \theta \Rightarrow r = \frac{3}{\cos \theta}$; $\frac{3}{\cos \theta} = 4 + 4 \cos \theta \Rightarrow 3 = 4 \cos \theta + 4 \cos^2 \theta \Rightarrow (2 \cos \theta + 3)(2 \cos \theta - 1) = 0$

$\Rightarrow \cos \theta = \frac{1}{2}$ or $\cos \theta = -\frac{3}{2} \Rightarrow \theta = \frac{\pi}{3}$ or $\frac{5\pi}{3}$ (the second equation has no solutions); $\tan \psi_2 = \frac{4(1 + \cos \theta)}{-4 \sin \theta}$

$= -\frac{1 + \cos \theta}{\sin \theta}$ is $-\sqrt{3}$ at $\frac{\pi}{3}$ and $\tan \psi_1 = \frac{3 \sec \theta}{3 \sec \theta \tan \theta} = \cot \theta$ is $\frac{1}{\sqrt{3}}$ at $\frac{\pi}{3}$. Then $\tan \beta$ is undefined since

$1 + \tan \psi_1 \tan \psi_2 = 1 + \left(\frac{1}{\sqrt{3}}\right)(-\sqrt{3}) = 0 \Rightarrow \beta = \frac{\pi}{2}$. Also, $\tan \psi_2\big|_{5\pi/3} = \sqrt{3}$ and $\tan \psi_1\big|_{5\pi/3} = -\frac{1}{\sqrt{3}}$

$\Rightarrow 1 + \tan \psi_1 \tan \psi_2 = 1 + \left(-\frac{1}{\sqrt{3}}\right)(\sqrt{3}) = 0 \Rightarrow \tan \beta$ is also undefined $\Rightarrow \beta = \frac{\pi}{2}$.

121. $\dfrac{1}{1-\cos\theta}=\dfrac{1}{1-\sin\theta}\Rightarrow 1-\cos\theta=1-\sin\theta\Rightarrow\cos\theta=\sin\theta\Rightarrow\theta=\dfrac{\pi}{4}$; $\tan\psi_1=\dfrac{\left(\dfrac{1}{1-\cos\theta}\right)}{\left[\dfrac{-\sin\theta}{(1-\cos\theta)^2}\right]}=\dfrac{1-\cos\theta}{-\sin\theta}$;

$\tan\psi_2=\dfrac{\left(\dfrac{1}{1-\sin\theta}\right)}{\left[\dfrac{\cos\theta}{(1-\sin\theta)^2}\right]}=\dfrac{1-\sin\theta}{\cos\theta}$. Thus at $\theta=\dfrac{\pi}{4}$, $\tan\psi_1=\dfrac{1-\cos\left(\dfrac{\pi}{4}\right)}{-\sin\left(\dfrac{\pi}{4}\right)}=1-\sqrt{2}$ and

$\tan\psi_2=\dfrac{1-\sin\left(\dfrac{\pi}{4}\right)}{\cos\left(\dfrac{\pi}{4}\right)}=\sqrt{2}-1$. Then $\tan\beta=\dfrac{\left(\sqrt{2}-1\right)-\left(1-\sqrt{2}\right)}{1+\left(\sqrt{2}-1\right)\left(1-\sqrt{2}\right)}=\dfrac{2\sqrt{2}-2}{2\sqrt{2}-2}=1\Rightarrow\beta=\dfrac{\pi}{4}$

123. (a) $\tan\alpha=\dfrac{r}{\left(\dfrac{dr}{d\theta}\right)}\Rightarrow\dfrac{dr}{r}=\dfrac{d\theta}{\tan\alpha}\Rightarrow\ln r=\dfrac{\theta}{\tan\alpha}+C$ (by integration) $\Rightarrow r=Be^{\theta/(\tan\alpha)}$ for some constant B;

$A=\dfrac{1}{2}\displaystyle\int_{\theta_1}^{\theta_2}B^2e^{2\theta/(\tan\alpha)}\,d\theta=\left[\dfrac{B^2(\tan\alpha)e^{2\theta/(\tan\alpha)}}{4}\right]_{\theta_1}^{\theta_2}=\dfrac{\tan\alpha}{4}\left[B^2e^{2\theta_2/(\tan\alpha)}-B^2e^{2\theta_1/(\tan\alpha)}\right]$

$=\dfrac{\tan\alpha}{4}\left(r_2^2-r_1^2\right)$ since $r_2^2=B^2e^{2\theta_2/(\tan\alpha)}$ and $r_1^2=B^2e^{2\theta_1/(\tan\alpha)}$; constant of proportionality $K=\dfrac{\tan\alpha}{4}$

(b) $\tan\alpha=\dfrac{r}{\left(\dfrac{dr}{d\theta}\right)}\Rightarrow\dfrac{dr}{d\theta}=\dfrac{r}{\tan\alpha}\Rightarrow\left(\dfrac{dr}{d\theta}\right)^2=\dfrac{r^2}{\tan^2\alpha}\Rightarrow r^2+\left(\dfrac{dr}{d\theta}\right)^2=r^2+\dfrac{r^2}{\tan^2\alpha}=r^2\left(\dfrac{\tan^2\alpha+1}{\tan^2\alpha}\right)$

$=r^2\left(\dfrac{\sec^2\alpha}{\tan^2\alpha}\right)\Rightarrow \text{Length}=\displaystyle\int_{\theta_1}^{\theta_2}r\left(\dfrac{\sec\alpha}{\tan\alpha}\right)d\theta=\int_{\theta_1}^{\theta_2}Be^{\theta/(\tan\alpha)}\cdot\dfrac{\sec\alpha}{\tan\alpha}\,d\theta=\left[B(\sec\alpha)e^{\theta/(\tan\alpha)}\right]_{\theta_1}^{\theta_2}$

$=(\sec\alpha)\left[Be^{\theta_2/(\tan\alpha)}-Be^{\theta_1(\tan\alpha)}\right]=K(r_2-r_1)$ where $K=\sec\alpha$ is the constant of proportionality

CHAPTER 9 ADDITIONAL EXERCISES–THEORY, EXAMPLES, APPLICATIONS

1. Directrix $x=3$ and focus $(4,0)\Rightarrow$ vertex is $\left(\dfrac{7}{2},0\right)$

$\Rightarrow p=\dfrac{1}{2}\Rightarrow$ the equation is $x-\dfrac{7}{2}=\dfrac{y^2}{2}$

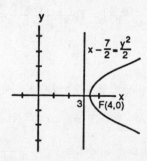

3. $x^2=4y\Rightarrow$ vertex is $(0,0)$ and $p=1\Rightarrow$ focus is $(0,1)$; thus the distance from $P(x,y)$ to the vertex is $\sqrt{x^2+y^2}$

and the distance from P to the focus is $\sqrt{x^2+(y-1)^2}\Rightarrow\sqrt{x^2+y^2}=2\sqrt{x^2+(y-1)^2}$

$\Rightarrow x^2+y^2=4\left[x^2+(y-1)^2\right]\Rightarrow x^2+y^2=4x^2+4y^2-8y+4\Rightarrow 3x^2+3y^2-8y+4=0$, which is a circle

5. Vertices are $(0, \pm 2) \Rightarrow a = 2;\ e = \frac{c}{a} \Rightarrow 0.5 = \frac{c}{2} \Rightarrow c = 1 \Rightarrow$ foci are $(0, \pm 1)$

7. Let the center of the hyperbola be $(0, y)$.

   (a) Directrix $y = -1$, focus $(0, -7)$ and $e = 2 \Rightarrow c = -\frac{a}{e} = 6 \Rightarrow \frac{a}{e} = c - 6 \Rightarrow a = 2c - 12$. Also $c = ae = 2a$

   $\Rightarrow a = 2(2a) - 12 \Rightarrow a = 4 \Rightarrow c = 8;\ y - (-1) = \frac{a}{e} = \frac{4}{2} = 2 \Rightarrow y = 1 \Rightarrow$ the center is $(0, 1);\ c^2 = a^2 + b^2$

   $\Rightarrow b^2 = c^2 - a^2 = 64 - 16 = 48$; therefore the equation is $\dfrac{(y-1)^2}{16} - \dfrac{x^2}{48} = 1$

   (b) $e = 5 \Rightarrow c - \frac{a}{e} = 6 \Rightarrow \frac{a}{e} = c - 6 \Rightarrow a = 5c - 30$. Also, $c = ae = 5a \Rightarrow a = 5(5a) - 30 \Rightarrow 24a = 30 \Rightarrow a = \frac{5}{4}$

   $\Rightarrow c = \frac{25}{4};\ y - (-1) = \frac{a}{e} = \frac{\left(\frac{5}{4}\right)}{5} = \frac{1}{4} \Rightarrow y = -\frac{3}{4} \Rightarrow$ the center is $\left(0, -\frac{3}{4}\right);\ c^2 = a^2 + b^2 \Rightarrow b^2 = c^2 - a^2$

   $= \frac{625}{16} - \frac{24}{16} = \frac{75}{2}$; therefore the equation is $\dfrac{\left(y + \frac{3}{4}\right)^2}{\left(\frac{25}{16}\right)} - \dfrac{x^2}{\left(\frac{75}{2}\right)} = 1$ or $\dfrac{16\left(y + \frac{3}{4}\right)^2}{25} - \dfrac{2x^2}{75} = 1$

9. $xy = 2 \Rightarrow x\dfrac{dy}{dx} + y = 0 \Rightarrow \dfrac{dy}{dx} = -\dfrac{y}{x};\ x^2 - y^2 = 3$

   $\Rightarrow 2x - 2y\dfrac{dy}{dx} = 0 \Rightarrow \dfrac{dy}{dx} = \dfrac{x}{y}$. If (x_0, y_0) is a point of

   intersection, then the product of the slopes is

   $\left(-\dfrac{y_0}{x_0}\right)\left(\dfrac{x_0}{y_0}\right) = -1 \Rightarrow$ the curves are orthogonal.

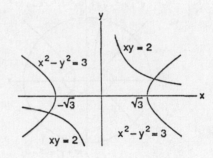

11. $2x^2 + 3y^2 = a^2 \Rightarrow 4x + 6y\dfrac{dy}{dx} = 0 \Rightarrow \dfrac{dy}{dx} = -\dfrac{2x}{3y};\ ky^2 = x^3$

   where k is a constant $\Rightarrow 2ky\dfrac{dy}{dx} = 3x^2 \Rightarrow \dfrac{dy}{dx} = \dfrac{3x^2}{2ky} = \dfrac{3x^2 y}{2ky^2}$

   $= \dfrac{3x^2 y}{2x^3}$ (since $ky^2 = x^3$). If (x_0, y_0) is a point of intersection

   then the product of the slopes is $\left(-\dfrac{2x_0}{3y_0}\right)\left(\dfrac{3x_0^2 y_0}{2x_0^3}\right) = -1$

   $\Rightarrow$ the curves are orthogonal at their points of intersection.

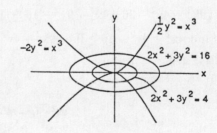

13. $y^2 = 4px \Rightarrow 2y\dfrac{dy}{dx} = 4p \Rightarrow \dfrac{dy}{dx} = \dfrac{2p}{y} \Rightarrow m_{\tan} = \dfrac{2p}{y_1}$ at

   $P(x_1, y_1) \Rightarrow$ the tangent line is $y - y_1 = \left(\dfrac{2p}{y_1}\right)(x - x_1)$

   and meets the axis of symmetry when $y = 0$

   $\Rightarrow -y_1 = \left(\dfrac{2p}{y_1}\right)(x - x_1) \Rightarrow -\dfrac{y_1^2}{2p} + x_1 = x \Rightarrow -\dfrac{4px_1}{2p} + x_1$

   $= x \Rightarrow x = -x_1$; that is, the tangent line meets the

   axis of symmetry x_1 units to the left of the vertex.

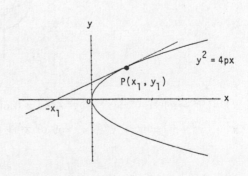

15. $xy = a^2$ is a hyperbola whose asymptotes are the x and y

axes; $xy = a^2 \Rightarrow y + x\frac{dy}{dx} = 0 \Rightarrow \frac{dy}{dx} = -\frac{y}{x}$. Let $P(x_1, y_1)$

be a point on the hyperbola $\Rightarrow m_{tan} = -\frac{y_1}{x_1} \Rightarrow$ the equation

of the tangent line is $y - y_1 = \left(-\frac{y_1}{x_1}\right)(x - x_1)$. The

tangent line intersects the coordinate axes to form the

triangle (see figure). If $x = 0$, then $y - y_1 = \left(-\frac{y_1}{x_1}\right)(0 - x_1)$

$\Rightarrow y - y_1 = y_1 \Rightarrow y = 2y_1$; if $y = 0$, then $-y_1 = \left(-\frac{y_1}{x_1}\right)(x - x_1)$

$\Rightarrow x_1 = x - x_1 \Rightarrow x = 2x_1$. Therefore the area is $A = \frac{1}{2}(2x_1)(2y_1) = 2x_1y_1 = 2a^2$.

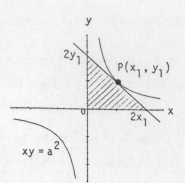

17.

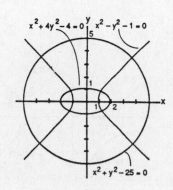

19.

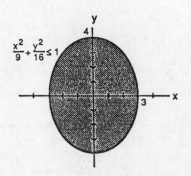

21. $\left(9x^2 + 4y^2 - 36\right)\left(4x^2 + 9y^2 - 16\right) \le 0 \Rightarrow 9x^2 + 4y^2 - 36 \le 0$

and $4x^2 + 9y^2 - 16 \ge 0$ or $9x^2 + 4y^2 - 36 \ge 0$ and

$4x^2 + 9y^2 - 16 \le 0$

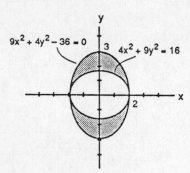

23. $x^4 - \left(y^2 - 9\right)^2 = 0 \Rightarrow x^2 - \left(y^2 - 9\right) = 0$ or

$x^2 + \left(y^2 - 9\right) = 0 \Rightarrow y^2 - x^2 = 9$ or $x^2 + y^2 = 9$

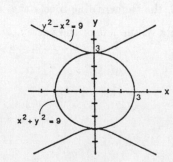

25. Arc PF = Arc AF since each is the distance rolled;

$$\angle PCF = \frac{\text{Arc PF}}{b} \Rightarrow \text{Arc PF} = b(\angle PCF); \; \theta = \frac{\text{Arc AF}}{a}$$

$$\Rightarrow \text{Arc AF} = a\theta \Rightarrow a\theta = b(\angle PCF) \Rightarrow \angle PCF = \left(\frac{a}{b}\right)\theta;$$

$$\angle OCB = \frac{\pi}{2} - \theta \text{ and } \angle OCB = \angle PCF - \angle PCE$$

$$= \angle PCF - \left(\frac{\pi}{2} - \alpha\right) = \left(\frac{a}{b}\right)\theta - \left(\frac{\pi}{2} - \alpha\right) \Rightarrow \frac{\pi}{2} - \theta$$

$$= \left(\frac{a}{b}\right)\theta - \left(\frac{\pi}{2} - \alpha\right) \Rightarrow \frac{\pi}{2} - \theta = \left(\frac{a}{b}\right)\theta - \frac{\pi}{2} + \alpha$$

$$\Rightarrow \alpha = \pi - \theta - \left(\frac{a}{b}\right)\theta \Rightarrow \alpha = \pi - \left(\frac{a+b}{b}\right)\theta.$$

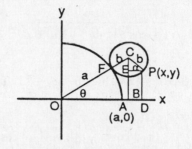

Now $x = OB + BD = OB + EP = (a+b) \cos\theta + b \cos\alpha = (a+b) \cos\theta + b \cos\left(\pi - \left(\frac{a+b}{b}\right)\theta\right)$

$$= (a+b) \cos\theta + b \cos\pi \cos\left(\left(\frac{a+b}{b}\right)\theta\right) + b \sin\pi \sin\left(\left(\frac{a+b}{b}\right)\theta\right) = (a+b) \cos\theta - b \cos\left(\left(\frac{a+b}{b}\right)\theta\right) \text{ and}$$

$$y = PD = CB - CE = (a+b) \sin\theta - b \sin\alpha = (a+b) \sin\theta - b \sin\left(\pi - \left(\frac{a+b}{b}\right)\theta\right)$$

$$= (a+b) \sin\theta - b \sin\pi \cos\left(\left(\frac{a+b}{b}\right)\theta\right) + b \cos\pi \sin\left(\left(\frac{a+b}{b}\right)\theta\right) = (a+b) \sin\theta - b \sin\left(\left(\frac{a+b}{b}\right)\theta\right);$$

therefore $x = (a+b) \cos\theta - b \cos\left(\left(\frac{a+b}{b}\right)\theta\right)$ and $y = (a+b) \sin\theta - b \sin\left(\left(\frac{a+b}{b}\right)\theta\right)$

27. $x = \frac{1-t^2}{1+t^2} \Rightarrow x^2 = \frac{(1-t^2)^2}{(1+t^2)^2}$ and $y = \frac{2t}{1+t^2} \Rightarrow y^2 = \frac{4t^2}{(1+t^2)^2}$

$$\Rightarrow x^2 + y^2 = \frac{(1-t^2)^2 + 4t^2}{(1+t^2)^2} = \frac{t^4 + 2t^2 + 1}{(1+t^2)^2} = \frac{(t^2+1)^2}{(1+t^2)^2} = 1;$$

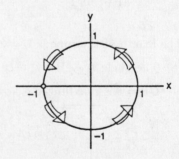

$y = 0 \Rightarrow \frac{2t}{1+t^2} = 0 \Rightarrow t = 0 \Rightarrow x = 1 \Rightarrow (-1, 0)$ is not

covered; $t = -1$ gives $(0, -1)$, $t = 0$ gives $(1, 0)$, and $t = 1$

gives $(0, 1)$. Note that as $t \to \pm\infty$, $x \to -1$ and $y \to 0$.

29. (a) $x = e^{2t} \cos t$ and $y = e^{2t} \sin t \Rightarrow x^2 + y^2 = e^{4t} \cos^2 t + e^{4t} \sin^2 t = e^{4t}$. Also $\frac{y}{x} = \frac{e^{2t} \sin t}{e^{2t} \cos t} = \tan t$

$$\Rightarrow t = \tan^{-1}\left(\frac{y}{x}\right) \Rightarrow x^2 + y^2 = e^{4 \tan^{-1}(y/x)} \text{ is the Cartesian equation. Since } r^2 = x^2 + y^2 \text{ and}$$

$\theta = \tan^{-1}\left(\frac{y}{x}\right)$, the polar equation is $r^2 = e^{4\theta}$ or $r = e^{2\theta}$ for $r > 0$

(b) $ds^2 = r^2 \, d\theta^2 + dr^2; \; r = e^{2\theta} \Rightarrow dr = 2e^{2\theta} \, d\theta$

$$\Rightarrow ds^2 = r^2 \, d\theta^2 + \left(2e^{2\theta} \, d\theta\right)^2 = \left(e^{2\theta}\right)^2 d\theta^2 + 4e^{4\theta} \, d\theta^2$$

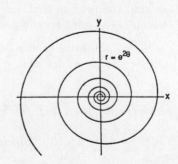

$$= 5e^{4\theta} \, d\theta^2 \Rightarrow ds = \sqrt{5}\, e^{2\theta} \, d\theta \Rightarrow L = \int_0^{2\pi} \sqrt{5}\, e^{2\theta} \, d\theta$$

$$= \left[\frac{\sqrt{5}\, e^{2\theta}}{2}\right]_0^{2\pi} = \frac{\sqrt{5}}{2}\left(e^{4\pi} - 1\right)$$

31. $r = 1 + \cos\theta$ and $S = \displaystyle\int 2\pi\rho\ ds$, where $\rho = y = r\sin\theta$; $ds = \sqrt{r^2\ d\theta^2 + dr^2}$

$= \sqrt{(1 + \cos\theta)^2\ d\theta^2 + \sin^2\theta\ d\theta^2}\ \sqrt{1 + 2\cos\theta + \cos^2\theta + \sin^2\theta}\ d\theta = \sqrt{2 + 2\cos\theta}\ d\theta = \sqrt{4\cos^2\left(\frac{\theta}{2}\right)}\ d\theta$

$= 2\cos\left(\frac{\theta}{2}\right)\ d\theta$ since $0 \le \theta \le \frac{\pi}{2}$. Then $S = \displaystyle\int_0^{\pi/2} 2\pi(r\sin\theta)\cdot 2\cos\left(\frac{\theta}{2}\right)\ d\theta = \int_0^{\pi/2} 4\pi(1 + \cos\theta)\cdot\sin\theta\ \cos\left(\frac{\theta}{2}\right)\ d\theta$

$= \displaystyle\int_0^{\pi/2} 4\pi\left[2\cos^2\left(\frac{\theta}{2}\right)\right]\left[2\sin\left(\frac{\theta}{2}\right)\cos\left(\frac{\theta}{2}\right)\cos\left(\frac{\theta}{2}\right)\right]\ d\theta = \int_0^{\pi/2} 16\pi\cos^4\left(\frac{\theta}{2}\right)\sin\left(\frac{\theta}{2}\right)\ d\theta = \left[\frac{-32\pi\ \cos^5\left(\frac{\theta}{2}\right)}{5}\right]_0^{\pi/2}$

$= \dfrac{(-32\pi)\left(\frac{\sqrt{2}}{2}\right)^5}{5} - \left(-\dfrac{32\pi}{5}\right) = \dfrac{32\pi - 4\pi\sqrt{2}}{5}$

33. $e = 2$ and $r\cos\theta = 2 \Rightarrow x = 2$ is the directrix $\Rightarrow k = 2$; the conic is a hyperbola with $r = \dfrac{ke}{1 + e\cos\theta}$

$\Rightarrow r = \dfrac{(2)(2)}{1 + 2\cos\theta} = \dfrac{4}{1 + 2\cos\theta}$

35. $e = \frac{1}{2}$ and $r\sin\theta = 2 \Rightarrow y = 2$ is the directrix $\Rightarrow k = 2$; the conic is an ellipse with $r = \dfrac{ke}{1 + e\sin\theta}$

$\Rightarrow r = \dfrac{2\left(\frac{1}{2}\right)}{1 + \left(\frac{1}{2}\right)\sin\theta} = \dfrac{2}{2 + \sin\theta}$

37. The length of the rope is $L = 2x + 2c + y \ge 8C$.

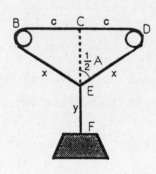

(a) The angle A ($\angle$BED) occurs when the distance

CF $= \ell$ is maximized. Now $\ell = \sqrt{x^2 - c^2} + y$

$\Rightarrow \ell = \sqrt{x^2 - c^2} + L - 2x - 2c$

$\Rightarrow \dfrac{d\ell}{dx} = \frac{1}{2}(x^2 - c^2)^{-1/2}(2x) - 2 = \dfrac{x}{\sqrt{x^2 - c^2}} - 2$.

Thus $\dfrac{d\ell}{dx} = 0 \Rightarrow \dfrac{x}{\sqrt{x^2 - c^2}} - 2 = 0 \Rightarrow x = 2\sqrt{x^2 - c^2}$

$\Rightarrow x^2 = 4x^2 - 4c^2 \Rightarrow 3x^2 = 4c^2 \Rightarrow \dfrac{c^2}{x^2} = \dfrac{3}{4} \Rightarrow \dfrac{c}{x} = \dfrac{\sqrt{3}}{2}$.

Since $\dfrac{c}{x} = \sin\dfrac{A}{2}$ we have $\sin\dfrac{A}{2} = \dfrac{\sqrt{3}}{2} \Rightarrow \dfrac{A}{2} = 60° \Rightarrow A = 120°$

(b) If the ring is fixed at E (i.e., y is held constant) and E is moved to the right, for example, the rope will slip around the pegs so that BE lengthens and DE becomes shorter $\Rightarrow$ BE + ED is always $2x = L - y - 2c$, which is constant $\Rightarrow$ the point E lies on an ellipse with the pegs as foci.

(c) Minimal potential energy occurs when the weight is at its lowest point $\Rightarrow$ E is at the intersection of the ellipse and its minor axis.

39. If the vertex is $(0, 0)$, then the focus is $(p, 0)$. Let $P(x, y)$ be the present position of the comet. Then

$\sqrt{(x - p)^2 + y^2} = 4\times 10^7$. Since $y^2 = 4px$ we have $\sqrt{(x - p)^2 + 4px} = 4\times 10^7 \Rightarrow (x - p)^2 + 4px = 16\times 10^{14}$.

Also, $x - p = 4 \times 10^7 \cos 60° = 2 \times 10^7 \Rightarrow x = p + 2 \times 10^7$. Therefore $\left(2 \times 10^7\right)^2 + 4p\left(p + 2 \times 10^7\right) = 16 \times 10^{14}$

$\Rightarrow 4 \times 10^{14} + 4p^2 + 8p \times 10^7 = 16 \times 10^{14} \Rightarrow 4p^2 + 8p \times 10^7 - 12 \times 10^{14} = 0 \Rightarrow p^2 + 2p \times 10^7 - 3 \times 10^{14} = 0$

$\Rightarrow \left(p + 3 \times 10^7\right)\left(p - 10^7\right) = 0 \Rightarrow p = -3 \times 10^7$ or $p = 10^7$. Since p is positive we obtain $p = 10^7$ miles.

41. $\cot \alpha = \dfrac{A - C}{B} = 0 \Rightarrow \alpha = 45°$ is the angle of rotation $\Rightarrow A' = \cos^2 45° + \cos 45° \sin 45° + \sin^2 45° = \dfrac{3}{2}$, $B' = 0$,

and $C' = \sin^2 45° - \sin 45° \cos 45° + \cos^2 45° = \dfrac{1}{2} \Rightarrow \dfrac{3}{2}x'^2 + \dfrac{1}{2}y'^2 = 1 \Rightarrow b = \sqrt{\dfrac{2}{3}}$ and $a = \sqrt{2} \Rightarrow c^2 = a^2 - b^2$

$= 2 - \dfrac{2}{3} = \dfrac{4}{3} \Rightarrow c = \dfrac{2}{\sqrt{3}}$. Therefore the eccentricity is $e = \dfrac{c}{a} = \dfrac{\left(\dfrac{2}{\sqrt{3}}\right)}{\sqrt{2}} = \sqrt{\dfrac{2}{3}} \approx 0.82$.

43. $\sqrt{x} + \sqrt{y} = 1 \Rightarrow x + 2\sqrt{xy} + y = 1 \Rightarrow 2\sqrt{xy} = 1 - (x + y) \Rightarrow 4xy = 1 - 2(x + y) + (x + y)^2$

$\Rightarrow 4xy = x^2 + 2xy + y^2 - 2x - 2y + 1 \Rightarrow x^2 - 2xy + y^2 - 2x - 2y + 1 = 0 \Rightarrow B^2 - 4AC = (-2)^2 - 4(1)(1) = 0$

$\Rightarrow$ the curve is part of a parabola

45. (a) The equation of a parabola with focus $(0, 0)$ and vertex $(a, 0)$ is $r = \dfrac{2a}{1 + \cos \theta}$ and rotating this parabola

through $\alpha = 45°$ gives $r = \dfrac{2a}{1 + \cos\left(\theta - \dfrac{\pi}{4}\right)}$.

(b) Foci at $(0, 0)$ and $(2, 0) \Rightarrow$ the center is $(1, 0) \Rightarrow a = 3$ and $c = 1$ since one vertex is at $(4, 0)$. Then $e = \dfrac{c}{a}$

$= \dfrac{1}{3}$. For ellipses with one focus at the origin and major axis along the x-axis we have $r = \dfrac{a\left(1 - e^2\right)}{1 - e \cos \theta}$

$= \dfrac{3\left(1 - \dfrac{1}{9}\right)}{1 - \left(\dfrac{1}{3}\right)\cos \theta} = \dfrac{8}{3 - \cos \theta}$.

(c) Center at $\left(2, \dfrac{\pi}{2}\right)$ and focus at $(0, 0) \Rightarrow c = 2$; center at $\left(2, \dfrac{\pi}{2}\right)$ and vertex at $\left(1, \dfrac{\pi}{2}\right) \Rightarrow a = 1$. Then $e = \dfrac{c}{a}$

$= \dfrac{2}{1} = 2$. Also $k = ae - \dfrac{a}{e} = (1)(2) - \dfrac{1}{2} = \dfrac{3}{2}$. Therefore $r = \dfrac{ke}{1 + e \sin \theta} = \dfrac{\left(\dfrac{3}{2}\right)(2)}{1 + 2 \sin \theta} = \dfrac{3}{1 + 2 \sin \theta}$.

47. Arc PT = Arc TO since each is the same distance rolled. Now Arc PT $= a(\angle TAP)$ and Arc TO $= a(\angle TBO)$

$\Rightarrow \angle TAP = \angle TBO$. Since $AP = a = BO$ we have that $\triangle ADP$ is congruent to $\triangle BCO \Rightarrow CO = DP \Rightarrow OP$ is

parallel to $AB \Rightarrow \angle TBO = \angle TAP = \theta$. Then OPDC is a square $\Rightarrow r = CD = AB - AD - CB = AB - 2CB$

$\Rightarrow r = 2a - 2a \cos \theta = 2a(1 - \cos \theta)$, which is the polar equation of a cardiod.

49.

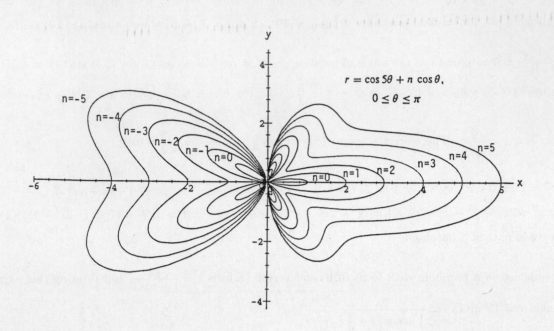

$$r = \cos 5\theta + n \cos \theta,$$
$$0 \leq \theta \leq \pi$$

NOTES: